# YEAST BIOTECHNOLOGY

TITLES OF RELATED INTEREST

*A biologist's advanced mathematics*
D. R. Causton

*Comparative plant ecology*
J. P. Grime, J. G. Hodgson & R. Hunt

*Chromosomes today vol 9*
A. Stahl, J. M. Luciani & A. M. Vagner-Capodano (eds.)

*Crop genetic resources*
J. H. W. Holden & J. T. Williams (eds.)

*The eukaryote genome in development and evolution*
B. John & G. Miklos

*Introduction to vegetation analysis*
D. R. Causton

*Light and plant growth*
J. W. Hart

*Lipids in plants and microbes*
J. L. Harwood & N. J. Russell

*The physical chemistry of membranes*
B. Silver

*Plant breeding systems*
A. J. Richards

# YEAST BIOTECHNOLOGY

Edited by
D.R.BERRY,
*Department of Bioscience & Biotechnology, University of Strathclyde*
I.RUSSELL, G.G.STEWART
*Labatt Brewing Co., London, Ontario*

London
ALLEN & UNWIN
Boston Sydney Wellington

Allen & Unwin, the academic imprint of
**Unwin Hyman Ltd,**
PO Box 18, Park Lane, Hemel Hempstead, Herts HP2 4TE, UK
40 Museum Street, London WC1A 1LU, UK
37/39 Queen Elizabeth Street, London SE1 2QB, UK

Allen & Unwin Inc.,
8 Winchester Place, Winchester, Mass 01890, USA

Allen & Unwin (Australia) Ltd,
8 Napier Street, North Sydney, NSW 2060, Australia

Allen & Unwin (New Zealand) Ltd in association with the
Port Nicholson Press Ltd,
60 Cambridge Terrace, Wellington, New Zealand

First published in 1987

**British Library Cataloguing in Publication Data**

Yeast biotechnology.
1. Biotechnology 2. Yeast
I. Berry, David R. II. Russell, I. III. Stewart, G. G.
660′.62 TP248.24
ISBN 0-04-574042-9

**Library of Congress Cataloging in Publication Data**

Yeast biotechnology.
Includes bibliographies and index.
1. Yeast fungi—Biotechnology. 2. Saccharomyces cerevisiae—Biotechnology.
I. Stewart, Graham, G., 1942– . II. Berry, David R. (David Richard)
III. Russell, Inge, 1947– .
TP248.27.Y43Y43 1987 660′.62 87-12583
ISBN 0-04-574042-9 (alk. paper)

Typeset in 10 on 11 point Times by Paston Press, Loddon, Norfolk
and printed in Great Britain by St. Edmondsbury Press, Suffolk

# *Preface*

Biotechnology is now established as a major area of technology, concerned with the 'application of biological organisms, systems or processes to manufacturing or service industries'. Although the exploitation of organisms by man is not new, many of the techniques which are stimulating the rapid advances in biotechnology have developed from recent scientific discoveries.

Throughout history man has, knowingly or not, been exploiting yeast in the production of alcoholic beverages and bread, and these processes still represent major biotechnological industries. The brewer's and baker's yeast *Saccharomyces cerevisiae* is, however, also a favoured organism for the production of many new biotechnological products. The successful production of both traditional and novel products from yeast requires an up-to-date knowledge and understanding of the developments in bioscience and biotechnology as they relate to the exploitation of yeast.

In this volume we have collected together a set of contributions to five areas which we believe are fundamental to all yeast biotechnological processes; isolation and identification; genetic manipulation, physiology and technology of growth, product formation, and product isolation and extraction. While it is impossible to cover this topic comprehensively in a single volume, we believe that each part contains essential information which is relevant to most yeast industries, and should be valuable to those who work with yeast either in a research or production capacity. While the book is largely concerned with *Saccharomyces cerevisiae*, reference to other yeasts is made where it is appropriate to give adequate coverage of a topic or process.

D. R. Berry, G. G. Stewart and I. Russell

# *Acknowledgements*

The editors would like to thank Elisabeth Berry, Karen Smith and Janice Riddell for their help and co-operation during the preparation of this volume.

We are grateful to the following individuals and organizations who have kindly given permission for the reproduction of copyright material (figure numbers in parentheses):

Pergamon Press, Canada (6.1, 6.2); Figure 6.3 reproduced from D. Fraenkel, Carbohydrate metabolism, in *The molecular biology of the yeast* Saccharomyces, J. N. Strathern, E. W. Jones & J. R. Broach (eds) by permission of Cold Spring Harbor Laboratories; Elsevier Science Publishers (6.4); Massey University, New Zealand (16.1, 16.2, 16.10, 16.14, 16.15); M. Pyke & Massey University (16.7).

# Contents

# *List of tables*

# List of contributors


Wilfred Niels Arnold, Department of Biochemistry, University of Kansas Medical Center, Kansas City, Kansas 66103, USA

J. P. Barford, Department of Chemical Engineering, University of Sydney, NSW 2006, Australia

D. R. Berry, Department of Bioscience and Biotechnology, Division of Applied Microbiology, University of Strathclyde, 204 George Street, Glasgow G1 1XW, UK

E. Alan Bevan, School of Biological Sciences, Queen Mary College (University of London), Mile End Road, London E1 4NS, UK

C. Brown, Department of Chemical and Process Engineering, University of Strathclyde, James Weir Building, Montrose Street, Glasgow G1 1XW, UK

B. L. A. Carter, Zymogenetics Inc., 2121 North 35th Street, Seattle, Washington, USA

Trevor G. Cartledge, Department of Life Sciences, Trent Polytechnic, Nottingham, UK

S. Doel, Botany School, University of Oxford, South Parks Road, Oxford, UK

A. R. Goodey, Delta Biotechnology, Castle Court, Castle Boulevard, Nottingham, UK

R. M. Jones, Labatt Brewing Company Limited, Production Research Department, 150 Simcoe Street, London, Ontario, Canada N6A 4M3

David King,* Biochemistry Division, Department of Biochemistry, University of Surrey, Guildford, Surrey GU2 5XH, UK

Barbara Kirsop, Curator, National Collection of Yeast Cultures, Food Research Institute, Colney Lane, Norwich NR4 7UA, UK

Marc-André Lachance, Department of Plant Sciences, University of Western Ontario, London, Ontario, Canada N6A 5B7

Vidar F. Larsen,† Biotechnology Department, Massey University, Palmerston North, New Zealand

Argyrios Margaritis, Department of Civil and Biochemical Engineering, Faculty of Engineering Science, University of Western Ontario, London, Ontario, Canada N6A 5B9

Fahar J. A. Merchant, Department of Chemical and Biochemical Engineering, Faculty of Engineering Science, University of Western Ontario, London, Ontario, Canada N6A 5B9

Diane J. Mitchell, School of Biological Sciences, Queen Mary College (University of London), Mile End Road, London E1 4NS, UK

C. J. Panchal, Labatt Brewing Company Limited, Production Research Department, 150 Simcoe Street, London, Ontario, Canada N6A 4M3

* Current address: CellTech Limited, 250 Batt Road, Slough, Berks SL1 4DX, UK

† Current address: Department of Chemical and Process Engineering, University of Strathclyde, Montrose Street, Glasgow G1 1XW, UK

J. R. Piggott, G. D. Searle and Co. Ltd, P.O. Box 53, Lane End Road, High Wycombe, Buckinghamshire HP12 4HL, UK
Knut Rosen, Danish Distillers Limited, Postboks 1738, 2300 Copenhagen S, Denmark
I. Russell, Labatt Brewing Company Limited, Production Research Department, 150 Simcoe Street, London, Ontario, Canada N6A 4M3
G. G. Stewart, Labatt Brewing Company Limited, Production Research Department, 150 Simcoe Street, London, Ontario, Canada N6A 4M3
D. C. Watson, Chivas Bros Distillers Limited, Keith, UK
M. E. E. Watson, G. D. Searle and Co. Ltd, P.O. Box 53, Lane End Road, High Wycombe, Buckinghamshire HP12 4HL, UK
Michael A. Winkler, Department of Chemical and Process Engineering, University of Surrey, Guildford, Surrey GU2 5XH, UK
Alan Wiseman, Biochemistry Division, Department of Biochemistry, University of Surrey, Guildford, Surrey GU2 5XH, UK

PART I

# Yeast identification and maintenance

# 1 *Maintenance of yeast cultures*

BARBARA KIRSOP

## 1.1 Introduction

The value of a production strain to a biotechnology company may be incalculable. Not only does it represent a considerable past investment in terms of screening, selection and strain-improvement programmes, but it is the sole means by which future production can be achieved. It is therefore important that the way in which stock strains can best be maintained for future use receives early attention, and is not considered only when problems arise during production.

It is clear that production will cease if a viable culture cannot be supplied, but it is less often recognized that a viable culture exhibiting atypical properties may cause substantial industrial problems. Production procedures are designed around expected levels of microbial performance, and deviation from normal behaviour can have serious and expensive consequences. Thus, if a brewing yeast loses its ability to flocculate at a particular stage in fermentation, separation problems arise; if an antibiotic-producing micro-organism no longer gives the levels of product to be expected in a given timespan, production schedules are upset. Not only, therefore, must cultures of production strains give standard levels of viability that will allow the use of standard inoculation levels, but they must exhibit no behavioural instability.

It is only fairly recently that the effect of preservation procedures on strain stability has been recognized, abnormality in culture performance having previously been attributed to changes in raw materials, to plant or management failure, to contamination or to mutation. It is now known that considerable changes in microbial activity can result from inappropriate culture maintenance procedures, and that the factors leading to loss of viability and instability are complex. Traditionally the maintenance of yeasts has not been considered to be a problem. Yeast cultures may survive periods of neglect and appear alive when subcultured, and they have consequently gained a reputation for robustness. However, yeasts are now used increasingly for novel biotechnological activities, and are subjected to closer monitoring than hitherto, and it has become recognized that cells surviving in an apparently active culture may be small in number and no longer typical. Failure to recognize sensitivity to preservation stresses in the past reflects the fact that culture handling and production procedures in the long-established biotechnological industries of brewing, baking and distilling selected in favour of the wanted strain. Thus, bottom cropping in lager fermentations eliminated the accumulation of top-fermenting subpopulations; conversely, accumulation of sedimenting strains in a top-fermenting yeast is prevented when surface-collection is practised.

When compared with other micro-organisms, yeasts have certain characteristics that predispose them to greater cellular damage in the course of maintenance. Higher fungi, for example, often form resistant spores at some stage during their life cycle that can better withstand the stresses of preservation procedures than vegetative cells; bacteria, on the other hand, are smaller than yeast cells and escape much of the damage believed to be caused by ice-crystal formation during freezing protocols. By contrast, yeasts do not produce spores that are significantly more resistant to environmental extremes, and their vegetative cells are of a size that makes them more susceptible to ice damage during cryopreservation or the freezing stage of freeze-drying. Over the years a number of methods have been developed for the successful maintenance of yeasts. Often these have arisen from empirical studies, the underlying mechanisms of cellular damage having been little understood, and many have been developed in the laboratories of the service culture collections where there exist opportunities to assess the suitability of methods for a wide range of genera.

In addition to the establishment of suitable preservation procedures, the successful management of an industrial culture collection depends upon careful documentation, quality control and culture distribution procedures. The national culture collections have experience in all activities necessary for the provision of a microbiological support service and, accordingly, the last section of this chapter gives a survey of the major yeast collections and the services they provide, together with recommendations for setting up and running an industrial culture collection.

## 1.2 Yeast preservation

Methods used for the preservation of yeasts (Kirsop & Snell 1984) should confer maximum survival and stability, and also be appropriate to the requirements and facilities of the laboratory. The methods commonly used by yeast laboratories fall into the following four categories:

(1) subculture, or active transfer;
(2) drying, or desiccation;
(3) freeze-drying; and
(4) freezing, or cryopreservation.

Each of these will be described in turn, consideration being given to the principles of the method, general methodology, survival levels, strain stability, advantages and disadvantages.

### *1.2.1 Subculture*

Subculturing, also termed active transfer or serial transfer, has been practised since the beginning of microbiology as a means of preventing the death of cultures. Micro-organisms can exist in the vegetative state only for limited periods if conditions are unsuitable for growth, and it follows that in order to extend the lifespan of cultures it is necessary either to ensure the continuing suitability of conditions or to encourage the formation of more-resistant cell types.

If good growth conditions are to be maintained, continual replenishment of appropriate nutritive medium is required, and other environmental factors, such as temperature and oxygen availability, must be provided at optimum levels. In the case of yeasts this means subculture from spent to fresh medium every few days, and this is clearly impracticable for the maintenance of large collections. Encouragement of the formation of resistant cells is also unlikely to succeed since, although yeasts produce spores, these are little more resistant to extremes of environment than the vegetative cells (Put & DeJong 1980). It follows that for subculturing to succeed growth rate must be reduced to practical levels. This can be achieved either by reducing the temperature of incubation to that at which cell division ceases, or by restricting the access to oxygen.

## Methodology

### MEDIUM

Any medium suitable for the growth of yeasts can be used. For a wide range of species media such as yeast morphology broth (Difco 0711-01-8) or yeast nitrogen base plus glucose (Difco 0392-15-9) are appropriate. Broth medium is preferable to agar, since ascospore formation occurs more readily on agar slopes with some strains, and this is unacceptable in the interests of strain stability.

### INOCULATION, INCUBATION AND STORAGE

Cultures are inoculated from the tube containing the spent medium to a tube of fresh medium, using an inoculation loop or Pasteur pipette for the purpose. A two-tube method is generally adopted, whereby one tube (the first to be inoculated) is reserved solely for inoculation of future stock cultures and is only opened once, the second tube being used for all other laboratory requirements. Strict aseptic conditions should be observed during inoculation, and laminar-flow cabinets should be used if laboratory conditions warrant. Efforts should be made to relieve the monotony of the procedure and the resultant errors that occur when subculturing large numbers of cultures. If bottles to be inoculated are arranged randomly on the work bench, a deliberate effort must be made to select the correct bottle, and the concentration of the operator must be sustained.

Cultures are usually incubated at temperatures between 20 and 30°C. Optimum growth temperatures of yeast cultures vary to some extent with species, but also between strains. For example, the optimum growth temperatures of two ale strains of *Saccharomyces cerevisiae* (NCYC 240 and NCYC 1026) are from 18.7 to 30.0°C and from 28.7 to 33.7°C, respectively. Incubation is continued until the cells reach the stationary phase of growth; this is usually 72 h after inoculation, but may be longer for strains of slow-growing genera, such as *Brettanomyces*. No information is available on whether yeasts grown as shaken or standing cultures survive better, but in many laboratories it will not be practicable to provide shaking facilities for large numbers of strains simultaneously.

Grown cultures are placed in refrigerators or constant-temperature rooms at 1–4°C, and generally can be stored for about 6 months without further attention. At the end of the storage period fresh cultures are prepared, using the unopened bottle of the duplicate cultures as inoculum.

### SURVIVAL LEVELS AND STRAIN STABILITY

Cultures appear to survive well by this method of maintenance. If problems occur, then the period between transfer may be reduced until a more successful protocol is found. General satisfaction with the method has not been based on numerical studies of percentage survival of the original population, but rather on an awareness that 'cultures' survive. However, von Rehberg (1978) carried out experiments to establish the precise extent to which cells survived, and on agar slope cultures found that as few as 10% of inoculated cells remained viable after a 6-month storage period. Culture collections 'lose' few cultures, but are aware that the growth rate of stored strains is often slower than normal. This may be attributed either to total loss of viability of a proportion of the original population or to sublethal damage to a larger number of cells.

Low survival levels are acceptable if surviving cells remain typical of the original population. Studies have been carried out to determine whether this is true of strains in the National Collection of Yeast Cultures (NCYC) and an examination of the morphological, physiological (Table 1.1) and industrial characters of a wide range of subcultured species showed considerable variation. Morphological change was characterized by loss of formation of ascospores, pseudomycelium and other characteristic features; physiologically, cultures acquired or lost the ability to metabolize carbon or nitrogen sources. Thus, in a study of about 600 strains following 10–25 years of storage, 46% of ascosporogenous strains had lost the ability to form spores, 25% of strains had acquired the ability to metabolize maltose, 50% of strains deposited with specific amino acid requirements had lost their nutritional markers. In addition, it was found that 10% of 300 brewing strains had changed over a 10-year period with regard to their ability to flocculate, a characteristic of significant industrial importance.

**Table 1.1** Variation in subcultured strains.

| Type of variation | Number of strains tested | Number of strains showing variation |
|---|---|---|
| maltose fermentation (gain) | 127 | 32 |
| maltose fermentation (loss) | 223 | 0 |
| maltose assimilation (gain) | 127 | 37 |
| galactose fermentation (gain) | 113 | 14 |
| galactose fermentation (loss) | 224 | 3 |
| galactose assimilation (gain) | 60 | 7 |
| amino acid requirement (loss) | 22 | 13 |

ADVANTAGES AND DISADVANTAGES

Experiences similar to those of the NCYC have been reported frequently for other micro-organisms (Pridham & Hesseltine 1975, Stanley 1979), and it is now generally recognized that maintenance by subculturing can lead to unacceptable levels of strain degeneration and should not be used for long-term storage. However, for short-term purposes the method has a number of advantages. It is cheap, simple, requires no sophisticated equipment and can be performed by relatively unskilled staff. However, operators must be persuaded of the importance of such a monotonous activity, and steps must be taken to make the work as interesting as possible. It must be accepted that occasional errors will still occur and effective quality control measures must be established to enable mislabelling or contamination to be detected (see Section 1.3.2).

The method is particularly versatile, since any combination of medium and incubation temperature can be used to meet the needs of the laboratory. Positive selection pressures can be applied for the maintenance of unstable mutants, stringent cultural conditions can be used for the maintenance of fastidious species and appropriate precursors can be added to the medium to retain high levels of specific metabolite production. The use of agar slopes rather than broth cultures is acceptable to meet distribution needs, but should not be used for storage of ascosporogenous species. The shelf-life of agar slope cultures may be extended by addition of sterile mineral oil to the bottle, so that the culture is covered completely and access to oxygen is restricted. This method has been used successfully for higher fungi (Kirsop & Snell 1984, Smith & Onions 1983).

### *1.2.2 Drying*

Drying is the traditional way of preserving natural substances containing water, and the method that has been used for many kinds of food has also been applied to the maintenance of micro-organisms. During the process water molecules are removed from the surface of the cell, the cell shrinks and the concentration of solutes in the cytoplasm increases. These physical processes have variable effects on the cells, and may result in loss of viability.

There are two methods of drying yeasts that have been used successfully in a number of laboratories. In one the cells are directly in contact with silica gel as a desiccant; in the other the cells are first placed on filter paper, and drying over silica gel subsequently takes place.

#### **Silica gel**

PREPARATION

Silica gel is placed in screw-cap containers, such as bijou, McCartney or Universal bottles, to a depth of about 1 cm. The gels are sterilized in a hot-air oven or an autoclave (with subsequent drying in an oven) and stored with the caps firmly tightened. A 5% solution of skimmed milk is prepared and sterilized for use as a suspending medium.

Cultures are grown on a suitable agar medium to reach stationary phase. Yeast morphology agar (Difco 0711-02-8) slopes are suitable for the purpose.

INOCULATION, STORAGE AND REVIVAL

Inoculation is preceded by cooling the suspending medium, sterile gels, pipettes and trays in a refrigerator for 24 h. A considerable amount of heat is generated on rehydration of the gels, and to prevent damage to the cells this should be dissipated as rapidly as possible by the use of chilled material for inoculation.

Inoculation is carried out by washing the growth from the agar slopes with the sterile milk to make a thick suspension. Drops of the suspension are transferred aseptically to the gels, which are shaken well to distribute the cells through the crystals. The gels should not be saturated; three or four drops from a Pasteur pipette are sufficient to inoculate a 1-cm depth of silica gel in a bijou bottle.

Inoculated gels are left at room temperature to dry for about 2 weeks, after which time they are transferred to airtight containers for storage at 4°C. Indicator silica gel is placed in the bottom of the container and replenished as needed.

Cultures are recovered by sprinkling a few gels aseptically on to appropriate growth medium (agar or broth) and incubating in the usual way.

SURVIVAL LEVELS AND STRAIN STABILITY

Survival of yeasts stored on silica gel has been good in some laboratories (Woods 1976) and poor in others (Bassel *et al.* 1977), success having been

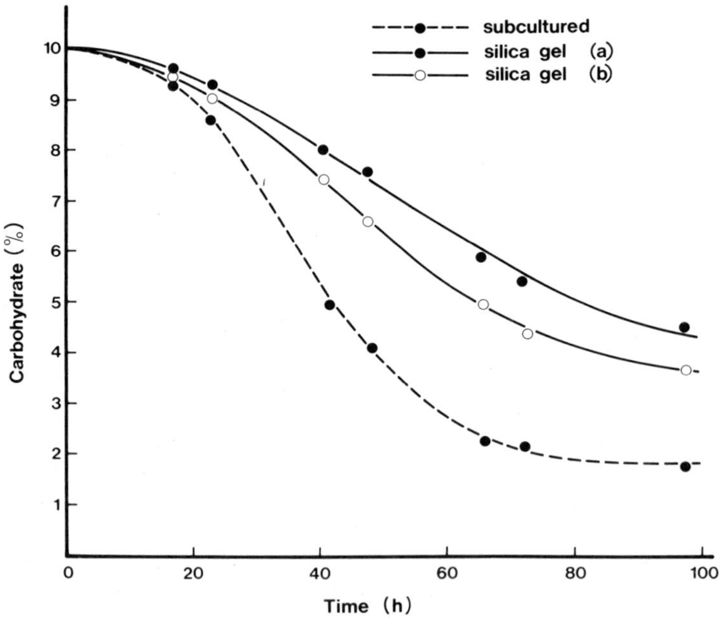


**Figure 1.1** Fermentation patterns of *Saccharomyces cerevisiae* NCYC 240, following drying on silica gel.

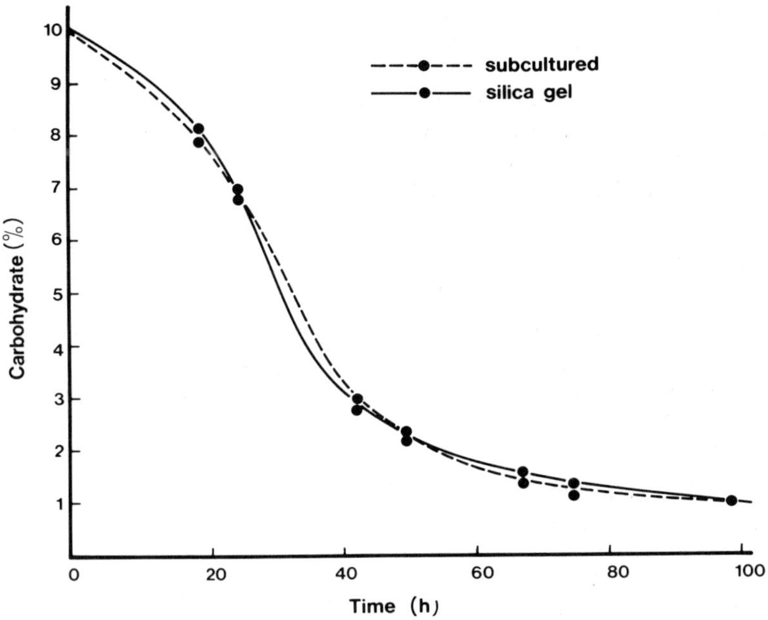


**Figure 1.2** Fermentation patterns of *Saccharomyces cerevisiae* NCYC 1245, following drying on silica gel.

reported more commonly for strains used in genetical studies. The NCYC explored the suitability of the method for representatives of 17 genera, and found that only 50% of strains could be recovered after 2 years of storage. Surviving strains varied considerably with regard to stability and, although small morphological changes were detected and carbon and nitrogen assimilation patterns remained stable, substantial changes took place in the fermentation behaviour of industrial strains. The damaging effects of desiccation were strain specific, NCYC 240 *Saccharomyces cerevisiae* showing large changes and NCYC 1245 *Saccharomyces cerevisiae* being unaffected (Figs 1.1 & 2).

ADVANTAGES AND DISADVANTAGES

It is clear from the above that the method has serious limitations with regard to survival and stability. However, it is equally clear that strains vary with regard to their sensitivity to desiccation, and if a limited number of cultures are to be maintained preliminary experiments can be carried out to establish the suitability of the method.

Requirements for equipment and skilled technical help are minimal, and the method is therefore cheap. It is particularly suitable for parts of the world where high ambient temperatures make freezing systems uneconomic, or where uncertain power supplies give rise to doubt in the use of systems depending upon refrigerators or freezers.

**Paper replica**

PREPARATION

Squares of filter paper are sterilized in tinfoil envelopes and stored, several in each envelope, for use. Proprietary tinned milk is used as the suspending medium. For inoculation, cultures are grown to stationary phase either on agar slopes or in Petri dishes using any general yeast-growth medium.

INOCULATION, STORAGE AND SURVIVAL

Heavy suspensions of cultures are prepared by mixing cells in drops of sterile milk placed in a sterile Petri dish. The filter-paper squares are immersed in the suspension aseptically and returned to the foil envelopes. The envelopes, folded once only, are placed in a desiccator to dry for 2 or 3 weeks. Once dry the envelopes are closed and placed in an airtight container at 4°C for storage.

Cultures are revived by placing a paper square into suitable medium and incubating.

SURVIVAL LEVELS AND STRAIN STABILITY

It is reported that about 99% of several hundred genetically marked strains of *Saccharomyces cerevisiae* have survived for periods of 3–6 years (Bassel *et al.* 1977). They are stable in respect of their genetic markers. Survival and stability of other genera and species is uncertain; poor results have been obtained with *Yarrowia lipolytica*.

ADVANTAGES AND DISADVANTAGES

The method has only been applied to a limited number of species, and generalizations cannot be made. For yeasts known to be resistant to desiccation the method has the same advantages as the silica-gel method. Additionally, the paper-replica method requires a very small amount of storage space and is particularly suitable for distribution purposes, since single squares can be wrapped individually in sterile foil and mailed in envelopes. Clearly, if strains falling into the category of 'infectious perishable biological substances' were to be distributed, then the appropriate postal and quarantine requirements would apply (see Section 1.3.3).

### *1.2.3 Freeze-drying*

Freeze-drying, also known as lyophilization, differs from the desiccation methods described above in that water is removed from frozen material by sublimation rather than from the liquid phase. Preliminary freezing may be carried out as a separate stage, the cells being frozen by immersion in an appropriate freezing mixture or in a mechanical freezer and then being transferred to a vacuum system for subsequent drying. More commonly freezing and drying occur as part of a continuous process using one of the commercial machines available for the purpose. Freeze-dryers may be of the centrifugal or shelf type, the latter being more appropriate for large-scale batch operations.

In the centrifugal system cells are mixed with a suitable suspending medium, and are inoculated into glass ampoules. These are placed in a circular holder in the vacuum chamber of the freeze-dryer. The ampoules are centrifuged during the initial stages of the process, both to provide a greater surface area for more efficient drying and to prevent frothing caused by the removal of dissolved gases. Freezing takes place by evaporation under vacuum, and subsequent drying is by sublimation, the water being removed from the atmosphere by the use of a desiccant or a refrigerated vapour trap. In the shelf system the cell suspension is placed in glass vials and is frozen by cooling the shelves of the freeze-dryer or by prefreezing in a mechanical freezer or freezing bath. Drying follows, and at the end of the process the vials are plugged automatically with rubber seals.

The centrifugal system is more appropriate for most purposes, since ampoules may be plugged with cotton wool to prevent both contamination and the formation of aerosols when the ampoules are opened. In addition, the sealed ampoules obtained by the centrifugal system have a longer shelf-life than the rubber-sealed vials of the shelf system, which are not completely impermeable to air and water. The shelf system has advantages if very large numbers of samples of a single strain are required, and it is appropriate for the preparation of vaccines and similar biological products.

There are a number of ways in which freeze-drying protocols may be adjusted to improve the final product, and the following parameters are known to affect survival levels:

- age of the inoculum
- the medium used for growth of the inoculum
- the oxygen availability during growth of the inoculum
- the size of the inoculum
- the suspending medium
- the length of primary and secondary drying
- the resuscitation procedures
- the intrinsic nature of the cells

It is clear that any part of the procedure that affects the composition of the cells may change their resistance or sensitivity to freeze-drying damage. The age of the inoculum is an important factor, and it has been found that with all species of yeast post-logarithmic cells survive better than younger cells. A number of species that show particularly low survival levels have become more resistant to damage following growth on low-nutrient medium before freeze-drying. Thus, the growth of *Brettanomyces* and other low-surviving strains is marginally improved (Table 1.2) following pre-growth on sodium acetate agar. In the interests of strain stability this procedure may be unsuitable for ascosporogenous strains, since the medium encourages ascospore formation and segregation during meiosis may lead to genetic change.

It has been shown that some strains may show different survival levels following pre-growth for a standard time with and without shaking (Table 1.3). Attempts to correlate the different lipid composition of high- and low-surviving strains with resistance or sensitivity to damage have so far been unsuccessful (Table 1.4). It would seem that changes in lipid composition cannot be isolated from possible changes in other cellular components, and further research is required to establish molecular relationships occurring during preservation procedures.

**Table 1.2** Survival of sensitive strains following growth on potassium acetate agar before freeze-drying.

| Genus | Number of strains | Mean viability (%) | |
|---|---|---|---|
| | | YM broth | Potassium acetate agar |
| *Saccharomyces* (low-surviving strains) | 8 | 0.01 | 0.8 |
| *Brettanomyces* | 8 | 0.9 | 4.8 |
| *Candida* (low-surviving strains) | 2 | 1.7 | 27.2 |

**Table 1.3** Survival of strains of *Saccharomyces cerevisiae* following growth with and without shaking before freeze-drying.

| Cultural treatment | Viability (%) | | | |
|---|---|---|---|---|
| | NCYC 1236 | NCYC 1245 | NCYC 0229 | NCYC 1062 |
| standing culture | 0.2 | 0.2 | 21.9 | 17.0 |
| shaken culture | 0.3 | 1.3 | 0.2 | 2.0 |

**Table 1.4** Fatty acid composition of high- and low-surviving strains of *Saccharomyces cerevisiae* (% total fatty acid).

| Fatty acid | High-surviving strains | | Low-surviving strains | |
|---|---|---|---|---|
| | NCYC 1062 | NCYC 1331 | NCYC 1236 | NCYC 739 |
| $C_8$–$C_{12}$ | 5 | 9 | 8 | 6 |
| $C_{14}$ | 1 | 2 | 3 | 0.7 |
| $C_{14:1}$ | 0.5 | 1 | 1 | 0.3 |
| $C_{16}$ | 23 | 28 | 26 | 15 |
| $C_{16:1}$ | 26.5 | 24 | 30 | 20 |
| $C_{18}$ | 7 | 6 | 7 | 5 |
| $C_{18:1}$ | 37 | 30 | 25 | 53 |

Inoculum size has been found to affect survival levels arithmetically, and if large numbers of viable cells are required for use concentrated suspensions should be prepared for inoculation into ampoules.

Much of the development work on freeze-drying has been concerned with determining the suspending medium conferring greatest protection on cells, and a number of substances are now considered appropriate (glucose–serum, inositol–serum, inositol broth, skimmed milk and sucrose).

Although it is possible to vary the drying time, reliable data are not available on the levels of residual moisture that give maximum survival and stability. It is generally considered that a moisture level of about 1% is desirable, but there is no published information on the long-term survival of different yeast species at different residual moisture levels.

It could be expected that stresses imposed during the rehydration of dried material would be similar in magnitude to those existing during freeze-drying, and that changes in resuscitation procedures would substantially affect survival levels. In practice it is difficult to separate the effects of freezing and drying from those of rehydration, but attempts to vary the composition of rehydration medium, and the temperature and time of rehydration, appear to have a limited effect on survival. Nevertheless, it is of interest that survival following rehydration in half-strength medium has sometimes led to marginally increased viability figures, suggesting that dilute medium may allow a more orderly reorganization of sublethally damaged cell components.

It is therefore clear that there are a number of changes that can be made to the freeze-drying protocol that may affect the stability of the product. It is nevertheless the experience of the NCYC that the overriding factor affecting survival of yeasts is the intrinsic, and as yet undefined, ability of individual strains to resist the imposed stresses of the process. Figure 1.3 shows survival

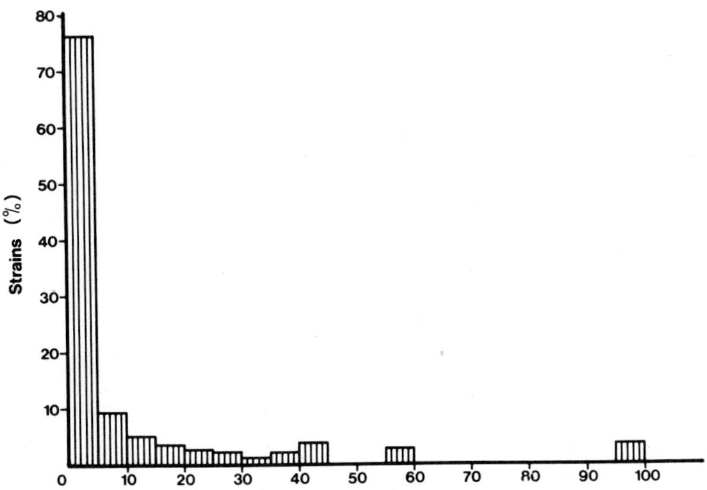


**Figure 1.3** Survival levels of different strains of *Saccharomyces cerevisiae* following freeze-drying.

levels of different strains of *Saccharomyces cerevisiae*. It has been shown that these levels of survival are consistent for individual strains and that, although survival can be affected marginally by fine tuning of the procedures, poor-surviving strains continue to survive poorly and resistant strains remain resistant.

For the preservation of a large number of strains from a wide range of species, it is clear that a standard method must be adopted that gives maximum viability and stability for most of the strains to be preserved. The method that is described below has been found to be suitable for most yeasts in the NCYC.

## Methodology

### PREPARATION

Glass ampoules are washed, dried, identified with filter-paper labels that have been numbered with non-toxic ink, plugged loosely with non-absorbent cotton wool, sterilized by autoclaving and dried in a hot-air oven.

### INOCULATION, STORAGE AND RECOVERY

The inoculum is prepared by mixing equal proportions of a 72-h YM broth (Difco 0711-01-8) suspension, adjusted if necessary to give a cell concentration of at least $10^6$ $ml^{-1}$, and sterile 7.5% glucose–horse serum. The yeast is grown as a standing culture without aeration, since for most strains these growth conditions give higher survival levels.

Ampoules are filled aseptically with 0.2 ml of the mixture, care being taken to deliver an accurate amount if viable counts are to be made. The filled ampoules are placed in the ampoule holder in such a way that on centrifugation the suspension will not obscure the label; the holder is then placed in the primary drying chamber of a centrifugal freeze-dryer. If the freeze-dryer uses a chemical desiccant, then this is placed in the desiccant trays just before the start of the run. When the ampoule holder is in place the chamber lid is replaced, and the centrifuge and vacuum pump are switched on. Primary drying continues for about 3 h, when the sample should appear completely dry.

At this stage the ampoules are removed and the cotton-wool plugs pushed down to just above the dried yeast wedges. The ampoules are then constricted in a gas–air flame and replaced on the manifold of the machine for secondary drying. The desiccant, if used, is replenished before the start of the secondary drying stage, which is continued overnight. A 3–4-h secondary drying period has been found equally suitable for many species.

The ampoules are sealed under vacuum and, if found to be satisfactory, stored in a cool (4°C), dark container. Cultures are rehydrated by opening the ampoules aseptically, resuspending the dried material in an appropriate growth medium and incubating at about 25°C for about 3 days. The growth rate is generally slower than normal following freeze-drying, and additional time should be allowed before discarding cultures as non-viable.

### SURVIVAL LEVELS AND STRAIN STABILITY

Tables 1.5 and 1.6 show average survival levels for ascosporogenous and anascosporogenous genera. These figures are low, and could suggest the selection of resistant biotypes from within the original population. However,

**Table 1.5** Survival of ascosporogenous yeasts following freeze-drying.

| Genus | Mean survival (%) | Standard deviation | S.E.M. | Number of strains |
|---|---|---|---|---|
| *Debaryomyces* | 15.04 | 18.2 | 5.5 | 12 |
| *Dekkera* | 0.35 | 0.13 | 0.10 | 2 |
| *Hanseniaspora* | 7.5 | 7.64 | 5.4 | 2 |
| *Hansenula* | 5.1 | 6.99 | 1.2 | 34 |
| *Kluyveromyces* | 9.12 | 19.29 | 3.78 | 26 |
| *Lipomyces* | 5.29 | 5.46 | 3.15 | 3 |
| *Metschnikowia* | 11.19 | 8.64 | 2.88 | 9 |
| *Pichia* | 9.10 | 15.54 | 2.63 | 35 |
| *Saccharomyces* | 5.16 | 10.80 | 0.45 | 580 |
| *Saccharomycodes* | 0.01 | 0.006 | 0.003 | 5 |
| *Saccharomycopsis* | 11.36 | 10.4 | 3.83 | 7 |
| *Schizosaccharomyces* | 7.04 | 10.55 | 3.99 | 7 |

repeated freeze-drying of the same strain failed to raise the level of survival, which would be expected if selection had occurred.

In view of the low survival levels, extensive tests have been carried out to establish whether surviving cells accurately represent the original population. Examination of a large number of strains before and after freeze-drying failed to show any change in the morphological, physiological or industrial characteristics. However, an increased level of respiratory-deficient mutants following freeze-drying of strains of *Saccharomyces cerevisiae* suggests that some cellular damage occurs during preservation.

The long-term survival levels of freeze-dried yeasts are generally very satisfactory, loss of viability being of the order of 1% per year. Nevertheless, batches of ampoules of individual strains can suddenly lose viability, and it is expedient to monitor survival levels on a routine basis if the loss of cultures is to be avoided.

**Table 1.6** Survival of anascosporogenous yeasts following freeze-drying.

| Genus | Mean survival (%) | Standard deviation | S.E.M. | Number of strains |
|---|---|---|---|---|
| *Brettanomyces* | 2.21 | 2.53 | 0.89 | 8 |
| *Bullera* | 3.7 | 3.05 | 1.76 | 3 |
| *Candida* | 12.95 | 19.1 | 2.45 | 61 |
| *Cryptococcus* | 21.48 | 27.49 | 8.69 | 10 |
| *Kloeckera* | 25.44 | 24.74 | 8.75 | 8 |
| *Rhodosporidium* | 4.40 | 2.47 | 1.75 | 2 |
| *Rhodotorula* | 11.75 | 22.05 | 4.50 | 24 |
| *Sporobolomyces* | 1.4 | 1.37 | 0.54 | 6 |
| *Torulopsis (Candida)* | 27.94 | 36.00 | 6.27 | 34 |
| *Trichosporon* | 8.27 | 11.56 | 4.72 | 6 |

ADVANTAGES AND DISADVANTAGES

The chief advantages of the method are the longevity of freeze-dried cultures and the undemanding storage requirements. Additionally, the method is particularly suitable for laboratories concerned with the distribution of yeast cultures, such as the service culture collections or the central laboratories of large industrial companies that supply production strains to subsidiary laboratories on a routine basis. If strain stability is of major importance, the method offers considerable advantages over subculturing and drying.

The disadvantages include the initial high capital cost of the equipment, the cost of training technicians to carry out the fairly skilled and labour-intensive process, and the poor back-up in developing countries for the replacement of spare parts of commercial equipment. In addition, there are industrial disadvantages in dependence upon a product with initially diminished activity, although recognition of this disadvantage can allow the establishment of appropriate schedules to compensate for slower growth rates.

### *1.2.4 Freezing*

The use of freezing for preserving cells is also known as cryopreservation, and was first applied in veterinary science for the storage of bull semen for the artificial insemination of cows. The birth of many generations of healthy calves bears witness to the success of the system, and it has since been used for storage of other cell types. Its use for yeast preservation has been examined by a number of workers, some of whom used a few strains of a single species (Wellman & Stewart 1973), others using a wider range of genera and species (Hubalek & Kockova-Kratochvilova 1978, Kirsop & Henry 1984).

Cells may be frozen by the use of an appropriate freezing mixture, a cooling bath or a mechanical refrigerator. As with freeze-drying, there are a number of factors that affect the survival of cells; these factors are often interrelated, and in practice it is difficult to dissociate their effects. Nevertheless, it is generally accepted that the following factors contribute to the success of the method:

- the age of the cells
- the cultural conditions (temperature, medium and oxygen level for pre-growth of the inoculum)
- the cryoprotectant in which the cells are cooled
- the cooling rate
- the storage temperature
- the warming rate and temperature
- the intrinsic properties of the cells

In general, the same pre-growth conditions give the highest survival levels, both in cryopreservation and in freeze-drying, so that stationary-phase cells and, to a lesser extent, growth on nutrient-deficient medium appear to be beneficial in both systems. Pre-growth at reduced temperatures is reported to increase survival levels of some strains (Morris 1981); conversely, cooling the inoculum before cryopreservation has been found to lower the survival levels of some strains. It has been found that for test strains examined by the NCYC pre-growth with unlimited access to oxygen (shaken cultures) leads to better survival than when standing cultures are used; this is in contrast with the effects of oxygen on the viability of freeze-dried cultures. Although all strains do not behave similarly in this respect and the observations are generalizations, NCYC protocols for cryopreservation and freeze-drying are based on these observations.

A number of cryoprotectants have been found appropriate for use with yeast cells, and those commonly used are glycerol (5, 10, 15 or 20%), dimethylsulphoxide (DMSO) or a combination of glycerol and DMSO. The effectiveness of different cryoprotectants depends upon their molar concentration and their ability to penetrate the cells. The cooling rate is probably the factor of greatest importance in the success of any cryopreservation programme, optimum cooling rates for different cell types varying from 1 to 1000°C $min^{-1}$. Yeast cells generally survive best when cooling rates of about 10°C $min^{-1}$ are used and, if a controlled cooling rate bath is available, cooling at the optimum rate can be achieved. However, few laboratories have this equipment, and an alternative method must be used. Fortunately, the disadvantages of uncontrolled, sub-optimal cooling rates can be overcome by adopting a two-step system by which cells are cooled to a temperature of between −20 and −30°C, are held at this temperature to allow the cells to dehydrate and are subsequently cooled to the temperature of storage, again at an uncontrolled cooling rate. It may be necessary to establish the optimum temperature and time for the holding operation, but the NCYC has found −30°C for 2 h to be a suitable regime for a wide range of yeasts. The success of this interrupted cooling system depends upon the formation during the initial cooling step of extracellular ice, which increases the osmolality of the suspending medium. This has the effect of removing water from the cells, causing them to shrink. After a holding period to allow an optimum level of shrinkage to take place, the cells may be further cooled to the storage temperature, eliminating further injury from intracellular ice formation that would occur in fully hydrated cells. This method is very simple to carry out in practice, and it has been applied to a wide range of cell types.

Although some workers store cells at temperatures above that of liquid nitrogen (−196°C), it is generally accepted that molecular activity continues to take place at temperatures above −130°C and that for long-term storage with maximum genetic stability temperatures below this level should be used. Many microbiologists store cultures in freezers at −20, −70 or −90°C and, while this is acceptable for short storage periods or when no alternative exists, for long-term preservation of important cultures a temperature below −130°C should be used. In contrast with the variable response of different cell types to different cooling rates, almost all cells show superior recovery rates following rapid thawing. Strict safety precautions should be observed in all methods in which liquid nitrogen is used. Protective clothing should be worn by operators, refrigerators should be kept in a well-ventilated room and glass ampoules, which can explode on thawing if liquid nitrogen has seeped into them, should not be used. The following miniaturized method (Kirsop & Snell 1984), using polypropylene straws, has been developed by the NCYC and used successfully for a wide range of yeast species.

PREPARATION

Coloured polypropylene drinking straws are cut into 2.5-cm lengths and sealed at one end by holding with tweezers and placing close to the flame of a Bunsen burner to allow the plastic to melt and form a seal. These small cryotubes are placed in a Petri dish and sterilized by autoclaving. A 10% glycerol solution is prepared and sterilized by filtration.

INOCULATION, STORAGE AND RECOVERY

The inoculum is prepared by mixing equal amounts of a 72-h YM broth (Difco 0711-01-8) culture, containing between $10^6$ and $10^7$ cells $ml^{-1}$, and the sterile glycerol solution. The final concentration of the cryoprotectant is thus 5%. Using a Pasteur pipette the straws, which may be colour coded, are filled aseptically with about 0.1 ml inoculum. The filled straws are sealed, again by holding close to a Bunsen flame, and placed in a 2-ml plastic ampoule. Five or six straws can be contained in each ampoule.

The ampoules are placed in a refrigerator or cooling bath at −30°C and held at this temperature for 2 h to allow dehydration of the cells to occur. At the end of this period the ampoules are removed and immersed directly into the liquid nitrogen at −196°C. Ampoules may be stored either on metal canes, several of which are placed in a metal cylinder, or in drawers using one of the inventory systems available commercially. The latter system allows easier retrieval of ampoules.

Cultures are recovered by removing a straw from the ampoule and thawing the contents rapidly in a water bath at 35°C. The straws are opened aseptically using sterile scissors, and the contents are removed with a Pasteur pipette and transferred to an appropriate growth medium.

SURVIVAL LEVELS AND STRAIN STABILITY

All reports of recovery rates following cryopreservation are greatly superior to those obtained by freeze-drying. The average percentage survival of NCYC

**Table 1.7** Survival of ascosporogenous yeasts in liquid nitrogen.

| Genus | Number of strains | Mean survival (%) | Standard deviation |
|---|---|---|---|
| *Debaryomyces* | 3 | 54.8 | 10.8 |
| *Guilliermondella* | 1 | 40.9 | — |
| *Hanseniaspora* | 1 | 65.0 | — |
| *Hansenula* | 7 | 51.4 | 35.3 |
| *Hyphopichia* | 1 | 123.0 | — |
| *Kluyveromyces* | 6 | 62.4 | 19.7 |
| *Lipomyces* | 7 | 68.7 | 34.6 |
| *Pichia* | 10 | 50.6 | 35.8 |
| *Saccharomyces* | 75 | 66.1 | 34.8 |
| *Schizosaccharomyces* | 2 | 103.0 | 71.0 |
| *Williopsis* | 1 | 26.6 | — |
| *Yarrowia* | 1 | 61.5 | — |
| *Zygosaccharomyces* | 2 | 59.4 | 37.2 |

**Table 1.8** Survival of anascosporogenous yeasts in liquid nitrogen.

| Genus | Number of strains | Mean survival (%) | Standard deviation |
|---|---|---|---|
| *Brettanomyces* | 4 | 29.2 | 17.5 |
| *Candida* | 25 | 70.4 | 31.7 |
| *Cryptococcus* | 3 | 85.4 | 17.9 |
| *Kloeckera* | 2 | 34.2 | 33.1 |
| *Rhodotorula* | 1 | 90.3 | — |
| *Sporobolomyces* | 5 | 62.0 | 30.5 |
| *Trichosporon* | 2 | 85.6 | 16.3 |
| Variation between 6 straws | | | |
| NCYC 1062 | | 71.8 | 11.7 |
| NCYC 1236 | | 78.7 | 23.8 |

strains preserved by the method described above are given in Tables 1.7 & 1.8. It can be seen that, even without establishing the optimum protocols for individual strains that would undoubtedly lead to higher survival levels, recovery rates are very satisfactory. Figures in excess of 100% recovery are sometimes obtained with clumping or chain-forming strains, since cooling to temperatures of −196°C may lead to cell separation.

Few reports of genetic instability in cryopreserved yeasts have so far been reported. Phenotypic and industrial characters have been shown to be unchanged following preservation (Wellman & Stewart 1973, Hubalek & Kockova-Kratochvilova 1978), and it has been established that yeast plasmids in strains of *Saccharomyces cerevisiae* and *Zygosaccharomyces bailii* are not lost (Kirsop & Henry 1984). Additionally, petite mutants, commonly recovered following freeze-drying, have been detected rarely in cultures of *Saccharomyces cerevisiae* recovered from liquid nitrogen (Wellman & Stewart 1973).

### ADVANTAGES AND DISADVANTAGES

There are major advantages to cryopreservation in terms of survival levels and genetic stability. Currently there is no method of yeast preservation that will yield higher viability figures or superior stability. However, substantial disadvantages exist in the practical execution of the method. As has been indicated above, the use of liquid nitrogen demands great care if accidents are to be avoided. Not only must steps be taken to avoid splashes and the explosion of glass containers on thawing, but proper ventilation must be assured because of the continuous evaporation of the liquid nitrogen. The evaporation necessitates regular replenishment of the refrigerators, and this can be an inconvenience during holiday periods. Additionally, the delivery of liquid nitrogen by the suppliers has been known to fail because of industrial action. This difficulty can be mitigated if storage tanks are large enough to cover long periods of use or if, as is the case with the National Collections, the laboratory can be put on a list for priority delivery. This disadvantage can be removed entirely if the possibility of making liquid nitrogen exists on site.

Contamination caused by the occasional leakage of liquid nitrogen into plastic ampoules is avoided by the use of sealed straws placed inside the ampoules, giving double security. Alternatively, plastic tubing may be placed over the screw caps of ampoules and heat-sealed to provide an impervious protective layer. The use of sealed glass ampoules or storage in the vapour phase would overcome the difficulty of leakage, but neither solution is to be recommended. The initial capital outlay for equipment is substantial, and in warm countries the cost of replenishment of the liquid nitrogen may be excessive.

The method is not convenient for distribution of cultures, since either recovery and regrowth must take place first or an expensive refrigerated distribution system must be employed.

FREEZING AT TEMPERATURES ABOVE −196°C

It has already been mentioned that long-term storage of important strains at temperatures above that of liquid nitrogen is ill-advised. However, for shorter storage periods, or for strains of ephemeral interest, storage in refrigerators at temperatures between −70 and −130°C is acceptable. The growth requirements, cryoprotectants and freezing protocols described for liquid nitrogen storage apply to storage at higher temperatures, and a number of suitable refrigerators are commercially available. Of particular interest is the recent introduction of a mechanical refrigerator operating at −130°C (White & Wharton 1985), a temperature at which molecular activity is virtually non-existent, and which therefore incorporates the major advantage of liquid-nitrogen storage. Clearly, the use of such a refrigerator eliminates many of the disadvantages inherent in dependence on the use of liquid nitrogen, but incorporates other disadvantages of high capital investment and dependence on a cheap and reliable power supply.

### *1.2.5 Summary of methods*

The relative advantages and disadvantages of the methods described are listed in Table 1.9. The choice of method will depend greatly on the circumstances of the laboratory, but the following criteria should be considered before deciding which method to adopt:

- the number of strains to be preserved
- the existing equipment
- the existing expertise
- the storage space
- the capital cost of purchasing new equipment
- the cost of training staff
- the required shelf-life
- the expected level of culture distribution
- the frequency of use of cultures
- the ambient temperature
- the reliability of services (electricity, liquid-nitrogen delivery, etc.)

**Table 1.9** Comparison of preservation methods. Rating (1, poor–5, good).

| Method | Survival level (% population) | Shelf-life (years) | Stability | Ease of revival | Expertise required | Space | Supply of cultures | Running costs | Capital outlay |
|---|---|---|---|---|---|---|---|---|---|
| SUBCULTURING | | | | | | | | | |
| broth | ? | 0.5 | 1 | 5 | 4 | 2 | 3 | 3 | 4 |
| agar slants | <10 | 1 | 1 | 5 | 4 | 2 | 3 | 3 | 4 |
| agar slants with oil overlay | ? | 2 | ? | 4 | 3 | 2 | 3 | 3 | 4 |
| DRYING | | | | | | | | | |
| paper replica (genetic strains of *Saccharomyces cerevisiae* only) | ? | 3–6 | 4 | 4 | 3 | 5 | 5 | 4 | 5 |
| silica gel | ? | 0–5 | variable | 5 | 3 | 2 | 3 | 4 | |
| FREEZE-DRYING | 0.01–30 | 5–30 | 4 | 2 | 1 | 4 | 5 | 4 | 2 |
| CRYOPRESERVATION | | | | | | | | | |
| ampoules | 20–100 | >5 | 4 | 3 | 2 | 3 | 3 | 2 | 2 |
| straws | 20–100 | >5 | 4 | 3 | 2 | 5 | 3 | 2 | 2 |

The final choice will ultimately depend upon what level of investment is justified in terms of past effort and the future returns that can be expected from the preserved strains. In view of the strain-specificity of sensitivity to environmental stresses, it is important that individual optimum protocols are established for valuable production strains. Additionally, reliance should not be placed on a single method.

At the NCYC, where each strain is considered to be of equal importance, but where numbers of strains preclude individual treatment, seed stocks are maintained in liquid nitrogen and freeze-dried cultures are prepared from these as back-up material and for distribution. Additionally a 'conservation collection' is maintained, whereby a sample of each strain is cryopreserved for the future at the time of deposit. Additionally, samples of each NCYC strain are kept in a separate location to guard against accidents at the collection laboratory. It is clearly wise to take similar precautions with strains of particular importance in biotechnology.

## 1.3 Collection management

In addition to selection of culture preservation methods, there are a number of other activities associated with running a culture collection that need consideration.

### *1.3.1 Documentation*

If a collection of yeasts has been established with the specific purpose of providing cultures on a regular basis, a great deal of reliance will be placed on this service. It is therefore important that proper records be kept, and the following documentation should be considered.

(a) Strain data

- accession number
- origin and history
- macromorphology (cultural appearance in broth or agar)
- microscopic appearance
- physiological properties
- industrial properties (fermentation characteristics, flocculence and product levels)

(b) Preservation requirements

- method (subculturing, drying, freeze-drying or cryopreservation)
- growth conditions for inoculum
- cryoprotectant or suspending medium
- number of samples
- intervals between preservation

(c) Preservation records

- date of preservation

- run data (date, inoculum, vacuum, time)
- viability figures, percentage survival
- number of ampoules or straws

(d) Quality control

- tests (morphological, physiological or industrial)
- test results

(e) Distribution records

- strain number
- date
- feedback (performance and future demand)

Records may be stored in notebooks or files, on punch cards or in computers (Painting *et al.* 1984). Although computers are particularly well-suited for storing records that need updating or searching, care should be taken not to use sophisticated methods where simple ones are better. The storage of strain data on a computer is appropriate, since it enables strains with particular combinations of characters to be selected. However, it is more convenient to keep details of freeze-drying runs in a notebook which can be kept near the freeze-dryer for ready reference.

A number of makes of microcomputer are suitable for data storage and retrieval. Specialist advice should be sought to select the best equipment for the needs of the laboratory and, in particular, capacity requirements should be estimated. If minicomputers are accessible to the culture collection, then storage capacity is not likely to be a problem and a large number of data base management programs can be used. The NCYC has used an Apple II microcomputer and Commercial software packages for customer records, stock control, invoicing and administration analysis. Eicon 2 8-inch hard disks were used to provide sufficient capacity for these records. Other systems have been used (Lawrence 1983) for handling smaller amounts of data.

For the storage and retrieval of strain data on each of 2000 strains in the NCYC, a VAX-VMS minicomputer and DATATRIEVE is used. This system enables 70–100 characters on each strain to be stored and searched, so that strains with specific combinations of characters can be found rapidly. Additionally, printouts of all known data on individual strains can be supplied to users on request.

In addition to data storage, computers can be used for a number of activities in the collection, such as accession procedures, preparation of catalogues or lists of strains and preservation protocols.

### *1.3.2 Quality control of culture purity, viability and stability*

Although laboratories may have the highest microbiological standards, mistakes will still occur and cultures will be mislabelled or become contaminated. It is essential that errors of this kind are detected and that steps are taken to remedy the situation before serious damage occurs at production level. The only way to detect errors is to monitor strain purity, viability and stability on a routine basis. Purity is best checked by spreading suitably diluted samples of each culture on to appropriate growth medium. Streak plates or spread plates are prepared, incubated until colonies are large enough to show distinctive characteristics, and examined. Errors will not be detected by this method unless proper records of the macromorphology of each strain are available for comparison. Photographs of colonies grown under standard conditions are useful, although it should be recognized that many different yeast strains have a similar colony appearance.

If impurity or contamination is suspected, then other stock cultures of the same strain should be examined to establish whether the whole batch of cultures is similarly affected. To guard against the total loss of strains, batches of new stocks should be prepared from known pure stocks well before the latter have run out and, in practice, stock holdings should be established at a level that ensures that difficulties of this kind do not arise. A computer may be used to keep records of stock levels and a program run routinely to find which strains need replenishing.

Quality control procedures must be established to monitor the stability of characteristics of cultures at predetermined intervals. If a preservation method is used that is known to limit or prevent genetic change, then check tests need only be carried out at infrequent intervals. If, conversely, a method is used that is known to result in strain drift, checks must be carried out more frequently or selection pressures applied to limit strain degeneration. It is commonly found necessary to carry out lengthy and expensive reselection procedures on production strains maintained by inferior preservation methods, and it is clearly to the advantage of industrial laboratories to use methods that confer maximum stability.

The NCYC is concerned with the long-term stability of the entire yeast genome, and so must use optimum preservation methods. Even so, it is felt necessary to check selected properties as well as viability levels and purity routinely. Diagnostic tests may be carried out on each strain or alternatively 'fingerprinting' methods, such as pyrolysis, gas–liquid chromatography, polyacrylamide gel-electrophoresis of whole cell proteins or DNA analysis following restriction digest, may be considered. In order to monitor the stability of genetically marked, genetically manipulated or plasmid-bearing strains, specialist expertise in genetical techniques is required.

### *1.3.3 Culture distribution*

If microbial cultures are to be distributed to other laboratories it is important that postal and quarantine regulations are followed. Although most species of yeasts that are important in biotechnology are considered non-pathogenic, there are strict regulations relating to the air-freight of infectious perishable biological substances (IPBS), particulars of which may be obtained from national postal headquarters or the World Federation for Culture Collections' Committee for Postal and Quarantine Regulations (Anon. 1980) (Table 1.10). Distribution of IPBS by air requires the use of packing materials and labels acceptable to the postal authorities and the International Air Transport Authorities, and despatch by registered mail. The responsibility of deciding

**Table 1.10** Organizations able to advise on cultures and culture collection activities.

*(1) Data Centres*
World Data Center for Collections of Microorganisms (WDC)
Director: Professor K. Komagata
World Data Center,
RIKEN,
2-1 Hirosawa, Wako,
Saitama 351-01,
Japan

Microbial Strain Data Network (MSDN)
Information Officer, MSDN: Ms E. Ross
12301 Parklawn Drive,
Rockville, MD 20852,
USA
(After September, 1987:
c/o Biotechnology Centre,
Cambridge University,
Cambridge, UK)

*(2) World Federation for Culture Collections*
Contact Address:
International Council of Scientific Unions,
51 Boulevard de Montmorency,
75016 Paris,
France
Tel.: (33.1) 45.25.03.29
Telex: ICSU 630553F
or
Data Centres (see 1 above)

*(3) European Culture Collections' Organisation (ECCO)*
Chairman: Dr M. Kocur
Czechoslovak Collection of Microorganisms,
J. E. Purkyne University,
Trida Obrancu Miru 10,
66243, Brno,
Czechoslovakia

Secretary: Mme Y. Cerisier
Collection Nationale de Cultures de Microorganismes,
Institut Pasteur,
25 rue du Docteur Roux,
75724 Paris Cedex 15,
France

*(4) National Federations*
Contact Addresses: National microbiological societies or data centres (see 1 above)
Australian Federation for Culture Collections (AFCC)
Chinese Central Committee for Culture Collections (CCCCC)
Japanese Federation for Culture Collections (JFCC)
UK Federation for Culture Collections (UKFCC)
US Federation for Culture Collections (USFCC)

*Continued*

**Table 1.10** *Continued.*

*(5) Microbial Resource Centres (MIRCENs)*
Contact Address: Dr E. J. DaSilva
Division of Scientific Research & Higher Education,
UNESCO,
7, Place de Fontenoy,
75700 Paris,
France

*(6) World Intellectual Property Organisation (WIPO)*
[For information on the Budapest Treaty on the International Recognition of the Deposit of Microorganisms for the Purposes of Patent Procedure]
34 Chemin des Colombettes,
Place des Nations,
1211 Geneva 2D,
Switzerland

which micro-organisms are IPBS rests with the distributor, and no accepted lists have been compiled. The National Collection of Yeast Cultures must err on the side of caution in this respect, and treats all yeast species other than baking, brewing, distilling and wine-making strains as IPBS, since it is not possible to be completely certain that all other species are non-infectious. It is no inconvenience to treat all yeasts as IPBS for postal purposes and to use accepted packaging and labels routinely. Unfortunately, a few national postal authorities have yet to reach agreement with IATA on postal requirements, and IPBS cultures can only be sent to these countries using private air-freight arrangements. This is an expensive way to distribute cultures, and it is hoped that pressure can be brought to bear on the few remaining postal authorities to accept the regulations already agreed by the great majority of countries.

Distribution of cultures by surface mail requires less-demanding postal regulations, and the supply of cultures to associate laboratories within a company is not usually a problem. Difficulties are more likely to arise with the distribution of bulk supplies of cultures where propagation facilities do not exist in subsidiary laboratories. It is usually not possible to use postal services for bulk distribution, because cultures must either be transported very rapidly or in a refrigerated condition, and it is generally found more convenient to hire private transport.

## 1.4 Service collections of yeasts

Biotechnology laboratories carry out a number of activities that can receive support from the service culture collections. As well as acting as resource centres for cultures and information, the collections can provide identification and preservation, safe-deposit and patent-deposit services, as well as functioning as training centres (Table 1.11). With the growth of biotechnology the demand for cultures, advice and other services has increased, and this section outlines ways in which the culture collections may provide help for industrial laboratories.

**Table 1.11** Collections* supplying yeasts and other services†.

| Acronym (for addresses see Table 1.12) | Cultures | Catalogue | Information | Identification | International Depository Authority (Budapest Treaty) | Preservation | Safe-deposit |
|---|---|---|---|---|---|---|---|
| ATCC | + | + | + | + | + | + | + |
| CBS (Yeast Division) | + | + | + | + | + | – | – |
| DSM | + | + | + | + | + | + | + |
| IFO | + | + | + | | + | | |
| NCPF (pathogens only) | + | | + | + | – | – | – |
| NCYC¶ | + | + | +‡ | + | + | + | + |
| NRRL | + | – | +‡ | +§ | + | – | – |
| YGSC | + (genetic strains of *Saccharomyces cerevisiae* only) | + | | – | – | – | – |

* For other collections see World Directory of Collections of Cultures of Microorganisms (McGowan & Skerman 1982).
† For charges, contact curators.
‡ Computer searches on strain data and printouts of properties of individual strains.
§ Some limitations, for details contact curator.
¶ On-line computer services for ordering, catalogue searches, electronic mail, probabilistic yeast identification system, bulletin board. For details contact the NCYC.

### *1.4.1 Culture and information centres*

The potential value of the 500 000 micro-organisms listed in the *World directory of collections of cultures of microorganisms* (McGowan & Skerman 1982) is incalculable, since the role microbes will play in the future cannot be judged. It is therefore important that these resource centres continue to be developed to serve the future requirements of microbiology. To this end, culture centres actively increase their holdings by collecting newly described type strains, strains studied and reported in the literature, and strains of industrial and genetic importance. In addition, and at their own discretion, collections accept strains offered for deposit and, as far as possible, act as safety nets for specialist collections that can no longer be supported for financial or political reasons, or because of changes in the direction of research interests.

The service collections preserve cultures in their care, using methods that provide maximum viability and stability. In addition, they supply cultures on demand and without restriction. Their prime function is to act as resource centres from which microbiologists in education, research and industry may obtain authenticated cultures together with help on such matters as strain selection, growth and preservation requirements, phenotypic characteristics or industrial properties. The supply of specific information on strains is helped greatly by the use of computers for data storage and retrieval, since printouts on individual strain properties can be provided with cultures. A database system similar to that in use at the NCYC is to be used for the other service collections in the UK, and eventually in Europe, to provide a central, on-line, microbial culture information service (MiCIS) housed in London at the Laboratory of the Government Chemist (Department of Trade and Industry, Cornwall House, Waterloo Road, London SE1 8XY). The scheme will enable users to find strains of any category of micro-organism [bacteria, fungi (including yeasts), algae, protozoa and animal cells], held in the UK and other European culture collections, searching either by name or by a specific character or combination of characters. Similar systems are being developed in other regions of the world, and plans are well developed to co-ordinate these activities through an international Microbial Strain Data Network (MSDN) (Krichevsky & Hill 1983), an organization sponsored by the World Federation for Culture Collections, the International Union of Microbiological Societies (IUMS) and the Committee on Data for Science and Technology (CODATA).

### *1.4.2 Identification*

The identification of micro-organisms is generally a specialized activity that needs experience of a kind not always available in biotechnology laboratories. It is appropriate that the service collections, which are daily concerned with establishing the authenticity of strains, should provide centralized identification services. The collections of yeasts providing a service of this kind are listed in Table 1.11. The NCYC has recently developed a computer-assisted identification system, based on a probabilistic matrix and using selected characters known to be reproducible and independent. This system is available to users throughout the world by use of computer networking systems (Packet

Switch Stream, PSS), and further information on this system may be obtained by direct application to the NCYC.

### *1.4.3 Preservation*

Laboratories do not always have equipment for freeze-drying or cryopreserving cultures, and yet require a supply of stable, active material for routine inoculation purposes. The service collections have such equipment, are experienced in culture preservation methods and many offer a preservation service (Table 1.11). Cultures are usually freeze-dried and returned to the depositor, but alternative arrangements can be made to suit the needs of the laboratory. This is a particularly valuable service for small- or medium-sized biotechnology companies which require a guaranteed supply of a production strain, but for which it would be uneconomic to develop in-house expertise.

### *1.4.4 Safe-deposit*

The regular supply of production strains, upon which much of a company's income may depend, places a great responsibility on the laboratory concerned with maintenance of the cultures. It is often felt advisable to deposit valuable micro-organisms with service collections, to guard against loss through error or accident. Collections providing this service guarantee to maintain strains on a strictly confidential basis, and samples of the culture are made available to the depositor only.

### *1.4.5 Patent depository*

If a process that involves the use of a micro-organism is to be patented, then the micro-organism must be deposited in a recognized depository. Under the terms of the Budapest Treaty for the Deposit of Microorganisms for Patent Purposes (Anonymous 1977), the governments of member states have designated a number of collections as International Depository Authorities (IDAs) (Table 1.11). A deposit in a single IDA enables depositors to apply for patents in a number of countries. Collections that are IDAs are able to help potential depositors comply with the regulations associated with the Budapest Treaty or with national patent laws. In addition the Patents Committee of the World Federation for Culture Collections (Table 1.10) has much experience in the procedures relating to the deposit of micro-organisms for patent purposes, and may be contacted for advice. Further information of a more general nature is available in Crespi (1982).

### *1.4.6 Education*

The service collections have experience in a number of areas, and they are concerned to make their technical knowledge available to others. Accordingly, scientists from the collections write articles and scientific papers, lecture, accept visiting workers to their laboratories for specific instruction and arrange training courses. These activities are organized either through the culture collections or through national or international federations for culture collections. The major federations for culture collections are listed in Table 1.10, and may be approached for advice.

### 1.4.7 *Collections, federations and committees*

Some of the major yeast collections, with acronyms, addresses, telephone and telex numbers are listed in Table 1.12. Table 1.11 shows the services provided by the collections. Other organizations from whom advice and information may be obtained are listed in Table 1.10.

**Table 1.12** Addresses, telephone and telex numbers of yeast culture collections listed in Table 1.11.

| Acronym | World Data Centre Number | Address | Telephone | Telex |
|---|---|---|---|---|
| ATCC | 1 | American Type Culture Collection, 12301 Parklawn Drive, Rockville, Maryland 20852, USA | 301–881–2600 | 908768 ATCCROVE |
| CBS Yeast Division | 133 | Centraalbureau voor Schimmelcultures, Yeast Division, Julianalaan 67A, 2628 BC Delft, The Netherlands | (015)782416 or 782394 | |
| DSM | 274 | Deutsche Sammlung von Mikroorganismen, Grisebachstrasse 8, D-3400 Göttingen, Federal Republic of Germany | 0551 393827 | |
| IFO | 191 | Institute of Fermentation, 17–85 Juso-honmachi 2-chome, Yodogawa-Ku, Osaka 532, Japan | | |
| NCPF | 184 | National Collection of Pathogenic Fungi, Central Public Health Laboratory, 61 Colindale Avenue, London NW9 5HT, UK | 01–200 4400 | |

*Continued*

**Table 1.12** *Continued.*

| Acronym | World Data Centre Number | Address | Telephone | Telex |
|---|---|---|---|---|
| NCYC | 169 | National Collection of Yeast Cultures, Institute of Food Research, Norwich Laboratory, Colney Lane, Norwich NR4 7UA, UK | 0603–56122 | 975453 FRINOR (Electronic Mail: DIALCOM and COMPASS) |
| NRRL | 97 | ARS Culture Collection, Northern Regional Research Center, 1815 North University Street, Peoria, Illinois 61604, USA | | 91065 20117 A/B NRRC PEO |
| YGSC | | Yeast Genetic Stock Center, Department of Biophysics & Medical Physics, University of California, Berkeley, California 94720, USA | 415–642–0815 | |

## References


Anonymous 1977. *Budapest Treaty on the International Recognition of the Deposit of Microorganisms for the Purposes of Patent Procedure.* Geneva, Switzerland: World Intellectual Property Organization.

Anonymous 1980. Committee on Postal and Quarantine Regulations. *World Federation for Culture Collections Newsletter* No. 7, 5–8.

Bassel, J., R. Contopoulou, R. Mortimer and S. Fogel 1977. *UK Federation for Culture Collections Newsletter* No. 4, 7.

Crespi, R. S. 1982. *Patenting in the biological sciences.* Chichester: Wiley.

Hubalek, Z. and A. Kockova-Kratochvilova 1978. The liquid nitrogen storage of yeasts. *Antonie van Leeuwenhoek* **44**, 229–41.

Kirsop, B. E. and J. E. Henry 1984. Development of a miniaturised cryopreservation method for the maintenance of a wide range of yeasts. *Cryoletters* **5**, 191–200.

Kirsop, B. E. and J. J. S. Snell (eds) 1984. *Maintenance of microorganisms. A manual of laboratory methods.* London: Academic Press.

Krichevsky, M. I. and L. R. Hill 1983. *Needs and specifications for an international microbial strain data network*. Proceedings of a Workshop, United Nations Environmental Programme, Brussels.

Lawrence, D. R. 1983. *Yeast differentiation and identification*. Proceedings of European Brewing Congress, 449–56.

McGowan, V. F. and V. B. D. Skerman (eds) 1982. *World directory of collections of cultures of microorganisms*. University of Queensland: World Data Center.

Morris, G. J. 1981. *Cryopreservation*. Cambridge: Institute of Terrestrial Ecology.

Painting, K., P. Foster and B. Kirsop 1984. The use of computers in a culture collection. *Binary* **1**, 7–9.

Pridham, T. G. and C. W. Hesseltine 1975. Culture collections and patent depositors. *Advances in Applied Microbiology* **19**, 1–21.

Put, H. M. C. and J. DeJong 1980. The heat resistance of selected yeasts causing spoilage of canned soft drinks and fruit products. In *Biology and activity of yeasts*, F. A. Skinner and R. R. Davenport (eds), 181–213. London: Academic Press.

von Rehberg, R. 1978. Kulturhaltung und Konservierung von Hefestämmen. *Die Branntweinwirtschaft* **118**(1), 1–6.

Smith, D. and A. H. S. Onions 1983. *The preservation and maintenance of living fungi*. Commonwealth Agricultural Bureaux.

Stanley, G. 1979. Variation in mixed and single strain starters as used in the preparation of fermented dairy products. In *The stability of industrial organisms*, B. E. Kirsop (ed.), 20–31. Commonwealth Agricultural Bureaux.

Wellman, A. M. and G. G. Stewart 1973. Storage of brewing yeasts by liquid nitrogen. *Applied Microbiology* **26**(4), 577–83.

White, W. and K. L. Wharton 1985. A cryogenic preservation system. *International Laboratory* (January–February), 58–69.

Woods, R. 1976. *UK Federation for Culture Collections Newsletter* No. 2, 5.

# 2 *Approaches to yeast identification*

MARC-ANDRÉ LACHANCE

## 2.1 Introduction

The identification of commercial and wild yeasts is of major concern in many industrial processes based on the activities of yeasts. Where the utilization and maintenance of a pure culture is of prime importance, as in the production of baker's yeast or wine yeasts, or as in brewing, then the ability to recognize wild and culture yeasts is essential. The aims of practical identification are different from those of yeast taxonomy. Biologists involved with yeast classification seek the eventual erection of a 'natural' system – one in which taxonomic assignments reflect evolutionary relationships between taxa. Whereas the fundamental unit of yeast classification is the species, an assemblage of strains presumed or shown to be highly related, the strain itself is the focus of an industrial process.

## 2.2 Yeast classification

The most widely accepted biological system of yeast classification at present largely follows the thoughts of Lodder and Kreger-van Rij (1952) combined with guidelines and methodologies proposed by Wickerham (1951). The two approaches merged with the publication, in 1970, of the well-known collaborative treatise under Lodder's editorship. An updated version of *The yeasts, a taxonomic study* (Kreger-van Rij 1984) has recently been published. In the currently accepted classification (Phaff *et al*. 1978, Kreger-van Rij 1984) yeasts are arranged hierarchically in subdivisions, families, subfamilies, genera and species. At the highest level identification is based on fundamental aspects of the yeasts' sexuality (Ascomycotina or Basidiomycotina), or the lack of it (Deuteromycotina). Lower categories tend to be defined on morphological differences. This is summarized in Table 2.1. Within each genus, species delineation usually corresponds to estimated discontinuities in the genetic relatedness of groups of yeasts. In practice yeast species are classified according to their nutritional abilities or requirements.

In the industrial laboratory yeasts are divided into categories meaningful to each specific industrial process. The strain used in production may be known as the 'culture yeast' and all others are referred to as 'wild yeasts'. Relevant methodologies must therefore allow the detection and recognition of unwanted yeasts. Wild yeasts are contaminants from the environment, or genetic variants

**Table 2.1** Major taxonomic categories of yeasts (Phaff *et al.* 1978, Kreger-van Rij 1984).

| | |
|---|---|
| subdivision Ascomycotina | |
| (yeasts which form ascospores) | |
| family Spermopthoraceae | |
| (ascospores are needle- or spindle-shaped) | e.g. *Metschnikowia* |
| family Saccharomycetaceae | |
| (ascospores are not needle- or spindle-shaped) | |
| subfamily Schizosaccharomycetoideae | |
| (division by fission) | e.g. *Schizosaccharomyces* |
| subfamily Nadsonioideae | |
| (bipolar budding) | e.g. *Hanseniaspora* |
| subfamily Lipomycetoideae | |
| (ascus is an external sac) | e.g. *Lipomyces* |
| subfamily Saccharomycetoideae | |
| (none of the above special features) | e.g. *Saccharomyces* |
| subdivision Basidiomycotina | |
| (yeasts which form basidiospores) | e.g. *Leucosporidium* |
| subdivision Deuteromycotina | |
| (yeasts without sexual spores) | |
| family Sporobolomycetaceae | |
| (ballistospores are formed) | e.g. *Sporobolomyces* |
| family Cryptococcaceae | |
| (ballistospores are not formed) | e.g. *Candida* |

of the cultured strain. The most problematic are yeasts endowed with physiological advantages over the industrial strain, resulting in a 'take over', or those able to spoil the product, even when present in small numbers.

Brewers are by far the most concerned with contamination problems. The yeasts most frequently encountered in breweries have been discussed in detail by Helbert (1982). *Brettanomyces* spp., by virtue of their ability to produce acetic acid from glucose, are known to impart a vinegary taste to beer. *Saccharomyces diastaticus*, a variant of *Saccharomyces cerevisiae*, produces amylases, causing overattenuation of wort or spoilage of finished beer. Other closely related variants may impart sulphury tastes or release diacetyl, a highly fragrant compound with an unpleasant taste.

Distillers and wine-makers are less preoccupied with wild yeasts, although there is now evidence that the yeast used to produce a wine has more effect on its flavour than was previously believed (Gallander 1982).

Baker's yeast manufacturers are most concerned with the presence of yeasts which may replace their culture strain because of their higher growth rate (Fowell 1967). They also fear interference of contaminants with the maltase activity and raising power of the yeast in dough (Goncharova *et al.* 1965). Pigmented yeasts may develop on yeast cakes, causing aesthetic spoilage after shipment to bakeries. The increased use of 'liquid yeast' is progressively solving that problem. Makers of baker's yeast are increasingly involved in the production of specialty yeasts aimed at wine-makers, distillers and amateur brewers. This new challenge increases the potential for cross-contamination of large culture vessels, and thus the need for methods able to discriminate between the various strains.

## 2.3 General identification methods

Most methods used in gathering information useful for yeast identification have been reviewed in detail by van der Walt and Yarrow (1984). The following discussion is a summary of the principal techniques.

### *2.3.1 Culture characterization*

Media used for the isolation and maintenance of yeast cultures were described in Chapter 1. Many of these are suitable for the observation of culture characteristics. Most taxonomy laboratories have adopted either malt extract agar (5–10°Balling) or YM agar as general purpose media. YM agar (Wickerham 1951), also known as MYGP agar (Ingledew & Casey 1982b), is available commercially in dehydrated form (Difco).

To obtain isolated colonies, yeasts from solid substrates should be suspended in small amounts of sterile water before being streaked on to a plate. The texture, colour and shape of yeast colonies is often characteristic of a genus, and should be noted carefully. Experienced laboratory workers rapidly recognize characteristics unique to familiar strains, a skill which can be of great value in quality control. Yeast growing in broth may also exhibit special characteristics such as the formation of hard flocs or dry creeping pellicles. The latter trait is diagnostic of many species of *Pichia* and *Debaryomyces*. For convenience these features may be observed while yeasts are assessed for glucose fermentation, in Durham tubes.

### *2.3.2 Morphological characterization*

Vegetative growth features are best observed in young (2–3-day) cultures grown in broth or on agar media. Cells from agar may be suspended in distilled water and mounted between slide and coverslip for microscopy. In this laboratory we normally mount up to ten cultures on a block of 1.5% agar cut out of a thin plate, using a bent inoculation wire to streak thin layers of cells. A 22-mm coverslip is carefully placed on top, and the mount is scanned in order, under phase-contrast oil immersion. Some phase-contrast microscopes cannot be adjusted to suit such a preparation, as the working distance of their phase condensor is too short. The preparation of agar-coated slides may alleviate this problem. The use of oil immersion (100× objective) may appear tedious to some, but its importance cannot be overemphasized. Essential but subtle features of yeast cells are easily overlooked at lower powers. Staining is usually not required unless specific organelles need be visualized. The reader is referred to Robinow's (1975) excellent treatment of this topic.

Vegetative features important for yeast identification include the cell shape and size, the mode of bud formation and the presence of unusual organelles. Shapes vary from nearly spherical to cylindrical. Slight variations in shape are usually not important, but certain yeasts may be assigned to a genus simply on the basis of their unusual cell morphology. The cells of *Dekkera* or *Brettanomyces*, for example, tend to acquire a distinct bullet (ogival) shape. Size also varies somewhat. Diploid cells of *Saccharomyces* are typically quite large (8 $\mu$m or more), and they are easily recognized by an experienced microscopist. The formation of buds by fission of large, cylindrical cells is diagnostic of the genus *Schizosaccharomyces*, while bipolar budding, which results in the accumulation of apiculate (lemon-shaped) cells, is typical of the subfamily Nadsonioideae.

Further morphological information may be gathered by inoculating poor media (e.g. corn meal agar or Difco) in such a way as to create an air gradient around the growing culture. The Dalmau plate method (Wickerham 1951) involves streaking a yeast on to agar and covering the streak with an alcohol-sterilized coverslip. Alternatively, the agar may be stabbed. These forms of inoculation result in the production of hyphae or pseudohyphae by yeasts having that potential. Direct microscopic observation of the plates at moderate magnification is done after 1 week or more of incubation. The diagnostic value of filamentous growth is limited in most instances.

The observation of ascospores usually requires growing yeast cultures on special sporulation media. These include media containing acetic acid as carbon source, vegetable juice media or media with a high nitrogen: carbohydrate ratio (e.g. Gorodkowa agar). Many sporulating yeasts exhibit their ability to do so on McClary's acetate agar (McClary *et al.* 1959) or on YM agar, and sometimes on both. The composition of McClary's medium is given in Table 2.2. Other formulations are given by Ingledew and Casey (1982a, b), by van der Walt and Yarrow (1984) and in Chapter 1. Ascospore numbers and shapes, the mode of ascus formation and ascus dehiscence at maturity are important in the identification of ascomycetous yeast genera. This is discussed at length in Kreger-van Rij's (1984) treatise. Basidiomycetous yeasts are frequently observed in their imperfect (asexual or anamorphic) form and, consequently, their sexual morphology is less useful in determinative activities.

**Table 2.2** Composition of McClary's acetate agar for yeast sporulation.

| | % |
|---|---|
| D-glucose | 0.10 |
| potassium chloride | 0.18 |
| yeast extract | 0.25 |
| sodium acetate trihydrate | 0.82 |
| agar | 1.50 |

### 2.3.3 *Physiological characterization*

The qualitative determination of fermentative abilities can be performed in Durham tubes, using 0.5% yeast extract and 2% sugar (4% raffinose; van der Walt & Yarrow 1984). Barnett *et al.* (1983) recommend 50 mmol $l^{-1}$ sugar. If time is not a significant factor, then glucose fermentation should be evaluated first, since it is a prerequisite for the fermentation of other sugars. It should be pointed out that the usefulness of fermentation spectra in yeast identification has decreased over the past few years.

Various approaches are used to determine the oxidative assimilatory abilities of yeasts. In general, carbon and nitrogen assimilation tests make use of Wickerham's (1951) synthetic media yeast nitrogen base and yeast carbon base, both of which are available commercially in dehydrated form. Certain researchers perform assimilation tests in liquid media contained in culture tubes, kept still or agitated by reciprocal or gyratory shaking. The tissue culture 'Rollordrum' (New Brunswick) is also used. Substantial variations can result from those methodological differences. In practice the choice of one method or the other depends upon convenience and specific objectives. Agitated cultures grow more rapidly. The preparation of media and inocula is described by van der Walt and Yarrow (1984), and by Barnett *et al.* (1983). The former authors recommend the addition of carbon sources at a concentration of 0.5% (1% for raffinose), whereas the latter suggest a level of 50 mmol $l^{-1}$. Nitrogen sources are used at about 5 mmol $l^{-1}$. Filter sterilization of concentrates of the basal media and carbon or nitrogen sources is recommended. In this laboratory and others, autoclaving has been found to be adequate, **provided that basal media and compounds to be tested are sterilized separately**.

Carbon and nitrogen utilization may be determined rapidly, for small numbers of yeasts, by auxanography (Beijerinck 1889). The yeast is suspended in basal agar medium previously melted and cooled to about 45°C, and the suspension is poured into Petri dishes. A few milligrams of each carbon compound (starch excepted) are added directly on the agar surface with a spatula, near the edge of a plate. Four, or perhaps five, compounds per plate can be tested. In most cases clear results are obtained within 2 or 3 days. Slowly utilized compounds cannot be evaluated adequately by this method (Wickerham 1951). Nitrogen compounds are best added as drops of 2–4% sterile solutions.

The disadvantages of auxanography are such that it cannot be used to perform all of the tests required for general identification. In addition, flocculent and filamentous yeasts perform poorly under those conditions. Nevertheless, Kirsop (1974) succeeded in sorting, by auxanography, 130 closely related brewing strains into 25 groups based on their growth responses on melezitose, methyl-glucoside, lactic acid, glycerol, ethanol and succinic acid. Huppert *et al.* (1975) have proposed a modification of the auxanographic technique in which an agar plate is seeded with a lawn of yeast, followed by the application of 12 carbon sources absorbed in paper discs together with basal medium. The results are obtained within 2 days, and an accuracy of 95% is observed in the characterization of clinically significant yeasts.

Convenient assimilation determinations may be performed with API strips (API System SA). Each strip is an alignment of plastic cupules, each containing a different dehydrated medium. Yeast suspensions are added, and the turbidity is assessed at suitable times after inoculation. Fermentative responses are evaluated by plugging appropriate cupules with wax. API 20C strips (20 tests) were designed for clinical yeasts. A strip with 50 tests (API 50CHL), marketed in Europe but not worldwide, has been studied and found to be adequate for the identification of brewery contaminants (Day & Meyling 1983). However, serious problems are encountered when strictly aerobic (B. Kirsop, pers. commun.) or highly flocculent (J. F. T. Spencer, pers. commun.) yeasts are examined. In addition, contaminated cupules are not easily detected.

**Replica plating**

When very large numbers of isolates are tested, replica plating may be a more suitable approach. A detailed description of the procedure used in this laboratory follows.

PREPARATION OF MEDIA

Basal media are prepared in bulk, using only 0.9 times the desired volume of water, and are dispensed into as many flasks as are required. Agar (1.5% of final volume) is added to each flask. Standard-quality agar is used for carbon assimilation, but the nitrogen growth tests require purified agar (e.g. Agar Noble or Difco). The tested compounds are dissolved separately, concentrated tenfold, in one-tenth of the final water volume. After autoclaving for 12–15 min at 121°C, the media are cooled to 50°C in a water bath, and the tested compounds are mixed aseptically into each flask before pouring plates. Each plate receives about 30 ml agar medium. Volatile compounds are spread on to the surface of appropriate plates 1 day before use. Additional yeast nitrogen base–glucose plates are prepared and incubated at different temperatures to determine minima and maxima. Cycloheximide (Actidione) resistance, salt and sugar tolerance, hydrolysis of gelatin, casein or lipids, and vitamin independence are also determined on plates containing suitable media (see van der Walt & Yarrow 1984).

INOCULATION OF PLATES

Replica plating may be accomplished with the velveteen pad technique (Shifrine *et al.* 1954), but the use of a multipoint inoculator offers better reproducibility and simplicity of operation. Several models are commercially available. A manual replication device is produced by the Institut Armand Frappier (Laval, Quebec, Canada), and is marketed by Diagnostic Equipment Inc. (St Paul, Minnesota, USA). In this and several other laboratories the electrically operated inoculator made and sold by Denley (Sussex, UK) has been found to be very satisfactory. Both replicators have floating inoculation needles which allow highly reproducible drops of yeast suspensions to be deposited on the agar surface. Young agar slant cultures are washed with sterile distilled water, and the suspensions are distributed in cups held in appropriate positions by a special rack. The replicator allows rapid inoculation of any number of plates. Glass or plastic Petri dishes may be used, but in the latter case it is essential to seal each plate with a strip of Parafilm, in order to prevent cross-feeding by vapours of ethanol, methanol or ammonia which may be released during growth. This does not seem to be a problem with glass dishes.

RECORDING GROWTH RESPONSES

Plates are read at defined intervals, H. J. Phaff (pers. commun.) recommends 4-, 8- and 18-day readings. The growth of colonies is evaluated in comparison with that obtained on carbon-free or nitrogen-free media as negative controls, and yeast nitrogen base–glucose or yeast carbon base–ammonium sulphate as positive controls. Differences in opacity, thickness and diameter of colonies are taken into account. Growth is recorded, each time, as negative, weak, +, ++ or +++. The final interpretation of results is as follows: if, after 18 days, growth is absent or very scant, the response is considered negative; a weak or + growth is considered a weak response; a ++ or +++ growth attained only at the third reading is considered a delayed response; a minimum growth of ++ at the first or second reading is considered a positive response.

Care must be exercised in the interpretation of growth on certain media. The hydrolysis of sucrose, raffinose, melibiose, trehalose, cellobiose, inulin or starch may cause the release of hexoses, which diffuse in agar and cross-feed other colonies. Cross-fed colonies are typically asymmetrical, and their early growth is scant. Methyl-glucoside hydrolysis gives rise to methanol, which diffuses very rapidly through agar; methanol utilizers therefore tend to exhibit symmetrical but delayed growth, which should not be recorded as positive. Vitamin cross-feeding is prevalent, and only early and strong growth on vitamin-free agar (using Difco yeast vitamin-free base) should be recorded as positive. When in doubt about any potentially false positive response a liquid assimilation test is prepared.

In spite of the possible problems described above, nutritional characterization of yeasts by replica plating is a convenient, rapid and reliable method which allows large numbers of yeast strains to be handled simultaneously. Contaminated plates or cultures are easily detected, and the growth of flocculent yeasts is easily evaluated, two additional advantages over procedures based on liquid media.

### 2.3.4 *Identification keys*

Once morphological and physiological data have been collected, published keys are used to determine the taxonomic identity of a yeast isolate. Most keys found in Kreger-van Rij (1984) are hierarchical, requiring that a generic assignment first be made. The genera are keyed largely on their important morphological features, including those of their sexual life cycles. For each genus there is a key to the species. Alternatively, many species may be keyed directly without any reference to life-cycle characteristics. Before a final specific assignment is made, it is essential that the results obtained for an unknown isolate be compared in their entirety with a standard description, and that they be in agreement. The 1984 edition of *The yeasts, a taxonomic study* suffers from one major disadvantage. The choice and numbers of growth responses vary between genera, and even between species. For this reason the 1970 edition should be kept handy.

Between the appearance of the books by Lodder (1970) and Kreger-van Rij (1984), Barnett *et al.* published a number of keys which are not taxonomically hierarchical, and which emphasize physiological responses over morphological features. The latest version (Barnett *et al.* 1983) is an invaluable source of information. Particularly pertinent to the industrial microbiologist are four identification keys which focus on yeasts associated with food, soft drinks, wine and brewing, respectively. In view of the abundance of nomenclatural ambiguities generated by recent advances in the natural classification of yeast species, the 'register of names' provided by these authors will be extremely useful to industrial laboratories. The keys are not without some disadvantages. They are strictly dichotomous, with the result that the probability of misidentification increases exponentially with the number of false or doubtful responses. The responses used to construct the keys were obtained in a single laboratory where assimilation tests are conducted in shaken tubes with liquid media. Results obtained with stationary tubes, API strips, replica plating or auxanography may (and do) vary somewhat. Finally, early in the keys are found some characters which are not normally conducted in every laboratory; for example, the determination of single-vitamin requirements. These and other problems have been discussed by Miranda (1984). In spite of these reservations laboratories where yeast identification is a frequent activity should benefit from the use of the keys, especially if the final taxonomic assignment is made by comparing the profile of each isolate with standard descriptions found in individual publications of new species, or in the treatise by Kreger-van Rij (1984).

The computer revolution is far from having had its full impact on yeast identification. The efforts of Barnett *et al.* were a first step towards the design of automatic identification systems, several of which are presently being explored. B. Kirsop (pers. commun.) and her colleagues, for example, are developing probabilistic identification programs which it is hoped will be made available over computer networks. Data from Barnett *et al.* (1983) and identification programs are available on microcomputer diskettes from Cambridge University Press. In this laboratory an inelegant but useful algorithm is used routinely to match properties of unknown strains with those of all published species. The degree to which any computerized yeast-identification system effectively achieves its intended purpose depends largely upon the quality of the data used to characterize both the reference strains and the strains being identified. Many recent taxonomic fusions have resulted in the creation of highly heterogeneous assemblages, introducing ambiguities which cannot be eluded, however sophisticated computer programs may be.

A special problem affecting the use of contemporary published keys or any other identification system, computerized or not, is that many industrially important *Saccharomyces* species have been reclassified extensively in the recent past. It is therefore suggested that the original monograph of Lodder and Kreger-van Rij (1952) be consulted when a more precise identification of *Saccharomyces* strains is to be achieved.

## 2.4 Identification of industrial contaminants

Quality control laboratories are concerned principally with the need to detect rapidly yeasts which may disrupt an industrial process or lessen the quality of the final product. The assignment of a valid taxonomic designation to each contaminant is usually not a priority, if desired at all. If an approximate identification is acceptable, then simplified schemes such as that proposed by Davenport (1981) may be considered. He suggests a primary separation of isolates based on colony types, followed by a secondary division according to habitats and cellular characteristics. Tertiary separation relies on selected physiological and morphological traits.

The identification of contaminant yeasts revolves around two major approaches. One is based on the formulation of media which discriminate between different strains. Elective media favour the growth of wild yeasts over that of culture yeasts, by supplying nutrients not utilized by the latter. Selective media contain specific inhibitors of culture yeasts and, it is hoped, allow growth of wild yeasts. Differential media support the growth of many yeast types, but they are formulated so as to create or accentuate differences in colony morphology or colour. The three terms are frequently interchanged in the literature.

The other major strategy is to recognize the presence of wild yeasts without the need to wait for their growth on culture media, which may take several days. It exploits specific biochemical properties of yeasts, revealed by physical–chemical analysis or immunochemical reactions, sometimes combined with microscopy.

### *2.4.1 Elective media*

Theoretically, once the growth characteristics of a cultured yeast have been determined, a medium which favours the exclusive growth of wild yeasts can be formulated. In practice, problems of cross-feeding restrict the efficiency of such media. Wiles (1953) suggested the use of five synthetic media, each containing a carbon source not utilized by brewer's yeast. Besides the impracticality of routinely plating yeast samples on that many media, important contaminants are easily overlooked (Fowell 1967).

Walters and Thiselton (1953) explored the utilization of lysine in several yeast strains, and found that many *Saccharomyces* species were unable to utilize this amino acid as sole nitrogen source. Morris and Eddy (1957) confirmed their findings, and proposed a slight modification of the lysine medium. The latter is available commercially in dehydrated form (Oxoid). According to the data tabulated by Barnett *et al.* (1983), only 55 yeast species out of 469 are unable to utilize lysine. The discriminatory power of this test to elicit the presence of non-*Saccharomyces* brewery contaminants is therefore immense. Lysine media are also useful in the detection and even the recognition of common wild yeasts in baker's yeast production (Fowell 1965). Kits containing a filtration membrane and a pad impregnated with lysine medium are available commercially (Nalge). They were found adequate, provided that the pads be supplemented with 0.5 ml additional lysine medium (Lin 1981).

### *2.4.2 Selective media*

One of the earliest attempts to inhibit selectively the growth of certain yeasts involved the addition of tartaric acid to culture media. Windish (1955) investigated this method and concluded that it is unreliable.

Cycloheximide (actidione) was initially added to Wallerstein Laboratory (WL) agar as a means of suppressing yeast growth for the determination of bacterial counts in beer (Green & Gray 1950). Green (1955) later observed that small amounts of cycloheximide resulted in the inhibition of brewing yeasts, while many wild yeasts developed into normal colonies. The usefulness of this approach has been challenged, as much variation was observed from one laboratory to another. Harris and Watson (1968) attributed this irreproducibility to the partial thermolability of the antibiotic. If the inhibitor is added after autoclaving, then its concentration can indeed be adjusted precisely to the minimal inhibitory level of the culture yeast, thus enabling detection of many wild strains.

Kozulis and Page (1968) recommended the addition of beer [25% (v/v)] to a medium containing 2.5% dehydrated tomato juice broth, 1.5% peptonized milk (both from Difco) and 1% glucose. The ingredients are autoclaved, and cycloheximide (1 mg $l^{-1}$) is added before pouring. The recovery of bacteria and wild yeast is said to exceed that observed on WL or other agar media. The formulation is called 'Universal Beer Agar'.

After testing several dyes Kato (1967) selected crystal violet as a specific inhibitor of brewer's yeast. After 48 h of incubation on 10.5°Balling wort agar containing 20 mg $l^{-1}$ crystal violet, non-brewing species of *Saccharomyces* formed colonies, while ale and lager yeasts (up to 50 000 cells) were suppressed.

Brenner *et al.* (1970) published a modification of the Schwarz 'differential' medium (SDM) for use in the detection of wild brewery yeasts. YM agar is supplemented with a mixture of basic fuchsin, sodium sulphite and dextrin, as shown in Table 2.3. All ingredients are mixed and boiled for 15 min before pouring plates. The medium is stored for 1 day at 30°C. Brewer's yeasts (ale or lager) are inhibited, but non-brewing strains of *Saccharomyces*, and species of *Hansenula*, *Candida* and *Rhodotorula*, grow well. The growth responses of yeasts on SDM have been compared with their serological classification according to Campbell's system (Brenner *et al.* 1972). In general, it was found that *Saccharomyces* strains in antigenic group A were usually inhibited, whereas the majority of other serotypes grew well. Several exceptions were also noted. The selectivity of SDM was also compared with that of lysine medium (Walters & Thiselton 1953), and it became evident that the two media complemented one another. The SDM appeared superior when dealing with yeasts closely related to brewing strains (now considered synonyms of *Saccharomyces cerevisiae*).

**Table 2.3** Composition of Schwarz differential medium (Brenner *et al.* 1970).

| | |
|---|---|
| fuchsin–sulphite mixture | |
| basic fuchsin | 4 g |
| sodium sulphite (anhydrous) | 25 g |
| dextrin | 1 g |
| medium | |
| YM agar | 4.1% |
| fuchsin–sulphite mixture | 0.35% |

**Table 2.4** Composition of Lin's medium for the detection of wild brewery strains of *Saccharomyces* (Lin 1974).

| | % |
|---|---|
| yeast extract | 0.4 |
| malt extract | 0.2 |
| peptone | 0.2 |
| D-glucose | 1.0 |
| dipotassium phosphate | 0.1 |
| ammonium chloride | 0.05 |
| crystal violet* | 0.04 |
| fuchsin–sulphite mixture† | 0.1 |
| agar | 2.0 |

* The minimum inhibitory concentration for the culture yeast should be used.
† See Table 2.3.

Lin (1974) proposed combining crystal violet and the fuchsin–sulphite mixture used in SDM in a single medium (Table 2.4). All ingredients are autoclaved together and stored at 5°C for 24 h before use. Liquid samples should be filtered over two Nuclepore membranes (N40CPR04700) which are then transferred to Lin's medium and lysine agar (Lin 1975). A modification of Lin's medium (Longley *et al.* 1980) has received extensive testing, and was not found to offer any distinct advantages (American Society of Brewing Chemists 1983).

Kish *et al.* (1983) have formulated a selective medium for the isolation of wine yeasts. A yeast nutrient agar not unlike YM is supplemented with 12% (v/v) ethanol and 150 mg $l^{-1}$ sodium metabisulphite at the time of pouring. Apiculate yeasts are suppressed.

### 2.4.3 *Differential media*

Czarnecki and van Engel (1959) observed that *Saccharomyces carlsbergensis* could form three distinct colony types on Wallerstein Laboratory (WL) nutrient agar. Normal colonies are white, as they fail to absorb brom cresol green from the medium. Respiratory-deficient clones develop into 'petite' (small) colonies. A third type, referred to as 'verdant' because of the dark green colour of the colonies, represents variants which in many cases produce larger levels of diacetyl in beer. Kot and Helbert (1977) isolated two verdant strains of *Saccharomyces uvarum*, and demonstrated that dark green colonies were respiratory-deficient. Moderate green verdants are normal in their respiratory process but, like the paler variant, they produce bad beer. Hall (1970) showed that wild yeasts are easily distinguished from ale yeasts when growing on that medium. The latter tend to form smooth and conical-to-hemispherical colonies of a medium–dark green coloration, whereas wild yeast colonies can be flat, rough, white, blue or yellowish-green.

Glucose–eosin–methylene blue agar (EMB) has been found useful to detect relatively low (1:5000) numbers of wild yeasts in baker's yeast (Fowell 1967). Most brewer's yeasts and wild yeasts are distinguished from baker's yeast by their colour. Some wild strains of *Saccharomyces carlsbergensis* appear identical to baker's yeast.

Nakatomi (1980) has obtained a patent for a medium composed of yeast nitrogen base, yeast extract, methionine, ponsol R and aniline blue. After incubation at 30°C, strains of *Saccharomyces cerevisiae* may be distinguished from one another by their blue or purple colorations.

### *2.4.4 Physical–chemical methods*

Wild brewery yeasts exhibit more heat tolerance than brewer's yeasts. Lund and Thygesen (1957) utilized that property to detect low (1:500 000) levels of contamination by heating yeast suspensions for 20 min at 50°C. After transfer to fresh broth the yeast mixture finds itself enriched in wild strains. Upon further transfer to an acetate medium, wild yeasts are reported to sporulate within 2 days, unlike brewer's yeast which take an additional day.

*Candida utilis*, a frequent contaminant of baker's yeast, can be identified in whole-yeast cakes by gas–liquid chromatography of fatty acids (Biacs 1979). The cells are washed, dried and ground in the presence of methanolic potassium hydroxide. Saponified lipids are then extracted with ethers, are esterified and the chromatographic profiles are examined for the presence of linoleic and linolenic acids. These compounds are present in *Candida utilis* and absent in baker's yeast, thus allowing the detection of infection by the wild yeast at levels of 5% or more. This approach was also applied to the brewing environment (Chen 1981). In general, brewer's yeasts tended to exhibit high ratios of saturated to unsaturated fatty acids and of stearic to oleic acids. The converse prevailed in wild yeasts. Wild yeasts also possessed unidentified fatty acids not present in culture yeast.

Industrial strains of *Saccharomyces cerevisiae* exhibit slight differences in the way they metabolize certain carbohydrates. Perry *et al.* (1983) made use of this property and succeeded in discriminating between baker's yeast, brewer's yeast and distiller's yeast by flow microcalorimetry in 16 h. They proposed the application of that method for characterization of industrial strains. The application of this kind of methodology for routine identification is doubtful.

Yeasts able to utilize lysine as a sole carbon source can be detected in brewery samples by incubating samples in the presence of $^{14}$C-lysine, followed by filtration, rinsing with a detergent solution and scintillation counting of the dried filter membrane (Bourgeois *et al.* 1973). Tens of cells may be detected within 9 h, fewer within 24 h. An analogous method was suggested by Box and Ferguson (1975) to detect the presence of any kind of yeast cells in very dilute samples. Labelled glucose was used, and comparable sensitivities were reported. Harrison *et al.* (1974) have explored radiotracer methods in parallel with approaches involving changes in pH during yeast growth. They have found the latter to be highly sensitive in the detection of low yeast numbers.

### 2.4.5 *Serology*

Campbell *et al.* have investigated the serology of *Saccharomyces* and related yeasts over a number of years. It was found initially that serology could discriminate between yeasts which are otherwise very similar, in particular between those used in brewing and the frequent contaminants of the brewing process (Campbell & Allan 1964, Campbell & Brudzinski 1966). *Saccharomyces cerevisiae* was found to differ antigenically from *Saccharomyces carlsbergensis*. The latter comprises two serotypes, one of which cross-reacts with *Saccharomyces ellipsoideus* and *Saccharomyces diastaticus*. Four serotypes (A, B, C and D, Table 2.5) were eventually defined (Campbell 1968) in such a way that yeasts now known to be genetically unrelated to *Saccharomyces cerevisiae* were rather accurately identified as such. Serotype B included yeasts now reclassified in the genus *Kluyveromyces*, and known to represent a distinct evolutionary lineage (Bicknell & Douglas 1970). Serotype D consisted of yeasts now regarded as members of the genus *Torulaspora*, and known to be genetically distinct from true *Saccharomyces* species as well (Price *et al.* 1978). Important for the brewer is the ability of antisera A and C to discriminate between brewing and other strains now suspected to be conspecific with *Saccharomyces cerevisiae* (see Kreger-van Rij 1984). Although taxonomists may treat them as the same entity, industry counts on the existence

**Table 2.5** Campbell's yeast serotypes (see text for references). Names in parentheses are the currently accepted designations following Kreger-van Rij (1984).

| Serotype | Eliciting antigen | Absorbing antigen |
|---|---|---|
| A | *Saccharomyces carlsbergensis* I (*Saccharomyces cerevisiae*) | *Saccharomyces ellipsoideus* (*Saccharomyces cerevisiae*) |
| B | *Saccharomyces fragilis* (*Kluyveromyces marxianus*) | *Saccharomyces rouxii* (*Zygosaccharomyces rouxii*) |
| C | *Saccharomyces inusitatus* or *Saccharomyces ellipsoideus* (*Saccharomyces cerevisiae*) | *Saccharomyces cerevisiae* |
| D | *Saccharomyces fermentati* or *Saccharomyces microellipsoides* (*Torulaspora delbrueckii*) | *Saccharomyces cerevisiae* |
| E | *Candida lipolytica* *Torulopsis anomala* (*Candida versatilis*) or *Schizosaccharomyces pombe* | |
| F | *Candida krusei* *Candida utilis* or *Hansenula* sp. | *Saccharomyces ellipsoideus* (*Saccharomyces cerevisiae*) |

of criteria to identify them either as cultured yeasts or as wild yeasts. Campbell (1971) later identified two more fundamental serotypes (E and F) in other non-*Saccharomyces* yeasts with known propensities to be industrial contaminants.

Campbell (1972) proposed the use of slide agglutination of colonies from any media, including special formulations used in breweries, to characterize rapidly yeasts based on his six serotypes. Using sonicated and washed yeast cell walls as antigens, Callejas *et al.* (1977) have confirmed the usefulness of serotype determinations in distinguishing contaminants from brewer's yeasts. Patents have been granted to Sapporo Breweries (1980a, b) for the use of absorbed antisera to detect wild brewery yeasts by slide agglutination. In one case five commercially available antisera for the identification of pathogenic *Candida* species were used. A detection rate of 75% was claimed.

Cowan (1977) has refined the discriminatory power of serological tests by means of immunoelectrophoresis. Nineteen antigens were demonstrated among 43 brewing strains. The application of this classification in breweries would depend on the eventual commercial availability of purified antibody preparations, each possessing one of the 19 specificities.

### 2.4.6 *Immunofluorescence microscopy*

Richards and Cowland (1967) succeeded in preparing a specific immunofluorescent staining procedure using mixed antisera elicited by *Saccharomyces pastorianus* and *Saccharomyces ellipsoideus*. Cross-reactivity with brewing yeast was eliminated by absorption. Richards (1969, 1970, 1972) investigated the technique further, using additional antisera, and reached the conclusion that the approach had little relevance in yeast taxonomy. On the other hand, its promise as a tool for the identification of brewery contaminants was obvious. The technique was subsequently adopted as a standard method by the Institute of Brewing (Pierce 1973). Yeasts are washed in buffer, placed on a microscope slide (3 million cells), dried and heat-fixed. Horse serum is added to prevent non-specific binding of rabbit antibodies. After rinsing, the specific rabbit antiserum is allowed to react, rinsed extensively and stained with anti-rabbit sheep immunoglobulin conjugated with fluorescein isothiocyanate (FITC). The slide is rinsed further, covered with mounting medium and a coverslip, sealed and observed by fluorescence microscopy.

Campbell (1968) investigated the possibility that his agglutinating sera might also be useful in detecting individual cells of wild yeast under the microscope. In spite of some ambiguities, his method showed much promise. Concerned with the possibility that immunofluorescent avidity of yeast cells may vary as a function of their physiological state, Cowland (1968) investigated the response of synchronous cultures to serological staining. Indeed, fluctuations were observed which paralleled the cell cycle to some extent. However, mature cultures were found to accept the stain with high efficiency.

More recently a method has been proposed which combines serological detection of contaminants with the determination of yeast cell viability in a single operation (Chilver *et al.* 1978). Antisera types A–F are pooled and absorbed extensively with the cultured yeast strain. Washed cells are treated with non-specific horse serum, reacted with the multivalent absorbed serum, stained immunologically with FITC–anti-rabbit serum conjugate, washed further and finally mixed with 0.01% methylene blue. The technique is reported to allow the detection of ten wild yeast cells in a million combined with an estimate of viability, in less than 3 h.

Bouix and Leveau (1980) proposed the reverse elicitation of an antiserum specific for *Saccharomyces carlsbergensis* and absorbed with *Saccharomyces uvarum*. Note that the presumed conspecificity of these two taxa (Barnett *et al.* 1983, Kreger-van Rij 1984) is still under debate (Vaughan-Martini & Kurtzman 1984). Samples are allowed to react with the antiserum and stained with FITC-labelled anti-rabbit serum. The fluorochrome Evans blue is added to stain unreacting yeast cells. Under the conditions specified, wild yeasts stain red, whereas brewing yeasts fluoresce in green. Forty yeast strains comprising 22 different non-*Saccharomyces* species and four varieties of non-brewing *Saccharomyces* were tested. Two strains exhibited high levels of green fluorescence, and ten more fluoresced weakly. The reported sensitivity was one wild yeast in 500–1000.

Hammond and Jones (1979) experienced some serious difficulties with immunofluorescence. Staining of a certain batch of pitching yeast with sera C (to detect wild *Saccharomyces*) and D (for *Brettanomyces*) indicated a massive (1000 cells per million) infection by a serotype C yeast. The contaminant was extensively characterized, and it was ultimately identified as a subtle variant of the cultured yeast, altered only in its ability to flocculate in the presence of calcium. The authors therefore concluded that immunofluorescence microscopy should be accompanied by other supportive tests.

### 2.4.7 *Special characterization techniques*

The flocculation characteristics of yeasts may be determined by the Tullo test (Gilliland 1951). Yeast is inoculated into 5 ml of wort and incubated at 25°C for 3 days. Non-flocculent yeasts are termed 'class I'. Classes II and III are defined by the late formation of small loose clumps or dense masses, respectively, whereas class IV represents yeasts that flocculate early during fermentation.

## 2.5 Concluding remarks

Yeast systematics may be approached from many directions, and for several different purposes. Substantial efforts are now channelled towards creating a classification system which would be an image, as faithful as possible, of the paths followed by yeasts to evolve into the present array of species. Natural yeast taxa do not necessarily lend themselves well to a convenient classification, because of one major component of the biological reality, variation. The use of recently developed molecular techniques by yeast taxonomists has not solved the problem of identification, and the development of rapid and clear-cut identification systems by brewers and other practically minded researchers has been of little use in evolutionary systematics. Nevertheless, research in this area should benefit from enhanced communication between the many researchers involved.

## Acknowledgements

Research support from the Natural Science and Engineering Council of Canada is gratefully acknowledged. I thank Dr G. G. Stewart for his encouragement and suggestions, Dr B. Kirsop for the information she communicated, and the Labatt's librarians for their assistance with bibliographical research.

## References

American Society of Brewing Chemists 1983. Report of Subcommittee on Microbiology. *Journal of the American Society of Brewing Chemists* **41**, 100–3.

Barnett, J. A., R. W. Payne and D. Yarrow 1983. *Yeasts: characteristics and identification*. Cambridge: Cambridge University Press.

Beijerinck, M. W. 1889. L'auxanographie, ou la méthode de l'hydro diffusion dans la gélatine appliquées aux recherches microbiologiques. *Arch. neerland. sci.* **23**, 367–72.

Biacs, P. A. 1979. Rapid method for the determination of wild yeast contamination in baker's yeast by lipid analysis. *Acta Aliment. Acad. Sci. Hung.* **8**, 57–67.

Bicknell, J. N. and H. C. Douglas 1970. Nucleic acid homologies among species of *Saccharomyces*. *Journal of Bacteriology* **101**, 505–12.

Bourgeois, C., P. Mafart and D. Thouvenot 1973. Méthode rapide de détection des contaminants dans la bière par marquage radioactif. *European Brewery Convention, Proceedings of the Congress, 14th, Salzburg* **1973**, 219–30.

Bouix, M. and J. Y. Leveau 1980. Detection of wild yeasts in brewing yeasts by double fluorescence. *MMBA Technical Quarterly* **17**, 163–6.

Box, T. C. and B. Ferguson 1975. Rapid estimation of small numbers of yeast by a radiotracer method. *American Society of Brewing Chemists Proceedings* **33**, 133–6.

Brenner, M. W., M. Krapiscak, H. Stern and W. P. Hsu 1970. A differential medium for detection of wild yeast in the brewery. *American Society of Brewing Chemists Proceedings* **1970**, 79–88.

Brenner, M. W., W. P. Egan and W. P. Hsu 1972. Relationship of antigenic properties of wild yeasts to growth on Schwarz differential medium. *American Society of Brewing Chemists Proceedings* **1972**, 75–8.

Callejas, C., M. A. Rodriguez and A. M. Cavales 1977. Use of immunological methods in the interspecies relationships of genus *Saccharomyces* and its possible application in detection of contaminating strains. *Journal of the American Society of Brewing Chemists* **35**, 12–15.

Campbell, I. 1968. Serological identification scheme for the genus *Saccharomyces*. *Journal of Applied Bacteriology* **31**, 515–24.

Campbell, I. 1971. Antigenic properties of yeasts of various genera. *Journal of Applied Bacteriology* **34**, 237–42.

Campbell, I. 1972. Simplified identification of yeasts by a serological technique. *Journal of the Institute of Brewing* **78**, 225–9.

Campbell, I. and A. M. Allan 1964. Antigenic analysis of *Saccharomyces cerevisiae*. *Journal of the Institute of Brewing* **70**, 316–20.

Campbell, I. and A. Brudzinski 1966. Serological studies on brewing yeasts. *Journal of the Institute of Brewing* **72**, 556–60.
Chen, E. C. H. 1981. Fatty acid profiles of some cultured and wild yeasts in brewery. *Journal of the American Society of Brewing Chemists* **39**, 117–24.
Chilver, M. J., J. Harrison and T. J. B. Webb 1978. Use of immunofluorescence and viability stains in quality control. *Journal of the American Society of Brewing Chemists* **36**, 13–18.
Cowland, T. W. 1968. Variation in the serological behaviour of yeasts of the genus *Saccharomyces* towards fluorescent antibody. *Journal of the Institute of Brewing* **74**, 457–64.
Cowan, D. 1977. The serological assessment of brewing yeasts. *Journal of the Institute of Brewing* **83**, 3.
Czarnecki, H. T. and E. L. van Engel 1959. The isolation and identification of diacetyl-producing brewer's yeast. *Brewer's Digest* **34**(3), 52–6.

Davenport, R. R. 1981. Yeasts and yeast-like organisms. In *Smith's introduction to industrial mycology*, A. H. S. Onions, D. Allsopp and H. O. W. Eggins (eds), 65–92. London: Edward Arnold.
Day, A. and J. C. Meyling 1983. Practical experience with rapid methods of detecting yeasts in breweries. *Journal of the Institute of Brewing* **89**, 204–6.

Fowell, R. R. 1965. The identification of wild yeast colonies on lysine agar. *Journal of Applied Bacteriology* **28**, 373–83.
Fowell, R. R. 1967. Infection control in yeast factories and breweries. *Process Biochemistry* **2**(12), 11–15.

Gallander, J. F. 1982. Influence of different wine yeasts on the quality of Vidal Blanc wines. *Developments in Industrial Microbiology* **23**, 123–9.
Gilliland, R. B. 1951. The flocculation characteristics of brewing yeasts during fermentation. *Proceedings, European Brewery Convention, Brighton* **1951**, 35–58.
Goncharova, L. A., N. N. Bocharova, Y. P. Kobrina and E. S. Zvigur 1965. Effect of wild yeasts on the yield and quality of baker's yeast. *Microbiology (USSR)* **34**, 157–62.
Green, S. R. 1955. A review of differential techniques in brewing microbiology. *Wallerstein Laboratory Communications* **18**, 239–51.
Green, S. R. and P. P. Gray 1950. A differential procedure applicable to bacteriological investigation in brewing. *Wallerstein Laboratory Communications* **13**, 357–66.

Hall, J. F. 1970. The use of Difco WLN agar for (a) the detection of wild yeast in the presence of *Sacch. cerevisiae* (b) demonstration of the instability of strains of *Sacch. carlsbergensis*. *Journal of the Institute of Brewing* **76**, 522–3.
Hammond, J. R. M. and M. Jones 1979. The immunofluorescence staining technique for the detection of wild yeasts – practical problems. *Journal of the Institute of Brewing* **85**, 26–30.
Harris, J. O. and W. Watson 1968. The use of controlled levels of actidione for brewing and nonbrewing yeast strain differentiation. *Journal of the Institute of Brewing* **74**, 286–90.
Harrison, J., T. J. B. Webb and P. A. Martin 1974. The rapid detection of brewery spoilage micro-organisms. *Journal of the Institute of Brewing* **80**, 390–8.
Helbert, J. R. 1982. Beer. In *Prescott and Dunn's industrial microbiology*, G. Reed (ed.), 403–67. New York: McGraw-Hill.
Huppert, M., G. Harper, S. H. Sun and V. Delanerolle 1975. Rapid methods for identification of yeasts. *Journal of Clinical Microbiology* **2**, 21–34.

Ingledew, W. M. and G. P. Casey 1982a. The use and understanding of media used in brewing mycology. I. Media for wild yeasts. *Brewer's Digest* **57**(3), 18–22.

Ingledew, W. M. and G. P. Casey 1982b. The use and understanding of media used in brewing mycology. II. Media for moulds/rapid methods. *Brewer's Digest* **57**(4), 22–6.

Kato, S. 1967. A new measurement of infectious wild yeasts in beer by means of crystal violet medium. *Bulletin of Brewing Science* **13**, 19–24.

Kirsop, B. 1974. Segregation of strains of brewing yeast by their carbon assimilation patterns. *Journal of the Institute of Brewing* **80**, 555–7.

Kish, S., R. Sharf and P. Margalith 1983. A note on a selective medium for wine yeasts. *Journal of Applied Bacteriology* **55**, 177–9.

Kot, E. J. and J. R. Helbert 1977. Two verdant types of *Saccharomyces uvarum*. *Journal of the American Society of Brewing Chemists* **35**, 25–8.

Kozulis, J. A. and H. E. Page 1968. A new universal beer agar medium for the enumeration of wort and beer microorganisms. *American Society of Brewing Chemists Proceedings* **1968**, 52–8.

Kreger-van Rij, N. J. W. 1984. *The yeasts, a taxonomic study*. Amsterdam: Elsevier.

Lin, Y. 1974. Detection of wild yeasts in the brewery. III. A new differential medium. *American Society of Brewing Chemists Proceedings* **1974**, 69–76.

Lin, Y. 1975. Detection of wild yeast in the brewery. VI. Use and effect of membrane filters. *American Society of Brewing Chemists Proceedings* **33**, 124–32.

Lin, Y. 1981. Evaluation of nalgene nutrient pad kits for wild yeast detection. *Brewer's Digest* **56**, 42–3.

Lodder, J. (ed.) 1970. *The yeasts, a taxonomic study*. Amsterdam: North–Holland.

Lodder, J. and N. J. W. Kreger-van Rij 1952. *The yeasts, a taxonomic study*. Amsterdam: North–Holland.

Longley, R. P., G. R. Edwards and S.A. Mathews 1980. Improved detection of wild *Saccharomyces*. *Journal of the American Society of Brewing Chemists* **38**, 18–22.

Lund, A. and P. Thygesen 1957. Detection of spore-forming wild yeasts. *Proceedings of the European Brewery Convention, Copenhagen* **1957**, 241–8.

McClary, D. O., W. L. Nulty and G. R. Miller 1959. Effect of potassium versus sodium in the sporulation of *Saccharomyces*. *Journal of Bacteriology* **78**, 362–8.

Miranda, M. 1984. Book review. *International Journal of Systemic Bacteriology* **34**, 512–13.

Morris, E. O. and A. A. Eddy 1957. Method for the measurement of wild yeast infection in pitching yeast. *Journal of the Institute of Brewing* **63**, 34–5.

Nakatomi, Y. 1980. Culture media for identification of yeasts. *Jpn. Kokai Tokkyo Koho* **80**, 50 898.

Perry, B. F., A. E. Beezer and R. J. Miles 1983. Characterization of commercial yeast strains by flow microcalorimetry. *Journal of Applied Bacteriology* **54**, 183–9.

Phaff, H. J., M. W. Miller and E. M. Mrak 1978. *The life of yeasts*. Cambridge, Massachusetts: Harvard University Press.

Pierce, J. S. 1973. Immunofluorescent staining technique for the detection of wild *Saccharomyces* contaminants in *Saccharomyces cerevisiae*. *Journal of the Institute of Brewing* **79**, 134–6.

Price, C. W., G. B. Fuson and H. J. Phaff 1978. Genome comparison in yeast systematics: delimitation of species within the genera *Schwanniomyces*, *Saccharomyces*, *Debaryomyces*, and *Pichia*. *Microbiological Reviews* **42**, 161–93.

Richards, M. 1969. The rapid detection of brewing contaminants belonging to the genus *Saccharomyces* – examination of lager yeasts. *Journal of the Institute of Brewing* **75**, 476–80.
Richards, M. 1970. Detection of yeast contaminants in pitching yeasts. *Wallerstein Laboratory Communications* **33**, 11–18.
Richards, M. 1972. Serology and yeast classification. *Antonie van Leeuwenhoek* **38**, 177–92.
Richards, M. and T. W. Cowland 1967. The rapid detection of brewery contaminants belonging to the genus *Saccharomyces* by a serological technique. *Journal of the Institute of Brewing* **73**, 552–8.
Robinow, C. F. 1975. The preparation of yeasts for light microscopy. In *Methods in cell biology*, Prescott, D. M. (ed.), 2–21. New York: Academic Press.

Sapporo Breweries, Ltd 1980a. Brewing yeast classification by pathogenic *Candida* factor antibodies. *Jpn. Kokai Tokkyo Koho* **80**, 68 298.
Sapporo Breweries, Ltd 1980b. Detection of contaminant yeasts in brewing by factor antibodies. *Jpn. Kokai Tokkyo Koho* **80**, 68 299.
Shifrine, M., H. J. Phaff and A. L. Demain 1954. Determination of carbon assimilation patterns of yeasts by replica plating. *Journal of Bacteriology* **68**, 28–35.

van der Walt, J. P. and D. Yarrow 1984. Methods for the isolation, maintenance, classification and identification of yeasts. In *The yeasts, a taxonomic study*, N. J. W. Kreger-van Rij (ed.), 45–104. Amsterdam: Elsevier.
Vaughan-Martini, A. and C. P. Kurtzman 1984. A reconsideration of relationships among species of the '*sensu strictu*' group I of the yeast genus *Saccharomyces* based upon nuclear DNA–DNA homologies. 6th International Symposium on Yeasts, Montpellier, France. Abst No. XIV-6-0.

Walters, L. S. and M. R. Thiselton 1953. Utilization of lysine by yeasts. *Journal of the Institute of Brewing* **59**, 401–4.
Wickerham, L. J. 1951. Taxonomy of yeasts. *USDA Technical Bulletin* **52**.
Wiles, A. E. 1953. Identification and significance of yeasts encountered in the brewery. *Journal of the Institute of Brewing* **59**, 265–84.
Windish, S. 1955. Uber den Nachweis wilder Hefen durch Kultur in weisaurer Wurze. *Proceedings of the European Brewery Convention, Baden-Baden* **1955**, 69.

PART II

# Genetic manipulation

# 3 *Classical genetic and protoplast fusion techniques in yeast*

R. M. JONES, I. RUSSELL
and G. G. STEWART

## 3.1 Introduction

The yeast genus *Saccharomyces* is without doubt one of the oldest 'plants' cultivated by man. It has recently progressed from being a 'workhorse' in the traditional food and beverage industries to the organism of choice for the 'gene manipulators'. This has resulted in a plethora of literature describing the genetics and biochemistry of laboratory strains of *Saccharomyces cerevisiae* and related species; however, there is still a general lack of knowledge regarding the genetics and biochemistry of industrial *Saccharomyces* strains. Non-*Saccharomyces* strains have not yet been studied as extensively, but interest in such genera as *Kluyveromyces* and *Saccharomycopsis* is increasing as researchers utilize these organisms for industrial purposes.

A variety of techniques are currently being employed to effect genetic changes in industrial yeast strains. These include classical hybridization, mutation, spheroplast (protoplast) fusion, electrofusion and transformation. The last of these techniques is dealt with in Chapter 4. Each of the techniques offers advantages and disadvantages, and thus a combination is often necessary to achieve the desired objective. Each technique will be reviewed and recent successes discussed.

## 3.2 Whole-cell hybridization

The study of yeast genetics was pioneered by Winge and his co-workers at the Carlsberg laboratory in Denmark. In 1935 they established the haploid–diploid life cycle (Winge 1935). Yeast can display two mating types, designated *a* and $\alpha$, which are manifested by the extracellular production of an *a* or an $\alpha$ mating pheromone. When *a* haploids are mixed with $\alpha$ haploids mating takes place and diploid zygotes are formed (Fig. 3.1). These diploids can be induced to sporulate forming four-spored asci. Ascus walls can be removed by the use of glucanase preparations such as snail gut enzyme. The four spores from each ascus can be isolated by use of a micromanipulator, induced to germinate and subsequently employed for further hybridization trials. Both haploid and diploid organisms can exist stably and undergo cell division via mitosis and budding.

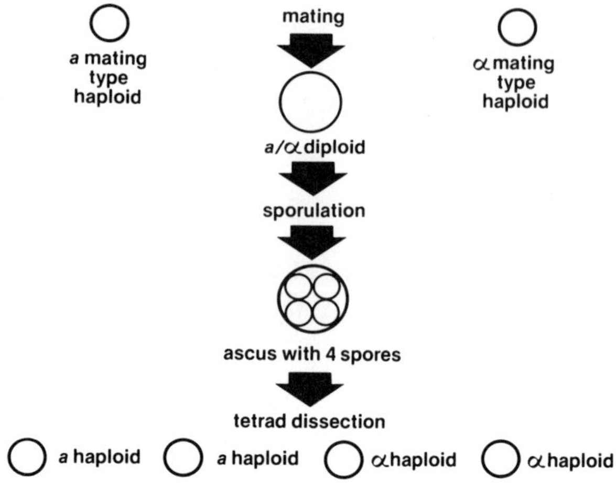


**Figure 3.1** Haploid–diploid life-cycle of *Saccharomyces* sp.

Although the technique of hybridization fell into disfavour for a number of years when recombinant DNA was thought to be the solution to all future gene manipulation requirements, it has again gradually come to be accepted as a very valuable technique. The production of a hybrid yeast by this technique has three prerequisites. First, it must be possible to induce sporulation in the parent strains; secondly, single spore isolates must be viable; and, thirdly, hybridization (mating) must take place when the spores or single-spore cultures are placed in contact with one another. Industrial strains have often showed little or no mating ability, poor sporulation and low spore viability. Research in recent years has shown that it is possible to increase sporulation ability in many of these industrial strains thus making them much more amenable to hybridization (Stewart & Russell 1986).

### *3.2.1 Induction of sporulation*

Traditionally, spore formation in diploid cultures is achieved simply by transferring them to a starvation medium such as potassium acetate. This results in the formation of four-spored asci. However, when attempting to sporulate industrial strains of *Saccharomyces* those characteristics which made them so suitable for production purposes (i.e. their genetic stability and reduced susceptibility to mutational forces) has resulted in low spore yields and low spore viability. This has presented difficulties in the hybridization of many industrial yeasts.

In recent studies in this laboratory (Bilinski *et al.* 1986a) a marked improvement in the sporogenic ability of a polyploid strain of *Saccharomyces cerevisiae* was observed on cultivation of cells in liquid rather than on agar culture media. Also, culture conditions for improved tetrad production were identified. These conditions are sporulation at 21°C instead of the usual 27°C and the substitution of glucose with acetate as the presporulation carbon source. A marked induction of ascus formation was obtained in sporulation medium employing exponential-phase acetate-grown rather than stationary-phase glucose-grown cells. In a strain of *Saccharomyces uvarum* (*carlsbergensis*), used in lager beer production, similar results were obtained except maximum ascus yields occurred when acetate-grown cells harvested from stationary phase were transferred to sporulation medium (Bilinski *et al.* 1986b). This increase in sporulation and tetrad formation has permitted the isolation of an array of segregants from ale- and lager-brewing strains.

### *3.2.2 Isolation of spores*

In order to facilitate the isolation of spores, they must be removed from the ascus wall. To digest the ascus, an inoculum from sporulation agar is placed in a test tube containing a 1:40 dilution of filter-sterilized snail enzyme (e.g. Glusulase from Endo Laboratories Inc., Garden City, New York, USA). This is a simple and rapid method of obtaining liberated spores for the purpose of tetrad dissection. There are numerous alternative methods employed when the objective is to obtain a large number of single spores, which involve heat treatment, paraffin treatment and electrophoresis to separate vegetative cells from spores.

The tetrads are dissected with the aid of a very fine glass needle mounted in a micromanipulator. The operation of this instrument requires considerable training and experience, but in time tetrads can be dissected with relative ease (Fowell 1969). The single spores are allowed to germinate on the agar surface on which they have been dissected. Tetrad analysis can then be employed to determine the genetic make-up of industrial strains. By reducing the strain to haploid units it is possible to determine the number of genes present for such characteristics as maltose uptake or flocculation. Then, depending upon what is desired, spores containing the appropriate genotype can be mated to each other to reconstruct the strain.

A technique that is often employed in the investigation of industrial strains is random spore analysis. This procedure is most often employed when the yeast strain under study exhibits a lack of tetrad formation or poor spore viability. This technique involves the removal of the ascus wall and subsequent agitation to release the spores. When plated on nutrient agar the haploids can be qualitatively distinguished by their relatively smaller colonial morphology.

### *3.2.3 Hybrid production*

Vegetative cells can be hybridized under the microscope by the use of a micromanipulator to mate single spores from the asci of two parental yeasts to form a zygote. This method is usually only employed for yeasts whose spores readily conjugate; however, many strains do not meet these requirements, in which case a mass-mating technique is applied. Although the mass-mating technique is less time consuming and matings are easily obtained, it is susceptible to error.

Hybridization has been used in industrial research laboratories to produce novel strains with superior fermentation capabilities. For example, typical

*Saccharomyces* yeast strains used for the production of industrial and fuel ethanol are incapable of dextrin and starch utilization. Therefore, the addition of commercial amylases to hydrolyse the starch is required, and this presents a significant expense for the industry. However, the yeast species *Saccharomyces diastaticus* is capable of its own glucoamylase production, thereby reducing the amount of exogenous enzyme addition required in alcoholic fermentations (Russell *et al.* 1986a). Results from this laboratory report on the improvement in ethanol production in a corn mash (Whitney *et al.* 1985) and a cassava mash (Russell *et al.* 1986b) using a diastatic yeast strain genetically modified by hybridization. The mass-mating technique is employed when it is not easily possible to obtain matings by direct hybridization. In mass-mating procedures high densities ($10^8$ cells $ml^{-1}$) of cells of opposite mating type are mixed and regenerated by plating on to complete medium. It is a less time-consuming method, but is subject to error in contrast with hybridization, where two cells of opposite mating type are observed to form a zygote directly under the microscope.

### *3.2.4 Rare mating*

Rare mating is a technique employed when normal hybridization procedures are ineffective. When non-mating strains are mixed at a high cell density a few true hybrids with fused nuclei form, and these can usually be selectively isolated (Tubb *et al.* 1981). A possible disadvantage of this method is that while incorporating nuclear genes into, for example, a brewing yeast, it is probable that many undesirable properties may also be inherited from the non-brewing partner. For example, in the construction of dextrin-fermenting brewer's yeast, the introduction of the *POF* gene (*P*henolic-*O*ff-*F*lavour) – which imparts the ability to decarboxylate ferulic acid to 4-vinyl guaiacol, giving beer a phenolic or clove-like off-flavour (Goodey & Tubb 1982) – made the hybrid unsuitable.

Rare mating can also be used to introduce cytoplasmic elements into one strain without transfer of nuclear genes from the other. The process, termed cytoduction (Zakharov & Yarovoy 1977), requires the presence of a specific nuclear gene mutation *Kar* (Conde & Fink 1976) which impairs nuclear fusion. Cytoduction can be used to achieve three main goals: (a) substitution of the mitochondrial genome; (b) transfer of double-stranded RNA species and (c) introduction of DNA plasmids.

When used in the substitution of the mitochondrial genome, it is possible to study the effects of these genetic elements on various cell functions. Mitochondrial substitution has been demonstrated to bring about variations in respiratory functions, cell surface activities (Wilkie & Nudd 1981) and various other strain characteristics (Conde & Mascort 1981, Russell *et al.* 1985). Rare mating has also been used to transfer 'zymocidal' or 'killer' factor from laboratory haploid strains to brewing yeast strains without altering the primary fermentation characteristics of the brewer's yeast strain (Young 1983, Hammond & Eckersley 1984, Russell & Stewart 1985). The 'killer' character of *Saccharomyces* spp. is determined by the presence of two species of cytoplasmically located dsRNA plasmids (Bevan & Somers 1969). The rare mating protocol

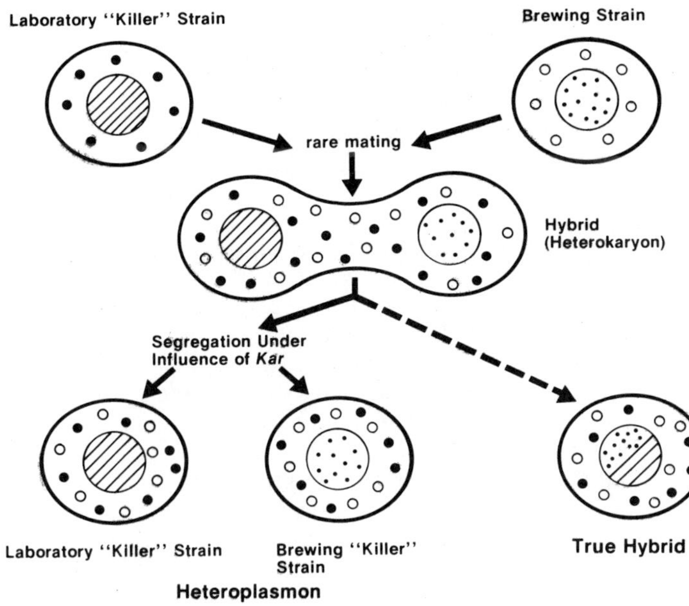


**Figure 3.2** Rare mating protocol to produce brewing strains with zymocidal ('killer') activity.

depicted in Figure 3.2 demonstrates the introduction of killer factor into a brewing yeast. Finally, rare mating used to introduce DNA plasmids has been successful in the introduction of specific genetic elements constructed by gene-cloning experiments (Kielland-Brandt 1979, Toh-e *et al.* 1980).

## 3.3 Mutation

Mutation is any change that alters the sequence of bases along the DNA molecule, thus modifying the genetic material. Chemical and physical treatments are used to induce mutation frequencies to detectable levels. One disadvantage of the approach is that mutagenesis is a destructive process and can cause gross rearrangement of the genome, resulting in numerous undesirable changes. In addition, mutagenesis is seldom employed with industrial food and beverage yeast strains. These strains are usually polyploid, and after mutagenic treatment the mutations usually do not reveal themselves because of the presence of non-mutated alleles. An important point in mutant screening is the rate of mutation. This depends upon the mutagen employed, the dose rate and the physiological conditions. For example, by changing the mutagen or the environment the dose can be adjusted to obtain the maximum mutation frequency. A moderate mutagen dose (i.e. one with a 20% survival rate) has been found advisable for obtaining superior strains (Alikanian 1970, Backus & Stauffer 1955). The chance of forming multiple mutations is reduced, which therefore increases the frequency of isolating strains differing at a single locus but which are otherwise isogenic (Sermonti 1979). Many applications of mutation have met with success, and these will be briefly reviewed.

### *3.3.1 Catabolite repression*

Glucose repression is the effect conferred by high concentrations of glucose on many strains of *Saccharomyces* yeasts, which renders various substrate-utilizing systems inactive. These repressing effects can be seen in a brewer's wort fermentation (Pfisterer *et al.* 1978, Stewart *et al.* 1985). Brewer's yeasts are normally capable of utilizing sucrose, glucose, fructose, maltose and maltotriose, in this sequence, leaving the dextrins unfermented. Only when approximately 50% of the wort glucose has been taken up by the yeast will maltose uptake begin (Phillips 1955, Stewart *et al.* 1979). Therefore, glucose repression is a major limiting factor in wort fermentation. The non-metabolizable glucose analogue 2-deoxyglucose (2-DOG) (Laires *et al.* 1983) was used to simulate conditions of catabolite repression by glucose, thus providing a means to isolate spontaneously arising mutants resistant to the repressive effect of glucose (Fig. 3.3).

A number of stable 2-DOG mutants were found to be capable of utilizing maltose in the presence of significant concentrations of glucose. For example, 2-DOG resistant mutants of *Saccharomyces cerevisiae* in a synthetic medium

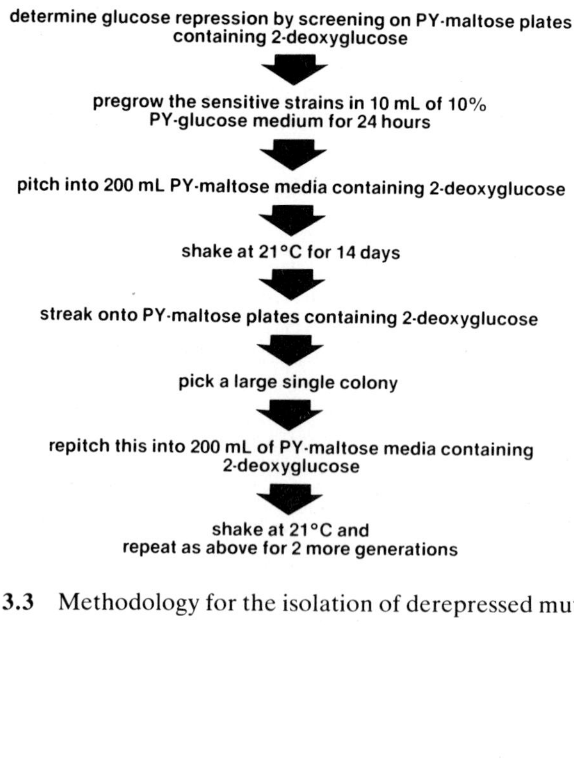


**Figure 3.3** Methodology for the isolation of derepressed mutants.

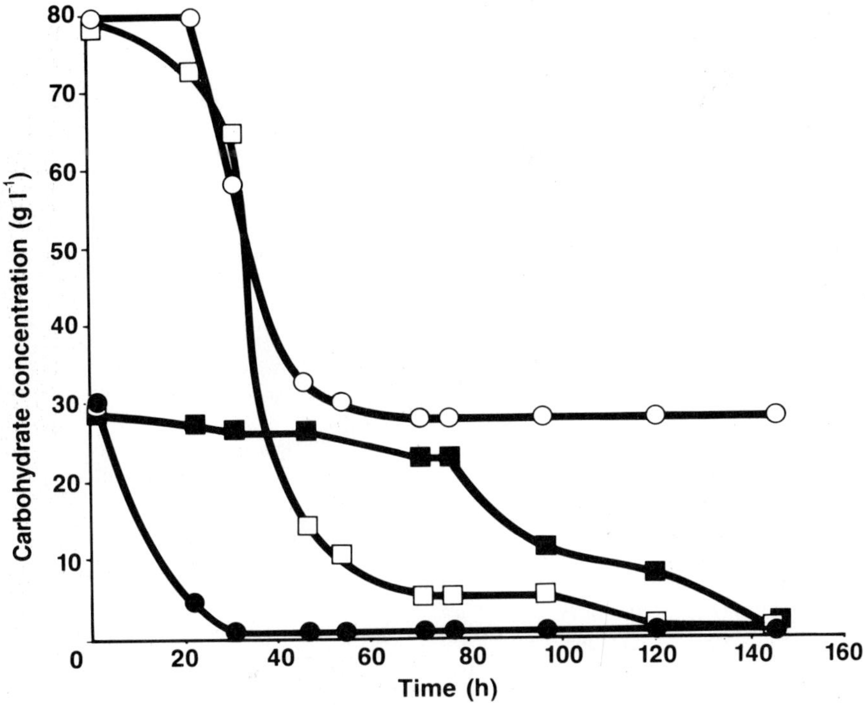


**Figure 3.4** Carbohydrate uptake in an 8% (w/v) maltose 3% (w/v) glucose peptone–yeast extract medium by polyploid *Saccharomyces cerevisiae* strain 154 and its derepressed mutant. Glucose uptake, parent (●—●) and mutant (■—■); maltose uptake, parent (○—○) and mutant (□—□).

containing both maltose 8% (w/v) and glucose 3% (w/v), were able to metabolize the maltose completely (Fig. 3.4) whereas in the same medium, maltose uptake was slow with the parental strain and only 60% complete when fermentation ceased. Fermentation and ethanol formation rates in 12°Plato wort were also increased in the 2-DOG mutants when compared with the parental strain (Fig. 3.5).

Bailey *et al.* (1982) have described a mutant allele in *Saccharomyces cerevisiae* that is characterized not only by 2-DOG resistance, but also by insensitivity to glucose repression for invertase production, maltose and galactose uptake, as well as the mitochondrial enzyme cytochrome *c*-oxidase. The levels of hexokinase activity towards both fructose and glucose were also approximately three times higher in the mutants. The molecular mechanisms by which catabolite repression is affected in yeast are still unclear, although many theories have been discussed (Boucherie 1985). Nevertheless, derepressed mutants of industrial yeast strains will be of significant economic importance, particularly with respect to increased control of fermentation rate.

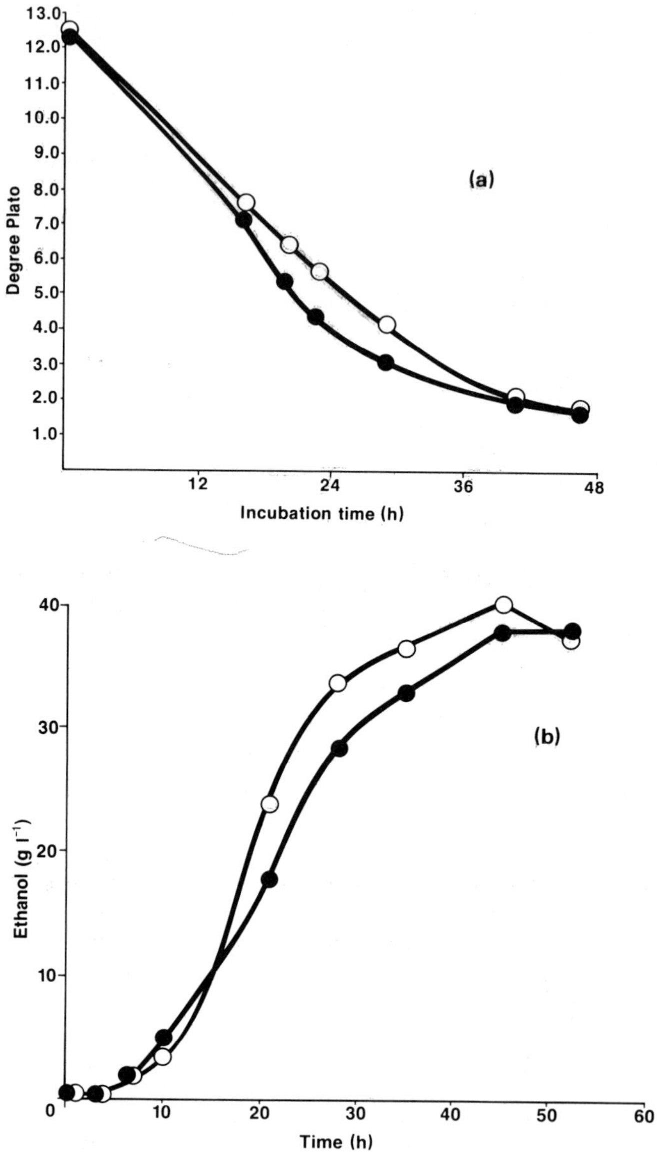


**Figure 3.5** (a) Wort fermentation by a polyploid *Saccharomyces cerevisiae* strain 154 (○—○) and its derepressed mutant (●—●); (b) ethanol production by a polyploid *Saccharomyces cerevisiae* strain 154 (●—●) and its derepressed mutant (○—○).

### 3.3.2 *Production of auxotrophic markers*

Mutation has been used to produce auxotrophic strains from commercial yeasts which are then useful as spheroplast fusion partners or as recipients for DNA transformation experiments. Due to the polyploid nature of industrial strains, the mutation often does not reveal itself, due to the presence of a number of non-mutated alleles, and only if the mutation is dominant is the phenotypic expression obtained. However, it has been possible to uncover recessive auxotrophic markers in brewing strains by exposure to benomyl, a commercially available fungicide (Delgado & Conde 1983, Bilinski *et al.* 1984) (DuPont Corp.) the active component of which has been reported to inhibit β-tubulin polymerization and to induce chromosome loss in *Saccharomyces* spp. (Wood 1982a, b). A number of auxotrophic brewing yeasts were isolated in this manner in this laboratory. The fermentation profiles of a threonine-requiring and of a lysine-requiring mutant of a *Saccharomyces cerevisiae* brewing strain in a 12.3°Plato wort (Fig. 3.6) demonstrated a reduced fermentation rate, even with additional amino acid supplementation, when compared with the wild type strain. Attempts at fusing these two strains proved unsuccessful.

Another method used to obtain auxotrophic markers in yeast employs the method of thymidylate starvation (Barclay & Little 1977). When yeast is grown in a medium containing aminopterin and sulphanilamide, deoxythymidine 5′-monophosphate (dTMP) biosynthesis is blocked. This results in cells undergoing thymineless death, unless they contain a specific mutation which results

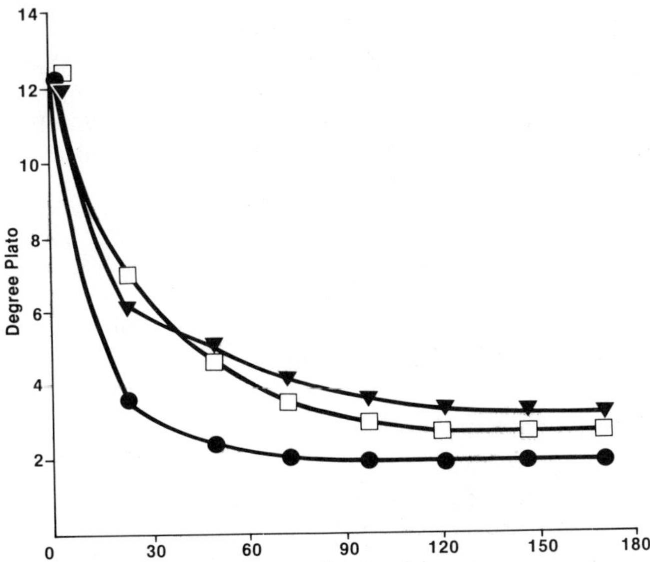


**Figure 3.6** Fermentation of a 12.3°Plato wort by a *Saccharomyces cerevisiae* polyploid yeast strain and its induced auxotrophic mutants: ●—●, wild type; □—□, lysine auxotroph; ▼—▼, threonine auxotroph.

in a block in RNA or protein synthesis. The addition of purines, pyrimidines or amino acids such as adenine, histidine or methionine can be used as an enrichment for such mutants; therefore, spontaneously occurring mutants in the population are at a selective advantage. In this manner mutants auxotrophic for methionine, lysine, leucine, histidine, tyrosine or uracil have been recovered. In this laboratory the method has been successfully employed and stable uracil-requiring mutants isolated from a polyploid ale-brewing yeast strain.

### *3.3.3 Induction of respiratory-deficient mutants*

Mitochondrial mutants lacking the ability to respire have become of interest due to their use as partners in spheroplast fusion experiments, for the increased production of ethanol under aerobic conditions (Esser *et al.* 1982), improved alcohol yield, productivity and growth rate of *Candida pseudotropicalis* (Moulin *et al.* 1982) and the improved final ethanol production by *Saccharomyces cerevisiae* from barley (Garcia-Acha & Grilione 1982).

Nuclear respiratory-deficient mutants (designated *pet*) are distinguished from cytoplasmic respiratory-deficient mutants (designated $\varrho^-$) using a special screening for temperature-sensitive lactate utilization (Schweizer *et al.* 1977). The temperature sensitivity is demonstrated by growth at 22°C, but not at 38°C. Respiratory-deficient mutants are characterized by their inability to oxidize non-fermentable substrates such as acetate, lactate, glycerol and ethanol. A number of methods have been found to induce respiratory mutations in yeast. These methods include the intercalating dyes acriflavine and ethidium bromide, nitrous acid, ultraviolet light, N-nitroso-N-methyl urethane and denaturing agents such as sodium dodecyl sulphate, guanidine–HCl, sodium lauryl sulphate and sodium persulphate (Bacila *et al.* 1978). Recently, it has been demonstrated that the formation of cytoplasmic respiratory-deficient mutants by acriflavine is completely inhibited by the addition of 4.0% methanol (Hamada *et al.* 1984).

It should also be noted that in many haploid, diploid and polyploid strains of *Saccharomyces* cytoplasmic respiratory-deficient formation is a spontaneous event not requiring the addition of a mutagen. Mutants can be selectively isolated using the triphenyl-tetrazolium chloride (TTC) overlay technique (Ogur *et al* 1957) in which respiratory sufficient colonies stain red and respiratory deficient colonies remain white.

### *3.3.4 Production of other selectable markers in yeast*

Mutagenesis of yeast is also used in the production of a variety of other selectable characteristics of industrial importance. For example, under normal conditions yeast cells are capable of the transport of thiamine (vitamin $B_1$), available in a brewer's wort, into the cell. Since vitamin $B_1$ is an important constituent for human nutrition it would be desirable for beer to contain at least the recommended dietary allowance. Using high doses of ultraviolet radiation treatment on haploid strains of *Saccharomyces cerevisiae* and *Saccharomyces uvarum* it has been possible to isolate a mutant capable of thiamine synthesis

(Silhankova 1985). Moreover, this mutant is also capable of thiamine excretion into the fermentation medium without any severe loss of the fermentation abilities of the yeast strain.

The carcinogenicity of various metals and metal ions to human health has been investigated by determining their mutagenic effect on nuclear and mitochondrial DNA in yeast (Singh 1984). Also of interest is the production of yeast mutants with markers for spheroplast fusion experiments such as resistance to chromium (Ono & Weng 1982), canavanine (Fogel *et al.* 1981), copper (Fogel & Welch 1982), Geneticin G418 (Jimenez & Davies 1980) and mercury (Ono & Sakamoto 1985).

## 3.4 Protoplast (spheroplast) fusion

When the walls of two cells of different mating types come into contact with each other they lyse at the point of contact and fuse. In the case of cells with the same mating type, or those with none at all, the wall does not lyse and fusion does not occur. Since many yeasts do not possess a mating type due to their polyploid or asexual nature, methods have been developed to overcome the barriers to conjugation. One such method involves the enzymatic removal of the yeast cell wall to produce protoplasts (spheroplasts). When protoplasted cells come into contact with each other, cell fusion between cells of either the same or no mating type may occur.

The basic method used by laboratories to fuse yeast protoplasts is universal. The first step involves the production of protoplasts or spheroplasts. A protoplast is defined as: 'the resulting form of a plant cell or microorganism when its cell wall has been mechanically or enzymatically removed. A spheroplast has all the properties of a protoplast but contains remnants of the cell wall' (Stewart 1981a). It is desirable that remnants of the cell wall remain, since protoplasts devoid of any cell wall exhibit poor cell-wall regeneration (Hinnen & Meyhack 1982). There are principally four methods used for the isolation of protoplasts (Villanueva & Garcia-Acha 1971). The first involves the use of mechanical or autolytic enzymes. The efficiency of autolytic enzymes in the formation of protoplasts can be greatly affected by the age, strain, culture conditions (Phaff 1963, Villaneuva 1966) and growth stage (Eddy & Williamson 1957). It has also been found that cells pretreated with mercapto-compounds such as $\beta$-mercaptoethanol or dithiothreitol tend to protoplast easier, since the glycosidic bonds are rendered more susceptible to enzymatic cleavage (Davies & Elvin 1964, Duell *et al.* 1964, Russell *et al.* 1973). The second method employs an enzyme to dissolve the cell wall. These can be: (a) snail enzymes isolated from the digestive juice of the snail *Helix pomatia* (i.e. glusulase); (b) microbial enzymes isolated from species such as *Bacillus*, *Streptomyces*, *Micromonospora* and, more specifically, zymolyase from *Arthrobacter luteus* (Kitamura & Yamamoto 1972) or (c) other enzyme preparations including proteolytic enzymes (i.e. trypsin and papain) (Eddy 1958), pancreatic lipase (Garcia-Acha *et al.* 1966) and phosphomannanase from *Bacillus circulans* (Nagasaki *et al.* 1966). The third method used for the production of spheroplasts involves the inhibition of cell-wall synthesis without affecting other reactions within the

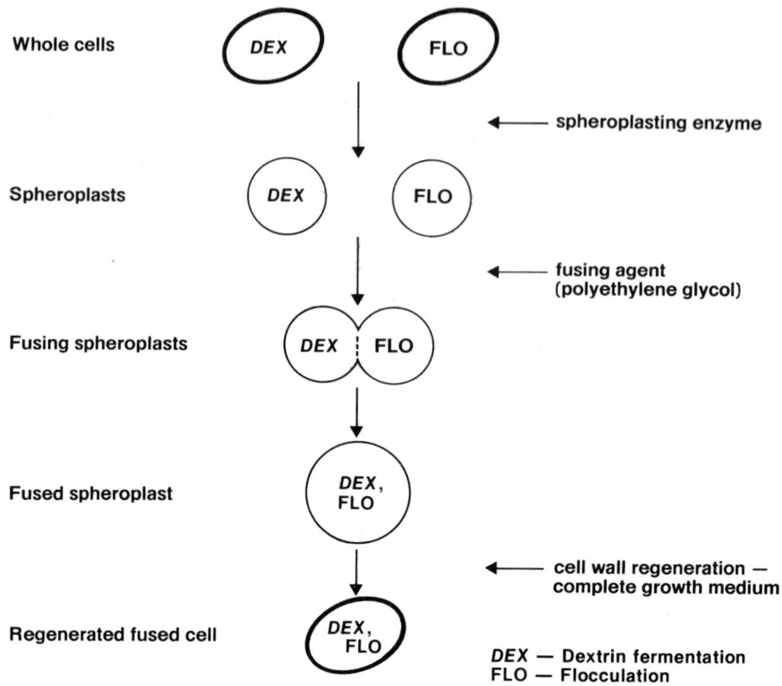


**Figure 3.7** Spheroplast fusion in yeast.

protoplasm. The result is called 'unbalanced growth', where cells are formed with abnormal or no cell walls. The fourth and final method involves the growth of cells on thickened guinea-pig serum, on which, after 7–14 days, abnormal fungal cells – reported to be protoplasts – were observed (Villanueva & Garcia-Acha 1971).

The methods most commonly used for yeast protoplast production are the use of snail enzymes and enzymes from *Arthrobacter luteus*. Once the cell wall has been removed, the cells are osmotically fragile and lysis will occur unless they receive external osmotic support. A variety of media have been used to stabilize protoplasts (Villanueva & Garcia-Acha 1971). Among these are the sugar alcohols (mannitol and sorbitol), salts (KCl, NaCl and $MgSO_4 \cdot 7H_2O$), and sugars (rhamnose, maltose, sucrose, xylose, fructose and sorbose). The concentrations of stabilizers required vary in different species, as the internal osmotic pressure differs.

Mutants with auxotrophic antibiotic resistance, temperature sensitivity, respiratory deficiency or morphological markers are employed for fusion (Fig. 3.7). After treatment, cells are plated in an agar medium such that only those with nutritional complementation (i.e. fusion products) are selected. The agar medium also allows for cell-wall regeneration. Controls are employed to detect reverse mutations or genetic transfer in ways other than by protoplast fusion. The presumed fusion products are then characterized using cytological, biochemical and genetic methods.

Spheroplast fusion of haploids of opposite mating type would generally achieve the same results as mating, except that direct mating would be significantly less time consuming than spheroplast fusion. However, when mating will not occur due to incompatibility or lack of mating type, spheroplast fusion achieves the desired result. The majority of fusion products recovered are stable and contain a full genetic complement from both parents.

### *3.4.1 Examples of protoplast fusion*

From the plethora of literature it is now obvious that the protoplast-fusion method has rapidly become a successful procedure for producing hybrids between yeast strains. A review of the successes to date is discussed in this section.

**Intrageneric fusions**

Among haploid laboratory strains of *Saccharomyces cerevisiae* the process of spheroplast fusion has been intensively investigated (Van Solingen & van der Plaat 1977, Yamamoto & Fukui 1977, Svoboda 1978). Cells of opposite and identical mating types have been fused using protoplast fusion techniques, constructing hybrids similar to those conducted by classical mating. The diploid hybrids received an equal genetic complement from both parents.

Recently, Terrance and Lipke (1981) reported on the use of sex pheromones to increase sexual agglutinability in *Saccharomyces cerevisiae*. Mating type *a*-cells synthesize and secrete a peptide pheromone *a*-factor, which affects cells of $\alpha$-mating type. Conversely $\alpha$-cells produce the peptide pheromone $\alpha$-factor which affects cells of *a*-mating type. When either mating type was treated with the appropriate sex pheromone, cell–cell binding was significantly increased. They also determined that the agglutination observed was similar in pheromone-treated and untreated cells.

Kao and Michayluk (1974) first suggested the use of polyethylene glycol (PEG) as an aid to increasing the fusion frequency of plant protoplasts. Protoplasts of fungi (Ferenczy *et al.* 1975, Anne & Peberdy 1976) and bacteria (Fodor & Alfoldi 1976, Schaeffer *et al.* 1976) have also been fused using PEG. Since PEG-induced fusion of yeast spheroplasts was first reported (Ferenczy *et al.* 1976, van Solingen & van der Plaat 1977), the method has grown in popularity. For example, the PEG-induced method of protoplast fusion has also been used successfully to construct stable diploids from identical mating type strains of *Schizosaccharomyces pombe* (Sipiczki & Ferenczy 1977b). Cells were protoplasted with enzymes from *Trichoderma viride* and mixed with PEG. The fusion frequency was 0.17%, producing cells with increased amounts of genetic material per cell unit.

There have also been reports of the successful fusion of *Rhodosporidium toruloides* protoplasts both of $\alpha$-mating type (Sipiczki & Ferenczy 1977a). Spheroplasts of haploid strains of identical mating type of the starch-fermenting strains of *Saccharomyces diastaticus* have been fused in the presence of $CaCl_2$ and PEG (Sakai *et al.* 1986). The resulting hybrids produced more ethanol and amylase than the parental strains. They therefore proposed the use of spheroplast fusion as a method of breeding potent starch-fermenting strains from haploid cultures.

Successful intraspecific fusions have also been described for diploids and haploids of many genera of yeast other than *Saccharomyces*, thus providing a useful tool into the studies of the genetic properties of organisms with either commercial or clinical significance. For example, the yeast *Candida albicans* is the most commonly encountered fungal pathogen in humans; however, since it lacks any sexual or asexual process of genetic recombination it was not possible to investigate variations in its genetic characteristics. Recently, it has been reported (Sarachek *et al.* 1981) that protoplasts of complementing auxotrophs of *Candida albicans* have been fused in the presence of PEG. The initial products isolated were slow-growing heterokaryons which stabilize to form cells harbouring a single nucleus. Spontaneous fusions between these cells occurred, producing rapidly growing, uninucleate hybrids of varying levels of ploidy. Thus, the technique of protoplast fusion has produced hybrids suitable for chromosomal or genetic analysis where in the past this has not been possible. A closely related yeast, *Candida tropicalis*, has also been successfully fused utilizing PEG (Fournier *et al.* 1977), producing unstable heterokaryons which readily dissociated into uninucleate hybrids of varying levels of ploidy as described with *Candida albicans*. Again, the heterokaryon nuclei fused under suitable conditions producing recombinants of different ploidy levels with various auxotrophic characteristics.

The first reported interspecific yeast hybrid fusion was between *Kluyveromyces lactis* and *Kluyveromyces fragilis* protoplasts (Whittaker & Leach 1978), and since then a variety of fusions of commercial importance have been cited. Fusions between two haploid strains of *Schwanniomyces alluvius*, an asexual yeast species recognized for its ability to hydrolyse starch and ferment dextrin, has resulted in the production of stable diploid recombinants (Wilson *et al.* 1982). Another species of the genus *Schwanniomyces castellii* is also capable of starch hydrolysis, due to the presence of an amylolytic enzyme system. The two species differ in the mechanism and effectiveness of their amylolytic enzyme systems (Sills & Stewart 1982), probably due to an additional starch-hydrolysing glucoamylase in *Schwanniomyces castellii* (Dhawale & Ingledew 1983b). Both yeasts are known to contain at least one indigenous $\alpha$-amylase. Using fusion techniques, stable diploid products between the two normally haploid *Schwanniomyces* species have been constructed (Dhawale & Ingledew 1983a). The goal of this work was to incorporate the amylolytic systems of both species, and thus produce a superior starch-hydrolysing yeast that could be employed industrially in fuel ethanol production.

The direct fermentation of D-xylose to ethanol is also of commercial importance. It has been reported to occur in several yeasts such as *Pachysolen*, *Candida*, *Kluyveromyces*, *Hansenula* and *Pichia* (Maleszka & Schneider 1982); however, the ethanol yield was commonly found to be too low to be economically significant. Results reported for a strain of *Candida shehatae* (du Preez & van der Walt 1983) demonstrated promising results from a 9% D-xylose fermentation, and protoplast fusion was utilized to increase the gene pool of this yeast in the hope of improving its fermentation ability (Johannsen *et al.* 1985). It was determined that diploid, triploid and tetraploid strains of *Candida shehatae* could be constructed by spheroplast fusion. Each contained a successive increase in mean cell length, width and volume compared with the parental haploid strain. It was also demonstrated that in a 9% D-xylose fermentation the rate of ethanol production was improved with increasing cell ploidy.

In an effort to overcome the inhibitory effect of high initial sugar concentrations on ethanol production by *Saccharomyces cerevisiae*, fusions have been performed with the osmotolerant yeast *Saccharomyces mellis* (Leamann & Margalith 1986). Resultant fusion products gave significantly improved fermentation efficiencies. Moreover, superior fermentation profiles were found compared with *Saccharomyces cerevisiae* at higher sugar concentrations. It was further shown that the hybrids accumulated less ethanol within the cell than *Saccharomyces mellis*, yet with higher tolerance to high sugar concentrations than *Saccharomyces cerevisiae*.

The techniques of protoplast fusion have also been applied in genetic studies involving a variety of other yeasts such as *Saccharomycopsis lipolytica* (Stahl 1978), *Hansenula polymorpha* (Savchenka & Kapul'tsevich 1980) and *Pichia guillermondii* (Bottcher *et al.* 1986) in which stable products have been obtained.

**Intergeneric fusions**

Many attempts have been made at intergeneric fusions with varying degrees of success being achieved. In the past, with the exception of *Candida utilis* for single-cell protein, non-*Saccharomyces* yeasts have been overlooked for industrial applications, due to their poor fermentation ability. However, intergeneric fusion is now providing a means of studying the characteristics of such yeasts after fusion with *Saccharomyces* strains. Some of the earlier successes involved the fusion of *Candida tropicalis* and *Saccharomycopsis fibuligera* (Provost *et al.* 1978), *Saccharomycopsis lipolytica* and *Candida guillermondii* (Spata & Weber 1980), *Saccharomyces cerevisiae* and *Saccharomycopsis lipolytica* (de van Broock *et al.* 1980) and *Saccharomyces cerevisiae* and *Schizosaccharomycopsis lipolytica* (Svoboda 1980).

More recently attempts have been made at fusing *Schwanniomyces alluvius* and *Saccharomyces uvarum* (*carlsbergensis*) (Dhawale & Ingledew 1983a) in an attempt to introduce the amylolytic enzyme system of *Schwanniomyces* into *Saccharomyces* strains. Unfortunately, the hybrids formed were unstable and spontaneously segregated into the original auxotrophic parent cultures. It has been postulated that, due to their genetic diversity (*Saccharomyces* species have 17 chromosomes while the number in *Schwanniomyces* is still in question), stable nuclear fusions between the two genomes will be difficult to obtain.

Successful fusions have been reported between *Saccharomyces cerevisiae*, *Saccharomyces diastaticus* and the following strains: *Pichia membranaefaciens*, *Hansenula capsulata*, *Saccharomyces rosei* and *Schizosaccharomyces pombe* (Spencer & Spencer 1982). Assimilation assays confirmed genomic and mitochondrial fusions between species.

Protoplast fusion of a sake yeast strain of *Saccharomyces cerevisiae* and a lactose-utilizing yeast strain *Kluyveromyces lactis* has resulted in stable hybrids demonstrating complementing sugar utilization and vitamin-requiring characteristics (Taya *et al.* 1984). In addition, the fusion product was capable of improved growth and ethanol production rates compared with the *Kluyveromyces lactis* parental strain in glucose and lactose media. *Kluyveromyces lactis* has also been fused with the hydrocarbon-utilizing yeast *Yarrowia* (*Saccharomycopsis*) *lipolytica* (Groves & Oliver 1984). Of the fusion products obtained, some were capable of erythritol and n-decane utilization similar to *Yarrowia lipolytica* in addition to the sucrose and galactose-assimilation capability of the *Kluyveromyces lactis* parent. The fusion of *Kluyveromyces fragilis*, a yeast strain capable of lactose fermentation, and *Saccharomyces cerevisiae*, a yeast with high ethanol tolerance, has also been undertaken (Farahnak *et al.* 1986). Fusants obtained were capable of assimilating lactose and producing in excess of 13% (v/v) ethanol.

Studies on DNA reassociation have recently been conducted on a number of interspecific and intergeneric fusion products. In one publication it was determined that an intergeneric fusion of *Zygosaccharomyces rouxii* and *Saccharomyces diastaticus* yielded hybrids bearing a higher degree of DNA reassociation with the *Saccharomyces diastaticus* than with that of the *Saccharomyces rouxii* parental strain (Spencer *et al.* 1985b).

**Commercial yeast fusions**

As previously discussed, commercial (industrial) strains of *Saccharomyces cerevisiae* employed for brewing, distilling, baking and fuel ethanol are generally polyploid and, consequently, do not possess a mating type. To circumvent the hybridization problems encountered, the technique of spheroplast fusion can be employed. However, other difficulties are often encountered. One such problem is the lack of genetic markers in many of the industrial yeasts. This can be overcome using spheroplast fusion in conjunction with many of the mutation methods described previously. The yeast is first mutated, by any number of methods, to obtain a derivative which contains a selectable marker. This strain is then used in subsequent fusion experiments.

Recent successful fusions with industrial or commercial implications include: the construction of a brewing yeast with amylolytic activity, thus being capable of converting dextrin and starch to ethanol by the fusion of *Saccharomyces cerevisiae* and *Saccharomyces diastaticus* (Freeman 1981, Stewart 1981b); a polyploid *Saccharomyces cerevisiae* capable of high ethanol production by fusion of a flocculent strain with saké yeasts (Seki *et al.* 1983); an industrial strain with improved osmotolerance by fusion of *Saccharomyces diastaticus* and *Saccharomyces rouxii* (Spencer *et al.* 1985a); and an industrial strain capable of cellobiose utilization and ethanol production by the fusion of *Saccharomyces cerevisiae* and *Zygosaccharomyces fermentatia* (Pina *et al.* 1986).

In all of these instances the spheroplast fusion technique has allowed the construction of industrially important hybrids containing new and unique fermentation or growth characteristics. Once constructed, numerous large-scale fermentation trials are required to investigate the yeasts' performance and end-product suitability.

## 3.5 Electrofusion

### *3.5.1 Theory*

Spheroplast fusion techniques have a number of limitations and these have been discussed in detail by Zimmerman and Vienken (1982):

(a) optimum fusion conditions vary between each species;
(b) the fusion process cannot be visually monitored under a microscope;
(c) cell number to be fused cannot be preselected;
(d) the fusion process often extends over a long time;
(e) a loss of intracellular material occurs, which may decrease hybrid viability;
(f) fusogenic compounds also affect cell viability; and
(g) hybrid yield is low.

In order to overcome these problems a novel fusion technique, electric field-induced fusion, has been proposed (Zimmermann 1982, 1983, Zimmermann *et al.* 1980). The technique is based on dielectrophoresis with subsequent electrofusion.

Dielectrophoresis is a process in which close membrane contact is established between cells by application of a non-uniform alternating field. In this situation the ions within the cell move with the field. If the frequency is increased sufficiently, then the ions cannot shift rapidly enough, become out of phase and an induced dipole results within the cell. Since one field of the chamber is made stronger than the other, the cells will move towards the stronger region. Mutual attraction occurs between cells, since each sees the other to be an electrode, and the result is the formation of cell 'chaining'. Once dielectrophoresis is complete (i.e. the cells are in close membrane contact) they are induced to fuse by the application of a d.c. pulse which causes membrane disruption. When sufficient voltage is applied, the ions on both sides of the touching membranes force pore formation in the membranes and cell cytoplasms are allowed to fuse. For a complete review on the theory and protocol involved in dielectrophoresis and electrophoresis see Zimmermann and Vienken (1982).

### *3.5.2 Examples of electrofusion*

Since electrofusion is based on a physical process common to all membranes, either biological or artificial, the method can be applied to a wide variety of systems. For example, some of the earlier successes obtained with this system involved the construction of large liposomes from synthetic and natural lipids by electrofusion (Buschl *et al.* 1982). Success of the method was also demonstrated by the fusion of mesophyll plant protoplasts of *Vicia faba* (Zimmermann & Scheurich 1981), eggs of the sea urchin *Paracentrotus lividus* (Richter *et al.* 1981), myeloma cells (Vienken *et al.* 1984), mouse hybridoma cells (Vienken & Zimmermann 1985), mammalian cells (Blangero & Teissie 1985) and chinese hamster with mouse cell lines (Finaz *et al.* 1984).

Electrofusion, although still in its infancy, has also recently been applied to yeast-cell fusions in an attempt to increase hybrid yields. Cells are pregrown and protoplasts produced in a similar manner to that for spheroplast fusion. It has been possible to fuse a respiratory-deficient diploid strain of *Saccharomyces cerevisiae* with a respiratory sufficient diploid of the same species and obtain stable diploid products (Halfmann *et al.* 1982). Further analysis revealed that the hybrids contained the chromosomal characteristics of only one strain, but a mixed cytoplasm. Therefore, electrofusion had resulted in plasmogamy but not in karyogamy. Hybrid frequency was reported as being 1–2 $\times 10^{-2}$ magnitudes higher than that obtained by the classical fusion method, although no data were reported (Halfmann *et al.* 1982).

The electrofusion of *Saccharomyces* yeast cells of identical mating types (Halfmann *et al.* 1983), resulted in both plasmogamy (i.e. haploid cultures) at a frequency of $5 \times 10^{-2}$ and karyogamy (i.e. diploid cultures) at a frequency of $5 \times 10^{-3}$. Cell fusion, with karyogamy occurring, is dependent upon the dielectrophoretic yield, which in turn is inversely related to the electrical conductivity of the medium (i.e. the addition of cations decreases the yield) (Lopez *et al.* 1984) and directly related to the a.c. voltage, cell concentration, time of exposure (Iglesias *et al.* 1984) and the addition of $Ca^{2+}$ and $Mg^{2+}$ salts and albumin (Schnettler & Zimmermann 1985). These improved conditions have recently enabled an improvement in haploid cell fusion of the same mating type (Schnettler *et al.* 1984) and the transformation of a haploid *Saccharomyces cerevisiae* yeast with plasmid YRp7 (Kaube *et al.* 1985).

As a means of promoting yeast cell fusion electrofusion has not yet been demonstrated to be a superior method of cell fusion, since the majority of the work is still at the haploid stage. Greater study will be required on method optimization before it can replace the current spheroplast-fusion method.

## 3.6 Conclusions

The classicial genetic- and spheroplast-fusion techniques reviewed in this chapter will continue to influence the type of yeast that can be produced. Further developments and improvements in cell-fusion techniques, in conjunction with improvement in polyploid yeast sporulation, hybridization and mutation, will open new avenues to the biotechnologist.

Strains will be specifically bred for such characteristics as extracellular amylase production, carbon catabolite derepression, killer activity, increased osmo- and ethanol-tolerance, and pentose, cellobiose and lactose utilization. It is hoped that a closer association between the on-going academic yeast research and industrial yeast research will accelerate the introduction of genetically modified yeast strains into full-scale production.

## Acknowledgements

The authors wish to acknowledge the helpful assistance of C. A. Bilinski, B. Blacklock, C. M. Crumplen, K. P. Farrell and C. J. Panchal in the preparation of this manuscript. Part of the research from this laboratory that is discussed was funded by IRAP and PILP contracts from the National Research Council of Canada, and these are gratefully acknowledged.

## References

Alikanian, S. I. 1970. Applied aspects of microbial genetics. *Current Topics in Microbiology and Immunology* **53**, 91–148.

Anne, J. and J. F. Peberdy 1976. Induced fusion of fungal protoplasts following treatment with polyethylene glycol. *Journal of General Microbiology* **92**, 413–17.

Bacila, M., A. W. Xavier and J. Horii 1978. Induction of respiration-deficient mutants of *Saccharomyces* and evaluation of their efficiency for ethanol production. In *Biochemistry and genetics of yeasts*, M. Bacila, B. L. Horecker and A. O. M. Stoppani (eds), 577–94. New York: Academic Press.

Backus, M. P. and J. P. Stauffer 1955. The production and selection of a family of strains in *Penicillium chrysogenum*. *Mycologia* **47**, 429–62.

Bailey, R. B., T. Benitez and A. Woodward 1982. *Saccharomyces cerevisiae* mutants resistant to catabolite repression: use in cheese whey hydrolysate fermentation. *Applied Environmental Microbiology* **44**, 631–9.

Barclay, B. J. and J. C. Little 1977. Selection of yeast auxotrophs by thymidylate starvation. *Journal of Bacteriology* **132**, 1036–7.

Bevan, E. A. and J. M. Somers 1969. Somatic segregation of the killer (K) and neutral (N) cytoplasmic genetic determinants in yeast. *Genetical Research* **14**, 71–7.

Bilinski, C. A., A. M. Sills and G. G. Stewart 1984. Morphological and genetic effects of benomyl on polyploid brewing yeasts: isolation of auxotrophic mutants. *Applied Environmental Microbiology* **48**, 813–17.

Bilinski, C. A., I. Russell and G. G. Stewart 1986a. Analysis of sporulation in brewer's yeast: induction of tetrad formation. *Journal of the Institute of Brewing* (in press).

Bilinski, C. A., I. Russell and G. G. Stewart 1986b. Physiological requirements for induction of sporulation in lager yeast. *Journal of the Institute of Brewing* (in press).

Blangero, C. and J. Teissie 1985. Ionic modulation of electrically induced fusion of mammalian cells. *Journal of Membrane Biology* **86**, 247–53.

Bottcher, F., D. Becher, U. Klinner, I. A. Samsonova and B. Schilowa 1986. Genetic structure of yeast hybrids constructed by protoplast fusion. In *Advances in protoplast research*, L. Ferenczy and G. L. Farkas (eds), 99–104. Oxford: Pergamon Press.

Boucherie, H. 1985. A study on the control of carbon catabolic repressed proteins in *Saccharomyces cerevisiae*. *Biochimica et Biophysica Acta* **825**, 360–4.

Buschl, R., H. Ringsdorf and U. Zimmermann 1982. Electric field-induced fusion of largc liposomes from natural and polymerizable lipids. *FEBS Letters* **150**, 38–42.

Conde Zurita, J. and G. R. Fink 1976. A mutant of *Saccharomyces cerevisiae* defective for nuclear fusion. *Proceedings of the National Academy of Sciences of the USA* **73**, 3651–5.

Conde Zurita, J. and S. J. L. Mascort 1981. Effect of the mitochondrial genome on the fermentation behaviour of brewing yeast. In *European Brewery Convention Proceedings of 18th Congress, Copenhagen*, 177–86. Oxford: IRL Press.

Davies, R. and P. A. Elvin 1964. The effect of $\beta$-mercaptoethanol on release of invertase and formation of protoplasts of *Saccharomyces fragilis*. *Biochemical Journal* **93**, 8P–9P.

de van Broock, M. R., M. F. Sierra and L. de Figueroa 1980. Intergeneric fusion of yeast protoplasts. In *Advances in protoplast research*, G. G. Stewart and I. Russell (eds), *Vth International Symposium on Yeasts*, London, Ontario, Canada, 20–25 July 1980, 165–70. Toronto: Pergamon Press.

Delgado, M. A. and J. Conde Zurita 1983. Benomyl as a tool for the parasexual genetic analyses of brewing yeast. In *European Brewery Convention Proceedings of the 19th Congress, London*, 465–71. Oxford: IRL Press.

Dhawale, M. R. and W. M. Ingledew 1983a. Interspecific protoplast fusion of *Schwanniomyces* yeasts. *Biotechnology Letters* **5**, 825–30.

Dhawale, M. R. and W. M. Ingledew 1983b. Starch hydrolysis by derepressed mutants of *Schwanniomyces castellii. Biotechnology Letters* **5**, 185–90.

du Preez, J. C. and J. P. van der Walt 1983. Fermentation of D-xylose to ethanol by a strain of *Candida shehatae. Biotechnology Letters* **5**, 357–62.

Duell, E. A., S. Inoue and M. F. Utter 1964. Isolation and properties of intact mitochondria from spheroplasts of yeasts. *Journal of Bacteriology* **88**, 1762–73.

Eddy, A. A. 1958. Presumed nuclear bodies isolated from yeast protoplasts using ultrasonics. *Proceedings of the Royal Society of London, Series B* **17**, 447–64.

Eddy, A. A. and D. H. Williamson 1957. A method of isolating protoplasts from yeast. *Nature* (London) **179**, 1252–3.

Esser, K., U. Schmidt and U. Stahl 1982. Ethanol and biomass production of wild strains and respiratory deficient mutants of *Saccharomyces cerevisiae* under anaerobic and aerobic conditions. *European Journal of Applied Microbiology and Biotechnology* **16**, 161–4.

Farahnak, F., T. Seki, D. D. Y. Ryu and D. Ogrydziak 1986. Construction of lactose-assimilating and high-ethanol-producing yeasts by protoplast fusion. *Applied Environmental Microbiology* **51**, 362–7.

Ferenczy, L., F. Kevei and M. Szegedi 1975. High frequency fusion of fungal protoplasts. *Experientia* **31**, 1028–30.

Ferenczy, L., F. Kevei, M. Szegedi, A. Franko and I. Rojck 1976. Factors affecting high-frequency fungal protoplast fusion. *Experientia* **32**, 1156–8.

Finaz, C., A. Lefevre and J. Teissie 1984. A new, highly efficient technique for generating somatic cell hybrids. *Experimental Cell Research* **150**, 477–82.

Fodor, K. and L. Alfoldi 1976. Fusion of protoplasts of *Bacillus megaterium. Proceedings of the National Academy of Sciences of the USA* **73**, 2147–50.

Fogel, S. and J. W. Welch 1982. Tandem gene amplification mediates copper resistance in yeast. *Proceedings of the National Academy of Sciences of the USA* **79**, 5342–6.

Fogel, S., R. K. Mortimer and K. Lusnak 1981. Mechanisms of meiotic gene conversion, or 'Wanderings on a foreign strand'. In *The molecular biology of the yeast* Saccharomyces, *life cycle and inheritance*, J. N. Strathern, E. W. Jones and J. R. Broach (eds), 289–320. New York: Cold Spring Harbor Laboratory.,

Fournier, P., A. Provost, C. Bourguignon and H. Heslot 1977. Recombination after protoplast fusion in the yeast *Candida tropicalis. Archives of Microbiology* **115**, 143–9.

Fowell, R. R. 1969. *The yeasts*, Vol. 1: *Sporulation and hybridization of yeasts*, A. H. Rose and J. S. Harrison (eds), 303–85. London: Academic Press.

Freeman, R. F. 1981. Construction of brewing yeasts for production of low carbohydrate beers. In *European Brewery Convention Proceedings of the 18th Congress, Copenhagen*, 497–504. Oxford: IRL Press.

Garcia-Acha, I. and P. Grilione 1982. Ethanol production characteristics for a respiratory deficient mutant yeast strain. *Transactions of the ASAE* **25**, 1396–9.

Garcia-Acha, I., F. Lopez-Belmonte and J. R. Villanueva 1966. Use of lipase streptozyme for isolation of protoplasts from fungal mycelium. *Journal of General Microbiology* **45**, 515–23.
Goodey, A. R. and R. S. Tubb 1982. Genetic and biochemical analysis of the ability of *Saccharomyces cerevisiae* to decarboxylate cinnamic acids. *Journal of General Microbiology* **128**, 2615–20.
Groves, D. P. and S. G. Oliver 1984. Formation of intergeneric hybrids of yeast by protoplast fusion of *Yarrowia* and *Kluyveromyces* species. *Current Genetics* **8**, 49–55.

Halfmann, H. J., W. Rocken, C. C. Emeis and U. Zimmermann 1982. Transfer of mitochondrial function into a cytoplasmic respiratory-deficient mutant of *Saccharomyces* yeast by electro-fusion. *Current Genetics* **6**, 25–8.
Halfmann, H. J., C. C. Emeis and U. Zimmermann 1983. Electro-fusion of haploid *Saccharomyces* yeast of identical mating type. *Archives of Microbiology* **134**, 1–4.
Hamada, H., M. Yamashita, M. Kojima and T. Mouta 1984. Effect of methanol on the induction of respiration-deficient mutants in yeast by acriflavine. *Chemical Pharmaceutical Bulletin* **32**, 623–7.
Hammond, J. R. M. and K. W. Eckersley 1984. Fermentation properties of brewing yeast with killer character. *Journal of the Institute of Brewing* **90**, 167–77.
Hinnen, A. and B. Meyhack 1982. Vectors for cloning in yeast. In *Current topics in microbiology and immunology*, Vol. 96, 101–17. New York: Springer-Verlag.

Iglesias, F. J., M. C. Lopez, C. Santamaria and A. Dominiquez 1984. Dielectrophoretic properties of yeast cells dividing by budding and by transversal fission. *Biochimica et Biophysica Acta* **804**, 221–9.

Jimenez, A. and J. Davies 1980. Expression of a transposable antibiotic resistance element in *Saccharomyces*. *Nature* (London) **287**, 869–71.
Johannsen, E., L. Eagle and G. Bredenhann 1985. Protoplast fusion used for the construction of presumptive polyploids of the D-xylose fermenting yeast *Candida shehatae*. *Current Genetics* **9**, 313–19.

Kao, K. N. and M. R. Michayluk 1974. A method for high frequency intergeneric fusion of plant protoplasts. *Planta* **115**, 355–67.
Kaube, I., E. Tamiya and H. Matsuoka 1985. Transformation of *Saccharomyces cerevisiae* spheroplasts by high electric pulse. *FEMS Letters* **182**, 90–4.
Kielland-Brandt, M. C., T. Nilsson-Tillgren, S. Holmberg, J. G. L. Petersen and B. A. Svenningsen 1979. Transformation of yeast without the use of foreign DNA. *Carlsberg Research Communications* **44**, 77–87.
Kitamura, K. and Y. Yamamoto 1972. Purification and properties of an enzyme, zymolyase, which lyses viable yeast cells. *Archives of Biochemistry and Biophysics* **153**, 403–6.

Laires, A., I. Spencer-Martins and N. van Uden 1983. Use of D-glucosamine and 2-deoxyglucose in the selective isolation of mutants of the yeast *Lipomyces starkeyi* derepressed for the production of extracellular endodextranase. *Zeitschrift für Allgemeine Mikrobiologie* **23**, 601–3.
Leamann, R. and D. Margalith 1986. Ethanol formation by hybrid yeasts. *Applied Microbiology and Biotechnology* **23**, 198–202.
Lopez, M. C., F. J. Iglesias, C. Santamaria and A. Dominiquez 1984. Dielectrophoretic behaviour of *Saccharomyces cerevisiae*: effect of cations and detergents. *FEMS Microbiology Letters* **24**, 149–52.

Maleszka, R. and H. Schneider 1982. Fermentation of D-xylose, xylitol, and D-xylulose by yeasts. *Canadian Journal of Microbiology* **28**, 360–3.
Moulin, G., H. Boze and P. Galzy 1982. Utilization of a respiratory-deficient mutant for alcohol production. *Journal of Fermentation Technology* **60**, 25–9.

Nagasaki, S., N. P. Neuman, P. Arnow, L. P. Schnable and J. O. Lampen 1966. An enzyme which degrades the wall of living yeasts. *Biochemical and Biophysical Research Communications* **25**, 158–64.

Ogur, M., R. St John and S. Nagai 1957. Tetrazolium overlay technique for population studies of respiration deficiency in yeast. *Science* **125**, 928–9.
Ono, B. I. and E. Sakamoto 1985. *Saccharomyces cerevisiae* strains sensitive to inorganic mercury. *Current Genetics* **10**, 179–85.
Ono, B. I. and M. Weng 1982. Chromium resistant mutants of the yeast *Saccharomyces cerevisiae*. *Current Genetics* **6**, 71–7.

Pfisterer, E. A., I. F. Garrison and R. A. McKee 1978. Brewing with syrups. *Technical Quarterly Master Brewers Association of the Americas* **15**, 59–63.
Phaff, H. J. 1963. Cell wall of yeasts. *Annual Review of Biochemistry* **17**, 15–30.
Phillips, A. Q. 1955. Utilization by yeasts of the carbohydrates of wort. *Journal of the Institute of Brewing* **61**, 122–6.
Pina, A., I. L. Calderon and T. Benitez 1986. Intergeneric hybrids of *Saccharomyces cerevisiae* and *Zygosaccharomyces fermentatia* obtained by protoplast fusion. *Applied Environmental Microbiology* **51**, 995–1003.
Provost, A., C. Bourguignon, P. Fournier, A. M. Ribet and H. Heslot 1978. Intergeneric hybridization in yeasts through protoplast fusion. *FEMS Microbiology Letters* **3**, 309–12.

Richter, H. P., P. Scheurich and U. Zimmermann 1981. Electric field-induced fusion of sea urchin eggs. *Development, Growth and Differentiation* **23**, 479–86.
Russell, I. and G. G. Stewart 1985. Valuable techniques in the genetic manipulation of industrial yeast strains. *Journal of the American Society of Brewing Chemists* **43**, 84–90.
Russell, I., I. F. Garrison and G. G. Stewart 1973. Studies on the formation of spheroplasts from stationary phase cells of *Saccharomyces cerevisiae*. *Journal of the Institute of Brewing* **79**, 48–54.
Russell, I., R. Jones and G. G. Stewart 1985. The influence of cytoplasmic control of the expression of nuclear genes in brewing yeasts. In *European Brewery Convention Proceedings of the 20th Congress, Helsinki*, 235–42. Oxford: IRL Press.
Russell, I., R. M. Jones and G. G. Stewart 1986a. The genetic modification of brewer's yeast and other industrial yeast strains. In *Biotechnology in food processing*, S. K. Harlander and T. P. Labuza (eds), 171–95. New Jersey: Noyes Publications.
Russell, I., C. M. Crumplen, R. M. Jones and G. G. Stewart 1986b. Efficiency of genetically engineered yeast in the production of ethanol from dextrinized cassava starch. *Biotechnology Letters* **8**, 169–74.

Sakai, T., K. Koo, K. Saitoh and T. Katsuragi 1986. Use of protoplast fusion for the development of rapid starch-fermenting strains of *Saccharomyces diastaticus*. *Agricultural Biology and Chemistry* **50**, 297–306.
Sarachek, A., D. D. Roads and R. H. Schwarzhoff 1981. Hybridization of *Candida albicans* through fusion of protoplasts. *Archives of Microbiology* **129**, 1–8.
Savchenka, G. V. and Y. G. Kapul'tsevich 1980. Hybridization of yeast from the genus

*Hansenula* by protoplast fusion. In *Advances in protoplast research*, L. Ferenczy and G. L. Farkas (eds), 125–30. Oxford: Pergamon Press.

Schaeffer, P., B. Cami and R. D. Hotchkiss 1976. Fusion of bacterial protoplasts. *Proceedings of the National Academy of Sciences of the USA* **73**, 151–215.

Schnettler, R. and U. Zimmermann 1985. Influence of the composition of the fusion medium on the yield of electrofused yeast hybrids. *FEMS Microbiology Letters* **27**, 195–8.

Schnettler, R., U. Zimmermann and C. C. Emeis 1984. Large-scale production of yeast hybrids by electrofusion. *FEMS Microbiology Letters* **24**, 81–5.

Schweizer, E., W. Pemmer, U. Holzner and H. W. Tahedl 1977. Controlled mitochondrial inactivation of temperature-sensitive *Saccharomyces cerevisiae* nuclear petite mutants. In *Mitochondria 1977: genetics and biogenesis of mitochondria*, W. Bandlow, R. J. Schweyen, D. Y. Thomas, K. Wolf and F. Kaudewitz (eds), 91–125. Berlin: de Gruyter.

Seki, T., S. Myoga, S. Limtong, S. Uedono, J. Kamnuanta and H. Taguchi 1983. Genetic construction of yeast strains for high ethanol production. *Biotechnology Letters* **5**, 351–6.

Sermonti, G. 1979. Mutation and microbial breeding. In *Genetics of industrial microorganisms*, O. K. Sebek and A. I. Laskin (eds), 10–14. Washington, DC: ASM.

Silhankova, L. 1985. Yeast mutants excreting vitamin $B_1$ and their use in the production of thiamine rich beers. *Journal of the Institute of Brewing* **91**, 78–81.

Sills, A. M. and G. G. Stewart 1982. Production of amylolytic enzymes by several yeast species. *Journal of the Institute of Brewing* **88**, 313–16.

Singh, I. 1984. Induction of gene conversion and reverse mutation by manganese sulphate and nickel sulphate in *Saccharomyces cerevisiae*. *Mutation Research* **137**, 47–9.

Sipiczki, M. and L. Ferenczy 1977a. Fusion of *Rhodosporidium* (*Rhodotorula*) protoplasts. *FEMS Letters* **2**, 203–5.

Sipiczki, M. and L. Ferenczy 1977b. Protoplast fusion of *Schizosaccharomyces pombe* auxotrophic mutants of identical mating type. *Molecular and General Genetics* **151**, 77–81.

Spata, L. and H. Weber 1980. A study on protoplast fusion and parasexual hybridization of alkane utilizing yeasts. In *Advances in protoplast research*, L. Ferenczy and G. L. Farkas (eds), 131–7. Oxford: Pergamon Press.

Spencer, J. F. T. and D. M. Spencer 1982. The use of mitochondrial mutants in the isolation of hybrids involving industrial yeast strains. III. Restoration of mitochondrial function in petites of industrial yeast strains by fusion with respiratory-competent protoplasts of other yeast species. *Current Genetics* **4**, 177–80.

Spencer, J. F. T., C. Bizeau, N. Reynolds and D. M. Spencer 1985a. The use of mitochondrial mutants in hybridization of industrial yeast strains. VI. Characterization of the hybrid, *Saccharomyces diastaticus* and *Saccharomyces rouxii*, obtained by protoplast fusion, and its behavior in simulated dough-raising tests. *Current Genetics* **9**, 649–52.

Spencer, J. F. T., D. M. Spencer, C. Bizeau, A. V. Martini and A. Martini 1985b. The use of mitochondrial mutants in hybridization of industrial yeast strains. V. Relative parental contributions to the genomes of interspecific and intergeneric yeast hybrids obtained by protoplast fusion, as determined by DNA reassociation. *Current Genetics* **9**, 623–5.

Stahl, U. 1978. Zygote formation and recombination between like mating types in the yeast *Saccharomycopsis lipolytica* by protoplast fusion. *Molecular and General Genetics* **160**, 111–13.

Stewart, G. G. 1981a. Biotechnology in the 1980s. *Brewer's Digest* **56**, 42–54.

Stewart, G. G. 1981b. The genetic manipulation of industrial yeast strains. *Annual Review of Microbiology* **37**, 121–43.

Stewart, G. G. and I. Russell 1986. One hundred years of yeast research and development in the brewing industry. *Journal of the Institute of Brewing* **92**, 537–58.

Stewart, G. G., J. A. Erratt, I. F. Garrison, T. E. Goring and I. F. Hancock 1979. Studies on the utilization of wort carbohydrates by brewers yeast strains. *Technical Quarterly Master Brewers Association of the Americas* **16**, 1–7.

Stewart, G. G., R. M. Jones and I. Russell 1985. The use of derepressed yeast mutants in the fermentation of brewery wort. In *European Brewery Convention Proceedings of the 20th Congress, Helsinki*, 243–50. Oxford: IRL Press.

Svoboda, A. 1978. Fusion of yeast protoplasts induced by polyethylene glycol. *Journal of General Microbiology* **109**, 167–75.

Svoboda, A. 1980. Intergeneric fusion of yeast protoplasts, *Saccharomyces cerevisiae* and *Schizosaccharomyces pombe*. In *Advances in protoplast research*, L. Ferenczy and G. L. Farkas (eds), 125–30. Oxford: Pergamon Press.

Taya, M., H. Honda and T. Kobayashi 1984. Lactose-utilizing hybrid strain derived from *Saccharomyces cerevisiae* and *Kluyveromyces lactis* by protoplast fusion. *Agricultural Biology and Chemistry* **48**, 2239–43.

Terrance, K. and P. N. Lipke 1981. Sexual agglutination in *Saccharomyces cerevisiae*. *Journal of Bacteriology* **148**, 889–96.

Toh-e, A., P. Guerry-Kopecko and R. B. Wickner 1980. A stable plasmid carrying the yeast LEU2 gene and containing only yeast DNA. *Journal of Bacteriology* **141**, 413–16.

Tubb, R. S., B. A. Searle, A. A. Goodey and A. J. P. Brown 1981. Rare mating and transformation for construction of novel brewing yeasts. In *European Brewery Convention Proceedings of the 18th Congress, Copenhagen*, 487–91. Oxford: IRL Press.

Van Solingen, D. and J. B. van der Plaat 1977. Fusion of yeast spheroplasts. *Journal of Bacteriology* **130**, 946–7.

Vienken, J. and U. Zimmermann 1985. An improved electrofusion technique for production of mouse hybridoma cell. *FEBS Letters* **182**, 278–80.

Vienken, J., U. Zimmermann, M. Fouchard and D. Zagury 1984. Electrofusion of myeloma cells on the single cell level. Fusion under sterile conditions without proteolytic enzyme treatment. *FEBS Letters* **163**, 54–6.

Villanueva, J. R. 1966. Protoplasts in fungi. In *The fungi*, Vol. 2, G. C. Ainsworth and A. S. Sussman (eds), 3–62. London: Academic Press.

Villanueva, J. R. and I. Garcia-Acha 1971. Production and use of fungal protoplasts. In *Methods in microbiology*, Vol. 4, C. Booth (ed.), 665–718. London: Academic Press.

Whitney, G. K., C. R. Murray, I. Russell and G. G. Stewart 1985. Potential cost savings for fuel ethanol production by employing a novel hybrid yeast strain. *Biotechnology Letters* **7**, 349–54.

Whittaker, P. A. and S. M. Leach 1978. Interspecific hybrid production between the yeasts *Kluyveromyces lactis* and *Kluyveromyces fragilis* by protoplast fusion. *FEMS Microbiology Letters* **4**, 31–4.

Wilkie, D. and R. C. Nudd 1981. Aspects of mitochondrial control of cell surface characteristics in *Saccharomyces cerevisiae*. In *Current developments in yeast research*, G. G. Stewart and I. Russell (eds), 345–9. Toronto: Pergamon Press.

Wilson, J. J., G. G. Khachatourians and W. M. Ingledew 1982. Protoplast fusion in the yeast *Schwanniomyces alluvius*. *Molecular and General Genetics* **186**, 95–100.

Winge, O. 1935. On haplophase and diplophase in some *Saccharomycetes. Comptes Rendus du Travail de la Laboratoire Carlsberg, Série de Physiologie* **21**, 77–111.

Wood, J. S. 1982a. Genetic effects of methyl benzimidazole-2-yl-carbamate on *Saccharomyces cerevisiae*. *Molecular and Cellular Biology* **2**, 1064–79.

Wood, J. S. 1982b. Mitotic chromosome loss induced by methyl benzimidazole-2-yl-carbamate as a rapid mapping method in *Saccharomyces cerevisiae*. *Molecular and Cellular Biology* **2**, 1080–7.

Yamamoto, M. and S. Fukui 1977. Fusion of yeast protoplasts. *Agricultural Biology and Chemistry* **41**, 1829–30.

Young, T. W. 1983. Brewing yeast with anti-contaminant properties. In *European Brewery Convention Proceedings of the 19th Congress, London*, 129–36. Oxford: IRL Press.

Zakharov, I. A. and B. Yarovoy 1977. Cytoduction as a new tool in studying the cytoplasmic heredity in yeast. *Molecular and Cellular Biochemistry* **14**, 15–18.

Zimmermann, U. 1982. Electric field-mediated fusion and related phenomena. *Biochimica et Biophysica Acta* **694**, 227–77.

Zimmermann, U. 1983. Electrofusion of cells: principles and industrial potential. *Trends in Biotechnology*, Vol. 1 (5), 149–55.

Zimmermann, U. and P. Scheurich 1981. High frequency fusion of plant protoplasts by electric fields. *Planta* **15**, 26–32.

Zimmermann, U. and J. Vienken 1982. Electric field-induced cell-to-cell fusion. *Journal of Membrane Biology* **67**, 165–82.

Zimmermann, U., J. Vienken, J. and P. Scheurich 1980. In *Biophysics of structure and mechanism*, Vol. 6, K. Gersonde (ed.), 86. Oxford: Springer-Verlag.

# 4 *Yeast plasmids and transformation*

C. J. PANCHAL

## 4.1 Introduction

The rapid advancements in understanding yeast biology made in the last 8–10 years have come about as a result of some major breakthroughs in molecular biology in the past two decades. These in turn have led to developments in yeast genetics that have helped to move the frontiers of yeast research in quantum jumps. Since yeasts such as *Saccharomyces cerevisiae* and *Saccharomyces uvarum* (*carlsbergensis*) constitute the group of micro-organisms most exploited commercially, it is not surprising that much research effort has been devoted to understanding them at the molecular level. The development of a transformation system in yeast (Hinnen *et al.* 1978) was a turning point in yeast genetics that has resulted in cloning of several yeast genes, and in a deeper understanding of the functions of many of them. The availability of shuttle vectors has greatly facilitated gene cloning, as well as the development of vector systems allowing heterologous gene expression in yeast.

The ability of shuttle vectors to replicate in yeast and in *Escherichia coli*, where they could be amplified readily, has been used to prepare gene banks in *Escherichia coli* from which genes of interest could be cloned in yeast. Several different types of yeast vector systems have been reported in the literature, and a recent review (Parent *et al.* 1985) lists the major ones grouped according to their mode of replication in yeast. The three main groups are: (a) yeast integrating plasmids (YIp); (b) yeast episomal plasmids (YEp); and (c) yeast replicating plasmids (YRp). Parent *et al.* (1985) have broadened these groups to include yeast centromeric plasmids (YCp), yeast linear plasmids (YLp), yeast promoter plasmids (YPp), yeast expression plasmids (YXp) and yeast hybrid plasmids (YHp). It is important to point out that in the construction of many of the above plasmid types, the availability of a resident yeast plasmid system (2-μm plasmid) has been of paramount significance. This chapter attempts to give a synopsis of the yeast 2-μm plasmid system and other commonly used plasmid systems in yeast (both laboratory strains and industrial strains).

## 4.2 The yeast 2-μm plasmid

Most strains of *Saccharomyces cerevisiae* and *Saccharomyces uvarum* (*carlsbergensis*) possess the covalently closed circular DNA plasmid called the

2-μm plasmid or 2-μm circle, since it has a monomeric contour length of approximately 2 μm (Cameron *et al.* 1977, Clark-Walker & Miklos 1974, Gubbins *et al.* 1977, Guerineau *et al.* 1971, Hollenberg *et al.* 1976, Livingston & Klein 1977). In addition, the plasmid has also been found in *Saccharomyces diastaticus* (Brown *et al.* 1981), *Saccharomyces italicus* (Gunge *et al.* 1981, Jazwinski & Edelman 1979) and *Torulopsis dattila* (Gunge 1983). Most polyploid brewing strains have also been found to possess the plasmid (Stewart *et al.* 1981). In most yeast strains the plasmid exists in a copy number of 50–100 per haploid genome.

The entire 2-μm plasmid from a laboratory strain of *Saccharomyces cerevisiae* has been sequenced (Hartley & Donelson 1980). The sequence has revealed a plasmid of 6318 bp and having identical inverted repeats of 599 bp which divide the plasmid into two unique sequences of 2774 bp (L region) and 2346 bp (S region). The origin of replication of the plasmid spans the junction between the larger unique region (L) and one of the inverted repeats. Recombination occurs readily between the two inverted repeat regions, resulting in an inversion of one unique region with respect to the other (Beggs 1978). Hence, 2-μm DNA, when isolated by CsCl-ethidium bromide centrifugation, is found to have a mixed population of equal amounts of two plasmids that differ in the orientation of one unique region with respect to the other; these forms have generally been termed the A form and the B form (Broach 1982). Intramolecular recombination is mediated by the product of a gene called *FLP* which has the longest open reading frame in the genome of 2-μm DNA (Broach 1982). The *FLP* gene is located on the S region of the DNA. The *FLP* function is *trans*-acting and hence a 2-μm (or 2-μm based) plasmid defective in the *FLP* region can replicate in a host yeast containing endogenous 2-μm plasmid (Broach & Hicks 1980).

Under normal growth conditions 2-μm plasmid behaves like a chromosomal replicon. Each molecule replicates only once during a cell cycle by the same cellular machinery that is used for chromosome replication (Zakian *et al.* 1979). Replication thus occurs at an early stage of S phase in the cell cycle, and the copy number for the cell is genetically controlled (Gerbaud & Guerneau 1980). However, it has been found that when a cell has a low plasmid copy number, the 2-μm DNA can override the cell-cycle control in order to replicate more than once in an attempt to stabilize the copy number. Moreover, a chromosomal mutation, *nib* was found to result in an elevated copy number of 2-μm DNA, and this was associated with promotion of rough 'nibbled' colonies exhibiting lethal sectoring (Holm 1982a, b).

The 2-μm plasmid amplification system consists of the proteins *REP*1 and *REP*2, which are *trans*-acting, and *REP*3, which is located a few hundred base pairs from *ori* the origin of replication and is only active in *cis* (Jayaram *et al.* 1983). The *REP* loci constitute only an amplification system and do not have the capacity to promote autonomous replication in yeast on their own, a function for which *ori* is required (Broach 1982, 1983).

The origin of replication spans a region extending from the middle of one of the inverted repeats 100 bp into the adjacent unique sequences (~3650–4000 in the A form and 650–1000 in the B form). This is the only site in the 2-μm molecule that promotes autonomous replication in yeast (Hartley & Donelson

1980). Hybrid plasmids containing 2-μm circle sequences bearing the origin of replication only are at low copy number in $cir^0$ strains and are lost rapidly during non-selective growth (Hollenberg 1982). However, the same plasmids in $cir^+$ strains are at high copy number and stable. The *trans*-acting products of *REP*1 and *REP*2 and the *cis*-acting product of *REP*3 are required for high copy number.

Jayaram *et al.* (1983, 1985) have proposed that the *REP* loci of 2-μm DNA comprise the amplification system which is activated when the plasmid copy number is low. Moreover, when hybrid plasmids are present in a yeast cell that also possesses 2-μm plasmid, some degree of incompatibility occurs. Jayaram *et al.* (1985) showed that in the presence of endogenous 2-μm plasmid the copy number of a hybrid plasmid CV20 decreases. On the other hand, Gerbaud and Guerineau (1980) have shown that the presence of hybrid 2-μm plasmids in a cell reduces the copy number of endogenous 2-μm plasmid. In fact, this phenomenon has been used to cure a cell of its endogenous 2-μm plasmid (Dobson *et al.* 1980).

In contrast with autonomously replicating plasmids (*ARS*, see below), where a high copy number may be associated with low stability, in the 2-μm based plasmids a positive correlation between copy number and stability exists.

### *4.2.1 Cellular location of 2-μm plasmid*

Initial reports on yeast 2-μm plasmid suggested that the plasmid was located in the cytoplasm (Cameron *et al.* 1977, Clark-Walker & Miklos 1974, Tabak 1977). This was based on: (a) non-Mendelian inheritance of the plasmid (Livingston 1977); (b) the absence in the nuclei (Clark-Walker & Miklos 1974); and (c) the presence of the plasmid in cell extracts presumed to be totally cytoplasmic, observed by electron microscopic analysis (Clark-Walker & Miklos 1974). However, recent biochemical and genetic evidence overwhelmingly points to a nuclear location of the 2-μm plasmid. The biochemical evidence includes the fact that 2-μm plasmid DNA replication strongly resembles chromosomal DNA replication, thus requiring the products of *CDC*4, *CDC*7 and *CDC*28 genes (Livingston & Kupfer 1977). Also, plasmid DNA synthesis is blocked by α-factor as is chromosomal DNA synthesis, but unlike that of the mitochondrial DNA synthesis (Livingston & Kupfer 1977, Newlon & Fangman 1975, Petes & Fangman 1973, Seligy *et al.* 1980).

The genetic evidence includes the fact that in cytoduction experiments involving the karyogamy-defective mutation *kar*1 (Conde & Fink 1976), plasmids were transmitted into cytoductants, always in association with the nuclear markers of their respective parental strains (Livingston 1977).

### *4.2.2 Two micron based chimeric plasmids*

The majority of the studies relating to plasmid-mediated transformation in yeast have been performed with chimeric plasmids containing some or all of the yeast 2-μm DNA, a part of the yeast chromosomal DNA (carrying functional genes such as *TRP*1, *LEU*2, *URA*3, etc.) and the *Escherichia coli* plasmid pBR322 (Beggs 1981). Such a plasmid would then be able to replicate in both yeast and in *Escherichia coli*, provided the 2-μm DNA segment contains the *ori* region. Yeast transformants are obtained at a high frequency, generally $10^3$–$10^5$ transformants $\mu g^{-1}$ DNA (Beggs 1981, Hollenberg 1982). However, under non-selective conditions the transformants are unstable and 10–30% of them may lose the plasmid after 15–20 generations (Futcher & Cox 1984, Hollenberg 1982).

A hybrid plasmid lacking the 2-μm *ori* segment, but containing an inverted repeat, can also propagate at moderately high copy number in yeast by recombination with the endogenous 2-μm plasmid. This would, of course, not be possible if transformation were carried out using a $cir^0$ recipient. Interestingly enough, in some cases intramolecular recombination with the chimeric plasmid may lead to loss of the pBR322 sequences, leaving a total yeast DNA plasmid. Dobson *et al.* (1980) reported that in some transformants obtained after introducing the chimeric plasmid pJDB219-13 (12.1 kbp) into a $cir^+$ recipient, loss of pMB9 DNA derived from *Escherichia coli* occurred, resulting in the formation of a total yeast DNA plasmid pYX-B (6.8 kbp). Replication of pYX-B in the transformant resulted in the loss of not only the chimeric pJDB219-13 plasmid, but also the resident 2-μm DNA.

The phenomenon has been successfully used to cure a strain of its 2-μm plasmid, something that has not been possible using physical or chemical methods (Dobson *et al.* 1980, Erhart & Hollenberg 1981). However, curing a strain of its 2-μm DNA using this method requires that the strain be auxotrophic for the particular marker gene carried by the chimeric plasmid (e.g. *LEU*2). Such auxotrophic markers are generally not present in most industrial strains, which are either polyploid or aneuploid (Stewart 1981). In order to circumvent this problem, Hollenberg and co-workers (Hollenberg 1982) constructed a hybrid plasmid carrying a transposable element Tn601 from *Escherichia coli* which has been found to confer resistance to kanamycin and gentamycin and Geneticin G418 in *Saccharomyces*. The antibiotic G418 is a very potent aminoglycoside and all species of *Saccharomyces*, including industrial strains, are very sensitive to small concentrations of the drug (Panchal *et al.* 1984). Hollenberg and co-workers (Hollenberg 1982) found that after 60 generations on non-selective medium most of the cells had lost the chimeric plasmid. However, about 5% of the transformants had also lost 2-μm DNA, and had therefore become $cir^0$.

## 4.3 Chromosomal *ARS*-containing plasmids

Autonomously Replicating Sequence (*ARS*) elements are DNA sequences which allow DNA molecules to transform yeast at a high frequency, giving rise to transformants in which the transforming DNA replicates autonomously (Williamson 1985). The *ARS* elements, of which there are about 400 in a haploid yeast genome (Chan & Tye 1980) have been proposed to be chromosomal origins of replication, although direct evidence for this hypothesis is lacking (Williamson 1985). Plasmids carrying the yeast *ARS* element (e.g. YRp7) transform yeast at high frequency, but the transformants are generally not stable, losing the plasmid quite readily (Struhl *et al.* 1979).

The plasmid YRp7 also carries the yeast *TRP*1 gene and pBR332 bacterial plasmid sequences. Transformants of a *trp*1 auxotroph occurs at a high frequency of around $10^3$–$10^4$ colonies $\mu g^{-1}$ DNA. The *ARS*1 sequences allow the plasmid to replicate autonomously at a high copy number. Loss of the plasmid during growth of the cells, particularly in non-selective media, is attributed to poor segregation, mainly biased towards the mother cell (Murray & Szostak 1983b). However, it has been found that linear *ARS*-bearing plasmids segregate equally to the mother and daughter cells. This is perhaps related to the attachment of replication intermediates to fixed nuclear sites during mitosis (Murray & Szostak 1983a, b). Plasmids bearing *ARS* can also be stabilized by *CEN* (centromere) sequences, which results in a drastic reduction of plasmid copy number, to around one (Murray & Szostak 1983a).

Besides *ARS*1, other *ARS* sequences from yeast that have been identified are *ARS*2 (near *ARG*4) (Hsiao & Carbon 1979), *ARS*3 (near *SUP*11) (Stinchcomb *et al.* 1981) and *ARS*4 (near *CEN*4) (Stinchcomb *et al.* 1981). Besides the yeast *ARS* elements, several heterologous *ARS* elements have also been cloned in yeast. These include sequences from *Xenopus* (Kearsey 1983), *Tetrahymena* (Kiss *et al.* 1981), *Trypanosoma brucei* (Davison & Thi 1982), *Cephalosporium acremonium* (Skatrud & Queener 1984), *Physarum polycephalum* (Wilhelm *et al.* 1986), *Drosophila melanogaster* (Marunouchi & Hosoya 1984), chloroplast DNA from *Petunia hybrida* (Overbeeke *et al.* 1984) and *Nicotiana tabacum* (Uchimiya *et al.* 1983) and human chromosomal DNA (Montiel *et al.* 1984). Many of the above *ARS* elements have been sequenced and homologies between them investigated. A consensus sequence based on the 2-μm DNA *ARS* and several other chromosomal ones is:

5′ T T T T A T A T T T A 3′.

However, some major differences are seen at bases 1, 7 and 11; it is nevertheless obvious that *ARS* elements are rich in adenine–thymine (A–T) base pairs. The functional significance of the consensus sequence is not yet clear (Williamson 1985).

## 4.4 All-yeast DNA vectors

The chimeric plasmid vector YRp7 has been used extensively for preparing gene banks for cloning genes of interest in yeast (Struhl *et al.* 1979). The plasmid carries the *ARS*1 sequences from yeast allowing autonomous replication in yeast, as well as the yeast *TRP*1 gene. Both the *TRP*1 and *ARS*1 genes are adjacent to each other on a 1.4-kb *Eco*RI DNA fragment. Zakian and Scott (1982) restricted YRp7 plasmid with *Eco*RI and *Bam*HI (which cuts into the pBR322 sequences), self-ligated the *TRP*1–*ARS*1 DNA fragment and transformed it into a recipient yeast *trp*1 strain. In this way a miniplasmid *TRP*1–*ARS*1, later known as YARp1, was constructed. Unlike YRp7, YRp1 was found to be extremely stable both mitotically and meiotically. Zakian and Scott (1982) also found that the plasmid propagated at a high copy number (100–200 copies per haploid cell) and needed the cell cycle functions coded for by *CDC*4,

*CDC*7, *CDC*8 and *CDC*28 genes, which are also required for chromosomal DNA replication. Although the plasmid has few restriction sites for insertion of heterologous DNA, the *Hind*III and *Eco*RI sites have been found to be appropriate for insertion of complete genes such as *URA*3 and the *GAL*1–10 promoter sequences (Fagan & Scott 1985). The larger plasmid, YARp11, carrying the *URA*3 gene has also been found to be very stable and in high copy number (100–150 copies per cell). The high stability of these miniplasmids YARp1 (1.4 kb) and YARp11 (2.6 kb) has been attributed to their small size, but it is still not entirely certain whether this is the only reason. Nevertheless, the *ARS*-based miniplasmids present an attractive alternative to chimeric plasmids for cloning genes of interest, particularly in industrial yeast strains (brewing, baking, etc.) where an all-yeast DNA vector would be suitable.

Beggs (1978) constructed a hybrid plasmid, pJDB219, which consisted of the entire yeast 2-μm DNA, the *Escherichia coli* Co1E1-related plasmid pMB9 and a 1.2 kb insert containing the yeast *LEU*2 gene. The *LEU*2 gene was linked through poly(dA:dT) sequences to the *Pst*1 site of the 2-μm DNA cloned in pMB9 at the *Eco*RI site in the shorter non-repeat segment of 2-μm DNA. Yeast cells transformed with pJDB219 stably maintained the plasmid in high copy number. However, the size of the plasmid (12.4 kb) and the large number of restriction sites on it make it unsuitable as a vector for cloning genes. Other vectors, smaller in size, have been derived from pJDB219. Toh-e *et al.* (1980) performed a *Hind*III endonuclease digestion of pJDB219 followed by self-ligation and transformation of a leucine auxotroph. Nucleic acid extracts from leucine transformants were analysed by gel-electrophoresis and found to contain both 2-μm DNA as well as the smaller 3.2-kb plasmid pSLe1, which was the self-ligated *Hind*III fragment of pJDB219. Toh-e *et al.* (1980) showed that the plasmid was maintained stably even in the presence of 2-μm DNA and a very low frequency of plasmid loss occurred in the transformants. They claimed that the plasmid was stably maintained even under non-selective conditions, and could be used as a vector for cloning genes in yeast. However, no reports using this plasmid as a cloning vector have yet appeared.

Kielland-Brandt *et al.* (1979, 1981) have described the construction of an all-yeast plasmid consisting of the yeast 2-μm DNA with an insertion of a yeast chromosomal sequence of 9400 base pairs containing the *HIS*4 gene. Using *kar*1-mediated cytoduction experiments, they have predicted the location of the plasmid to be nuclear, like the 2-μm plasmid. However, the plasmid maintains a very low copy number in transformants and, under non-selective conditions, rapid loss of the plasmid occurred. Due to its low copy number the plasmid was barely detectable in agarose gels stained with ethidium bromide, and was thus ligated to a cosmid vector in order to obtain a large copy number to enable restriction analysis. By hybridization with a labelled probe consisting of the *HIS*4 gene, Holmberg *et al.* (1979a, b) showed that the total-yeast DNA plasmid, pC504, was 15.7 kb in size and that the 2-μm part of the plasmid occurred in both the L and the S forms.

To increase the stability of the yeast plasmid, Kielland-Brandt *et al.* (1979) attempted integration of the plasmid into the chromosome of a *his* auxotroph. Partially stable *His*$^+$ colonies were obtained by crossing a *His* transformant containing the *HIS*-carrying plasmid against a *his* auxotroph in the presence of azide. Restriction digestion followed by agarose gel-electrophoresis and Southern (1975) hybridization with radioactive *His* probes of the total DNA from one of these partially stable isolates indicated that the 2-μm DNA of the plasmid was located between the wild type *HIS* gene and the mutant *his* gene. When this segment of DNA (approximately 36 kbp) was inserted into a cosmid vector and used to transform a *his* auxotroph, the transformants showed that the cosmid contained two wild type *HIS* genes, on either side of the 2-μm DNA, while the mutant *his*2–24 gene was absent. Kielland-Brandt *et al.* (1979) predicted that this could be due to a recombination event at the inverted repeat region of the 2-μm DNA sequences linked to *HIS*4 sequences. They have recommended this approach for obtaining induced chromosomal integrations of a plasmid containing a full copy of 2-μm DNA, thereby obtaining chromosomal fragments which are inverted repeats around the region of 2-μm DNA sequences.

## 4.5 Minichromosomes

Carbon and Clarke (1984) have described the minichromosome as a combination of a DNA segment fulfilling a centromeric function, a replicating site and one or more genes, including regulatory genes and structural genes, which are to be expressed by the transformed host cell. Such minichromosomes can be prepared stepwise by insertion of various DNA segments into a circular plasmid. Minichromosomes containing the yeast *CEN*3 element were constructed and found to mimic, to a large extent, the larger parental chromosomes (Clarke & Carbon 1980). They were stable both in mitosis and through the first and second division of meiosis. However, unlike chromosomes, minichromosomes do not always pair during replication. This does not prevent them from segregating during meiosis. The mitotic stability of *CEN*3-containing minichromosomes makes them suitable as cloning vectors. However, a chromosomal replicator is required on the plasmid for the centromere to be functional.

The centromere in yeast chromosomes provides the site for attachment of microtubules, and for regulating the proper chromosomal movement during mitosis and meiosis. The centromere regions of *CEN*3 and *CEN*11 have been cloned and sequenced (Fitzgerald-Hayes *et al.* 1982a, b). While the sequences of the two centromeres differed greatly, an extremely A–T-rich core region of about 90 bp was conserved in both *CEN*3 and *CEN*11. This may be the sequence organization required for microtubule binding (Gunge 1983).

## 4.6 Integrating plasmids

Transformation of yeast cells by plasmids can occur by integration of the plasmid into a chromosome by homologous recombination. This integration event results in an integrated copy of the plasmid vector sequences flanked by direct repeats of the homologous region on the plasmid and the chromosome. Indeed, in the first report on transformation of yeast Hinnen *et al.* (1978) reported that the plasmid pYeLeu10 carrying the yeast *LEU*2 gene integrated at the homologous *leu*2 region in the chromosome, since the plasmid lacked sequences allowing autonomous replication in the yeast cell. However, not all transformants result from integration of the plasmid. It was reported that in about 20% of the transformants the homologous DNA on the chromosome had been replaced by corresponding plasmid sequences without integration of the plasmid, these usually arise either via gene conversion or by a double-crossover event. In the case of pYeLeu10, Hinnen *et al.* (1978) reported that integration of the plasmid occurred not only at the *leu*2 region, but also at several other sites on different chromosomes. This unusual transformation event has been attributed to the presence of sequences of the yeast transposable element Ty and its flanking $\delta$ sequences which are present in the vicinity of the *LEU*2 gene. Since Ty1 elements are present in at least 30 copies in a haploid yeast cell (Boeke *et al.* 1985), integration of pYeLeu10 at several sites in the yeast genome is likely.

Plasmid sequences integrated into the chromosome by recombination (Campbell type) can be unstable, since they can be quite readily looped out of the genome by the reversal of the integrative recombination process. However, the transformation frequencies of integrative plasmids can be increased greatly by linearizing the plasmid, particularly with a restriction enzyme that cuts within the yeast gene to be integrated. Linearized plasmids with double-strand breaks are highly recombinogenic and give transformation frequencies 5–20 times higher than those obtained with uncut plasmids (Imai *et al.* 1983, Orr-Weaver *et al.* 1981).

Transformation frequencies with linear plasmids have also been increased by incorporating ribosomal DNA (rDNA) sequences into the plasmid, since about 140 copies of tandemly repeated rDNA sites are located on yeast Chromosome XII, providing multiple sites for integration (Orr-Weaver *et al.* 1981).

As a general rule, the larger the integrant is, then the higher the frequency of transformation. This integration at the *ura*3 locus usually occurs at a low frequency, because most plasmids carry a 1.1-kb *URA*3 gene, which is small compared with other marker genes such as *LEU*2 or *HIS*4. Thus, with a vector carrying the *URA*3 gene, integrations occur mostly at the location of choice provided the homology of the DNA segment to be integrated is at least 1000 bp to the region in the chromosome (Orr-Weaver & Szostak 1983).

Using the integrative transformation method, Rothstein (1983) developed a method to disrupt yeast genes in a single step. The method involves disrupting the cloned gene of interest on a plasmid vector by inserting a marker gene (e.g. *HIS*3 or *URA*3) using a restriction enzyme that cleaves within the gene of interest. One thus obtains a selectable marker gene flanked by the two ends of the disrupted gene. The *in vitro* disrupted gene is then liberated from the bacterial plasmid sequences by restriction enzyme digestion. It is important that the linear fragment contain sequences homologous to the gene of interest located in the chromosome. The linear DNA fragment is used to transform an auxotrophic (*his*3 or *ura*3) strain that possesses the gene of interest. Integration of the DNA occurs, resulting in the chromosomal gene being replaced by a disrupted gene (as well as the marker gene). Using the same approach an entire gene or gene fragment can be deleted from the chromosome. The method can thus be used, to determine whether a gene is essential.

### *4.6.1 Linear DNA integration and other linear plasmids*

Orr-Weaver *et al.* (1981) showed clearly the recombinogenic capability of several integrating plasmids that were linearized with a restriction enzyme. Integration into the yeast chromosome is known to occur by homologous recombination. Circular plasmids integrate at a very low frequency, mostly by a single reciprocal crossover event. Orr-Weaver *et al.* (1981) found that restriction digestion within the homologous region greatly enhanced efficiency of integration. Interestingly, they also found that when two restriction cuts were made with the same homologous sequence, thus deleting an internal segment of DNA, integration of such gapped plasmids still occurred, with the deleted part being replaced by chromosomal DNA sequences. It was further observed that the gene *RAD*52 was required for integration of linear plasmids (by homologous recombination), but not for integration of circular plasmids (by reciprocal exchange).

Since integration of linear DNA into the yeast chromosome requires sequence homology between the plasmid and the chromosomal DNA, integration of heterologous DNA would pose a problem. Zamir *et al.* (1981) overcame the potential problem by co-transforming heterologous DNA (*Nif* genes of *Klebsiella pneumoniae*) with two plasmids. One plasmid contained a marker that was selectable in yeast and provided the necessary sequence homology with the yeast chromosomal DNA. Independent recombination events between the yeast DNA sequences and between the bacterial antibiotic resistance sequences common to the two plasmids eventually led to the integration of the heterologous *Nif* DNA into the yeast genome. However, the recipient yeast strain contained an integrated copy of the bacterial plasmid pBR313, and this had a sequence homology of over 1 kb with the plasmid carrying the *Nif* genes. Compton *et al.* (1982) further extended the experiments by integrating *Nif* genes into a yeast strain that contained no bacterial sequences. In fact, they repeated the experiment with plasmids containing other yeast sequences and foreign DNA sequences, and concluded that the co-transformation procedure applied to any yeast selectable marker and to any non-homologous sequence that could be cloned into a bacterial plasmid. A generalized scheme was proposed whereby three events occurred in sequential order: (1) primary integration, which occurred by homologous recombination between the yeast sequence present on the selectable plasmid and in the genome; (2) secondary integrations, which occurred as a result of selectable plasmid sequences within the genome providing potential sites for homologous recombination with both the selectable and non-selectable plasmids; and (3) multiple integrations, which led to a wide range of possible tandem and interspersed repeats of the two plasmids within the genome of the recipient.

Dani and Zakian (1983) constructed linear yeast plasmids with and without centromeres by ligating the termini from the extrachromosomal rDNA of the ciliated protozoan *Tetrahymena thermophila* as 'telomeres'. These linear plasmids transformed yeast at high frequency and were maintained as linear extrachromosomal molecules during mitotic growth. They found that linear plasmids containing centromere *CEN* sequences were also transmitted through meiosis, segregating mainly 2+ : 2− at the first meiotic division. These plasmids lacking the *CEN* sequences had mitotic stability comparable with circular plasmids. Thus, although *Tetrahymena* rDNA termini provided telomere function during meiosis, they had no effect on the mitotic stability of autonomously replicating plasmids. Moreover, linear plasmids containing the termini and *CEN* sequences were three- to four-times less stable mitotically than circular plasmids containing *CEN* sequences. It was proposed that linearization of a *CEN* plasmid had a direct detrimental effect on its mitotic stability, perhaps due to spatial constraints on the positions of centromeres and telomeres. Zakian *et al.* (1985) also showed that the *RAD*52 gene product which is required for most yeast mitotic and meiotic recombination events is not required for either the creation of new telomeres on ciliate termini or for the efficient maintenance of linear DNAs. They added synthetic tracts of $5'[C_{(1-3)}A]3'$ DNA to the ciliate termini of the linear plasmids, and found no detrimental effect on plasmid stability.

In a similar set of experiments, Szostak and Blackburn (1982) used a linear plasmid with *Tetrahymena* termini to clone a yeast chromosomal telomere. This yeast telomere has been sequenced and shown to consist of an irregular repeated sequence of the form $5'[C_{(1-3)}A]3'$. They also observed that after introduction of linear plasmid with the *Tetrahymena* termini into yeast cells, the ends of the plasmid grew longer by 100 to several hundred base pairs. This event occurred only when the DNA was introduced by transformation. Sequence analysis of the extra terminal ends showed presence of the same $5'[C_{(1-3)}A]3'$ sequences.

A unique set of linear plasmids has recently been described in two strains of the yeast *Kluyveromyces lactis*. Gunge *et al.* (1981) described the presence of two double-stranded linear DNA species, pGKL1 (of 8.9 kb) and pGKL2 (of 13.4 kb), in 50–100 copies per haploid cell in *Kluyveromyces lactis* IFO1267. Both plasmids stably coexist in the yeast, and show cytoplasmic inheritance. The smaller plasmid has been shown to be associated with production of a 'killer' protein toxin and immunity to it, while the larger plasmid is required for the maintenance of the smaller plasmid. Sequence analysis of the terminal ends of the plasmids showed that pGKL1 had A–T-rich inverted repeats of 202 bp while pGKL2 had 182 bp of repeats. The ends have also been found to be tightly associated to proteins, perhaps providing protection from nuclease digestion (Hishinuma *et al.* 1984, Kikuchi *et al.* 1984, Stark *et al.* 1984).

The toxin produced by *Kluyveromyces lactis* IFO1267 is secreted into the medium and is lethal to a wide variety of yeasts, including strains of *Kluyveromyces*, *Saccharomyces*, *Candida* and *Torulopsis* (Sugisaki *et al.* 1984). This is in contrast with the *Saccharomyces cerevisiae* toxin, which is lethal to *Saccharomyces* strains only. Moreover, the toxin is a large protein of molecular weight over 100 000, having three subunits. The mode of action of the toxin is not completely understood, but inhibition of adenylate cyclase by the toxin has been shown (Sugisaki *et al.* 1983). Interestingly, the plasmids have been transferred to *Saccharomyces cerevisiae* both by protoplast fusion and by transformation, and have been found to be functional (Gunge & Sakaguchi 1981, Gunge *et al.* 1982). The linear plasmids were found to coexist with the *Saccharomyces cerevisiae* 2-μm plasmids, and toxin secretion was observed. In fact, in both *Saccharomyces cerevisiae* and another strain of *Kluyveromyces* –

*Kluyveromyces fragilis* – toxin secretion, for some reason, was found to be enhanced (Sugisaki *et al.* 1985).

Several attempts have been made by many researchers to use the linear plasmids as cloning vehicles, without much success. It is known that the smaller plasmid codes for four open-reading frames (Sor & Fukuhara 1985, Stark *et al.* 1984), and that the central part of the linear plasmid can function as an autonomously replicating sequence on a circular plasmid in *Saccharomyces cerevisiae* (Gunge & Yamane 1984, Thompson & Oliver 1984). Attempts to construct circular plasmids containing a functional 'killer' gene have also been unsuccessful.

Gunge *et al.* (1982) also found that the plasmids were maintained stably only in $\varrho^0$ (rho$^0$) cells of *Saccharomyces cerevisiae*, and when mitochondrial DNA was introduced via cytoduction the linear plasmids became unstable and were often lost (Gunge & Yamane 1984). This would suggest incompatibility between the plasmids and mitochondrial DNA. The reason for such incompatibility is not clear.

## 4.7 Plasmids carrying dominant selectable markers

As noted above, transformations of laboratory yeast strains generally involve complementation of an auxotrophic marker by a plasmid carrying the fully functional gene that is defective in the recipient. The genes coding for enzymes in the amino acid biosynthetic pathway, such as *LEU2* or *TRP1*, are commonly used for such purposes, since it is fairly easy to obtain haploid auxotrophs for leucine and tryptophan biosynthesis. However, for industrial yeast strains this is not the case. Yeasts used in the brewing, baking and wine industries are usually polyploid in their genetic make-up, and it is very difficult to obtain stable auxotrophs of such strains without altering the physiological and fermentative characteristics for which these yeasts have been selected (Stewart *et al.* 1982). Transformation of such strains therefore has to be approached from a different angle, involving the use of dominant selectable markers. The gene imparting resistance or tolerance to the metal-ion copper constitutes such a marker.

The gene imparting resistance to high levels of copper ($>0.5$ mmol $l^{-1}$), *CUP1*, has been cloned and characterized by Fogel and Welch (1982) and by Butt *et al.* (1984a). The gene codes for a copper-chelating metallothionine protein rich in cysteine residues. Fogel and Welch (1982) showed that the cloned DNA fragment contained one or more tandemly repeated DNA sequences with a unit length of 1.95 kb. The repeat units were found to be tandemly arranged in the yeast chromosome, and copper-resistant yeast strains were found to contain as many as 15 reiterated repeats. Butt *et al.* (1984b) determined the structure of the yeast copper-binding protein by DNA sequence analysis of the *CUP1* gene. They found that the 6573-dalton protein was induced by copper and by zinc, cadmium and gold. Butt *et al.* (1984a, b) obtained definitive proof that the 5′-upstream or 5′-untranslatable, or both, DNA sequence of the *CUP1* gene was responsive to copper regulation by fusing the cloned DNA fragment with the *Escherichia coli galK* gene. A 50-fold increase in *galK* expression upon addition of $CuSO_4$ to the medium was obtained when a plasmid containing a single copy of *CUP*1 gene was used. Thus, copper inducibility of the copper-metallothionine (Cu-MT) protein was retained on an autonomously replicating plasmid in yeast. Butt *et al.* (1984a, b) conclude from their studies that the Cu-MT promoter would be an excellent system for the expression of heterologous gene products in yeast. Henderson *et al.* (1985) also cloned the *CUP*1 gene independently and used it for transformation of brewing industrial yeast strains. Eight brewing strains that were extremely sensitive to copper were chosen as recipients. They were all transformed with the plasmid pET13.1 which carried the *CUP*1 gene. The transformants were able to grow on agar plates containing 1.0 mmol $l^{-1}$ $CuSO_4$. However, the plasmid was unstable in the transformants, and resistance to copper was quickly lost when the strains were grown in non-selective media. Meaden and Tubb (1985) further performed wort fermentations with one of the transformants as well as the parent strain. They showed that in both a small-scale set-up and in the pilot brewery, fermentation profiles of the transformant and the parent strains were almost indistinguishable. They state that no differences in the resulting beers were found by a taste panel. Meaden and Tubb (1985) also constructed a single plasmid containing the *CUP*1 gene and the *DEX* gene coding for extracellular amyloglucosidase (Meaden *et al.* 1985). The *DEX–CUP*1 plasmid was used to transform a brewing strain. One of the transformants was subjected to wort fermentations, and was found to utilize 26–30% of the carbohydrate which is ordinarily unfermentable.

### 4.7.1 *Use of antibiotic-resistance selection*

In bacterial transformations use is commonly made of plasmids carrying genes conferring resistance to certain antibiotics. Thus, plasmid pBR322 carries genes for resistance to ampicillin and tetracycline. The use of antibiotic resistance as a selection for plasmid-mediated transformations of yeast has been hampered by the availability of antibiotics that are lethal to yeasts. However, a few exceptions do exist. One of the more useful ones has been Geneticin G418, an aminoglycoside that has been found to be active against many procaryotes and eucaryotes, including plant and animal cells. The antibiotic is structurally related to Kanamycin and gentamycin. The bacterial transposon Tn601 (or Tn903) codes for the enzyme aminoglycoside phosphotransferase-3′(I), which inactivates G418. Jimenez and Davies (1980) demonstrated that yeast transformed with a chimeric plasmid carrying Tn601 becomes resistant to G418 at concentrations greater than 150 $\mu g\ ml^{-1}$. Hollenberg (1982) also reported on transformation of strain AH22 to G418 resistance using plasmids carrying the Tn601 transposon. Webster and Dickson (1983) used the Kanamycin resistance gene of Tn903 to transform a strain of *Saccharomyces cerevisiae* by direct selection on yeast extract–peptone–dextrose (YEPD) plates containing G418. They observed that salts in yeast nitrogen-base media could inactivate the antibiotic and give high background values.

Webster and Dickson (1983) successfully used the antibiotic selection to clone autonomously replicating sequences (*ARS*) from the lactose-utilizing yeast *Kluyveromyces lactis*.

Zhu *et al.* (1985) have described a dominant selection system for transformation of *Saccharomyces cerevisiae* using a mouse dihydrofolate reductase (*dhfr*) gene. A multi-copy 2-μm derived plasmid carrying the *dhfr* gene imparts to the yeast cell resistance to the anti-metabolite methotrexate. Zhu *et al.* (1985) constructed a variety of yeast–*Escherichia coli* chimeric plasmids carrying the *dhfr* gene as well as the yeast *URA*3 gene, the yeast cytochrome C (*CYC*1) promoter and terminator regions and different segments of the 2-μm plasmid. When the plasmids contained the 2-μm origin of replication together with *REP*1, *REP*2 and *REP*3 genes, no endogenous 2-μm functions were required. Transformants selected on plates containing increasing concentrations of methotrexate showed a 25% increase in copy number of the plasmid. However, a more pronounced amplification was seen when a plasmid containing only the origin of replication region (2.2-kb *Eco*RI fragment) of the 2-μm DNA was used.

Without selection, the recombinant plasmid was present in low copy number (25 copies $cell^{-1}$), while at maximal selection pressure (8 mg $ml^{-1}$ sulphanilamide + 0.2 mg $ml^{-1}$ methotrexate) a much higher copy number was obtained. Moreover, a twofold increase in the endogenous 2-μm DNA was also observed, perhaps due to the requirement of *REP*1 and *REP*2 gene products by the recombinant plasmid.

Using the *CYC*1 promoter, Zhu *et al.* (1985) obtained moderate levels of expression of the *dhfr* cDNA. However, when the cDNA was fused without the 5′-untranslated sequence to the stronger actin promoter, increased levels of expression were obtained and the plasmid showed a better transforming efficiency. Zhu *et al.* (1985) attribute this to either the fact that using *dhfr* DNA the transformation efficiency is dependent upon the strength of the promoter or to the structure and size of the plasmid vector. (The actin promoter containing plasmid carried 2-μm DNA sequences and no *ARS*1 DNA while the *CYC*1 promoter containing plasmid carried both 2-μm and *ARS*1 DNA sequences.)

Using this dominant selection method, Zhu *et al.* (1985) obtained high-level expression of human interleukin-2 cDNA (IL-2 cDNA) in yeast. The cDNA was fused to the yeast triose phosphate isomerase (TPI) promoter on a plasmid carrying the *dhfr* gene. Again, by increasing the selective pressure (concentrations of sulphanilamide and methotrexate) a marked increase in levels of the protein (up to 5% of total protein) was obtained.

A similar strategy has also been used with other drugs (J. Ernst, pers. communs), such as the use of the *HYQ* gene, imparting resistance to the drug hygromycin B (Gritz & Davies 1983) and the *NEO* gene imparting resistance to the aminoglycoside Geneticin G418.

## 4.8 Expression vectors

It has been shown quite convincingly that heterologous eukaryotic genes can be expressed in *Saccharomyces cerevisiae* by placing them under the control of various yeast gene 'promoters' (Brake *et al.* 1984, Hitzeman *et al.* 1983, Kramer *et al.* 1984, Smith *et al.* 1985). The yields of the heterologous proteins are determined by the 'strength' of the promoter. Thus, the yeast gene for phosphoglycerate kinase (*PGK*) has a promoter that can direct the expression of heterologous genes with greater than 500-fold increase in efficiency over the *TRP*1 promoter. Mellor *et al.* (1985) have used the *PGK* promoter fragment to direct the expression of human interferon-$\alpha$-2 (IFN$\alpha$2) on a high copy number plasmid in yeast. Yields of 1–3% of the total yeast protein as IFN$\alpha$2 were obtained. The highest yields of any heterologous product in yeast have been around 5% of total cell protein using an efficient promoter and a 2-μm based plasmid vector. Theoretically, however, a higher yield should be possible since the 2-μm based plasmids are usually at a copy number varying from 20 to 100 $cell^{-1}$.

In cloning genes of industrial significance, it is useful and economical to have the gene product(s) secreted into the medium. The secretion of proteins by *Saccharomyces cerevisiae* occurs by a mechanism very similar to that observed in mammalian cells. Schekman and collaborators (Schekman & Novick 1982, Novick *et al.* 1981) have shown that the yeast secretory pathway involves a series of membrane-bound structures that mediate the transfer of exported proteins from their site of synthesis at the endoplasmic reticulum to their site of release at the plasma membrane. As with other organisms, the secreted proteins are synthesized as large precursors including 'signal' sequences which are subsequently cleaved by a membrane bound protease to yield the mature gene products.

There have been published reports on the secretion of heterologous proteins from yeast using the signal sequences of the mammalian proteins being secreted (Hitzeman *et al.* 1983). However, such signal sequences do not generally permit more than about 20% of the protein synthesized to be secreted. Moreover, it has been reported that with such signal sequences proteolytic cleavage seems to be imprecise. Because of such findings considerable effort has been devoted to developing plasmids carrying signal sequences from yeast genes coding for secretory proteins such as the enzymes invertase, acid phosphatase, melibiase, glucoamylase and proteins like $\alpha$-factor pheromone and killer toxin. The structural genes involved in the synthesis of these proteins have been cloned and extensively characterized.

The gene *SUC*2 responsible for synthesis of invertase (which is mostly secreted into the periplasmic space) has been cloned by Carlson and Botstein (1982). The *SUC*2 gene codes for two differentially regulated mRNAs and two forms of the sucrose-hydrolysing enzyme invertase. The secreted, glycosylated form of the enzyme is required for the extracellular hydrolysis of sucrose, and is regulated by glucose. The intracellular, non-glycosylated invertase has no known function and is produced constitutively in the cell. The *SUC*2 gene has been cloned and sequenced (Taussig & Carlson 1983). The two forms of invertases are encoded by two differently regulated mRNAs which differ only at the 5′-ends (Carlson *et al.* 1983). A 1.9-kb RNA, which is glucose-repressible, encodes a signal peptide-containing precursor to the secreted invertase. However, the constitutively synthesized 1.8-kb RNA does not encode a complete signal sequence, and hence the translation product remains intracellular (Carlson & Botstein 1982).

Since the *SUC2* gene carries a strong promoter as well as a signal sequence for secretion, it has been found to be useful in the construction of hybrid DNA sequences by fusing heterologous genes of interest with the promoter or signal sequence region, or both, of the *SUC2* gene. Thus, secretion of calf prochymosin (the inactive precursor of chymosin or renin) has been achieved (Smith *et al.* 1985) by fusing in translational reading frame, the *SUC2* transcriptional promoter region, the coding region for the amino-terminal secretion signal and codons for the first 11 amino-acid residues of mature invertase with a prochymosin-encoding DNA fragment. The fused genes were placed on a plasmid vector and transformed into an appropriate recipient yeast strain for expression. Experiments involving fusion of other heterologous genes with the *SUC2* promoter have also been reported (Emr *et al.* 1983). Since depletion of glucose in the medium would activate the *SUC2* promoter, the use of *SUC2* promoter signal sequence system for expression and secretion of heterologous proteins such as glucoamylase is very appealing.

## 4.9 Transformation

The most commonly used procedure for transformation of yeast has been the one that was used for the first reported transformation of yeast (Hinnen *et al.* 1978). Recipient yeast cells are first converted into spheroplasts by removing the cell wall enzymatically. The enzymes – mostly glucanases – are obtained commercially, generally as crude enzyme mixtures. Several different types of enzymes or enzyme systems have been reported in the literature. Hinnen *et al.* (1978) used Glusulase (Endo Laboratory, USA) which is derived from snail gut. A similar enzyme mixture Helicase (Industrie Priologique Française, France) was used by Beggs (1978), while the enzyme lyticase isolated from *Oerskovia xanthineolytica* by Scott and Schekman (1980) has been found to be very effective for preparing clean yeast spheroplasts. Kirin Breweries (Japan) and Miles Laboratories (USA) have been the major suppliers of the enzyme mixture known as Zymolyase, isolated from *Arthrobacter luteus*, while Novo has been marketing the enzyme Novozyme 234 which has been found to be effective at low concentrations. With all enzyme treatments pretreatment of cells with thiol reagents such as $\beta$-mercaptoethanol and dithiothreitol or sulphur amino acids (e.g. cysteine) is required to obtain high spheroplast-formation frequencies (Hinnen & Meyhack 1982, Dowhanick *et al.* 1984). For effective spheroplast formation it is essential that yeast cells be in mid-exponential phase of growth. With some enzymes, such as Zymolyase, enzymatic treatment has to be timed accurately to avoid overdigestion and hence a drop in spheroplast regeneration frequency. This is particularly the case with Zymolyase. Cell-wall digestion is generally monitored by checking the osmotic stability of the spheroplasts in distilled water and observing lysis under the microscope.

Osmotically sensitive spheroplasts are suspended in a sorbitol buffer (1.0–1.2 mol $l^{-1}$), treated with $Ca^{2+}$, and plasmid DNA is added to the mixture. After a short incubation period the polymer polyethylene glycol (20% of PEG 4000) is added. The function of PEG is not clear, but it is believed to be involved in facilitating entry of DNA into the yeast. After an incubation period of about 20 min the PEG is usually removed by centrifugation and washing. The spheroplasts are gently mixed in regeneration agar (selective medium + sorbitol + 3% agar) and plated on regeneration plates. Several dilutions are plated, as are controls (no plasmid and no sorbitol – to measure the frequency of spheroplast formation). An incubation period of 3–5 days is generally required for transformants to come up. Frequencies of transformation are strain dependent and can vary from $1 \times 10^2$ to $1 \times 10^5$ transformants $\mu g^{-1}$ DNA (Gingold 1984, Struhl 1983).

A technique similar to that used for *Escherichia coli* transformation was recently described by Ito *et al.* (1983), and has gained considerable popularity. The method is simple and does not require spheroplasting of the yeast cells. The cells are subjected to treatment with alkali cations (e.g. 0.1 mol $l^{-1}$ lithium acetate) together with PEG, after being mixed with DNA. After an incubation period of 30 min they are subjected to heat shock at 42°C for 5 min, followed by cooling and plating on selective media. Transformation frequencies are generally low ($1–10^3$ $\mu g^{-1}$ DNA) and strain dependent. While spheroplast transformations can be performed with as little as 1 μg of DNA, whole-cell transformations generally require about 20 μg of DNA, often with heterologous carrier DNA present. Nevertheless, the technique is simple and rapid, and transformants usually come up in less than 3 days. A modification of the technique using only $CaCl_2$ and no lithium acetate or PEG has also been reported, but a much smaller transformation frequency is obtained with such a modification (Iimura *et al.* 1983). Another modification involving treatment of intact cells with only PEG has also been reported, where transformation frequencies of up to 140 $\mu g^{-1}$ DNA were obtained (Yamakawa *et al.* 1985).

Suzuki *et al.* (1983) reported successful transformations of yeast strains with linear DNA plasmids which were circularized *in vivo* by the host yeast ligase system. The chimeric plasmid pJDB219 carrying the yeast *LEU2* gene and the 2-μm origin of replication was initially restricted with *Hind*III, producing four differently sized fragments. The 3.2-kb fragment was found to contain the *LEU2* gene as well as the 2-μm origin of replication, but no bacterial DNA (Toh-e *et al.* 1980). This fragment (pSLe1) could transform a *leu2* strain to leucine prototrophy at a high frequency and was recovered from the transformants as a circular plasmid, thus confirming *in vivo* ligation. An equimolar mixture of *Hind*III cut pSLe1 and *Hind*III linearized YIp5 (a chimeric plasmid carrying the yeast *URA3* gene but no origin of replication) was used to transform a yeast strain to leucine and uracil prototrophy. A frequency of transformation of $1–5 \times 10^3$ transformants $\mu g^{-1}$ DNA was obtained. Four different size plasmids were seen when the transformants were analysed for plasmid DNA content. However, each of the plasmids could transform a yeast auxotroph to leucine and uracil prototrophy at a high frequency ($10^3–10^4$ transformants $\mu g^{-1}$ DNA). All of the four plasmids were found to be circular, again confirming *in vivo* ligation. Such *in vivo* ligation was also observed when the plasmids were restricted with an enzyme (e.g. *Hpa*I) that resulted in flush ends rather than cohesive ends. In fact Suzuki *et al.* (1983) claimed that *in vivo* ligation even occurred when one plasmid was restricted with *Hind*III and the other with *Eco*RI.

## 4.10 Summary

The development of a plasmid vector system for *Saccharomyces* yeasts has come a long way in the past 15 years. Use of *Escherichia coli*–yeast chimeric plasmids for cloning yeast genes or genes of interest into yeast is now almost routine. While much attention has been focused on such chimeric plasmids, a few studies devoted to total-yeast DNA plasmids have also been carried out. These have been mainly in the context of using plasmids for food and beverage purposes. While increasing attention is paid to such industrial strains to produce novel products (e.g. proteins with pharmaceutical value), the development of appropriate vector systems for such purposes has also progressed rapidly. The development of plasmids with dominant selectable markers and the use of integrating vectors are cases in point. With the use of plasmids in industrial yeast strains, concern has also been expressed over the stability of such plasmids in large-scale fermentations. While there are no instances of yeast plasmids propagating with 100% stability (except perhaps for 2-μm DNA in diploids and polyploids), major insights into functions involved in plasmid stability have been made. The study of stability of linear plasmids with telomeric ends will undoubtedly lead to an understanding of how yeast chromosomes behave during mitosis and meiosis.

As mentioned at the beginning of this chapter, progress in yeast genetics has come about rapidly, in part due to development of certain techniques which have opened up a whole new world of science. The ability to carry out transformation of yeast, most certainly, was one such example. The recently reported technique to perform karyotyping yeast DNA using Orthogonal Field Alternation Gel (OFAGE; Carle & Olson 1985, Carle *et al.* 1986, Schwartz & Cantor 1984) promises to be another one. The potentials for cloning genes and for following the course of events during genetic manipulations of yeasts (especially industrial yeasts) are enormous. Yeast genetics is certainly in for some interesting times in the foreseeable future.

## Acknowledgements

The author wishes to acknowledge, with gratitude, his colleagues Dr G. G. Stewart, I. Russell, Dr C. A. Bilinski, L. Bast and R. M. Jones for their helpful discussions and suggestions during the preparation of the manuscript.

## References

Beggs, J. D. 1978. Transformation of yeast by a replicating hybrid plasmid. *Nature* (London) **275**, 104–9.

Beggs, J. D. 1981. Multi-copy yeast plasmid vectors. In *Molecular genetics in yeast. Alfred Benzon symposium*, Vol. 16, D. von Wettstein, J. Friis, M. Kielland-Brandt and A. Stenderup (eds), 383–95. Copenhagen: Munksgaard.

Boeke, J. D., D. J. Garfinkel, C. A. Styles and G. R. Fink 1985. Ty elements transpose through an RNA intermediate. *Cell* **40**, 491–500.

Brake, A. J., J. P. Merryweather, D. G. Coit, U. A. Heberlain, F. R. Masiarz, G. T. Mullenbach, M. S. Urdea, P. Valenzuela and P. J. Barr 1984. Alpha-factor directed synthesis and secretion of mature foreign proteins in *Saccharomyces cerevisiae*. *Proceedings of the National Academy of Sciences of the USA* **81**, 4642–6.

Broach, J. R. 1982. The yeast plasmid 2 μ circle. In *The molecular biology of the yeast* Saccharomyces, *life cycle and inheritance*, J. N. Strathern, E. W. Jones and J. R. Broach (eds), 445–70. New York: Cold Spring Harbor Laboratories.

Broach, J. R. 1983. Construction of high copy yeast vectors using 2 μm circle sequences. In *Methods in enzymology* **101**, 307–25. New York: Academic Press.

Broach, J. R. and J. B. Hicks 1980. Replication and recombination functions associated with the yeast plasmid, 2 μ circle. *Cell* **21**, 501–8.

Brown, A. J. P., A. R. Goodey and R. S. Tubb 1981. Interstrain transfer of the 2 μm DNA plasmid of *Saccharomyces* by cytoduction. *Journal of the Institute of Brewing* **87**, 234–8.

Butt, T. R., E. Sternberg, J. Herd and S. T. Crooke 1984a. Cloning and expression of a yeast copper metallothionein gene. *Gene* **27**, 23–33.

Butt, T. R., E. J. Sternberg, J. A. Gorman, P. Clark, D. Hamer, M. Rosenberg and S. T. Crooke 1984b. Copper metallothionein of yeast, structure of the gene, and regulation of expression. *Proceedings of the National Academy of Sciences of the USA* **81**, 3332–6.

Cameron, J. R., P. Philippson and R. W. Davis 1977. Analysis of chromosomal integration and deletions of yeast plasmids. *Nucleic Acids Research* **4**, 1429–48.

Carbon, J. and L. B. Clarke 1984. Construction and use of functional minichromosomes. US Patent 4 464 472.

Carle, G. F. and M. V. Olson 1985. An electrophoretic karyotype for yeast. *Proceedings of the National Academy of Sciences of the USA* **82**, 3756–60.

Carle, G. F., M. Frank and M. V. Olson 1986. Electrophoretic separations of large DNA molecules by periodic inversion of the electric field. *Science* **232**, 65–8.

Carlson, M. and D. Botstein 1982. Two differentially regulated mRNAs with different 5′ ends encode secreted and intracellular forms of yeast invertase. *Cell* **28**, 145–54.

Carlson, M., R. Taussig, S. Kustu and D. Botstein 1983. The secreted form of invertase in *Saccharomyces cerevisiae* is synthesized from mRNA encoding a signal sequence. *Molecular and Cellular Biology* **3**, 439–47.

Chan, C. S. M. and B-K. Tye 1980. Autonomously replicating sequences in *Saccharomyces cerevisiae*. *Proceedings of the National Academy of Sciences of the USA* **77**, 6329–33.

Clarke, L. and J. Carbon 1980. Isolation of a yeast centromere and construction of functional small circular chromosomes, *Nature* (London) **287**, 504–9.

Clark-Walker, G. D. and G. L. G. Miklos 1974. Localization and quantification of circular DNA in yeast. *European Journal of Biochemistry* **41**, 359–65.

Compton, J. L., A. Zamir and A. A. Szalay 1982. Insertion of non-homologous DNA into the yeast genome mediated by homologous recombination with co-transforming plasmid. *Molecular and General Genetics* **188**, 44–50.

Conde, J. and G. R. Fink 1976. A mutation of *Saccharomyces cerevisiae* defective for nuclear fusion. *Proceedings of the National Academy of Sciences of the USA* **73**, 3651–5.

Dani, G. M. and V. A. Zakian 1983. Mitotic and meiotic stability of linear plasmids in yeast. *Proceedings of the National Academy of Sciences of the USA* **80**, 3406–10.

Davison, J. and V. H. Thi 1982. The *Trypanosoma brucei* maxi-circle DNA contains *ars* elements active in *Saccharomyces cerevisiae*. *Current Genetics* **6**, 19–20.

Dobson, M. J., A. B. Futcher and B. S. Cox 1980. Loss of 2 μm DNA from *Saccharomyces cerevisiae* transformed with the chimeric plasmid pJDB219. *Current Genetics* **2**, 201–5.

Dowhanick, T. M., C. J. Panchal and G. G. Stewart 1984. Effects of pretreatments with sulhydryls and alkylating agents on spheroplast formation from *Schwanniomyces* species. *Canadian Journal of Microbiology* **30**, 368–74.

Emr, S. D., R. Schekman, M. C. Flessel and J. Thorner 1983. An MFalpha$_1$-*SUC2* (alpha factor-invertase) gene fusion for study of protein localization and gene expression in yeast. *Proceedings of the National Academy of Sciences of the USA* **80**, 7080–4.

Erhart, E. and C. P. Hollenberg 1981. Curing of *Saccharomyces cerevisiae* 2 μm DNA by transformation. *Current Genetics* **3**, 83–9.

Fagan, M. C. and J. F. Scott 1985. New vectors for construction of recombinant high-copy-number yeast acentric ring plasmids. *Gene* **40**(2–3), 217–29.

Fitzgerald-Hayes, M., I. M. Buhler, T. Cooper and J. Carbon 1982a. Isolation and subcloning analysis of functional centromere DNA (*CEN*11) from *Saccharomyces cerevisiae* chromosome XI. *Molecular and Cellular Biology* **2**, 82–7.

Fitzgerald-Hayes, M., L. Clarke and J. Carbon 1982b. Nucleotide sequence comparisons and functional analysis of yeast centromere DNAs. *Cell* **29**, 235–44.

Fogel, S. and J. W. Welch 1982. Tandem gene amplification mediates copper resistance in yeast. *Proceedings of the National Academy of Sciences of the USA* **79**, 5342–6.

Futcher, A. B. and B. S. Cox 1984. Copy number and the stability of 2 μm circle-based artificial plasmids of *Saccharomyces cerevisiae*. *Journal of Bacteriology* **157**, 283–90.

Gerbaud, C. and M. Guerneau 1980. 2 μm Plasmid copy number in different yeast strains and repartition of endogenous and 2 μm chimeric plasmids in transformed strains. *Current Genetics* **1**, 219–28.

Gingold, E. B. 1984. Yeast transformation. In *Methods in molecular biology*. Vol. 2: *Nucleic acids*, J. M. Walker (ed.), 251–5. Clifton, New Jersey: Humana Press.

Gritz, L. and J. Davies 1983. Plasmid encoded hygromycin B resistance: the sequence of hygromycin B phosphotransferase gene and its expression in *Escherichia coli* and *Saccharomyces cerevisiae*. *Gene* **25**, 179–88.

Gubbins, E. J., C. S. Newlon, M. D. Kann and J. E. Donelson 1977. Sequence organization and expression of a yeast plasmid. *Gene* **1**, 185–207.

Guerineau, M., C. Grandchamp, C. Paoletti and P. P. Slonimski 1971. Characterization of a new class of circular DNA molecules in yeast. *Biochemical and Biophysical Research Communications* **42**, 550–7.

Gunge, N. 1983. Yeast–DNA plasmids. *Annual Review of Microbiology* **37**, 253–76.

Gunge, N. and K. Sakaguchi 1981. Intergeneric transfer of deoxyribonucleic acid killer plasmids, pGKL1 and pGKL2, from *Kluyveromyces lactis* into *Saccharomyces cerevisiae* by cell fusion. *Journal of Bacteriology* **147**, 155–60.

Gunge, N. and C. Yamane 1984. *Kluyveromyces lactis* linear killer plasmids contain *ARS* elements functional in heterologous yeasts. *12th International Conference on Yeast Genetics and Molecular Biology Abstracts*, 81, Edinburgh, Scotland.

Gunge, N., A. Tamaru, F. Ozawa and K. Sakaguchi 1981. Isolation and characterization of linear deoxyribonucleic acid plasmids from *Kluyveromyces lactis* and the plasmid-associated killer character. *Journal of Bacteriology* **145**, 382–90.

Gunge, N., K. Murata and K. Sakaguchi 1982. Transformation of *Saccharomyces cerevisiae* with linear DNA killer plasmids from *Kluyveromyces lactis*. *Journal of Bacteriology* **151**, 462–4.

Hartley, J. L. and J. E. Donelson 1980. Nucleotide sequence of the yeast plasmid. *Nature* (London) **286**, 860–5.

Henderson, C. A., B. S. Cox and R. S. Tubb 1985. The transformation of brewing yeasts with a plasmid containing the gene for copper resistance. *Current Genetics* **9**, 133–8.

Hinnen, A. and B. Meyhack 1982. Vectors for cloning in yeast. In *Current Topics in Microbiology and Immunology*, Vol. 96, W. Henle, P. H. Hofschneider, J. Koprowski, F. Melchers and R. Rott (eds), 101–17. New York: Springer-Verlag.

Hinnen, A., J. B. Hicks and G. R. Fink 1978. Transformation of yeast. *Proceedings of the National Academy of Sciences of the USA* **75**, 1929–33.

Hishinuma, F., K. Nakamura, K. Hirai, R. Nishizawa, N. Gunge, T. Maeda 1984. Cloning and nucleotide sequences of the linear DNA killer plasmids from yeast. *Nucleic Acids Research* **12**(19), 7581–97.

Hitzeman, R. A., D. W. Leung, L. J. Perry, W. J. Kohr, H. L. Levine and D. V. Goeddel 1983. Secretion of human interferons by yeast. *Science* **219**, 620–5.

Hollenberg, C. P. 1982. Cloning with 2 μm DNA vectors and the expression of foreign genes in *Saccharomyces*. In *Current Topics in Microbiology and Immunology*, Vol. 96, W. Henle, P. H. Hofschneider, J. Koprowski, F. Melchers and R. Rott (eds), 119–44. New York: Springer-Verlag.

Hollenberg, C. P., A. Degelman, B. Kusterman-Kuhn and H. D. Royer 1976. Characterization of 2 μm DNA of *Saccharomyces cerevisiae* by restriction fragment analysis and integration in an *Escherichia coli* plasmid. *Proceedings of the National Academy of Sciences of the USA* **73**, 2072–6.

Holm, C. 1982a. Clonal lethality caused by the yeast plasmid 2 μm DNA. *Cell* **29**, 585–94.

Holm, C. 1982b. Sensitivity to the yeast plasmid 2 μm DNA is conferred by the nuclear allele *nib*1. *Molecular and Cellular Biology* **2**(8), 985–92.

Holmberg, S., J. G. L. Petersen, T. Nilsson-Tillgren and M.C. Kielland-Brandt 1979a. Molecular characterization of a *Saccharomyces* plasmid containing the *HIS*4 gene. *Carlsberg Research Communications* **44**, 269–82.

Holmberg, S., M. C. Kielland-Brandt, T. Nilsson-Tillgren and J. G. L. Petersen 1979b. Molecular characterization of three *his*4 deletion mutants in *Saccharomyces cerevisiae*. *Carlsberg Research Communications* **44**, 77–87.

Hsiao, C-L. and J. Carbon 1979. High-frequency transformation of yeast by plasmids containing the cloned *ARG*4 gene. *Proceedings of the National Academy of Sciences of the USA* **76**, 3829–33.

Iimura, Y., K. Gotoh, K. Ouchi and T. Nishiya 1983. Yeast transformation without the spheroplasting process. *Agricultural Biology and Chemistry* **47**(4), 897–901.

Imai, Y., K. Suzuki, I. Yamashita and S. Fukui 1983. High-frequency transformation of *Saccharomyces cerevisiae* with linear deoxyribonucleic acid. *Agricultural Biology and Chemistry* **47**(4), 915–18.

Ito, H., Y. Fukuda, K. Murata and A. Kimura 1983. Transformation of intact yeast cells treated with alkali cations. *Journal of Bacteriology* **153**, 163–8.

Jayaram, M., Y. Y. Li and J. R. Broach 1983. The yeast plasmid 2 μm circle encodes components required for its high copy propagation. *Cell* **34**, 95–104.

Jayaram, M., A. Sutton and J. R. Broach 1985. Properties of *REP*3, a *cis*-acting locus required for stable propagation of the *Saccharomyces cerevisiae* plasmid 2 μm circle. *Molecular and Cellular Biology* **5**, 2466–75.

Jazwinski, S. M. and G. M. Edelman 1979. Replication *in vitro* of the 2 μm DNA plasmids of yeast. *Proceedings of the National Academy of Sciences of the USA* **76**, 1223–7.

Jimenez, A. and J. Davies 1980. Expression of a transposable antibiotic resistance element in *Saccharomyces*. *Nature* (London) **287**, 869–71.

Kearsey, S. E. 1983. Analysis of sequences conferring autonomous replication in baker's yeast. *EMBO Journal* **2**, 1571–7.

Kielland-Brandt, M. C., T. Nilsson-Tillgren, S. Holmberg, J. G. L. Petersen and B. A. Svenningsen 1979. Transformation of yeast without the use of foreign DNA. *Carlsberg Research Communications* **44**, 77–87.

Kielland-Brandt, M. C., T. Nilsson-Tillgren, J. G. L. Petersen and S. Holmberg 1981. Transformation in yeast without the involvement of bacterial plasmids. In *Molecular genetics in yeast, Alfred Benzon symposium*, Vol. 16, D. von Wettstein, J. Friis, M. Kielland-Brandt and A. Stenderup (eds), 369–82. Copenhagen: Munksgaard.

Kikuchi, Y., K. Hirai and F. Hishinuma 1984. The yeast linear DNA killer plasmids, pGKL1 and pGKL2, possess terminally attached proteins. *Nucleic Acids Research* **12**(14), 5685–92.

Kiss, G. B., A. A. Amin and R. E. Pearlman 1981. Two separate regions of the extrachromosomal ribosomal deoxyribonucleic acid of *Tetrahymena thermophila* enable autonomous replication of plasmids in *Saccharomyces cerevisiae*. *Molecular and Cellular Biology* **1**, 535–43.

Kramer, R. A., T. M. DeChiara, M. D. Schaber and S. Hilliker 1984. Regulated expression of a human interferon gene in yeast: control by phosphate concentration or temperature. *Proceedings of the National Academy of Sciences of the USA* **81**, 367–70.

Livingston, D. M. 1977. Inheritance of 2 μm DNA plasmid from *Saccharomyces*. *Genetics* **86**, 73–84.

Livingston, D. M. and H. L. Klein 1977. Deoxyribonucleic acid sequence organization of a yeast plasmid. *Journal of Bacteriology* **129**, 472–81.

Livingston, D. M. and D. M. Kupfer 1977. Control of *Saccharomyces cerevisiae* 2 μm DNA replication by cell division cycle genes that control nuclear DNA replication. *Journal of Molecular Biology* **116**, 249–60.

Marunouchi, T. and H. Hosoya 1984. Isolation of an autonomously replicating sequence (*ARS*) from satellite DNA of *Drosophila melanogaster*. *Molecular and General Genetics* **196**, 258–65.

Meaden, P. G. and R. S. Tubb 1985. A plasmid vector system for the genetic manipulation of brewing strains. *European Brewery Convention Proceedings of the 20th Congress, Helsinki*, 219. Oxford: IRL Press.

Meaden, P., K. Ogden, H. Bussey and R. S. Tubb 1985. A *DEX* gene conferring production of extracellular amyloglucosidase on yeast. *Gene* **34**, 325–34.

Mellor, J., M. J. Dobson, N. A. Roberts, A. J. Kingsman and S. M. Kingsman 1985. Factors affecting heterologous gene expression in *Saccharomyces cerevisiae*. *Gene* **33**, 215–26.

Montiel, J. F., C. J. Norbury, M. F. Tuite, M. J. Dobson, J. S. Mills, A. J. Kingsman and S. M. Kingsman 1984. Characterization of human chromosomal DNA sequences which replicate autonomously in *Saccharomyces cerevisiae*. *Nucleic Acids Research* **12**(2), 1049–68.

Murray, A. W. and J. W. Szostak 1983a. Construction of artificial chromosomes in yeast. *Nature* (London) **305**, 189–93.

Murray, A. W. and J. W. Szostak 1983b. Pedigree analysis of plasmid segregation in yeast. *Cell* **34**, 961–70.

Newlon, C. S. and W. L. Fangman 1975. Mitochondrial DNA synthesis in cell cycle mutants of *Saccharomyces cerevisiae*. *Cell* **5**, 423–8.
Novick, P., S. Ferro and R. Schekman 1981. Order of events in the yeast secretory pathway. *Cell* **25**, 461–9.

Overbeeke, N., M. A. Haring, H. John, J. Nijkamp and Ad. J. Kool 1984. Cloning of *Petunia hybrida* chloroplast DNA sequences capable of autonomous replication in yeast. *Plant Molecular Biology* **3**, 235–41.
Orr-Weaver, T. L., J. W. Szostak and R. J. Rothstein 1981. Yeast transformation: a model system for the study of recombination. *Proceedings of the National Academy of Sciences of the USA* **78**(10), 6354–8.
Orr-Weaver, T. L. and J. W. Szostak 1983. Yeast recombination: the association between double-strand gap repair and crossing-over. *Proceedings of the National Academy of Sciences of the USA* **80**, 4417–21.

Panchal, C. J., G. K. Whitney and G. G. Stewart 1984. Susceptibility of *Saccharomyces* spp. to the aminoglycoside antibiotic G418. *Applied Environmental Microbiology* **47**(5), 1164–6.
Parent, S. A., C. M. Fenimore and K. A. Bostian 1985. Vector systems for the expression, analysis and cloning of DNA sequences in *Saccharomyces cerevisiae*. *Yeast* **1**, 83–138.
Petes, T. D. and X. Fangman 1973. Preferential synthesis of yeast mitochondrial DNA in $\alpha$-factor arrested cells. *Biochemical and Biophysical Research Communications* **55**, 603–9.

Rothstein, R. J. 1983. One-step gene disruption in yeast. In *Methods in enzymology*, Vol. 101, R. Wu, L. Grossman and K. Moldave (eds), 202–11. New York: Academic Press.

Schekman, R. and P. Novick 1982. The secretory process and yeast cell surface assembly. In *Molecular biology of the yeast* Saccharomyces: *metabolism and gene expression*, J. N. Strathern, E. W. Jones and J. R. Broach (eds), 361–93. New York: Cold Spring Harbor Laboratories.
Schwartz, D. C. and C. R. Cantor 1984. Separation of yeast chromosome-sized DNA by pulsed field gradient gel electrophoresis. *Cell* **27**, 67–75.
Scott, J. and R. Schekman 1980. Lyticase: endoglucanase and protease activities that act together in yeast cell lysis. *Journal of Bacteriology* **142**, 414–23.
Seligy, V. L., D. Y. Thomas and B. L. A. Miki 1980. *Saccharomyces cerevisiae* plasmids, Scp or 2 μm: intracellular distribution, stability and nucleosomal-like packaging. *Nucleic Acids Research* **8**, 3371–91.
Skatrud, P. L. and S. W. Queener 1984. Cloning of a DNA fragment from *Cephalosporium acremonium* which functions as an autonomous replication sequence in yeast. *Current Genetics* **8**, 155–63.
Smith, R. A., M. J. Duncan and D. T. Moir 1985. Heterologous protein secretion from yeast. *Science* **229**, 1219–24.
Sor, F. and H. Fukuhara 1985. Structure of a linear plasmid of the yeast *Kluyveromyces lactis*; compact organization of the killer genome. *Current Genetics* **9**, 147–55.
Southern, E. M. 1975. Detection of specific sequences among DNA fragments separated by gel electrophoresis. *Journal of Molecular Biology* **98**, 503–17.
Stark, M. J. R., A. J. Mileham, M. A. Romanos and A. Boyd 1984. Nucleotide sequence and transcription analysis of a linear DNA plasmid associated with the killer character of the yeast *Kluyveromyces lactis*. *Nucleic Acids Research* **12**(15), 6011–30.

Stewart, G. G. 1981. The genetic manipulation of industrial yeast strains. *Canadian Journal of Microbiology* **27**, 973–90.

Stewart, G. G., I. Russell and C. J. Panchal 1981. The genetic manipulation of industrial yeast strains. In *Current developments in yeast research*, G. G. Stewart and I. Russell (eds), 17–24. Toronto: Pergamon Press,

Stewart, G. G., C. J. Panchal, I. Russell and A. M. Sills 1982. Advances in ethanol from sugars and starch. In *Ethanol from biomass*, H. E. Duckworth and E. A. Thompson (eds), 4–57. Toronto: Royal Society of Canada.

Stinchcomb, D. T., C. Mann, E. Selker and R. W. Davis 1981. DNA sequences that allow the replication and segregation of yeast chromosomes. *ICN–UCLA Symposium on Molecular and Cellular Biology* **22**, 473–88.

Struhl, K. 1983. The new yeast genetics. *Nature* (London) **305**, 391–7.

Struhl, K., D. T. Stinchcomb, S. Scherer and R. W. Davis 1979. High frequency transformation of yeast: autonomous replication of hybrid DNA molecules. *Proceedings of the National Academy of Sciences of the USA* **76**, 1035–9.

Sugisaki, Y., N. Gunge, K. Sakaguchi, M. Yamasaki and G. Tamaru 1983. *Kluyveromyces lactis* killer toxin inhibits adenylate cyclase of sensitive yeast cell. *Nature* (London) **304**, 464–6.

Sugisaki, Y., N. Gunge, K. Sakaguchi, M. Yamasaki and G. Tamaru 1984. Characterization of a novel killer toxin encoded by a double-stranded linear DNA plasmid of *Kluyveromyces lactis*. *European Journal of Biochemistry* **141**, 241–5.

Sugisaki, Y., N. Gunge, K. Sakaguchi, M. Yamasaki and G. Tamaru 1985. Transfer of DNA killer plasmids from *Kluyveromyces lactis* to *Kluyveromyces fragilis* and *Candida pseudotropicalis*. *Journal of Bacteriology* **164**(3), 1373–5.

Suzuki, K., Y. Imai, I. Yamashita and S. Fukui 1983. *In vivo* ligation of linear DNA molecules to circular forms in the yeast *Saccharomyces cerevisiae*. *Journal of Bacteriology* **155**(2), 747–54.

Szostak, J. W. and E. H. Blackburn 1982. Cloning yeast telomeres on linear plasmid vectors. *Cell* **29**, 245–55.

Tabak, H. F. 1977. Absence of 2 μm DNA sequences in *Saccharomyces cerevisiae* Y379-5D. *FEBS Letters* **84**, 67–70.

Taussig, R. and M. Carlson 1983. Nucleotide sequence of the yeast *SUC2* gene for invertase. *Nucleic Acids Research* **11**, 1943–54.

Thompson, A. and S. G. Oliver 1984. Isolation and characterization of autonomously replicating sequences from the K1 killer DNA plasmid of *Kluyveromyces lactis*. *12th International Conference on Yeast Genetics and Molecular Biology, Abstracts*, 82. Edinburgh, Scotland.

Toh-e, A., P. Guerry-Kopecko and R. B. Wickner 1980. A stable plasmid carrying the yeast *LEU2* gene and containing only yeast deoxyribonucleic acid. *Journal of Bacteriology* **141**(1), 413–16.

Uchimiya, H., T. Ohtani, T. Ohgawara, H. Harada, M. Sugita and M. Sugiura 1983. Molecular cloning of tobacco chromosomal and chloroplast DNA segments capable of replication in yeast. *Molecular and General Genetics* **1972**, 1–4.

Webster, T. D. and R. C. Dickson 1983. Direct selection of *Saccharomyces cerevisiae* resistant to the antibiotic G418 following transformation with a DNA vector carrying the kanamycin-resistance gene of Tn903. *Gene* **26**, 243–52.

Wilhelm, M. L., M. F. Demouveau and F. X. Wilhelm 1986. Identification of an autonomously replicating sequence near a histone gene of *Physarum polycephalum*. *Current Genetics* **10**, 459–62.

Williamson, D. H. 1985. The yeast *ARS* element, six years on: a progress report. *Yeast* **1**(1), 1–14.

Yamakawa, M., F. Hishinuma and N. Gunge 1985. Intact cell transformation of *Saccharomyces cerevisiae* by polyethylene glycol. *Agricultural Biology and Chemistry* **49**(3), 869–71.

Zakian, V. A. and J. F. Scott 1982. Construction, replication and chromatin structure of *TRP*1RI circle, a multiple-copy synthetic plasmid derived from *Saccharomyces cerevisiae* chromosomal DNA. *Molecular and Cellular Biology* **2**(3), 221–32.

Zakian, V. A., B. J. Brewer and W. L. Fangman 1979. Replication of each copy of the yeast 2 micron DNA plasmid occurs during the S phase. *Cell* **17**, 923–34.

Zakian, V. A., H. M. Blanton and G. M. Dani 1985. Formation and stability of linear plasmids in a recombination deficient strain of yeast. *Current Genetics* **9**, 441–5.

Zamir, A., C. V. Maina, G. R. Fink and A. A. Szalay 1981. Stable chromosomal integration of the entire nitrogen fixation gene cluster from *Klebsiella pneumoniae* in yeast. *Proceedings of the National Academy of Sciences of the USA* **78**, 3496–500.

Zhu, J., R. Contreras, D. Gheysen, J. Ernst and W. Fiers 1985. A system for dominant transformation and plasmid amplification in *Saccharomyces cerevisiae*. *Biotechnology* **3**, 451–6.

# 5 *dsRNA killer systems in yeast*

DIANE J. MITCHELL
and E. ALAN BEVAN

## 5.1 Introduction

Killer yeasts, first identified by Bevan and Makower (1963), secrete proteinaceous toxins which kill other, sensitive yeasts. It is now more than a decade since the cytoplasmic determinants associated with the killing phenomenon were shown to be encapsidated dsRNA molecules. It was originally demonstrated that only two major species of dsRNA molecule were involved – the L and M dsRNAs. Since then a plethora of literature has emerged identifying or reporting the existence of a whole range of dsRNA species. Some of these have been named, and are, in decreasing size order, XL, $L_{BC}$ and $L_A$ (co- or closely migrating species), T, W and M. Despite this apparent increase in the complexity of the system, there is evidence for only one L genome, $L_A$, and the M genome being involved in the killer phenomenon. No evidence links the 'other' species, XL, $L_{BC}$, T and W with the killer system. However, since these species have not, unlike the 5.9-kb encapsidated dsRNA intermediates of the yeast Ty elements, been implicated in any other system, they will be considered here. The involvement of $L_A$ and M genomes is beyond doubt. $L_A$ dsRNAs encode their own major capsid polypeptide, and that of M dsRNA. M dsRNAs encode the killer toxin and immunity function.

*Saccharomyces cerevisiae* has numerous other extrachromosomal elements, some of which are well characterized, others of which are not. None of these elements, [rho], [CIR] (2-μm DNA), [URE.3], [psi], [MAL] or [20S RNA] bear any relationship to the L or M genomes (reviewed in Mitchell & Bevan 1983).

As will be seen later, some 30 or more host nuclear genes have been shown to be involved in the maintenance or expression of the M dsRNAs. Some of these genes have also been shown to be involved in the maintenance of the $L_A$ but not the XL, $L_{BC}$, T or W species.

The secretion of toxin by killer cells is normally associated with (own-) toxin-specific immunity. There are thus clear analogies between toxin-producing strains of yeast and colicin-producing strains of bacteria. Different strains of yeast secrete toxins with differing specificities, and which cross-react. On only three such cross-reacting groups has any significant amount of research been undertaken: the K1, K2 and K3 groups.

This chapter will consider primarily the K1 system, on which the major impetus of research has been directed. Where appropriate, analogies with the

K2 and K3 systems will be made. It is beyond the scope of this chapter to consider two other killer systems; that of the DNA plasmid-encoded toxin-producing *Kluyveromyces* species and the dsRNA *Ustilago* killer system. For research in these areas the reader is directed to some of the more recent literature (Sor & Fukuhara 1985, Hishinuma *et al.* 1984, Kikuchi *et al.* 1984, Stark *et al.* 1984, Sugisaki *et al.* 1984) and (Koltin 1977, Koltin & Kandel 1978, Koltin *et al.* 1978, Widgerson & Koltin 1982), respectively.

## 5.2 Frequency and distribution in nature

Killer yeasts exist amongst laboratory strains and natural isolates of *Saccharomyces cerevisiae* and other species and genera. The phenomenon is widespread, but much more information is available concerning *Saccharomyces cerevisiae* than other 'analogous' systems.

Only a minority of *Saccharomyces cerevisiae* cultures have both L and M dsRNA, and are killers. However, an overwhelming majority of laboratory cultures, at least, carry L genomes and are therefore sensitives (Fink & Styles 1972). Killer yeasts have been identified amongst natural isolates of *Saccharomyces cerevisiae*. Maule and Thomas (1973) identified a killer yeast as a brewery contaminant; Naumov *et al.* (1973) found killer wine-fermentation yeasts; and Imamura *et al.* (1974) found killer yeasts as contaminants in saké brewing.

Extensive surveys of yeasts representative of many genera and species have been conducted (Middelbeek *et al.* 1979, 1980c, Nomoto *et al.* 1984, Pfeiffer & Radler 1984, Philliskirk & Young 1975, Radler *et al.* 1985, Rosini 1985, Stumm *et al.* 1977). Table 5.1 shows a range of genera and species amongst which 'killers' have been identified. The inhibitory substances or toxins which such strains produce may be compared either structurally or functionally. For example, the colicins produced by colicinogenic bacteria have been compared functionally by determining their activity against various resistant mutants (Davies & Reeves 1975). Comparisons between the toxic substances produced by the wide range of killer yeasts have been largely functional, and have been made on the basis of cross-reactivity (Pfeiffer & Radler 1984, Middelbeek *et al.* 1980b, Rogers & Bevan 1978, Rosini 1985, Stumm *et al.* 1977, Young & Yagiu 1978). The fact that killer strains are cross-reactive may generally be taken to imply toxin differences. The converse is not necessarily true. At this point it is considered to be appropriate to note that in the most widely studied system, the K1 system of *Saccharomyces cerevisiae*, both toxin and toxin-specific immunity functions are determined by the M dsRNA genome (see Section 5.7.2). However, resistance to toxin can also be conferred by host genes. *kre*1, *kre*2 and *kre*3 (nuclear) mutations when introduced into K1 sensitives confer resistance not only to K1 toxin, but also to a broad range of other toxin types (Al-Aidroos & Bussey 1978, Bussey *et al.* 1979).

On the basis of toxin specificities (e.g. temperature stability, pH optima and effect of proteolytic enzymes) ten different groups, K1–K10, were identified by Young and Yagiu (1978). Rogers and Bevan (1978) identified four different groups, three of which compare directly with three of those of Young and Yagiu

**Table 5.1** Occurrence of killer yeasts.

| Genera and Species | Reference |
|---|---|
| *Candida albicans* | Mitchell (1974) |
| *crusei* | Pfeiffer and Radler (1984) |
| *valida* | Philliskirk and Young (1975) |
| *Cryptococcus laurentiae* | Middelbeek *et al.* (1980c) |
| *Debaryomyces vanriji* | Philliskirk and Young (1975) |
| *Hansenula anomola* | Philliskirk and Young (1975) |
| *mrakii* | Philliskirk and Young (1975) |
| *subpellicosa* | Philliskirk and Young (1975) |
| *Kluyveromyces drosophilarum* | Philliskirk and Young (1975) |
| *fragilis* | Philliskirk and Young (1975) |
| *Pichia fermentans* | Pfeiffer and Radler (1984) |
| *kluyveri* | Philliskirk and Young (1975) |
| *membranaefaciens* | Philliskirk and Young (1975) |
| **Saccharomyces (capensis) cerevisiae* | Philliskirk and Young (1975) |
| *cerevisiae* | Bevan and Makower (1963) |
| *diastaticus* | Philliskirk and Young (1975) |
| *uvarum* | Philliskirk and Young (1975) |
| *Torulopsis glabrata* | Bussey and Skipper (1975) |

* Only isolates of this genera have been shown to possess dsRNA, associated with the killer phenotype.

(1978) and the fourth has been renamed K11 by Wickner (1979). Pfeiffer and Radler (1984) identified seven different groups on the basis of cross-reactivity and toxin specificities. Two of these correlated with the K1 and K2 groups and the remaining five, two of which were *Saccharomyces cerevisiae*, did not correlate with any group. No direct comparisons have been made between killers identified in the other surveys and those of groups K1–K11. Thus, it remains entirely feasible that more groups than these 11 actually exist.

Several points are worthy of mention here. dsRNA, associated with a killer phenotype, has only been demonstrated for strains of *Saccharomyces* species. Killers of this genus have been assigned exclusively to one of three groups K1, K2 or K3, other genera belonging to one of the remaining groups K4–K11. The K1 killers are typified by having a 1.9-kb M genome and pH optimum of 4.6 for toxin activity; K2 killers have a 1.7-kb M genome and a pH optimum of 4.2 for toxin activity; K3 killers have a 1.5-kb M genome and a pH optimum of 4.6 for toxin activity (Bevan *et al.* 1973, Philliskirk & Young 1975, Rogers & Bevan 1978, Young & Yagiu 1978). Three further isolates of toxin-producing *Saccharomyces cerevisiae* strains exist which may not fall into any of the K1, K2 or K3 groups. Extremera *et al.* (1982) have identified a killer strain of *Saccharomyces cerevisiae*, induced by nitrosoguanidine treatment, which has a 2.0-kb M genome and kills isolates belonging to the K1 or K2 groups. They called this strain a 'K3' killer, but it is clearly different – on the basis of M genome size – from the K3 group identified above. Of the two *Saccharomyces cerevisiae* (non-K1 or non-K2) killers identified by Pfeiffer and Radler (1984) one, strain 42297, has not been characterized further, but the other, strain 28, has been shown to have a particularly stable toxin, active at a pH optimum of 5.8. Neither of these two strains has been examined for dsRNA content, but presuming they do indeed carry dsRNA, then strain 28 at least must belong to yet another group.

M dsRNA genomes, associated with killer strains belonging to groups K1, K2 and K3, are readily cured by elevated temperatures, cycloheximide or 5-fluorouracil (Fink & Styles 1972, Mitchell *et al.* 1976b, Nesterova *et al.* 1981, Wickner 1974a, Young & Yagiu 1978, authors' unpubl. obs.). No dsRNA has been detected in the non-*Saccharomyces* species assigned to groups K4–K11, neither could such species be cured of their killer activity by elevated temperatures or cycloheximide (Young & Yagiu 1978). Thus, the basis of the killing phenomenon in the non-*Saccharomyces* yeasts assigned to groups K4–K11 may be entirely different from that of the *Saccharomyces* yeasts.

Polonelli *et al.* (1983) and Morace *et al.* (1984) undertook epidemiological surveys of the sensitivity of *Candida albicans* and other pathogenic yeast isolates to 'killer' yeasts. Their recorded data include some 24 species of yeast which have a positive 'killer' effect. Species not included in Table 5.1 are: *Candida guilliermondii*, *Candida* (*Yarrowia*) *lipolytica*, *Candida maltosa* (*sake*), *Hansenula bimundalis*, *Hansenula californica*, *Hansenula canadensis*, *Hansenula dimennae*, *Hansenula fabiani*, *Hansenula holstii*, *Hansenula petersonii*, *Pichia farinosa*, *Pichia guillermondii*, *Pichia ohmeri*, *Pichia spartinaiae*, *Rhodotorula glutinis*, *Rhodotorula rubra* and *Trichosporon capitatum*. These authors scored a killer effect as positive on the basis of either an 'inhibitory effect' or a 'bluish' (methylene blue media) 'zone of coloured cells', or both. Thus, the killing effects noticed could have been due to either a true dsRNA encoded killer toxin or to some other genome-encoded toxin, or even to an interstrain interaction caused by, for example, mating-type pheromones. Strains such as these, and those already assigned to groups K4–K11, are worthy of further investigation in order to elucidate the basis of their interstrain inhibitory effects. This would be particularly pertinent in the case of *Torulopsis glabrata* killers (K4; Young & Yagiu 1978). Isolates of *Torulopsis glabrata* with killing ability were discovered by Bussey and Skipper (1975, 1976). No dsRNA or linear DNA plasmids have been detected, although the appropriate extraction procedures were followed (Young & Yagiu 1978, Sriprakash & Batum 1984), nor is the killer trait cured by cycloheximide, heat, acridine orange or ethidium bromide (Young & Yagiu 1978). Sriprakash and Batum (1984) undertook cytoduction between killer and non-killer strains of *Torulopsis glabrata* and, although they clearly established transfer of mitochondrial DNA, they were unable to demonstrate transfer of the killer trait. Killer and non-killer protoplasts were fused and then treated with benlate, which induces chromosomal non-disjunction. The protoplasts were killer, implying dominance, and following benlate treatment the killer trait segregated, implying a chromosomal location, but independently of the four auxotrophic markers used in the analysis. These authors concluded that this was evidence for a possible chromosomal location of the *Torulopsis glabrata* killer determinant. The evidence would be even stronger if linkage to a chromosomal gene could be established during such non-disjunction analysis.

## 5.3 Killer toxins

A diffusible product of killer cells (Bevan & Makower 1963) with a protein moiety (Bussey 1972, Woods & Bevan 1968) was first shown to be responsible for the killing reaction. It is now known that – in the case of K1 strains at least, and presumably for K2 and K3 strains also – the M genome encodes the killer toxin and killer-immunity determinant (see Section 5.7.2).

### *5.3.1 Properties and structural analysis of K1 toxins*

Type K1 toxin is a non-glycosylated extracellular (secreted) protein with an estimated molecular weight of approximately 11 500 (Palfree & Bussey 1979, Bussey *et al.* 1983b). Toxin is heat labile, being stable only around 22°C and active only within a narrow pH range (4.2–4.8) close to its pI (4.5). It is irreversibly inactivated at pH values in excess of 6.5, and has a half-life in culture media of around 60 min (Bussey *et al.* 1983a). This lability of toxin, following secretion, is analogous to the situation in *Escherichia coli* for the degradation of colicin E4 (Pugsley 1983). Killer toxin is thought to be degraded by a protease in wild type strains. This is supported by the observation that a protease inhibitor, phenylmethyl sulphonylfluoride (PMSF), increases the stability of the secreted toxin and the amount of secreted toxin present in the medium. The presence of the recessive allele *ski*5 gives rise to a similar phenotype. It results not only in the oversecretion of toxin, but also in the oversecretion of other low-molecular weight proteins and probably also $\alpha$-factor pheromone. It is thus concluded that the *ski*5 mutation exerts its effects via a defect in a cell-surface PMSF-inhibited protease (Bussey *et al.* 1983a).

It has been concluded that a considerable fraction of toxin, 70–90%, is inactive. Separation of active from inactive toxin has been achieved by controlled-pore glycerated glass columns (Palfree & Bussey 1979) and by 1,6$\beta$-D-glucan affinity (Hutchins & Bussey 1983), although no difference between the two components could be detected by gel analysis (Bussey *et al.* 1983b) or by antisera (Bostian *et al.* 1980b).

As mentioned above, the molecular weight of toxin, as identified by SDS–polyacrylamide gel-electrophoresis, was put at about 11 000 (Palfree & Bussey 1979, Bostian *et al.* 1980b). However, its apparent size in phosphate-buffered gels is only about 9 kda (Bostian *et al.* 1984). Much of the reported work on K1 toxin has involved studies on two strains or their derivatives. These strains are a standard laboratory strain K12-1 and a superkiller, T158c, which has normal amounts of a more stable toxin (Palfree & Bussey 1979). The toxin of strain K12-1 has recently been shown to consist of two polypeptide components, $\alpha$ (9.5 kda) and $\beta$ (9.0 kda), which may just be resolved by reducing SDS–polyacrylamide gel-electrophoresis. Strain M5c, a derivative of T158c, has similarly been shown to possess two components, with $\beta$ having the same apparent size as from strain K12-1 but with $\alpha$ migrating slightly faster (Bostian *et al.* 1984). N-terminal sequence analysis on the $\alpha$- and $\beta$-components of toxin from K12-1 has shown that they are widely separated within the cDNA sequence of $M_1$ (Fig. 5.1). Secreted toxin from both strains has been detected as 18–20-kda dimers, the separate components being disulphide-linked and assumed to be $\alpha\beta$ dimers (Bostian *et al.* 1984). Sequence analysis of cDNA clones from K12-1 (Bostian *et al.* 1984, Tipper & Bostian 1984) and from a derivative of T158c (Skipper *et al.* 1984) show that the $\beta$-component sequences are identical and that the $\alpha$-component sequences vary, resulting in changes in amino acids at two different sites (Fig. 5.1). As noted by Bostian *et al.* (1984), the relatively high hydrophobic and charged-amino acid contents of the $\alpha$- and $\beta$-subunits, respectively, is worth noting. Toxin has a pI of 4.5, its pH of maximum stability (Bussey 1981), and all of its negative charge is apparently due to the excess of acidic residues in the $\beta$-component. Hydropathicity analysis (Kyte & Doolittle 1982) suggests that killer toxin could be functionally organized like the abrin and ricin class of toxins (Olsnes & Pihl 1973), in which the domains for receptor binding and toxin reside on separate disulphide-bonded polypeptide chains (Bostian *et al.* 1984).

Analysis of the $\beta$-component fails to reveal any significant stretches of hydrophobic residues. However, in the $\alpha$-component two regions of hydrophobic residues, separated by a hydrophilic region encoding four acidic residues, are found at positions 72–91 and 112–123 in the gene sequence (Fig. 5.1). These hydrophobic regions each have a high potential for forming the membrane-spanning regions of an integral membrane protein. On the basis of this evidence, Bostian *et al.* (1984) concluded that the $\alpha$-component may be the toxin domain, with the core stretching from residue 72–123, while the $\beta$-component could be the (1,6$\beta$-D-glucan cell wall) receptor-binding domain.

### 5.3.2 *Mode of action of K1 toxins and immunity*

For some time it has been known that killer toxins bind rapidly, irreversibly and in an energy-independent manner to receptors in the cell wall (Al-Aidroos & Bussey 1978, Bussey *et al.* 1973). Evidence from competition studies with pure polysaccharides, from chemical and enzymic degradation, from cell-wall fractionation and from mutants defective in receptor-binding suggests that 1,6$\beta$-D-glucan acts as toxin cell-wall receptor (Hutchins & Bussey 1983). These cell-wall receptors are found in large numbers in both sensitive and killer strains, whereas receptor-defective mutants (*kre*1 and *kre*2) have markedly reduced cell-wall binding. *kre*1 resistant mutants have been shown to be defective in some aspect of 1,6$\beta$-D-glucan synthesis, but no information is available concerning *kre*2 mutants (Al-Aidroos & Bussey 1978, Bussey *et al.* 1973, Hutchins & Bussey 1983).

Bussey (1981) has suggested that the primary site of action of the toxin is the plasma membrane. An energy-dependent process, required subsequent to cell-wall binding, is presumably involved in transfer of toxin from the cell wall to the plasma membrane (Skipper & Bussey 1977). Protoplasts of both *kre*1 and *kre*2 mutants are sensitive to toxin, whereas protoplasts of killer cells retain their immunity (Bussey *et al.* 1973). Cells and protoplasts of *kre*3 mutants, pppGAAAAAUAAAGAA AUG ACG AAG CCA ACC CAA GUA UUA GUU AGA UCC GUC AGU AUA UUA UUU UUC AUC ACA UUA CUA CAC CUA GUC GUA GCG 90

Met Thr Lys Pro Thr Gln Val Leu Val Arg Ser Val Ser Ile Leu Phe Phe Ile Thr Leu Leu His Leu Val Val Ala

δ →

CUG AAC GAU GUG GCC GGU CCU GCA GAA ACA GCA CCA GUG UCA UUA CUA CCU CGU GAA GCG CCG UGG UAU GAC AAG AUC UGG GAA GUA AAA 180

Leu Asn Asp Val Ala Gly Pro Ala Glu Thr Ala Pro Val Ser Leu Leu Pro Arg Glu Ala Pro Trp Tyr Asp Lys Ile Trp Glu Val Lys

α Toxin →

GAU UGG CUA UUA CAG CGU GCC ACA GAU GGC AAU UGG GGC AAG UCG AUC ACC UGG GGU UCA UUC GUA GCG AGC GAU GCA GGU GUA GUA AUC 270

Asp Trp Leu Leu Gln Arg Ala Thr Asp Gly Asn Trp Gly Lys Ser Ile Thr Trp Gly Ser Phe Val Ala Ser Asp Ala Gly Val Val Ile

UUU GGU AUC AAU GUG UGU AAG AAC UGC GUG GGU GAG CGU AAG GAU GAU *AGC AGU ACG GAC UGC GGC AAG CAA ACA CUU GCU UUA CUA GUC 360

Phe Gly Ile Asn Val Cys Lys Asn Cys Val Gly Glu Arg Lys Asp Asp Ser Ser Thr Asp Cys Gly Lys Gln Thr Leu Ala Leu Leu Val

AGC AUU UUU GUA GCA GUU *GCA UCC GGC CAU CAU CUU AUA UGG GGU GGU AAU AGG CCG GUG UCG CAG UCA GAU CCU AAU GGC GCU ACC GUU 450

Ser Ile Phe Val Ala Val Ala Ser Gly His His Leu Ile Trp Gly Gly Asn Arg Pro Val Ser Gln Ser Asp Pro Asn Gly Ala Thr Val

γ immunity? →

GCU CGU CGU GAC AUU UCU ACU GUC GCA GAC GGG GAU AUU CCA CUG GAC UUU AGU GCG UUG AAC GAC AUA UUA *AAC GAA CAU GGU AUU AGU 540

Ala Arg Arg Asp Ile Ser Thr Val Ala Asp Gly Asp Ile Pro Leu Asp Phe Ser Ala Leu Asn Asp Ile Leu Asn Glu His Gly Ile Ser

AUA CUC CCA GCU AAC GCA UCA CAA UAU GUC AAA AGA UCA GAC ACA GCC GAA CAC ACG ACA AGU UUU GUA GUG ACC AAC AAC UAC ACU UCU 630

Ile Leu Pro Ala [Asn Ala Ser] Gln Tyr Val Lys Arg Ser Asp Thr Ala Glu His Thr Thr Ser Phe Val Val Thr Asn [Asn Tyr Thr] Ser

720

UUG CAU ACC GAC CUG AUU CAU CAU GGU AAU GGA ACA UAU ACC ACG UUU ACC ACA CCU CAC AUU CCA GCA GUG GCC AAG CGU UAU GUU UAU

Leu His Thr Asp Leu Ile His His Gly [Asn Gly Thr] Tyr Thr Thr Phe Thr Thr Pro His Ile Pro Ala Val Ala Lys Arg Tyr Val Tyr

β toxin →

CCU AUG UGC GAG CAU GGU AUC AAG GCC UCA UAC UGU AUG GCC CUU AAU GAU GCC AUG GUG UCG GCU AAU GGU AAC CUG UAU GGA CUA GCA

Pro Met Cys Glu His Gly Ile Lys Ala Ser Tyr Cys Met Ala Leu Asn Asp Ala Met Val Ser Ala Asn Gly Asn Leu Tyr Gly Leu Ala 900

GAA AAG CUG UUU AGU GAG GAU GAG GGA CAA UGG GAG ACG AAU UAC UAU AAA UUG UAU UGG AGU ACU GGC CAG UGG AUA AUG UCG AUG AAG

Glu Lys Leu Phe Ser Glu Asp Glu Gly Gln Trp Glu Thr Asn Tyr Tyr Lys Leu Tyr Trp 960 Ser Thr Gly Gln Trp Ile Met Ser Met Lys

UUU AUU GAG GAA AGU AUU GAU AAC GCC AAU AAU GAC UUU GAA GGC UGU GAC ACA GGC CAC UAG GGCAUCGUGUCUGACC

Phe Ile Glu Glu Ser Ile Asp Asn Ala Asn Asn Asp Phe Glu Gly Cys Asp Thr Gly His Stop

**Figure 5.1** Nucleotide sequence of the $M_1$ dsRNA preprotoxin gene. The sequence of the coding strand of strain T158c (K1, superkiller) is shown and was derived from RNA sequencing (bases 1–174; Thiele *et al*. 1982, 1984b) and DNA sequencing of cDNA clones (bases 42–1077; Skipper *et al*. 1984). Sequencing of cDNA clones derived from strain K12-1 (bases 111–1010; Bostian *et al*. 1984) revealed minor differences shown at the bottom. N-terminal positions of $\delta$ leader peptide, $\alpha$- and $\beta$-toxin components and putative $\gamma$-immunity regions are arrowed, and possible Asn-linked glycosylation residues are boxed. Sequence is shown only up to residue 980, after which divergence, between that for T158c and K12-1, occurs. K12-1 sequence (Bostian *et al*. 1984) differs at four residues which result in two amino acid differences within the $\alpha$-toxin component. These are indicated by * and are:

| 320–322 | 380–382 | 494–496 | 524–526 |
|---|---|---|---|
| AUC | ACA | CUC | AAU |
| Ile | Thr | [Leu] | [Asn] |
| | | (silent) | (silent) |

which have normal cell wall binding, are resistant to toxin, despite apparently normal cell-wall binding (Al-Aidroos & Bussey 1978, Bussey cited in Tipper & Bostian 1984), so *kre*3 may be a pleiotropic mutation. The existence of *kre*3 suggests a membrane receptor.

The cDNA coding sequence located between the $\alpha$ and $\beta$ toxin components has been called $\gamma$. It has been suggested that $\gamma$ has a potential role in immunity (Bostian *et al.* 1984), although this has yet to be established with certainty. cDNA clones, containing the identified 948 bp (316 aa) open-reading frame ($\delta$ N-terminal leader, $\alpha$-component, $\gamma$- and $\beta$-component) expressed in yeast under the control of the yeast ADH1 promoter actually express immunity in addition to toxin (Lolle *et al.* 1984). This may be taken as evidence for the immunity component residing in this reading frame, and further supports the hypothesis that $\gamma$ may be this component.

Subsequent to cell wall binding, there is an energy-dependent process which results in membrane damage (Skipper & Bussey 1977). Since *in vitro* experiments have shown that killer toxin from *Pichia kluyveri* inserts into a phospholipid bilayer system, forming ion-permeable channels (Kagan 1983), it is thought likely that a similar situation occurs *in vivo* (Bussey *et al.* 1982). Colicin K also acts by forming voltage-dependent channels in phospholipid bilayer membranes (Schlein *et al.* 1978).

There is an initial lag period, dependent upon culture conditions, of some 40 min before the events leading to cell death become apparent and irreversible. After the (reversible) lag period there is a co-ordinate cessation of macromolecule synthesis (Bussey & Sherman 1973) with a coincident increase in permeability of the plasma membrane to protons, resulting in, for example, potassium-ion efflux. Finally, cell death is identified by leakage of metabolic pools such as ATP (Bussey & Sherman 1973, Skipper & Bussey 1977). Damaged cells shrink in volume, but they do not lyse nor, since macromolecule leakage does not occur, do they contain large pores (Bussey 1972).

### *5.3.3 K2 and other killer toxins*

Toxins, from a K2 strain (NCYC399) and from the unclassified strain 28, have recently been purified and analysed for amino acid content (Pfeiffer & Radler 1984). Although some similarities exist between the toxins of NCYC399 and strain 28, they both clearly differ from the published (Palfree & Bussey 1979, Bostian *et al.* 1984) amino acid content of K1 toxin. No N-terminal sequence analysis has yet been done on the NCYC399 K2 toxin or that of strain 28. Such an analysis or cDNA cloning followed by DNA sequence analysis should yield more information. Neither of these toxins has been examined by reducing SDS–polyacrylamide gel-analysis, which would identify possible $\alpha$- and $\beta$-components as are found in K1 toxin. However, examination by non-reducing SDS–polyacrylamide gel-analysis places the molecular weights of the toxins both from strain 28 and from K2 killers (NCYC399 & NCYC738) at about 16 000 (Pfeiffer & Radler 1984) compared with the toxin of K1 killers, run under similar conditions, which has a molecular weight of about 11 500 (Palfree and Bussey 1979).

A wild yeast contaminant was found in a saké yeast, of K1 immunity (Bussey 1981), brewing mash (Imamura *et al.* 1974). Properties of the wild yeast killer toxin and its effects on sensitive *Saccharomyces cerevisiae* cells are very similar to those described above for K1 toxin (Imamura *et al.* 1974, Kotani *et al.* 1977). However, two additional features were described. Calcium ions were shown to reverse the effects of killer toxin during the reversible transient (lag) phase by preventing leakage of ATP from the cells. Kotani *et al.* (1977) also demonstrated that addition of ADP to the culture medium enhanced the effects of the killer toxin by reducing the length of the transient phase. The mechanism for this was not elucidated.

The production, purification and properties of toxin from a *Pichia kluyveri* isolate, which kills K1 killers, have been studied in some detail (Middelbeek *et al.* 1979, 1980a, d, e, Stumm *et al.* 1977). The toxin has a reported molecular weight of around 19 000, it may be glycosylated and is most stable at its optimal pH values close to its pI of 4.3. It was shown that binding to cell-wall receptors was energy independent and was a reversible phase in the killing process. Recovery could be effected by inactivating bound toxin at this stage.

Kagan (1983) has reported that the toxin from *Pichia kluyveri* forms ion-permeable channels in artificial phospholipid bilayer membranes. These channels show a relative lack of ion selectivity, consistent with the *in vivo* physiology, and have approximately the right conductance to account for the *in vivo* $K^+$ efflux data of Skipper & Bussey (1977) and Middelbeek *et al.* (1980d) (Kagan 1983). The experiments of Middelbeek *et al.* (1980a) showed that toxin-bound cells could be 'rescued' by the addition of physiological potassium and hydrogen ions to the medium. This suggests that the lethal effect of toxin is the loss of intracellular potassium ions and the decrease in intracellular pH. Similar rescue experiments have been performed with *Escherichia coli*, in which it was demonstrated that the leakage of potassium and magnesium ions seemed to be the lethal effects of colicins K and E1 (Kopecky *et al.* 1975), which have also been shown to be channel formers (Schlein *et al.* 1978).

Thus, it seems that there are strong resemblances between killer toxins and bacterial colicins. As noted by Kagan (1983), both are encoded by extra-chromosomal genomes and both induce relatively non-selective ion-permeable channels in the plasma membrane. Immunity components are also implicated in both systems.

### *5.3.4 dsRNA genomes and killer phenotypes*

Several different species of dsRNA have been found in various isolates of K1, K2 and K3 *Saccharomyces* strains. Differentiation between these species has been made on the basis of size, and they are thus identified by a letter. In decreasing size order (see Table 5.2) they are: XL, L, T, W and M. Various subspecies of the L class of molecule have been demonstrated. These are identified by subscript letters, with or without a preceding number. The latter is dependent upon whether their presence is restricted to a particular killing group; for example, $L_{BC}$ is found in K1 and K2 strains, whereas $L_{1A}$ is found only in K1 strains. A full consideration of each of these dsRNA species may be

**Table 5.2** dsRNA species and relationship to killer system.

| Group | dsRNA | Size (kb) | VLP | Gene products | Relevance to killer system |
|---|---|---|---|---|---|
| K1 | XL | >4.7 | ? | ? | ? |
| | $L_{BC}$ ($L_B$, $L_C$, $L_{BC}$) | 4.7* | ScV-$L_{BC}$ | 82 kda capsid pp | ? |
| | $L_{1A}$ | 4.7* | ScV-$L_{1A}$ | 88 kda capsid pp | required for $M_1$ maintenance [has capsid-associated RNA polymerase activity] [88 kda capsid pp also encapsidates $M_1$] |
| | T | 2.7 } | none | ? | ? |
| | W | 2.25 } | | | |
| | $M_1$ | 1.9 | ScV-$M_1$ | K1 toxin<br>K1 immunity | determines K1 killer (toxin or resistance) phenotype |
| K2 | $L_{BC}$ ($L_B$, $L_C$, $L_{BC}$) | . . . . . . . . . . . . . . exactly as for $L_{BC}$ in K1 group . . . . . . . . . . . . . . | | | |
| | $L_{2A}$ | 4.7† | ScV-$L_{2A}$ | 84 kda capsid pp | required for $M_2$ maintenance |
| | minor species | 3.0‡ | ? | ? | ? |
| | $M_2$ | 1.7 | ScV-$M_2$ | K2 toxin<br>K2 immunity | determines K2 killer (toxin or resistance) phenotype |
| K3 (one strain: NCYC 761) | $L_{3A}$ | 4.7† | | major capsid pp | presumably acts like $L_{1A}$ and $L_{2A}$ in being required for $M_3$ maintenance |
| | $M_3$ | 1.5 | | K3 toxin?<br>K3 immunity? | determines K3 killer (toxin or resistance) phenotype |

* Under certain conditions $L_{BC}$ and $L_{1A}$ may be separated by gel electrophoresis (El-Sherbeini *et al.* 1983) – see text.
† Minor differences in molecular weights of these L species have been reported (Young & Yagiu 1978).
‡ These species were seen to migrate between L and M genomes (Young & Yagiu 1978). They may be similar to T and W found in strains of K1 group.

found together with references to the original literature in subsequent sections. It is the purpose of this section merely to correlate their presence or absence with the various killer phenotypes.

Killer phenotype testing is done according to established methods (Somers & Bevan 1969) on media containing 0.03% methylene blue and buffered to the optimal pH for toxin activity. The incubation of killer cells, streaked across a background lawn of sensitive cells, results in a clear zone of (growth) inhibition surrounded by a ring of dead (blue-stained) sensitive cells. Killer toxin is responsible for this effect and own-toxin immunity confers resistance on the killer cells themselves.

**K1 killers**

Strains exhibiting K1 killer phenotype have a requirement for $L_{1A}$ and $M_1$ genomes (Bevan *et al.* 1973, El-Sherbeini *et al.* 1983, Sommer & Wickner 1982a). $L_{1A}$ has been shown to encode the major capsid polypeptide for its own encapsidation, as well as that of $M_1$. $M_1$ has been shown to encode the K1 killer toxin and immunity (resistance) function (Bostian *et al.* 1980b, 1984, Skipper *et al.* 1984, Lolle *et al.* 1984).

$L_{BC}$ is found in most K1 killers. No requirement for its presence has ever been demonstrated and, indeed, some killer strains lack $L_{BC}$ (El-Sherbeini *et al.* 1983, Sommer & Wickner 1982a). Thus, $L_{BC}$ probably bears no relationship to the killer system.

XL, T or W may be present or absent in K1 killers and they also, like $L_{BC}$, have no established role in this system (Wesolowski & Wickner 1984).

**K2 killers**

Killer strains of the K2 group, like those of K1, require an $L_A$ and an M species, being $L_{2A}$ and $M_2$. Similarly $L_{2A}$ encodes the major capsid polypeptide which encapsidates both species, $L_{2A}$ and $M_2$, and $M_2$ determines the K2 killer phenotype (El-Sherbeini *et al.* 1984, Hannig & Leibowitz 1985, Young & Yagiu 1978).

$L_{BC}$ is also found in some K2 killers, but again no requirement for its presence or role in the system has been demonstrated. Minor species, which may be equivalent to T and W in K1 killers, have been found in some K2 killers (Young & Yagiu 1978).

**K3 killers**

Only one isolate exhibiting K3 killer phenotype has yet been identified (Young & Yagiu 1978). It is NCYC761, and possesses $L_{3A}$ and $M_3$. It apparently does not have an $L_{BC}$ species, nor have any other, minor, species been identified. The L genome of NCYC761 has been identified as $L_{3A}$ on the following basis.

(a) It encodes its own major capsid polypeptide and that of $M_3$ (R. Jewers, this laboratory, unpubl. obs.).
(b) It may be heat-cured (J. Wooldridge, this laboratory, unpubl. obs.), like $L_{1A}$ and $L_{2A}$.
(c) It shows partial homology, in Northern hybridization analysis, to $L_{2A}$ and $L_{1A}$, with hybridization being stronger to $L_{2A}$ than $L_{1A}$ (J. Wooldridge, unpubl. obs.).

The $M_3$ genome also cross-hybridizes to $M_2$ but not $M_1$ (J. Wooldridge, unpubl. obs.).

**Sensitives**

Sensitive strains, which neither produce toxin nor are resistant to it, are thus killed by killer strains and, in general, they lack an M genome. The only exceptions to the latter are certain sensitive strains which possess an altered (mutant) M genome (see Section 5.7.4).

Sensitive strains may possess any of the possible combinations of L genomes according to their group, i.e. K1 sensitives may have $L_{BC}$ and $L_{1A}$, $L_{BC}$ alone, $L_{1A}$ alone or, indeed, neither subspecies.

A sensitive strain (only K1 group examined to date), regardless of its L or M content, may have both T and W, only W, or neither species (Wesolowski & Wickner 1984). This further strengthens the observation made earlier that neither T or W dsRNA species may have a role to play in the killer system.

## 5.4 ScV particles

Adler and Mackenzie (1972) were the first to demonstrate the presence of isometric 40-nm particles, isolated from cell-free extracts, in yeast. However, no correlation was made with either dsRNA or the killer phenotype. dsRNA-associated 33–35-nm spherical particles were later reported by Buck *et al.* (1973), but again no correlation was made with the killer phenotype. The first critical experiments, which demonstrated that viruses containing dsRNA were associated with the killer phenotype, were made by Herring and Bevan (1974) and were later confirmed by Adler *et al.* (1976). Isometric 39-nm particles were correlated with the presence of a dsRNA species already shown to be associated with the killer trait (Berry & Bevan 1972, Bevan *et al.* 1973, Vodkin & Fink 1973).

Viral particles, sometimes referred to as virus-like particles (VLPs), found in *Saccharomyces cerevisiae* are designated ScV (see Table 5.2). Further differentiation between these particles is made according to the encapsidated dsRNA genome; ScV-$L_{1A}$, ScV-$L_{2A}$, etc. Amongst strains of the K1 killer system a number of different dsRNAs have been found. Major species are L ($L_{1A}$ and $L_{1BC}$) and M ($M_1$ and mutant M genomes). Minor species are XL, T and W. None of the latter species has been shown to be associated with particles, despite attempts (at least for T and W) to determine such an association (Wesolowski & Wickner 1984). All of the major, L and M, species have been shown to be separately encapsidated as icosahedral particles, which are indistinguishable from one another under the electron microscope. The evidence, both genetic and biochemical, for separate encapsidation is overwhelming and has been reviewed elsewhere (Mitchell & Bevan 1983).

ScV-L particles differ from ScV-M particles in their physical properties such that they may be separated on sucrose gradients (Herring & Bevan 1974). The known physical properties of the ScV particles are summarised in Table 5.3. Bruenn (1980) has made various predictions as to the physical properties of

**Table 5.3** Physical properties of ScV particles.

| ScV particle | Size (nm) | Density CsCl ($g\,cm^{-3}$) | $S_{20,w}$ | $A_{260}/A_{280}$ | Reference |
|---|---|---|---|---|---|
| ScV-L | 33–41 | 1.40–1.42 | 150–162 | 1.4–1.6 | Adler *et al.* (1976)<br>Buck *et al.* (1973)<br>Herring and Bevan (1974)<br>Hopper *et al.* (1977)<br>Kane *et al.* (1979)<br>Pietras and Bruenn (1976) |
| ScV-M | 39 | 1.36–1.38 | ? | 1.2 | Herring and Bevan (1974)<br>Kane *et al.* (1979)<br>Pietras and Bruenn (1976) |
| ScV-M (S3)* | ? | 1.36–1.38 | ? | 1.3 | Kane *et al.* (1979)<br>Pietras and Bruenn (1976) |
| ScV-empty | ? | 1.29–1.32 | 110–1.32 | 0.75–0.95 | Adler *et al.* (1976)<br>Buck *et al.* (1973)<br>Pietras and Bruenn (1976) |

* These are the particles associated with the mutant M (internal deletion) genomes of suppressive sensitive mutants (Somers 1973, Vodkin *et al.* 1974, Sweeney *et al.* 1976, Thiele *et al.* 1984a, b).

these particles. He estimated the numbers of L-P1 (capsid) polypeptides in ScV-L particles to be 120, extrapolated this same figure to ScV-M particles, and calculated an expected density for ScV-M particles of 1.35–1.36. The latter is in agreement with observed data (Kane *et al.* 1979, Pietras & Bruenn 1976); however, it has now been reported that there are in fact only 60 molecules of capsid polypeptide per particle (Reilly *et al.* 1984). Bruenn (1980) further calculated the particle sizes of ScV-L to be $13.6 \times 10^6$ da (22.8% RNA) and of ScV-M to be $11.7 \times 10^6$ da (10.3% RNA).

Although the ScV particles are unrelated immunologically to the dsRNA viruses of *Aspergillus foetidus*, *Aspergillus niger* and *Penicillium stoloniferum* (Buck *et al.* 1973), their physical properties (reviewed in Saksena & Lemke 1978) do appear to be similar:

(a) same particle size range: 30–40 nm;
(b) $S_{20,w}$ in same range: 140–170;
(c) CsCl densities in same range: 1.3–1.45 $g\,cm^{-3}$;
(d) molecular weights of particles similar: $13 \times 10^6$ da;
(e) all dsRNA species separately encapsidated;
(f) molecular weights of component dsRNA genomes in same range: $0.06–3 \times 10^6$ da; and
(g) RNA content similar: 10–15 per cent.

**Table 5.4** Encapsidation of K1 dsRNA genomes, major capsid polypeptides and capsid-associated polymerase activities.

| Particle | dsRNA genome | Major capsid polypeptide | Capsid-associated polymerase activity |
|---|---|---|---|
| ? | XL | ? | ? |
| ScV-$L_{BC}$ | $L_{BC}$ ($L_B$ or $L_C$) | $L_{BC}$-P1 (82 kda) | yes |
| ScV-$L_{1A}$ | $L_{1A}$ | $L_{1A}$-P1 (88 kda) | yes, but different activity from $L_{BC}$ polymerase |
| none | T, W | ? | ? |
| ScV-$M_1$ | $M_1$ | $L_{1A}$-P1 (88 kda) | yes, same activity as $L_{1A}$ polymerase |

ScV-$M_1$ and ScV-$L_{1A}$ share the same major capsid polypeptide, $L_{1A}$-P1 (88 kda) (Bostian *et al.* 1980b, El-Sherbeini *et al.* 1984, Harris 1978, Hopper *et al.* 1977) and capsid-associated polymerase activity (Brennan *et al.* 1981a, b, Bruenn *et al.* 1980, Hannig & Leibowitz 1985, Herring & Bevan 1977, Welsh & Leibowitz 1980, Welsh *et al.* 1980). ScV-$L_{1BC}$ has a different (82 kda) major capsid polypeptide, $L_{1BC}$-P1, and a capsid associated RNA polymerase of different activity to that of ScV-$L_{1A}$ and ScV-$M_1$. Thiele *et al.* (1984a) demonstrated that ScV-$L_{BC}$ ($L_C$) particles have a much lower density than ScV-$L_{1A}$ particles, and hence were able to isolate them separately. As has been noted by El-Sherbeini *et al.* (1984), this low density may in part reflect proteolytic damage during isolation. The main features of capsid polypeptides and capsid-associated polymerase activities are summarized in Table 5.4 (but see Section 5.6.2).

Other VLPs have been found in yeast. They are spherical to ovoid in shape, about 60 nm in diameter and associated with Ty RNA (Mellor *et al.* 1985, Garfinkel *et al.* 1985). Since the presence of Ty-VLPs is dependent specifically upon the transposition-induced condition, they have been concluded (Garfinkel *et al.* 1985) to be clearly different from the ScV-killer particles.

## 5.5 dsRNA transcription and replication

Yeast (ScV) dsRNA is thought to replicate in two discrete steps; the replication of dsRNA to produce single-stranded intermediates (ds → ss) and the doubling-up reaction (ss → ds). Each of these steps will be considered, separately, in the following subsections.

Much of the work on transcription and replication of the dsRNAs was reported before the differentiation of L genomes as $L_A$ or $L_{BC}$. Most of the strains used in this work have subsequently been analysed with respect to L dsRNA content. Thus, wherever possible, the L genomes used in the earlier work will be identified.

*5.5.1 ds → ss and analysis of ss transcripts*

Purified ScV-L and ScV-M particles possess capsid-associated, *in vitro*, RNA polymerase activity which results in full-length single-stranded transcripts that have messenger activity (Brennan *et al.* 1981a, b, Bruenn *et al.* 1980, Hannig & Leibowitz 1985, Herring & Bevan 1977, Welsh & Leibowitz 1980, Welsh *et al.* 1980). Strains possessing $L_{1A}$ alone or $L_{BC}$ ($L_B$ or $L_C$) alone both exhibit such capsid-associated activity (Sommer & Wickner 1982a, Thiele *et al.* 1984a). The polymerase activity associated with ScV-$L_{1A}$ differs in its properties (pH optimum, sensitivity to salt concentrations and UMP incorporation kinetics) from that associated with ScV-$L_{BC}$ ($L_C$) (Thiele *et al.* 1984a).

Herring and Bevan (1977) showed that the ssRNA product of $L_{1A}$ *in vitro* polymerization seemed to be a transcript of a *single* strand of the dsRNA. Bostian *et al.* (1983a) worked with the same ($L_{1A}$) strain and, using *in vitro* translation, demonstrated that much of the full-length $L_{1A}$ ssRNA found *in vivo* had mRNA activity. Therefore, it is by definition plus-stranded. However, these authors could not exclude the possible existence of *in vivo* minus-stranded transcripts. Further evidence that L ssRNA molecules are plus-stranded comes from the work of Sclafani and Fangman (1984). These authors cloned a 360-bp fragment of L cDNA into an M13 vector. Single-stranded progeny were identified and separated according to whether they carried the plus or minus polarity strand of L cDNA. Labelling and Northern blotting to total cellular ssRNA from *cdc*7 arrested cells (i.e. rich in L ssRNA) resulted in hybridization of minus- but not plus-strand clones. Thus, the L ssRNA present in the total cellular extracts was plus-stranded and the estimate of possible minus-strand L ssRNA content was <0.5%.

The *in vivo* transcripts m (full length 1.9 kb) and $m_a$ (sub-genomic, 1.2 kb) of $M_1$ dsRNA have mRNA activity (Bostian *et al.* 1983a). Hybridization, to separated plus and minus strands of $M_1$ dsRNA, confirms the positive polarity of these transcripts. The full-length *in vitro* $M_2$ transcript, $m_2$, has similarly been shown to be plus-stranded (Hannig & Leibowitz 1985).

Thus, from both *in vivo* and *in vitro* evidence, it seems that the full-length ssRNA molecules transcribed from L and M dsRNAs probably function both as mRNAs and as replicative intermediates.

Polyadenylation of L transcripts probably does not occur. However, the evidence is conflicting. Haylock and Bevan (1981) found that full-length *in vivo* $L_{1A}$ transcripts did not bind to oligo(dT)-cellulose. Bostian *et al.* (1983a) reported that full-length *in vivo* $L_{1A}$ transcripts did bind to poly(U)-sepharose. An examination of *in vitro* $L_{1A}$ transcripts and separated strands of $L_{1A}$ dsRNA revealed that neither bound to oligo(dT)-cellulose (Hannig *et al.* 1984). If the *in vivo* transcripts from $L_{1A}$ have either short 3′-terminal or internal poly(A) tracts, a possibility noted by Hannig *et al.* (1984), then the conflicting data would be reconciled since it is known that poly(U)-sepharose will bind molecules with shorter poly(A) tracts than will oligo(dT)-cellulose (Hunter & Garrels 1977). Bostian *et al.* (1983a) found, in addition to the full-length $L_{1A}$ transcripts l (4.7 kb), partial length transcripts $l_a$ (2.3 kb). Both these *in vivo* transcripts had full messenger activity but $l_a$, unlike l, did not bind to poly(U)-sepharose.

Terminal polyadenylation of $M_1$ transcripts clearly does not occur. Early evidence, presumptive of polyadenylation, came from Bostian *et al.* (1983a). They reported that m and $m_a$ *in vivo* transcripts bound tightly to poly(U)-sepharose. Hannig *et al.* (1984) showed that although full-length *in vitro* transcripts of $M_1$ bound, only the plus strands of separated $M_1$ dsRNA had affinity for oligo(dT)-cellulose. Sequence analysis of both the plus genomic strand and the *in vitro* transcripts failed to reveal any 3′-terminal poly(A). Hannig *et al.* (1984) concluded that binding is due to the 200-bp AU-rich internal region seen in $M_1$ dsRNA (Fried & Fink 1978), with the plus strand being A-rich and the minus strand being U-rich. When these authors tested the oligo(dT)-cellulose affinity of both strands of S3 dsRNA, an $M_1$ deletion mutant which lacks the 200-bp AU-rich region plus flanking sequences (Fried & Fink 1978, Thiele *et al.* 1984b), they found that neither strand bound, thus strengthening their hypothesis. Sequence analysis of approximately 87% of S3 also failed to reveal any poly(A) tracts on S3 dsRNA (Thiele *et al.* 1984b). Such sequence analysis has revealed potential polyadenylation recognition sequences (Fitzgerald & Shenk 1981, Montell *et al.* 1983) at the 3′-termini of the plus strands: 5′-AAGAAA-3′ (positions 9–13) for $M_1$ (Hannig *et al.* 1984); 5′-AAUAAA-3′ (positions 161–166) for S3 (Thiele *et al.* 1984b). However, as noted by these authors, it seems that these sequences do not function in the polyadenylation of either the *in vitro* synthesized or *in vivo* m and s (S3) transcripts, and the polyadenylate observed is template-encoded for both $M_1$ and S3 *in vivo* and *in vitro* transcripts.

$M_2$ dsRNA has also been shown to possess an AU-rich region with a polyadenylate-like tract on the plus genomic strand and the $m_2$ *in vitro* transcript (Hannig & Leibowitz 1985).

The lack of terminal polyadenylation makes the situation analogous to reovirus where both the *in vivo* (Stolzfus *et al.* 1973) and *in vitro* (Antczak *et al.* 1982) transcripts have been shown to lack 3′-terminal poly(A) tracts. However, in this respect a lack of polyadenylation makes it unusual for a eukaryotic mRNA. Yeast mRNAs have been demonstrated to have poly(A) tails, although these are on average shorter than the poly(A) tails found in higher eukaryotes (McLaughlin *et al.* 1973).

The significance of the full, l and m, and partial, $l_a$ and $m_a$, plus-stranded *in vivo* transcripts of $L_{1A}$ and M dsRNAs, respectively, is unclear (Bostian *et al.* 1983a, Hannig *et al.* 1984). All of these transcripts have *in vitro* messenger activity, and it has been suggested by these authors that the full-length l and m transcripts may act solely as replicative intermediates, whereas the partial-length $l_a$ and $m_a$ transcripts may be the *in vivo* messages. There is no evidence yet to support or deny this hypothesis.

As noted above, the transcripts of the L and M dsRNA species are full-length, faithful copies of the genomic dsRNAs. They are plus-stranded, initiate at the AU-rich termini and terminate with the addition of a single 3′-terminal purine. Only A has been identified *in vivo*, although *in vitro* additional purines are A (53%) or G (38%) (Bruenn & Brennan 1980). There is an eight-base sequence at the 3′-termini of the minus strands, being $_{OH}ACUU^{A}_{U}UU^{A}_{U}$, which is common to $L_{1A}$, $L_{2A}$, $M_1$, $M_2$ and S3 dsRNAs (see Fig. 5.2). This sequence is postulated to be the *in vivo* transcriptase recognition site (Brennan *et al.*

| dsRNA | strand | (AU-rich terminus) sequence (C-rich terminus) |
|---|---|---|
| $L_{1A}$ | + | pppGAAAAAUUUUUAAAUUCAUAUAACUCCCAUGCUAAGA..[ ].....................UAGGGAAUUACCCAUAUGCA$_{OH}$ |
| | - | $_{OH}$ACUUUUUA..... .....GGUAUACG$_{PPP}$ |
| $L_{2A}$ | + | pppGAAUAAUUUGAAUAUUC......................[ ].......................GCUUAUACACAUAUGCA$_{OH}$ |
| | - | $_{OH}$ACUUAUUA..... .....UGUAUACG$_{PPP}$ |
| $L_{BC}$ | + | pppGAAUUUUUCCACUCGCACAC...................[ ]........................... .....GCA$_{OH}$ CG$_{PPP}$ |
| | - | $_{OH}$ACUUAAAA..... .....GCG$_{OH}$ CG$_{PPP}$ |
| $M_1$ | + | pppGAAAAAUAAAGAAAUGACGAAGCCAACCCAAGUAUUA..[ ]..........UAAUGCAACAGCAUAGAAGAAACACACAUCA$_{OH}$ |
| | - | $_{OH}$ACUUUUUA..... .....GUGUGUAG$_{PPP}$ |
| $M_1$(S3) | + | pppGAAAAAUAAAGAAAUGACGAAGCCAACCCAAGUAUUA..[ ].UAGCCUGGUUUAAUGCAACAGCAUAGAAGAAACACACAUCA$_{OH}$ |
| | - | $_{OH}$ACUUUUUA..... .....UGUGUAG$_{PPP}$ |
| $M_2$ | + | pppGAAAAAAUGAAAGAGACUACCACCAGC............[ ].....UGGACGCACCGAAAUUAUGAGGUACAUUUACCUAGCA$_{OH}$ |
| | - | $_{OH}$ACUUUUUA..... .....AUGGAUCG$_{PPP}$ |

**Figure 5.2** Terminal sequences of L and M dsRNA genomes. Plus (+) and minus (−) strands, AUG initiation codons (followed by open-reading frames) and putative eight base (3′ minus-strand) polymerase recognition sequences are all identified.

1981b). $L_{BC}$, which it has already been noted has a capsid-associated RNA polymerase with different properties from that associated with $L_{1A}$ (Thiele *et al.* 1984a), also differs at residues 6 and 7 within this putative transcriptase recognition sequence. The eight-base sequence for $L_{BC}$ is $_{OH}$ACUUAAAA (Field *et al.* 1982).

Potential ribosome binding sites have been identified for M but not for L genomes (Hannig *et al.* 1984, Thiele & Leibowitz 1982, Thiele *et al.* 1984a).

The L and M dsRNAs are uncapped, both 5′-ends of all L and M species being $_{PPP}G_{P}$ (Bruenn & Keitz 1976). There is no published evidence for any of the transcripts being capped or uncapped. However, Bobek and Brennan have unpublished work cited by Bruenn (1980) in which it is stated that the product of the *in vitro* reaction with ScV RNA polymerase is uncapped: 'the product' is unspecified. If the 5′-termini of the mRNAs (transcripts) are uncapped, then this is in contrast with the blocked 5′-termini of the plus genomic strands and transcripts of reovirus (Chow & Shatkin 1975, Furuichi *et al.* 1975a, b, Miura *et al.* 1974). Capping also occurs in higher eukaryotic mRNAs (Adams & Cory 1975) and yeast mRNAs (Sripati *et al.* 1976), which are generally either 7MeG$_{PPP}$G$_{P}$ or 7MeA$_{PPP}$A$_{P}$, and this seems to be of primary importance in the initiation of translation. If, indeed, the 5′-termini of the dsRNA transcripts are uncapped, then the potential 18S and 5.8S ribosomal RNA binding sites, just upstream of the AUG initiation codon in $M_1$ (Hannig *et al.* 1984) and in $M_2$ (Hannig & Leibowitz 1985) may be significant. No such potential ribosome binding sites have been identified in $L_{1A}$ sequences (Thiele *et al.* 1984a). Furthermore, AUG flanking nucleotides of the $L_A$ dsRNA, and $l_a$ transcript, do not resemble those found adjacent to the majority of functional eukaryotic initiation codons (Kozak 1981).

Three modes of replication could exist: semi-conservative (all, random, molecules); semi-conservative (small replicating population); or conservative. Density-shift experiments, in which cells are grown first in 'heavy' ($^{13}C^{15}N$) and then 'light' ($^{12}C^{14}N$) media, have yielded data which tend to support conservative synthesis of transcripts and thus a conservative mode of replication. In a 'random' semi-conservative mode of replication the majority of dsRNA molecules would be expected to be of hybrid density after a shift from heavy to light medium. In a conservative mode of replication no hybrid molecules would be expected; before doubling the molecules would be fully dense (HH) and after doubling they would be fully light (LL). Such density-shift experiments have been reported for L dsRNA (Newman *et al.* 1981, Sclafani & Fangman 1984) and for M dsRNA (Sclafani & Fangman 1984). Newman *et al.* (1981) reported up to 18% of L dsRNA molecules being [HL] but, with better resolution of the density classes, Sclafani and Fangman (1984) found no more than 2% of L or M dsRNA molecules being [HL]. These results tend to disprove a 'random' semi-conservative replication mode and support a conservative mode, but do not rule out a semi-conservative mode in which only a very small proportion of the molecules act as masters in many rounds of semi-conservative replication.

The results from *in vitro* experiments (Welsh & Leibowitz 1980, Welsh *et al.* 1980) suggest that the newly synthesized single-strand molecules are displaced from their parent capsids, as in reovirus (reviewed in Silverstein *et al.* 1976).

### 5.5.2 *ss → ds*

All of the available evidence points to the conclusion that the full length l and m plus-stranded transcripts act as replicative intermediates whether or not they act as *in vivo* messages. Second-strand synthesis then must initiate at their C-rich 3′-termini.

The RNA polymerase activity necessary for doubling the ss to dsRNA has been demonstrated for ScV-$L_{1A}$. It has been shown to be associated with particles, isolated from log-phase cells, that sediment more slowly than particles containing ds $L_{1A}$. The end product of the *in vitro* reaction is full-length $L_{1A}$ dsRNA (Bevan & Herring 1976).

There are sequences, just before the $-AUGCA_{OH}$ 3′-termini of the plus-strand of $L_{1A}$ dsRNA and $l_a$ transcript, which could form secondary stem and loop structures (Brennan *et al.* 1981b, Thiele *et al.* 1984a). Brennan *et al.* (1981b) suggested that these may be replicase recognition sites. Such stem and loop structures exist in the genomic RNAs of alfalfa mosaic virus and are bound by capsids as an apparent prerequisite for replication initiation (Koper-Zwarthoff & Bol 1980). These secondary structural features of the 3′ plus-strands of $L_{1A}$ are not found in the 3′ plus-strands of $M_1$ dsRNAs (including that of S3 '$M_1$' dsRNA) (Thiele & Leibowitz 1982, Thiele *et al.* 1984b). Thus, the significance of these sequences in $L_{1A}$ is unclear.

There are some interesting regions of homology between $M_1$ and $M_2$ dsRNAs towards their 3′ plus-strand termini (Fig. 5.2). The most significant stretch is a 10-bp conserved sequence at positions (starting from the 3′ plus-strand terminus) 105–114 (Hannig & Leibowitz 1985). The functional significance of these sequences is unknown, and they are absent from the 3′ plus-strand terminus of $L_{1A}$ (Thiele *et al.* 1984a).

Comparisons, then, between 3′ plus-strand (C-rich) terminal sequences of L and M genomes reveal little in the way of homology (Fig. 5.2). Proposed capsid–replicase interaction sites found in $L_{1A}$ are absent from M genomes, and conserved sequences seen in various M genomes are absent from $L_{1A}$.

### 5.5.3 *Replication and the cell cycle*

Presumably ScV particle replication is controlled somehow in line with host-cell division such that intracellular numbers are limited. A mechanism for particle copy number must exist to ensure both host and viral survival, and is indeed indicated by the strain-specific, even under different physiological conditions, copy number observations of Oliver *et al.* (1977).

Clare and Oliver (1979) have shown that L synthesis is co-ordinated with host-cell protein synthesis, and that during nitrogen starvation ScV-L replication is subordinate to the requirements of the host cell. Extensive degradation of pre-existing L dsRNA occurs, although ScV-L particles are never reduced to zero levels. Higher levels of ScV-L are also found in ethanol-grown cells as opposed to glucose-grown cells (Oliver *et al.* 1977). These authors concluded that there was a relationship between mitochondrial metabolism and L dsRNA replication, although the mechanism for this was not elucidated.

Radioactive labelling experiments of size-separated cells have shown that the dsRNA is replicated continuously, at a rate which increases exponentially throughout the cell cycle (Newman *et al.* 1981). Other workers (Bostian *et al.* 1983a) support continuous replication, but Zakian *et al.* (1981) suggest a pause in L dsRNA replication during S phase. Cells in G1 phase of cell cycle replicate these molecules but accumulate them when G1 is arrested by either $\alpha$-factor or by a *cdc*7 mutation. Cells cease to replicate L dsRNA once they enter the S phase. Sclafani and Fangman (1984) suggested that if the plus-strand ss transcripts are replicative intermediates, then their levels might also be controlled by the cell cycle. These authors measured levels of ssRNA (from both L and M) in $\alpha$-factor or *cdc*7 G1-arrested cells and in S-phase cells. They found that L ssRNA accumulation is modulated in both the G1 phase and the S phase in parallel with L dsRNA accumulation. Whereas M ssRNA and dsRNA were similarly affected in G1-arrested cells, they were differentially affected once the G1 block was removed and the cells had entered S phase: the accumulation of ssRNA continued during S phase to the extent that its copy number approached that of its ds counterpart. It had previously been concluded that some aspect of L and M dsRNA replication must be different, because of the multiplicity of *MAK* gene products required for M but not L dsRNA maintenance (Wickner 1981). The differential control of M ssRNA and L ssRNA metabolism during S phase supports such a conclusion.

### *5.5.4 Proposed sequence of events in replication*

To summarize, there is now very strong evidence to support a conservative mode of ScV-dsRNA replication, like that employed by reovirus (reviewed in Silverstein *et al.* 1976). The proposed sequence of events is as follows, viz. The minus strands of parental dsRNA duplexes are transcribed, by capsid-associated polymerase, to produce full-length plus-stranded ssRNA intermediates. These are displaced from parental duplexes and released from their parental VLPs. Capsid-associated polymerase with replicase activity then completes the process by doubling-up the ssRNAs to yield progeny duplexes. The point at which progeny duplexes become encapsidated has not been established, but capsid binding is probably intimately associated with initiation of replication as is the situation in alfalfa mosaic virus (Koper-Zwarthoff & Bol 1980).

## 5.6 L dsRNA species

Early genetic studies, involving the K1 killer strain K7, suggested the existence of differences between L genomes (Mitchell *et al.* 1976a). More recently, several laboratories (El-Sherbeini *et al.* 1983, Field *et al.* 1982, Sommer & Wickner 1982a) have documented genetic and biochemical evidence for the coexistence of various forms of L. These L dsRNA species all have identical, or nearly identical, sizes of about 4.7 kb. The literature has been very confusing as regards the nomenclature of these species, and warrants clarification. We are using here the nomenclature agreed with other workers and as adopted in El-Sherbeini *et al.* (1984). Table 5.5 defines the L genomes with the subscript

**Table 5.5** L dsRNA gene products.

| dsRNA | *In vivo* products | | Products of *in vitro* translation | | Molecular weight (kda) | Comments |
|---|---|---|---|---|---|---|
| | Old* | New* | Old* | New* | | |
| $L_{1A}$ | ScV-P1 | $VL_{1A}$-P1 | L-P1 | $L_{1A}P1$ | 88 | major capsid pp for $L_{1A}$ and $M_1$ genomes: |
| | | | L-P4 | | 75 | *in vitro* only product: L-P1 fragment |
| $L_{BC}$ ($L_B$, $L_C$ or $L_{BC}$) | ScV-P2 | $VL_{BC}$-P1 | L-P2 | $L_{BC}$-1 | 82 | major capsid pp for $L_{BC}$ (K1 and K2 strains) |
| | ScV-P3 | | L-P3 | $L_{BC}$-P2 | 78 | unknown: identity of *in vivo* and *in vitro* products not proven |
| $L_{2A}$ | | $VL_{2A}$-P1 | L-P5 | $L_{2A}$-P1 | 84 | major capsid pp for $L_{2A}$ (and ?$M_2$) genomes |
| | | | L-P6 | | 82 | unknown |
| | | | L-P7 | | 67 | unknown |

* Old and new refer to the nomenclature. The new nomenclature is consistent with that used for L genomes.

number denoting the group, i.e. K1, K2 or K3, and the following subscript letter denoting the sub-species.

El-Sherbeini *et al.* (1983) demonstrated that most killer strains possessed two forms of L – 'L1' and 'L2' – which were, following a rapid isolation procedure, separable by gel-electrophoresis. Two strains, K7 and K12, were shown to contain only the faster-migrating, 'L2', species, all other strains tested having both 'L1' and 'L2'.

Field *et al.* (1982) identified a major species, '$L_1$', in K1 strains and '$L_2$' in K2 strains, and a minor species, '$L_a$', found in both K1 and K2 strains. Terminal sequence analysis differentiated between '$L_1$' and '$L_a$'.

Sommer and Wickner (1982a) identified three L species, called 'L-A', 'L-B' and 'L-C'. $T_1$ fingerprints of all three species were different, but the results of liquid hybridization suggested that 'L-B' and 'L-C' were about 50% related, although neither showed homology to 'L-A'.

Since both El-Sherbeini *et al.* (1983) and Sommer and Wickner (1982a) used the same strain K7, which has only one form of L, 'L2' is equivalent to 'L-A'. This species is now called $L_{1A}$ and is presumably also equivalent to '$L_1$' of Field *et al.* (1982). 'L1' has subsequently been identified with 'L-BC' (El-Sherbeini *et al.* 1984) and is now called $L_{BC}$, or $L_B$ or $L_C$ where they have been differentiated. $L_{BC}$ is found in K1 and K2 strains, and hence has no subscript number to denote group; it is presumably equivalent to '$L_a$' of Field *et al.* (1982).

### 5.6.1 *Sequence analysis and genome comparisons*

Bruenn and Brennan (1980) observed terminal sequence heterogeneity amongst L dsRNAs before the identification of 'alternative' L species. The cumulative data from such terminal sequencing that has been done (Brennan *et al.* 1981b, Bruenn & Brennan 1980, Field *et al.* 1982, Hannig & Leibowitz 1985, Thiele *et al.* 1982, 1984a, b) is represented in Figure 5.2. The limited data available suggest little sequence conservation, between $L_A$ and $L_{BC}$, beyond the first nine or ten residues. There is some evidence to suggest the presence of a 170-bp long internal inverted repeat in $L_{1A}$. As such this would be the longest known non-terminal inverted repeat in a viral dsRNA (Bruenn *et al.* 1985). Hybridization analyses and the results from $T_1$ fingerprinting (Bobek *et al.* 1982, El-Sherbeini *et al.* 1984, Field *et al.* 1982, Sommer & Wickner 1982a) have revealed no internal homology between $L_{1A}$ and $L_{BC}$. However, liquid hybridization suggests that $L_B$ and $L_C$ are about 50% related (Sommer & Wickner 1982a).

$T_1$ fingerprint analysis distinguishes between the three major classes of L, $L_{1A}$, $L_B$ and $L_C$ (Sommer & Wickner 1982a). It is not altogether surprising that $L_B$ and $L_C$ should show different fingerprints, since it has been calculated (Sommer & Wickner 1982a) that as little as 12% sequence divergence between them could result in non-identical fingerprints.

Solution hybridization experiments demonstrated that about 50% of $L_B$ formed RNAase-resistant hybrids with $L_C$, although $L_{1A}$ formed resistant hybrids with neither $L_B$ nor $L_C$ (Sommer & Wickner 1982a).

Hybrid plasmids have been constructed (Bobek *et al.* 1982) containing cDNA synthesized from $L_{1A}$ *in vitro* transcripts. Nick-translated probes were used in Northern blots to search for homology with $L_{BC}$ or $L_{2A}$, or both. One probe, derived from a 325-bp insert, mapping very close to the 5′ AU-rich end of the $L_{1A}$ transcript, hybridized well to $L_{1A}$, showed considerable homology to $L_{2A}$ (Bobek *et al.* 1982) but failed to show any significant homology to $L_{BC}$ (Field *et al.* 1982). Two other probes, representing internal $L_{1A}$ sequences starting at least 585 bp in from the 5′ terminus of the transcript, failed to hybridize significantly with $L_{2A}$ (Bobek *et al.* 1982). One of the latter two probes, derived from a 280-bp insert, also failed to hybridize to $L_{BC}$ (Field *et al.* 1982).

Probes, representative of all of $L_{1A}$ and all of $L_{1BC}$, were made separately by kinase labelling of the respective denatured L genomes (El-Sherbeini *et al.* 1984). These probes were used, in Northern blots, against total dsRNA prepared from strains of varying constitution with respect to $L_{1A}$, $L_B$, $L_C$, $L_{BC}$, $L_{2A}$ and $M_1$ or $M_2$ content. The lack of homology between $L_{1A}$ and $L_B$, $L_C$ or $L_{BC}$ was confirmed. The similarity between $L_B$ and $L_C$ was also confirmed, the $L_{BC}$ probe hybridizing well and indistinguishably to $L_B$, $L_C$ and $L_{BC}$ species. $L_{1A}$ cross-hybridized with $L_{2A}$ to a limited extent. Densitometric measurements led to an estimate for this partial cross-hybridization of about 25% (El-Sherbeini *et al.* 1984).

The conclusions from all of these L genome comparison studies may be summarized as follows.

(a) $L_B$ and $L_C$ are markedly similar. Although they show different $T_1$ fingerprints, this may be accounted for by as little as 12% sequence divergence. Solution hybridization experiments suggest that $L_B$ and $L_C$ are about 50% related, but they are indistinguishable on the basis of Northern blot hybridizations. $L_B$ and $L_C$ are therefore very related, and may be variants of the same species.

(b) $L_{1A}$ is seen to be unrelated to $L_{BC}$ ($L_B$, $L_C$ or $L_{BC}$) by $T_1$ fingerprint analysis, solution hybridization and Northern blot hybridization.

(c) $L_{2A}$ may be somewhat related to $L_{1A}$. Kinase-labelled probes made from $L_{1A}$ dsRNA hybridized to about 25% with $L_{2A}$. cDNA probes hybridized well at the 5′ AU-rich terminus, but internal probes failed to hybridize. $L_{1A}$ and $L_{2A}$ therefore have terminal sequence homology, but probably lack much significant internal homology.

### 5.6.2 *Gene products*

Both *in vivo* products and the products of *in vitro* translation of the various L genomes are shown in Table 5.5. This table also lists both the new and old product designations, and confirms the identity of *in vivo* with *in vitro* products where it has been established.

The only confirmed gene products of the various L genomes are their respective major capsid polypeptides. The experiments which led to the identification of the *in vitro* translation products with the *in vivo* (particle) associated polypeptides are the work of several groups of collaborators, and will be discussed in some detail below. L genomes have sufficient coding capacity to accommodate (and there is some circumstantial evidence for their encoding) an RNA-dependent RNA polymerase, but this has not been proven.

$L_{1A}$ encodes the major, 88-kda, capsid polypeptide, $VL_{1A}$-P1, which encapsidates $L_{1A}$ and $M_1$ genomes (Bostian *et al.* 1980a). The identity of the *in vitro* $L_{1A}$ translation product, $L_{1A}$-P1, with the *in vivo* polypeptide, $VL_{1A}$-P1, was confirmed by Hopper *et al.* (1977) on the basis of cross-reactivity with anticapsid ($VL_{1A}$-P1) IgG; by their co-migration in gels and by their exhibiting identical peptide maps using the technique of Cleveland *et al.* (1977). L-P4 has been deduced to be a premature termination product of $L_{1A}$ translation, lacking about 15% of the $L_{1A}$-P1 (L-P1) C-terminal sequence (El-Sherbeini *et al.* 1984). Both L-P4 and $L_{1A}$-P1 were precipitated by anticapsid ($VL_{1A}$-P1) IgG, and both gave identical terminal hydrolysis products, derived with similar kinetics from partial hydrolysis products, in peptide-mapping experiments (El-Sherbeini *et al.* 1984).

$L_B$, $L_C$ and $L_{BC}$ each encode the major $VL_{BC}$-P1 82-kda capsid polypeptide. The *in vitro* $L_{BC}$ translation product $L_{BC}$-P1 co-migrates in gels with the *in vivo* $VL_{BC}$-P1 polypeptide. Furthermore, co-migration of the stained V8 protease fragments of $VL_{BC}$-P1 and the autoradiographically detected fragments of $L_{BC}$-P1 was demonstrated, thus confirming the identity of $VL_{BC}$-P1 with $L_{BC}$-P1. The minor $L_{BC}$ *in vitro* translation product, $L_{BC}$-P2, was unrelated to $L_{BC}$-P1 by peptide mapping (El-Sherbeini *et al.* 1984).

Two groups of collaborators have reported the existence of light particles as opposed to the heavy $L_A$-containing VLPs. Thiele *et al.* (1984a) reported that $L_C$ VLPs have a very low density with three proteins (76, 68 and 65 kda) uniquely associated with them. El-Sherbeini *et al.* (1984) showed that $VL_{BC}$-P1 is extremely protease sensitive, and this may render $L_{BC}$ VLPs particularly susceptible to proteolytic damage during preparation. This would perhaps explain the 'odd' protein pattern and low density of such particles seen by Thiele *et al.* (1984a). A second group of workers (Long *et al.* 1982) demonstrated that low-density particles from a K1 killer strain were much more heat labile than the denser particles. These authors concluded that the low-density particles were ScV-$M_1$, but others (Tipper & Bostian 1984) have suggested that it is much more likely that they were damaged forms of ScV-$L_{BC}$.

$L_{2A}$ encodes the major $VL_{2A}$-P1 84-kda capsid polypeptide. Peptide-mapping experiments of El-Sherbeini *et al.* (1984) confirmed the identity of the $L_{2A}$-P1 (L-P5) *in vitro* product with $VL_{2A}$-P1 *in vivo* capsid polypeptide. Two other translation products of $L_{2A}$, L-P6 and L-P7 of 82 kda and 67 kda, respectively, were also seen. Although L-P6, at 82 kda, was shown to co-migrate with the $L_{BC}$ product $L_{BC}$-P1 (≡ $VL_{BC}$-P1, major $VL_{BC}$ capsid polypeptide) it was seen in the translation of dsRNA from a K2 strain reported as lacking $L_{BC}$. The identity and significance of L-P6 and L-P7, and their relationship to $L_{2A}$-P1, are therefore unknown.

Comparisons between the various capsid polypeptides and the 'minor' translation products have been made (El-Sherbeini *et al.* 1984). $VL_{1A}$-P1, $VL_{BC}$-P1 and $VL_{2A}$-P1 all differ in size and have marked differences in their peptide maps. Anti-capsid ($VL_{1A}$-P1) IgG precipitated $L_{1A}$-P1 (and L-P4, its premature translation product) but none of the products of translation of $L_{BC}$. The same anti-capsid also precipitated *in vivo* capsid $VL_{1A}$-P1 but not $VL_{BC}$-P1. Thus, $VL_{1A}$-P1 and $VL_{BC}$-P1 are antigenically unrelated. It was shown previously (Bobek *et al.* 1983) that partial cross-reaction occurs between

ScV-$L_{2A}$ protein and anti-V$L_{1A}$-P1-capsid serum. Thus, as discussed above, not only do $L_{1A}$ and $L_{2A}$ share some sequence homology, but their capsid polypeptides are somewhat related antigenically.

Thiele *et al.* (1984a) have shown that RNA polymerase activity associated with ScV-$L_{1A}$ particles has quite different characteristics from that associated with ScV-$L_C$ particles. The activities differ in their UMP incorporation kinetics, pH optima for activity and response to various salt concentrations. It thus seems likely that $L_{1A}$ and $L_{BC}$ particles not only have unique capsid proteins, but that their associated polymerase activities may also be quite distinct. Although there is sufficient coding capacity on the various L genomes to encode a polymerase, this L-encoded function has not yet been demonstrated. One possible candidate for an RNA polymerase could be the unique minor *in vitro* translation product of $L_{BC}$, $L_{BC}$-P2 (L-P3) which may correspond to the identically sized *in vivo* protein ScV-P3. However, it still remains possible that ScV-P3 is a contaminant or a non-antigenic fragment of the capsid polypeptide V$L_{1A}$-P1 (El-Sherbeini *et al.* 1984).

To summarize, identified gene products are as follows.

$L_{1A}$ encodes its own major capsid polypeptide, the 88-kda V$L_{1A}$-P1.
$L_{2A}$ encodes its own major capsid polypeptide, the 84-kda V$L_{2A}$-P1.
$L_{BC}$ ($L_B$, $L_C$ and $L_{BC}$) encodes its own major capsid polypeptide, the 82-kda V$L_{BC}$-P1.

V$L_{1A}$-P1, V$L_{2A}$-P1 and V$L_{BC}$-P1 are all different in size and have marked differences in their peptide maps. V$L_{1A}$-P1 is also antigenically unrelated to V$L_{BC}$-P1, but some relatedness to V$L_{2A}$-P1 has been demonstrated. Finally, although it is considered likely that these L genomes actually encode their own RNA polymerase, this has not yet been confirmed.

### *5.6.3 Variant and mutant genomes*

**$L_{BC}$ variants**

$L_{BC}$ genomes are found in K1 and K2 isolates, and only two subspecies, $L_B$ and $L_C$, have been found. They may only be clearly differentiated by $T_1$ fingerprint mapping (Sommer & Wickner 1982a) and no functional differences have been detected: $L_{BC}$ (undifferentiated), $L_B$ and $L_C$ all encode the same gene product, the 82-kda V$L_{BC}$-P1 capsid polypeptide (El-Sherbeini *et al.* 1984).

**$L_A$ variants**

$L_A$ variants have all been defined on a largely functional basis. Since they differentially affect the continued presence of $M_1$ or $M_2$, or both, in a particular genetic background, i.e. *SK*1, it is appropriate here to discuss the exclusion of $M_2$ by $M_1$ and to define the *SK*1 and *ski* genes.

Mating K1 strains with K2 strains results in the loss of $M_2$ but retention of $M_1$ (Naumov & Naumova 1973). Originally it was suggested that the exclusion of $M_2$ was caused by the $L_{1A}$ species (Field *et al.* 1982). The loss of exclusion ability has now been correlated with the loss of $M_1$ and not $L_{1A}$. Furthermore, when *mak* segregants ($L_{1A}$ but no $M_1$) from several crosses were crossed with K2 strains no exclusion of $M_2$ occurred (Wickner 1983b). Thus, $M_1$ is responsible for the exclusion of $M_2$ and as suggested by Wickner (1983a) this exclusion may operate via competition for any of the various components of the (host supplied) replication machinery. 'M3' (not K3 group, see Section 5.2) is also excluded from cells with $M_1$, but $M_2$ and '$M_3$' appear to be able to coexist stably (Martin *et al.* 1981).

*ski* (superkiller) mutants were originally identified amongst K1 killers which had enhanced killer activity and elevated $M_1$ levels (Toh-e *et al.* 1978, Toh-e & Wickner 1980, Ridley *et al.* 1984). *ski* mutations also result in enhanced $M_2$ copy number (Ridley *et al.* 1984). Recently it has been shown that the *SK*1 gene products lower the copy number of $L_{BC}$, $L_A$, $M_1$ and $M_2$ (Ball *et al.* 1984). A further point of interest is that when K1 strains are cured of $M_1$ then $L_A$ ($L_{1A\text{-}HN}$), but not $L_{BC}$, copy number is increased (Ridley *et al.* 1984).

**Table 5.6** Stability of $L_A$ and M dsRNA genomes when introduced, in various combinations, into *SK*1 genetic backgrounds. This table represents compiled information from Ball *et al.* 1984, Hannig *et al.* (1985), Ridley *et al.* (1984), Sommer and Wickner (1984), Wickner (1980, 1983b) and Wickner and Toh-e (1982).

| $L_A$ variants | Stability of $M_1$ or $M_2$, separately, with the listed $L_A$ variants | Comments |
|---|---|---|
| $L_A$ | $M_1$ lost | presence of $L_{A\text{-}null}$ detectable only by hybridization |
| $L_{A\text{-}E}$ | $M_1$ lost<br>$M_2$ lost | |
| $L_{A\text{-}HE}$ | $M_1$ stable<br>$M_2$ lost | |
| $L_{2A\text{-}H}$ | $M_1$ stable<br>$M_2$ stable | <br>presence of $M_2$ represses $L_{2A\text{-}H}$ copy number by 2- to 3-fold |
| $L_{1A\text{-}HN}$ | $M_1$ stable<br>$M_2$ stable† | presence of $M_1$ represses $L_{1A\text{-}HN}$ copy by 5- to 10-fold |
| $L_{A\text{-}E}$ }<br>$L_{2A\text{-}H}$ } | $M_1$ stable or lost<br>$M_2$ lost | if $M_1$ stable, then poor expression; in addition, $L_{2A\text{-}H}$ copy number drops by at least 15-fold; no degradation of pre-existing $M_2$ molecules occurs |
| $L_{A\text{-}E}$ }<br>$L_{2A\text{-}HR}$* } | $M_2$ stable | no reduction in $L_{2A\text{-}HR}$ copy number |
| $L_{A\text{-}E}$ }<br>$L_{1A\text{-}HN}$ } | $M_1$ stable<br>$M_2$ stable | |
| $L_{2A\text{-}H}$ }<br>$L_{1A\text{-}HN}$ } | $M_1$ stable<br>$M_2$ stable | <br>$L_{2A\text{-}H}$ copy number drops by at least 15-fold |

* $L_{2A\text{-}HR}$ are mutants, derived from $L_{2A\text{-}H}$, which prevent exclusion of $M_2$ by $L_{A\text{-}E}$.
† If *MKT* and not *mkt* is present, see text.

Variants of $L_A$ genomes have been identified. They were originally classed as unidentified cytoplasmic determinants and were referred to as [HOK] (helper of $M_1$), [EXL] (exclusion of $M_2$) and [NEX] (non-exclusion, synergistic effect to [EXL]) (Wickner 1980, Wickner & Toh-e 1982). [HOK] (El-Sherbeini *et al.* 1983, Sommer & Wickner 1982a, b), [NEX] and [EXL] (Sommer & Wickner 1982a, b) have now been shown to be functional variants of $L_A$ genomes. Thus, these variants are differentiated by additional subscript letters, i.e. H for [HOK], N for [NEX] and E for [EXL]. For example, an $L_{1A}$ genome with [HOK] and [NEX] activity is identified as $L_{1A\text{-}HN}$.

$L_{1A\text{-}HN}$, $L_{A\text{-}HE}$ and $L_{A\text{-}E}$ probably share more than 99% sequence homology. Their *in vivo* VLP protein patterns were shown to be essentially the same, and only minor differences were seen by $T_1$ fingerprint analysis (Sommer & Wickner 1982a). No other structural or sequence analysis has been reported, and the variant $L_A$ genomes are thus defined functionally.

These variant $L_A$ genomes are either only required or are only seen to exert their effects in a *SK*1 genetic background. There is only one exception to this, and that is the instability of $M_1$ in a particular, *ski*2, strain 46B (Ridley *et al.* 1984). Strain 46B has barely detectable levels of an $L_A$ genome, '$L_{A\text{-}null}$', which lacks any [HOK], [NEX] or [EXL] functions (Wickner, cited in Hannig *et al.* 1985) and the low $L_A$ content may, in turn, account for the instability of the $M_1$ genome.

Each of the $L_A$ variants exerts either a positive or negative effect on the stability $M_1$ or $M_2$, or both, and, indeed, on each other. These effects have been tabulated (Table 5.6).

### $L_{1A\text{-}HN}$ AND $L_{2A\text{-}H}$

In a *SK*1 genetic background, M genomes require the [HOK] function of $L_A$ (see Table 5.6) i.e. $M_1$ needs $L_{1A\text{-}HN}$ (or $L_{2A\text{-}H}$) and $M_2$ needs $L_{2A\text{-}H}$ (or $L_{1A\text{-}HN}$) (Hannig *et al.* 1985, Ridley *et al.* 1984, Wickner & Toh-e 1982). In a *ski* genetic background M genomes have no such requirement for [HOK], and can replicate in the presence of $L_A$ variants which do not supply this function, e.g. $L_{A\text{-}E}$ or $L_{A\text{-}null}$ (Ridley *et al.* 1984, Wickner & Toh-e 1982).

The exact role of the [HOK] function is unclear. Since $L_{1A}$ is known to encode the coat protein itself and $M_1$, and similarly $L_{2A}$ for itself and $M_2$ (Bostian *et al.* 1980a, El-Sherbeini *et al.* 1984, Hopper *et al.* 1977), it has generally been assumed that [HOK] is that function. However, an alternative possibility exists: Ball *et al.* (1984) have suggested that $L_A$ ($L_{1A\text{-}HN}$), which is present in much larger copy number than $M_1$, may prevent the reduction of $M_1$ copy numbers to zero levels by soaking up most of the (repressive) *SK*1 gene products.

### $L_{A\text{-}E}$, $L_{A\text{-}HE}$ AND $L_{A\text{-}HR}$

Exclusion [EXL] activity is found associated with an $L_A$ genome of unknown K1 and K2 derivation, being $L_{A\text{-}E}$. (Variant $L_{A\text{-}HE}$ also exists.) The synergistic, non-exclusion activity [NEX] is found associated with the normal $L_{1A}$ genome, $L_{1A\text{-}HN}$. Exclusion [EXL] and non-exclusion [NEX] were originally defined by the observation that [EXL] excludes $M_2$ from strains that lack [NEX], but not from strains that have [NEX] (Sommer & Wickner 1982a, Wickner 1980).

Restating this in molecular terms, if $L_{A\text{-}E}$ is introduced into a strain carrying $L_{2A\text{-}H}$ and $M_2$, then $M_2$ is *excluded*, whereas if $L_{A\text{-}E}$ is introduced into a strain carrying $L_{1A\text{-}HN}$, then $M_2$ is *not excluded* (Hannig *et al.* 1985).

Not only does $L_{A\text{-}E}$ eliminate $M_2$, it also dramatically lowers the copy number of $L_{2A\text{-}H}$ (Hannig *et al.* 1985). It is thought that these two phenomena may be related, since mutants of ($L_{2A\text{-}H}$ $M_2$) strains, resistant to $M_2$ exclusion by $L_{A\text{-}E}$, were shown by Hannig *et al.* (1985) to have altered $L_{2A\text{-}H}$ ($L_{2A\text{-}HR}$) genomes whose copy number was not lowered by $L_{A\text{-}E}$. These authors therefore concluded that $M_2$ is excluded by the $L_{A\text{-}E}$ enforced copy number reduction in $L_{2A\text{-}H}$ (needed for $M_2$ maintenance in a *SK*1 background). Further support for this comes from Sommer and Wickner (1984), who have shown that exclusion of $M_2$ by $L_{A\text{-}E}$ (and by $M_1$) is not a result of degradation of pre-existing molecules.

The situation with regard to $M_1$ exclusion in *SK*1 backgrounds is less certain (see Table 5.6). As is the case for $M_2$, when $L_{A\text{-}E}$ is introduced into a strain carrying $L_{1A\text{-}HN}$, then $M_1$ is not excluded. However, when $L_{A\text{-}E}$ is introduced into a strain carrying $L_{2A\text{-}H}$, then the $M_1$ may be lost, or may be stable but poorly expressed (Hannig *et al.* 1985). One of the parents of the diploids from which $M_1$ was lost was derived from strain 46B. As noted above, strain 46B carries an extremely low copy number of an 'odd' $L_A$, $L_{A\text{-}null}$, and furthermore is extremely unstable for $M_1$ even in a *ski*2 genetic background. Thus, it may be that the $L_{A\text{-}null}$ from strain 46B is contributing to the instability of $M_1$ in derived strains.

When $L_{A\text{-}E}$ is the only L genome present, neither $M_1$ nor $M_2$ are maintained (see Table 5.6).

It is possible that the reason for non-exclusion of $M_2$ (or $M_1$) by $L_{A\text{-}E}$ in the presence of $L_{1A\text{-}HN}$ may simply be that $L_{A\text{-}E}$ does not lower $L_{1A\text{-}HN}$ copy number as it does $L_{2A\text{-}H}$, but this has not been tested. If this were the case, then the $L_{2A\text{-}HR}$ mutants (resistant to $M_2$ exclusion by $L_{A\text{-}E}$) may be somewhat analogous to $L_{1A\text{-}HN}$, since it has already been shown that $L_{2A\text{-}HR}$ copy number is unaffected by $L_{A\text{-}E}$ (see Table 5.6) (Hannig *et al.* 1985). Some support for the notion that $L_{1A\text{-}HN}$ may be a more successful competitor than $L_{2A\text{-}H}$ for replication components comes from a study of copy numbers in mixed $L_{2A\text{-}H}$ and $L_{1A\text{-}HN}$ diploids. In such diploids $L_{2A\text{-}H}$ copy number is reduced to levels which are no longer detectable, i.e. less than 50 ng (Hannig *et al.* 1985). Studying copy numbers of $L_{1A\text{-}HN}$ and $L_{A\text{-}E}$ in a mixed diploid should resolve this issue.

### EXCLUSION BY $L_{1A\text{-}HN}$

$L_{1A\text{-}HN}$ excludes $M_2$ in a strain carrying the recessive alleles, *mkt*1 or *mkt*2, of two nuclear maintenance genes (Wickner 1980, 1983b). In an *mkt*1 host this exclusion occurs at 30°C but not at 20°C (Ridley *et al.* 1984). These authors also showed that recessive alleles of six *SK*1 and *ski* genes – *ski*2, *ski*3, *ski*4, *ski*6, *ski*7 and *ski*8 – suppress this $M_2$ exclusion, increase M copy number at 30°C and confer cold-sensitivity. It is already known that the host *LTS*5 (≡ *MAK*6) product is required both for $M_1$ maintenance (Wickner *et al.* 1982) and for low-temperature growth (Singh & Manney 1974). Ridley *et al.* (1984) proposed that the elevated M copy number in *ski* hosts diverts the *LTS*5 product away from the host and to the M replication process. They further suggested that

$L_{1A\text{-}HN}$ and $M_2$ may compete for the *MKT*1 gene product, with $M_2$ having a lower affinity, and that the lower copy number noted for $L_{2A\text{-}H}$ would leave sufficient product, even in *mkt*1 strains for $M_2$. If, as they suggest, the *ski* mutants are de-repressed for *MK*1 product, then that could account for non-exclusion of $M_2$ by $L_{1A\text{-}HN}$ in *ski* hosts. One prediction of this hypothesis (Ridley *et al.* 1984) is that elimination of $M_2$ should increase the copy number of $L_{1A\text{-}HN}$. Such an increase in $M_2$ upon elimination of $L_{2A\text{-}H}$, and an increase in $M_1$ upon elimination of $L_{1A\text{-}HN}$, has been observed (see Table 5.6) (Ball *et al.* 1984). Ridley *et al.*(1984) also reported the effects of the dominant allele of one gene (*MKS*50) and the recessive alleles of two other genes (*mks*1 and *mks*2) in suppressing $M_2$ exclusion by $L_{1A\text{-}HN}$. However, these mutants exhibited no other characteristics of the *ski* mutations.

**Conclusions**

L genomes exist in copy numbers per cell, for some strains, estimated at thousands (Oliver *et al.* 1977). Normal M copy numbers are around 100–1000 $cell^{-1}$ (Brewer & Fangman 1980, Ridley & Wickner 1983). It has already been noted that the presence of one L genome in a cell affects the copy numbers of other L genomes (see Table 5.6), and that the loss of M genomes results in an increase in L copy number (Ball *et al.* 1984). These authors have also shown that *ski* mutations dramatically increase L and M copy numbers. Thus, in the repressive *SK*1 environment, the basis of the exclusion phenomenon and stability or lack of stability of M genomes may be a result of competition for some common components in dsRNA replication; with some variant genomes being more successful than others. Exclusion by, for example, competition for common replication factors is well documented for prokaryotic phage and plasmids.

### *5.6.4 Interactions with host genes*

**$L_{BC}$**

Cytoduction experiments suggest that both $L_B$ and $L_C$ require a host to be *CL*0. Inexplicably, meiotic analysis confirms this result only for $L_B$ and not for $L_C$. The complex *clo* chromosomal defect is still unidentified, and has no effect on the maintenance of $L_A$, $M_1$ or $M_2$ species (Wesolowski & Wickner 1984).

The only other chromosomal loci known to affect $L_{BC}$ are the *SK*1 and *ski* genes. The *SK*1 gene products (*SK*1 2, 3, 4, 6, 7 and 8) lower the copy numbers of $L_{BC}$, $L_{1A\text{-}HN}$, $M_1$ and $M_2$ (Ball *et al.* 1984).

**$L_A$**

Table 5.7 lists the host genes involved in the maintenance of $L_A$ or $M_1$, or both, but not $L_{BC}$, genomes. In addition to the effects of the *SK*1 and *ski* loci (see above) three other genes are known to affect $L_1$ ($L_{1A\text{-}HN}$) maintenance: the products of *MAK*3, *MAK*10 and *PET*18 (*PES*1) are all necessary (El-Sherbeini *et al.* 1983, Sommer & Wickner 1982a).

The complex *PET*18 locus has recently been cloned. Sequence analysis has revealed the presence of four open-reading frames, and from this and the results of complementation analysis it has been concluded that the function of

**Table 5.7** K1 host genes involved in the maintenance of $L_{1A}$ ($L_{1A\text{-}HN}$) and $M_1$ genomes.

| Mutant allele | Equivalent locus | Host-cell function affected, and comments | dsRNA lost in presence of mutation | Suppressed by *ski*1 | Suppressed by *ski*2–*ski*4† | Suppressed by [Kil-$b_1$] |
|---|---|---|---|---|---|---|
| *mak*16 | ? | ? | $M_1$ | – | – | +‡ |
| *mak*3 | ? | ? | | | | |
| *mak*10 | ? | loss of functional mit DNA bypasses mutation* | $L_{1A}$ $M_1$ | + | – | – |
| *pet*18 | [*pes*1] | complex locus containing multiple genes: also causes loss of mit DNA | | | | |
| *mak*1 | ? | deficient in DNA topoisomerase activity | | | | |
| *mak*6 | *lts*5 | unable to grow at low temperature, ?defective ribosomes | | | | |
| *mak*8 | *tcm*1 | loss of trichodermin resistance and ribosomal protein L3 | | | | |
| *spe*2 | — | deficient in adenosylmethionine decarboxylase (spermine and spermidine synthesis) | $M_1$ | | | |
| *spe*10 | — | deficient in ornithine decarboxylase (putrescine synthesis) | | +§ | +§ | +/–¶ |
| *mak*4, 5<br>*mak*7<br>*mak*9<br>*mak*11–28 | ? | ? | $M_1$ | | | |

* Antibiotic-induced respiratory deficiency does not have same effect (Wickner 1977).
† *ski*5 has been considered earlier (Section 5.3) and *ski*6–*ski*8 have not yet been tested for such suppression.
‡ [Kil-$b_1$] bypass of *mak*16 was shown by cytoduction (Ball *et al*. 1984). This is in contrast with the results obtained in meiosis (Toh-e & Wickner 1980).
§ Not all tested.
¶ Not all tested. Of mutants tested for bypass by [Kil-$b_1$] some are + (meiotic and cytoduction analysis); some are – (meiotic) but + (cytoduction); and some are – (meiotic and cytoduction).

this locus, concerning maintenance of L and $M_1$, is controlled by a separate gene(s) from that concerning the maintenance of mitochondrial DNA (Toh-e & Sahashi 1985).

Mutations in many of the other ($M_1$ maintenance) *MAK* genes tested do not detectably affect maintenance or copy number of $L_{1A}$ (or $L_{BC}$) (Ball *et al.* 1984, Wickner & Toh-e 1982). *ski*1 but not *ski*2, *ski*3 or *ski*4 mutations suppress *mak*3, *mak*10 and *pet*18 defects (Toh-e & Wickner 1980).

No complete analysis of the requirements of the variant $L_A$ genomes for the different host genes (Table 5.7) has been done. However, it is known that both $L_{A\text{-}HE}$ and $L_{A\text{-}E}$ require *MAK*3, but that whereas $L_{A\text{-}E}$ has a requirement for *MAK*27, $L_{A\text{-}HE}$ does not (Wickner 1983b).

## 5.7 M dsRNA species

### *5.7.1 Sequence analysis*

Terminal sequence analysis of $M_1$, $M_1$(S3) and $M_2$ dsRNAs has already been considered (Section 5.5) and data are presented in Figure 5.2.

A 1.1-kb cDNA derived from the AU-rich terminus of $M_1$ dsRNA from a derivative of the super-killer T158c has been cloned and sequenced (Skipper *et al.* 1984). cDNA from an *in vivo* transcript of $M_1$ dsRNA from strain K12-1 has also been cloned and sequenced (Bostian *et al.* 1984). There are only two coding differences between these two sequences, giving rise to amino acid differences within the $\alpha$-toxin component. In both sequences, a single open-reading frame exists, commencing with an AUG at bases 14–16, coding for 316 amino acid residues (34.7 kda) and terminating with an UAG amber codon at bases 963–965 (see Figs 5.1 & 5.3).

No other open-reading frames that would encode a protein in excess of 51 amino acids exist within the first 1.1 kb of $M_1$ on either strand (Skipper *et al.* 1984). The last 100 bp of this 1.1-kb sequence is the first half of the 200-bp AU-rich bubble (Fried & Fink 1978, Skipper *et al.* 1984) which is followed by some 650 bp ending at the 3′ (plus-strand) terminus of $M_1$. 3′-Terminal sequence analysis revealed no open-reading frame in either strand of approximately the last 300 bp (Thiele *et al.* 1982, 1984b). Thus, only the remaining 350 bp could conceivably be a candidate for encoding any other, as yet unidentified, $M_1$ product.

The first 210 bases of the 3′-terminus of the negative strand of $M_2$ dsRNA have been sequenced (Hannig & Leibowitz 1985). An AUG at positions 7–9 (see Fig. 5.2) of the positive strand begins an open-reading frame predicted to encode the $M_2$ toxin precursor. As noted by these authors, no significant homology exists between the N-terminal amino acid sequences encoded by $M_1$ and $M_2$, respectively.

### *5.7.2 Gene products*

That $M_1$ genomes are responsible for toxin and immunity (resistance) phenotypes has long been established. Mitotic and meiotic analyses (Bevan *et al.*

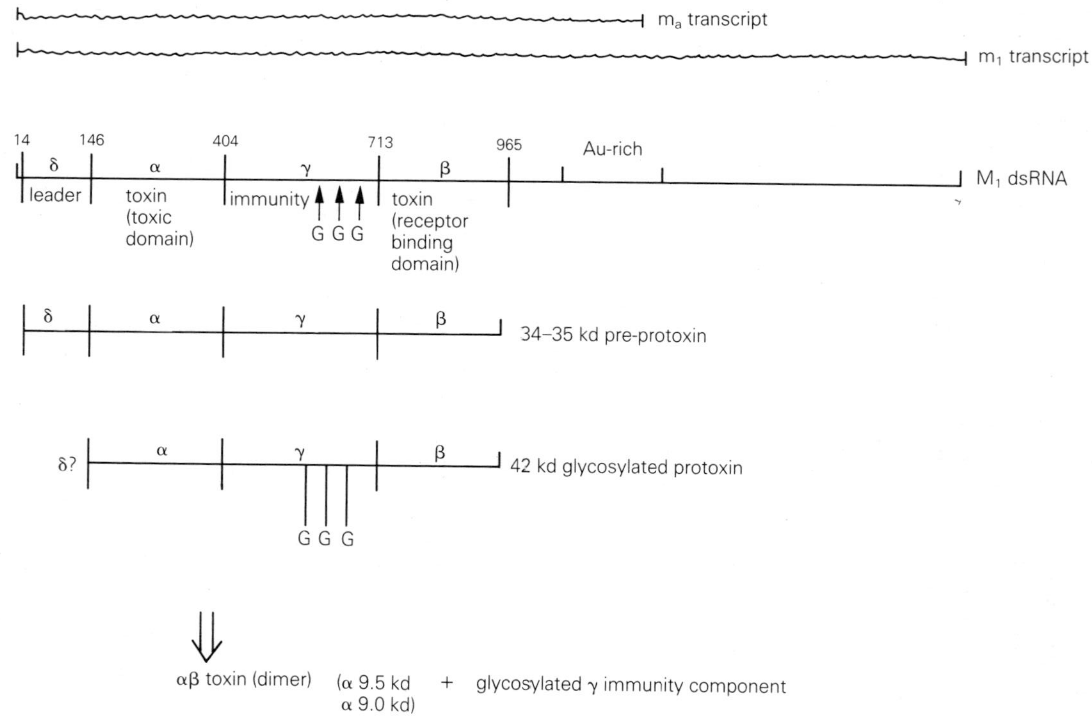


**Figure 5.3** $M_1$ dsRNA: gene products. Full ($m_1$) and partial-length ($m_a$) transcripts initiate at the 5′ (plus-strand) terminus of $M_1$ dsRNA. The primary product of translation is the non-glycosylated pre-protoxin, which undergoes glycosylation (G) and processing to yield an $\alpha$–$\beta$ dimeric toxin and a glycosylated $\gamma$ product, the putative immunity protein.

1973), curing of M genomes (Fink & Styles 1972, Mitchell *et al.* 1976b, Wickner 1974a) and analysis of mutant M genomes (Bevan & Somers 1969, Bussey *et al.* 1982) all provide evidence for M being the determinant for both toxin and immunity.

Direct expression in yeast of cDNA $M_1$ clones (Skipper 1983, Skipper *et al.* 1984) confirmed that both toxin and immunity functions are contained within the identified single open-reading frame (Lolle *et al.* 1984). The $\alpha$- and $\beta$-toxin components, together with the suggested $\delta$-immunity sequence (Bostian *et al.* 1984) which were all considered earlier (Section 5.3) are identified in Figures 5.1 and 5.3.

The established open-reading frame (AUG at 14–16 to UAG at 963–965) is contained within both the full length, m, and sub-genomic, $m_a$ (see Section 5.5) ssRNA *in vivo* molecules both of which have full *in vitro* messenger activity (see Fig. 5.3).

A study of *in vitro* translation products and *in vivo* intermediates suggests that toxin and immunity components share common precursor molecules (see Fig. 5.3) (Bussey 1984). Since much of the literature has been reviewed in detail elsewhere (Tipper & Bostian 1984), the following is just a summary of the main post-transcriptional events leading to the secretion of killer toxin.

(1) Synthesis of a primary translation product, the 34–35-kda pre-protoxin. Pre-protoxin has $\delta$, $\alpha$, $\gamma$ and $\beta$ sequences, is not glycosylated and has been detected by *in vitro* translation of M1 dsRNA (Bostian *et al.* 1980b, Bussey *et al.* 1983b).

(2) Pre-protoxin is then glycosylated to a 42-kda protoxin which most probably has an intact $\delta$ leader sequence. Glycosylated protoxin has been detected *in vivo* (Bussey *et al.* 1982, 1983b, Bostian *et al.* 1983b) and glycosylation is thought to occur at three putative [Asn-(X)-Thr/Ser] sites within $\gamma$ as indicated (Figs 5.1 & 5.3) (Bostian *et al.* 1984, Skipper *et al.* 1984). It is thought that the hydrophobic N-terminal region within $\delta$ (32–34, val to 92–94, leu) contains a sequence which directs the mature molecule into the correct glycosylation pathway (Lolle *et al.* 1984). A similar hydrophobic region occurs within the N-terminal region of $M_2$ (Hannig & Leibowitz 1985).

(3) Further processing (proteolytic cleavage) yields a non-glycosylated $\alpha$–$\beta$ dimeric toxin (see Section 5.3) and, presumably, a glycosylated $\gamma$ product, the putative immunity protein (Bostian *et al.* 1984, Lolle *et al.* 1984).

This sequence of events is summarized in Figure 5.3.

Most secreted proteins in yeast are found in the cell wall (actually, in the periplasmic space between the plasma membrane and the cell wall) or both in the cell wall and in the medium. Furthermore, most are glycosylated (for a review, see Schekman & Novick 1982). Secretion of toxin (Bussey *et al.* 1983b), in common with other secreted proteins of yeast, occurs via the normal yeast secretory pathway as defined by the *sec* mutants (Novick & Schekman, 1979, Novick *et al.* 1980, 1981). $\alpha$-Factor and toxin bear similarities. Both are unlike other yeast-secreted proteins in being found primarily in the medium, in being non-glycosylated but in having glycosylated precursors (see Schekman &

Novick 1982). Secretion of insulin-like hormones involves the removal of an N-terminal peptide and excision of an internal 'C' peptide leaving a disulphide-linked dimeric product (Chan *et al.* 1979). This is analogous to the steps involved in release of mature toxin, except that in the case of killer toxin, the internal peptides ($\gamma$ region) are glycosylated and thought to be the non-secreted immunity proteins (Bostian *et al.* 1984).

Toxin and immunity factor are the only known gene products of M genomes.

### 5.7.3 *Interactions with host genes*

The dominant alleles of 30, unclustered, genes are required for the maintenance of $M_1$ dsRNA (Cohn *et al.* 1978, Guerry-Kopecko & Wickner 1980, Mitchell *et al.* 1976a, Tyagi *et al.* 1984, Wickner 1974b, 1978, Wickner & Leibowitz 1976a, 1977, 1979). Table 5.7 lists the mutant alleles together with their host cell defects (where known). Their ability to be suppressed or bypassed by the *ski* mutations or the mutant M genome [*kil-b*$_1$], or both, is also indicated. Three *MAK/mak* genes have recently been identified with host genes of known function. These are *MAK*1 (≡gene for yeast topoisomerase) (Thrash *et al.* 1984), *MAK*6 (≡*LTS*5) (Ridley *et al.* 1984) and *MAK*8 (≡ *TCM*1) (Wickner *et al.* 1982) (see Table 5.7). A dominant unstable mutation, *KRB*1, has been isolated. *KRB*1, which is located on a new centromere, was isolated as a suppressor of *mak*7 defect, but has also been shown to suppress the *pet*18 mutation (Wickner *et al.* 1983).

$M_2$ dsRNA was shown first to require the *MAK*10 function (Naumov & Naumova 1973). It is now also known to require *MAK*3 and *PET*18; *MAK*1, *MAK*8, *MAK*16, *MAK*27, *SPE*2 and *SPE*10 (Tyagi *et al.* 1984, Wickner 1983a). $M_2$ dsRNA is not maintained in the presence of the normal $L_{1A}$ genome, $L_{1A\text{-}HN}$ as opposed to $L_{2A\text{-}H}$ (normal $L_{2A}$ genome). The effects of five nuclear genes in suppressing this exclusion have already been considered [Section 5.6.3(b)(iii)].

As already discussed (Section 5.6.4), *SK*1 gene products lower the copy numbers of $L_{BC}$, $L_{1A\text{-}HN}$, $M_1$ and $M_2$ (Ball *et al.* 1984). The effects of some of the *ski* mutations, in suppressing or by-passing M dsRNA requirements for host maintenance genes, are shown in Table 5.7.

Aside from the *sec* or *ski* loci, three other nuclear genes, which affect the expression of $M_1$ dsRNA, have been described. *kex*1 and *kex*2 (killer expression) both make glycosylated protoxin which is apparently not processed correctly and both fail to secrete toxin whilst immunity remains unaffected (Wickner & Leibowitz 1976b, Bussey *et al.* 1983b). *kex*2 is a pleiotropic mutation resulting, amongst other things, in a failure to secrete functional $\alpha$-factor (Bussey *et al.* 1983b). The third mutation, *rex*1, is largely uncharacterized. *rex*1 mutants result in a loss of immunity function while toxin is still secreted, and they therefore have a 'suicide' phenotype (Wickner 1974b).

### 5.7.4 *Variant and mutant genomes*

$M_2$ and $M_3$ dsRNAs, like $M_1$, presumably each encode toxins and immunity determinants of their respective specificities (see Table 5.2 and Section 5.3) and may thus be considered to be variant genomes. The exclusion of $M_2$ dsRNA by $M_1$ has been discussed earlier (Section 5.6.3).

Mutant M genomes have only so far been identified for $M_1$ dsRNA. Most of the known $M_1$ mutants and their main features are outlined in Table 5.8. One mutant [Kil-$d_1$] is not included in Table 5.8: very little is known about this mutant, other than the fact that it is apparently maintained only in a/$\alpha$ diploid cells (Wickner 1976).

**[Kil-$sk_1$]: superkiller mutant**

The toxin and immunity gene sequence of the mutant [Kil-$sk_1$] $M_1$ genome (derived from strain T158c) is shown, together with that of a wild type [Kil-$k_1$] $M_1$ genome (strain K12-1) in Figure 5.1. Sequence analysis (Skipper *et al.* 1984) reveals only two coding changes, both within the $\alpha$-toxin component, which result in the production of a more stable toxin. This increased stability may be caused by reduced sensitivity to the *SK*15 encoded cell-surface protease (Bussey *et al.* 1983a; see Section 5.3.1).

**[Kil-$n_1$]: neutral mutants**

Neutral mutants were first discovered by Somers and Bevan (1969). N1 and N2 are two such mutants (Table 5.8), both of which retain immunity but fail to secrete toxin. The basis for the mutant phenotype is assumed to be a mutation in $M_1$ but the nature and site of lesion are unknown.

Bussey *et al.* (1982) obtained additional neutral mutants by ethyl methane sulphate (EMS) mutagenesis of a derivative of the superkiller strain T158c. Four such mutants NLP1, NLP3, NLP7 and NLP10 are shown in Table 5.8. Gross changes in M genome are seen only in one mutant, NLP3, which inexplicably has two M genomes, one of normal size and one somewhat smaller. NLP3, like NLP7 and NLP10, secretes a normal-size 'toxin' protein. Both of the last two mutants probably have mutations within the $\alpha$-toxin component (see Fig. 5.1) since they retain functional $\beta$, the putative cell-wall binding receptor domain (Bussey *et al.* 1982). NLP1 is probably a point mutation within the $\beta$-domain since Bussey *et al.* (1982) have shown that it is readily reverted by EMS, it retains immunity yet produces smaller than wild-type pre-protoxin and protoxin.

No sequence analysis is yet available for these mutants.

**[Kil-$i_1$]: suicide mutants**

Immunity minus, toxin producing strains were first described by Vodkin *et al.* (1974) and Sweeney *et al.* (1976). Following EMS mutagenesis of wild-type killers, Bussey *et al.* (1982) have similarly isolated suicide mutants, designated SLP2, SLP4 and SLP6 (Table 5.8). SLP2 is the only one of the three to have an altered sized, smaller, M genome. However, the putative deletion probably resides outside the toxin and immunity gene since normal sized pre-protoxin is produced. Thus, the defect, resulting in immunity deficiency, like SLP4 and SLP6, probably resides within the $\gamma$-domain (see Fig. 5.1). All three of these mutants retain some immunity, with the most sensitive mutant, SLP2, still requiring some 100 times the toxin concentration needed for the same kill as with its isonuclear, M-cured, sensitive strain (Bussey *et al.* 1982).

**Table 5.8** $M_1$ dsRNA mutant genomes.

| Killing (K) and immunity (R) phenotype | $M_1$ genotype and designation | Mutant* | Pre-protoxin (kda) | Protoxin (kda) | Secreted protein (kda) | M dsRNA (kb) | Nature and site of lesion | Comments |
|---|---|---|---|---|---|---|---|---|
| $K^+ R^+$ | [Kil-$k_1$] wild type | — | 34–36 | 42(G) | 11 | 1.85 | no lesion | K12-1 is a wt K1 killer: for sequence see Figure 5.1 |
| $K^{++} R^+$ | [Kil-$sk_1$] superkiller | (strain T158c) | 34–36 | 42(G) | 11 | 1.85 | two mis-sense changes in $\alpha$: 320–322 (ile $\rightarrow$ ser) 380–382 (thr $\rightarrow$ ala) | mutant produces same amount of a more stable toxin: sequence given in Figure 5.1 |
| $K^- R^+$ | [Kil-$n_1$] neutral | N1 | 34–36 | 42(G)† | none | 1.85 | ? | possibly defective in protein processing |
| | | N2 | 34–36 | 42(G)† | none | 1.85 | ? | possibly defective in toxin secretion |
| | | NLP1 | 30 | 39(G)† | none | 1.85 | ?nonsense ($\beta$) | possibly a nonsense mutation, but not suppressed by nonsense suppressors |
| | | NLP3 | | | 11† | 1.85 + <1.85 | ? | |
| | | NLP7 | | | 11† | 1.85 | ?mis-sense ($\alpha$) | retained receptor binding ($\beta$) |
| | | NLP10 | | | 11† | 1.85 | ?mis-sense ($\alpha$) | retained receptor binding ($\beta$) |
| $K^+ R^-$ | [Kil-$i_1$] suicide | SLP2 | 34–36 | | | <1.85 | ?deletion outside reading frame + ?mis-sense ($\gamma$) | retained partial immunity SLP2 < SLP4 < SLP6 but even SLP $\gg M_1^-$ sensitive |
| | | SLP4 | | | | 1.85 | ?mis-sense ($\gamma$) | |
| | | SLP6 | | | | 1.85 | ?mis-sense ($\gamma$) | |
| $K^- R^-$ | [Kil-$s_1$] suppressive | S3 | 8.0 | 6.4(−) | | 0.72 | internal 1.2 kb deletion: sequence changes at residue 230 | 'new' stop codon at 281–283 |
| | | S1 S4 | 8.0 | 6.4(−) | | | | both derived by further rearrangement of S3 |

* All mutants, except N1 and N2 of unknown origin, were derived from T158c [Kil-$sk_1$] superkiller. (G), glycosylated.
† Reacts with antitoxin.

**[Kil-$s_1$]: suppressive sensitive mutants**

Mutant $M_1$ genomes, defective in toxin production and immunity, have large internal deletions and interfere (suppress) the replication of non-mutant $M_1$ genomes (Ridley & Wickner 1983, Somers 1973, Sweeney *et al.* 1976, Tzen *et al.* 1974, Vodkin *et al.* 1974). Three such suppressive sensitive mutants S1, S3 and S4, lack the internal AU-rich region (Fried & Fink 1978) but retain both termini. Some 300 bases of both 3′-termini of S3 dsRNA have been sequenced (Thiele *et al.* 1984b). The 3′-terminus of the negative strand is identical to that of $M_1$ dsRNA up to nucleotide 229, from which point there is marked divergence. This position agrees precisely with the previously determined start of the deletion as estimated by heteroduplex analysis (Fried & Fink 1978). The 3′-termini of the positive strands of S3 dsRNA and $M_1$ are identical.

*In vitro* translation of suppressive-sensitive [Kil-$s_1$] dsRNAs results in the synthesis of a small (premature termination) polypeptide of about 8 kda which is immunologically cross-reactive with the mature killer toxin (Bostian *et al.* 1983b). The premature termination probably occurs at the recently identified termination codon at 281–283 (Thiele *et al.* 1984b). The mutant genomes of S3 are encapsidated in VLPs with major protein components identical to those of $L_{1A}$ and $M_1$ (Thiele *et al.* 1984a). Suppressive-sensitive mutants are thus somewhat analogous to the defective-interfering animal virus genomes (reviewed by Huang 1977, Perrault 1981).

**[Kil-$b_1$]: bypass mutant**

[Kil-$b_1$], a presumed mutant $M_1$ genome, with a superkiller phenotype, was first isolated by Toh-e and Wickner (1980) and shown to bypass the requirement for some, but not all, of the host maintenance genes (see Table 5.7). Bypass of the various *mak* mutations was examined both by meiotic analysis (Toh-e & Wickner 1980) and following cytoduction (Ball *et al.* 1984). Both analyses showed bypass of *mak*4, *mak*7 and *mak*17, but not *mak*3 or *mak*10. On meiotic analysis neither *pet*18 nor *mak*21 was by-passed by [Kil-$b_1$], cytoduction analysis not having been done. In addition cytoduction analysis showed bypass of *mak*1, *mak*8, *mak*13 and *mak*22. Strangely, *mak*16 and *mak*12 were not bypassed by [Kil-$b_1$] in meiotic analysis but were bypassed following cytoduction.

It seems that requirement for, or bypass of, nuclear genes depends upon the particular M genome and on its copy number. Ball *et al.* (1984) examined the copy number of M genomes, derived from three different killers, in isonuclear backgrounds. They showed that strain 1368 [Kil-$b_1$] had twice as much M dsRNA as strain 18 [Kil-$k_1$], which in turn had twice as much M dsRNA as strain A364A [Kil-$k_1$]. Whereas A364A (lowest M copy number, standard killer) required the full complement of *MAK* alleles, strain 18 (intermediate M copy number and killing) showed 'weak' bypass of some of the *mak* alleles and strain 1368 (high [Kil-$b_1$] M copy number and super-killer phenotype), as considered above, showed bypass of many *mak* alleles. This relationship between super-killing, $M_1$ copy number and ability to bypass *mak* mutations is also correlated with the ability to repress $L_{1A}$ copy number (Ball *et al.* 1984). This is worthy of further investigation to determine whether there is any basis for this relationship.

**[Kil-Sd$_1$]: SK1-dependent mutant**
Mitotic segregants, in which the $M_1$ genome had become dependent on a *ski* mutation, were isolated from *ski* killer strains. The *ski*-dependent mutation was assumed to be a mutation in the M genome which was thus designated [Kil-sk$_1$] (Toh-e & Wickner 1979). It is now known that although the $M_1$ in [Kil-sd$_1$] strains may indeed be mutant, the phenomenon may be more simply explained by the host lacking $L_{1A\text{-}HN}$ but having $L_{A\text{-}E}$ and a wild-type $M_1$ genome. Thus, in a *ski* host the $M_1$ genome would be maintained, whereas in a *SK*1 host it would be lost (Ridley *et al.* 1984, see Table 5.6).

## 5.8 Practical applications

### *5.8.1 Cloning: cDNA secretion vectors*

Our comprehension of many cellular processes in the yeast *Saccharomyces cerevisiae* has increased dramatically over the past decade, due largely to the application of recombinant DNA technology. The reader is referred to several recent reviews covering some of the more basic and some of the more sophisticated techniques and systems available (Botstein & Davis 1982, Broach *et al.* 1983, Hinnen & Meyhack 1982, Parent *et al.* 1985, Struhl 1983).

Amongst the wide range of vectors available (see review by Parent *et al.* 1985) are many which are useful for homologous or heterologous protein synthesis, but only a few which may be utilized for protein secretion. The latter are plasmids containing coding sequences which are able to direct post-translational processing and protein secretion.

Most secreted proteins in yeast are actually found in the periplasmic space rather than the medium (see review by Schekman & Novick 1982). Another problem common to most yeast-secreted proteins such as acid phosphatase and invertase is that core glycosylation followed by a massive elongation of the carbohydrate chains occurs (Esmon *et al.* 1981).

Interest in developing vector systems based on M dsRNA, cDNA, sequences lies in the fact that toxin, like $\alpha$-factor, is non-glycosylated (although it has a glycosylated precursor) and is secreted into the medium. $\alpha$-Factor sequences have already been used to construct expression vectors which direct post-translational processing and protein secretion (Brake *et al.* 1984, Bitter *et al.* 1984, Carlson & Botstein 1982, Emr *et al.* 1983, Singh *et al.* 1984).

cDNA of the toxin and immunity (pre-protoxin) gene of $M_1$ dsRNA has been expressed under the control of both the *PHO*5 promoter with its own N-terminal secretion leader peptide (Hanes *et al.* cited in Tipper & Bostian 1984) and the *ADH*1 promoter (Lolle *et al.* 1984). In one of the latter clones, pADH·10A, the hydrophobic N-terminal region within $\delta$ (see Fig. 5.1 and Section 5.7.2) was largely intact; immunity was expressed and active toxin secreted. Plasmids such as these could be used or developed for the insertion of heterologous genes at various positions within the pre-protoxin gene in order to achieve proteolytic processing and secretion of the gene products. One such target site for insertions would be the $\gamma/\beta$ boundary (see Fig. 5.1), a presumed processing site (Tipper & Bostian 1984).

### 5.8.2 *Cloning: dsRNA vectors*

None of the already constructed pre-protoxin cDNA clones has authentic $M_1$ dsRNA-derived termini (Bostian *et al.* 1984, Skipper *et al.* 1984). If clones containing authentic termini could be constructed, then mRNA transcripts might prime the synthesis of dsRNA and thence be encapsidated. The host cell would necessarily have to be wild-type for the maintenance genes and, presumably, contain an appropriate $L_A$ dsRNA species (see Sections 5.6.3 & 5.7.3). In a somewhat analogous situation, a phage $Q\beta$ clone was shown to produce actual phage progeny when the clone was introduced into an *Escherichia coli* host (Taniguchi *et al.* 1978).

One advantage of using expression vectors based on M dsRNA would be the very high copy number of about 100–1000 $cell^{-1}$, which is increased even further in *ski* genetic backgrounds (see Section 5.6.3).

### 5.8.3 *Strain manipulations*

Having read this chapter it will have become apparent to the reader that the majority of strain manipulations involving the killer dsRNA genomes have been based on formal genetic approaches, i.e. meiotic and mitotic analyses, cytoduction, curing and mutagenesis. More recently, transformation systems have been developed utilizing plasmids in which cDNA copies of the dsRNA genomes are carried (see above).

No infective cycle has ever been demonstrated, and reports of extracellular *de novo* infection (Lhoas 1972, Nesterova 1974) have never been firmly established. Recently, however, there has been a report that ScV-$M_1$ and ScV-$M_2$ particles are capable of extracellular transmission. El-Sherbeini and Bostian (1985) have co-infected M dsRNA-negative hosts with ScV-M particles and a yeast plasmid (YEp24) carrying an auxotrophic (*URA*3) gene. The infected clones were stable killers and maintained normal levels of M dsRNA. As noted by these authors, this technique could have widespread applications in studying host–viral interactions.

### 5.8.4 *Manipulation of industrial yeasts*

Brewer's yeast strains are often polyploid in nature and, although this polyploidy leads to a high degree of genetic stability, it means that such strains sporulate infrequently and with low spore viability (Stewart *et al.* 1977). Thus, standard genetic analysis is extremely difficult. Furthermore, brewing strains have been selected over long periods for the possession of a wide range of attributes, some of which are well defined genetically, e.g. the *MAL* genes and ability to ferment maltose and maltotriose (Stewart *et al.* 1977), and some of which are ill-defined, e.g. failure to produce 'off-flavours'. Any standard genetic manipulation, e.g. crossing or production of cell-fusion products, may result in the acquisition of some new attribute, but would almost certainly result in the loss of others, due to the mixing of genotypes. Thus, attention has recently been focused on the application of recombinant DNA technology to industrial yeasts.

In recombinant DNA work some positive selection is required for obtaining transformants and stably maintaining the vectors. Industrial strains are polyploid and prototrophic, so auxotrophic markers cannot be used as a basis for selection. Recent attention has been directed toward utilizing such selection techniques as drug- or metal-resistance (Henderson 1983, Kaster *et al.* 1984, Webster & Dickson 1983). There are obvious practical problems in adopting such techniques for use in industrial fermentations.

The pre-protoxin gene, as a cDNA copy in a plasmid, could be a useful selection marker for the transformation of industrial yeasts. It is possible that killer toxins are produced under brewing conditions, since wild-killer yeasts have often been noted to contaminate and spoil such fermentations (see Section 5.2). Bussey and Meaden (1985) have actually tested the feasibility of such an approach. Using plasmids carrying (expressed) cDNA copies of the pre-protoxin gene these authors have shown that transformants are self-selected, with the evidence being strongly presumptive for toxin being the selective agent. Furthermore, the immune and toxin-producing transformants were stably maintained in the absence of any other selection.

### *5.8.5 Epidemiological typing*

Initial interest in pathogenic yeasts was based upon the premise that pathogenicity might in some way be related to the presence or absence of the dsRNA genomes. No such correlation was observed (Mitchell 1974). The killer system has now been adopted as a means of strain differentiation amongst pathogenic yeasts. Classification on the basis of sensitivity to toxic effects of different killer yeasts was first made on isolates of *Candida albicans* (Polonelli *et al.* 1983) and was later extended to include a wide range of pathogenic yeasts (Morace *et al.* 1984). The killer system proved to be valid as an epidemiological marker even where other characteristics were heterogeneous and failed to permit categorization.

## 5.9 Concluding remarks

The past decade has seen an explosion in research effort directed towards the elucidation of the killer system. With each publication one problem is seemingly solved but another, new, problem is posed. Despite the obvious progress there are still major areas unsolved. For example, very little is known about the synthesis of ds molecules from ss, or the putative involvement of capsid proteins in replication. Also, although it has always been assumed that the L genomes encode RNA polymerase, there is no *a priori* reason why this should be so: RNA polymerase could well be host-encoded.

Notwithstanding the wealth of information published on the 30-plus maintenance genes, very few of them have identified products or known host functions. The relationship between *MAK* and *SK*1 genes has not been elucidated, nor has the role they have to play in determining L and M copy numbers. What also is the basis for coexistence or exclusion of the various genomes? Is it just that some genomes are replicated preferentially? If so, is this a result of, for example, the possession of consensus versus non-consensus sequences for replicase and capsid binding? The establishment of such binding regions, comparisons of binding affinities and comparative sequence analysis of these regions may help to solve this problem.

The potential usefulness of killer 'genes' to yeast biotechnologists, is now beginning to be recognized. For example, there is the possibility of developing a stable, high-copy number dsRNA cloning system. cDNA copies of the toxin-immunity gene have already been cloned into plasmids. These have the possibility of being developed into sophisticated secretion vectors, with the advantage of true export of foreign products without the disadvantage of possible glycosylation. The toxin-immunity gene could itself be used as a selective marker for the transformation of polyploid industrial yeasts: this would be of enormous benefit where the utilization of drug or metal resistance are practical infeasibilities. Finally, the acquisition of killer genomes by brewing yeasts could render them less susceptible to contamination by wild yeasts.

## Acknowledgements

The authors wish to thank Vivienne Winrow, Jocelyn Wooldridge and Michael Daly for their help with the proofreading of this manuscript. We should also like to express our gratitude to Vera Lamport for expert secretarial assistance.

## References

Adams, J. M. and S. Cory 1975. Modified nucleotides and bizarre 5′-termini in mouse myeloma mRNA. *Nature New Biology* **255**, 28–33.

Adler, J. and D. W. R. Mackenzie 1972. *Abstracts of the Annual Meeting of the American Society of Microbiology* 8.

Adler, J., H. A. Wood and R. F. Bozarth 1976. Virus-like particles from killer, neutral and sensitive strains of *Saccharomyces cerevisiae*. *Journal of Virology* **17**, 472–6.

Al-Aidroos, K. and H. Bussey 1978. Chromosomal mutants of *Saccharomyces cerevisiae* affecting the cell wall binding site for killer factor. *Canadian Journal of Microbiology* **24**, 228–37.

Antczak, J. B., R. Chmelo, D. J. Pickup and W. K. Joklik 1982. Sequences at both termini of the 10 genes of Reovirus serotype 3 (strain Dearing). *Virology* **121**, 307–19.

Ball, S. G., C. Tirtiaux and R. B. Wickner 1984. Genetic control of L-A and L-(BC) dsRNA copy number in killer systems of *Saccharomyces cerevisiae*. *Genetics* **107**, 199–217.

Berry, E. A. and E. A. Bevan 1972. A new species of double-stranded RNA from yeast. *Nature* (London) **239**, 279–80.

Bevan, E. A. and A. J. Herring 1976. The killer character in yeast: preliminary studies of virus-like particle replication. In *Genetics, biogenesis and bioenergetics of mitochondria*, R. J. Bandelow, D. Y. Schweyen, K. Thomas, K. Wolf and F. Kaudewitz (eds), 153–62. Berlin, New York: Walter de Gruyter.

Bevan, E. A. and M. Makower 1963. The physiological basis of the killer character in yeast. *Proceedings of the XIth International Congress on Genetics* **1**, 202–3 (abstr.).

Bevan, E. A. and J. M. Somers 1969. Somatic segregation of the killer (k) and neutral (n) cytoplasmic genetic determinants in yeast. *Genetical Research Cambridge* **14**, 71–7.

Bevan, E. A., A. J. Herring and D. J. Mitchell 1973. Preliminary characterisation of two species of dsRNA in yeast and their relationship to the 'killer' character. *Nature* (London) **245**, 81–6.

Bitter, G. A., K. K. Chen, A. R. Banks and P.-H. Lai 1984. Secretion of foreign proteins from *Saccharomyces cerevisiae* directed by $\alpha$-factor gene fusions. *Proceedings of the National Academy of Sciences of the USA* **81** (17), 5330–4.

Bobek, L. A., J. A. Bruenn, L. J. Field and K. W. Gross 1982. Cloning of cDNA to a yeast viral double-stranded RNA and comparison of three viral RNAs. *Gene* **19**, 225–30.

Bobek, L. A., J. A. Bruenn, L. J. Field, J. D. Reilly and K. W. Gross 1983. Cloning of cDNAs homologous to a yeast viral dsRNA. In *Double-stranded RNA viruses*, R. W. Compans and D. H. L. Bishop (eds), 451–6. New York: Elsevier.

Bostian, K. A., J. A. Sturgeon and D. J. Tipper 1980a. Encapsidation of yeast killer double-stranded ribonucleic acids: dependence of M on L. *Journal of Bacteriology* **143** (1), 463–70.

Bostian, K. A., J. E. Hopper, D. T. Rogers and D. J. Tipper 1980b. Translational analysis of the killer-associated virus-like particle dsRNA genome of *S. cerevisiae*: M dsRNA encodes toxin. *Cell* **19**, 403–14.

Bostian, K. A., S. Jayachandran and D. J. Tipper 1983a. A glycosylated protoxin in killer yeast: models for its structure and maturation. *Cell* **32**, 169–80.

Bostian, K., V. E. Burn, S. Jayachandran and D. J. Tipper 1983b. Yeast killer dsRNA plasmids are transcribed *in vivo* to produce full and partial-length plus-stranded RNAs. *Nucleic Acids Research* **4** (11), 1077–97.

Bostian, K. A., Q. Elliot, H. Bussey, V. Burn, A. Smith and D. J. Tipper 1984. Sequence of the preprotoxin dsRNA gene of type 1 killer yeast: multiple processing events produce a two-component toxin. *Cell* **36**, 741–51.

Botstein, D. and R. W. Davis 1982. Principles and practice of recombinant DNA research with yeast. In *The molecular biology of the yeast* Saccharomyces. *II. Metabolism and gene expression*, J. N. Strathern, E. W. Jones and J. R. Broach (eds), 607–36. New York: Cold Spring Harbor Laboratory.

Brake, A. J., J. P. Merryweather, D. G. Coit, U. A. Heberlein, F. R. Valenzuela and P. J. Barr 1984. $\alpha$-Factor-directed synthesis and secretion of mature foreign proteins in *Saccharomyces cerevisiae*. *Proceedings of the National Academy of Sciences of the USA* **81** (15), 4642–6.

Brennan, V. E., L. A. Bobek and J. A. Bruenn 1981a. Yeast dsRNA viral transcriptase pulse products: identification of the transcript strand. *Nucleic Acids Research* **10**, 5049–60.

Brennan, V. E., L. Field, P. Cizdziel and J. A. Bruenn 1981b. Sequence of the 3′ ends of yeast dsRNA: proposed transcriptase and replicase initiation sites. *Nucleic Acids Research* **9**, 4007–21.

Brewer, B. J. and W. L. Fangman 1980. Professional inclusion of extrachromosomal genetic elements in yeast meiotic spores. *Proceedings of the National Academy of Sciences of the USA* **77** (9), 5380–4.

Broach, J. R., Y.-Y. Li, L.-C. C. Wu and M. Jayaram 1983. Vectors for high-level, inducible expression of cloned genes in yeast. In *Experimental manipulation of gene expression*, M. Inouye (ed.), 83–117. New York: Academic Press.

Bruenn, J. A. 1980. Virus-like particles of yeast. *Annual Review of Microbiology* **34**, 49–68.

Bruenn, J. A. and V. E. Brennan 1980. Yeast viral double-stranded RNAs have heterogeneous 3′ termini. *Cell* **19**, 923–33.

Bruenn, J. and B. Keitz 1976. The 5′ ends of yeast killer factor RNAs are pppGp. *Nucleic Acids Research* **3**, 2427–37.
Bruenn, J., L. Bobek, V. Brennan and W. Held 1980. Yeast viral double-stranded RNAs have heterogeneous 3′ termini. *Cell* **19**, 923–33.
Bruenn, J., K. Madura, A. Siegel, Z. Miner and M. Lee 1985. Long internal inverted repeat in a yeast viral double-stranded RNA. *Nucleic Acids Research* **13**, 1575–91.
Buck, K. W., P. Lloas and B. K. Street 1973. Virus particles in yeast. *Biochemical Society Transactions* **1**, 1141–2.
Bussey, H. 1972. Effects of yeast killer factor on sensitive cells. *Nature New Biology* **235**, 73–5.
Bussey, H. 1981. Physiology of killer factor in yeast. In *Advances in microbial physiology*, A. H. Rose and G. Morris (eds), 93–122. New York: Academic Press.
Bussey, H. 1984. Yeast killer toxin and an immunity protein are processed from a composite precursor. *Microbiological Science* **1**, 62–6.
Bussey, H. and P. Meaden 1985. Selection and stability of yeast transformants expressing cDNA of an M1 killer toxin-immunity gene. *Current Genetics* **9**, 285–91.
Bussey, H. and D. Sherman 1973. Yeast killer factor: ATP leakage and coordinate inhibition of macromolecular synthesis in sensitive cells. *Biochimica et Biophysica Acta* **298**, 868–75.
Bussey, H. and N. Skipper 1975. Membrane-mediated killing of *Saccharomyces cerevisiae* by glycoproteins from *Torulopsis glabrata. Journal of Bacteriology* **124**, 476–83.
Bussey, H. and N. Skipper 1976. Killing of *Torulopsis glabrata* by *Saccharomyces cerevisiae* killer factor. *Antimicrobial Agents and Chemotherapy* **9** (2), 352–4.
Bussey, H., D. Sherman and J. M. Somers 1973. Action of yeast killer factor: a resistant mutant with sensitive spheroplasts. *Journal of Bacteriology* **113**, 1193–7.
Bussey, H., D. Saville, K. Hutchins and R. G. E. Palfree 1979. Binding of yeast killer toxin to a cell wall receptor on sensitive *Saccharomyces cerevisiae. Journal of Bacteriology* **140**, 888–92.
Bussey, H., W. Sacks, D. Galley and D. Saville 1982. Yeast killer plasmid mutations affecting toxin secretion and activity and toxin immunity function. *Molecular and Cellular Biology* **2**, 346–54.
Bussey, H., O. Steinmetz and D. Saville 1983a. Protein secretion in yeast: two chromosomal mutants that oversecrete killer toxin in *Saccharomyces cerevisiae. Current Genetics* **7**, 449–56.
Bussey, H., D. Saville, D. Greene, D. J. Tipper and K. A. Bostian 1983b. Secretion of yeast killer toxin: processing of the glycosylated precursor. *Molecular and Cellular Biology* **3**, 1362–70.

Carlson, M. and D. Botstein 1982. Two differentially regulated mRNAs with different 5′ ends encode secreted and intracellular forms of yeast invertase. *Cell* **28**, 145–54.
Chan, S. J., C. Patzelt, J. R. Duguid, P. Quinn, A. Labrecque, B. Noyes, P. Keim, R. L. Heinrikson and D. F. Steiner 1979. Precursors in the biosynthesis of insulin and other peptide hormones. In *From gene to protein*, T. R. Russel, H. Bren, H. Faber and J. Schultz (eds), 361–78. New York: Academic Press.
Chow, N. and A. J. Shatkin 1975. Blocked and unblocked 5′ termini in reovirus genome RNA. *Journal of Virology* **15**, 1057–64.
Clare, J. J. and S. G. Oliver 1979. The regulation of RNA synthesis in yeast IV [synthesis of double-stranded RNA]. *Molecular and General Genetics* **171**, 161–6.
Cleveland, D. W., S. G. Fisher, M. W. Kirschner and U. K. Laemmli 1977. Peptide mapping by limited proteolysis in sodium dodecyl sulfate and analysis by gel electrophoresis. *Journal of Biological Chemistry* **252**, 1102–6.
Cohn, M. S., C. W. Tabor, H. Tabor and R. B. Wickner 1978. Spermidine or spermine

requirement for killer double-stranded RNA plasmid replication in yeast. *Journal of Biological Chemistry* **253**, 5225–7.

Davies, J. K. and P. Reeves 1975. Genetics of resistance to colicins in *Escherichia coli* K-12: cross-resistance among colicins of group B. *Journal of Bacteriology* **123**, 96–101.

El-Sherbeini, M. and K. A. Bostian 1985. Infection of yeast protoplasts with the K1 and K2 dsRNA viruses. *Abstracts of Cold Spring Harbor Symposium on the Molecular Biology of Yeast*, 364.

El-Sherbeini, M., E. A. Bevan and D. J. Mitchell 1983. Two biochemically and genetically different forms of L dsRNA of *Saccharomyces cerevisiae* exist: one form, $L_2$, is correlated with the [HOK] plasmid. *Current Genetics* **7**, 63–8.

El-Sherbeini, M., D. J. Tipper, D. J. Mitchell and K. A. Bostian 1984. Virus-like particle capsid proteins encoded by different L double-stranded RNAs of *Saccharomyces cerevisiae*: their roles in maintenance of M double-stranded killer plasmids. *Molecular and Cellular Biology* **4**, 2818–27.

Emr, S. D., R. Scheckman, M. C. Flessel and J. Thorner 1983. An *MF*$\alpha$*1–SUC2* ($\alpha$-factor-invertase) gene fusion for study of protein localisation and gene expression in yeast. *Proceedings of the National Academy of Sciences of the USA* **80**, 7080–4.

Esmon, B., P. Novick and R. Schekman 1981. Compartmentalized assembly of oligosaccharides on exported glycoproteins in yeast. *Cell* **25**, 451–60.

Extremera, A. L., I. Martin and E. Montoya 1982. A new killer toxin produced by *Saccharomyces cerevisiae*. *Current Genetics* **5**, 17–19.

Field, L. J., L. A. Bobek, V. E. Brennan, J. D. Reilly and J. A. Bruenn 1982. There are at least two yeast viral double-stranded RNAs of the same size: an explanation for viral exclusion. *Cell* **31**, 193–200.

Fink, G. R. and C. A. Styles 1972. Curing of a killer factor in *Saccharomyces cerevisiae*. *Proceedings of the National Academy of Sciences of the USA* **69**, 2846–9.

Fitzgerald, M. and T. Shenk 1981. The sequence 5′-AAUAAA-3′ forms part of the recognition site for polyadenylation of late SV40 mRNAs. *Cell* **24**, 251–60.

Fried, H. M. and G. R. Fink 1978. Electron microscopic heteroduplex analysis of 'killer' double-stranded RNA species from yeast. *Proceedings of the National Academy of Sciences of the USA* **75**, 4224–8.

Furuichi, Y., S. Muthukrishnan and A. J. Shatkin 1975a. 5′-Terminal $m^7G(5')$-ppp(5′)Gmp *in vivo*: identification in reovirus genome. *Proceedings of the National Academy of Sciences of the USA* **72**, 742–5.

Furuichi, Y., M. Morgan, S. Muthukrishnan and A. J. Shatkin 1975b. Reovirus messenger RNA contains a methylated, blocked 5′-terminal structure, $M^7G(5')$-ppp(5′)GmpCp. *Proceedings of the National Academy of Sciences of the USA* **72**, 362–6.

Garfinkel, D. J., J. D. Boeke and G. R. Fink 1985. Ty element transposition: reverse transcriptase and virus-like particles. *Cell* **42**, 507–17.

Guerry-Kopecko, P. and R. B. Wickner 1980. Isolation and characterization of temperature-sensitive *mak* mutants of *Saccharomyces cerevisiae*. *Journal of Bacteriology* **144**, 1113–18.

Hannig, E. M. and M. J. Leibowitz 1985. Structure and expression of the M2 genomic segment of a type 2 killer virus of yeast. *Nucleic Acids Research* **13**, 4379–400.

Hannig, E. M., D. J. Thiele and M. J. Leibowitz 1984. *Saccharomyces cerevisiae* killer

virus transcripts contain template-coded polyadenylate tracts. *Molecular and Cellular Biology* **4**, 101–9.
Hannig, E. M., M. Leibowitz and R. B. Wickner 1985. On the mechanism of exclusion of $M_2$ double-stranded RNA in *Saccharomyces cerevisiae*. *Yeast* **1**, 57–65.
Harris, M. S. 1978. Virus-like particles and double-stranded RNA from killer and nonkiller strains of *Saccharomyces cerevisiae*. *Microbios* **21**, 161–76.
Haylock, R. W. and E. A. Bevan 1981. Characterisation of the LdsRNA encoded mRNA of yeast. *Current Genetics* **4**, 181–6.
Henderson, R. C. A. 1983. D.Phil. thesis, Botany School, Oxford.
Herring, A. J. and E. A. Bevan 1974. Virus-like particles associated with the double-stranded RNA species found in killer and sensitive strains of the yeast *Saccharomyces cerevisiae*. *Journal of General Virology* **22**, 387–94.
Herring, A. J. and E. A. Bevan 1977. Yeast virus-like particles possess a capsid-associated single-stranded RNA polymerase. *Nature* (London) **268**, 464–6.
Hinnen, A. and B. Meyhack 1982. Vectors for cloning in yeast. *Current Topics in Microbiology and Immunology* **96**, 101–17.
Hishinuma, F., K. Nakamura, K. Hirai, R. Nishizawa, N. Gunge and T. Maeda 1984. Cloning and nucleotide sequences of the linear DNA killer plasmids from yeast. *Nucleic Acids Research* **12**, 7581–97.
Hopper, J. E., K. A. Bostian, L. B. Rowe and D. J. Tipper 1977. Translation of the L-species dsRNA genome of the killer-associated virus-like particles of *Saccharomyces cerevisiae*. *Journal of Biological Chemistry* **252**, 9010–17.
Huang, A. S. 1977. Viral pathogenesis and molecular biology. *Bacteriological Reviews* **41**, 811–21.
Hunter, T. and J. I. Garrels 1977. Characterisation of the mRNAs for $\alpha$, $\beta$ and $\gamma$ actin. *Cell* **12**, 767–81.
Hutchins, K. and H. Bussey 1983. Cell wall receptor for yeast killer toxin: involvement of $(1 \rightarrow 6)$-$\beta$-D-Glucan. *Journal of Bacteriology* **154**, 161–9.

Imamura, T., M. Kawamoto and Y. Takaoka 1974. Isolation and characterisation of killer-resistant mutants of *Sake* yeast. *Journal of Fermentation Technology* **52** (5), 300–5.

Kagan, B. L. 1983. Mode of action of yeast killer toxins: channel formation in lipid bilayer membranes. *Nature* (London) **302**, 709–11.
Kane, W., D. Pietras and J. Bruenn 1979. The evolution of defective-interfering dsRNAs of the yeast killer virus. *Journal of Virology* **32**, 692–6.
Kaster, K. R., S. G. Burgett and T. D. Ingolia 1984. Hygromycin B resistance as dominant selectable marker in yeast. *Current Genetics* **8**, 353–8.
Kikuchi, Y., K. Hirai and F. Hishinuma 1984. The yeast linear DNA killer plasmids, pGKL1 and pGKL2, possess terminally attached proteins. *Nucleic Acids Research* **12**, 5685–92.
Koltin, Y. 1977. Virus-like particles in *Ustilago maydis*: mutants with partial genomes. *Genetics* **86**, 527–34.
Koltin, Y. and J. S. Kandel 1978. Killer phenomenon in *Ustilago maydis*: the organization of the viral genome. *Genetics* **88**, 267–76.
Koltin, Y., I. Mayer and R. Steinlauf 1978. Killer phenomenon in *Ustilago maydis*: mapping and viral functions. *Molecular and General Genetics* **166**, 181–6.
Kopecky, A. L., C. P. Copeland and J. E. Lusk 1975. Viability of *Escherichia coli* treated with colicin K. *Proceedings of the National Academy of Sciences of the USA* **72**, 4631–4.
Koper-Zwarthoff, F. C. and J. F. Bol 1980. Nucleotide sequence of the putative

recognition site for coat protein in the RNAs of alfalfa mosaic virus and tobacco streak virus. *Nucleic Acids Research* **8**, 3307–18.

Kotani, H., A. Shinmyo and T. Enatsu 1977. Killer toxin for sake yeast: properties and effects of adenosine 5′-diphosphate and calcium ions on killing action. *Journal of Bacteriology* **129**, 640–50.

Kozak, M. 1981. Possible role of the flanking nucleotides in the recognition of the AUG initiator codon by eukaryotic ribosomes. *Nucleic Acids Research* **9**, 5233–52.

Kyte, J. and R. F. Doolittle 1982. A simple method for displaying the hydropathic character of a protein. *Journal of Molecular Biology* **157**, 105–32.

Lhoas, P. 1972. Electron micrographs of cells of *Saccharomyces cerevisiae* infected with double-stranded RNA virus from *Aspergillus niger* and *Penicillium stoloniferum*. *Nature: New Biology* **236**, 86–7.

Lolle, S., N. Skipper, H. Bussey and D. Y. Thomas 1984. The expression of cDNA clones of yeast M1 double-stranded RNA in yeast confers both killer and immunity phenotypes. *EMBO Journal* **3** (6), 1383–7.

Long, M. S., B. L. Brizzard and S. R. De-Kloet 1982. Selective inhibition of yeast killer specific double-stranded RNA transcription *in vitro*. *Biochemical and Biophysical Research Communications* **108** (1), 31–5.

Martin, I., A. L. Extremera and E. Montoya 1981. Interactions between pairs of plasmids related to killer character in yeast. *Microbios Letters* **18**, 81–6.

Maule, A. P. and P. D. Thomas 1973. Strains of yeast lethal to brewery yeasts. *Journal of the Institute of Brewing* **79**, 137–41.

McLaughlin, C. S., J. R. Warner, M. Edmonds, H. Nakazato and M. H. Vaughan 1973. Polyadenylic acid sequences in yeast messenger ribonucleic acid. *Journal of Biological Chemistry* **248**, 1466–71.

Mellor, J., M. H. Malim, K. Gull, M. F. Tuite, S. McCready, T. Dibbayawan, S. M. Kingsman and A. J. Kingsman 1985. Reverse transcriptase activity and Ty RNA are associated with virus-like particles in yeast. *Nature* (London) **318**, 583–6.

Middelbeek, E. J., J. M. H. Hermans and C. Stumm 1979. Production, purification and properties of a *Pichia kluyveri* killer toxin. *Antonie van Leeuwenhoek Journal of Microbiology and Serology* **45**, 437–50.

Middelbeek, E. J., A. Q. H. Crutzen and G. D. Vogels 1980a. Effects of potassium and sodium ions on the killing action of a *Pichia kluyveri* toxin in cells of *Saccharomyces cerevisiae*. *Antimicrobial Agents and Chemotherapy* **18**, 519–24.

Middelbeek, E. J., J. M. H. Hermans, C. Stumm and H. L. Muytjens 1980b. High incidence of sensitivity to yeast killer toxins among *Candida* and *Torulopsis* isolates of human origin. *Antimicrobial Agents and Chemotherapy* **17**, 350–4.

Middelbeek, E. J., J. G. W. H. Peters, C. Stumm and G. D. Vogels 1980c. Properties of a *Cryptococcus laurentiae* killer toxin and conditional killing effect of the toxin on *Cryptococcus albidus*. *FEMS Microbiology Letters* **9**, 81–4.

Middelbeek, E. J., C. Stumm and G. D. Vogels 1980d. Effects of a *Pichia kluyveri* killer toxin on sensitive cells. *Antonie van Leeuwenhoek Journal of Microbiology and Serology* **46**, 205–20.

Middelbeek, E. J., H. H. A. M. van der Laar, J. M. H. Hermans, C. Stumm and G. D. Vogels 1980e. Physiological conditions affecting the stability of *S. cerevisiae* to a *Pichia kluyveri* killer toxin and energy requirement for toxin action. *Antonie van Leeuwenhoek Journal of Microbiology and Serology* **46**, 483–97.

Mitchell, D. J. 1974. The correlation between nucleic acid differences and the killer phenotypes in yeast. Ph.D. thesis, University of London.

Mitchell, D. J. and E. A. Bevan 1983. ScV 'killer' viruses in yeast. In *Yeast genetics (fundamental and applied aspects)*, J. F. T. Spencer, D. M. Spencer and A. R. W. Smith (eds), 371–419. New York: Springer-Verlag.

Mitchell, D. J., A. J. Herring and E. A. Bevan 1976a. The genetic control of dsRNA virus-like particles associated with *Saccharomyces cerevisiae* killer yeast. *Heredity* **37** (1), 129–34.

Mitchell, D. J., A. J. Herring and E. A. Bevan 1976b. Virus uptake in yeasts. In *Microbial and plant protoplasts*, J. F. Peberdy, A. H. Rose, H. J. Rogers and E. C. Cocking (eds), 91–105. London: Academic Press.

Miura, K. I., K. Watanake and M. Sugura 1974. 5′-terminal nucleotide sequences of the double-stranded RNA of silkworm cytoplasmic polyhedrosis virus. *Journal of Molecular Biology* **86**, 31–48.

Montell, C., E. F. Fisher, M. H. Caruthers and A. J. Berk 1983. Inhibition of RNA cleavage but not polyadenylation by a point mutation in mRNA 3′ consensus sequence AAUAAA. *Nature* (London) **305**, 600–5.

Morace, G., C. Archibusacchi, M. Sesito and L. Polonelli 1984. Strain differentiation of pathogenic yeasts by the killer system. *Mycopathologia* **84** (2–3), 81–6.

Naumov, G. I. and T. I. Naumova 1973. Comparative genetics of yeast. XIII. Comparative study of killer strains of *Saccharomyces* from different collections. *Genetika* **9**, 140–5.

Naumov, G. I., L. V. Tyurina, N. I. Bur'Yan and T. I. Naumova 1973. Wine-making, an ecological niche of type $K_2$ killer *Saccharomyces*. *Biologicheskie Nauki* **16**, 103–7.

Nesterova, G. F. 1974. Infection of germinating spores of *Saccharomyces cerevisiae* with cytoplasmic factor k. *Genetika* **10**, 78–82.

Nesterova, G. F., L. V. Semykina and A. A. Filatov 1981. A study of the killer plasmid mutants isolated by treatment with 5-fluorouracil. *Genetika* **17**, 391–8.

Newman, A. M., S. G. Elliott, C. S. McLaughlin, P. A. Sutherland and R. C. Warner 1981. Replication of double-stranded RNA of the virus-like particles in *Saccharomyces cerevisiae*. *Journal of Virology* **38**, 263–71.

Nomoto, H., K. Kitano, T. Shimazaki, K. Kodama and S. Hara 1984. Distribution of killer yeasts in the genus *Hansenula*. *Agricultural and Biological Chemistry* **48** (3), 807–9.

Novick, P. and R. Schekman 1979. Secretion and cell-surface growth are blocked in a temperature-sensitive mutant of *Saccharomyces cerevisiae*. *Proceedings of the National Academy of Sciences of the USA* **76**, 1858–62.

Novick, P., C. Field and R. Schekman 1980. Identification of 23 complementation groups required for post-translational events in the yeast secretory pathway. *Cell* **21**, 205–10.

Novick, P., S. Ferro and R. Schekman 1981. Order of events in the yeast secretory pathway. *Cell* **25**, 461–9.

Oliver, S. G., S. J. McCready, C. Holm, P. A. Sutherland, C. S. McLaughlin and B. S. Cox 1977. Biochemical and physiological studies of the yeast virus-like particle. *Journal of Bacteriology* **130**, 1303–9.

Olsnes, S. and A. Pihl 1973. Isolation and properties of abrin, a toxic protein inhibiting protein synthesis. Evidence for different biological functions of its two constituent-peptide side chains. *European Journal of Biochemistry* **35**, 179–85.

Palfree, R. G. E. and H. Bussey 1979. Yeast killer toxin: purification and characterisation of the protein toxin from *Saccharomyces cerevisiae*. *European Journal of Biochemistry* **93**, 487–93.

Parent, S. A., C. M. Feinmore and K. A. Bostian 1985. Vector systems for the expression, analysis and cloning of DNA sequences in *S. cerevisiae*. *Yeast* **1**, 83–138.

Perrault, J. 1981. Origin and replication of defective interfering particles. *Current Topics in Microbiology and Immunology* **93**, 151–207.

Pfeiffer, P. and F. Radler 1984. Comparison of the killer toxin of several yeasts and the purification of a toxin of type $K_2$. *Archives of Microbiology* **137** (4), 357–61.

Philliskirk, G. and T. W. Young 1975. The occurrence of killer character in yeasts of various genera. *Antonie van Leeuwenhoek Journal of Microbiology and Serology* **41**, 147–51.

Pietras, D. F. and J. A. Bruenn 1976. The molecular biology of yeast killer factor. *International Journal of Biochemistry* **7**, 173–9.

Polonelli, L., C. Archibusacchi, M. Sesito and G. Morace 1983. Killer system: a simple method for differentiating *Candida albicans* strains. *Journal of Clinical Microbiology* **17**, 774–80.

Pugsley, A. P. 1983. Colicin E4-CT9 is proteolytically degraded after discharge from producing cells in liquid cultures. *Journal of General Microbiology* **129**, 833–40.

Radler, F., P. Pfeiffer and M. Dennert 1985. Killer toxins in new isolates of the yeasts *Hanseniaspora uvarum* and *Pichia kluyveri*. *FEMS Microbiology Letters* **29** (3), 269–72.

Reilly, J. D., J. Bruenn and W. Held 1984. The capsid polypeptides of the yeast viruses. *Biochemical and Biophysical Research Communications* **121**, 619–25.

Ridley, S. P. and R. B. Wickner 1983. Defective interference in the killer system of *Saccharomyces cerevisiae*. *Journal of Virology* **45**, 800–12.

Ridley, S. P., S. S. Sommer and R. B. Wickner 1984. Super killer mutations in *Saccharomyces cerevisiae* suppress exclusion of M-2 double-stranded RNA by L-A-HN and confer cold sensitivity in the presence of M and L-A-HN. *Molecular and Cellular Biology* **4**, 761–70.

Rogers, D. and E. A. Bevan 1978. Group classification of killer yeasts based on cross-reactions between strains of different species and origin. *Journal of General Microbiology* **105**, 199–202.

Rosini, G. 1985. Interaction between killer strains of *Hansenula anomala* var. *anomala* and *Saccharomyces cerevisiae* yeast species. *Canadian Journal of Microbiology* **31**, 300–2.

Saksena, K. N. and P. A. Lemke 1978. Viruses in fungi. In *Comprehensive virology*, H. Fraenkel-Conrat and R. R. Wagner (eds), Vol. 12, 102–43. New York: Plenum.

Schekman, R. and P. Novick 1982. The secretory process and yeast cell-surface assembly. In *The molecular biology of the yeast* Saccharomyces. *II. Metabolism and gene expression*, J. N. Strathern, E. W. Jones and J. R. Broach (eds), 361–93. New York: Cold Spring Harbor Laboratory.

Schlein, S. J., B. L. Kagan and A. Finkelstein 1978. Colicin K acts by forming voltage-dependent channels in phospholipid bilayer membranes. *Nature* (London) **276**, 159–63.

Sclafani, R. A. and W. L. Fangman 1984. Conservative replication of double-stranded RNA in *Saccharomyces cerevisiae* by displacement of progeny single strands. *Molecular and Cellular Biology* **4**, 1618–26.

Silverstein, S. C., J. K. Christman and G. Acs 1976. The reovirus replicative cycle. *Annual Review of Biochemistry* **45**, 375–408.

Singh, A. and T. R. Manney 1974. Genetic analysis of mutations affecting growth of *Saccharomyces cerevisiae* at low temperature. *Genetics* **77**, 651–9.

Singh, A., J. M. Logovoy, W. J. Kohr and L. J. Perry 1984. Synthesis, secretion and processing of $\alpha$-factor-interferon fusion proteins in yeast. *Nucleic Acids Research* **12**, 8927–38.

Skipper, N. 1983. Synthesis of double-stranded cDNA transcript of the killer toxin-coding region of the yeast M1 double-stranded RNA. *Biochemical and Biophysical Research Communications* **114**, 518–25.

Skipper, N. and H. Bussey 1977. Mode of action of yeast toxins: energy requirement for *Saccharomyces cerevisiae* killer toxin. *Journal of Bacteriology* **129**, 668–77.

Skipper, N., D. Y. Thomas and P. C. K. Lau 1984. Cloning and sequencing of the preprotoxin-coding region of the yeast M1 double-stranded RNA. *EMBO Journal* **3** (1), 107–11.

Somers, J. M. 1973. Isolation of suppressive sensitive mutants from killer and neutral strains of *Saccharomyces cerevisiae*. *Genetics* **74**, 571–9.

Somers, J. M. and E. A. Bevan 1969. The inheritance of the killer character in yeast. *Genetic Research Cambridge* **13**, 17–83.

Sommer, S. S. and R. B. Wickner 1982a. Yeast L dsRNA consists of at least three distinct RNAs; evidence that the non-Mendelian genes [HOK], [NEX] and [EXL] are on one of these dsRNAs. *Cell* **31**, 429–41.

Sommer, S. S. and R. B. Wickner 1982b. Co-curing of plasmids affecting killer double-stranded RNAs of *Saccharomyces cerevisiae*: [HOK], [NEX], and the abundance of L are related and further evidence that $M_1$ requires L. *Journal of Bacteriology* **150**, 545–51.

Sommer, S. S. and R. B. Wickner 1984. Double-stranded RNAs that encode killer toxins in *Saccharomyces cerevisiae*: unstable size of M double-stranded RNA and inhibition of $M_2$ replication by $M_1$. *Molecular and Cellular Biology* **4**, 1747–53.

Sor, F. and H. Fukuhara 1985. Structure of a linear plasmid of the yeast *Kluyveromyces lactis*; compact organization of the killer genome. *Current Genetics* **9**, 147–55.

Sripati, C. E., Y. Groner and J. R. Warner 1976. Methylated, blocked 5′ termini of yeast mRNA. *Journal of Biological Chemistry* **251**, 2898–904.

Sriprakash, K. S. and C. Batum 1984. Possible chromosomal location for the killer determinant in *Torulopsis glabrata*. *Current Genetics* **8**, 115–19.

Stark, M. J. R., A. J. Mileham, M. A. Romanos and A. Boyd 1984. Nucleotide sequence and transcription analysis of a linear DNA plasmid associated with the killer character of the yeast *Kluyveromyces lactis*. *Nucleic Acids Research* **12**, 6011–30.

Stewart, G. G., T. E. Goring and I. Russell 1977. Can a genetically manipulated yeast strain produce palatable beer? *Journal of the American Society of Brewing Chemistry* **35**, 168–78.

Stoltzfus, C. M., A. J. Shatkin and A. K. Banerjee 1973. Absence of polyadenylic acid from reovirus messenger ribonucleic acid. *Journal of Biological Chemistry* **248**, 7993–8.

Struhl, K. 1983. The new yeast genetics. *Nature* (London) **305**, 391–7.

Stumm, C., J. M. H. Hermans, E. J. Middelbeek, A. F. Croes and G. J. M. L. de Vries 1977. Killer-sensitive relationships in yeasts from natural habitats. *Antonie van Leeuwenhoek Journal of Microbiology and Serology* **43**, 125–8.

Sugisaki, Y., N. Gunge, K. Sakaguchi, M. Yamasaki and G. Tamura 1984. Characterisation of a novel killer toxin encoded by a double-stranded linear DNA plasmid of *Kluyveromyces lactis*. *European Journal of Biochemistry* **141**, 241–5.

Sweeney, T. K., A. Tate and G. R. Fink 1976. A study of the transmission and structure of double-stranded RNAs associated with the killer phenomenon in *Saccharomyces cerevisiae*. *Genetics* **84**, 28–42.

Taniguchi, T., M. Palmieri and C. Weissmann 1978. Q$\beta$ DNA-containing hybrid plasmids giving rise to Q$\beta$ phage formation in the bacterial host. *Nature* (London) **274**, 223–8.

Thiele, D. J. and M. J. Leibowitz 1982. Structural and functional analysis of separated strands of killer double-stranded RNA of yeast. *Nucleic Acids Research* **10**, 6903–18.

Thiele, D. J., R. W. Wang and M. J. Leibowitz 1982. Separation and sequence of the 3′ termini of M double-stranded RNA from killer yeast. *Nucleic Acids Research* **10**, 1661–78.

Thiele, D. J., E. M. Hannig and M. J. Leibowitz 1984a. Multiple L double-stranded RNA species of *Saccharomyces cerevisiae*: evidence for separate encapsidation. *Molecular and Cellular Biology* **4**, 92–100.

Thiele, D. J., E. M. Hannig and M. J. Leibowitz 1984b. Genome structure and expression of a defective interfering mutant of the killer virus of yeast. *Virology* **137**, 20–31.

Thrash, C., K. Voelkel, S. di Nardo and R. Sternglanz 1984. Identification of *Saccharomyces cerevisiae* mutants deficient in DNA topoisomerase 1 activity. *Journal of Biological Chemistry* **259**, 1375–7.

Tipper, D. J. and K. A. Bostian 1984. Double-stranded ribonucleic acid killer systems in yeasts. *Microbiological Reviews* **48** (2), 125–56.

Toh-e, A. and Y. Sahashi 1985. The *PET*18 locus of *Saccharomyces cerevisiae*: a complex locus containing multiple genes. *Yeast* **1**, 159–71.

Toh-e, A. and R. B. Wickner 1979. A mutant killer plasmid whose replication depends on a chromosomal 'superkiller' mutation. *Genetics* **91**, 673–82.

Toh-e, A. and R. B. Wickner 1980. 'Superkiller' mutations suppress chromosomal mutations affecting double-stranded RNA killer plasmid replication in *Saccharomyces cerevisiae*. *Proceedings of the National Academy of Sciences of the USA* **77**, 527–30.

Toh-e, A., P. Guerry and R. B. Wickner 1978. Chromosomal superkiller mutants of *Saccharomyces cerevisiae*. *Journal of Bacteriology* **136**, 1002–7.

Tyagi, A. K., R. B. Wickner, C. W. Tabor and H. Tabor 1984. Specificity of polyamine requirements for the replication and maintenance of different double-stranded RNA plasmids in *Saccharomyces cerevisiae*. *Proceedings of the National Academy of Sciences of the USA* **81**, 1149–53.

Tzen, J. C., J. M. Sauers and D. J. Mitchell 1974. A dsRNA analysis of suppressive sensitive mutants of 'killer' *Saccharomyces cerevisiae*. *Heredity* **33**, 132 (abstr.).

Vodkin, M. H. and G. R. Fink 1973. A nucleic acid associated with a killer strain of yeast. *Proceedings of the National Academy of Sciences of the USA* **70**, 1069–72.

Vodkin, M., F. Katterman and G. R. Fink 1974. Yeast killer mutants with altered double-stranded ribonucleic acid. *Journal of Bacteriology* **117**, 681–6.

Webster, T. D. and R. C. Dickson 1983. Direct selection of *Saccharomyces cerevisiae* resistant to the antibiotic G418 following transformation with a DNA vector carrying the kanamycin resistance gene of Tn903. *Gene* **26**, 243–52.

Welsh, J. D. and M. J. Leibowitz 1980. Transcription of killer virion double-stranded RNA. *Nucleic Acids Research* **8**, 2365–75.

Welsh, J. D., M. J. Leibowitz and R. B. Wickner 1980. Virion DNA-dependent RNA polymerase from *Saccharomyces cerevisiae*. *Nucleic Acids Research* **8**, 2349–63.

Wesolowski, M. and R. B. Wickner 1984. Two new double-stranded RNA molecules showing non-Mendelian inheritance and heat inducibility in *Saccharomyces cerevisiae*. *Molecular and Cellular Biology* **4**, 181–7.

Wickner, R. B. 1974a. 'Killer character' of *Saccharomyces cerevisiae*: curing by growth at elevated temperature. *Journal of Bacteriology* **117**, 1356–7.

Wickner, R. B. 1974b. Chromosomal and nonchromosomal mutations affecting the 'killer character' of *Saccharomyces cerevisiae*. *Genetics* **76**, 423–32.

Wickner, R. B. 1976. Mutants of the killer plasmid of *Saccharomyces cerevisiae* dependent on chromosomal diploidy for expression and maintenance. *Genetics* **82**, 273–85.

Wickner, R. B. 1977. Deletion of mitochondrial DNA bypassing a chromosomal gene needed for maintenance of the killer plasmid of yeast. *Genetics* **87**, 441–52.

Wickner, R. B. 1978. Twenty-six chromosomal genes needed to maintain the killer double-stranded RNA plasmid of *Saccharomyces cerevisiae*. *Genetics* **88**, 419–25.

Wickner, R. B. 1979. The killer double-stranded RNA plasmids of yeast. *Plasmid* **2**, 303–22.

Wickner, R. B. 1980. Plasmids controlling exclusion of the $K_2$ killer double-stranded RNA plasmid of yeast. *Cell* **21**, 217–26.

Wickner, R. B. 1981. Killer systems in *Saccharomyces cerevisiae*. In *Molecular biology of the yeast* Saccharomyces cerevisiae: *life cycle and inheritance*, J. N. Strathern, E. W. Jones and J. R. Broach (eds), 415–44. New York: Cold Spring Harbor Laboratory.

Wickner, R. B. 1983a. Genetic control of replication of the double-stranded RNA segments of the killer systems in *Saccharomyces cerevisiae*. *Archives of Biochemistry and Biophysics* **222** (1), 1–11.

Wickner, R. B. 1983b. Killer systems in *Saccharomyces cerevisiae*: 3 distinct modes of exclusion of M2 double-stranded RNA by 3 species of double-stranded RNA, M1, L-A-E and L-A-HN. *Molecular and Cellular Biology* **3**, 654–61.

Wickner, R. B. and M. J. Leibowitz 1976a. Chromosomal genes essential for replication of a double-stranded RNA plasmid of *Saccharomyces cerevisiae*: the killer character of yeast. *Journal of Molecular Biology* **105**, 427–43.

Wickner, R. B. and M. J. Leibowitz 1976b. Two chromosomal genes required for killing expression in killer strains of *Saccharomyces cerevisiae*. *Genetics* **82**, 429–42.

Wickner, R. B. and M. J. Leibowitz 1977. Dominant chromosomal mutation by-passing chromosomal genes needed for killer RNA plasmid replication in yeast. *Genetics* **87**, 453–69.

Wickner, R. B. and M. J. Leibowitz 1979. *mak* mutants of yeast: mapping and characterisation. *Journal of Bacteriology* **140**, 154–60.

Wickner, R. B. and A. Toh-e 1982. [HOK], a new yeast non-Mendelian trait, enables a replication-defective killer plasmid to be maintained. *Genetics* **100**, 159–74.

Wickner, R. B., F. Boutelet and F. Hilger 1983. Evidence for a new chromosome in *Saccharomyces cerevisiae*. *Molecular and Cellular Biology* **3**, 415–20.

Wickner, R. B., S. P. Ridley, H. M. Fried and S. G. Ball 1982. Ribosomal protein L3 is involved in replication or maintenance of the killer double-stranded RNA genome of *Saccharomyces cerevisiae*. *Proceedings of the National Academy of Sciences of the USA* **79**, 4706–8.

Widgerson, M. and Y. Koltin 1982. Dual toxin specifications and the exclusion relations among the *Ustilago* dsRNA viruses. *Current Genetics* **5**, 127–36.

Woods, D. R. and E. A. Bevan 1968. Studies on the nature of the killer factor produced by *Saccharomyces cerevisiae*. *Journal of General Microbiology* **51**, 115–26.

Young, T. W. and M. Yagiu 1978. A comparison of the killer character in different yeasts and its classification. *Antonie van Leeuwenhoek Journal of Microbiology and Serology* **44**, 59–77.

Zakian, V. A., D. W. Wagner and W. L. Fangman 1981. Yeast L-dsRNA is synthesised during the G1 phase but not the S phase of the cell cycle. *Molecular and Cellular Biology* **1**, 673–9.

PART III

# Growth of yeast

# 6 *Physiology of yeast growth*

D. R. BERRY and C. BROWN

## 6.1 Introduction

Studies on the physiology and chemistry of yeast growth are as old as the disciplines themselves. The demonstration of the role of yeast in alcoholic fermentations by Pasteur in 1876, the conversion of glucose to ethanol by cell-free extracts of yeast by Buchner in 1897 and the studies on the regulation of respiration by Warburg in 1926, are indicative of the central role of yeast studies in the development of biochemistry (Rose 1977). Many of these studies were carried out because of the importance of yeast in the production of alcoholic beverages. However, as early as 1860 the need for aeration to obtain high yields of yeast for baking was recognized. This led to the development of the Vienna process and a specialized baker's yeast industry. Since then yeast has been grown extensively both anaerobically, to produce ethanol, and aerobically, to produce baker's yeast and yeast biomass for use as food and feed products. It is probably this ability of *Saccharomyces cerevisiae* to grow equally vigorously in both anaerobic and aerobic conditions which gives it such a unique physiology which, in spite of extensive studies over the past century, are still far from understood.

The growth of any microbial organism can be considered as a continuous process of interaction between the characteristics of the organism and the environmental conditions in which it is being grown. Because of the unique genotype of different species of micro-organisms, they react to environmental stimuli in different ways. In this chapter the reaction of the yeast *Saccharomyces cerevisiae* to growth in different media is considered, and an attempt is made to examine how these reactions are controlled by the biochemical properties of the cell and how they influence the design of different fermentation processes. The concentration on *Saccharomyces cerevisiae* is, perhaps, inevitable since it is commercially the most important yeast and has also been most extensively studied. However, other industrially important yeasts will also be referred to where they illustrate different physiological characteristics.

Although the growth requirements of yeast are not complex – they can readily be grown in defined media – the mode of yeast growth is extremely sensitive to changes in the growth conditions. In particular, the nature and concentration of the carbon substrate and the availability of oxygen are critical in determining the yield of cells and the nature of the products of the fermentation. The importance of oxygen and glucose in the regulation of the growth characteristics in yeast have been recognized for many years. The reduction in the rate of glycolysis when yeast is grown in the presence of oxygen was reported by Pasteur, and has been referred to as the Pasteur effect, and the repression of mitochondrial function when yeast is grown in high levels of glucose has been referred to as the Crabtree effect, the glucose effect or catabolite repression (Fiechter *et al.* 1981). However, the effects of both oxygen and glucose are both varied and complex, and dependent upon the physiological state of the cell, so it is probably more helpful to consider the effects of different environmental parameters and their interaction, rather than attempting to fit these responses into defined effects. It is clear from the work of Wales *et al.* (1980) on the control of mitochondrial development that at least four different states can be recognized with respect to the oxygen and carbohydrate status of the growth conditions; namely aerobic catabolite-repressed and derepressed states, and anaerobic catabolite-repressed and derepressed states. These different states exhibit different levels of mitochondrial enzymes, but also have different nutritional requirements and different growth characteristics.

## 6.2 Yeast nutrition

In common with all micro-organisms, in order to grow yeasts must obtain from their environment all of the substances that they require for the synthesis of their cellular materials, and the provision of energy to drive these biosynthetic processes. The culture medium used must therefore contain all the essential elements for growth, in proportions similar to those occurring in the yeast biomass (Table 6.1).

This section contains a brief synopsis of the main nutritional requirements of yeast, in particular *Saccharomyces cerevisiae*, and the interactions between these nutrients. Several reviews dealing with yeast nutrition have been published by Suomalainen and Oura (1971), Jones *et al.* (1981) and recently by Jones and Greenfield (1984). Additional information to that given below and literature pertaining to many of the specific examples used may be found in these reviews.

### *6.2.1 Carbon*

Organic compounds such as glucose have a dual nutritional role: as a source of carbon and as a source of energy. In combination primarily with hydrogen, oxygen and nitrogen, they make up the major structural units of yeast; approximately 50% of yeast (dry weight) is carbon. The most widely utilized carbon sources are the sugars, though the particular sugars which may be assimilated are variable within species and strain. Yeasts normally metabolize the same sugars aerobically as they ferment. Table 6.2 shows the carbon sources capable of being utilized by *Saccharomyces cerevisiae*. Hydrogen and oxygen are also provided by carbohydrates. Industrially impure sources of carbohydrates are often employed; these contain more than one form of carbon source. For example, beet molasses contains glucose, fructose, sucrose and raffinose. The uptake of carbon sources is regulated by catabolite repression (Section 6.5.2).

**Table 6.1** Average elemental composition of *Saccharomyces* species (adapted from Jones & Greenfield 1984).

| Element | Average value [g (100 g)$^{-1}$ dry weight] |
|---|---|
| potassium | 2.2 |
| phosphorus | 1.6 |
| sulphur | $300 \times 10^{-3}$ |
| magnesium | $270 \times 10^{-3}$ |
| sodium | $60 \times 10^{-3}$ |
| calcium | $50 \times 10^{-3}$ |
| barium | $15 \times 10^{-3}$ |
| zinc | $12 \times 10^{-3}$ |
| iron | $10 \times 10^{-3}$ |
| copper | $5 \times 10^{-3}$ |
| manganese | $3 \times 10^{-3}$ |
| cobalt | $0.5 \times 10^{-3}$ |
| nickel | $250 \times 10^{-6}$ |
| arsenic | $180 \times 10^{-6}$ |
| lead | $150 \times 10^{-6}$ |
| iodine | $125 \times 10^{-6}$ |
| molybdate | $7 \times 10^{-6}$ |
| boron | $5 \times 10^{-6}$ |
| aluminium | $1 \times 10^{-6}$ |
| chromium | $10 \times 10^{-26}$ |
| vanadium | $5 \times 10^{-26}$ |

*Notes*

(a) Particularly in the case of the lesser trace elements, few data are available, hence the values given above serve only as an approximation.

(b) Elemental composition from yeast growth under fermentative and respiratory conditions were included in the data.

**Table 6.2** Carbon sources utilized by *Saccharomyces cerevisiae* (adapted Barnett *et al*. 1983).

| | |
|---|---|
| D-glucose | sucrose |
| D-galactose | raffinose |
| mannose | trehalose |
| fructose | deoxyribose |
| melibiose | maltotriose |
| melezitose | starch(?) |
| glycerol | D-glucitol |
| D-mannitol | DL-lactate |
| ethanol | succinate |
| D-glucono-1-5-lactone | Me-α-D-glucoside |

### *6.2.2 Nitrogen*

Nitrogen is the next most important constituent of a growth medium after carbon, in terms of quantitative requirement. In the reduced form, nitrogen is used in organic combinations in amino acids, nucleotides and certain vitamins. A wide range of substances are able to serve as nitrogen sources.

(a) *Inorganic nitrogen sources.* Ammonium nitrogen in the form of $NH_4NO_3$, $NH_4Cl$ and $(NH_4)_2SO_4$. The last of these is the favoured source, as it provides the dual nutritional benefit as a source of sulphur. It is commonly employed as a nitrogen source in industrial fermentations.

(b) *Amino acids.* As shown in Table 6.3, most amino acids can serve as a sole source of nitrogen, although the maximum growth rate that they can promote varies with the particular amino acid. Basic amino acids, such as asparagine, on transamination provide a carbon skeleton which can be used in biosynthesis, while hydrophobic amino acids, such as valine, have their carbon skeletons excreted after transamination.

(c) *Mixtures of amino acids.* Casein hydrolysate, which provides a mixture of amino acids, generally allows a higher maximum specific growth rate and cell yield than any single amino acid. This is also true of wort, where amino nitrogen constitutes the largest fraction of available nitrogen, although $NH_4^+$ has been shown to inhibit amino-acid uptake (Cooper 1982a).

(d) *Peptides.* A limited number of dipeptides and tripeptides can be utilized by yeast. Polypeptides have also been reported to be taken up by yeast from wort. Peptide uptake is limited by molecular size; they are cleaved by intracellular peptidases (Becker *et al.* 1971).

**Table 6.3** Relative maximum specific growth rates of yeast on nitrogen sources. The exact order is strain-dependent (adapted from Cooper, 1982a).

| | |
|---|---|
| (highest $\mu_{max}$) | |
| glutamine | valine |
| ammonia | leucine |
| asparagine | proline |
| glutamate | tyrosine |
| serine | isoleucine |
| arginine | threonine |
| urea | tryptophan |
| alanine | allantoate |
| phenylalanine | methionine |
| ornithine | histidine |
| citrulline | glycine |
| allantoin | |
| | (lowest $\mu_{max}$) |

NB: Lysine and cysteine not determined, but can also serve as a source of nitrogen.

(e) *Urea*. Urea provides a good nitrogen source, although addition of a biotin supplement is required. Due to nitrogen suppression (see below) urea utilization via suppression of the urea transport system is decreased when a better nitrogen source is available.
(f) *Purines and pyrimidines*. Purines and pyrimidines are able to provide sole sources of nitrogen, they are actively transported into the cell. Allantoin, a product of adenine and guanine catabolism in many organisms, can also be utilized.

If one or more of the above nitrogen sources are present together, as often occurs with industrial media, then they are not all utilized at the same rate. Good nitrogen sources will be chosen over poor ones (see Table 6.3). Cooper proposed that to be a good nitrogen source the compound must:

(a) be rapidly taken up into the cell;
(b) undergo conversion once in the cell with the minimum intervening steps to form glutamate or ammonia, or both; and
(c) have no toxic side effects on the cell.

Yeast selectively degrades the best nitrogen sources first; this is thought to occur by means of repression and metabolite compartmentation (Cooper & Sumrada 1983). Hence, the presence of a good nitrogen source (e.g. glutamate) in the medium results in the exclusion of a poorer one (e.g. proline) from the cells.

### 6.2.3 *Phosphorus*

Phosphorus is assimilated only in the form of the dihydrogen phosphate (orthophosphate) ion, $H_2PO_4^-$. The major role of phosphorus is as a component of sugar phosphates, nucleic acid and nucleoside di- or triphosphates. It is also found as condensed inorganic phosphates in the form of linear polymers, the polyphosphates, important in the regulation of cellular metabolism and in providing a reservoir of phosphates and energy (Kulaev & Vagabov 1983). The intracellular concentration of $PO_4^{2-}$ regulates the synthesis of lipids and carbohydrates.

Both $KH_2PO_4$ and $Na_2HPO_4$ can serve as sources of phosphorus. The $K^+$ ion has been shown to stimulate phosphate uptake, since it is co-transported with $PO_4^{2-}$ into the cell (Schmidt *et al.* 1949). Hence $KH_2PO_4$ serves as a better $PO_4$ source than $Na_2HPO_4$, particularly in the absence of another source of $K^+$. An additional feature of phosphorus metabolism in yeast is that under phosphorus limitation the cells secrete a non-specific phosphomonoesterase enzyme, acid phosphatase, to the cell surface, allowing the cells to obtain phosphorus from phosphate esters (Schurr & Yagil 1971).

### 6.2.4 *Sulphur*

Sixty per cent of cellular sulphur is incorporated into S-containing amino acids in proteins. Of the remainder, 5% is osmotically free inorganic sulphate, the rest is in the form of bound sulphate or free amino acids (Maw 1963). Sulphur is also found in certain vitamins (e.g. biotin) and other sulphydryl compounds.

Inorganic sulphur in the form of sulphate ion, $SO_4^{2-}$ is the most commonly used sulphur source in media in the form of $K_2SO_4$ or, more frequently, $(NH_4)_2SO_4$. The active uptake of sulphate is correlated with the availability of homocysteine synthase which brings about its incorporation to form cysteine (Breton & Surdin-Kerjan 1977). *Saccharomyces* species are also able to utilize sulphite and thiosulphate (Maw 1960). Only methionine amongst the sulphur-containing amino acids can be used as a sole source of sulphur; in fact, it allows greater and more rapid growth than $SO_4^{2-}$ (Hough 1971).

### 6.2.5 *Potassium*

The large quantity of potassium found in the analysis of the elemental composition of yeast (Table 6.1) reflects a central physiological role for this ion. The important functions of potassium are the following.

(a) As an internal divalent cation regulator, for each divalent metal ion taken up $2K^+$ are excreted (Fuhrmann & Rothstein 1968).
(b) At low pH it stimulates fermentation and respiration. This has been shown to be due to elevated levels of NADPH, ADP and inorganic phosphate (Pêna 1981).
(c) It acts as an effector on a variety of enzymes such as pyruvate kinase, aldolase, aldehyde dehydrogenase and membrane ATPase (see Maiorella *et al*. 1984).
(d) As a structural component of RNA.

Sources of potassium include KCl, $K_2HPO_4$ and $KH_2PO_4$. Only half of the quantity of potassium is taken up by respiring yeast compared with fermenting yeast (see Jones & Greenfield 1984).

### 6.2.6 *Magnesium*

Magnesium is involved with many functions important to yeast metabolism:

(a) It is required for the proper functioning of over 100 enzymes as a cofactor or effector. As such it participates in the following important areas of metabolism:
   (i) activation of glycolytic enzymes;
   (ii) stimulation of essential fatty acid synthesis;
   (iii) regulation of cellular ionic levels;
   (iv) activation of membrane ATPase; and
   (v) along with $K^+$ involved with phosphate uptake.
(b) It has a structural role and is bound to ribosomes, cell membranes and nucleic acids. $MgCl_2$ and $MgSO_4$ are two of the magnesium sources commonly used. The growth-enhancing effect of magnesium is dependent on the presence of $Zn^{2+}$, which must also be added. Magnesium deficiencies result in the synthesis of acetate rather than ethanol.

### 6.2.7 *Calcium*

Calcium is not an essential growth requirement, but has been found to have stimulatory effects on *Saccharomyces* species. It is taken up only during the growth phase, and is incorporated into newly synthesized cell wall and plasma membranes. It is in the latter that it exerts its stimulatory effects by buffering the cell membrane against an adverse environment. Calcium is also involved as an effector for calmodulin and pyruvate kinase, as well as being involved in the budding and flocculation processes (see Maiorella *et al.* 1984). It can be provided in synthetic media in the form of $CaCl_2 \cdot 6H_2O$.

### 6.2.8 *Chlorine*

Chlorine has been reported as being either essential or stimulatory to growth. The metabolic functions that it is involved with are not well understood, but it is thought to act at antiports moving counter to the positive-ion direction.

### 6.2.9 *Sodium*

Sodium has no known metabolic role in yeast growth, although it may be required in trace amounts. The intracellular level is normally maintained at 10% of the extracellular level, resulting in an increased maintenance requirement under hypertonic conditions (this results in growth inhibition) (Maiorella *et al.* 1984).

### 6.2.10 *Barium*

Barium is thought to be involved in cytosolic metabolism. It stimulates growth at low concentrations.

### 6.2.11 *Zinc*

Zinc has a number of roles in yeast metabolism:

(a) It is an essential cofactor for a variety of enzymes, e.g. aldolase, and is essential for glycolysis.
(b) It is required for the synthesis of some vitamins, e.g. riboflavin.
(c) It acts to stimulate the uptake of the maltose and maltotriose sugars.

### 6.2.12 *Iron*

Iron as $Fe^{2+}$ or $Fe^{3+}$, if added, does not increase cell yield, but is required in trace amounts as it is found in the active site of many yeast proteins (the haem-enzymes).

### 6.2.13 *Manganese*

Manganese ($Mn^{2+}$) is found to have a number of effects on yeast metabolism. Addition of manganese results in:

(a) a higher cell-nitrogen content;
(b) increased protein synthesis; and
(c) an increase in cell yield.

### *6.2.14 Other ions*

Copper, cobalt, nickel, arsenic, lead, iodine, molybdate, boron, aluminium, chromium and vanadium are found in the elemental composition of yeast. However, the only reported effect of arsenic, lead, aluminium, chromate and vanadium are to inhibit growth and fermentation. Some ions, such as copper, may be added in synthetic media but only in trace amounts, however, more commonly they may be assumed to be present as impurities, even in AnalaR reagents.

### *6.2.15 Growth factors*

In addition to the nutrient element requirements detailed above, growth factors are also required either as an absolute requirement or in promotion of growth and optimal cell yield. Growth factor requirement varies amongst species and strains, as does the actual concentrations required. Table 6.4 shows the vitamin requirements common to most strains of *Saccharomyces cerevisiae*.

**Table 6.4** Growth factor requirements of *Saccharomyces cerevisiae* (strain variable, see below). (Values given are those of Oura 1974.)

| | Requirement ($mg\,l^{-1}$) |
|---|---|
| inositol | 125 |
| Ca-pantothenate | 6.25 |
| pyridoxine HCl | 6.25 |
| thiamine HCl | 5 |
| nicotinic acid | 5 |
| biotin | 1 |

**Inositol**

Inositol plays an important role in the lipid composition of the cell, which is known to be a critical factor in maintaining the physiological state of the cell membrane (Henry 1983). On being taken up inositol is converted into phosphotidylinositol, which serves as the precursor to the synthesis of inositol-containing lipids, such as triphosphoinositides (Steiner & Lester 1972). Only the *meso* form is taken up and utilized (Nikawa *et al*. 1982). Inositol plays a key role in regulating the rate of fermentation. Yeasts do not normally have an absolute requirement for inositol, since biosynthesis of inositol occurs in the cyotoplasm of yeast (Culbertson *et al*. 1976), so the cells are capable of growing in inositol-free media.

**Pantothenate**

Most *Saccharomyces* species have an essential requirement for pantothenate (Suomalainen & Oura 1971). In synthetic media it is normally provided as calcium pantothenate, and in the yeast cell it is integrated into coenzyme A. Coenzyme A is involved in *keto* acid oxidation reactions and other reactions such as fatty-acid metabolism which involve the transfer of an acyl group (Stryer 1981).

**Pyridoxine (vitamin $B_6$)**

On being taken up it is converted to pyridoxal phosphate and pyridoxamine phosphate, which serve as coenzymes involved in deamination and decarboxylation of amino acids (Stryer 1981). It is also known to play an important role in the metabolism of polyamines, carbohydrates and nucleic acids.

**Thiamin (vitamin $B_1$)**

The effect of thiamine deficiency or imbalance on yeast physiology has been extensively studied by Kamihara and Nakamura (1982). It has been shown to influence several areas of yeast metabolism including:

(a) respiratory competence;
(b) fatty acid, sterol and lipid formation; and
(c) glycolysis and ethanol production.

**Nicotinic acid**

Synthesis of nicotinic acid by the cell is inhibited if nicotinic acid is added to the medium, thus releasing the cell of a biosynthetic load and stimulating growth (Suomalainen & Oura 1971). Nicotinic acid is required for $NAD^+$ and $NADP^+$, which are essential for ATP generation and reductive biosynthesis, respectively (Stryer 1981).

**Biotin**

The low quantity of biotin added to synthetic media and as a supplement in complex media does not reflect the major importancc of this vitamin, which is essential to yeast growth. Under low- ($<25$ ng ml$^{-1}$) biotin conditions an active transport system operates to take up the biotin (Rogers & Lichstein 1969). It is involved in several areas of yeast metabolism including:

(a) carboxylation of pyruvate;
(b) synthesis of pyridine nucleotides;
(c) nucleic-acid synthesis;
(d) formation of purine and pyrimidine bases;
(e) protein synthesis;
(f) polysaccharide synthesis; and
(g) fatty-acid synthesis.

(From Oura and Suomalainen 1982.)

Biotin, unlike other yeast growth factors, is deficient in complex industrial media, such as malt wort, and therefore it is commonly added to promote growth.

**Other vitamin supplements**

Folic acid, *para*-aminobenzoic acid and riboflavin are also occasionally added to media, though normally all can be synthesized by the yeast. Only a few strains of *Saccharomyces* spp. show an absolute requirement for these growth factors.

**Anaerobic growth factor requirements**

The synthesis of unsaturated fatty acids and ergosterol in yeast occurs only in the presence of oxygen. Therefore, strictly anaerobically grown cells require both unsaturated fatty acid and sterol (Andreasen & Stier 1954).

**Interactions**

As has been shown above, the function and source of each nutrient element and growth factor has been, in most cases, well defined for yeast growth. However, the nutritional content of the complex media commonly used in industry is not so well defined. There are three main difficulties which arise:

(a) interactions occurring between nutrients;
(b) inhibitory effects of medium components; and
(c) the presence of complexing agents.

A variety of complex interactions exist between different elements, resulting in variations in the optimal concentration for growth for one particular nutrient. This is illustrated by the effect of interactions which occur between concentrations of $(NH_4)_2SO_4$ and $KH_2PO_4$ on the maximum specific growth rate of yeast which has been investigated by Nagamune *et al.* (1981) (see Fig. 6.1).

As the concentration of glucose normally determines the final biomass concentration, the requirement for other nutrients and vitamins is related to

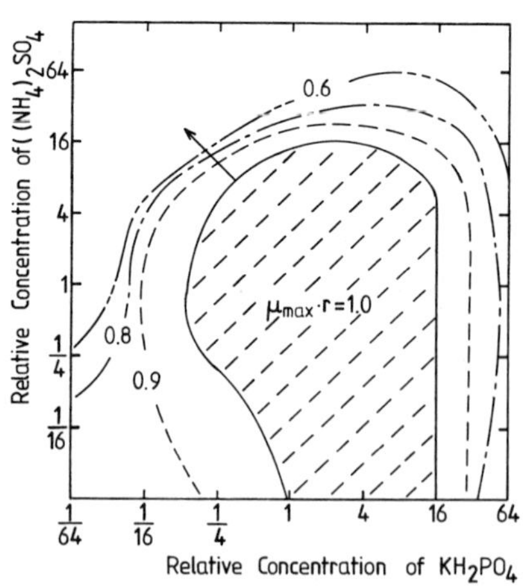


**Figure 6.1** Effect of interaction between $(NH_4)_2SO_4$ and $KH_2PO_4$ concentrations on the maximum growth rate of yeast cells (Nagamune *et al.* 1981).

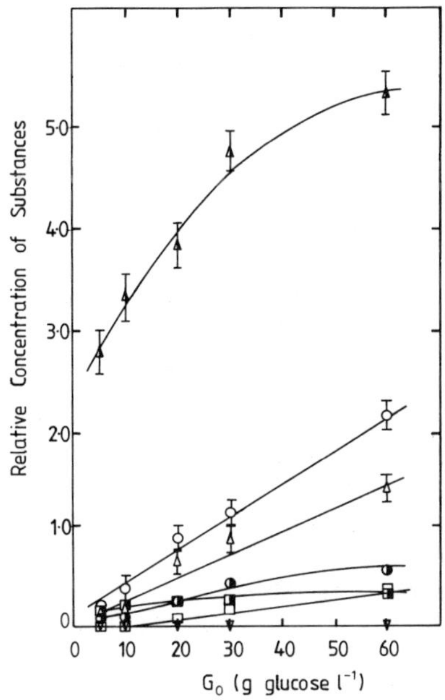


**Figure 6.2** Relative concentrations of nutrients required for yeast growth on different concentrations of glucose (Nagamune *et al.* 1981). ○, $Fe^{++}$, $Zn^{++}$, $Cu^{++}$; △, vitamins; □, $MgSO_4$; ▽, $CaCl_2$; ◑, asparagine; ▲, $(NH_4)_2SO_4$; ◨, $KH_2PO_4$.

the glucose concentration. By altering nutrients and vitamin concentrations using Hayduck's synthetic medium, Nagamune *et al.* determined their interaction on yeast growth and cell yield. Interestingly, variations in the relative vitamin concentration resulted in the largest changes in $\mu_{max}$ and cell yield. From their experiments they determined the minimal essential component-to-glucose concentration ratios necessary to maintain optimal physiological activities of *Saccharomyces cerevisiae* (Fig. 6.2).

**Inhibition**

If the concentration of a nutrient in a medium exceeds that required for growth and metabolism, it frequently results in inhibition or even cessation of yeast growth. This is often observed with several inorganic nutrients. This is illustrated by copper which, at a concentration of 1.5 mmol $l^{-1}$ stimulates growth but at 10 mmol $l^{-1}$ inhibits growth. The inhibitory effects are also affected by interaction between ions. For example, $Zn^{2+}$ can reduce $Cu^{2+}$ toxicity (Maiorella *et al.* 1984).

**Complexing**

In complex industrial media nutrients are prevented from being freely available for microbial utilization by the chelating, sequestering and adsorbing action of various materials such as humic acids (Jones & Greenfield 1984). However, these effects may be advantageous in removing potentially inhibitory concentrations of nutrients.

Useful information on the formulation of synthetic media and techniques to avoid problems of precipitation during medium development can be obtained from Oura (1974) and Hutner (1972). Formulation on an industrial scale is dealt with by Zabriskie *et al.* (1980) in *Trader's guide to fermentation media formulation*.

## 6.3 Substrate assimilation

### *6.3.1 Carbohydrates*

The nature of the carbon substrate and the characteristics of the transport mechanism responsible for its assimilation are of fundamental importance in understanding the regulation of the metabolism of the microbial cell. Different yeasts can grow on a wide variety of both carbohydrate and non-carbohydrate substrates (see Chs 9 and 10). However, it is not possible to consider the physiology of yeast growth without some discussion of the mechanisms which exist for the assimilation of low-molecular weight sugars.

Three mechanisms for the assimilation of solutes into cells have been described; simple chemical diffusion, facilitated diffusion and active transport. Chemical diffusion is not very important in the assimilation of carbohydrates, although it has been implicated in the assimilation of polyols (Fiechter *et al.* 1981). Both facilitated diffusion and active transport involve the use of a membrane-associated carrier. Since these carriers are proteins, the rate of uptake is sensitive to temperature, is capable of being saturated at higher substrate concentrations and can be competitively inhibited by substrates which have structures sufficiently similar to bind to the active site of the carrier. The main difference between facilitated diffusion and active transport is that whereas a direct input of energy is essential for active uptake, allowing the substrate to be assimilated against a concentration gradient, no direct energy input is involved in facilitated diffusion (Cooper 1982b).

Facilitated diffusion appears to be the most important mechanism for the uptake of monosaccharides such as glucose, fructose and mannose, whereas disaccharides such as maltose are assimilated by active transport. Whereas maltose can be concentrated in the cell to 15× the extracellular concentration, free glucose is not normally detected in the cell, even under conditions of rapid glucose assimilation. When glucose is being assimilated rapidly, glucose 6-phosphate rather than free glucose accumulates in the cell. Three enzymes have been identified in yeast – hexokinase PI, hexokinase PII and glucokinase – which have the capability of phosphorylating glucose, but which differ in their affinity for fructose. The difference in the function of these three enzymes is not clear, since the presence of any one of them permits growth on glucose. Since the loss of any one or two of these enzymes does not directly affect the uptake of labelled glucose it was considered by Franzusoft and Cirillo (1982) that none of them was directly involved in the uptake process. This supports the model of glucose assimilation which involves the entry of glucose into the cell by facilitated diffusion and the maintenance of the diffusion gradient by the rapid removal of free glucose in the cell by phosphorylation. This model is also favoured by the high levels of hexokinase activity which occur in baker's yeast. However, it has been difficult to distinguish between the rapid phosphorylation of glucose after entering into the cell and direct phosphorylation of glucose during the actual uptake process. If glucose is phosphorylated after entry into the cell, then the low level of free glucose indicates that it is the transport process which is rate limiting during the metabolism of glucose. The available evidence suggests that the uptake of most monosaccharides is carried out by the single constitutive facilitated diffusion system which is responsible for glucose uptake. However, different sugars have different affinities for this transport system (Table 6.5a) such that sugars with a low affinity for the carrier, e.g. arabinose and xylose, can act as competitive inhibitors of glucose uptake in a similar manner to analogues of glucose, such as 2-deoxyglucose.

**Table 6.5(a)** Half-saturation constants of the hexose-uptake system (data from Kotyk & Janecek, 1975).

| Sugar | $K_1$ (mmol $l^{-1}$) |
|---|---|
| D-glucose | 5 |
| D-fructose | 17 |
| D-mannose | 27 |
| 2-deoxy-D-glucose | 4 |
| L-sorbose | 21 |
| D-galactose | 35 |
| 1,5-anhydroglucitol | 50 |
| L-glucose | 100 |
| D-arabinose | 115 |
| D-xylose | 130 |
| D-allose | 225 |
| 3-O-methyl-D-glucose | 250 |
| L-xylose | 400 |
| D-ribose | 600 |

**Table 6.5(b)** Carbohydrate uptake systems in *Saccharomyces cerevisiae* (from Cooper 1982b).

| Carbohydrate(s) | Mode of uptake | Control of uptake | Gene(s) |
|---|---|---|---|
| glucose, fructose, mannose | facilitated | constitutive | — |
| galactose | facilitated | inducible | *GAL2* |
| sucrose | active | — | — |
| maltose | active | inducible | — |
| $\alpha$-methylglucoside | facilitated | constitutive | *MGL2*, *MGL4* |
| trehalose | unclear | unknown | |
| acyclic polyols | unclear | unknown | |
| acetate, lactate, glycerol, ethanol | unknown | unknown | |

Although galactose can enter the cell by the constitutive facilitated diffusion system, the rate of entry is very low. In the presence of D-galactose a second inducible facilitated diffusion system develops which has a higher affinity for galactose (Kou & Cirillo 1970) and which is subject to repression in the presence of glucose (Matern & Holzer 1977) (Table 6.5b).

Studies on the uptake of disaccharides in yeast are complicated by the fact that yeast produces enzymes such as invertase and $\beta$-galactosidase which hydrolyse disaccharides to their component monosaccharide sugars, which are subsequently assimilated. However, in the case of sucrose evidence has recently been presented that an active transport system for sucrose accumulation exists in yeast (Santos *et al.* 1982). Such active transport systems have been demonstrated for maltose uptake. This transport system is inducible by maltose and is 150× more active in cells grown on maltose than in glucose-grown cells. The system concentrates maltose up to 15-fold, and is inhibited by inhibitors of aerobic respiration such as sodium fluoride and DNP. The accumulation of maltose by this system is associated with an increase in pH of the medium, suggesting that a proton is co-transported through the membrane with each molecule of maltose assimilated.

A similar transport system has been described for $\alpha$-methyl glucoside, which is capable of concentrating $\alpha$-methyl glucoside up to 150× more than the external concentration. This transport system can be induced by $\alpha$-methyl glucoside or the non-metabolisable analogue, $\alpha$-thioethyl D-glucosepyranoside (TEG). However, it is interesting to note that $\alpha$-methyl glucoside can also be assimilated by the facilitated diffusion system, but not by the maltose uptake system.

The mechanisms of uptake of other carbon substrates, such as trehalose, polyols, acetate, lactate, glycerol and ethanol, are not known (Cooper 1982b). This is regrettable, since many of these compounds play a central role in the control of yeast metabolism (Table 6.5b).

### *6.3.2 Nitrogen*

Since yeast is capable of assimilating a variety of nitrogen sources, including ammonia, urea, allantoin, amino acids, bases and some small peptides, it is not surprising that complex regulatory mechanisms operate to control the assimilation of nitrogenous substrates. The assimilation of urea, allantoin and certain amino acids has been extensively studied and has been recently reviewed by Cooper (1982a). Enzymes responsible for the breakdown of organic nitrogen sources such as arginase and allantoinase, which are involved in the breakdown of arginine and allantoin, are subject to dual control. They are induced by arginine or allantoin, respectively – but are also subject to repression by ammonia and other readily assimilated nitrogen sources such as asparagine and glutamine. The induction of the catabolic enzymes is very rapid. An increase in the level of arginase can be detected within 5 min and a 50-fold increase in level can occur very rapidly.

The preferential use of the most useful nitrogen sources has generally been considered to be controlled by a repression phenomenon referred to as nitrogen catabolite repression. The addition of asparagine to yeast growing in the presence of arginine can lead to the repression of the formation of several enzymes involved in the breakdown of allantoin, arginine and urea, and the loss of mRNA produced from the corresponding genes. The loss of biosynthetic ability occurs very rapidly, having a half-life of 3 min, however, this repressive effect only influences the formation of new enzyme and has no effect on existing enzymes. The loss of existing enzymes occurs much more slowly, and 10% of the induced activity of one of these enzymes, urea amidolyase, has been reported to persist after 6 h or three generations. This indicates that enzyme loss occurs by dilution-out during cell division (Cooper & Sumrada 1983). This slow rate of response is inadequate to explain the much more rapid inhibition of uptake of amino acids. Cooper has recently presented evidence that the ability of yeast to take up allantoin and proline is repressed very rapidly in the presence of a good nitrogen source, and he proposes that the mechanism of rapid nitrogen catabolite repression is more concerned with the transport of nitrogenous molecules into the cell than the repression of the formation of enzymes for their catabolism.

However, evidence has been presented that the control of amino-acid assimilation can vary at different growth rates. Watson (1977) has studied the uptake of glutamate and proline by yeast growing in continuous culture on a medium containing 1 mmol $l^{-1}$ proline and 1 mmol $l^{-1}$ glutamate. At a dilution rate of 0.3 $h^{-1}$ glutamate and proline were assimilated equally; however, when the dilution rate was increased to 0.35 $h^{-1}$ 90% of the proline remained in the medium and glutamate was assimilated preferentially. The total amino acid pool was measured during these experiments, and showed a 10% increase at the higher growth rate.

These results may have implications for the growth of genetically engineered yeast which have auxotrophic requirements for certain amino acids. Addition of necessary amino acids to the growth medium may not always lead to their assimilation if other more-readily assimilable nitrogen sources are present.

## 6.4 Modes of yeast growth

When *Saccharomyces cerevisiae* is inoculated into an aerobic batch culture containing a high sugar concentration a diphasic growth curve is obtained. After an initial lag phase exponential growth occurs, during which the glucose is fermented to ethanol. During this phase aerobic respiration is repressed. When the glucose has been metabolized a lag phase occurs before aerobic growth continues during which the ethanol is metabolized to carbon dioxide and water. During this lag phase the RQ drops to 1.0 or less, and mitochondrial function develops. Such a yeast has been referred to by Fiechter as a glucose-sensitive yeast (Fiechter *et al.* 1981).

Not all yeasts exhibit this growth characteristic and some appear to be insensitive to high concentrations of glucose. These yeasts which include species of *Candida*, *Rhodotorula*, *Trichosporon*, *Pichia*, *Torulopsis* and *Hansenula* do not produce ethanol when grown on high-glucose media in aerobic conditions. They respire aerobically and give a high yield of biomass, hence the use of *Candida* for the production of fodder yeast. However, they are still capable of fermenting glucose to ethanol in anaerobic conditions or conditions of reduced aeration. These yeasts have been referred to as glucose-insensitive yeasts.

*Schizosaccharomyces pombe* represents a third mode of respiratory activity in yeast. It is glucose sensitive, so ferments glucose to ethanol in glucose-repressed conditions. However, it is incapable of growth on ethanol as the sole carbon substrate, so growth ceases at the end of the ethanol-fermentation phase, even in aerobic conditions.

These three types of yeasts also exhibit different growth characteristics in continuous culture. Glucose-insensitive yeasts exhibit classical kinetics of an organism in continuous culture. The biomass concentration remains at a constant high level until the dilution rate exceeds $\mu_{max}$ when washout occurs. However, glucose-sensitive yeasts exhibit different kinetics. At lower dilution rates the biomass concentration remains at a higher level, corresponding to a high biomass yield, and the RQ indicates that aerobic respiration is occurring; but at higher dilution rates the RQ rises, ethanol begins to accumulate in the medium and the yield of biomass drops to a lower level. However, washout does not occur, and steady states at higher dilution rates can be established during which the yeast exhibits ethanol fermentation as well as aerobic respiration. Glucose repression can be induced by high levels of glucose or by a high influx of glucose, even when the concentration of glucose is low (10–100 mg $l^{-1}$) (van Dijken & Scheffers 1984). However, the dilution rate at which glucose repression starts is influenced not only by the concentration of the glucose in the inlet medium, but also by other medium components. Low levels of iron and copper can cause glucose repression to occur at low dilution rates, so concentrations of 100 mg $l^{-1}$ $FeCl_3 \cdot 6H_2O$ and 0.79 mg $l^{-1}$ $CuSO_4 \cdot 5H_2O$ have been recommended in order to prevent this occurring (Fiechter *et al.* 1981). If yeast is inoculated into defined medium in strictly controlled anaerobic conditions (Fiechter 1977), then growth does not occur. Growth only occurs if trace levels of oxygen are present or if small quantities of ergosterol and unsaturated fatty acids are added to the medium (Andreasen & Stier 1954). Nicotinic acid has also been reported to be required in strictly anaerobic conditions (Shatzman 1975). In such conditions the major fermentation product is ethanol, but glycerol and pyruvate can also be produced in anaerobic chemostat culture (Oura 1977). When anaerobically grown yeast is used as an inoculum in beer fermentations the requirement for unsaturated fatty acids and sterols is met partly by the presence of these compounds in the wort and partly by blowing air or oxygen through the wort (Aries & Kirsop 1977). Studies on the levels of sterols in yeast by Aries and Kirsop showed that saturation of the wort with air led to an increase in sterol concentration up to 10 mg $g^{-1}$ dry weight yeast, but it never increased beyond this level. If the inoculum level of a beer fermentation is low, 0.5 g $l^{-1}$, then insufficient sterol is produced if aeration is restricted to the initial period of the fermentation in order to achieve maximum levels of biomass, and the fermentation then becomes oxygen limited. This did not occur when high levels of inoculum, 5 g $l^{-1}$, were used. It was also reported that strains differed in their requirement for oxygen in these conditions. In anaerobic conditions the absence of a cytochrome system to oxidize NADH results in a requirement for other hydrogen-acceptor molecules if glycolysis is to continue. Normally this is satisfied by the transfer of hydrogen from NAD to acetaldehyde to form ethanol. However, in the absence of available acetaldehyde, due to the diversion of metabolic intermediates into biosynthetic reactions, glyceraldehyde may be reduced to glycerol. The maintenance of redox balance is of critical importance in the anaerobic growth of yeast.

## 6.5 Yeast metabolism

### *6.5.1 Intermediary metabolism*

*Saccharomyces cerevisiae*, in common with most other micro-organisms, can produce the necessary enzymes to metabolize simple sugars such as glucose and fructose to carbon dioxide and water by the process of aerobic respiration. In conditions of good aeration and low carbohydrate concentration the glucose is metabolized by the Emden Meyerhof Parnas (EMP) pathway to pyruvate, which is then oxidized by the tricarboxylic acid cycle to carbon dioxide and water (Fig. 6.3) (Fraenkel 1982). These key metabolic pathways serve three functions in the cell: (a) the generation of ATP; (b) the generation of reducing power in the form of NADH; and (c) the formation of intermediates for the biosynthesis of cellular materials. In aerobic conditions it has been reported that only 70% of the glucose is converted to pyruvate via the EMP pathway, the remainder being metabolized by the pentose phosphate pathway (PPP or HMP) which plays a major role in producing reducing power in the form of NADPH, required for most biosynthetic reactions. It also plays a role in the biosynthesis and metabolism of pentose sugars such as xylose, which are present in large amounts in hydrolysates of hemicelluloses.

Most of the ATP generation during aerobic respiration occurs during the oxidation of acetate to carbon dioxide and water via the tricarboxylic acid cycle, and involves oxidative phosphorylation during the transfer of electrons from reduced NADH to oxygen. However, in anaerobic conditions or in conditions of high glucose flux the repression of mitochondrial function can occur and ATP cannot be generated by this route. Under these conditions the only ATP produced is that generated at the 1,3-diphosphoglycerate to 3-phosphoglycerate and the phosphoenol pyruvate to pyruvate steps in glycolysis. Since ATP is used to phosphorylate glucose, this represents a net gain of two ATP molecules for each glucose molecule metabolized.

During vigorous growth yeast maintains an energy charge of between 0.7 and 1.0 (Franco *et al.* 1984a). However, it has been reported that yeast can survive at an energy charge value of 0.15 (Ball & Atkinson 1975), which is unusually low. An additional problem which arises in conditions which repress mitochondrial function is that NADH cannot be oxidized by the cytochrome system. Thus, NADH generated in the glyceraldehyde, 3-phosphate to 1,3-diphosphoglycerate step can only be oxidized by passing on the hydrogen atoms to other substrates. In yeast this is normally acetaldehyde, resulting in the generation of ethanol. However, under certain conditions this can occur as a result of the reduction of glyceraldehyde phosphate to glycerol phosphate. The

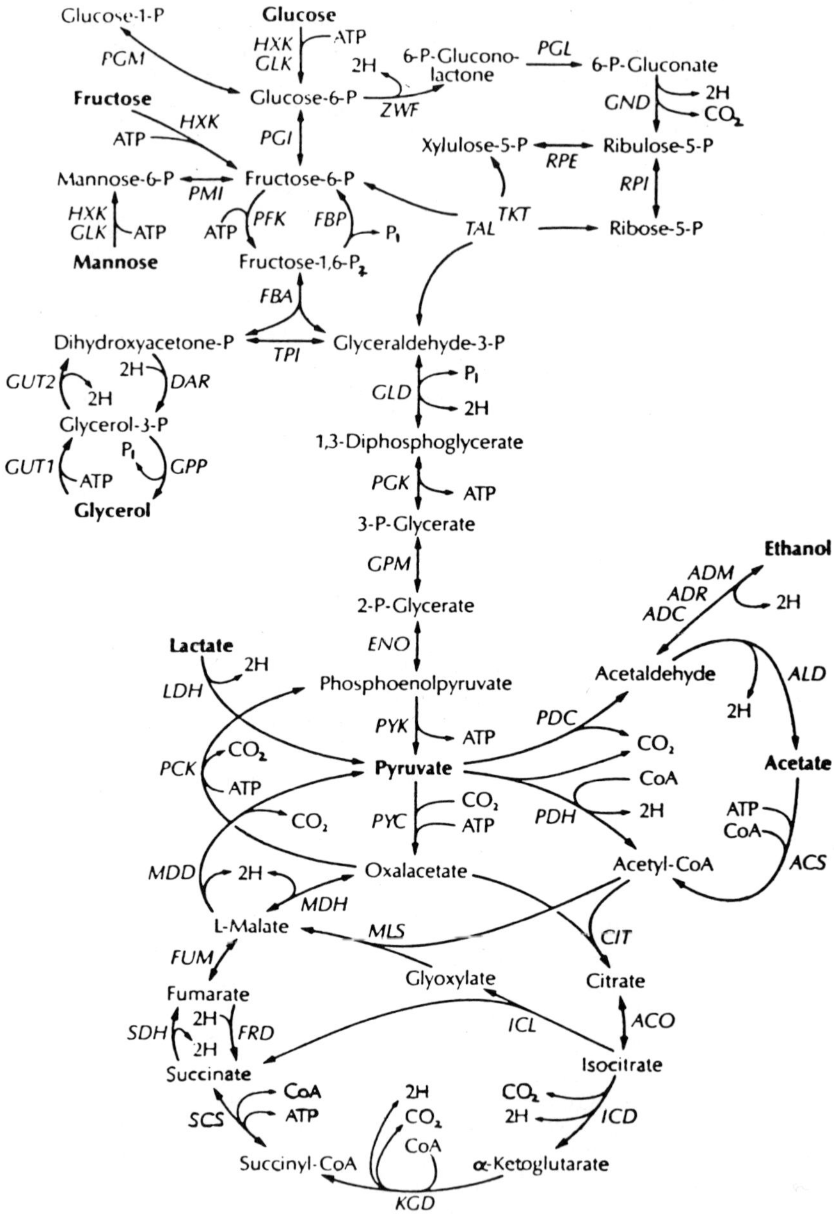

Glucose-1-P
Glucose
HXK
GLK
ATP
PGM
2H
6-P-Glucono-
lactone
PGL
6-P-Gluconate
Fructose
Glucose-6-P
ZWF
GND
2H
CO2
HXK
ATP
PGI
Xylulose-5-P
RPE
Ribulose-5-P
Mannose-6-P
PMI
Fructose-6-P
RPI
HXK
GLK
ATP
ATP
PFK
FBP
P1
TAL
TKT
Ribose-5-P
Mannose
Fructose-1,6-P2
FBA
Dihydroxyacetone-P
TPI
Glyceraldehyde-3-P
GUT2
2H
2H
DAR
GLD
P1
2H
Glycerol-3-P
1,3-Diphosphoglycerate
GUT1
P1
ATP
GPP
PGK
ATP
Glycerol
3-P-Glycerate
Ethanol
GPM
ADM
ADR
ADC
2H
2-P-Glycerate
Lactate
ENO
2H
LDH
Phosphoenolpyruvate
Acetaldehyde
ALD
2H
PDC
PYK
ATP
CO2
PCK
CO2
ATP
Pyruvate
CoA
Acetate
CO2
CO2
PYC
ATP
PDH
2H
ATP
CoA
MDD
2H
Oxalacetate
Acetyl-CoA
ACS
MDH
L-Malate
MLS
CIT
FUM
Glyoxylate
Citrate
Fumarate
ACO
SDH
2H
2H
FRD
ICL
Isocitrate
Succinate
CoA
ATP
2H
CO2
CoA
CO2
2H
ICD
SCS
Succinyl-CoA
α-Ketoglutarate
KGD

**Figure 6.3** Pathways of yeast intermediary metabolism. Gene symbols are in italics. Substances in boldface type are commonly used as carbon sources for growth. Many of the reactions are incompletely described. For example, for kinases, ATP is indicated as a reactant, but ADP is not shown as a product. The existence, function or direction of several of the indicated reactions is speculative (see text). ($P_i$) Inorganic phosphate: 2H indicates a dehydrogenase reaction (from Fraenkel 1982).

*Enzymes and genes of yeast intermediary metabolism*

| | |
|---|---|
| *ACO*$^{h}$ | aconitate hydratase, 4.2.1.3$^{h}$ |
| (*ACS*) | acetyl-CoA synthetase, 6.2.1.1 |
| *ADCADMADR* | alcohol dehydrogenase, 1.1.1.1 |
| (*ALD*) | aldehyde dehydrogenase (NADP), 1.2.1.5 |
| *CIT* | citrate synthase, 4.1.3.7 |
| (*DAR*) | dihydroxyacetone-P reductase (glycerol-3-P dehydrogenase), 1.1.1.94 |
| (*ENO*) | enolase, 4.2.1.11 |
| (*FBA*) | fructose-bisphosphate aldolase, 4.1.2.13 |
| (*FBP*) | fructose bisphosphatase, 3.1.3.11 |
| (*FRD*) | fumarate reductase, 1.3.99.1 |
| (*FUM*) | fumarase, 4.2.1.2 |
| (*GLD*) | glyceraldehyde-3-P dehydrogenase, 1.2.1.12 |
| *GLK* | glucokinase, 2.7.1.2 |
| (*GND*) | 6-phosphoglucanate dehydrogenase, 1.1.1.44 |
| *GPM* | phosphoglycerate mutase, 2.7.5.3 |
| (*GPP*) | glycerol-1-phosphatase, 3.1.3.21 |
| *GUT1* | glycerol kinase, 2.7.1.30 |
| *GUT2* | glycerol-3-P dehydrogenase, 1.1.99.5 |
| *HXK1*, *HXK2* | hexokinase; Bernard (1975) |
| (*ICD1*) | isocitrate dehydrogenase ($NAD^+$), 1.1.1.42 |
| (*ICD2*) | isocitrate dehydrogenase ($NADP^+$), 1.1.1.43 |
| *ICL* | isocitrate lyase, 4.1.3.1; Olson (1959) |
| *KGD* | α-ketoglutarate dehydrogenase, 1.2.4.2 |
| (*LDH*) | lactate dehydrogenase, 1.1.2.3 (cytochrome $b_2$) |
| (*MDD*) | malate dehydrogenase (decarboxylating), 1.1.1.40 |
| *MDH* | malate dehydrogenase, 1.1.1.37 |
| (*MLS*) | malate synthase, 1.1.3.2 |
| *PCK* | phosphoenolpyruvate carboxykinase, 4.1.1.49 |
| *PDC* | pyruvate decarboxylase, 4.1.1.1 |
| (*PDH*) | pyruvate dehydrogenase, 1.2.4.1 |
| *PFK* | phosphofructokinase, 2.7.1.11 |
| *PGI* | phosphoglucose isomerase, 5.3.1.9 |
| *PGK* | phosphoglycerate kinase, 2.7.2.3 |
| (*PGL*) | 6-phosphogluconolactonase, 3.1.1.31 |
| *PGM1*, *PGM2* | phosphoglucomutase, 2.7.5.1 |
| *PMI* | phosphomannose isomerase, 5.3.1.8 |
| (*PYC*) | pyruvate carboxylase, 6.4.1.1 |
| *PYK* | pyruvate kinase, 2.7.1.40 |
| (*RPE*) | ribulose-5-P 3-epimerase |
| (*RPI*) | ribose-5-phosphate isomerase, 5.3.1.6 |
| *SDH* | succinate dehydrogenase, 1.3.99.1 |
| (*SCS*) | succinyl-CoA synthetase, 6.2.1.4 |
| (*TAL*) | transaldolase, 2.2.1.2 |
| (*TKT*) | transketolase, 2.2.1.1 |
| *TPI* | triosephosphate isomerase, 5.3.1.1 |
| *ZWF* | glucose-6-phosphate dehydrogenase, 1.1.1.49 |

repression of mitochondrial function also affects the functioning of the glyoxylate cycle, which is a metabolic pathway playing a key role in generating carbon skeletons for TCA cycle acids and for amino-acid biosynthesis. In aerobic conditions *Saccharomyces cerevisiae* can be grown on ethanol or acetate as the sole carbon source. Acetate has been used as the carbon source in presporulation medium by several workers (Roth & Halvorson 1969). Under these conditions the TCA cycle functions normally, but the role of the EMP pathway is reversed and it functions to convert acetate into the hexose and pentose molecules required for cellular biosynthesis. This is achieved by the process of gluconeogenesis. Most of the steps in the EMP pathway are reversible, but the fructose 6-phosphate to fructose 1,6-diphosphate and the phosphoenol pyruvate to pyruvate steps are not reversible using the same enzymes. Whereas ATP and kinase enzymes are involved in the phosphorylation reactions, phosphatases are involved in the removal of the phosphate groups. During growth on ethanol or acetate two carbon compounds are converted into three carbon compounds by carboxylation reactions. These are dependent upon the availability of carbon dioxide in the growth medium; however, little information appears to be available on the effects of carbon-dioxide concentration on the activity of these reactions during yeast growth.

### *6.5.2 Regulation of metabolism*

The reduced rate of glycolysis observed when *Saccharomyces cerevisiae* is grown aerobically has traditionally been interpreted in terms of the much higher level of ATP generated during aerobic respiration by mitochondrial oxidative phosphorylation than by substrate-level phosphorylation, which is the only method of generating ATP in fermentative conditions. However, intensive biochemical studies have indicated that such a mechanism is backed up by other biochemical control mechanisms. This control can involve several different levels of regulation. High levels of citrate and ATP can inhibit phosphofructokinase activity, possibly leading to a reduced rate of glycolysis, and glucose 6-phosphate accumulation has been reported to suppress the glucose-uptake mechanism (Sols *et al.* 1971).

Enzyme inhibition is probably important in the short-term response of yeast to changed growth conditions, whereas changes in the level of key enzymes are important in the longer-term adaptation of yeast to changes. This may involve the induction of respiratory enzymes by oxygen, the repression of enzymes by glucose, catabolite repression or the inactivation of enzymes in the presence of glucose, or both, a phenomenon known as catabolite inactivation (Fraenkel 1982).

Repression of mitochondrial formation by glucose and other readily metabolizable sugars such as maltose has been extensively studied (Polakis & Bartley 1965, Fiechter *et al.* 1981). A detailed study of the effect of glucose and oxygen on the levels of mitochondrial enzymes has been carried out by Wales *et al.* (1980) (Table 6.6). Under aerobic conditions the activities of TCA enzymes were reduced by between 69% and 89% by growth on glucose at 10% concentration when compared with controls on 0.8% glucose. The same enzymes were even further repressed by growth in anaerobic conditions and

**Table 6.6** Specific activities of the tricarboxylic acid cycle and related enzymes in cell-free extracts of *Saccharomyces carlsbergensis* grown under four different conditions (from Wales *et al.* 1980).

| | Enzyme specific activity ($nmol^{-1} min^{-1} mg^{-1}$ protein) | | | |
|---|---|---|---|---|
| Enzyme | Aerobic, glucose-derepressed | Aerobic, glucose-repressed | Anaerobic, glucose-derepressed | Anaerobic, glucose-repressed |
| citrate synthase | 128 | 13(11) | 7(5) | 44(34) |
| aconitase | 221 | 49(22) | 19(9) | 1(0.3) |
| NADP-linked isocitrate dehydrogenase | 94 | 35(37) | 10(11) | 14(15) |
| 2-oxoglutarate decarboxylase | 27 | 4(16) | 0.3(0) | 0.3(0) |
| succinate dehydrogenase | 44 | 6(14) | 1(2) | 1(3) |
| fumarase | 510 | 98(19) | 2(0.3) | 3(0.5) |
| malate dehydrogenase | 4120 | 1630(40) | 175(4) | 155(3) |
| alcohol dehydrogenase | 1380 | 1680(122) | 93(7) | 1040(76) |
| L-lactate dehydrogenase | 148 | 23(15) | 5(3) | 3(2) |
| fumarate reductase | 73 | 125(170) | 12(16) | 8(11) |
| isocitrate lyase | ND | ND | ND | ND |
| malate synthase | ND | ND | ND | ND |

ND, not detectable.
Figures in parentheses show the specific activities expressed as a percentage of the corresponding activity for growth under aerobic, glucose-derepressed conditions.

2-oxoglutarate decarboxylase was absent altogether in anaerobic conditions. The levels of most TCA enzymes were no lower in glucose repressed anaerobically grown cells than in glucose-derepressed anaerobic cells, and one enzyme, citrate synthetase, was actually higher.

The role of oxygen in the induction of mitochondrial structure and formation is not understood. This induction does not appear to be dependent on the need for unsaturated fatty acids and sterols, since the supplying of these does not lead to normal mitochondrial development (Rogers & Stewart 1973, Oura 1974). Furthermore, even under completely anaerobic conditions the level of respiratory enzymes is not zero.

The formation of cytochrome C oxidase during the adaptation of anaerobically grown cells to aerobic conditions was studied by Jenkins *et al.* (1984), whose results showed that within 10 min cytochrome C oxidase activity was associated with a membrane fraction, but that these membranes had a different density from mature mitochondria. However, most of the cytochrome C oxidase activity was associated with the mitochondrial fraction after 3 h of adaptation.

The level of cytochrome in yeast can vary markedly under normal growth conditions. High cytochrome levels have been reported during the later stages of batch fermentation in *Candida utilis*, but it has been reported that the level of cytochromes in well-controlled steady-state cultures was relatively low (Cail *et al.* 1979). However, much higher levels were observed by the same workers in less well-controlled fermentations in which glucose was added manually at intervals. In these experiments the highest level of cytochrome was observed at a DOT of 1% air saturation. Addition of pulses of ethanol had a similar stimulatory effect on the level of cytochromes. These results were interpreted in terms of the regulation of the level of cytochromes as a response to changes in the ADP/ATP value in the cell (Cail *et al.* 1979).

Studies on another glucose-insensitive yeast, *Candida tropicalis*, indicated that the ratio of enzyme activities in glucose repressed and derepressed conditions was only 1–1.5 for most mitochondrial enzymes (quoted in Fiechter *et al.* 1981).

It is clear that there is a relationship between the different growth characteristics of glucose-sensitive yeasts, such as *Saccharomyces cerevisiae*, and glucose-insensitive yeasts, such as *Candida tropicalis*, and the intracellular control of mitochondrial enzyme levels. The repression of mitochondrial function can be considered as a special adaptation for growth in fermentative conditions.

Catabolite repression has been implicated in the control of other cellular functions in addition to mitochondrial development. The uptake and metabolism of a wide range of sugars including maltose, lactose and melibiose is repressed by glucose in many yeasts, and sporulation by diploid yeast is also subject to glucose repression in *Saccharomyces cerevisiae*. However, some enzyme activities are at their highest in conditions of catabolite repression. The level of ADH 1, the isozyme of alcohol dehydrogenase which is active in the fermentation of glucose to ethanol, is present at high levels in catabolite-repressed conditions but low in conditions which favour ethanol oxidation and gluconeogenesis (Eltayeb & Berry 1982).

Although catabolite repression has been the subject of many studies, its mechanism is far from clear and it is still not known whether glucose or some other product of glucose catabolism is the key intermediate in regulation. Several different techniques have recently been developed to isolate mutants which are not subject to catabolite repression. Some of these affect the catabolite repression of a single enzyme, while others affect several pathways (Fraenkel 1982). The initial steps in glucose metabolism, the uptake and phosphorylation of glucose by hexokinase have been implicated by several studies. Yeast contains three enzymes which are involved in the phosphorylation of glucose; hexokinase PI and PII, which are capable of phosphorylating both glucose and fructose, and glucokinase, which is only active against glucose.

Strains of *Saccharomyces cerevisiae* have been isolated (*hex*1, *hxk*2) which lack the genes responsible for hexokinase PI and PII and which do not exhibit catabolite repression. Restoration of the hexokinase PII gene by transformation restores hexokinase activity and also carbon catabolite repression (Fröhlich *et al.* 1984). It appears that hexokinase PII plays a central role in the control of catabolite repression, and that it is an enzyme with two functional domains – one for its catalytic function, i.e. the phosphorylation of glucose, and one for its regulatory function in carbon catabolite repression (Entian & Fröhlich 1984). Catabolite-derepressed strains of *Schwanniomyces occidentalis* have also been reported to have reduced hexokinase activity (McCann & Barnett 1984).

However, other catabolite derepressed mutants (*grr*1, *hex*2) have been reported which have elevated levels of hexokinase (Bailey & Woodward 1984, Zimmerman & Scheel 1977). All of these mutations have broad phenotypic effects, and their presence results in the derepression of cytoplasmic wall enzymes such as invertase, maltase and galactokinase, as well as mitochondrial enzymes such as cytochrome C oxidase and succinic dehydrogenase. Of particular interest to the brewer is that several of the catabolite-derepressed mutants isolated, e.g. *grr*1 (Bailey & Woodward 1984) and *CCR* $80^r$ (Ciriacy 1978) are dominant, and the strain *grr*1 was actually isolated in a diploid strain by its resistance to 2-deoxyglucose.

The *CCR*$80^r$ mutant exhibited a very slow aerobic growth rate, which indicated a major disruption of either glycolytic flux or sugar uptake. However, since the strain exhibited derepression even when maltose was used as the carbon source, catabolite derepression in this strain cannot be attributed to a defect in hexose uptake.

A common control mechanism of catabolite repression was indicated by the identification of a gene *SNF*1 (sucrose non-fermenting) which is essential for sucrose utilization. This gene was found to be essential for growth on galactose, maltose, melibiose and several non-fermentable carbon substrates, all of which are subject to catabolite repression. The *snf*1 mutant strain also exhibits a reduced growth rate on glucose (Celenza & Carlson 1984). The *SNF*1 gene, which has now been cloned, appears to be essential for the formation of the 1.9-kb mRNA which encodes for the signal peptide containing precursor of secreted invertase.

Further evidence for the nature for the control of catabolite repression comes from studies on the gene for alcohol dehydrogenase II, the glucose-repressible isoenzyme which is active during growth on ethanol. The expression of ADHII, the gene responsible for this enzyme, is subject to control by a second gene, ADR1. Mutations in ADR1 can result in the constitutive formation of ADHII even in the presence of glucose (Ciriacy 1979). The ADHII gene has been cloned and investigated by deletion analysis to identify those sequences of the ADHII gene which are involved in the regulation of its expression. It has been demonstrated that derepression of ADHII requires positive activation, which is mediated through a sequence of bases between 216 and 257 base pairs upstream from the start side for ADHII transcription (Beier *et al.* 1985).

By analogy with bacterial systems, many attempts have been made to implicate cAMP in the regulation of catabolite repression in yeast. Inverse relationships between glucose levels and cAMP levels in *Schizosaccharomyces pombe* (Schlanderer & Dellweg 1974) and in *Saccharomyces cerevisiae* (Sy & Richter 1972) have been reported. However, these results were not confirmed by Foury and Goffeau (1975).

Studies using mutants which were defective in adenyl cyclase suggested that cAMP was not involved in the control of catabolite repression (Matsumoto *et al.* 1983). Much stronger evidence is available that cAMP is involved in the phosphorylation of enzymes by cAMP-dependent protein kinases. A cAMP-dependent protein kinase which phosphorylates fructose 1,6-diphosphatase, a key regulatory enzyme in glycolysis, has been described (Wingender-Drissen 1983). If such an enzyme is involved in the control of catabolite repression, it may be that it can be replaced by a cAMP-independent protein kinase in cAMP-deficient strains. Both types of protein kinase have been reported in yeast (Wingender-Drissen & Becker 1983).

### 6.5.3 *Levels of intermediates*

In a recent study on the growth of baker's yeast, the levels of a wide range of glycolytic and tricarboxylic acid cycle intermediates were monitored in yeast grown in different conditions (Franco *et al.* 1984a). In this study three different media – a control medium, a low-nitrogen medium and a low-phosphate medium – were used at a range of dilution rates in a glucose-limited continuous culture. It was clear from the results that, even in continuous culture when the medium composition, temperature, aeration and pH were controlled, intermediate levels could vary with growth rate. In general the levels of most glycolytic intermediates, e.g. glucose 6-phosphate, fructose 6-phosphate, fructose 1,6-diphosphate, phosphoenol pyruvate, 1-phosphoglycerate and pyruvate, were high at low dilution rates, decreased as the dilution rate increased, then increased again as the dilution rate approached 0.2 $h^{-1}$. At this dilution rate glucose accumulated in the medium, ethanol production occurred and the yield of yeast decreased. The levels of citrate, oxalacetate and glyoxylate showed a different pattern; they increased with dilution rate before the onset of fermentative metabolism, then decreased. It was apparent that the levels of pyruvate and other glycolytic intermediates increased during the onset of fermentative growth; however, high levels of these intermediates occurred at dilution rates which did not induce fermentative metabolism. So it is unlikely that their accumulation is the key mechanism controlling the onset of fermentation. The transition to fermentation occurred at around $D = 0.16\ h^{-1}$. However, significant levels of aerobic respiration as indicated by the levels of TCA intermediates continued at higher dilution rates. These results indicate that stable growth states exist, during which both aerobic and fermentative respiration can occur. The transition to fermentation was not accompanied by a dramatic change in the level of most intermediates.

The effects of depletion of nitrogen and phosphate were complex, but the levels of glucose 6-phosphate, fructose 6-phosphate, pyruvate and citrate in particular were high in low-nitrogen media, and the level of glyoxylate was stimulated by low-phosphate medium.

Similar studies were carried out on a batch fermentation. In this experiment the initial growth period lasted for 12 h and was followed by a 4-h lag period before the onset of an aerobic growth phase using ethanol as a substrate (Franco *et al.* 1984b). In these experiments a reduction in most of the glycolytic intermediates occurred during the lag phase, but the levels of 6-phosphogluconate, oxalacetate, citrate and glyoxylate all increased. However, several of the glycolytic intermediates increased again during the second log phase of growth, before decreasing again during the stationary phase. In both of these studies it was apparent that the levels of intermediates can be a sensitive indicator of metabolic state, and that by monitoring these many more physiological states can be identified than are apparent by more-routine techniques of monitoring fermentations. The levels of glucose 6-phosphate, 6-phosphogluconate, pyruvate, citrate and glyoxylate were identified as being the most valuable indicators of metabolic state.

In these experiments a transition to fermentative metabolism occurred at a dilution rate above 0.16 $h^{-1}$. Such transitions have been reported by other workers (Fiechter *et al.* 1981). However, other studies have been reported in which the specific rate of oxygen uptake remained constant up to a dilution rate of 0.6 $h^{-1}$ (Barford & Hall 1979). In these experiments the RQ rose from 1.0 to >2.0 between $D = 0.3$ and $D = 0.6$ $h^{-1}$, and the yield of yeast cells dropped from 5 to <2 g $l^{-1}$ over the same period. However, washout did not occur at these high dilution rates.

It is appreciated that monitoring intracellular intermediates involves difficult and time-consuming extraction procedures, and so may not be appropriate for the routine monitoring of fermentations, but it is possible that the levels of extracellular intermediates which can be monitored much more readily can be used as indicators. Studies on the levels of intermediates excreted during growth in continuous culture showed that they were influenced by growth rate and the nutritional status of the medium.

The technique of isotachophoresis has recently been employed to investigate the level of a wide range of intermediates in yeast (Ryder *et al.* 1983). It may be that further development of this technique could offer a new and rapid method for monitoring the levels of cellular intermediates during fermentation.

### 6.5.4 *Redox balances and the metabolism of pentoses*

The role of reduced pyridine nucleotides in the control of yeast metabolism is well established. The ability of the cell to re-oxidize NADH is an essential feature of anaerobic fermentation (Sols *et al.* 1971). However, the importance of NADPH demand in the metabolic control of the yeast cell has been less extensively studied.

Since the amount of NADPH required for a given biosynthetic reaction is constant, it is possible to calculate the amount of NADPH required to produce a given weight of biomass from cellular intermediates, such as amino acids, fatty acids, nucleotides and sugar phosphates. A theoretical analysis of NADPH production has been published (Bruinenberg *et al.* 1983a, b) in which it was calculated that 930.5 mmol NADPH were required to produce 100 g yeast when growing on glucose with ammonia as the nitrogen source. This corresponds to the metabolism of at least 2% of the available glucose via the hexose monophosphate pathway, the main pathway for NADPH generation in yeast, when ammonia is supplied as the nitrogen source. When nitrate is the nitrogen source, as it can be for *Candida utilis*, the NADPH required to reduce nitrate to ammonium increases the percentage of glucose required to provide NADPH to 20%. The demand for NADPH also alters when different carbon sources are supplied. When xylose is the carbon source there is an extra requirement for NADPH in *Candida utilis* for the conversion of xylose to xylulose before it is metabolized further. However, when gluconate is the carbon source, NADPH is generated when it is metabolized in the hexose monophosphate pathway, so the overall requirement for NADPH during biomass formation is low. During growth of *Candida utilis* on acetate the level of $NADP^+$-dependent acetaldehyde dehydrogenase was found to increase, indicating that this may be important in NADPH generation (Bruinenberg *et al*. 1983c).

Many yeasts which can assimilate and metabolize xylose in aerobic conditions are unable to ferment it to ethanol in anaerobic conditions, e.g. *Candida utilis*, or can only ferment it at very slow rates, e.g. *Pachysolen tannophilus*. Studies on *Candida utilis* (Bruinenberg *et al*. 1983c) showed that failure to ferment xylose in anaerobic conditions was not due to the loss of fermentative capacity, since ethanol production commenced immediately upon the addition of glucose, but was more concerned with the initial steps in the assimilation of xylose. Growth and ethanol formation on a xylose medium by *Candida utilis* can be induced by the addition of xylose isomerase, which converts xylose to xylulose, which can then be fermented by *Candida utilis*. In *Candida utilis* the direct conversion of xylose to xylulose by xylose isomerase does not occur. The formation of xylulose occurs via a two-step reaction in which xylose is reduced to xylitol by xylose reductase, followed by the subsequent oxidation of xylitol to xylulose by xylitol dehydrogenase. The xylose reductase step requires NADPH, which must be generated by the hexose monophosphate pathway, whereas the xylitol reductase step generates NADH, which cannot be used either directly or indirectly (i.e. via a transhydrogenase) to provide the NADPH required by the xylose reductase (see Fig. 6.4). The fermentation comes to a halt unless the NADH can be re-oxidized. This can be achieved by the addition of acetoin, which can act as a hydrogen acceptor since *Candida utilis* contains an NADH-dependent acetoin reductase.

However, a different situation exists in *Pachysolen tannophilus* and *Pichia stipitis*, two yeasts which are capable of fermenting xylose to ethanol in anaerobic conditions (Bruinenberg *et al*. 1984). In both of these species the

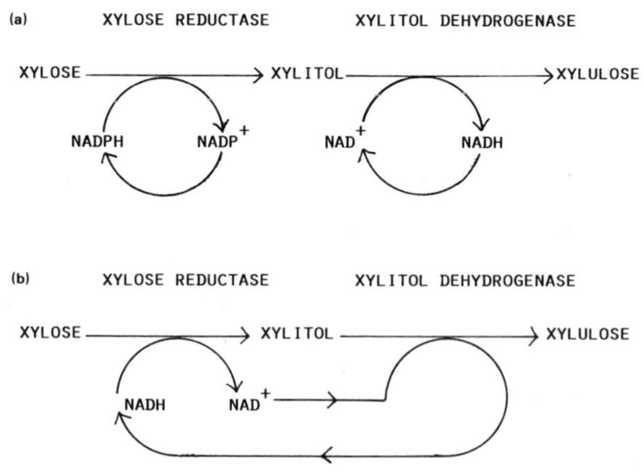


**Figure 6.4** Metabolism of xylose (a) in yeasts which are unable to ferment xylose anaerobically and (b) in *Pichia stipitis*, which can ferment xylose anaerobically (from van Dijken & Scheffers 1984).

xylose reductase and the xylitol dehydrogenase enzymes can both utilize NAD and NADP as co-factors. This means that the reduced pyridine nucleotides generated in one step can be utilized in the other. Since the rate of fermentation in *Pichia stipitis* is much higher than in *Pachysolen tannophilus*, and the level of the NADH activity of the xylose reductase is much higher in *Pichia* than in *Pachysolen*, it seems likely that it is the NADH which is used predominantly for the reduction of xylose to xylitol. Whereas the *Pichia* can ferment xylose rapidly in anaerobic conditions, *Pachysolen* requires small quantities of oxygen to achieve rapid rates of alcohol production. However, neither species can grow on xylose in the absence of oxygen, probably reflecting the disturbance of the redox balance during these fermentations.

The main pathway for the metabolism of xylose and xylulose in yeast has usually been considered to be the pentose phosphate pathway. This has been investigated by Evans and Ratledge (1984), who observed that the molar yield of $C_2$ units from xylose was higher than could be accounted for by the metabolism of all of the xylose via the pentose phosphate pathway. The investigation showed that phosphoketolase, an enzyme which catalyses the cleavage of xylulose 5-phosphate, and which is involved in pentose metabolism in heterofermentative lactobacilli, was present in most of the 25 yeasts investigated, and the yields obtained on xylose in 15 of these indicated that the phosphoketolase played a key role in the metabolism of xylose. Using this pathway, in which one xylulose phosphate molecule is cleaved to glyceraldehyde 3-phosphate and acetyl phosphate, two molecules of ethanol can be derived from one pentose molecule, compared with 1.67 $C_2$ molecules if all of the xylose is metabolized via the pentose phosphate pathway.

## 6.6 Glycerol formation

Low levels of glycerol production occur in all yeast alcoholic fermentations. If the availability of acetaldehyde is restricted either by the use of inhibitors, such as bisulphite, or by the diversion of intermediates into other biosynthetic pathways, then it is not available as a substrate for re-oxidation of NADH generated during glycolysis. Under these conditions the NADH is oxidized by the reduction of dihydroxyacetone phosphate to glycerol phosphate, which is then converted to glycerol. This pathway has been exploited to produce glycerol by fermentation. In these processes in which ethanol production is repressed by the addition of sulphite the yield is low (around 20%) and extraction is difficult, so the process is not presently economic. However, the maintenance of redox balance is not the only reason that yeast produces glycerol, since glycerol can be produced by aerobically grown yeast under certain conditions. The addition of high concentrations of NaCl, e.g. 0.44 mol $l^{-1}$ to the growth medium results in a marked stimulation of glycerol production by yeast, irrespective of whether the yeast is growing aerobically or anaerobically. This effect is associated with a loss of cell yield and a decrease in ethanol production in fermentative conditions. This effect of hypertonic concentrations of NaCl has been attributed to the formation of glycerol as an osmotic agent to facilitate the retention of cellular water, since it can be stimulated by the addition of a wide variety of salts and is not restricted to NaCl (Maiorella *et al.* 1984). Glycerol : ethanol ratios of more than 6 : 1 have been achieved in yeast fermentations upon the addition of high concentrations of KCl. Since common medium components such as potassium phosphate and ammonium phosphate have such an effect, it is clear that this phenomenon should be considered when high yields of ethanol are required.

## 6.7 RQ as a guide to metabolic state

When yeast is grown aerobically on low levels of glucose which do not repress mitochondrial function the RQ of the fermentation is unity, that is, $QCO_2$ (mmol $CO_2$ produced $g^{-1}$ dry weight of yeast $h^{-1}$) is equal to $QO_2$ (mmol $O_2$ consumed $g^{-1}$ dry weight yeast $h^{-1}$). However, the RQ of yeast growth frequently deviates from unity. In catabolite-repressed conditions the amount of carbon dioxide derived from the metabolism of glucose directly to ethanol can be calculated as $QCO_2 - QO_2$. Under these conditions RQ values of greater than unity are obtained (Rickard & Hogan 1978a). However, values of less than unity are obtained when yeast is grown on ethanol or acetate as the sole carbon source. Some of the ethanol or acetate is converted into oxalacetate via the glyoxylate cycle. This involves the consumption of 3.5 mol oxygen $mol^{-1}$ ethanol but no release of carbon dioxide, so the RQ for this step is zero. Under experimental conditions RQ values as low as 0.2 have been observed in yeast cells growing on ethanol (Rickard & Hogan 1978b).

Mass spectrometry has recently been used to study the regulation of respiration in yeast. The simultaneous measurement of oxygen and carbon dioxide levels in suspensions of yeast were carried out in a 7-ml vessel using a quadrupole mass spectrometer (Lloyd *et al.* 1983a). Using these data the $K_m$ value for oxygen during respiration was obtained by plotting the reciprocal of oxygen concentration against the reciprocal of the oxygen consumption rate, and was found to be between 1 and 2 $\mu mol\ l^{-1}\ O_2$. The inhibitory effect of oxygen on $CO_2$ evolution (one aspect of the Pasteur effect) was measured by plotting the reciprocal of the oxygen concentration against the reciprocal of the glycolytic $CO_2$ evolution in glucose-repressed cells. The $K_i$ for oxygen for this reaction was found to be 6–14 $\mu mol\ l^{-1}$. It is not clear why this difference in affinity for oxygen should occur if the Pasteur effect is a response to increased ATP generation during aerobic respiration.

Using similar techniques, Lloyd *et al.* (1983b) showed that the addition of ammonium ions to washed yeast cells led to a stimulation in the rate of glycolytic $CO_2$ production. This stimulation could not be attributed either to the uncoupling of oxidative phosphorylation or to a Pasteur effect, since both these effects could be demonstrated to be independent of the ammonium effect.

## 6.8 Ethanol toxicity

Recent interest in the development of processes to produce ethanol on a scale and at a price which would allow it to be an alternative fuel to hydrocarbons has stimulated a great deal of interest in the effect of ethanol on the yeast cell. It has been known for many years that increased ethanol concentrations can cause a decrease in yeast viability, a reduction in yeast growth and a reduction in the rate of ethanol production. However, in spite of a considerable amount of work, the effect of ethanol on the yeast cell is far from understood (Ingram & Bukkte 1984). The phenomenon of alcohol toxicity does not appear to be restricted to ethanol, since other alcohols such as the higher alcohols produced by yeast itself are also inhibitory. Not all alcohols are equally toxic; secondary alcohols have been reported to be less toxic than primary alcohols, and long-chain alcohols more inhibitory than those containing fewer carbon atoms. The differential toxicity of different alcohols has been related to their lipid solubility (Leao & van Uden 1982). The higher the lipid solubility, then the stronger the inhibitory effect.

One of the difficulties of working in this area has been that of measuring the rate of uptake and excretion of ethanol and its intracellular concentration.

Several groups have obtained evidence that the concentration of ethanol in the cell can exceed the extracellular concentration, so that studies involving the addition of ethanol to yeast cultures do not necessarily give any guide to the true inhibitory concentration in the cell. This infers an active transport system which can maintain a concentration gradient across the cell membrane (Slapack *et al.* 1987). However, the high lipid solubility of ethanol and the rapid rate of transport have convinced other workers that transport is purely by diffusion. This view has recently been supported by workers using radioactive ethanol (Guijarro & Lagunas 1984).

Efforts to elucidate the primary site of action of ethanol have been restricted by the fact that such a wide range of cellular functions are affected by ethanol. Slapack *et al.* (1987) have produced a list of 13 different effects on microbial cells, including inhibition of solute uptake, denaturation and inhibition of glycolytic enzymes, uncoupling of oxidative phosphorylation and the induction of petite yeast strains.

While it is apparent that ethanol does inhibit certain glycolytic enzymes, e.g. glucokinase and glycerophosphate dehydrogenase (Miller *et al.* 1982), and that this may influence fermentation rates (particularly at ethanol concentrations above 16%), most of the evidence points to the cell membrane being the primary site of action for ethanol in the yeast cell. Brown *et al.* (1981) noted that growth and viability had different inhibition constants for ethanol. Many attempts have been made to improve the tolerance of yeast to ethanol. Increasing the level of unsaturated fatty acids and sterols has been reported to increase the alcohol tolerance of yeast (Thomas *et al.* 1978). However, attempts to isolate strains with increased ethanol tolerance have met with only limited success. The use of feedback selection techniques in continuous culture led to the isolation of strains which were able to ferment glucose in the presence of 10% ethanol at twice the rate of the parent strain (Brown & Oliver 1982). This suggests that this is a valuable technique for isolating strains of yeast with improved alcohol tolerance.

## 6.9 Storage carbohydrates

Both glycogen and trehalose can accumulate in certain conditions as storage materials in yeast. It has been reported that up to 40% of the dry weight of the cell can occur as storage carbohydrates, representing the consumption of up to 3% of the carbohydrates supplied in a beer fermentation (Quain & Tubb 1982). The differences in the role of trehalose and glycogen in the yeast cell are not clear. In general, the levels of both carbohydrates rise and fall in parallel, although the level of glycogen is normally higher than the level of trehalose in growing cells. Trehalose levels are normally higher in aerobically grown cells than in anaerobically grown cells, and also during ascospore formation, trehalose being the special storage material for ascospore germination. The levels of storage carbohydrates tend to increase in conditions of nitrogen limitation (Patel & Ingledew 1973) and at low dilution rates in continuous culture (Kuenzi & Fiechter 1972).

The pathways involved in the biosynthesis and mobilization of both glycogen and trehalose are known (Fig. 6.5), but the factors influencing glycogen and

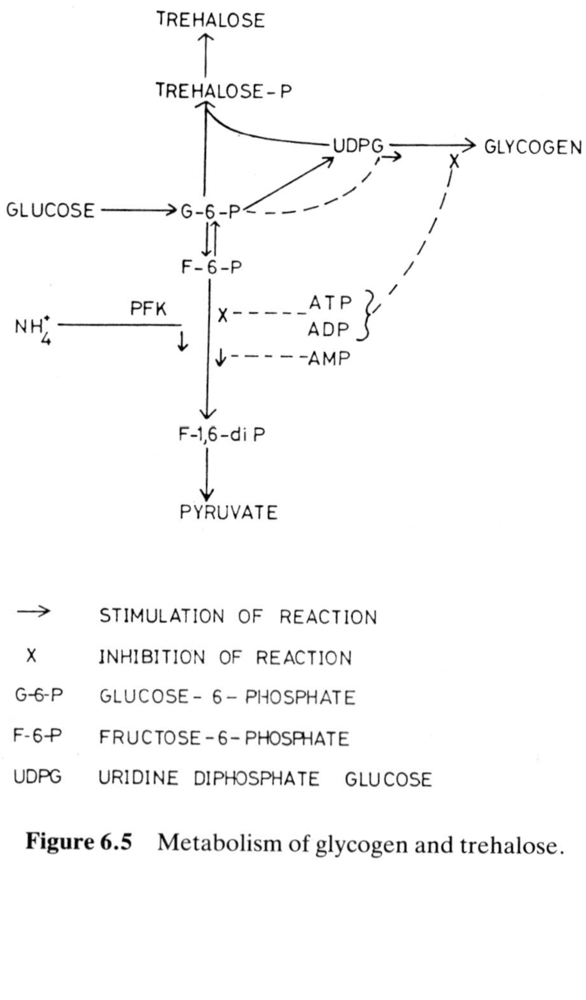


**Figure 6.5** Metabolism of glycogen and trehalose.

trehalose levels are far from understood. The biosynthetic pathways of both trehalose and glycogen are similar, both being produced from UDP glucose. The rates of metabolism of trehalose and glycogen may be more important in controlling their levels than their rates of biosynthesis. Both appear to have two mechanisms for mobilization. Glycogen can be metabolized either by glycogen phosphorylase or by the action of $\alpha$-glucosidase enzymes which have been described as occurring in sporulating yeast by Colonna and Magee (1978). Glycogen phosphorylase exists in two forms, a and b, but cyclic AMP has not been implicated in the phosphorylation of the phosphorylated form of glycogen phosphorylase B, as it is in higher organisms. Trehalose is hydrolysed to glucose by the enzyme trehalase. In the yeast two forms of trehalase appear to exist: a non-regulated trehalase, which has an acid pH optimum and is not phosphorylated, and a regulated trehalase, which has a neutral pH optimum and is phosphorylated. Evidence has been presented that cyclic AMP-dependent protein kinases are involved in the formation of the active form of trehalase which is phosphorylated (Thevelein 1984).

The level of reserve carbohydrates varies during different physiological conditions. In general the level of reserve carbohydrates decreases during periods of starvation, but the rapid depletion of glycogen has been reported when cells have been inoculated into fresh medium (Panek 1963) and in cells undergoing respiratory adaptation (Quain & Haslam 1979). A rapid decline in the level of glycogen has been reported by Quain *et al.* (1981), when yeast was inoculated into fresh beer worts. In these studies evidence was presented that this decreasing glycogen was associated with a period of rapid lipid synthesis and provided an energy source for this lipid synthesis. The level of glycogen has been monitored in beer fermentations (Patel & Ingledew 1973), and was found to increase during the period of sugar assimilation and then to decline slowly towards the end of the fermentation. The level of glycogen in pitching yeast is critical (Murray *et al.* 1984). Only yeast with a high level of glycogen (27.7%) gave a high level of yeast in suspension and a rapid rate of attenuation in a 16°C lager fermentation. Storage of yeast at 15°C in aerobic conditions leads to a decrease in the level of glycogen and a decrease in yeast performance.

## 6.10 Growth of genetically engineered yeast

The development of cloning techniques which permit the alteration of yeast genotype and the introduction of genetic information to permit the production of non-yeast proteins has led to the prospect of a new range of products which will require experience of yeast growth for their production. These include the peptide hormones, e.g. interferon (Hitzeman *et al.* 1981, 1983) and novel vaccines (Valenzuela *et al.* 1985).

Although most of the recent work has been concerned with *Saccharomyces cerevisiae*, work is proceeding in several other yeasts such as *Schizosaccharomyces pombe* (Beach & Nurse 1981), *Kluyveromyces* (Das & Hollenberg 1982) and others (Gunge 1983).

The growth of genetically engineered yeast and the development of new fermentation processes present new challenges for the physiologist, since it is necessary to grow yeasts whose characteristics have been determined more by the needs of the molecular geneticist than by the optimum requirements for fermentation.

The design of fermentations which involve the use of genetically engineered organisms has been considered by Pitt (1984), who listed 11 parameters which must be considered before the final decision on plant design and process operation can be made.

(a) Product demand. Has market research supplied information on this?
(b) State of operation. From (a) the scale of operation, with respect to size and number of fermentations, can be assessed assuming a threshold product titre for economic operation.
(c) Fermenter biomass concentration. This may be required at maximum possible concentration, or at a submaximal concentration as in the case of non-growth associated product formation.
(d) Production route.
(e) Cost of medium formulation. Given the product value and process unit costs, is this a significant cost factor in the overall process?
(f) Auxotrophic requirements of the host cell.
(g) Precursor requirements for product synthesis.
(h) Specific nutrient levels, especially for any limiting nutrient, desired to obtain the preferred physiological state of the host–vector system.
(i) Interfering substrate components. Removal or minimization of those interfering with product stability on downstream processing.
(j) Requirement for chemical selection pressure.
(k) Commercial availabilities of the various substrate components at the required grade for the duration of the production phase.

The cost of media in relation to that of product value and process unit costs is likely to have the greatest bearing on medium formulation. In the case of low-volume high-value products it has been estimated that in a recombinant bacterial DNA fermentation from a single fermentation run of 10 000 litres, 5 $\times 10^6$ doses (assuming that one dose per person represents 1 mg protein) of a typical human pharmaccutical product would be obtained (Pickett & Haddock 1983). Pickett and Haddock concluded that low fermenter volume requirements are therefore likely to be the norm, and hence the use of a chemically defined or semisynthetic medium at production scale may be economically viable. A synthetic medium provides a means of overcoming the problem commonly encountered with plasmid stability in cultures of recombinant yeast.

Plasmid stability refers to the efficiency of a plasmid in replicating and segregating to daughter cells during cell division. Under non-selective conditions plasmid-minus cells may arise as progeny of plasmid-containing cells. Due to their higher growth rate, these plasmid-minus cells may outgrow the plasmid-containing cells under the same growth conditions, so plasmid instability results in a 'diluting out' effect of plasmid-containing cells and an ever-increasing proportion of non-product forming cells in the total biomass. Geneticists can normally overcome this problem in the laboratory by the positive selection for the plasmid in yeast, containing the cloned gene of interest. For example, many yeast plasmids, such as YEp13, contain the *LEU2* gene (Broach 1983). When a *leu2* auxotrophic strain is transformed with such a plasmid it becomes prototrophic with respect to leucine. When the transformed strain is subsequently inoculated into a leucine-minus growth medium it will grow. Any plasmid-minus progeny will be leucine auxotrophs, being unable to grow in the leucine-minus medium to any degree. Similar results can be obtained by incorporating genes for other essential steps in yeast metabolism, e.g. triose phosphate isomerase into plasmids, or by using resistance genes to select for plasmid stability, e.g. copper resistance (Fogel & Welch 1982).

Plasmid stability in yeast growing in continuous culture has been investigated by Walmsley *et al.* (1983), who developed equations for predicting the kinetics of appearance of $leu^-$ cells in chemostat culture in conditions where the plasmid was considered to be at a selective disadvantage, a selective advantage or selectively neutral. Fermentations were carried out using different plasmids, and results were obtained which fitted closely the kinetics predicted by this model. Techniques for evaluating the stability of plasmids in small populations using flow-cytometry cell sorting have been described and may be valuable for further investigations in this area (Hjortso *et al.* 1985).

An advantage of synthetic media is that it is possible to add everything required for yeast growth minus the auxotrophic requirement used for plasmid selection. Synthetic medium without tryptophan as a selective pressure on the plasmid has been used in the commercial development of Hepatitis B surface antigen in yeast using 380-litre fermenters (Carty *et al.* 1984).

An additional problem that is encountered is that the host strains commonly used by geneticists have multiple auxotrophic requirements. Strains such as *2A* and *leu 2-3 leu 2-112* will be preferred as production strains, eliminating the cost of supplement addition to media, the double-leu mutant ensuring that the role of back-mutation is negligible.

Another useful feature of a synthetic medium is that it allows a degree of freedom in choosing sources of nutrient elements, both the form supplied and the concentration, thus allowing flexibility in the chemical and biochemical environment. Chemical and biochemical influences in the fermentation can both result in inactivation of protein or polypeptide products secreted by yeast and have a profound impact on the ease of protein recovery and fractionation during downstream processing (Dunnill 1983). The positive impact of media composition on recombinant DNA processes in yeast appears to be largely unexplored, though the potential is demonstrated by Franklin *et al.* (1984), who have shown that elevated NaCl concentrations cause a marked increase in extracellular secretion of proteins such as invertase.

The effects of harbouring high-copy number chimaeric plasmids and their encoded products on yeast physiology and nutrition are as yet, unknown. However in *Escherichia coli* the presence of a drug resistance plasmid, RPI, resulted in an increase in requirement for $K^+$, $Mg^{2+}$, $Fe^{2+}$ and $PO_4^{3-}$ (Klemperer *et al.* 1979).

Nutrient concentrations in media can affect the expression of certain promoter sequences inserted to increase the expression of the desired protein in genetic constructions. Acid phosphatase promoter, when present on a plasmid in yeast, increased the concentration of acid phosphatase six-fold under phosphate limitation (Brierley *et al.* 1984). The constitutive glycolysis promoters, such as phosphoglycerate kinase (PGK), will have highest activity in continuous culture. The maximal expression of insulin using PGK was found to occur during exponential growth, though maximum secretion occurred near the stationary phase (Hitzeman *et al.* 1983). The inducible promoters now available, such as galactokinase (*GAL*1), overcome possible problems of reduction in growth rate of transformed cells containing constitutive promoter sequences, due to overproduction of proteins which do not contribute to cell metabolism. The use of inducible promoter sequences will favour fed-batch production of recombinant yeast to obtain a high biomass before induction by substrate or by a temperature shift (Koltz 1983). The mode of operation of the process will affect the nutrient concentrations when they are added.

If chemically defined media were the rule, rather than the exception, in the yeast fermentation industries, then plasmid stability and other problems detailed above might not represent the stumbling blocks that they do to the commercial exploitation of genetically manipulated yeasts. In reality, only in the small-volume fermentations with high-value products is use of synthetic media possible. While recent publicity of the success of yeast molecular biology has been given to the production of high-value products such as interferons (Hitzeman *et al.* 1983), a steadily increasing number of cloned-gene functions of interest to the traditional yeast industries has begun. Genetic manipulation offers potential advantages to the yeast industries, by increasing production efficiencies or process economics, or both, via novel substrate utilization of low-cost raw materials (Stewart 1981).

The media used in the yeast industries, e.g. in the production of alcohol, are relatively low-value materials, such as sulphite waste liquor. Such media are complex, containing a variety of nutrient mixtures, and in addition are often subject to variability in composition (Suomalainen & Oura 1971, Maiorella *et al.* 1984). To exclude a particular auxotrophic requirement (e.g. leucine) from these complex media for the purposes of plasmid selection would not normally be an economically viable proposition. Other methods of achieving plasmid stability other than environmental selection will therefore have to be found.

## References

Andreasen, A. and T. Stier 1954. Anaerobic nutrition of *Saccharomyces cerevisiae*. II Unsaturated fatty acid requirement for growth in defined medium. *Journal of Cellular and Comparative Physiology* **43**, 27–32.

Aries, V. and B. H. Kirsop 1977. Sterol synthesis in relation to growth and fermentation by brewing yeasts inoculated at different concentrations. *Journal of the Institute of Brewing* **83**, 220–3.

Bailey, R. B. and A. Woodward 1984. Isolation and characterisation of a pleiotropic glucose repression resistant mutant of *Saccharomyces cerevisiae*. *Molecular and General Genetics* **193**, 507–12.

Ball, W. J. and D. E. Atkinson 1975. Adenylate change in *Saccharomyces cerevisiae* during starvation. *Journal of Bacteriology* **121**, 975–82.

Barford, J. P. and R. J. Hall 1979. An examination of the Crabtree effect in *Saccharomyces cerevisiae*, the role of respiratory adaptation. *Journal of General Microbiology* **114**, 267–75.

Barnett, J. A., R. W. Payne and D. Yarrow 1983. *Yeasts, characteristics and identification.* Cambridge: Cambridge University Press.

Beach, D. and P. Nurse 1981. High frequency transformation of the fission yeast *Schizosaccharomyces pombe. Nature* (London) **290**, 140–9.

Becker, J. M., F. Naider and E. Katchalski 1971. Peptide utilisation in yeast. Studies on methionine and lysine auxotrophs of *S. cerevisiae. Biochimica et Biophysica Acta* **291**, 388–97.

Beier, P. R., A. Slepziewski and E. T. Young 1985. Deletion analysis identifies a region, upstream of the ADH2 gene of *Saccharomyces cerevisiae*, which is required for ADR1-mediated derepression. *Molecular and Cellular Biology* **5**, 1743–9.

Breton, A. and Y. Surdin-Kerjan 1977. Sulphate uptake in *S. cerevisiae*: biochemical and genetic study. *Journal of Bacteriology* **132**, 224–32.

Brierley, R. A., D. Dibrasio, J. E. Miller, K. M. McNagny, E. A. Whiteside, J. Ma, J. F. Lemont and P. Saw 1984. Continuous fermentation of plasmid containing *Saccharomyces cerevisiae. Abstracts. Annual Meeting of the American Society of Microbiologists* 192.

Broach, J. R. 1983. Construction of high copy yeast vectors using 2 μm circle sequences. In *Methods in enzymology*, R. Wu, L. Grossman and K. Moldane (eds), Vol. 101, 307–25. New York: Academic Press.

Brown, S. W. and S. G. Oliver 1982. Isolation of ethanol tolerant mutants of yeast by continuous selection. *European Journal of Applied Microbiology and Biotechnology* **16**, 119–22.

Brown, S. W., S. C. Oliver, D. E. F. Harrison and R. C. Righelato 1981. Ethanol inhibition of yeast growth and fermentation: differences in the magnitude and complexity of the effect. *European Journal of Applied Microbiology and Biotechnology* **11**, 151–5.

Bruinenberg, P. M., J. P. van Dijken and W. A. Scheffers 1983a. A theoretical analysis of NADPH production and consumption in yeasts. *Journal of General Microbiology* **129**, 953–64.

Bruinenberg, P. M., J. P. van Dijken and W. A. Scheffers 1983b. An enzymic analysis of NADPH production and consumption in *Candida utilis. Journal of General Microbiology* **129**, 265–71.

Bruinenberg, P. M., P. H. M. de Bot, J. P. van Dijken and W. A. Scheffers 1983c. The role of redox-balance in anaerobic fermentations of xylose by yeasts. *European Journal of Applied Microbiology and Biotechnology* **18**, 287–92.

Bruinenberg, P. M., P. H. M. de Bot, J. P. van Dijken and W. A. Scheffers 1984. NADH-linked aldose reductase: the key to anaerobic alcoholic fermentation of xylose by yeast. *Journal of Applied Microbiology and Biotechnology* **19**, 256–60.

Cail, R., P. A. D. Rickard and P. L. Rogers 1979. Maximisation of *Candida utilis* cytochromes by pulse additions of ethanol to continuous culture. *Biotechnology and Bioengineering* **21**, 803–20.

Carty, C. E., F. X. Kouach, W. J. McAteer and R. Z. Maigetter 1984. Hepatitis B surface antigen production during fermentation by a recombinant yeast. *Abstracts. Annual Meeting of the Society for Microbiology*, 194.

Celenza, J. L. and M. Carlson 1984. Cloning and genetic mapping of SNF1, a gene required for expression of glucose-repressible genes in *Saccharomyces cerevisiae. Molecular and Cellular Biology* **4**, 49–53.

Ciriacy, M. 1978. A yeast mutant with glucose-resistant formation of mitochondrial enzymes. *Molecular and General Genetics* **129**, 329–35.

Ciriacy, M. 1979. Isolation and characterisation of further *cis*- and *trans*-acting regulator elements involved in the synthesis of glucose repressible alcohol dehydrogenase (ADH II). *Molecular and General Genetics* **176**, 427–31.

Colonna, W. J. and P. T. Magee 1978. Glycogenolytic enzymes in sporulating yeast. *Journal of Bacteriology* **134**, 844–53.

Cooper, T. G. 1982a. Nitrogen metabolism in *Saccharomyces cerevisiae*. In *The molecular biology of the yeast* Saccharomyces, J. N. Strathern, E. W. Jones and J. R. Broach (eds), 39–99. Cold Spring Harbor Monograph Series. New York: Cold Spring Harbor Laboratories.

Cooper, T. G. 1982b. Transport in *Saccharomyces cerevisiae*. In *The molecular biology of the yeast* Saccharomyces, J. N. Strathern, E. W. Jones and J. R. Broach (eds), 399–462. Cold Spring Harbor Monograph Series. New York: Cold Spring Harbor Laboratories.

Cooper, T. G. and R. A. Sumrada 1983. What is the function of nitrogen catabolite repression in *Saccharomyces cerevisiae*? *Journal of Bacteriology* **155**, 623–7.

Culbertson, M., T. Donahue and S. Henry 1976. Control of inositol biosynthesis in *Saccharomyces cerevisiae*. I. Properties of a repressible enzyme system in extracts of wild type ($1no^+$) cells. *Journal of Bacteriology* **126**, 232–42.

Das, S. and C. Hollenberg 1982. A high frequency transformation system for the yeast *Kluyveromyces lactis*. *Current Genetics* **6**, 123–8.

Dunnill, P. 1983. Trends in downstream processing of proteins and enzymes. *Process Biochemistry* (9 October), 9–13.

Eltayeb, Y. and D. R. Berry 1982. Studies on levels of isozymes of alcohol dehydrogenase during growth and sporulation of *Saccharomyces cerevisiae*. *Transactions of the British Mycological Society* **79**, 247–51.

Entian, K. D. and K. U. Fröhlich 1984. *Saccharomyces cerevisiae* mutants provide evidence of hexokinase PII as a bifunctional enzyme with catalytic and regulatory domains for triggering carbon catabolite repression. *Journal of Bacteriology* **158**, 29–35.

Evans, T. C. and C. Ratledge 1984. Induction of xylulase 5-phosphate phosphoketolase in a variety of yeasts grown on D-xylose: the key to efficient xylose metabolism. *Archives of Microbiology* **139**, 48–52.

Fiechter, A. 1977. Continuous cultivation of yeasts. In *Methods in cell biology II*, D. M. Prescott (ed.), 97–130. New York: Academic Press.

Fiechter, A., G. F. Fuhrmann and O. Kapelli 1981. Regulation of glucose metabolism in growing yeast cells. *Advances in Microbial Physiology* **22**, 123–83.

Fogel, S. and J. W. Welch 1982. Tandem gene amplification mediates copper resistance in yeast. *Proceedings of the National Academy of Sciences of the USA* **79**, 5342–6.

Foury, F. and A. Goffeau 1975. Stimulation of active uptake of nucleosides and amino acids by c-AMP in the yeast *Schizosaccharomyces pombe*. *Journal of Biological Chemistry* **250**, 2354–62.

Fraenkel, D. G. 1982. Carbohydrate metabolism. In *The molecular biology of the yeast* Saccharomyces, J. N. Strathern, E. W. Jones and J. R. Broach (eds), 38–57. Cold Spring Harbor Monographs. New York: Cold Spring Harbor Laboratories.

Franco, C. M. M., J. E. Smith and D. R. Berry 1984a. Effect of nitrogen and phosphate on the levels of intermediates in bakers yeast grown in continuous culture. *Journal of General Microbiology* **130**, 2465–72.

Franco, C. M. M., J. E. Smith and D. R. Berry 1984b. Changes in intermediate levels during batch culture of *Saccharomyces cerevisiae*. *Biotechnology Letters* **6** (12), 803–8.

Franklin, T. A., G. W. Malaney and R. D. Tanner 1984. The effect of salt and aeration rate on the secretion of proteins from growing bakers yeast *Saccharomyces cerevisiae*. In *6th Symposium on biotechnology for fuels and chemicals. Biotechnology and bioengineering*, C. P. Scotte (ed.), Vol. 14, 467–77. New York: John Wiley.

Franzusoft, A. and V. P. Cirillo 1982. Uptake and phosphorylation of 2-deoxy-D-glucose by wild type and single kinase strain of *Saccharomyces cerevisiae*. *Biochimica et Biophysica Acta* **689**, 295–304.

Fröhlich, K. U., K. D. Entian and D. Mecke 1984. Cloning and restriction analysis of the hexokinase gene of the yeast *Saccharomyces cerevisiae*. *Molecular and General Genetics* **194**, 144–8.

Fuhrmann, G. F. and A. Rothstein 1968. The transport of $Zn^{++6}$, and $Co^{++}$ and $Ni^{++}$ into yeast cells. *Biochimica et Biophysica Acta* **163**, 325–30.

Guijarro, J. M. and R. Lagunas 1984. *Saccharomyces cerevisiae* does not accumulate ethanol against a concentration gradient. *Journal of Bacteriology* **160**, 874–8.

Gunge, N. 1983. Yeast DNA plasmids. *Annual Review of Microbiology* **37**, 253–76.

Henry, S. A. 1983. Membrane lipids of yeast: biochemical and genetic studies. In *The molecular biology of the yeast* Saccharomyces, J. Strathern, E. Jones and J. Broach (eds), 101–58. New York: Cold Spring Harbor Laboratories.

Hitzeman, R. A., F. E. Hagie, H. L. Levine, D. V. Goeddel, G. Ammerer and B. D. Hall 1981. Expression of a human gene for interferon in yeast. *Nature* (London) **293**, 717–22.

Hitzeman, R. A., D. W. Leung, L. T. Perry, W. J. Kohr, H. L. Levine and D. W. Goeddel 1983. Secretion of human interferons by yeast. *Science* **21**, 620–5.

Hjortso, M. A., K. E. Dennis and J. E. Bailey 1985. Quantitative characterisation of plasmid instability in *Saccharomyces cerevisiae* using flow cytometry cell sorting. *Biotechnology Letters* **7**, 21–4.

Hough, J. S. 1971. *Malting and brewing science*. London: Chapman and Hall.

Hutner, S. H. 1972. Inorganic nutrition. *Annual Review of Microbiology* **26**, 313–46.

Ingram, L. A. and T. M. Bukkte 1984. Effect of alcohols on microorganisms. *Advances in Microbial Physiology* **25**, 253–303.

Jenkins, R. O., T. G. Cartledge and D. Lloyd 1984. Respiratory adaptation of anaerobically grown *Saccharomyces uvarum*. Changes in distribution of enzymes. *Journal of General Microbiology* **130**, 2809–16.

Jones, R. P. and P. F. Greenfield 1984. A review of yeast ionic nutrition, Part 1: Growth and fermentation requirements. *Process Biochemistry* (April), 48–60.

Jones, R. P., N. Pamment and P. F. Greenfield 1981. Alcohol fermentation by yeasts. The effect of environmental and other variables. *Process Biochemistry* (April/May), 42.

Kamihara, T. and I. Nakamura 1982. Regulation of respiration and its related metabolism by vitamin $B_1$ and vitamin $B_6$ in *Saccharomyces* yeasts. *Advances in Biochemical Engineering and Biotechnology* **29**, 35–82.

Klemperer, R. M., N. T. Ismail and M. R. Brown 1979. Effects of R plasmid Rp1 on the nutritional requirements of *E. coli* in batch culture. *Journal of General Microbiology* **115**, 325–31.

Koltz, L. C. 1983. Overproduction of proteins in recombinant organisms. In *Biochemical engineering III. Annals of the New York Academy of Sciences* **413**, 1–11.

Kotyk, A. and K. Janecek 1975. *Cell membrane transport*. New York: Plenum Press.

Kou, S. C. and V. P. Cirillo 1970. Galactose transport in *Saccharomyces cerevisiae*, III Characterisation of galactose uptake in transferase-less cells. Evidence against transport associated phosphorylation. *Journal of Bacteriology* **103**, 679–85.
Kuenzi, M. T. and A. Fiechter 1972. Regulation of carbohydrate composition of *Saccharomyces cerevisiae* under growth limitation. *Archiv für Mikrobiologie* **84**, 254–65.
Kulaev, I. S. and V. M. Vagabov 1983. Polyphosphate metabolism in microorganisms. In *Advances in microbial physiology*, A. H. Rose, T. G. Morris and D. W. Tempest (eds), Vol. 24, 83–174.

Leao, C. and N. van Uden 1982. Effect of ethanol and other alcohols on the kinetics and the activation parameters of thermal deaths in *Saccharomyces cerevisiae*. *Biotechnology and Bioengineering* **24**, 1581–90.
Lloyd, D., B. Kristensen and H. Degn 1983a. Glycolysis and respiration in yeast. The Pasteur effect studied by mass spectrometry. *Biochemical Journal* **212**, 749–54.
Lloyd, D., B. Kristensen and H. Degn 1983b. Glycolysis and respiration in yeast: the effect of ammonium ions studied by mass spectrometry. *Journal of General Microbiology* **129**, 2125–7.

McCann, K. and J. A. Barnett 1984. Starch utilisation by yeasts: mutants resistant to carbon catabolite repression. *Current Genetics* **8**, 525–30.
Maiorella, B. L., H. W. Blanch and C. R. Wilke 1984. Feed component inhibition of ethanolic fermentations by *Saccharomyces cerevisiae*. *Biotechnology and Bioengineering* **26**, 1155–66.
Matern, H. and H. Holzer 1977. Catabolite inactivation of the galactose uptake system in yeast. *Journal of Biological Chemistry* **252**, 6399–402.
Matsumoto, K., I. Uno, T. Ishikawa and Y. Oshima 1983. cAMP may not be involved in catabolite repression in *Saccharomyces cerevisiae*: evidence from mutants unable to synthesise it. *Journal of Bacteriology* **156**, 898–900.
Maw, G. A. 1960. Utilisation of sulphur compounds by brewers yeast. *Journal of the Institute of Brewing* **66**, 162–7.
Maw, G. A. 1963. Sulphur amino acid uptake by yeast. *Journal of General Microbiology* **31**, 247–59.
Miller, D. G., K. Griffiths-Smith, E. Algar and R. K. Scopes 1982. Activity and stability of glycolytic enzymes in the presence of ethanol. *Biotechnology Letters* **4**, 601–6.
Murray, C. R., T. Barich and D. Taylor 1984. The effect of yeast storage conditions on subsequent fermentations. *Master Brewers Association of the Americas Technical Quarterly* **21**, 189–94.

Nagamune, T., I. Indo and I. Inone 1981. The effects of medium composition on yeast physiological activities for ethanol production. In *Advances in biotechnology. II. Fuels, chemicals, foods and waste treatment*, M. Moo-Yong (ed.), 219–24. Oxford: Pergamon Press.
Nikawa, J. I., I. Nagumo and S. Yamashita 1982. Myo-inositol transport in *Saccharomyces cerevisiae*. *Journal of Bacteriology* **150**, 441–6.

Oura, E. 1974. Effect of aeration intensity on the biochemical composition of bakers yeast. I Factors affecting the type of metabolism. *Biotechnology and Bioengineering* **16**, 1197–212.
Oura, E. 1977. Reaction products of yeast fermentations. *Process Biochemistry* (April), 19–21.
Oura, E. and H. Suomalainen 1982. Biotin-active compounds: their existence in

nature and biotin requirements of yeast. *Journal of the Institute of Brewing* **88**, 299–308.

Panek, A. D. 1963. Function of trehalose in bakers yeast, *Saccharomyces cerevisiae. Archives of Biochemistry and Biophysics* **100**, 422–5.

Patel, G. B. and W. M. Ingledew 1973. Internal carbohydrates of *S. carlsbergensis* during commercial lager brewing. *Journal of the Institute of Brewing* **79**, 392–6.

Pêna, A. 1981. The transport of cations in yeasts; characteristics and postulated mechanisms. In *Advances in biotechnology, current developments in yeast research*, G. G. Stewart and I. Russell (eds), 395–410. Toronto: Pergamon Press.

Pickett, A. and B. Haddock 1983. Fermentation of recombinant organisms – problems and perspectives. *Processes in Biotechnology* **83**, 337–42.

Pitt, D. E. 1984. Fermentation parameters affected by the physiological characteristics of recombinant organisms. *Proceedings of Australian Conference on Biotechnology, University of Queensland*, Vol. 3, 3–18.

Polakis, E. S. and W. Bartley 1965. Changes in the enzyme activities of *Saccharomyces cerevisiae* during aerobic growth on different carbon sources. *Biochemical Journal* **97**, 284–97.

Quain, D. E. and J. M. Haslam 1979. Changes in glucose 6-phosphate and storage carbohydrate during catabolite derepression in *Saccharomyces cerevisiae. Journal of General Microbiology* **113**, 195–8.

Quain, D. E. and R. S. Tubb 1982. Importance of glycogen in brewing yeast. *Technical Quarterly of the Master Brewers Association of the Americas* **19**, 29–33.

Quain, D. E., P. A. Thurston and R. S. Tubb 1981. The structural and storage carbohydrates of *Saccharomyces cerevisiae*; changes during fermentation of wort and a role for glycogen catabolism in lipid biosynthesis. *Journal of the Institute of Brewing* **87**, 108–11.

Rickard, P. A. D. and C. B. J. Hogan 1978a. Effect of glucose on the activity and synthesis of fermentative and respiratory pathways of *Saccharomyces* sp. *Biotechnology and Bioengineering* **20**, 1105–10.

Rickard, P. A. D. and C. B. J. Hogan 1978b. Effect of the glyoxylic acid cycle on the respiratory quotient of *Saccharomyces* sp. *Biotechnology and Bioengineering* **20**, 1111–15.

Rogers, T. D. and H. L. Lichstein 1969. Regulation of biotin transport in *S. cerevisiae. Journal of Bacteriology* **100**, 565–72.

Rogers, P. J. and P. R. Stewart 1973. Mitochondrial and peroxisomal contribution to the energy metabolism of *Saccharomyces cerevisiae* in continuous culture. *Journal of General Microbiology* **79**, 205–17.

Rose, A. H. 1977. History and scientific basis of alcoholic beverage production. In *Economic microbiology*, Vol. 1, A. H. Rose (ed.), 1–41. New York: Academic Press.

Roth, R. and H. O. Halvorson 1969. Sporulation in yeast harvested in logarithmic growth. *Journal of Bacteriology* **98**, 831–2.

Ryder, D. S., D. R. Woods, M. Castian and C. A. Masschelein 1983. Glycolytic flux in lager yeast fermentations: application of isotachophoresis and bioluminescence for measurement of cellular intermediates. *Journal of the American Society of Brewing Chemists* **41**, 125–32.

Santos, E. L., M. V. Rodriguez, W. Elorza and R. Sentandreu 1982. Uptake of sucrose transport by *Saccharomyces cerevisiae. Archives of Biochemistry and Biophysics* **216**, 652–60.

Schlanderer, G. and H. Dellweg 1974. Cyclic AMP and catabolite repression in yeasts. In *Schizosaccharomyces pombe* glucose lowers both intracellular adenosine 3-5 monophosphate levels and the activity of catabolite sensitive enzymes. *European Journal of Biochemistry* **49**, 305–19.

Schmidt, G., L. Hecht and S. J. Thannhauser 1949. The effect of potassium ions on the absorption of orthophosphate and the formation of metaphosphate by bakers yeast. *Journal of Biological Chemistry* **178**, 733–42.

Schurr, A. and E. Yagil 1971. Regulation and characterisation of acid and alkaline phosphatase in yeast. *Journal of General Microbiology* **65**, 291–303.

Shatzman, H. 1975. Anaerobes Wachstum von *Saccharomyces cerevisiae*. Thesis 5504, ETH Zurich.

Slapack, G. E., I. Russell and G. G. Stewart 1987. Thermophilic bacteria and thermotolerant yeasts for ethanol production. *CRC Critical Reviews in Biotechnology* (in press).

Sols, A., C. Gancedo and G. Delafuente 1971. Energy yielding metabolism in yeast. In *The yeasts*, Vol. 2, A. H. Rose and J. S. Harrison (eds), 271–367. New York: Academic Press.

Steiner, S. and R. Lester 1972. Metabolism of diphosphoinositide and triphosphoinositide in *Saccharomyces cerevisiae*. *Biochimica et Biophysica Acta* **260**, 82–7.

Stewart, G. A. 1981. The genetic manipulation of industrial yeast strains. *Canadian Journal of Microbiology* **27**, 973–90.

Stryer, L. 1981. *Biochemistry*. San Francisco: W. H. Freeman.

Suomalainen, H. and E. Oura 1971. Yeast nutrition and solute uptake. In *The yeasts*, Vol. 2, A. H. Rose and J. S. Harrison (eds), 3–74. London: Academic Press.

Sy, J. and D. Richter 1972. Content of cyclic 3′5 adenosine monophosphate and adenyl cylase in yeasts at various growth conditions. *Biochemistry* **11**, 2788–91.

Thevelein, J. M. 1984. Regulation of trehalose mobilisation in fungi. *Microbiological Reviews* **48**, 42–59.

Thomas, D. S., J. A. Hossack and A. H. Rose 1978. Sterol and phospholipid composition influence ethanol tolerance in *Saccharomyces cerevisiae*. *Archives of Microbiology* **117**, 239–45.

Valenzuela, P., D. Coit, M. A. Medina-Selby, C. H. Kuo, G. Van Nest, R. L. Burke, P. Bull, M. S. Urdea and P. V. Graves 1985. Antigen engineering in yeast: synthesis and assembly of hybrid hepatitis B surface antigen – Herpes simplex 1 gD particles. *Biotechnology* **3**, 323–6.

van Dijken, J. P. and W. A. Scheffers 1984. Studies on alcoholic fermentation in yeasts. *Progress in Industrial Microbiology* **20**, 497–506.

Wales, D. S., T. G. Cartledge and D. Lloyd 1980. Effects of glucose repression and anaerobiosis on the activities and subcellular distribution of tricarboxylic acid cycle and associated enzymes in *Saccharomyces carlsbergensis*. *Journal of General Microbiology* **116**, 93–8.

Walmsley, R. M., D. G. Gardner and S. G. Oliver 1983. Stability of a cloned gene in yeast growth in chemostat culture. *Molecular and General Genetics* **192**, 361–5.

Watson, T. G. 1977. Inhibition of proline utilisation by glutamate during steady state growth of *Saccharomyces cerevisiae*. *Journal of General Microbiology* **103**, 123–6.

Wingender-Drissen, R. 1983. Yeast cAMP dependent protein kinase. *FEBS Letters* **163**, 28–32.

Wingender-Drissen, R. and J. U. Becker 1983. Regulation of yeast phosphorylase by phosphorylase kinase and cAMP-dependent protein kinase. *FEBS Letters* **163**, 33–6.

Zabriskie, D. W., W. B. Armiger, D. H. Phillips and P. A. Albana 1980. *Traders guide to fermentation media formulation*. Traders Oil Mill Co: Fort Worth, Texas.
Zimmerman, F. D. and I. Scheel 1977. Mutants of *Saccharomyces cerevisiae* resistant to carbon catabolite repression. *Molecular and General Genetics* **154**, 75–82.

# 7 *The technology of aerobic yeast growth*

J. P. BARFORD

## 7.1 Introduction

This chapter is concerned with the selection and operation of different types of fermentation apparatus for producing yeast products. The importance of such considerations as the biochemical pathways and microbial physiology of particular yeast strains, scale of operation, value of product, degree of sterility, nature of product and whether it is produced intracellularly or extracellularly on the selection of fermentation techniques will be discussed. Reference will be made to different industries which grow yeast aerobically, e.g. baker's yeast, the production of fodder (feed) yeast, yeast enzymes and other intracellular yeast components and genetically engineered products.

Extensive reviews of baker's yeast production, fodder (feed) yeast production and yeast enzymes and other intracellular yeast components have been regularly published (Aunstrup 1983, Aunstrup *et al.* 1979, Boing 1983, Demain 1978, Gaden & Humphrey 1977, Laskin 1977, Litchfield 1979, Nakao 1979, Peppler 1977, 1983, Reed 1982, Reed & Peppler 1973, Rehm & Reed 1983, Tannenbaum 1968) and it is not the intention of this work to rework these reviews. Relevant sections will be referenced for the readers' information.

An attempt was made to obtain up-to-date information in the four industrial production areas relevant: baker's yeast, fodder (feed) yeast, yeast enzymes and other intracellular yeast components, and genetically engineered products. This involved directly contacting over 100 major companies in the USA, Europe, Japan and Australasia; companies which included both major producers of yeast products, designers, manufacturers and suppliers of fermentation equipment to the yeast industry. Consistent with the experience of Sittig (1982), who attempted to gather realistic information about the latest state of fermentation reactors used in industry, the replies were generally scarce and uninformative. Attempts to follow up replies with specific questions generally resulted in statements such as the information was 'commercially sensitive', 'proprietary' or subject to 'secrecy agreements' between user and manufacturer or supplier. The few notable exceptions provide whatever up-to-date data is contained in this work. As a result of the varying degree of response obtained for the major products [baker's yeast, fodder (feed) yeast, yeast enzymes and other intracellular yeast components, and genetically engineered products], they will be treated in varying depth. In addition, some areas are the subject of separate chapters in this book (e.g. genetic manipulations). The author's own research interests – the control of sugar utilization by aerobically grown yeasts

(Barford 1984a, b, Barford & Hall 1978, 1981, Barford *et al.* 1981), the microbial physiology of yeasts (Barford & Hall 1976, Hall & Barford 1981, Schwartzkoff & Barford 1981) and yeast fermenter design (Prince & Barford 1982a, b, c, 1984) – provide the basis for the discussion of these aspects of baker's yeast and fodder (feed) yeast production and their important effect on the selection of fermentation techniques.

## 7.2 Baker's yeast

### *7.2.1 Introductory remarks*

Many excellent and up-to-date reviews exist concerning the production of baker's yeast (Reed 1982, Rehm & Reed 1983) and it is not therefore necessary to cover the general production techniques in detail. Rather, it is intended to illustrate how many factors have influenced the development of baker's yeast technology – notably the biochemistry and microbial physiology of the yeast *Saccharomyces cerevisiae*, the nature of the product (that is, the requirement for closely controlled metabolic activity characterized by the ability to undertake rapid fermentation in doughs of high osmotic strength and to maintain storage stability and viability), the production level and value of the product and the fermentation sterility requirements. It will be shown that each of these considerations has led to the choice of a fed-batch fermentation technique for baker's yeast production.

Two aspects are treated in rather more detail than others. These are the extent to which the biochemistry and microbial physiology of *Saccharomyces cerevisiae* has influenced baker's yeast technology, and approaches to aeration and mixing design as a component of fermenter design. These are chosen as they represent a level of detail, not presented in other reviews, of aspects which are of considerable importance in the industrial production of a good quality baker's yeast.

### *7.2.2 Factors affecting selection of fermentation techniques*

#### Major biochemical pathways and microbial physiology

The major biochemical pathways for sugar utilization by aerobically grown baker's yeast, *Saccharomyces cerevisiae*, are illustrated in Figure 7.1. Two major uses are evident in addition to biosynthesis; energy generation (by glycolysis or respiration) and carbohydrate storage (directly from glucose or by gluconeogenesis from excreted ethanol).

For energy production, sugar is taken up by the cell and transformed into pyruvate (glycolysis). The resulting pyruvate may be excreted as ethanol (fermentation) or further transformed to carbon dioxide and water using oxygen (respiration). Under conditions of excess sugar and oxygen the maximum possible rate of fermentation in *Saccharomyces cerevisiae* is a faster rate than the maximum possible rate of respiration. However, fermentation is a less-efficient process energetically. Consequently, the two major pathways

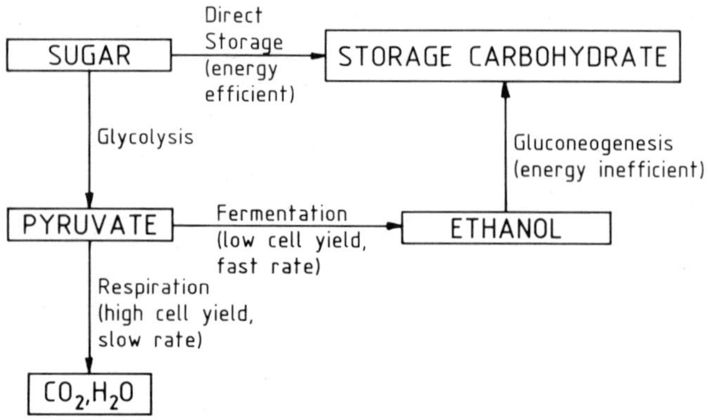


**Figure 7.1** Basic metabolic pathways of *Saccharomyces cerevisiae*.

consist of a rapid, energetically inefficient pathway (glycolysis and subsequently fermentation) and a slower, energetically efficient pathway.

Similarly, carbohydrate storage may occur by two major pathways; storage directly from sugar or gluconeogenesis from excreted ethanol. The two major carbohydrate storage products are trehalose and glycogen. Storage directly from sugar is a relatively simple polymerization step and is energy efficient compared with the process of gluconeogenesis, where two carbon units must be formed into complex carbohydrates.

Differences in the maximum rates at which these pathways can operate and their energy efficiency have a significant effect both on a single-cell and on an overall cell-culture level.

The overall metabolism of *Saccharomyces cerevisiae* is best illustrated by the growth of the strain in continuous culture under fully aerated conditions and sugar limitation. An ideal schematic diagram is given in Figure 7.2. The diagram may be conveniently considered in two parts; those dilution rates (or specific growth rates) below a critical dilution rate and those above a critical dilution rate. Below the critical dilution (S) rate, the metabolism is characterized by complete sugar utilization, no ethanol (E) excretion, high biomass : substrate yield ($y_{xs}$). During this range of dilution rates, the rate of sugar uptake by the organism is less than the maximum rates of both fermentation and respiration (due to restricted sugar availability). Consequently, the sugar is fully respired and this results in the predominant metabolic characteristics of complete sugar utilization, no ethanol excretion and high biomass : substrate yield. Also, during this range of dilution rates the specific oxygen uptake rate, in mmol oxygen $g^{-1}$ biomass $h^{-1}$ ($QO_2$), increases approximately linearly with dilution rate, indicating that the maximum respiratory capacity of the organism has not yet been reached (due to restricted sugar availability).

Above the critical dilution rate different metabolic characteristics are evident; incomplete sugar utilization, ethanol excretion and low biomass : substrate yield. During this range of dilution rates the rate of sugar uptake by the organism exceeds either the maximum rate of respiration as reflected by the maximum specific oxygen uptake rate $QO_2$(max) or the prevailing rate of respiration ($QO_2$). This is due to the increasing sugar availability with increasing dilution rate and the low maximum respiratory capacity of this organism. Consequently, not all of the sugar taken up by the organism can be respired, and significant ethanol excretion occurs. Hence, during this range of dilution rates the metabolism is predominantly fermentative.

The actual rate of respiration during this period, as indicated by the specific oxygen uptake rate, is generally considered to decrease, indicating respiratory repression (Dellweg *et al.* 1977, Meyenburg 1969). This is illustrated by specific oxygen uptake rate (see Fig. 7.2). These experimental findings have been questioned (Barford 1984a, Barford & Hall 1979, Barford *et al.* 1981) and a specific oxygen uptake rate profile such as that illustrated (see Fig. 7.2) is suggested as being the actual response when adequate adaptation times are allowed. Irrespective of the outcome of this issue, the actual rate of respiration is considerably less than the rate of fermentation. Since fermentation predominates, the observed metabolic characteristics of ethanol excretion and low biomass : substrate yield result.

Such an overall culture profile introduces an interesting consideration into the industrial manufacture of baker's yeast. Clearly, at low dilution (growth) rates a greater yeast yield per unit of sugar will be obtained. However, the fermentative activity of this yeast may be reduced. Alternatively, at high dilution (growth) rates a much reduced (approximately 0.15–0.2 g biomass $g^{-1}$ sugar, compared with approximately 0.5 g biomass $g^{-1}$ sugar at low dilution rates) yeast yield is obtained, but with possibly much higher fermentative activity due to the predominantly fermentative metabolism at these dilution rates. If it is remembered that the baker pays for yeast activity (i.e. the ability to produce gas in the dough, or rapid fermentation), then the trade-off is not completely obvious. It is known that in the baking process very little net growth

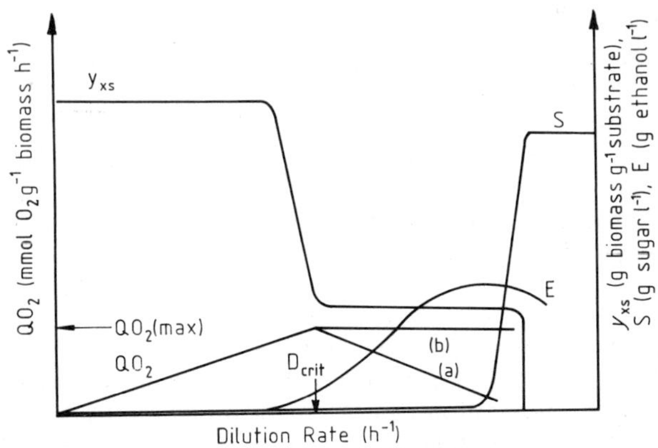


**Figure 7.2** Continuous culture of aerobically grown *Saccharomyces cerevisiae* with sugar as the limiting substrate (ideal schematic). [(a) and (b) indicate different responses observed experimentally for the specific oxygen uptake rate at dilution rates greater than the critical dilution rate.]

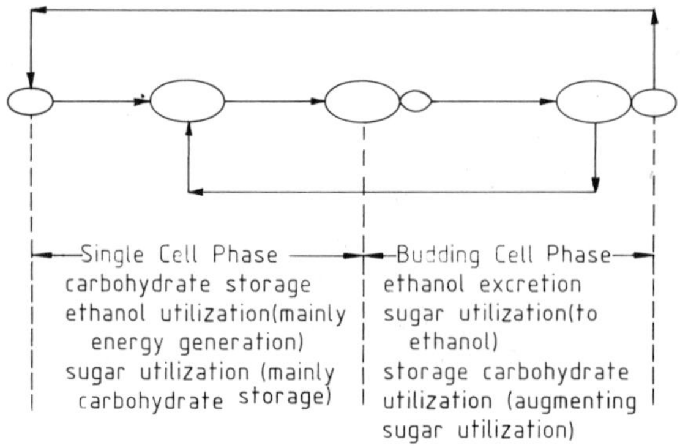


**Figure 7.3** Cell cycle organization for *Saccharomyces cerevisiae* (schematic).

of yeast occurs and that, consequently, the yeast inoculum level is high (the number of yeast cells per gram of dough exceeds that of a starting beer or wine fermentation by a factor of about 50) (Reed 1982). The level of glycolytic enzymes is (fortuituously) largely constitutive in yeast (Beck & von Meyenburg 1968) and, consequently, the growth rates which favour high biomass : substrate yield are favoured when consideration of net fermentative capacity (that is, fermentative capacity per cell × cell yield from substrate) is important.

The organization of the cell cycle of *Saccharomyces cerevisiae* also has an influence on the culture conditions prevailing in baker's yeast production. Cell cycle variations may be studied experimentally (Barford & Hall 1976, Meyenburg 1969) and by computer simulation (Hall & Barford 1981) amongst other techniques. Figure 7.3 is a schematic diagram of the cell cycle of *Saccharomyces cerevisiae*, the single-cell phase and the budding-cell phase. During the budding phase the processes of DNA replication and nuclear and cell division occur. This occurs in an approximately constant period (Barford & Hall 1976). For these processes to occur in approximately constant time, considerable use of storage carbohydrate sources (notably trehalose and glycogen) occurs, thus ensuring a high degree of independence of these events from the environmental conditions prevailing. During this period rapid use of the carbohydrate sources is undertaken and, consequently, the energy metabolism is predominantly fermentative with ethanol excretion. The extent to which storage carbohydrate is used, compared with the direct utilization of sugar, is not clear, although enough storage carbohydrate exists to complete the budding phase in the absence of external sugars. In the presence of an external sugar *Saccharomyces cerevisiae* will utilize the sugar to some extent, augmenting the utilization of storage carbohydrates for energy production. The unbudded phase of the cell cycle of *Saccharomyces cerevisiae* is predominantly used for the building up of storage carbohydrate reserves for later use during the budding period. However, the method by which the organism utilizes the two substrates available, viz. sugar and ethanol, has an important role to play in the efficient production of biomass. By utilizing the incoming sugar predominantly for carbohydrate storage (an energy-efficient pathway), it is able to utilize ethanol for providing the maintenance energy required for the unbudded cell phase (another energy-efficient process), only requiring storage carbohydrates to be made from ethanol (an energy-inefficient process) under exceptional circumstances. Consequently, the carbon source is utilized with greatest energy efficiency, given the complex organization of the cell cycle.

The growth rate of the micro-organism affects the predominance of either the unbudded-cell or budding-cell phase. Since the budding-cell phase is constant (at about 70–80 min), the culture is predominantly budding cells at the fastest growth rates, where cell cycle times are in the order of 90 min. At these growth rates (e.g. batch growth) the metabolic characteristics are rapid fermentation of sugars to ethanol and low biomass : substrate yield. At slow growth rates or long cell cycle times, the culture contains a substantial proportion of unbudded cells and the metabolic characteristics of such a culture are complete respiration, little ethanol excretion and high biomass : substrate yield. Hence, the organization of the cell cycle of *Saccharomyces cerevisiae* also favours the lower growth rates for efficient biomass production. As discussed in detail later, this is normally achieved using a fed-batch process rather than a continuous process. In addition, at lower growth rates the cell content of storage carbohydrates, e.g. trehalose, will be higher due to this cell cycle organization – an important consideration with respect to the subsequent storage stability of the yeast.

**Nature of product, scale of production, value of product and degree of sterility**

In addition to the basic biochemical pathways and microbial physiology of *Saccharomyces cerevisiae* influencing the selection of the fermentation techniques employed, the nature and value of the product also affects this selection.

Baker's yeast consists of whole-yeast cells of a closely controlled metabolic activity which is characterized by the ability to undertake rapid fermentation in doughs of high osmotic pressure. They are selected strains of the yeast *Saccharomyces cerevisiae* which, in addition to the properties exhibited in doughs, are chosen for their vigorous growth, high yields on molasses and storage stability and viability. Storage stability and viability is an important aspect of baker's yeast production technology, since long shelf-life is essential for consumer yeasts as well as transported yeasts. Batch fermentation at a number of sites reduces the requirement for extensive refrigerated transport and refrigerated storage at the bakery site, compared with centralized continuous culture installations (assuming baker's yeast quality to be equivalent). Storage stability and viability has been correlated with levels of various cellular components – storage carbohydrates, in general, and trehalose in particular (Bocharova *et al.* 1976, Panek 1975), as well as other components such as nitrogen and phosphate level. Compressed yeasts with a nitrogen content of above 8% (based on dry weight) have a shorter shelf-life than yeast with nitrogen levels of 7% or less (Reed 1982). As discussed above, the organization of the cell cycle has an influence on the levels of storage carbohydrates (e.g. trehalose) contained in baker's yeast with desirable high levels being favoured by lower growth rates. Low growth rates may be achieved most simply by either a fed-batch fermentation technique or a continuous culture fermentation technique.

The fed-batch fermentation technique is chosen for baker's yeast for a number of reasons in addition to any consideration regarding storage stability (which is satisfied by such a fermentation technique). First, there is very little growth of the baker's yeast in the dough. Reed (1982) has estimated that in 1–5 h of bread dough fermentation, yeast cell counts were $275 \times 10^6$ cells $ml^{-1}$ at the start of the fermentation and $300 \times 10^6$ cells $ml^{-1}$ at the end. Consequently, contaminants (normally bacteria with less osmotolerance than yeast) are of no consequence in the dough unless a major contamination has occurred during production. The lack of necessity for absolute sterility during baker's yeast production has two major consequences. First, it allows the more simple fermentation technique of batch culture to be used. The fermentation vessel need not be pressurized, which lowers the cost of production. In addition, it allows a variety of fermentation peripheral equipment to be used which may involve considerable risk if strict sterility (e.g. in continuous culture) were necessary. These include internal cooling coils, aeration systems where the fermenter contacts are recirculated through pumps, heat exchangers and external air-distribution systems. Downstream processing is also considerably simplified when strict sterility is not a requirement. It has been estimated that the total bacterial count is usually between $10^4$ and $10^9$ cells $g^{-1}$ baker's yeast (Reed 1982). Continuous cultivation may then be seen to confer additional operational difficulties, e.g. sterility requirements for both the fermentation vessel and fermentation peripherals, which are reduced when batch or fed-batch cultivation is employed. In addition, under continuous culture techniques the composition of the cell is constant and, although some cellular components (e.g. trehalose) may be at an approximately suitable cell level to ensure storage stability under such growth conditions, not all important cellular components may be at suitable levels for the subsequent storage and dough fermentation stages. Fed-batch fermentation techniques offer more flexibility in the manipulation of cellular composition.

Baker's yeast is an example of a relatively large production of a low-value product. Estimates of the worldwide baker's yeast production have been given by Peppler (1983). The 1980 production estimates totalled $1574 \times 10^6$ kg yeast (28–32% dry matter) per annum worldwide. Of this, North and Central America produced $363 \times 10^6$ kg yeast per annum. The current production in the USA has been estimated by Reed (pers. comm.) to be $200–227 \times 10^6$ kg per annum. The price of baker's yeast is approximately US \$2.0–2.2 $kg^{-1}$ yeast, of which about 30% is estimated to be the production cost. At such production levels continuous culture techniques are economically feasible, but the added complexity of operation (with respect to sterility requirements and the achievement of appropriate yeast composition), coupled with the transportation costs and storage of using a large central facility, have resulted in the use of a large number of fed-batch fermenters at a variety of locations. These fermenters are of the order of 100–200 $m^3$ in size, with an estimated output of up to 10 000 kg yeast per batch. Active dry yeast has not replaced compressed yeast in most wholesale bakery operations, and markets where its storage stability is advantageous (e.g. consumer markets) are only small markets. Consequently, active dry yeast production has not had any significant effect on the selection of the fermentation technique employed.

### 7.2.3 *Baker's yeast production*

**General outline**

As previously stated, excellent and up-to-date reviews concerning the production of baker's yeast have been published recently (Reed 1982, Rehm & Reed 1983), and it is not intended to cover the general production techniques in detail here. A brief outline is summarized from Reed (1982).

Baker's yeast (*Saccharomyces cerevisiae*) fermentations are carried out under highly aerobic conditions with incremental feeding of molasses. Molasses is received at about 80°Brix, with a fermentable sugar concentration of 50–55%. It is diluted (to about 40°Brix) and clarified for use in batch fermentation. Nitrogen sources used are ammonia, ammonium sulphate and urea; phosphorus sources are orthophosphates or phosphoric acid. Mineral (magnesium and trace minerals) and vitamin (biotin and thiamin) supplements are added. The fermentation is carried out at a pH of between 4 and 6, at a temperature of 30°C for periods of up to 24 h. When the fermentation is completed (yeast solids of the order of 40–60 kg $l^{-1}$) the yeast cells are concentrated by centrifugation to a yeast cream of 15–20% solids. The cream is cooled and then filtered using either a filter press or a rotary vacuum filter. The yeast press cake is then extruded and sold as compressed baker's yeast in blocks packaged in wax paper (0.5–2.5-kg blocks) or it is crumbled and sold in bulk. In the production of active dry yeast the press cake is extruded in the form of fine strands. These are dried in tunnel driers on endless steel wash belts, in rotary louvre driers or in air-lift driers. Active dry yeast contains 90–95% yeast solids compared with approximately 30% in compressed yeast. It has a longer shelf-life but less bake activity.

**Fermentation apparatus**

For the reasons previously discussed, baker's yeast is produced in batch fermenters. The final fermentation vessel is generally of the order of 100–200 $m^3$ in size, and is fabricated from stainless steel or other corrosion-resistant materials. The fermenter is equipped with an aeration system, cooling system, sugar and nutrient metering system, sampling ports, and foam- and pH-controlling devices.

Efficient aeration is a crucial aspect of baker's yeast production, and will be discussed here in detail. It is important to appreciate at the outset that simply ensuring sufficient aeration, i.e. rate of oxygen transfer to the fermentation medium greater than rate of oxygen utilization by the biomass, is not the only measure of the adequacy of an aeration system, since in many cases the aeration system is also used for mixing of the fermenter contents. Effective aeration does not necessarily imply complete mixing. Incomplete mixing is significant in baker's yeast production, since the product relies on a carefully controlled composition and activity, which in turn requires the adequate mixing of all nutrients within the fermenter. Incomplete mixing can result in reduction in yield and changes in microbial composition and activity. As a result, some discussion of the approaches to the quantification of mixing on production-scale fermenters is included here.

Generally, three major types of aeration systems are used in the baker's yeast industry:

(a) air spargers without mechanical agitation;
(b) agitated vessels (conventional stirred bioreactors); and
(c) proprietary aeration systems (e.g. Vogelbusch, Frings and Hoechst).

Air-sparged vessels are most common in the USA and Australasia. Considerable application of various proprietary aeration systems has occurred in Europe, Japan, Africa and South America. Agitated vessels can be used to assist distribution of air bubbles from an air sparger consisting of a network of perforated tubes, or may provide virtually the complete power input for creating and distributing the air bubbles.

Air spargers without mechanical agitation consist of a horizontal tube with side tubes provided with a large number of small holes. Rosen (1977) describes a 'typical' system consisting of a horizontal tube with 24 side tubes provided with 30 000 holes of diameter 1.5 mm. The provision of a large number of small holes does not necessarily guarantee good mass transfer, since small bubbles formed at tiny orifices may eventually coalesce to form large bubbles in a medium if there is inadequate mixing or ineffective surfactants, or both, present. Power input from the gas phase is also important. If the power input from the gas phase is insufficient to generate turbulence in the liquid phase, then the bubble size will be that of the bubbles formed at the orifice and will probably increase with liquid height in the fermenter, due to bubble coalescence. However, if the liquid is in turbulent motion bubble break-up will also occur, and coalescence and break-up equilibrium will determine the mean bubble size. Correlations for these phenomena are available in the excellent publications of Moo-Young and Blanch (1981) and Yoshida (1982). The gas-flow distribution and the height : diameter ratio of the fermenter will also then have a considerable effect on the extent of liquid turbulent motion for a given gas phase power input. As will be discussed later, although such correlations provide a theoretical basis and starting point for engineering calculations, differences in theoretical properties of fermentation liquids from 'ideal' experimental conditions and scale-up considerations must be taken into account with respect to the application of any correlation to a particular fermentation process.

The provision of air to air-sparged systems is generally of the order of 0.8–1.4 volumes of air per fermenter volume per minute (Reed 1984). Such figures are not meaningful for any comparative purposes (i.e. different fermenter geometry and different aeration systems), since they give no information regarding bubble size, gas hold-up, actual dissolved oxygen tension, etc. Specific energy consumption is of the order of 150 Wh $kg^{-1}$ yeast (28% dry solids).

Agitated vessels are used by some companies (e.g. Anhauser-Busch, USA). The major reasons for using these systems are to attempt to promote more-extensive mixing of the liquid and gas phases at larger scale, and to ensure effective power input by turbulent liquid motion. As previously stated, good mixing is essential in baker's yeast production. In small tanks, provided the impeller speed is above a critical value, both the liquid and gas phases are considered to be perfectly mixed. For large tanks this is often not the case industrially (Oosterhuis & Kossen 1983). The actual behaviour of the two phases should be established, since assumptions of either well-mixed or plug-flow behaviour of the gas phase may lead to gas absorption rate predictions which differ by a significant order of magnitude (Moo-Young & Blanch 1981). Modelling of both liquid- and gas-phase behaviour has been carried out at both laboratory (see Moo-Young & Blanch 1981) and industrial scale (Oosterhuis & Kossen 1983). Such approaches, as discussed later, are essential for scientifically based scale-up.

The major proprietary aeration systems are typified by those of Vogelbusch, Frings and Hoechst. The basis of the Vogelbusch aeration system is the development of a special centrifugal pump suitable for pumping homogeneous gas and liquid mixtures of a specific density of 0.3–0.95 at a hydraulic efficiency of 0.65–0.75, depending upon the application (Schreier 1975). The hydraulic part consists of a multistage impeller with a spiral casing. The gas separated in the pump is removed from the impeller by a system of boreholes, and is removed from the pump by a special constriction in the pump casing. The rear of the impeller is equipped with a special blading (separation stage) which, in conjunction with a specific casing cover design and a throttling flag in the pressure pipe, ensures that only gas without liquid leaves the pump. Fermentation medium (liquid and gas) is pumped from the bioreactor using this pump for circulation of reaction liquid and simultaneous discharging of a controlled part of the waste air. A pressure line is then used to supply an aeration installation with the pump exit reaction liquid. A tube heat exchanger is installed in the pump pressure line to remove the heat evolved during fermentation. The aeration installation involves injection of air into the medium, generally induced by pumping the medium past air inlet holes (Hoehne 1975). In the Vogelbusch system the aeration installation is placed at the top of a long shaft, the outlet of which is on the level of the surface of the liquid to be aerated. The air flow is self-priming. The air sucked in is finely dispersed, and a largely homogeneous mixed jet leaves the shaft end. The aerated jet penetrates nearly to the bottom of the fermentation vessel by virtue of its energy content. This causes intense turbulence in the fermentation vessel. Fine bubbles are then injected into the base of the fermenter and are exposed to fields of turbulence (created by the air jet trail) as they move to the liquid surface. In this way, Vogelbusch claim that the system has the conditions conducive to high mass transfer, viz. high turbulence and constant renewal of the contact surface, fine dispersion of gas, long residence time of gas in the medium and homogeneity of reactor contents (i.e. good mixing). The size of circulating pumps and overflow shafts are determined according to the oxygen transfer rate required with larger fermenters (up to 1000 $m^3$) having a multiple pump and shaft arrangement. Oxygen transfer rates of 8–12 (max) kg oxygen $m^{-3}$ reactor volume $h^{-1}$ (250–350 mol $m^3$ reactor volume $h^{-1}$) are claimed to be possible with an optimum specific energy consumption of the order of 0.45 kWh $kg^{-1}$ oxygen transferred.

Actual operating specific energy consumption is often of the order of 0.6 kWh $kg^{-1}$ oxygen transferred. However, it must be noted that oxygen transfer rate power consumption is highly dependent on the character and concentration of the substrate, and performance data for any particular application may deviate from these generalized figures. Fermenters with considerably simpler air-sparging systems have lower oxygen transfer rates but higher usable fermenter volume than those in which air bubbles are finely dispersed. Clearly, comparisons of oxygen transfer rates between these two types of aeration systems have to be made on an equivalent basis to be valid. Vogelbusch systems have been installed in a number of countries, including Yugoslavia (40, 80 and 120 $m^3$), Japan (100 $m^3$), Austria, Canada, Zaire and Bulgaria (50 and 110 $m^3$).

The Frings aeration system (the Friborator) was originally developed for the microbial production of vinegar and was subsequently used for baker's yeast production. Hoechst have modified this aeration system to allow for sterile operation (this being not strictly necessary for vinegar and baker's yeast production). The modified fermenter also contains a draught tube to promote a loop-type movement of the liquid flow pattern (Sittig 1982). The Frings aerator consists of a motor shaft which drives a turbine, at the base of which is a stator. The motor shaft enters the fermentation tank through the bottom. It is sealed against the bottom by a mechanical seal. The turbine of the Friborator is shaped such that it freely aspirates atmospheric air down to many metres below the liquid surface. The air is mixed intimately with the liquid by means of a stator (which also distributes the gas and liquid mixture uniformly across a defined cross section of the fermenter). The motor speed largely determines the amount of air aspirated, although there is a small decrease with increasing liquid height. The power consumption of the motor also increases with increasing liquid height. The continuous alteration of motor speed using a frequency alternator is useful in the baker's yeast industry, where varying quantities of air are required at different times of the batch process. To use energy most efficiently the tank height is limited to about 5 m. Above this height a combination of the Friborator and a blower is used. Part of the energy required for the pressurization of the air is supplied by the blower. The motor for the Friborator may then be reduced in size. Power consumption for the combined Friborator and blower is less than the self-aspirating system, because the turbine needs only to aspirate against a lower, and always constant, liquid level. The turbine must aspirate the gas to mix it into the liquid properly, and the air pressure must be at a level which does not lead to flooding of the turbine. The maximum specific energy consumption of this system is claimed to be 200 Wh $kg^{-1}$ yeast (28% dry solids) for a typical self-aspirating system (fermenter height 7.5 m with final operating level 4–4.5 m) compared with 150 Wh $kg^{-1}$ yeast (28% dry solids) for a typical blower supported system (fermenter height 8–9 m with final operating level 5.5–6 m). This compares with the Vogelbusch system, whose operating specific energy consumption figures of approximately 0.58 kWh $kg^{-1}$ dry matter is equivalent to 162 Wh $kg^{-1}$ yeast (28% dry solids). Frings aeration systems have been used extensively in Germany for baker's yeast production.

Most detailed design information relating to aeration systems and actual performance data from operating plants is generally considered to be secret by major baker's yeast companies. The design methods available to baker's yeast companies consist of empirical correlations contained in the literature combined with actual operating experience from plants and laboratory work at varying scales. It is important to appreciate, in the first instance, that empirical correlations relating overall mass transfer rates directly to major process variables exist, but only facilitate design calculations. A wealth of practical information under plant operating conditions exists within companies, and provides a significant (if not dominating) input to aerator design. There are a number of reasons why available correlations are only used as a starting point in design calculations. Most correlations available for biochemical systems are generally based on chemical engineering design approaches to mass transfer. Many important considerations unique to biochemical systems restrict the application of such approaches in their current form. The two major limitations are: fermentation broth characteristics and scale-up methodologies.

Fermentation broth characteristics may be generalized as the effects of surfactants (e.g. antifoams and end-products of fermentation) limiting substrate depletion, electrolyte concentration, buffering capacity, biomass concentration, etc., on the rheological properties of the fermentation medium and their effect on bubble size, mass-transfer coefficient, gas : liquid interfacial area and, most importantly, bubble coalescence. Many excellent sources of correlations exist (e.g. Moo-Young & Blanch 1981, Yoshida 1982) for such major process variables as:

(a) the behaviour of bubbles in the presence of surfactants (e.g. effect on bubble size, gas hold-up, mean bubble size and bubble size distribution, $k_L$ variations);
(b) the effect of biomass concentration on the mass transfer coefficient, $k_L$ and the interfacial area, $a$;
(c) effect of electrolyte concentration on the mass transfer coefficient, $k_L$ and the interfacial area, $a$;
(d) power requirements for aeration of both unsparged and sparged fermenters; and
(e) effect of superficial gas velocity on the mass transfer coefficient, gas hold-up, interfacial area, etc.

Given the extent to which correlations exist, it is important to appreciate the reasons for their limited use, in general. The major reason for this is that most work carried out to obtain experimental correlations is performed on media which do not accurately reflect the complexity of the actual fermentation medium, or on a scale or fermenter geometry which differs from the actual fermentation system which is to be designed. While it is understandable that variations in actual fermentation media and scales of operation are substantial and anything but an exhaustive experimental programme could not make any impact on such a range of variables, the fact remains that many correlations bear little relevance to the actual operating fermenter, and provide only a first estimate of important design parameters. For example, the effect of surfactants are often studied experimentally by the addition of soluble surfactants, e.g. sodium lauryl sulphate and bovine serum albumin and are generally found to result in a decrease in the mass transfer coefficient, $k_L$. In addition, the behaviour of bubbles can be observed to change significantly with surfactant addition (e.g. the rise of large bubbles through a swarm of fine bubbles and the coalescence of large bubbles with fine bubbles). Any attempt to use such experimental data on an actual fermentation process is extremely difficult, because the addition of surfactants to the system or the excretion or removal, or both, of substances with surfactant properties by the growing cell will be a dynamic process, not a well-defined static process as in the experimental case. In addition, considerable variations often exist between industrially utilized medium (e.g. cane or beet molasses) from batch to batch with respect to concentrations of a variety of medium components, some with surfactant properties. Rheological properties may vary significantly. This makes the application of generalized correlations difficult.

The scale of operation also presents a vital area for consideration, since actual fermentation systems are often orders of magnitude larger than the experimental vessel in which the correlations have been formulated. If it is attempted to utilize correlations directly, then it would generally be considered appropriate to maintain geometric similarity on scale-up and to keep a given criterion constant. If a given criterion is kept constant, then other variables change due to the laws governing heat, mass and momentum transport and their dependence on fermenter geometry and operating conditions (Oldshue 1966). This often results in the assumed controlling variable, on which the scale-up was based, no longer being the controlling variable, but some other variable becoming controlling. Even if geometric similarity in scale-up is not maintained and some other variable such as power per unit volume, $k_L$, $a$, etc., is kept constant, these restrictions are not eliminated.

Given the limitations of design correlations, both in the nature of the experimental procedures used to obtain them and the limitations of scale-up methodologies, how does a company approach large-scale aeration design for fermenters producing baker's yeast? The approach used by ICI in the design of their single-cell protein process (Scott 1983) may be considered to be the idealized approach from small scale to large scale. This involved a number of identifiable steps.

(1) The relationships between growth, product formation and the external environment are established for the micro-organisms of interest, using laboratory fermenters. These relationships are independent of scale provided the small-scale equipment used is selected and designed to ensure any artefacts, e.g. poor mixings, are not introduced into the results.

(2) Based on these data, an economic analysis is undertaken and scale-up targets for product concentration, productivity, yield and product purity can be set. Heat and mass transfer constraints on the large plant are taken into account.

(3) Experiments are then conducted on various scales. Typically the minimum-sized fermenter for carrying out these experiments is approximately 1 $m^3$, due to such considerations as higher surface : volume ratio in smaller vessels, relative sizes of probes to other vessel internals, etc.

These experiments determine the relationships between equipment and operating variables and environmental conditions. The values of these variables that give the desired environmental conditions, and hence the required process result, are then ideally obtained. Scale-up is then carried out. If the required process performance is not achieved on scaling-up, then further work is necessary to establish the environmental variable which is changing with scale and affecting the process. Modifications are then made either by altering product–growth kinetics by strain improvement, changing growth conditions or modifying the equipment, whichever is the most economical. A comprehensive discussion of scale-up criteria is given by Pace (1984).

It is not a simple matter to obtain sufficient information to determine to what extent these ideal scale-up procedures are incorporated into commercial baker's yeast plant design. Certainly the major metabolic parameters (e.g. specific growth rate, biomass-to-substrate yield, specific oxygen uptake rate) are generally available for a number of yeast strains of interest under controlled fermentation conditions, as well as information on less objective, but widely accepted, tests such as the ability of such strains to produce gas in various doughs. Such data are necessary, since the characteristics required of compressed yeast strains and active dry yeast strains differ as well as the characteristics required of commercial strains for varying types of dough (e.g. sweet breads). Aeration studies on a number of different scales are carried out by major companies who design their own aeration systems. The extent to which these studies are monitored varies considerably. It should be noted again that efficient aeration does not ensure efficient mixing, and that efficient mixing is an essential requirement for good yeast quality. Yeast yield, activity in the dough and quality will all deteriorate if the mixing of nutrients is not adequate in the large-scale fermenter. Consequently, pilot-scale experimentation where oxygen uptake is monitored, for example, using either a fixed oxygen electrode within the fermentation vessel or exit gas-stream oxygen concentration analyses, provides information relating to the overall specific oxygen uptake rate and whether the fermentation is oxygen-limited, but gives no indication of the extent to which effective mixing within the fermenter exists. As a result the measured performance of a particular yeast strain in such a system with respect to its major metabolic variables, e.g. specific growth rate, biomass-to-substrate yield, specific oxygen uptake rate performance in dough tests and its keeping qualities, may bear little relationship to its performance on a larger scale. Despite this consideration, most developmental work by yeast companies is limited to such approaches if that level of sophistication is involved. Also, it is important to note that ascribed variations between yeast strains may be accentuated by such methodological limitations. The experimental techniques required to quantify the oxygen transfer capacity of a production-scale bioreactor and to quantify the extent of the mixing obtained exist, and have been elegantly described by Oosterhuis and Kossen (1983). It involves a combination of experimental measurements with the use of compartment models which quantify the mixing in the fermenter of both gas and liquid fractions and impose varying degrees of mixing upon them. In the work described, an oxygen probe was mounted on a cable in the reactor. The cable was fixed by pulleys at the bottom and top of the vessel. As a result, it was possible to position the electrode at different heights in the reactor and measure the oxygen concentration profile in a production facility. In this case dissolved oxygen concentration was used as an indication of the extent of mixing within the fermenter. The bioreactor was a stirred tank fitted with two six-bladed Rushton-type turbine impellers with a liquid volume of up to 25 $m^3$. These workers noted that oxygen concentration increases at the lower stirrer level, has a low concentration in the space between the stirrers and decreases sharply in the upper part of the fermenter, above the upper stirrer. This behaviour was modelled using a two-compartment model in which the reactor consisted of two parts, a bubble zone and a well-mixed zone. For each zone different oxygen coefficients are estimated using empirical correlations. The model thus developed was successful in predicting refinement with respect to the degree of coalescence of the liquid phase which is required to improve the accuracy of the model. Even without this modification, such a model is more accurate than the simple correlations obtained in the literature, which generally overestimate the oxygen transfer capacity of the reactor. Such methodologies, in the author's opinion, are the basis for a much more rational design approach than either the use of correlations or limited experimental data alone, since they attempt to confront the major problem in translating laboratory results into design of production-scale fermenters, namely scale-up or quantifying the degree of heterogeneity on a large scale. A further approach used by these authors is the scale-down approach (Oosterhuis *et al.* 1983) where, after measurements are made on a production scale, attempts are made to reproduce the overall experimental observations experimentally on a small scale using modified fermenters or operating conditions. A full discussion of such methods is beyond the scope of this work.

The design approaches used by companies supplying proprietary aeration systems are generally not described in detail. However, pilot-plant work is generally involved, since detailed specification of the substrate source and concentration is generally required before data on oxygen transfer, biomass-to-substrate yields, electric power demand, fermenter volume, productivity, etc., are provided.

In either case the alternative to a well-based laboratory or pilot study, incorporating a range of experimental measurements and some mathematical modelling of the fermentation bioreactor, is overdesign (both in terms of the size of the fermenter and the energy requirements for a particular biomass production) and variable yeast quality. Approaches which incorporate the latest methodologies are likely to result in more reliable and cost-effective design.

## 7.3 Fodder (feed) yeast

### *7.3.1 Introductory remarks*

Unlike baker's yeast, fodder (feed) yeast does not have the same stringent requirements for metabolic activity under particular environmental conditions.

It is simply a protein source, and competes with other conventional protein sources on that basis. A very extensive literature exists on the range of raw materials and range of micro-organisms capable of being used for single-cell protein production. A review of this literature is beyond the scope of this work. Reviews are available in the literature (Gaden & Humphrey 1977, Laskin 1977, Litchfield 1979, Rehm & Reed 1983, Tannenbaum 1968, Zanetti 1984). As in the previous section, it is intended to illustrate how many factors have influenced the development of fodder (feed) yeast technology – notably the biochemistry and microbiology of the yeast *Candida utilis*, the production level, nature and value of product, and the fermentation sterility requirements. It will be shown that each of those considerations has led to the choice of a continuous, large-scale fermentation technique for single-cell protein production.

The use of single-cell protein as both an animal feed supplement (whole cells) and human food supplement (use of both whole cells and cell components with particular functional properties) will be discussed. Data on production levels and costs (limited) are presented.

### *7.3.2 Factors affecting selection of fermentation techniques*

**Major metabolic pathways and microbial physiology**

The major biochemical pathways for sugar utilization by aerobically grown fodder (feed) yeast, *Candida utilis*, are illustrated in Figure 7.4. Significant differences exist between this organism and *Saccharomyces cerevisiae*. The major difference is that the respiratory capacity of *Candida utilis* is much greater (that is, it has a much faster maximum rate). The maximum respiratory rate of *Saccharomyces cerevisiae* strains is equivalent to a specific oxygen uptake rate of approximately 100–150 ml oxygen $g^{-1}$ biomass $h^{-1}$, while the maximum respiratory capacity of *Candida utilis* is generally equivalent to a specific oxygen uptake rate of at least 250 ml oxygen $g^{-1}$ biomass $h^{-1}$.

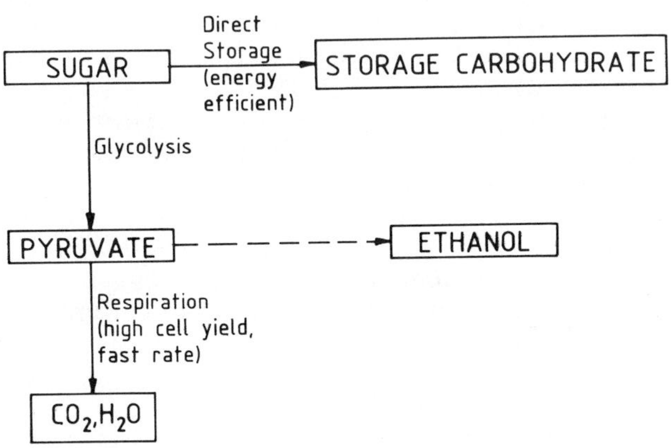


**Figure 7.4** Basic metabolic pathways of *Candida utilis*.

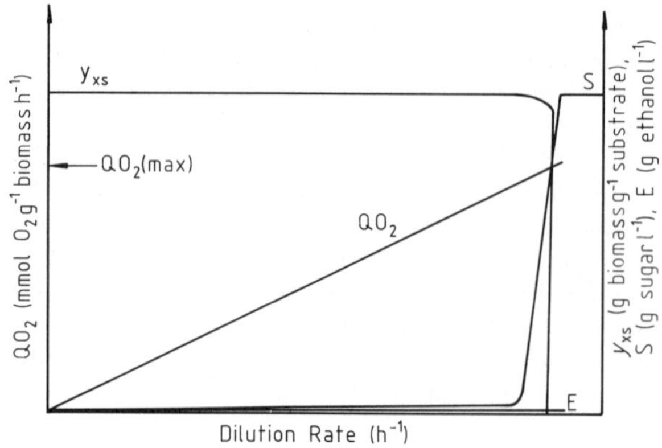


**Figure 7.5** Continuous culture of aerobically grown *Candida utilis* with sugar as the limiting substrate (ideal schematic).

Consequently, the energy metabolism of *Candida utilis* is completely respiratory, since the rate of sugar uptake by the cell does not exceed the respiratory capacity of the organism as it does in *Saccharomyces cerevisiae*. The metabolism is characterized by a large oxygen uptake demand. Only when the organism is poorly adapted to the carbon source (e.g. in batch growth with poor inoculum preparation) or the culture is oxygen limited (and hence the respiratory capacity of the organism is 'artificially' reduced) is a small ethanol excretion possible. However, this is not the usual culture condition. An additional metabolic feature of *Candida utilis* is that it does not require vitamin supplementation to the medium, unlike *Saccharomyces cerevisiae*, since it manufactures its own from the sugar source and nutrients. As such, it is a much more versatile organism and can grow on a range of cheap carbon sources. The role of storage carbohydrates in the metabolism of *Candida utilis* is poorly studied compared with *Saccharomyces cerevisiae*. Surprisingly, very few well-defined experiments have been performed (most tending to be on poorly defined industrial substrates). The most complete study indicated the presence of stored carbohydrates in *Candida utilis* but not to the same extent as *Saccharomyces cerevisiae* (Schwartzkoff & Rogers 1982). Whether they play any crucial role in the cell cycle, as is the case with *Saccharomyces cerevisiae*, is unclear. As previously stated, storage carbohydrates are often kept in reserve by the cell to ensure complete nuclear division in the case of disruption of external energy sources.

The growth of a strain of *Candida utilis* in continuous culture under fully aerated conditions and sugar limitation is shown as an ideal schematic in Figure 7.5. Again, notable differences are evident compared with the metabolism of *Saccharomyces cerevisiae* (Fig. 7.2). The cell yield is high for the full range of specific growth rates less than the maximum specific growth rate. This is indicative of complete respiration of the substrate, and is consistent with no ethanol excretion and complete sugar utilization. The major metabolic indicator of the extent of respiration, the specific oxygen uptake rate, increases linearly with dilution rate indicating a direct dependence on the sugar uptake rate of the cell which, in turn, is dependent on the rate of sugar addition to the continuous culture. That is, the rate of sugar uptake does not exceed the maximum respiratory capacity of the cell and is determined by the rate of sugar utilization. At the maximum specific growth rate, maximum sugar utilization rate is achieved, and this sets the maximum respiratory capacity [$QO_2$(max)] possible. This is unlike *Saccharomyces cerevisiae*, where the maximum rate of respiration is less than the maximum rate of fermentation, presumably set by some physical limitation in the mitochondrion. In addition, unlike *Saccharomyces cerevisiae* the overall metabolism *per se* does not affect the choice of operating conditions, since the biomass-to-substrate yield is high over all operating dilution rates. However, this does not imply that there are no constraints, since other factors have a significant influence on the dilution rate range likely to be used commercially.

As stated above, experimental work with *Candida utilis* under defined experimental conditions has been limited. Most experimental work has been carried out with industrial wastes of a poorly defined nature under conditions of poor environmental control. Schwartzkoff and Rogers (1982) reported the variation of cell composition in continuous culture under defined experimental conditions, and showed variations in protein content, nucleic acid content and storage carbohydrate content with dilution rate. Figure 7.6 is a schematic diagram of the continuous culture of *Candida utilis*, including the metabolic variables of practical commercial interest: cell yield, protein content of cell, oxygen requirements, heat evolution and dilution rate. At low dilution rates it can be seen that cell protein content is highest (due to the lower content of other major cell components, e.g. nucleic acids). However, the biomass yield (g biomass $g^{-1}$ substrate) is lower, due to the maintenance requirements of the organism. At these low dilution rates heat evolution and oxygen requirements are low. At high dilution rates cell protein content is lower (due to the increased

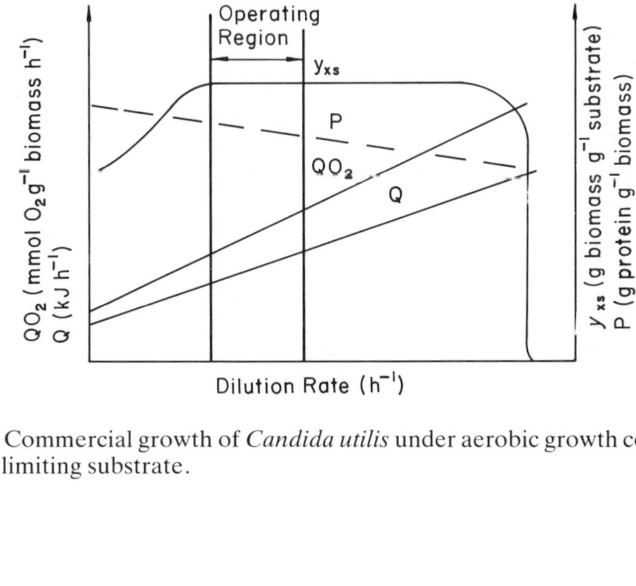


**Figure 7.6** Commercial growth of *Candida utilis* under aerobic growth conditions with sugar as the limiting substrate.

content of other major cell components, e.g. nucleic acids and storage carbohydrates). At these high dilution rates biomass yield is high, due to the decreasing influence of maintenance energy requirements. However, at these high dilution rates oxygen requirements and heat evolution are very significant and provide significant practical engineering problems. As a result of these considerations the operating region most suitable industrially is in the region 0.1–0.2 $h^{-1}$, where there is the combination of high biomass yield, high cell protein content and economically practical oxygen requirement and heat evolution.

**Nature of product, scale of production, value of product and degree of sterility**

As discussed in the previous section, the biochemistry and microbial physiology of the major yeast used for single-cell protein production leads to the conclusion that low growth rates are desirable for highest protein content, biomass-to-substrate yield and oxygen supply and heat removal requirements that are industrially practical using simple fermentation technology.

Single-cell protein, to be commercially successful, needs to be produced on a very large scale since it is a product of low market value. This coupled with the fact that single-cell protein has few of the constraints of baker's yeast with respect to stringent metabolic activity and cellular content of a range of intracellular components, leads to the general conclusion that continuous culture techniques are likely to be economically advantageous. Using continuous-culture techniques all of the requirements resulting from the consideration of the biochemistry and microbial physiology of the micro-organism can be met.

However, the stringent sterility requirements of continuous culture introduce some practical fermenter considerations. Aeration and cooling systems must not only be chosen with their primary aim in mind, but also to their susceptibility to microbial contamination and ease of cleaning, both routinely and after a process infection. In general, simplicity of design (for example, few moving parts) are sought in such cases. Single-cell protein technology has undoubtedly significantly contributed to the rapid development of continuous fermentation methodology. In the area of fermentation technology developed for the single-cell protein industry, the ICI air-lift fermenter typifies the development of a simple, continuous fermentation system. In contrast with the conventional stirred-tank fermenter systems, air-lift systems have no moving parts, and the only power requirement is the power required to deliver the air through the air-sparging system, using air compressors. There is no mechanical agitation, and the air bubbles forced through the sparger system are responsible for the induced turbulent mixing and mass transfer. The presence of a draft tube system inside the air-lift fermenter stabilizes the liquid circulation patterns and increases the interfacial area between gas and liquid. The main advantages of this system are the low shear and low energy requirements, as well as its simplicity of construction (aseptic seals are not required around rotating shafts) (Margaritis & Wallace 1984). As discussed in Section 7.3.3, a number of different fermenter types of varying operational complexity have been used for commercial single-cell protein production. However, limited operational data exist to compare their performance. Fermenter designs with simple, continuous operation are clearly desirable, providing oxygen transfer requirements, heat removal requirements and economic considerations are favourable.

The scale of operation and value of single-cell products produced are varied as is the nature of the product. The largest commercial single-cell protein facilities are the ICI plant in England (a bacterial process) using *Methylophilus methylotrophus* and a methanol substrate and the Rumanian plant in Curtea de Arges (a yeast process using *Candida parafinica*). Both processes are continuous, with a production level of approximately $60 \times 10^6$ kg per year. The ICI process uses an air-lift bioreactor ($1.5 \times 10^6$ litres) while the Rumanian plant uses four bubble-column tower bioreactors, each with a volume of $1.26 \times 10^6$ litres (Zanetti 1984). This scale of operation restricts the type of fermenters that are practical for such applications (as discussed in Section 7.3.3).

Many laboratory and pilot-plant studies have been reviewed by Litchfield (1979). Commercial applications are more limited, the major ones being (in addition to the ICI and Rumanian plants) Amoco (USA), Hoechst-Ude (FDR), Cellulose Attisholtz (Switzerland), Fromageries-Bel (France), and Mitsubishi, Kohjin, Sanyo-Kokusaku, Juyo and Kanegafuchi (Japan). In addition, significant single-cell protein production facilities exist in the USSR. Plant capacities vary from small ($<5 \times 10^6$ kg per year – Hoechst-Ude, Sanyo-Kokusaku, Juyo, Kanegafuchi) to medium ($5–10 \times 10^6$ kg per year – Amoco, Cellulose Attisholtz, Fromageries-Bel, Kohjin) to large ($>10 \times 10^6$ kg per year – USSR, Mitsubishi). Most products are used for animal feed, except for those of Amoco, Hoechst-Ude and Fromageries-Bel. Products of Amoco (Torutein range) are used in seasoning carriers, luncheon meat emulsifiers, salad dressings, pasta dough conditioners for pizzas, etc., of Hoechst-Ude, as a functional isolate and, of Fromageries-Bel (Protilac) as a dough conditioner for puff pastries, croissants and vol-au-vents. The nature of the product has a limited effect on the choice of the overall fermentation process. It may in some circumstances eliminate the necessity for a fermentation step, such as when spent brewer's yeast is used. About $4.5 \times 10^6$ kg per year of spent brewer's yeast is either used as fodder (feed) yeast (after drying) or dried together with brewer's spent grains (Reed 1984). Where fermentation is involved, even though the fermentation stage is not generally affected by the nature of the product, the subsequent downstream processing stage may need to be more elaborate if particular yeast components are required to be isolated for particular physicochemical properties (e.g. water- and fat-binding, emulsion stability, dispersibility, gel formation, whippability and thickening).

### *7.3.3 Fodder (feed) yeast production*

**General outline**

The method of fodder (feed) yeast production from sugar sources is similar to that for baker's yeast, except for the utilization of continuous-culture techniques, and will not be discussed in detail here except for some comment on the fermenter design and aeration systems utilized.

**Fermentation apparatus**

As discussed in previous sections, fodder (feed) yeast is produced generally by continuous fermentation. The fermentation vessel is sized using the desired production level and the substrate concentration, since the specific growth rate of micro-organisms is generally limited to a defined region for optimum cellular protein level and biomass-to-substrate yield. Fermenters of 1500–2000 $m^3$ are not uncommon, making them an order of magnitude larger than comparative baker's yeast fermenters. Consequently, the aeration systems and heat removal systems are more sophisticated for single-cell protein bioreactors. The major difference from baker's yeast bioreactors is that the proprietary aeration systems used in baker's yeast production (e.g. Frings, Vogelbusch and Hoechst) and stirred vessels with aeration are not practical past a scale of about 250 $m^3$, and consequently have limited application. The Vogelbusch aeration system has been used for single-cell protein production up to this stage (50–220 $m^3$) in Yugoslavia, Brazil, the Netherlands, Cuba and Iraq. This is a further example where scale of operation has an influence on the selection of the fermentation apparatus. While it is feasible to use multiple vessels and incorporate such proprietary aeration systems, such a solution is not generally economically attractive. In addition, the requirement for strict sterility has found such systems lacking. Hoechst have attempted to modify the Frings aeration system to overcome such limitations, but the restriction of scale of operation still remains.

The two aeration methods used for large-scale single-cell protein production are then the ICI air-lift fermenter (previously described) and the Rumanian plant aeration system which consists of bubble columns. The distribution of air through spargers at this stage generally involves computer design and control. Certainly, the previous discussion relating to aeration and mixing with respect to baker's yeast is pertinent here.

## 7.4 Yeast enzymes and other intracellular yeast components

### *7.4.1 Introductory remarks*

A considerable range of yeast enzymes and other intracellular yeast components are produced commercially. However, up-to-date figures of production levels and international pricing are not generally available. Generally such products may be considered low-volume products compared with baker's yeast and fodder (feed) yeast, but with higher unit value and an extremely wide range of product applications; the last of these characteristics contributes to the difficulty of obtaining relevant production figures. The market is also far more variable than for baker's yeast.

Yeast enzymes may be broadly divided into two categories: those with significant use in the food industry and those which may be categorized as fine chemicals and are largely for laboratory use. Invertase and lactase are the major enzymes in the first category, although some production of lipases exists. Invertase is produced from *Saccharomyces cerevisiae* and is mostly used for increasing the sugar concentration of liqueur chocolates, ice creams, jelly, candy, bonbons, nougat, shortcake and artificial honey, and as a wetting agent in fruit wine, etc. Lactase is produced from *Kluyveromyces fragilis* and is mostly used for hydrolysing milk lactose in a range of milk, cheese and whey products. Lipase is produced from *Candida lipolytica*, and is mostly used to improve food quality by oil and fat decomposition, e.g. prevention of acid odour of fat in chocolate, development of egg whippability by fat decomposition of egg white, and fixation of odour or flavour to sharply odoured milk products. Enzymes which may be categorized as fine chemicals and are largely used in laboratory applications are alcohol dehydrogenase, aldehyde dehydrogenase, glucose 6-phosphate dehydrogenase, $\alpha$-glucosidase and hexokinase.

Other intracellular yeast components may be broadly subdivided into three categories: whole-yeast cell extracts, nucleosides, and health and therapeutic compounds. Whole-yeast cell extracts consist of hydrolysed spent brewer's yeast (or baker's yeast) and are mostly used as fermentation media nutrients and food flavourings. Nucleotides and nucleosides are produced by enzymic hydrolysis of yeast RNA (usually from *Candida utilis*). Some production of 5′-IMP and 5′-GMP is carried out by fermentation, but this is largely by bacteria except for a small proportion produced by the yeast *Candida tropicalis*. Although yeasts are a good source of vitamins, those produced by fermentation are generally produced by other microbial species. Some riboflavin is produced by *Candida flareri* commercially. Some therapeutic compounds, e.g. CDP-choline (from *Saccharomyces carlsbergensis*) are also produced by yeast. Both of the latter applications are limited.

As previously stated, production figures and figures relating to the value of these products are limited. The total production of enzymes from all microbial sources were given by Peppler (1979) as US$150–175 million per year. Allowing for moderate annual expansion, this would be equivalent to a net present value of the order of US$250–300 million per year at a product price range of the order of US$10 000–100 000 $kg^{-1}$ enzyme. These enzymes are produced in USA, Japan and Europe. By contrast, there is no large-scale production of intracellular compounds such as nucleotides in the USA (although some are under consideration and several companies produce fine chemicals, e.g. nucleotides, enzymes, co-factors, etc., for laboratory use). Nucleotide and nucleoside production is dominated by Japan. The level of nucleotide production in Japan was estimated by Yamada (1977) to be of the order of 215 000 kg per year, although this should not be considered a current or exhaustive figure.

### 7.4.2 *Factors affecting selection of fermentation techniques*

As discussed with respect to baker's yeast and fodder (feed) yeast production, the major factors affecting the selection of fermentation techniques are the scale of operation, value of product, degree of sterility and the nature of the product (particularly whether it is produced intra- or extracellularly). The industrial production of yeast enzymes and other intracellular yeast components may be distinguished from baker's yeast and single-cell protein production in two major ways: they are high-value products of comparatively low volume, and the fermentation process is largely secondary to extensive downstream processing processes. Downstream processing will not be discussed in detail, although the basic processes used will be referred to. Details of the process steps involved are well covered in the works of Peppler (1979, 1983).

Since yeast enzymes are relatively low-volume products, it is difficult to justify the development and construction of specialized equipment for their production. If a source of spent yeast is available from which enzymes may be extracted, then the fermentation stage may be circumvented altogether, but this is unlikely for all but very low volume intracellular enzymes, e.g. some fine chemical yeast enzymes used in specialized laboratory applications. Certainly, for the production of extracellular enzymes where the enzymes are excreted into the fermentation liquor by the micro-organism, e.g. invertase from yeast, it is unlikely that a ready source of enzyme could be obtained without a fermentation stage. Given these considerations, the fermenter is chosen on the basis of the smallest, most flexible unit – thereby minimizing capital costs and having the option of using it for other fermentation processes. This generally results in the choice of a stirred tank with aeration, operated in a batch manner.

Although the fermenter is therefore similar to that used in the baker's yeast industry, several differences would exist: notably its smaller size and greater attention to strictly sterile operation. Many enzyme fermentations are especially vulnerable to microbial contamination, since they are generally of batch times of the order of 30–150 h. Some enzyme productions are carried out in a fed-batch mode of operation, generally because of the presence of catabolite repression.

The nature of the product has a significant effect on the subsequent processing steps. The two most important considerations are whether the enzyme is produced intra- or extracellularly and whether the enzyme is sold as a crude extract, in commercial grade or as a high-purity product. The first task in the recovery of an extracellular enzyme is to remove cells and other suspended material from the broth. This is normally achieved by centrifugation or vacuum filtration. Intracellular enzymes are extracted from the microbial cell mass and, consequently, extraction is normally achieved by centrifugation which gives easier access to the microbial cells. Intracellular enzymes may be extracted by autolysis or physical disruption methods (e.g. homogenizer, bead mill, etc.). When the cells have been broken the enzyme can be purified in the same way as an extracellular enzyme, but the process is usually more difficult, due to the content of cell debris and nucleic acids from the broken cells. Purification methods then depend upon whether a crude extract, commercial grade extract or highly purified form of enzyme is needed. Methodologies are described in many reviews (Aunstrup *et al.* 1979, Aunstrup 1983, Boing 1982, Peppler 1979, 1983) and will not be treated in detail here.

Whole-cell yeast extracts are generally made from autolysed spent brewer's yeast, and are usually used unpurified. The nature of this product eliminates the need for a specific fermentation or separation stage.

The other intracellular yeast components of significant commercial interest are the nucleotides and nucleosides, and to a limited extent vitamins. Nucleotides and nucleosides are of interest mainly due to their ability to enhance the flavour of food, although they have potential and actual (limited) medical applications (Demain 1978, Nakao 1979). The industrial production of 5′-IMP and 5′-GMP is at present carried out by both enzymic hydrolysis of

RNA and by direct fermentation, and is largely dominated by Japan. The source of RNA for the enzymic hydrolysis may be spent brewer's yeast (as it could be for vitamin production), but there is also manufacture of yeast specifically for nucleotide and nucleoside production. *Candida utilis* is most widely used for this purpose because of its ease of growth on cheap carbon sources, non-fastidious nutrient requirements and generally low DNA content. Generally, RNA is extracted from yeasts by a hot, dilute alkaline method, after which the RNA is dried and powdered. The scale of production of these yeast components would generally be relatively low and dictate batch growth, where both ease of cultivation and highest RNA content (at the maximum specific growth rate) would be economically attractive. The process would be carried out with strict sterility, since the compounds would later be used in the food industry. Production of nucleotides and nucleosides directly by fermentation is carried out by the yeast *Candida tropicalis*.

## 7.5 Genetically engineered products

### *7.5.1 Introductory remarks*

Genetic engineering or recombinant DNA technology may be broadly defined as the rearrangement of the genetic machinery of living things to achieve desired metabolic conversions. Such metabolic conversions may improve the production of particular components already produced by the micro-organism, or may involve the production of biochemicals completely foreign to the micro-organism.

In this section genetically engineering techniques applied to yeast, work aimed at particular aspects of strain improvements and the range of genetically engineered products from yeast currently being developed will be discussed. It is not intended that the coverage be comprehensive, but rather that the major principles of the techniques and examples of their application are presented. Detailed presentations of methods of genetic manipulation and non-yeast peptide product formation appears elsewhere in this book.

It should be noted that the current status of genetically engineered products from yeast is predominantly at the research, rather than production, stage, and that levels of expression currently achieved are low compared with genetically engineered products from bacteria, where up to 15–20% of the cells' protein may be expressed as the genetically engineered product (e.g. insulin).

Potentially, genetically engineered products using yeast are attractive from a number of viewpoints. Prokaryotic systems are not necessarily suitable for all purposes, e.g. proteins that require glycosylation or other types of modification. In addition, unlike *Escherichia coli* (the bacterium extensively used for genetically engineered products), *Saccharomyces cerevisiae* has no pathogenic relationship with man, and proteins made in yeast may therefore be more acceptable as human pharmaceuticals or human food products. Yeast is a eukaryotic micro-organism which has most of the genetic and microbiological advantages which have promoted the extensive use of *Escherichia coli*.

### 7.5.2 *Genetic techniques*

A range of methods are used in the genetic manipulation of yeasts. These may broadly be divided into traditional technologies, which involve mutagenesis and mating and sporulation, and newer technologies such as spheroplast–protoplast fusion and the use of host–vector systems utilizing various transformation methods. These methodologies are treated in detail elsewhere in this book, and will not be discussed in detail here. However, it should be noted that each of these methods can be applied for the manipulation of yeasts for commercial use. Much depends on what is hoped to be achieved as to the efficiency of a particular method for that application. For example, mutation is an adequate technique where it is desired to isolate temperature-resistant or ethanol-resistant mutants, since there is a positive selection process involved. However, it is tedious to isolate, for example, auxotrophs from a polyploid strain because there is no positive selection. Similarly, although cell-fusion techniques offer a relatively simple approach (theoretically) to introducing properties of one yeast strain into another, it has been shown to be unsuccessful due to the instability of fusion products, and the use of host–vector systems with transformation would appear to be the most promising approach for this application at present.

### 7.5.3 *Production of genetically engineered products*

This section will concentrate on a brief discussion of the applications of the newer technologies, namely spheroplast–protoplast fusion and the use of host–vector systems utilizing transformation methods. This is not intended to imply that the traditional methodologies of mutation and mating and sporulation are not used in commercial yeast strain development. However, the newer methods have the potential for greatly increasing the range of products produced by yeast, and modifying the well-known metabolic characteristics of traditional yeast strains.

The method of spheroplast–protoplast fusion, in theory, circumvents ploidy and mating type barriers. In practice this has not been shown to be the case; with only the use of closely related species (e.g. *Saccharomyces cerevisiae*, *Saccharomyces diastaticus* and *Saccharomyces carlsbergensis*) resulting in stable fusion products. The following are typical examples of the application of this method.

(a) Introduction of flocculence (Figueroa *et al.* 1984, Russell & Stewart 1979).
(b) Introduction of maltose or maltotriose utilization ability (Panchal *et al.* 1984).
(c) Introduction of melibiose or dextrin utilization ability (Stewart *et al.* 1982).
(d) Introduction of flocculence and ethanol tolerance (Seki *et al.* 1983).
(e) Amplification of maltose utilization rate by increased gene dosage (Mowshowitz 1979).
(f) Introduction of temperature tolerance.

The major advantage of spheroplast–protoplast fusion is that it is not necessary to know the precise DNA segment that is associated with the desired metabolic characteristic. In the examples given, successful fusion products are easily isolated, since there exists a positive selection marker (e.g. flocculence, ethanol tolerance, etc.). The major disadvantage of the spheroplast–protoplast fusion method is that it is a 'shotgun approach'. Since the whole genome of both fusion partners becomes integrated in an ill-defined manner, the fusion product (although successful in terms of the primary selection marker, e.g. flocculence, ethanol tolerance, etc.) is nearly always significantly different from both original fusion partners. Depending on the industrial process involved, this may have an undesirable result. Panchal *et al.* (1984), for example, reports fusion experiments with brewing yeast strains which achieve their primary aim of introducing flocculence into stable fusion products. However, significant differences in the flavour of beers resulting from the use of the fusion products (e.g. phenolic off-flavours, excessively heavy sulphur or ester formation) rendered the 'successful' fusion products commercially non-viable for beer production. The baker's yeast industry and wine industry would have similar considerations, with respect to flavour, about the possible introduction of yeast fusion products. In some industries such considerations may be less important. For example, flocculent yeast strains have been shown to carry out rapid and operationally simple continuous conversion of sugars to power ethanol (Prince & Barford 1982a, b, c). Provided successful yeast fusion products for the introduction of flocculence did not produce flavour compounds (or other end-products) in quantities that significantly affect the ethanol yield from the sugar based substrate, the presence of such compounds would be of no consequence. Commercially, it should be noted that most successful fusion studies have involved the use of haploid strains and may bear limited relevance to industrial utilization where polyploid strains are generally used due to their higher stability.

Fusions amongst unrelated strains have been unsuccessful, despite their obvious commercial significance. Examples of desired inter-species fusions are *Saccharomyces–Kluyveromyces*, with the aim of introducing lactose utilization ability (lactase) into *Saccharomyces* (Stewart 1981, Panchal *et al.* 1984), and *Saccharomyces–Schwanniomyces*, with the aim of introducing starch-utilization ability ($\alpha$-amylase and glucoamylase) into *Saccharomyces* (Panchal *et al.* 1984). Although fusion products have been isolated, they have been found to be unstable on propagation.

The use of host–vector systems utilizing various transformation methods has received much attention in the recent literature. In this method an attempt is made to incorporate limited DNA segments coding for the desired product with sufficient expression signals to ensure its expression. Amongst others, proteins such as mature leukocyte interferon (Hitzeman *et al.* 1981), interferon–$\gamma$-cDNA and hepatitis B surface antigen (Derynck *et al.* 1983), human interferon ($\gamma 1$, $\gamma 2$, $\gamma$; Hitzeman *et al.* 1983), $\beta$-isopropylamate dehydrogenase (Beggs 1978), bacterial $\beta$-lactamase (Breunig *et al.* 1982), chymosin (Mellor *et al.* 1983), hygromycin B resistance (Gritz & Davies 1983) and bacterial xylose isomerase (Ho *et al.* 1983) have been expressed in yeast, although the levels of expression are generally low. Plasmids used for recombinant DNA work involving yeasts are most commonly chimeric plasmids which have incorporated bacterial plasmids (e.g. pBR 322) with the yeast 2-$\mu$m plasmid together with yeast chromosomal DNA fragments (e.g. TRP1, SUC2, LEU2, ADH1, URA3, HIS3, ARG4, PGK and CYC1) which serve as expression vectors. The bacterial plasmids may be modified plasmids incorporating some marker (e.g. pMB9, derivative of the COLE1 plasmid incorporating tetracycline resistance) (Beggs 1978). It should be noted that yeast *Escherichia coli* hybrids used for commercial production of food products would be subjected to strict containment procedures, which may affect their commercial viability. In addition to production of different products, work on gene insertion (e.g. insertion of 'killer factor' into commercial wine yeast production (Universal Foods) and deletion of various undesired flavours from yeast is being undertaken. The potential for the production of high-value products (e.g. interferon) using yeast is still largely at the research and development stage. However, the possibility that yeast may be able to undertake a range of protein modifications previously largely restricted to higher cells is likely to ensure continued extensive research activity. Increased expression level in yeast is important for any commercial development. The complexity of the control of protein expression in yeasts compared with bacteria may prove a difficult hurdle, and significant use of bacteria as hosts is likely to continue to be more practical commercially in the near future.

### 7.5.4 *Factors affecting selection of fermentation techniques*

As discussed above, most work in the area of genetically engineered products is at the research and development stage rather than at the commercial production stage. Typically, fermentations are carried out in small (approximately 10-litre) stirred fermentation vessels with standard environmental controls. Cell Tech, for example, are investigating a range of genetically engineered yeast products such as calf prochymosin at this scale.

As discussed with baker's yeast, fodder (feed) yeast, yeast enzymes and other intracellular yeast components, factors such as scale of operation, value of product, degree of sterility, nature of product and metabolic and physiological considerations will have a significant bearing on the selection of the fermentation techniques. In this regard it is envisaged that genetic manipulations of yeast may be divided into two broad areas: manufacture of new genetically engineered products of high value and manipulation of currently used brewing, baking and wine yeast strains to improve quality (flavour enhancement and off-flavour deletion), improve reliability (insertion of killer factor) and increase available substrate range.

Genetically engineered products of high value would probably be of small volume and require sterile operation (particularly if for human or pharmaceutical application), and would consequently probably be batch fermentations. The scale of the operation would largely be a function of the expression levels achieved, given the relatively small market. The nature of the product (that is, whether it is produced intra- or extracellularly) would affect the subsequent downstream processing (see Section 7.4). The operating conditions of the batch fermenter would also probably be affected by the nature of the product. For example, if its production was subject to glucose repression, a fed-batch system similar to baker's yeast would probably result. If its production was cell-cycle dependent, certain growth rate regimes may be more favourable than others. If the product was intracellular, then the production of other extracellular end products would be of less importance to subsequent purification than if the product was extracellular where, desirably, little or no other end product should be produced (necessitating respiratory growth).

Manipulation of currently used brewing, baking and wine yeast strains to improve product quality, reliability or substrate range utilized is unlikely to change the fermentation practices now adopted commercially.

## Acknowledgements

The author would like to thank the following for their valuable assistance in preparing this manuscript: Dr J. Friend, Dr G. Pace, Mr L. Treu, Mr W. Da Costa and Miss J. Orlowski.

## References

Aunstrup, K. 1983. Enzymes of industrial interest – traditional products. *Annual Reports on Fermentation Processes* **6**, 175–201.

Aunstrup, K., O. Andersen, E. A. Falch and K. N. Tage 1979. Production of microbial enzymes. In *Microbial technology*, H. J. Peppler and D. Perlman (eds), Vol. 2, 292–311. New York: Academic Press.

Barford, J. P. 1984a. The control of respiration and fermentation in yeasts – a unified approach. *7th International Special Symposium on Yeasts*, New Delhi (in press).

Barford, J. P. 1984b. The control of fermentation and respiration in yeasts. *6th Australian Biotechnology Conference, Brisbane*, 335–44.

Barford, J. P. and R. J. Hall 1976. Estimation of the length of cell cycle phases from an asynchronous culture of *S. cerevisiae*. *Experimental Cell Research* **102**, 276–84.

Barford, J. P. and R. J. Hall 1979. An examination of the crabtree effect in *Saccharomyces cerevisiae* – the role of respiratory adaptation. *Journal of General Microbiology* **106**, 267–75.

Barford, J. P. and R. J. Hall 1981. A mathematical model for the aerobic growth of *Saccharomyces cerevisiae* with a saturated respiratory capacity. *Biotechnology and Bioengineering* **23**, 1763–96.

Barford, J. P., P. M. Jeffery and R. J. Hall 1981. The crabtree effect in *Saccharomyces cerevisiae* – primary control mechanism or transient? *Advances in Biotechnology* **1**, 255–60.

Beck, C. and H. K. von Meyenburg 1968. Enzyme pattern and aerobic growth of *Saccharomyces cerevisiae* under various degrees of glucose limitation. *Journal of Bacteriology* **96**, 479–86.

Beggs, J. D. 1978. Transformation of yeast by replicant hybrid plasmid. *Nature* (London) **275**, 104–9.

Bocharova, N. N., V. G. Chernysh and V. P. Ozerova 1976. Keeping characteristics of pressed yeast. *Khlebopek Konditer Promst* **8**, 37–8.

Boing, J. T. P. 1983. Enzyme production. In *Industrial microbiology*, G. Reed (ed.), 634–708. Westport: AVI.

Breunig, K. G., V. Mackendonski and C. P. Hollenberg 1982. Transcription of the bacterial β-lactamase gene in *Saccharomyces cerevisiae*. *Gene* **20**, 1–10.

Dellweg, H., H. K. Bronn and W. Hartmeier 1977. Respiration rates of growing and fermenting yeast. *Kem Kemi* **4**(12), 611–15.

Demain, A. L. 1978. Production of nucleosides and nucleotides by microorganisms. In *Economic microbiology*, A. H. Rose (ed.), Vol. 2, 187–209. New York: Academic Press.

Derynck, R., A. Singh and D. V. Goeddel 1983. Expression of the human interferon-γ cDNA in yeast. *Nucleic Acids Research* **11**(6), 1819–37.

Figueroa, de L. I., M. F. de Richard and M. R. de van Broock 1984. Interspecific protoplast fusion of baker's yeast *S. cerevisiae* and *S. diastaticus*. *Biotechnology Letters* **6**(4), 269–74.

Gaden, E. L. and A. E. Humphrey 1977. *Single cell protein from renewable resources*. New York: Wiley.

Gritz, L. and J. Davies 1983. Plasmid encoded hygromycin B resistance: the sequence of hygromycin b phosphotransferase gene and its expression in *Escherichia coli* and *Saccharomyces cerevisiae*. *Gene* **25**, 179–88.

Hall, R. J. and J. P. Barford 1981. A simulation of the integration of the internal energy metabolism and the cell cycle of *Saccharomyces cerevisiae*. *Biotechnology and Bioengineering* **23**, 1763–96.

Hitzeman, R. A., F. E. Hagie, H. L. Levine, D. V. Goeddel, G. Ammerer and B. J. Hall 1981. Expression of a human gene for interferon in yeast. *Nature* (London) **293**, 717–22.

Hitzeman, R. A., D. W. Leung, L. J. Perry, W. J. Kohr, H. L. Levine and D. V. Goeddel 1983. Secretion of human interferons by yeast. *Science* **219**, 620–5.

Ho, N. W. Y., P. S. Jervis, S. Rosenfeld, J. J. Huang and S. T. Tsao 1983. Expression of the *E. coli* xylose isomerase gene by a yeast promoter. *Biotechnology and Bioengineering Symposia* **13**, 245–50.

Hoehne, R. 1975. Belüftungssystein für Grossfernter. *Branntweinwirtschaft* **115**(22), 400–1.

Laskin, A. I. 1977. Single cell protein. *Annual Reports on Fermentation Processes* **1**, 151–80.

Litchfield, H. J. 1979. Production of single cell protein for use in food or feed. In *Microbial technology*, H. J. Peppler and D. Perlman (eds), Vol. 2, 93–156. New York: Academic Press.

Margaritis, A. and J. B. Wallace 1984. Novel bioreactor systems and their applications. *Biotechnology* **1**(5), 447–53.

Mellor, J., M. J. Dobson, N. A. Roberts, M. F. Tuite, J. S. Emtage, S. White, P. A. Lowe, T. Patel, A. J. Kingsman and S. M. Kingsman 1983. Efficient synthesis of enzymically active chymosin in *Saccharomyces cerevisiae*. *Gene* **24**, 1–14.

Meyenburg, von H. K. 1969. Katabolit-repression und der Sprossungszyklus von *Saccharomyces cerevisiae*. Ph.D. thesis, Zürich.

Moo-Young, M. and H. W. Blanch 1981. Design of biochemical reactors: mass transfer criteria for simple and complex systems. *Advances in Biochemical Engineering* **19**, 1–69.

Mowshowitz, D. B. 1979. Gene dosage effects on the synthesis of maltose in yeast. *Journal of Bacteriology* **137**, 1200–7.

Nakao, Y. 1979. Microbial production of nucleosides and nucleotides. In *Microbial technology*, H. J. Peppler and D. Perlman (eds), Vol. 2, 312–55. New York: Academic Press.

Oldshue, J. Y. 1966. Fermentation mixing scale-up techniques. *Biotechnology and Bioengineering* **8**, 3–24.

Oosterhuis, N. M. G. and N. F. Kossen 1983. Oxygen transfer in a production scale bioreactor. *Chemical Engineering Research and Design* **61**(5), 308–12.

Oosterhuis, N. M. G., V. M. Graesbeek, A. P. C. Olivier and N. W. F. Kossen 1983. Scale-down aspects of the gluconic acid fermentation. *Biotechnology Letters* **5**, 141–6.

Pace, G. W. 1984. Scale-up criteria for fermentation systems. Biotechnology Training Course Notes, University of Queensland.

Panchal, C. J., I. Russell, A. M. Sills and G. G. Stewart 1984. Genetic manipulation of brewing and related yeast strains. *Food Technology* **99**, 1068–111.

Panek, A. D. 1975. Trehalose synthesis during starvation of baker's yeast. *European Journal of Applied Microbiology* **2**(1), 39–46.

Peppler, H. J. 1977. Production of yeast and yeast products. In *Microbial technology*, H. J. Peppler and D. Perlman (eds), Vol. 2, 157–87. New York: Academic Press.

Peppler, H. J. 1983. Ventures in yeast utilisation. *Annual Reports on Fermentation Processes* **6**, 237–51.

Prince, I. G. and J. P. Barford 1982a. Continuous tower fermentation for power alcohol production. *Biotechnology Letters* **4**(4), 265–8.

Prince, I. G. and J. P. Barford 1982b. Tower fermentation of sugar cane juice. *Biotechnology Letters* **4**(7), 469–74.

Prince, I. G. and J. P. Barford 1982c. Induced flocculation of yeasts for use in the tower fermenter. *Biotechnology Letters* **4**(10), 621–6.

Prince, I. G. and J. P. Barford 1984. Flow and kinetic parameter estimation in a heterogeneous continuous fermenter. *5th Australian Biotechnology Conference, Brisbane*, 293–500.

Reed, G. 1982. Production of baker's yeast. In *Industrial microbiology*, G. Reed (ed.), 593–633. Westport: AVI.

Reed, G. and H. J. Peppler 1973. *Yeast technology*. Westport: AVI.

Rehm, H. J. and G. Reed 1983. *Biotechnology*, Vol. 3. Berlin: Springer-Verlag.

Rosen, K. 1977. Production of baker's yeast. *Process Biochemistry* **12**(3), 10–12.

Russell, I. and G. G. Stewart 1979. Spheroplast fusion of brewer's yeast strains. *Journal of the Institute of Brewing* **85**, 95–8.

Schreier, K. 1975. Neuer Hochleistungsfermenter nach dem Tauchstrahlverfahren. *Chemiker-Zeitung* **99**(7), 328–31.

Schwartzkoff, C. L. and J. P. Barford 1981. Macromolecular composition of yeast as a function of life cycle. *Advances in Biotechnology* **2**, 387–92.

Schwartzkoff, C. L. and P. L. Rogers 1982. Glycogen synthesis by glucose limited *Candida utilis*. *Journal of General Microbiology* **128**, 1635–8.

Scott, R. 1983. Design and evaluation of experiments to provide scale-up information for the ICI single cell protein process. *Proceedings International Conference on Commercial Applications and Implications of Biotechnology* 235–48.

Seki, T., S. Myoga, S. Limtong, S. Vedono, J. Kumnvata and H. Taguchi 1983. Genetic construction of yeast strains for higher ethanol production. *Biotechnology Letters* **5**(5), 351–6.

Sittig, W. 1982. The present state of fermentation reactions. *Journal of Chemical and Technical Biotechnology* **32**, 47–58.

Stewart, G. G. 1981. The genetic manipulation of industrial yeast strains. *Canadian Journal of Microbiology* **27**, 973–90.

Stewart, G. G., I. Russell and J. Panchal 1982. The genetics of alcohol metabolism in yeast. *Brewing and Distilling International* **12**(1), 23–36.

Tannenbaum, S. 1968. Factors in the processing of single cell protein. In *Single cell protein*, R. I. Mateles and S. R. Tannenbaum (eds), 343–52. Cambridge, Massachusetts: MIT Press.

Yamada, K. 1977. Biotechnology report: recent advances in industrial fermentation in Japan. *Biotechnology and Bioengineering* **19**, 1563–621.

Yoshida, F. 1982. Aeration and mixing fermentation. *Annual Reports on Fermentation Processes* **5**, 1–34.

Zanetti, R. 1984. Breathing new life into single-cell protein. *Chemical Engineer* **91**, 18–21.

# 8 *The technology of anaerobic yeast growth*

ARGYRIOS MARGARITIS
and FAHAR J. A. MERCHANT

## 8.1 Introduction

The ability of yeast to ferment sugars to ethanol has been commercially exploited by man for several thousand years. For most of this period the role of yeast anaerobiosis has been to produce potable ethanol, whereas the biological production of ethanol for use as a fuel or chemical feedstock can be traced back to as recently as the beginning of this century. Although ethanol is the largest (by volume) fermentation product today, the technology for its production owes little to the strides that biotechnology has made over the past 30 years (Righelato 1980). Thus, the innate conservatism of the brewing industry and direct economic competition from the catalytic ethylene-hydration route has, until recently, largely precluded any changes from established batch fermentation systems. These processes are slow and inefficient, and have been aptly described as 'a lot of water containing a dash of catalyst in large expensive fermenters' (Atkinson *et al.* 1980). The need for highly efficient continuous bioreactor systems has therefore become widely apparent, and it is quite likely that, in the near future, such processes will be commonly employed, both in the brewing industry and, especially, for the industrial production of fuel-grade ethanol.

In this chapter the various conventional and novel bioreactor systems employed for the anaerobic growth of yeast cells will be described. In doing so we shall point out the rationale for using different bioreactor systems, their efficiency (or otherwise), and the influence of both economics and scale on the design of fermentation processes. However, the production of industrial and potable ethanol will be discussed separately, since the process objectives and the economic constraints are not similar. For instance, any change in the fermentation process may be implemented in the alcoholic beverage industry only if the major flavour characteristics remain essentially unchanged. The problems facing the fermentation technologist in the beverage sector are therefore much more complex than those encountered when fuel ethanol production is the objective (Kirsop 1982).

## 8.2 Ethanol production using freely suspended yeast cells

The production of ethanol via anaerobic yeast growth may be achieved by using either free or immobilized cell systems. In this chapter an immobilized yeast-cell system will be defined as any system in which yeast cells are confined within a bioreactor thereby allowing their economical reuse. The term 'economical' has been used to exclude processes in which yeast cells are recovered and reused by employing techniques such as centrifugation and microfiltration which may introduce high capital and operating costs. On the other hand, flocculent yeast cells are categorized as being 'immobilized', since their recovery can be accomplished with relative ease by employing static settling tanks (Abbot 1977).

Having distinguished between the two fermentation processes, free- and immobilized-cell systems, we shall now evaluate the various free-cell systems that have been employed for the production of industrial ethanol.

### *8.2.1 Conventional batch fermentation*

Batch fermentation is the most common process used by industry for producing ethanol. In this system a large volume of the substrate [10–20% (w/v) of sugar supplemented with nutrients] is fed to an enclosed or open vessel with or without mechanical agitation, and inoculated with a 5–10% (v/v) inoculum of actively growing yeast. Yeast growth and ethanol production is allowed to proceed until maximum yields have been obtained. At this point the fermentation broth is transferred into a holding tank, the vessel is cleaned and sterilized, and a new cycle is initiated. A complete cycle time may vary from 36 to 72 h. Although this mode of fermentation is mechanically simple and reliable, it is inefficient and slow. Thus, in a batch fermentation, the lag and growth phases of the yeast represent a major drawback to more rapid fermentation, and the downtime associated with the cyclic emptying, cleaning and refilling of the bioreactor also adds to the overall low ethanol productivity (quantity of ethanol formed per unit volume in unit time) of 1.8–2.5 $g\ l^{-1}\ h^{-1}$ (Rose 1976).

There have, of course, been many efforts to secure faster batch fermentations. For instance, in several traditional processes a vessel in which fermentation is active is partially emptied into a second vessel and the first is replenished with fresh medium (Stewart 1974). In this way the lag period is reduced or even eliminated in replenished vessels. Alternatively, at the end of a fermentation, the yeast may be separated from the broth by centrifugation and recycled to the next batch. This system, known as the Melle–Boinot process, was initially developed in the 1940s (Boinot & Boige 1948). More recently, Nishizawa *et al.* (1984) demonstrated the use of microfiltration, instead of centrifugation, to concentrate yeast cells for the repeated batch production of ethanol. However, all batch systems are associated with high capital and operating costs. The continual start-up and shutdown nature of these systems makes it difficult to automate and, therefore, high labour costs are associated.

### *8.2.2 Continuous fermentation*

It is well known that the most substantial improvements over traditional batch processes result when the fermentation is made continuous. Continuous systems, with their attendant preparation and product-separation equipment, are generally smaller than those used in equivalent batch systems and, therefore, result in lower capital costs. In addition, since continuous processes are much more amenable to automation, substantial improvements in the efficiency of the process and product quality can be achieved, and subsequently higher productivities and lower operating costs ensue. Furthermore, unlike batch fermentation, the downtime between cycles is eliminated resulting in greater total production. Continuous processes have been extensively applied for the production of industrial ethanol in the USSR since the 1960s (Hospodka 1966), and are only recently being developed in some Western countries (Lyons 1984, Guidoboni 1984). Before proceeding any further, we must note that although ethanol fermentation is an anaerobic process, trace amounts of oxygen are required in continuous ethanol fermentation to maintain yeast cell viability and consequently high rates of ethanol production (Ryu *et al.* 1984a).

**Continuous stirred-tank bioreactor system**

The continuous stirred-tank bioreactor (CSTBR) is mechanically very simple and similar in construction to a batch fermenter, except that feed and overflow devices are incorporated. The fundamental difference lies in the fact that the contents of the vessel are at a steady state and the system needs to be well mixed to avoid short circuiting of feed between the inlet and outlet. Using an adapted yeast strain, Cysewski and Wilke (1976) achieved a cell density of 12 g dry wt $l^{-1}$ and an ethanol productivity of 7 g $l^{-1}$ $h^{-1}$, which is approximately three times higher than that obtained in a batch fermentation system. However, due to the homogeneity of the vessel contents, the exit conditions tend to reflect the physiological and biochemical environment in which the yeast cells operate. The major limitation of the CSTBR therefore arises from ethanol inhibition.

**Multistage stirred-tank bioreactor system**

In this system the overflow from the first bioreactor is fed into the second bioreactor, and so on. This arrangement has been found to be favourable since the inhibitory effects of ethanol are reduced and a higher product concentration can be achieved. Thus, in a series-CST bioreactor system (also known as the cascade process), all stages except the final one operate at an ethanol concentration which is not inhibitory to the yeast cells. This results in an overall increase in ethanol productivity when compared to a single stage CSTBR (Maiorella *et al.* 1981). Ghose and Tyagi (1979a) have shown that the productivity in a two-stage CST bioreactor system is 2.3 times higher than that of a single stage CSTBR. The cascade process (e.g. the Vogelbusch and Technipetrol processes) which may consist of up to 12 CST bioreactors arranged in series, has therefore been applied for the large-scale production of ethanol (Hospodka 1966, Guidoboni 1984).

Single- and multistage CST bioreactors do, however, have definite productivity limitations determined by the growth characteristics of the yeast strain employed. For instance, in a single-stage CSTBR operating at steady state the specific growth rate equals the dilution rate of the system. However, if the dilution rate exceeds the maximum specific growth rate, then yeast cells are purged from the system and cell washout occurs.

**Cell recycle bioreactor systems**

In these systems, non-flocculent yeast cells released in the product stream are recovered by either centrifugation or microfiltration, and are reintroduced into the CSTBR. Extremely high yeast cell concentrations of up to 100 g (dry weight) $l^{-1}$ can be maintained, enabling the bioreactor to operate at higher dilution rates, and consequently, ethanol productivities approaching 100 g $l^{-1}$ $h^{-1}$ have been achieved (Ghose & Tyagi 1979b, Mehaia & Cheryan 1984, Nishizawa *et al.* 1983, Cysewski & Wilke 1977). A number of large-scale, continuous, single- and multi-stage ethanol fermentation plants have therefore recently introduced yeast-cell recycle by centrifugation into their processes (Guidoboni 1984, Panushka 1983, Rosen 1978). However, the use of the centrifuge for cell recovery does increase the capital cost, requires considerable maintenance and added supervision, elevates energy requirements, and is associated with an increased risk of contamination. These disadvantages have been shown to be more than offset by the great increase in ethanol productivity and a consequent reduction in bioreactor volume (Maiorella *et al.* 1981). Furthermore, the use of either centrifugation or microfiltration is attractive, as they do not rely on the flocculating ability of the yeast for separation from the broth.

**Selective ethanol removal fermentation systems**

Although the cell density can be adequately increased by cell recycle, the inhibitory effect of ethanol becomes the limiting factor when concentrated substrates are employed. Processes in which ethanol is selectively removed during fermentation would therefore appear to be beneficial.

Vacuum operation of the CSTBR makes use of the high volatility of ethanol, essentially to boil-off the alcohol as it is formed, at temperatures compatible with yeast function. Ramalingham and Finn (1977) demonstrated that a 50% (w/v) sugar feed could be fermented under a vacuum of 32 mm Hg (30°C). In an independent study Cysewski and Wilke (1977) showed that the vacuum-CSTBR in conjunction with cell recycle permitted the use of a concentrated sugar solution [33.4% (w/v)] as feed, and ethanol productivities of up to 82 g $l^{-1}$ $h^{-1}$ were obtained. Furthermore, a reduction in distillation requirements resulted due to the formation of a highly concentrated ethanol product.

However, the operation of a vacuum fermentation is associated with the following potential disadvantages.

(a) The accumulation of toxic non-volatile feed components and fermentation by-products in the medium necessitates a continuous 'bleed' of the fermentation broth.
(b) Carbon dioxide produced during fermentation has to be compressed up to atmospheric pressure resulting in high energy requirements.
(c) Pure oxygen is required to maintain yeast-cell viability.
(d) The likelihood of bioreactor contamination and shutdown is increased.

Flash fermentation is another technique that has been tested and for which high ethanol productivities have also been claimed. A typical system consists of a CSTBR maintained under atmospheric pressure, connected to an external

**Table 8.1** Ethanol production by *Saccharomyces cerevisiae* using various free-cell bioreactor systems.

| Bioreactor system | Substrate/ reducing sugars concentration ($g\,l^{-1}$) | Dilution rate ($h^{-1}$) | Feed sugars utilized (%) | Biomass concentration [g (dry wt) $l^{-1}$] | Volumetric ethanol productivity ($g\,l^{-1}\,h^{-1}$) | Reference |
|---|---|---|---|---|---|---|
| simple batch | glucose/100 | — | NA | 5.0 | 1.8–2.5 | Rose (1976) |
| repeated batch – cell recovery by centrifugation | glucose/150 | — | NA | NA | 3.0 | Nishizawa *et al*. (1984) |
| repeated batch – cell recovery by microfiltration | glucose/200 | — | NA | 36.0 | 6.4 | Nishizawa *et al*. (1984) |
| simple CST bioreactor | glucose/100 | 0.17 | NA | 12.0 | 7.0 | Cysewski and Wilke (1976) |
| CST bioreactor + cell recycle by centrifugation | glucose/100 | 0.68 | 98 | 50.0 | 29.0 | Cysewski and Wilke (1972) |
| CST bioreactor + cell recycle by centrifugation | bagasse hydrolysate/150 | 0.50 | 83 | 42.0 | 32.0 | Ghose and Tyagi (1979b) |
| CST bioreactor + cell recycle by microfiltration | glucose/150 | 0.58 | 83 | 85.0 | 27.0 | Nishizawa *et al*. (1983) |
| vacuum CST bioreactor | glucose/334 | 0.26 | 99 | 50.0 | 40.0 | Cysewski and Wilke (1972) |
| vacuum CST bioreactor + cell recycle | glucose/334 | 0.23 | 99 | 124.0 | 82.0 | Cysewski and Wilke (1972) |
| CST bioreactor + cell recycle by microfiltration | glucose/100 | 2.0 | 100 | 100.0 | 100.0 | Mehaia and Cheryan (1984) |

NA, not available.

vessel under vacuum in which ethanol is flashed off (Maiorella *et al.* 1981, Roffler *et al.* 1984, Haraldson & Rosen 1982a, b). In this process, some of the operating difficulties encountered in the vacuferm process are overcome. However, as in the vacuum fermentation, a bleed of fermented broth is continually withdrawn. Thus, at comparable ethanol productivities and energy requirements, the flash fermentation may appear to be more attractive than direct vacuum fermentation. Fully integrated fermentation–distillation processes with cell recycle have also been designed and are currently being used for the pilot-scale (Atpal process) and commercial-scale (Biostil process) production of ethanol (Guidoboni 1984, Garlick 1983). Other techniques of *in situ* ethanol recovery have included the use of liquid–liquid extraction (Larsson & Mattiasson 1984) and adsorption (Lencki *et al.* 1983, Wang *et al.* 1981) coupled to fermentation.

From Table 8.1 it is apparent that with the use of continuous systems coupled to cell recycle or ethanol extraction, or both, substantial improvements in the efficiency of the fermentation result.

## 8.3 Ethanol production using immobilized yeast cells

Historically, the first attempts to produce ethanol continuously employed immobilized yeast cells rather than free-cell systems. In 1892 Delbruck proposed a continuous fermentation system in which yeast cells were retained within a porous cylinder whereas the fermenting substrate flowed past and diffused through the wall of the cylinder (Delbruck 1892). In 1899 Barbet devised an apparatus within which a culture of yeast was supported on a packing material (Barbet 1899). However, these early attempts were impractical owing to their primitive methods of vessel design and construction, and the failures to guard these processes against microbial contamination. Nevertheless, they established the fact that the maintenance of high cell densities was essential to achieve rapid fermentation (Portno 1970).

With developments in cell-immobilization techniques concurrently with novel bioreactor design, recent attempts in the production of both fuel and potable alcohol have met with more success and have been reviewed by Linko and Linko (1984), Margaritis and Merchant (1984) and Margaritis and Wallace (1984). In this section we shall evaluate a variety of techniques that have been used for the immobilization of yeast cells and subsequently used for ethanol production. These techniques may be grouped under four separate headings: (a) containment of yeast cells within membrane bioreactors; (b) attachment of yeast cells to solid supports; (c) entrapment of yeast cells in polymeric gels; and (d) flocculation of yeast cells.

In selecting a suitable technique for live-cell immobilization a number of criteria need to be considered. Primarily, the procedure should be mild enough to ensure retention of yeast-cell viability, at the same time being able to achieve and maintain a high concentration of yeast cells. Furthermore, the immobilized cell system must be capable of reactivation, if necessary, and be stable for prolonged periods. Finally, the technique should be simple, relatively inexpensive and capable of scale-up. Thus, a suitable technique may confer all or some of the advantages listed below to a fermentation system. The advantages of immobilized cell systems include the following:

(a) The possibility of much greater cell concentrations within the bioreactor, facilitating faster fermentation rates.
(b) Operation at high dilution rates without fear of yeast-cell washout.
(c) A reduced risk of microbial contamination due to the combined effects of (a) and (b).
(d) Less susceptibility to the effects of inhibitory compounds and nutrient depletion.
(e) The *in situ* removal of the cells from the product stream eliminates the cost of a centrifuge normally used as the first step in product recovery, thus reducing energy and capital cost requirements.
(f) As opposed to free-cell CSTBR systems, most immobilized-cell bioreactors can be operated in the plug-flow mode. One can therefore take advantage of the intrinsic 'kinetic' superiority of the latter bioreactor system.

### *8.3.1 Membrane-contained yeast-cell bioreactor systems*

The use of membrane-contained microbial cells predates the recent emergence of interest in cell immobilization. In these systems yeast cells may be confined within a bioreactor by incorporating either porous media, microfilters or dialysis membranes in a bioreactor.

In 1969 Schultz and Gerhardt published an extensive review on the design, theory and experimental evaluation of dialysis culture of micro-organisms. Such systems are composed of a dialysate circuit and a bioreactor circuit, either of which may be operated in a continuous or batch mode. Using a simple continuous dialysis system, Kyung and Gerhardt (1984) maintained an active culture of yeast cells. By operating the bioreactor circuit without an effluent the yeast cells were essentially 'immobilized'. The substrate, glucose, entered the bioreactor circuit from the dialysate circuit by diffusing through a dialysis membrane, whereas the product, ethanol, diffused back in the opposite direction and was recovered in an overflow. Although this system serves to introduce sterile feed into the bioreactor and maintains a high yeast cell density [up to 57 g (dry weight) $l^{-1}$], it is limited by the gradual fouling of the membrane and the inherently very slow process of diffusion of both nutrients and metabolic products, resulting in low ethanol productivities of only 5.8 g $l^{-1}$ $h^{-1}$ and low utilization of the substrate.

However, the problem of membrane fouling and the limitation due to slow diffusion rates can be overcome by microfiltration, as in the rotorfermenter which essentially consists of a rotating microporous membrane enclosed within a stationary fermentation vessel. The rotorfermenter has been described in detail and characterized by Margaritis (1975) and Margaritis and Wilke (1978a, b). In the rotorfermenter the cell-free filtrate along with the metabolic products are continuously removed by filtration through the rotating microporous membrane, whereas the growing yeast cells are retained within the bioreactor. Furthermore, since the membrane is rotated at high speeds, a strong centrifugal force develops at the membrane surface, which keeps the membrane clean and prevents fouling. High cell densities [30.9 g (dry weight) $l^{-1}$] and ethanol productivities (27.3 g $l^{-1}$ $h^{-1}$) have been achieved using a pilot-scale rotorfermenter (Taniguchi *et al.* 1983). However, more work is needed to optimize the performance characteristics of the rotorfermenter further.

High rates of filtration in the absence of membrane plugging has also been demonstrated with the use of filter-bioreactors. These bioreactors are equipped with a sintered stainless-steel plate (which may be coated with a thin layer of diatomaceous earth) located at the bottom of the vessel (Taniguchi *et al.* 1983, Dostalek & Haggstorm 1982). Membrane plugging can be prevented by adequate mechanical agitation near the filter surface or by back-flushing the membrane and consequently allowing high filtration rates to be maintained. Although high cell densities and ethanol productivities have been achieved, the main disadvantage of filter-bioreactors is the limited filtration area.

A new approach for the continuous production of ethanol has been recently developed using hollow-fibre bioreactors (Ghose & Tyagi 1979, Mehaia & Cheryan 1984, Inloes *et al.* 1983). These bioreactors essentially consist of a cylindrical bundle of a large number of hollow fibres sealed into a surrounding shell effectively separating the outside of the fibre from the fluid flowing inside the fibres. Yeast cells are inoculated into the shell side, and the nutrients are supplied and products removed by a single-pass perfusion of the fibre lumens. Although high cell densities and ethanol productivities have been achieved with hollow-fibre bioreactors, a number of problems have been encountered with prolonged use, due to the accumulation of $CO_2$ in the shell side (Inloes *et al.* 1983).

Inloes *et al.* (1983) observed that high pressures due to $CO_2$ accumulation resulted in the collapse of the hollow fibres and hindered the flow of the liquid medium through the fibre. Furthermore, high levels of $CO_2$ can adversely affect the metabolism of yeast cells (Jones & Greenfield 1982). Therefore, the efficient removal of $CO_2$ becomes imperative if hollow-fibre bioreactors are to be used for the continuous production of ethanol. Also, these bioreactors are complex and expensive, and the metabolic activity of the yeast cells is limited by low nutrient and product transfer rates across the membrane (Vick Roy *et al.* 1983). As is the case with other mechanically contained bioreactor systems, the feedstream must be pre-filtered to avoid clogging of the membranes and fibres. This will generally prevent the use of insoluble reactants, which are often less expensive.

The performance characteristics of various mechanically contained yeast-cell bioreactor systems are summarized in Table 8.2, indicating that high ethanol productivities may be achieved. However, due to the inherent limitations of such systems it is unlikely that these bioreactors will be used for the industrial production of ethanol, especially in view of the recent developments in other methods of yeast-cell immobilization.

### 8.3.2 *Ethanol production using attached yeast-cell bioreactor systems*

The attachment of yeast cells to inert supports can be accomplished either by adsorbing yeast cells to inert supports (by virtue of electrostatic interactions) or by attaching yeast cells to supports via covalent bonds.

**Table 8.2** The continuous production of ethanol by membrane-contained yeast-cell bioreactor systems.

| Bioreactor system | Substrate/ concentration ($g\,l^{-1}$) | Dilution rate ($h^{-1}$) | Feed sugars utilized (%) | Biomass concentration [g (dry weight) $l^{-1}$] | Volumetric ethanol productivity ($g\,l^{-1}\,h^{-1}$) | Reference |
|---|---|---|---|---|---|---|
| simple continuous dialysis bioreactor | glucose/200 | 0.092 | 78.1 | 56.9 | 5.81 | Kyung and Gerhardt (1984) |
| Rotorfermenter | glucose/102 | 0.58 | 99 | 30.9 | 27.3 | Kyung and Gerhardt (1984), Margaritis and Wilke (1972), Margaritis and Wilke (1978a, b) |
| filter-bioreactor | glucose/100 | 0.60 | 96 | 60 | 26.0 | Taniguchi *et al.* (1983) |
| filter-bioreactor | glucose/100 | 0.65 | 100 | 33 | 26.0 | Dostalek and Haggstorm (1982) |
| hollow-fibre bioreactor | glucose/100 | 0.25 | 85 | 100 | 10 | Mehaia and Cheryan (1984a, b) |
| hollow-fibre bioreactor | glucose/89 | 2.17 | 28 | 87.5 | 26.0 | Inloes *et al.* (1983) |

**Adsorption of yeast cells**

Adsorption was amongst the first methods used for the immobilization of yeast cells (Barbet 1899). The basis of cell adsorption is characterized by the electrostatic interaction between the support and yeast cells. However, the number of factors listed in Table 8.3 that govern the extent and strength of cell fixation are numerous and complex (Kolot 1980). Therefore, in selecting a suitable support for the production of ethanol by adsorbed yeast cells, the support material must have a high affinity for the yeast strain to be employed under the environmental conditions present within the bioreactor.

On this basis a number of supports have been utilized for the adsorption of yeast cells, and subsequently for the continuous production of ethanol. In most of these cases, adsorption is accomplished by circulating a concentrated suspension of yeast cells through the bioreactor for several hours. It is evident from the data summarized in Table 8.4 that in such bioreactor systems a high biomass concentration [up to 27 g (dry weight) $l^{-1}$ bioreactor volume] can be achieved (Daugulis *et al.* 1981), resulting in elevated ethanol productivities (62.0 g $l^{-1}$ $h^{-1}$). Furthermore, the long-term operational stability of at least 75 days (Ghose & Bandyopadhyay 1980) makes these processes comparable with, and in some cases superior to, the CSTBR with cell recycle. More recently, a potentially viable process which removes ethanol continuously (by liquid–liquid extraction) during fermentation (using adsorbed yeast cell bioreactor) has been devised, enabling feeds of high sugar content to be completely fermented (Minier & Goma 1982). Thus, the use of adsorbed yeast cells for the continuous production of ethanol employs a cheap and simple method of retaining high cell densities without the use of the expensive separation and recycle systems used conventionally. Also, the immobilization procedures do maintain the yeast cells in a viable state and do not require the use of toxic chemicals to induce adsorption.

However, the use of adsorbed yeast-cell bioreactor systems on an industrial scale may be limited by two major disadvantages. First, the amount of yeast cells that can be adsorbed by per unit weight of support is limited by the surface area of the carrier. Secondly, the operational stability of the bioreactor system

**Table 8.3** Factors affecting the adsorption of yeast cells.

| Properties of the inert support | Yeast cell characteristics | Conditions within the bioreactor |
|---|---|---|
| (1) particle size | (1) yeast strain | (1) pH |
| (2) porosity of support | (2) age of yeast cells | (2) temperature |
| (3) charge on support | (3) charge on cell surface | (3) feed flow rate |
| (4) carrier composition | (4) cell wall composition | (4) composition of culture medium |
| | (5) cell size | |

**Table 8.4** Continuous ethanol production by adsorbed yeast-cell bioreactor systems.

| Support | Bioreactor configuration | Substrate/ concentration ($g l^{-1}$) | Feed sugars utilized (%) | Biomass concentration* [g (dry weight) $l^{-1}$] | Volumetric ethanol† productivity ($g l^{-1} h^{-1}$) | Reference |
|---|---|---|---|---|---|---|
| beech wood chips | vertical PBBR | glucose/100 | 83.5 | 11.6–14.4 | 14.86 | Gianetto *et al.* (1983) |
| beech wood chips | vertical PBBR | glucose/150 | 75.5 | 26.3 | 16.0 | Specchia *et al.* (1983) |
| treated beech wood chips | vertical PBBR | glucose/130 | 96.9 | 0.18‡ | 21.8 | Moo-Young *et al.* (1980) |
| wood chips | vertical PBBR | glucose/130 | 73 | 15.5 | 25.2 | Gencer and Mutharasan (1981) |
| wood chips | vertical PBBR | glucose/164 | 82 | 24.6 | 39.1 | Gencer and Mutharasan (1983) |
| wood chips | vertical PBBR | glucose/300 | NA | ≃27.0 | 2.0 | Ryu *et al.* (1982) |
| wood chips | multistage, agitated | glucose/300 | NA | 14.0 | 4.3 | Ryu *et al.* (1982) |
| sawdust | vertical PBBR | glucose/100 | 70–90 | NA | 60 | Michaux *et al.* (1982) |
| unspecified inert support | vertical PBBR | molasses and reducing sugars/197 | 74.6 | 20.7 | 24.9 | Ghose and Bandyopadhyay (1980) |
| unspecified inert support | | molasses and reducing sugars/150 | 96 | 0.132‡ | 28.6 | Tyagi and Ghose (1982) |
| ion-exchange resin | vertical PBBR | glucose/120 | 84.2 | 27.2 | 52.0 | Daugulis *et al.* (1981) |
| porous brick particles | vertical PBBR (fermentation coupled to ethanol extraction) | glucose/400 (with dodecanol as extractant) | 100 | NA | 1.03 | Minier and Goma (1982) |
| $PSP_s$-spheres of stainless-steel knitted mesh | FBBR | 24.8 | 66.8 | 100 | 20.4 | Black *et al.* (1984) |
| $BSP_s$-cubes of reticulated polyester foam | CBBR | 46.0 | 97 | 150 | 21.6 | Black *et al.* (1984) |

* Based on bioreactor working volume; † based on bioreactor void volume; ‡ g (dry weight) $g^{-1}$ (dry weight) support. PBBR, packed-bed bioreactor; FBBR, fluidized-bed bioreactor; CBBR, circulating-bed bioreactor; $BSP_s$, biomass support particles; NA, not available.

is restricted by the rate of desorption of cells from the support (Minier & Goma 1982). This problem is especially severe when changes in the pH or the ionic strength occur (Michaux *et al.* 1982), or when cells are sheared from the carrier surface by virtue of rapid fluid flow (Gianetto *et al.* 1983, Moo-Young *et al.* 1980) and high turbulence created by $CO_2$ gas bubbles. However, these disadvantages may be partially resolved by the use of specially designed support particles as described by Atkinson *et al.* (1979).

These support particles are essentially an interconnecting void within an open network of matrix support material (e.g. stainless-steel knitted mesh, reticulated polyester foam, etc.). Due to the quiescent conditions which exist within such porous structures, the accumulation of yeast cells readily occurs, and the porous support eventually becomes filled with cells. This mode of immobilization provides an environment within which yeast cells are protected from the high external liquid shear, and where weak adhesion or flocculation is sufficient to retain the cells. The abrasive forces of interparticle and particle–wall collisions, together with liquid shear, serve to remove any excess cells which grow beyond the protected environment within these support particles. A constant population of well-defined yeast flocs is, therefore, maintained within the bioreactor.

A liquid–fluidized bed bioreactor (FBBR) and a gas stirred–circulating bed bioreactor (CBBR) were used to immobilize yeast cells within stainless-steel knitted mesh and reticulated polyester foam, respectively (Black 1984). In both cases, despite the diffusion limitation of the system, high ethanol productivities ($21.6\ g\ l^{-1}\ h^{-1}$) were attained. Extended operation of the bioreactors for up to 113 days was achieved with no loss of immobilized biomass or of cell viability.

**Attachment of yeast cells to solid supports by covalent binding**

The attachment of yeast cells to activated supports (e.g. gelatin-coated distillation packing) via cross-linking agents (e.g. glutaraldehyde) has also been attempted. By adsorbing gelatin to an inert support, a reactive base for glutaraldehyde is provided, thus allowing a covalent link to be formed between the yeast cells and the gelatin-coated support. This method has been employed to immobilize viable yeast cells, and was subsequently used for the continuous production of ethanol (Griffith & Compere 1976, Sitton & Gaddy 1980, Amin *et al.* 1983). High ethanol productivities of up to $20\ g\ l^{-1}\ h^{-1}$ (Table 8.5) and sustained operation for several months has been possible with these bioreactor systems.

However, certain limitations tend to restrict the use of covalently attached yeast-cell systems. For instance, the toxic effects of some cross-linking agents adversely affect the metabolic activity of yeast cells (Navarro & Durand 1977, Bandyopadhyay & Ghose 1982). Furthermore, in covalently attached yeast-cell systems the biomass loading (dry weight of cells bound per gram of dry support) is limited by the availability of binding sites on the support surface. However, if macroporous supports are used the number of covalent binding sites can be increased, thereby significantly improving the biomass loading capacity of the support (Dias *et al.* 1982).

Despite these disadvantages, the immobilization of yeast cells to a support via irreversible covalent bonds does produce a system free of the diffusion

**Table 8.5** Continuous ethanol production by covalently attached yeast-cell bioreactor systems.

| Immobilization technique | Bioreactor configuration | Substrate/ concentration ($g\,l^{-1}$) | Feed sugars utilized (%) | Volumetric ethanol productivity* ($g\,l^{-1}\,h^{-1}$) | Reference |
|---|---|---|---|---|---|
| *Saccharomyces cerevisiae* cells cross-linked via glutaraldehyde to supports coated with gelatin–polyelectrolyte mixture | vertical PBBR | glucose/300 | ≈100 | >20 | Griffith and Compere (1976) |
| *Saccharomyces cerevisiae* cells cross-linked via glutaraldehyde to gelatin-coated Raschig rings | vertical PBBR | glucose/30 | 99.4 | 4.24 | Sitton and Gaddy (1980) |
| *Saccharomyces bayanus* cells cross-linked via glutaraldehyde to gelatin- and agar-coated pebbles | vertical PBBR | glucose/150 | 93 | 20† | Amin *et al.* (1983) |

* Based on bioreactor void volume; † based on total bioreactor working volume. PBBR, packed-bed bioreactor.

limitations present in entrapped cell systems, and without the problems of cell desorption commonly encountered in adsorbed yeast-cell bioreactor systems. Thus, although the production of ethanol by covalently attached yeast-cell bioreactor systems is infrequently used, the development of macroporous supports and milder cross-linking agents may in future favour its use in place of adsorption.

### *8.3.3 Entrapped yeast-cell bioreactor systems*

Fundamentally, the immobilization of cells by entrapment is based on the inclusion of cells within a rigid polymeric matrix. The entrapment of live cells and their subsequent growth within such supports was first demonstrated by Updike *et al.* (1969). As shown by Margaritis and Merchant (1984), entrapment has since become a common method for the immobilization of viable yeast cells, and has been extensively evaluated for the continuous production of ethanol. A number of different polymers have been used for yeast-cell entrapment and subsequently for the continuous production of ethanol. The techniques used to induce gelation are listed in Table 8.6.

**Entrapment of yeast cells in alginate gels**

Alginate is a glycuronan consisting of residues of D-mannuronic acid and L-guluronic acid arranged in a blockwise fashion along the polymer chain. In the presence of multivalent cations, gel formation occurs. The

**Table 8.6** Methods used to entrap viable yeast cells for the continuous production of ethanol.

| Entrapment technique | Procedure used to induce gelation | Examples |
|---|---|---|
| (1) entrapment in ionic polymeric networks; cross-linking of linear polyanions | use of multivalent carbons, e.g. $Ca^{2+}$, $Ba^{2+}$, $Sr^{2+}$, $Al^{3+}$ | calcium alginate, aluminium alginate |
| (2) entrapment by precipitations of polymers | change in pH, temperature, solvent, etc. | carrageenan, gelatin, pectin, agar |
| (3) entrapment within covalent polymeric matrix | (a) cross-linking co-polymerization | polyacrylamide |
| | (b) polycondensation of prepolymers | epoxide |
| | (c) irradiation with X-rays | irradiated polyethylene-glycol dimethacrylate |
| (4) entrapment within photo-crosslinkable resins | illumination with 300–400-nm wavelength light | polyethylene glycolic photo-crosslinkable resins |

three-dimensional gel network formed is biochemically inert, and yeast cells can be trapped in the interstitial spaces of the gel. The immobilization procedure is attractive primarily because of its simplicity. It is achieved by dropping a mixture of sodium alginate solution [2–4% (w/v)] and the yeast-cell suspension into a calcium chloride solution. The pellets so formed are highly porous (Klein *et al.* 1983), readily diffusible to low-molecular weight compounds (Tanaka *et al.* 1984), and mechanically stable (Cheetham 1979).

The continuous production of ethanol using alginate-entrapped yeast cells was first reported by Kierstan and Bucke (1977). In this system cells of *Saccharomyces cerevisiae* were entrapped in Ca-alginate fibres packed in a vertical column. Since then, the use of Ca-alginate beads has been preferred. Furthermore, a number of different novel bioreactor configurations have been designed (Table 8.7) enabling ethanol productivities of up to 46 $g\,l^{-1}\,h^{-1}$ (based on total bioreactor volume) to be achieved (Shiotani & Yamane 1981). Several other modifications to the original entrapment technique have also been made. These include the following.

(a) Incorporation of magnetic material within Ca-alginate beads to facilitate retrieval of immobilized cells from batch or repeat batch systems, especially when highly viscous media or media containing particulate matter are used (Larsson & Mosbach 1979).
(b) Coentrapment of micrometre-sized silica or sand particles within Ca-alginate beads to enhance gel strength, density and internal surface adhesion (Chotani & Constantinides 1984, Fang *et al.* 1983).
(c) Incorporation of certain sterols and unsaturated fatty acids within Ca-alginate beads to maintain yeast-cell viability during fermentation (Oda *et al.* 1983, Nagashima *et al.* 1984).
(d) Partial drying of beads at room temperature to reduce bead size and increase mechanical stability of alginate beads (Linko & Linko 1984, Linko *et al.* 1983). Reduction in bead size has also shown to improve substantially the kinetics of ethanol production (Ryu *et al.* 1984), due to increases in mass transfer rates in and out of the beads.
(e) Use of higher alginate concentration [up to 8–10% (w/v)] to enhance the mechanical stability of the matrix (Linko & Linko 1982).

High rates of ethanol production have been maintained for prolonged periods under conditions of periodic aeration or nutrient addition, or both, to regenerate the activity of entrapped yeast cells (Lee *et al.* 1983, Williams & Munnecke 1981). Furthermore, alginate-entrapped cells of *Saccharomyces cerevisiae* have also been shown to ferment efficiently a variety of commercially available sugars such as sugar cane molasses (Linko & Linko 1981), waste sulphite liquor (Linko *et al.* 1983), glucose corn syrup (Ryu *et al.* 1984b) and starch syrup (Linko *et al.* 1984). The microenvironmental conditions within the Ca-alginate matrix have been implicated in the change in metabolic properties of entrapped yeast cells. According to established metabolic pathways for freely suspended yeast cells, a theoretical yield of 51% ethanol from glucose is possible. However, entrapped yeast cells have been found to operate with substantially improved yields (Holoberg & Margalith 1981). Mattiasson and Hagerdahl (1982) have shown that such changes are due to a decrease in

**Table 8.7** Alginate-entrapped yeast-cell systems used for the continuous production of ethanol.

| Immobilized yeast cell system | Bioreactor configuration | Substrate/ concentration ($g\,l^{-1}$) | Feed sugars utilized (%) | Volumetric ethanol productivity* ($g\,l^{-1}\,h^{-1}$) | Reference |
|---|---|---|---|---|---|
| *Saccharomyces cerevisiae* entrapped in Ca-alginate fibres | vertical PBBR | glucose/100 | 90 | 4.3† | Kierstan and Bucke (1977) |
| *Saccharomyces cerevisiae* entrapped in Ca-alginate beads | horizontal PBBR | glucose/196 | 86 | 46.0† | Shiotani and Yamane (1981) |
| *Saccharomyces cerevisiae* entrapped in Ca-alginate beads | vertical PBBR | glucose/150 | 99.3 | 37.8 | Lee *et al.* (1983) |
| *Saccharomyces cerevisiae* entrapped in Ca-alginate beads | vertical PBBR | glucose/100 | 95 | 25.0 | Cho *et al.* (1981) |
| *Saccharomyces cerevisiae* entrapped in Ca-alginate beads | FBBR | glucose/100 | 94 | 32.2 | Cho *et al.* (1981) |
| *Saccharomyces cerevisiae* entrapped in Ca-alginate beads | vertical PBBR | glucose/127 | 100 | 30.0 | Williams and Munnecke (1981) |
| *Saccharomyces cerevisiae* entrapped in Ca-alginate beads | CSTBR | glucose/100 | 97.3 | 11.3 | Merchant (1981) |
| *Saccharomyces cerevisiae* entrapped in Ca-alginate beads | FBBR | glucose/100 | 90 | 65.0 | Margaritis *et al.* (1983b) |
| *Saccharomyces cerevisiae* entrapped in Ca-alginate beads | vertical PBBR | sugar cane molasses/175 (reducing sugars) | 100 | 19–22‡ | Linko and Linko (1981) |

**Table 8.7** Continued.

| Immobilized yeast cell system | Bioreactor configuration | Substrate/ concentration ($g\,l^{-1}$) | Feed sugars utilized (%) | Volumetric ethanol productivity* ($g\,l^{-1}\,h^{-1}$) | Reference |
|---|---|---|---|---|---|
| *Saccharomyces cerevisiae* entrapped in partially dried Ca-alginate beads | vertical PBBR | sugar cane molasses/175 (reducing sugars) | 100 | 39–43‡ | Linko *et al.* (1983) |
| *Saccharomyces cerevisiae* co-entrapped with micrometre sized silica in Ca-alginate beads | CFBR | glucose/205 | 90 | 37.8 | Chotani and Constantinides (1984) |
| *Saccharomyces formosensis* co-entrapped with sand particles in Ca-alginate beads | vertical SCBR | glucose/92 | 88 | 60§ | Fang *et al.* (1983) |
| *Saccharomyces cerevisiae* montrachet and *Saccharomyces formosensis* co-entrapped in Al-alginate beads | 3 stage FBBR | cane juice/180 (reducing sugars) | 95 | 20 | Fukushima and Hatakeyama (1983) |
| *Saccharomyces cerevisiae* co-entrapped with sterols and unsaturated fatty acids in Ca-alginate beads | 2 stage FBBR | sugarcane molasses/140 (reducing sugars) | >90 | 50 | |

* Based on bioreactor void volume; † based on total bioreactor working volume; ‡ based on biocatalyst bed volume excluding space taken up with $CO_2$; § based on liquid volume.

PBBR, packed-bed bioreactor; FBBR, fluidized-bed bioreactor; CSTBR, continuous stirred-tank bioreactor; CFBR, cross-flow bioreactor; SCBR, segmented-column bioreactor.

the water activity within the microenvironment where the yeast cells are entrapped.

From the above studies it is evident that alginate-entrapped yeast cells are extremely versatile and highly efficient biocatalysts. It is therefore not surprising that their use at the pilot-plant level, as shown by Margaritis *et al.* (1983), has already shown promise for the future full-scale industrial production of ethanol. Most notably, recent work carried out in Japan (Oda *et al.* 1983, Nagashima *et al.* 1984) employing Ca-alginate-entrapped cells of *Saccharomyces cerevisiae* has been operated efficiently for several months. The system used for these purposes included a two-stage and also a three-stage fluidized-bed bioreactor with a total capacity of 4 $m^3$. The Ca-alginate beads containing over 250 g yeast cell (dry weight) $l^{-1}$ gel and occupying 50–60% of the bioreactor volume, were kept in a fluidized state by jets of air and the native $CO_2$ of the process. The system was efficiently used for producing 2400 litres of 99.5% (v/v) ethanol $day^{-1}$ from diluted sugar cane molasses. The need for sterilizing the sugar feed was found not to be important since periodic contamination was effectively removed by the use of bactericidal agents or dilute sulphuric acid, or both. Alternatively, the use of 0.3 g $l^{-1}$ of sodium azide (a respiratory inhibitor) in the feed has also been suggested to restrict contaminants from entering the bioreactor (Hahn-Hagerdal & Mattiasson 1982a) and at the same time prevent the shift towards aerobic metabolism in the presence of oxygen (Hahn-Hagerdal & Mattiasson 1982b).

The major disadvantage of using Ca-alginate pellets is that, in the presence of calcium chelating agents such as phosphates, gel disruption readily occurs (Cheetham 1979). Since the presence of phosphate in the medium is essential in maintaining the integrity of yeast cell function (Jones *et al.* 1981), techniques have been developed to prepare phosphate-stable alginate beads without affecting yeast-cell viability. This has included the treatment of Ca-alginate beads with high-molecular weight cationic polymers, such as polyethylene amine and cross-linking agents (Veliky & Williams 1981, Birnbaum *et al.* 1981). Alternatively, the use of more-efficient gel-inducing cations, other than $Ca^{2+}$, has also been suggested (Paul & Vignais 1980). Thus, highly stable Al-alginate beads entrapping a mixture of *Saccharomyces cerevisiae* montrachet (wine yeast) and an ethanol tolerant strain of *Saccharomyces formosensis* were used to produce ethanol continuously from non-sterile cane juice (Fukushima & Hatakeyama 1983). A FBBR composed of three vertically connected rhomboid-shaped units was used, in which the entrapped cells were fluidized by introducing either $CO_2$ or $N_2$ into the bioreactor. Prolonged and efficient operation of such a system was proven in a 150-litre pilot-scale bioreactor (Fukushima & Hanai 1982).

Based on the excellent performance characteristics (Table 8.7) and long-term operational stability of alginate-entrapped yeast-cell bioreactor systems even on scale-up, it is quite likely that new ethanol plants will employ systems similar to those described above.

### Entrapment of yeast cells in other natural gels

In addition to alginate, a number of other naturally occurring polymers (carrageenan, agar, pectin and gelatin) have been employed to entrap viable yeast cells. Like alginate, carrageenan is a low-cost carbohydrate derived from seaweed, and can be readily induced to gel under mild conditions.

The use of carrageenan entrapped viable yeast cells was first attempted by Chibata and his colleagues in Japan (Wada *et al.* 1979). In their study two different entrapment techniques were evaluated. In the first case a dilute suspension of yeast cells was mixed with a carrageenan solution and gel beads prepared by dropping the mixture into a KCl solution at 20°C. After incubating the carrageenan beads for 2–3 days in a 'nutrient-rich' medium, the number of viable yeast cells within the gel matrix increased by 1000-fold. Thus, immobilization was essentially achieved by growing the yeast cells within the gel, in which case the cells were largely distributed at the periphery of the bead (Wada *et al.* 1980a). In the second method a concentrated yeast cell paste was mixed with the carrageenan solution before gel induction. In this case the yeast cells were evenly distributed within the gel matrix. When comparing the two entrapment methods, Wada *et al.* (1979) observed that the ethanol productivity of the immobilized growing cell system was at least three times higher than that of the latter system at an equivalent cell concentration within the bead. This may be attributed to the greater availability of the substrate and nutrients in the former case, since yeast cells were largely localized near the gel surface. Immobilized growing yeast cells were subsequently utilized for the continuous production of ethanol using a vertical PBBR (Wada *et al.* 1980b). By gradually increasing the glucose concentration in the feed (from 100 to 250 $g\,l^{-1}$) a high concentration of ethanol (110 $g\,l^{-1}$) in the effluent stream could be maintained for up to 2 months at the highest substrate concentration (Mitani *et al.* 1984).

Various modifications to the original entrapment procedure have been made and employed for the continuous production of ethanol (Table 8.8). Wang and Hettwer (1982) devised a technique for preparing highly porous carrageenan beads. This was achieved by coentrapping tricalcium phosphate crystals and *Saccharomyces cerevisiae* cells in carrageenan beads followed by crystal dissolution at low pH. Due to a reduction in the internal mass transfer resistance, with the use of porous beads, the rate of ethanol production was substantially improved (Wang & Hettwer 1982). Similarly, entrapped yeast cells were subsequently employed in a repeated batch-fermentation process coupled with an ethanol extraction system using activated carbon (Lee & Wang 1982) enabling complete utilization of the concentrated glucose medium.

Although the carrageenan matrix is mechanically stable, preparations of greater stability have been obtained either by treating the carrageenan gel with polyacrylamide (Kuu & Polack 1983) or by co-entrapping celite and locust bean gum (Mitani *et al.* 1984a, b) within the matrix. Both of these systems have been successfully applied for the continuous production of ethanol over prolonged periods without appreciably affecting yeast-cell viability.

Cells of *Saccharomyces uvarum* entrapped in porous gelatin beads have also been successfully used for the continuous production of ethanol from sugar cane molasses (Sivaraman *et al.* 1982) and sweet sorghum juice (Mohite & Sivaraman 1984). The porous gelatin beads were prepared by dissolving the Ca-alginate component of the composite alginate–gelatin beads using phosphate buffer. These workers reported extremely high ethanol productivities at high feed flow rates, enabling the system to be operated without previous

**Table 8.8** Continuous ethanol production by yeast cells entrapped in naturally occurring polymers.

| Immobilized yeast cell system | Bioreactor configuration | Substrate/ concentration ($g\,l^{-1}$) | Feed sugars utilized (%) | Volumetric ethanol productivity* ($g\,l^{-1}\,h^{-1}$) | Reference |
|---|---|---|---|---|---|
| *Saccharomyces carlsbergensis* grown within K-carrageenan beads | vertical PBBR | glucose/100 | 100 | 92.6 | Wada *et al.* (1980b) |
| *Saccharomyces cerevisiae* grown within K-carrageenan beads | vertical PBBR | glucose/250 | 95 | 43.7† | Wada *et al.* (1981) |
| *Saccharomyces cerevisiae* co-entrapped with locust bean gum and celite in K-carrageenan gel slabs | FPBR | glucose/200 | 95 | 48.0† | Mitani *et al.* (1984a) |
| *Saccharomyces bayanus* grown within K-carrageenan beads | vertical PBBR | glucose/175 | 82 | 5.3† | Amin and Verachtert (1982) |
| K-carrageenan entrapped cells of *Saccharomyces bayanus* coated on small stones | vertical PBBR | glucose/150 | 98 | 69.8 | Amin *et al.* (1984) |
| *Saccharomyces uvarum* entrapped within porous gelatin beads | vertical PBBR | sugar cane molasses/100 (fermentable sugars) | 95 | 144.0 | Sivaraman *et al.* (1982) |
| *Saccharomyces uvarum* entrapped within porous gelatin beads | vertical PBBR | sweet sorghum juice/115 (fermentable sugars) | 100 | 86.0 | Mohite and Sivaraman (1984) |
| *Saccharomyces cerevisiae* entrapped in pectin beads | horizontal PBBR | sugar cane molasses/160 (reducing sugars) | 88 | 40 | Navarro *et al.* (1983), Navarro *et al.* (1984) |

* Based on bioreactor void volume; † based on gel volume.
PBBR, packed-bed bioreactor; FPBR, filter-press bioreactor.

sterilization of the feed (Varma *et al.* 1984). Day and Sarkar (1982) similarly observed that they could efficiently ferment non-sterile sugar cane molasses for up to 30 weeks in a vertical PBBR containing agar-entrapped yeast cells.

The use of naturally occurring polymers as matrices for viable yeast-cell entrapment therefore appears to be desirable in view of the mildness of the technique, the possibility of preparing porous spherical particles and the capability of long-term operation without significant gel disruption.

**Entrapment of yeast cells in synthetic polymers**

Amongst the synthetic polymers tested, the use of polyacrylamide as a cell entrapment matrix has been most popular. Entrapment can be achieved by the polymerization of an aqueous solution of acrylamide monomers in which micro-organisms are suspended. The polymeric gel containing microbial cells can then be granulated for use as a column packing. This method of immobilization has been employed to entrap viable yeast cells (Shiotani & Yamani 1981, Holoberg & Margalith 1981, Siess & Divies 1981, Furusaki *et al.* 1983, Burrill *et al.* 1983) and subsequently used for the continuous production of ethanol (Table 8.9).

When comparing two different matrices for yeast-cell immobilization, Shiotani and Yamane (1981) observed that although yeast cells grew in both alginate and polyacrylamide gels, the alginate-entrapped yeast-cell system was much more efficient with respect to ethanol production. Furthermore, entrapment within polyacrylamide gels has been shown to render up to 80% of the yeast cells in a non-viable state (Siess & Divies 1981). Thus, due to the toxicity of polyacrylamide, modifications to the original entrapment procedure have been developed for yeast-cell immobilization (Pines & Freeman 1982). Substantial enhancement in the survival of yeast cells, even in the presence of high ethanol concentrations [22% (v/v)] resulted, indicating that the modified technique was mild and the gel had a protective effect.

Polyacrylamide gel particles are usually of irregular shapes and sizes. In columns, these particles tend to pack irregularly, causing uneven fluid flow and the development of relatively high flow-induced pressure drops. These factors therefore necessitate the use of sectionalized columns or several columns in series. Furthermore, in well-mixed bioreactors (CSTBR, FBBR, etc.) the polyacrylamide gel granules are easily abraded. To alleviate some of these problems, Kuu and Polack (1983) treated agar or carrageenan beads with polyacrylamide to form mechanically stable, uniform spherical particles containing viable yeast cells which were capable of ethanol production in a vertical PBBR for more than 2 months.

The entrapment of viable yeast cells by radiation-induced polymerization (Fujimura & Kaetsu 1981) and also within porous epoxy beads (Klein & Kreesdorf 1982) has been attempted and evaluated for the production of ethanol. However, the most significant development in the immobilization of viable yeast cells has been the use of photo-crosslinkable resins as entrapment matrices.

Polyethylene glycolic photo-crosslinkable resins can be readily induced to gel when illuminated with light of wavelength 300–400 nm, and the resulting

**Table 8.9** Continuous ethanol production by yeast cells entrapped in synthetic polymers.

| Immobilized yeast cell system | Bioreactor configuration | Substrate/ concentration ($g\,l^{-1}$) | Feed sugars utilized (%) | Volumetric ethanol productivity* ($g\,l^{-1}\,h^{-1}$) | Reference |
|---|---|---|---|---|---|
| *Saccharomyces cerevisiae* entrapped in polyacrylamide gel | horizontal PBBR | glucose/196 | >85 | 18.7† | Shiotani and Yamane (1981) |
| *Saccharomyces cerevisiae* entrapped in polyacrylamide gel | CSTBR | glucose/ 180–200 | NA | 15.5 | Siess and Divies (1981) |
| *Saccharomyces formosensis* entrapped in polyacrylamide gel | CSTBR | glucose/92 | >92 | 32.0 | Furusaki *et al.* (1983) |
| *Saccharomyces formosensis* grown in polyacrylamide gel | CSTBR | glucose/92 | >92 | 92.0 | Furusaki *et al.* (1983) |
| *Saccharomyces cerevisiae* entrapped in polyacrylamide–hydrazide cross-linked by glyoxal | vertical PBBR | glucose/200 | >90 | 58.0† | Pines and Freeman (1982) |
| *Saccharomyces cerevisiae* entrapped in polyacrylamide treated agar beads | vertical PBBR | glucose/100 | 95 | 50.5‡ | Kuu and Polack (1983) |
| *Saccharomyces* sp. entrapped in photo-crosslinkable resins | 3-stage PFRBR | sugar cane molasses/170 (reducing sugars) | >95 | 11.0† | Oda *et al.* (1983) |

* Based on bioreactor void volume; † based on total bioreactor working volume; ‡ based on gel volume.

PBBR, packed-bed bioreactor; CSTBR, continuous stirred-tank bioreactor; PFRBR, parallel-flow rectangular bioreactor. NA, not available.

matrix possesses excellent characteristics, especially with regard to its mechanical strength (Oda *et al.* 1983). With the help of an automatic immobilization unit, sheets (1 mm thick and 50 cm wide) of entrapped yeast cells can be produced at the rate of up to 5 m length $min^{-1}$ under aseptic conditions. These sheets, when arranged vertically (1 mm apart) in a parallel-flow rectangular bioreactor, could effectively produce ethanol from a non-sterile sugar cane molasses feed. By operating the bioreactor at a low-feed pH and periodic addition of appropriate chemicals, sludge removal (using phosphate solution) and bacterial contamination were effectively controlled (Oda *et al.* 1983). Based on the performance characteristics on a bench scale (10 litres ethanol $day^{-1}$), a demonstration plant capable of producing 3.8 million litres of ethanol per year was constructed by the Japan Gas Company. The latter system was operated continuously for at least 13 months and industrial-scale plants capable of producing up to 38 000 000 litres ethanol per year are currently under construction in the USA (Anonymous 1984).

**Advantages and limitations of entrapped yeast-cell systems**

Systems employing entrapped yeast cells possess excellent properties, especially with respect to the overall performance of the bioreactor and long-term operational stability even on scale-up. Since high biomass loadings can be attained by yeast-cell entrapment, high ethanol productivities result. Furthermore, unlike adsorbed-cell systems, which are limited by comparatively lower biomass loadings and lower feed flow rates, entrapped-cell systems can be operated at extremely high dilution rates without any cell washout occurring, and cell loss is not affected by the turbulence caused by $CO_2$ evolution. However, the biomass loading that can be used in entrapped-cell systems is limited by the significant decrease in gel strength at high cell loadings (Krouwel *et al.* 1982). Thus, the biomass loading within the gel has to be optimized such that a high ethanol productivity is achieved by a mechanically stable cell-entrapment system.

Entrapped-cell systems are also subjected to mass transfer limitations imposed by the additional diffusion barrier created by the support matrix. This is especially true when low substrate concentrations or high-molecular weight substrates are used. Additionally, high levels of the toxic product, ethanol, may acculumate within the gel and thereby reduce the efficiency of the system. However, with the use of small-diameter, open, porous beads, substantial improvement in the mass transfer of substrate and product, into and out of the matrix, may be respectively achieved.

In ethanol fermentation systems, $CO_2$ gas is also a major product. This gaseous product is characterized by its relatively low solubility in aqueous media and, therefore, the diffusion of the gaseous product out of the matrix can be rate-limiting (Krouwel & Kossen 1980). Thus, if the diffusion of $CO_2$ out of the matrix is slower than its production, $CO_2$ will accumulate and be saturated within the matrix. Eventually, gas bubbles may be formed within the matrix and cause disruption of the gel matrix. This phenomenon may therefore set severe restrictions on the applicability of entrapped-cell systems for ethanol production. However, recent studies have shown that mechanically weak and brittle matrices (e.g. agar) are disrupted easily, whereas stronger and elastic matrices such as calcium alginate, can withstand high pressure differences without gel disruption (Krouwel & Kossen 1980, Krouwel & Kossen 1981).

The accumulation of $CO_2$ within the entrapment matrix may not only cause gel disruption, but also adversely affect the metabolic properties of the entrapped yeast cells (Jones & Greenfield 1982). Thus, by employing small, porous beads, combined with the use of mechanically strong and flexible matrices, the undesirable effects due to $CO_2$ accumulation and mass-transfer limitations may be minimized in entrapped yeast-cell systems.

### 8.3.4 *Flocculent yeast-cell bioreactor systems*

The flocculent nature of certain yeast cells has provided a means of maintaining a high cell concentration within the bioreactor without the need for inert support materials. Yeast flocculation is therefore the simplest natural mode of cell immobilization. Although high cell densities of non-flocculent yeast cells have been maintained in CST bioreactors coupled with cell recycle by centrifugation, the need for expensive centrifugal units can be eliminated if flocculent strains are employed.

The use of a single-stage CSTBR coupled with an external gravity settler resulted in high cell densities of the flocculent yeast strain in the bioreactor and consequently high rates of ethanol production have been reported (Del Rosario *et al.* 1979, Raminez & Boudarel 1983). This mode of operation forms the basis of the Alcon and Tate & Lyle processes, both of which have been developed at the pilot-scale level (Guidoboni 1984). In the former case a yeast cell density of up to 45 g (dry weight) $l^{-1}$ can be achieved (Fricker 1982), whereas in the latter case the use of an ethanol-tolerant, flocculent yeast strain allows a high ethanol concentration to be maintained in the effluent stream (Guidoboni 1984).

Tower fermentation is an alternative means of ethanol production using flocculent cells, and has been tried for large-scale beverage production. Designed for the brewing industry by the APV Co. in the early 1960s, the major features were described by Royston (1966) and subsequently by Greenshields and Smith (1974). Essentially, the tower bioreactor (TBR) consists of a vertical cylindrical column with a conical base through which the medium is fed. Fermentation proceeds as the medium rises through the dense floc of yeast cells. A gas-separation device incorporated at the top of the column allows a volume free from turbulence, therefore enabling yeast settling to occur. The advantages of the TBR lie in the simplicity of construction and mode of operation. No agitation is required to keep the yeast cells in suspension, nor is any external yeast separation and recycling device required to achieve the high cell density necessary for high rates of ethanol production. More recently, therefore, the TBR has been employed to produce ethanol for non-beverage uses, not only with naturally flocculent yeast (Prince & Barford 1982a, b, Jones *et al.* 1984), but also with a non-flocculent yeast which was induced to flocculate under the physicochemical conditions within the bioreactor (Prince & Barford 1982c).

The TBR does, however, have some features which may restrict its applications. For instance, in order to retain a constant yeast cell concentration within the TBR, a dynamic balance has to be achieved between cell growth, attrition and elutrition (Atkinson 1980). Consequently the TBR can only be operated within a narrower range of dilution rates without cell washout occurring, compared with other immobilized yeast-cell bioreactors (Brohan & McLoughlin 1984). Furthermore, floc break-up and flotation of yeast cells due to $CO_2$ production occurs, and the TBR therefore behaves midway between a plug-flow and a fluidized-bed bioreactor (Comberbach & Bu'Lock 1984). These factors impose severe design problems in kinetic modelling, location of instrumentation and process control, especially on scale-up. In addition, the operating characteristics are sensitive to the distribution of different particle characteristics through the bioreactor volume, and may vary unpredictably. An ideal situation would be to combine the cell-retention characteristics of the low-shear TBR with the mixing characteristics of the CSTBR (Smart 1984). This was achieved by using a gas-lift TBR for the continuous production of ethanol using a highly flocculent strain of *Saccharomyces uvarum* (Comberbach and Bu'Lock 1984). The gas-lift TBR consists essentially of an external settler for yeast recycle and the flocculated yeast particles for a fluidized bed which is maintained in a well-mixed suspension by a gas-lift using the evolved $CO_2$ gas (Comberbach & Bu'Lock 1984, Smart 1984, Limtong *et al.* 1984). Satisfactory operation has been maintained for prolonged periods with yeast concentrations in excess of 100 g (dry weight) $l^{-1}$ at retention times of less than 0.4 h, enabling ethanol productivities of up to 45 g $l^{-1}$ $h^{-1}$ to be achieved (Comberbach & Bu'Lock 1984).

The only mechanical input in the gas-lift TBR is the gas compressor outside the bioreactor volume. Based on the performance characteristics of such a system, a computer controlled two-stage system (Bu'Lock *et al.* 1984) and also a 100-litre fully automated pilot-scale bioreactor (Smart 1984) have been recently designed.

A variation of the gas-lift TBR, known as the Inter-Loop process, has been developed by Hoechst-Uhde (Guidoboni 1984). The essential feature of this partly aerobic process is that the aerated loop bioreactor (TBR containing a single draft tube) is connected in series to two sedimentation tanks used for recycle of the flocculent yeast (Faust *et al.* 1983). Furthermore, unlike the gas-lift TBR, the yeast particles in the latter system are fluidized by air instead of $CO_2$ gas. A demonstration unit capable of producing up to 15 000 litres ethanol day$^{-1}$ has been successfully operated for several months (Faust *et al.* 1983).

A vertical PBBR has also been tested at the laboratory scale for the continuous production of ethanol using flocculent cells of the fission yeast *Schizosaccharomyces pombe* (Hsiao *et al.* 1983). Thus, a wide variety of bioreactor configurations may be employed with the 'self-immobilized' flocculent yeast cells for the efficient production of ethanol. However, these systems are restricted to the use of naturally flocculent yeast strains. Recent attempts to induce cell aggregation in non-flocculent yeast strains either by altering the physicochemical environment within the bioreactors (Genon *et al.* 1983, Weeks *et al.* 1982), or alternatively by genetic manipulation of the desired yeast strains (Russell & Stewart 1979) have been successful. One may therefore anticipate a wider variety of yeast strains to be successfully employed in the above bioreactor systems.

### *8.3.5 Economic considerations*

Sugars (glucose, sucrose and fructose) needed for the production of ethanol may be derived from three major classes of raw materials, namely, sugar-containing feedstocks, starchy feed materials and celluloses. Most commonly sucrose, derived from sugar cane, and starch, derived from corn, are the two major feedstocks currently employed for the industrial production of ethanol in Brazil and the USA, respectively. However, these raw materials are expensive, and therefore dictate the overall high price of ethanol. Recent economic evaluations have shown that the cost of the raw material may contribute up to 70% of the final cost of ethanol (Maiorella *et al.* 1984, Luong & Tseng 1984). Thus, with the use of highly efficient bioreactors such as immobilized yeast-cell systems, substantial reductions can only be achieved in so far as the overall plant costs are concerned, which in turn would result in reduced product cost. For instance, based on a $10^9$ litre year$^{-1}$ plant capacity and using molasses feed, the use of a TBR may reduce the cost of the product by up to 8% compared with the use of a conventional batch system (Maiorella *et al.* 1984). This is attributed largely to a seven-fold reduction in bioreactor volume and consequently a 50% reduction in the capital cost of the bioreactor and auxiliary equipment. Fukushima and Hanai (1982) have shown that for a 100 000 litre day$^{-1}$ facility, using an alginate entrapped yeast-cell bioreactor system, a 30% reduction in the capital cost can be achieved compared with the Melle–Boinot process.

Substantial reductions in the final cost of ethanol can however be only achieved with the use of cheaper sugar sources such as:

(a) lactose derived from whey (waste product of the cheese industry);
(b) inulin-type polyfructans derived from Jerusalem artichokes (carbohydrate-rich plant capable of growing on marginal land); and
(c) mono- and disaccharides (glucose, xylose and cellobiose) derived from lignocellulosic wastes.

The bioconversion of these carbohydrates to ethanol using immobilized yeast-cell systems are currently being examined in several laboratories using, mostly, laboratory-scale substrates concurrently with highly efficient immobilized-cell bioreactors. One may anticipate further reductions in the overall cost of ethanol (Luong & Tseng 1984) (Table 8.10).

### *8.3.6 Bioreactor design considerations*

A wide variety of immobilized yeast-cell bioreactors have been tested at the laboratory, pilot and commercial scale for the continuous production of ethanol (Table 8.11). The operational characteristics of mechanically contained and flocculated yeast-cell bioreactors have been adequately discussed in the previous sections. We shall therefore only consider some of the common bioreactor systems employed with attached and entrapped yeast-cell systems.

**Table 8.10** Ethanol production from non-conventional carbohydrate raw materials using immobilized yeast-cell system.

| Fermentation system and immobilization technique | Reference |
|---|---|
| (a) Direct fermentation of lactose to ethanol | |
| (i) *Kluyveromyces fragi* cells entrapped in either polyacrylamide, carrageenan or Ca-alginate | King and Zall (1983a, b) |
| (ii) *Kluyveromyces marxianus* cells entrapped in Ca-alginate beads | Marwaha and Kennedy (1984a, b) |
| (iii) *Saccharomyces fragilis* cells entrapped in Ca-alginate beads | Linko and Linko (1984), Linko *et al*. (1981) |
| (iv) *Saccharomyces carlsbergensis* attached to pumice stone | Novais *et al*. (1983) |
| (v) *Saccharomyces fragi* adsorbed on cellulose acetate particles | Chen and Zall (1982) |
| (vi) mechanical containment of *Kluyveromyces fragilis* in a hollow-fibre bioreactor | Mehaia and Cheryan (1984c) |
| (b) Simultaneous lactose hydrolysis and ethanol fermentation | |
| (i) $\beta$-galactosidase–alginate complex co-immobilized with *Saccharomyces cerevisiae* in Ca-alginate beads | Hagerdal (1980), Hahn-Hagerdal (1982) |
| (ii) $\beta$-galactosidase covalently attached to yeast cell wall | Hartmeier (1984) |
| (c) Ethanol production from Jerusalem artichoke extracts by inulinase producing yeast strains (*Kluyveromyces fragilis*, *Kluyveromyces marxianus*, *Debaryomyces polymorphus*) entrapped in Ca-alginate beads | Margaritis and Bajpai (1981, 1982, 1983), Margaritis *et al*. (1983a) |
| (d) Ethanol production from xylose by a xylose utilizing yeast strain, *Pachysolen tannophilus*, entrapped in Ca-alginate beads | Maleszka *et al*. (1981), Slininger *et al*. (1982) |
| (e) Enzymatic conversion of xylose to xylulose by immobilized glucose isomerase coupled in series to ethanol production by Ca-alginate entrapped *Saccharomyces cerevisiae* | Chiang *et al*. (1982), Suihko and Poutanen (1984) |
| (f) Simultaneous cellobiose hydrolysis and ethanol fermentation | |
| (i) $\beta$-glucosidase–alginate complex co-immobilized with *Saccharomyces cerevisiae* in Ca-alginate beads | Hagerdal (1980), Hahn-Hagerdal (1984), Hagerdal and Mosbach (1979) |
| (ii) sepharose immobilized $\beta$-glucosidase co-entrapped with *Saccharomyces cerevisiae* in Ca-alginate beads | Kierstan *et al*. (1982) |
| (iii) $\beta$-glucosidase covalently attached to yeast cell wall | Hartmeier (1984) |

*Continued*

**Table 8.10** Continued.

| Fermentation system and immobilization technique | Reference |
|---|---|
| (g) Simultaneous saccharification and fermentation of cellulose to ethanol | |
| (i) cellulase enzymes covalently attached to yeast cell wall | Hartmeier (1981a) |
| (ii) cellulase enzymes (freely suspended) and *Saccharomyces uvarum* entrapped in porous gelatin beads | Deshpande *et al.* (1983) |

**Table 8.11** Bioreactor designs used with immobilized yeast cells for the production of ethanol.

(a) Mechanically contained yeast cells
- (i) simple dialysis bioreactor
- (ii) Rotorfermenter*
- (iii) filter bioreactor
- (iv) hollow-fibre bioreactor

(b) Attached yeast cells
- (i) vertical packed-bed bioreactor
- (ii) multistage stirred-tank bioreactor
- (iii) fluidized-bed bioreactor with or without draft tubes*
- (iv) circulating-bed bioreactor

(c) Entrapped yeast cells
- (i) vertical packed-bed bioreactor
- (ii) horizontal packed-bed bioreactor
- (iii) single- and multistage fluidized-bed bioreactors with or without draft tubes*
- (iv) stirred-tank bioreactor
- (v) cross-flow bioreactor
- (vi) filter-press bioreactor
- (vii) parallel-flow rectangular bioreactor*

(d) Flocculent yeast cells
- (i) stirred-tank bioreactor with external settler*
- (ii) tower bioreactor*
- (iii) gas-lift tower bioreactor*
- (iv) aerated loop bioreactor*
- (v) vertical packed-bed bioreactor

* Pilot- and/or commercial-scale bioreactors.

The most common bioreactor used in preliminary laboratory-scale studies has been the vertical PBBR. The choice of this configuration is obvious, due to the simplicity of construction and desirable kinetic properties. Thus, in a PBBR, plug-flow characteristics restrict the inhibitory effects of the product ethanol to only the small final portion of the bioreactor (Yamane & Shimizu 1982), and therefore high rates of ethanol production may be readily achieved. However, a number of factors tend to limit the use of the vertical PBBR, especially when scale-up is considered. These limitations include:

(a) $CO_2$ gas hold-up within the bioreactor resulting in reduced bioreactor efficiency (Cho *et al*. 1982) and unsteady temperature distribution across the bioreactor length (Ghose & Bandyopadhyay 1982);
(b) inability to use feed materials containing particulate matter due to plugging and fouling problems (Emery & Cardoso 1978);
(c) blockage due to released yeast cells (Sitton & Gaddy 1980); and
(d) the combined effects of (a), (b) and (c) result in extremely high pressure drops and, consequently, deformation of entrapment matrices (Shiotane & Yamane 1981).

The problems associated with $CO_2$ hold-up have, however, been partially resolved with the use of a horizontal PBBR (Shiotane & Yamane 1981). Alternatively, single- or multistage stirred-tank bioreactors may be used to alleviate the above problems. However, with STBR, limitations due to ethanol inhibition in a single-stage system, high power requirements and shear effects near the impeller region causing gel disruption and cell desorption have largely restricted its wider use with immobilized yeast-cell systems (Lee *et al*. 1983).

It appears that the most desirable bioreactor configuration for ethanol production by immobilized cells is the FBBR. In this system immobilized biocatalyst particles are suspended and agitated by the upward flow of fluid through bed. The flow is maintained at a rate sufficient to expand the bed of immobilized yeast cells and cause their suspension and agitation, but not so great as to carry particles from the confines of the bioreactor. Based on these operational characteristics, the use of an FBBR for ethanol production offers unique advantages (Margaritis & Wallace 1984, Allen *et al*. 1979) which include the following:

(a) foreign particles in the fluid readily pass through the bed and therefore do not allow plugging to occur;
(b) gas can be introduced (e.g. oxygen for yeast-cell regeneration) and $CO_2$ can be readily removed;
(c) low pressure drops and power requirements;
(d) reduced shear effect compared with the STBR; and
(e) good mass- and heat-transfer characteristics which are inherent in fluidized systems.

Because of these desirable properties, fluidized-bed bioreactors have been successfully tested for the industrial-scale production of ethanol using immobilized yeast-cell systems.

## 8.4 The production of alcoholic beverages using different bioreactor systems

In this section we describe the different bioreactor types used for the production of alcoholic beverages.

### *8.4.1 Batch fermentation for beer production*

Conventionally, beer was produced by fermentation using a suitable yeast strain in large, shallow, rectangular, open tanks in which the yeast rose to the top at the completion of attenuation. The yeast was then skimmed off, pressed and stored in the cold for subsequent re-use. The yeast-free beer was then chilled by a complex system of pipes distributed within the vessel. Up to 48 h was required for chilling, giving a complete cycle time, from start of fermentation to racking to cellars, of between 5 and 7 days. The subsequent cleaning of the open vessel was extremely labour-intensive and required close supervision and laboratory checks to ensure reasonable sterility. Furthermore, these large tanks occupied a large floor area and, since they were open vessels, they had no protection against airborne contaminants.

#### Asahi tanks

Enclosed fermentation systems have obvious microbiological advantages and the use of Asahi tanks and various cylindrical vessels are such examples. Asahi tanks have a vertical shell and a flat bottom which slopes downwards to the outlet (Lindsay & Larson 1975). Attemperation for fermenting beer is provided by cooling jackets mounted on the walls of the tanks. Experience with these vessels showed that the beer aged in these tanks was similar to conventionally lagered beers (Takayanagi & Harada 1968). In addition, the installation and operating costs of Asahi tanks was approximately 50% lower than the conventional system (Takayanagi & Harada 1968).

#### Small cylindrico-conical vessels

Today, however, the use of cylindrico-conical vessels (CCV) is more popular (Hoggan 1977). This configuration was first described by Nathan (1930), and gained widespread popularity following the pioneering work of Shardlow and Thompson (1971), Shardlow (1972) and Ulenberg *et al.* (1972). The vessel essentially consists of a metal cylinder with a flat or convex top and a coned base having a cone angle of between 75° and 120°. This vessel design leads to some distinct advantages not commonly encountered in the conventional vessel. Thus, the $CO_2$ rising from the depths of the cone sets up powerful convection currents, facilitating both heat and mass transfer, and consequently a more vigorous and rapid fermentation results than that achieved in the shallow rectangular vessels. Furthermore, at the end of the fermentation the settled yeast can be readily removed from the bottom of the steeped cone, and the beer can be subsequently aged in the same vessel, allowing greater flexibility since no intermediate transfer is necessary. These vessels are also versatile in the sense that the production of both lager and ale is possible, resulting in lower capital costs.

However, the fast rate of fermentation in CCV leads to practical difficulties of controlling foam formation, which occupies a large volume of the bioreactor. In addition, only yeasts which have a property of sedimentation can be advantageously used for fermentation. Furthermore, due to the height of the tank (21 m), the $CO_2$ content of the resulting beer is greater than that obtained conventionally, due to the hydrostatic pressure effects in the former case. The undesirable effects of $CO_2$ on yeast physiology have also been recently reviewed (Jones & Greenfield 1982). However, to date, most investigators who have used the CCV have reported that the beer produced in such systems is comparable with that obtained conventionally (Shardlow 1972, Ulenberg *et al.* 1972).

**The Uni-Tank process**

An extension of the CCV is the Uni-process employing the Uni-Tank, which consists of a thin-walled vertical cylinder (height: tank diameter ratio close to 1.0), with a dome or conical top and a conical (angle of cone is 155°) bottom (Lindsay & Larson 1975). Cooling jackets are built as integral parts of the tank wall, and provisions are made to inject $CO_2$ into the beer. The incline of the cone is sufficient to permit easy removal of yeast after fermentation without transferring the beer. The cooling jackets are sized to remove heat generated during fermentation, and also for cooling the beer after fermentation. Thus, the Uni-Tank also offers versatility, since it can be used as the bioreactor, ageing tank and also as a finishing tank. Extensive analysis has shown that the beer produced by the Uni-process is equal or superior in quality to beer produced in conventional tanks (Knudsen & Larson 1974). Furthermore, the capital and operating costs of the Uni-process are substantially lower than conventional tanks (Knudsen & Vacano 1973). More recently, stirred fermentations have also gained interest, due to the faster rates of reaction (Pajunen & Makinen 1975).

**Sphero-conical tanks**

In 1972, Knudsen and Vacano reported that the use of spherical tanks would minimize the cost of construction materials and also labour, due to the advantages offered by the lower surface area-to-volume ratio. However, one of the main disadvantages of this otherwise ideal system is the difficulty in recovery of the yeast. This problem has been solved by adding a conical bottom to the sphere as in the sphero-conical vessels (SCV). The advantages incurred with the use of SCV are several (Posada 1978, Martin *et al.* 1975) and include the following:

(a) the SCV can be operated at high pressures, enabling adequate $CO_2$ saturation in the beer during the lagering process, and thereby eliminating the need for further $CO_2$ treatment;
(b) the presence of the steep cone facilitates easy and efficient removal of the yeast cells;
(c) due to the spherical shape, the surfaces to be cleaned are minimized; and
(d) the spherical shape of the top minimizes foam formation, and facilitates the cooling process.

Thus, the versatility of the SCV is a consequence of their geometry which has allowed both lagering and fermentation to occur in the same tank, and the cost of construction is lower than that of the CCV (Martin *et al.* 1975).

### *8.4.2 Continuous production of alcoholic beverages*

The brewing industry has maintained interest in continuous fermentation for more than 70 years (Portno 1979, Purssell & Smith 1968, Hough & Button 1972) because of several distinct advantages, some of which have already been listed (Section 8.2.2). In addition, lower product cost can be envisaged in continuous systems due to reduced beer losses, improved hop utilization, higher alcohol yield and reduced detergent usage. Nevertheless, the conversion from batch to continuous systems has not been readily applied in the brewing industry, for the following reasons:

(a) It is difficult to maintain production-scale continuous fermentation free from contamination by both bacteria and wild yeast.
(b) Continuous fermentation requires round-the-clock engineering and scientific support (Portno 1979).
(c) Continuous fermentation as practised cannot reproduce the full cycle of changing metabolic patterns which characterize batch fermentation. Consequently, beer produced in a continuous system will not have the same flavour as that produced by more-conventional batch means (Stewart 1977).
(d) The risk of culture yeast altering its behaviour because of spontaneous mutation is demonstrably greater in continuous than in batch fermentation systems. For example, the level of respiratory-deficient mutants is nearly always greater in continuous than in batch systems. Killer yeast strains have also been detected in production-scale continuous systems (Maule & Thomas 1973).
(e) In a brewing operation where a number of different beers are produced, continuous fermentation does not allow as much flexibility as the batch process, since long runs of a single product are produced at a constant rate in the former case.

However, despite these disadvantages there are two continuous systems being commercially used today, namely the multistage and the APV tower fermentation processes (Kirsop 1982, Baxter 1976).

#### The multistage system

Depending on the product, the multistage system may consist of three or four tanks connected in series. In the four-stage system the first three vessels are equipped with stirrers. In the first vessel, which is known as the yeast propagation vessel, sterile high-gravity wort is added to the yeast and the contents aerated. Approximately 90% of the fermentation is achieved in the primary fermentation vessel (second vessel), and fermentation is completed in the third vessel. The fourth vessel is used as a chilling vessel, and has a conical bottom to facilitate yeast separation. All types of beer, including strong ales, have been produced using the multistage system, and the taste of the product has been found to be similar to that obtained using conventional batch fermentation (Hough *et al.* 1976). The multistage continuous system is currently used commercially in New Zealand for beer production.

**Tower fermentation**

In the APV continuous tower (described in Section 8.3.4) the wort advances non-directionally up the tower containing a high population of flocculent yeast cells (Bruens 1966, Klopper *et al.* 1966, Royston 1966, Ault *et al.* 1969, Smith & Greenshields 1973, Seddon 1975). However, the actual pattern of yeast distribution along the length of the tower does vary with the fermentation conditions. Thus, at low flow rates and particularly with a feed of low sugar content, the yeast cells settle heavily at the base of the bioreactor, forming a quasi-solid plug through which the feed permeates, fermenting as it does so. At higher flow rates, or at higher feed densities, a zone of maximum yeast concentration is formed midway in the tower, with a lower yeast concentration above and below. At maximum throughput the zone of maximum yeast concentration moves progressively upwards. The tower bioreactor is then on the borderline of instability and further increases in throughput or feed strength lead to loss of yeast from the bioreactor, resulting in a decrease in the fermentation efficiency. This represents the most serious disadvantage of the tower fermentation system.

Once in operation the TBR may take up to 3 months to achieve the product consistency desired, but steady-state operation can continue for a year or even longer (Ault *et al.* 1969). Under steady-state conditions the rates of fermentation can be very rapid, and complete attenuation may be achieved at residence times of 2–6 h (Smith & Greenshields 1973, Seddon 1975), depending on the wort concentration and the degree of fermentation. In conventional systems up to 40% of the bittering substances are lost during fermentation, whereas in the continuous TBR, only 10–15% loss occurs. Thus, due to faster fermentation rates and reduced hop losses, substantial cost savings may be envisaged. Furthermore, problems associated with yeast mutation and contamination are less frequently observed (Ault *et al.* 1969). More recently the use of tower fermentation for the production of both wine and cider has been demonstrated (Wiesenberger 1978). At present the tower fermentation system is used commercially only for the production of vinegar.

**Immobilized yeast-cell bioreactors**

In addition to the use of flocculent yeast cells in tower fermentation systems, several other immobilization techniques have been evaluated for the production of alcoholic beverages (Table 8.12). In most cases the organoleptic qualities of the fermented beverage have been undesirable and have therefore precluded production at the commercial scale. However, by incorporating a secondary treatment stage, beer of improved quality may be obtained (Baker & Kirsop 1973b). This has been recently achieved at the pilot-plant level in China using a 6000-litre bioreactor containing 20 kg alginate-entrapped yeast cells (Chiou 1979). Thus, with further process optimization and development of appropriate treatment stages, use of immobilized yeast cells for the

**Table 8.12** Continuous production of alcoholic beverages using immobilized yeast-cell systems.

| Fermentation system and immobilization technique | Reference |
|---|---|
| (a) Production of beer from wort | |
| (i) yeast cells retained in a porous cylinder | Delbruck (1892) |
| (ii) bed of yeast cells held between two porous plates | Berdelle-Hilge (1973) |
| (iii) yeast–kieselguhr mixture deposited on vertical filters | Narziss and Hellich (1971, 1972) |
| (iv) yeast-kieselguhr mixture held between two porous plates | Baker and Kirsop (1973a) |
| (v) yeast–diatomaceous earth mixture held between two filters | Grinbergs *et al.* (1977) |
| (vi) yeast cell adsorbed on to sterile cotton wool | Barbet (1899) |
| (vii) yeast cells adsorbed on to PVC, brick particles, diatoms, etc. | Corrieu *et al.* (1976), Moll (1977) |
| (viii) yeast cells entrapped in Ca-alginate beads | Linko and Linko (1981), Chiou (1979), White and Portno (1977), Masschelein and Francotte (1983), Pardonora *et al.* (1982) |
| (b) Production of low-carbohydrate beer | |
| (i) amyloglucosidase co-immobilized on surface of live yeast cells | Hough and Lyons (1972), Dellweg and Hartmeir (1983) |
| (ii) amyloglucosidase co-entrapped with yeast cells in alginate | Godfredsen *et al.* (1981) |
| (c) Production of wine | |
| (i) yeast cells entrapped in polyacrylamide gel | Divies (1977) |
| (ii) flocculent strains of wine yeast in slant tubes | Wick and Popper (1977) |
| (iii) pepsin co-immobilized on surface of wine yeast cells | Hartmeier (1981b) |

continuous production of alcoholic beverages may in future be an attractive alternative, especially in view of the potential advantages of such systems. However, more research is needed at the basic level before any realistic assessment can be made on the potential importance of alcoholic beverage production using immobilized cell systems.

## 8.5 Conclusions

The major developments that have occurred in the brewing industry with respect to the use of fermentation vessels, have been the adoption of cylindrico-conical vessels and Uni-Tank processes instead of the conventional rectangular tanks. These developments have, however, been restricted to the use of batch fermentation systems. Considerable effort was directed to the continuous production of beer in the 1950s and 1960s, and today there is only one continuous commercial process for the production of beer in New Zealand, where the multistage system is used. More recently, progress in immobilized-cell and enzyme technology has rejuvenated interest in the continuous production of alcoholic beverages. In the potable alcohol industry, consumer acceptability of the product is of prime concern, and since the type of fermentation system has a profound effect on the taste and quality of the beverage produced, the use of immobilized yeast-cell systems may only find favour in the production of some types of beverages. Further fundamental studies and pilot-scale testing will have to be carried out to evaluate fully the advantages and limitations of immobilized yeast-cell systems in the continuous production of alcoholic beverages.

In the production of industrial-grade ethanol the subtle complexities normally associated with beverage production are minor. Therefore, the use of highly efficient continuous fermentation systems at an industrial scale appear to be more promising. For instance, recent developments in the use of immobilized yeast cells in highly efficient bioreactors, such as fluidized bioreactor systems, have already made a significant impact on the biotechnology of ethanol production. In the future one might therefore anticipate the use of such bioreactor systems on a wider scale, especially in newer ethanol plants.

Developments in the selection of appropriate yeast strains capable of producing ethanol from a wider variety of cheap carbohydrate raw materials will contribute substantially to the overall economics of ethanol production. With advances in the genetic manipulation of yeasts, selection of ethanol and osmotolerant yeast strains capable of growing at thermophilic temperatures will also contribute to the economics of the process. However, these developments will make further demands on the design of novel bioreactors, and associated equipment and control systems.

As the range of products capable of being produced by yeasts increases due to advances in recombinant-DNA technology, the demands on suitable bioreactor systems for future large-scale processes employing yeast cells will be substantial. In this respect novel, highly efficient bioreactor systems will play a dominant role in future industrial-scale biotechnological processes. A clear understanding of the technology of anaerobic yeast growth and their advantages and limitations, especially on scale-up, therefore becomes imperative if these systems are to be employed commercially.

## Acknowledgements

One of the authors (A. Margaritis) wishes to acknowledge the financial support received from the Natural Sciences and Engineering Research Council (NSERC) of Canada, and Imperial Oil Ltd, Toronto, Canada.

## References

Abbot, B. J. 1977. Immobilized cells. In *Annual reports on fermentation processes*, D. Perlman (ed.), Vol. 1, 205–33.

Allen, B. R., R. W. Coughlin and M. Charles 1979. Fluidized-bed enzyme reactors. *Annals of the New York Academy of Sciences* **326**, 105–17.

Amin, G. and H. Verachtert 1982. Comparative study of ethanol production by immobilized-cell systems using *Zymomonas mobilis* and *Saccharomyces bayanus*. *European Journal of Applied Microbiology and Biotechnology* **14**, 59–63.

Amin, G., E. Van den Eynde and H. Verachtert 1983. Determination of by-products formed during ethanolic fermentation using batch and immobilized cell systems of *Zymomonas mobilis* and *Saccharomyces bayanus*. *European Journal of Applied Microbiology and Biotechnology* **18**, 1–5.

Amin, G., P. Standaert and H. Verachtert 1984. Effect of metabolic inhibitors on the alcoholic fermentation by several yeasts in batch or in immobilized cell systems. *Applied Microbiology and Biotechnology* **19**, 91–9.

Anonymous 1984. IRAS will commercialize JGC's continuous fermentation process. *Biomass Digest* **6**(1), 1–2.

Atkinson, B., G. M. Black, P. J. S. Lewis and A. Pinches 1979. Biological particles of given size, shape and density for use in biological reactors. *Biotechnology and Bioengineering* **21**, 193–200.

Atkinson, B., G. M. Black and A. Pinches 1980. Process intensification using cell support systems. *Process Biochemistry* (May), 24–32.

Ault, R. G., A. N. Hampton, R. Newton and R. H. Roberts 1969. Biological and biochemical aspects of tower fermentation. *Journal of the Institute of Brewing* **75**, 260–77.

Baker, D. A. and B. H. Kirsop 1973a. Rapid beer production and conditioning using a plug fermenter. *Journal of the Institute of Brewing* **79**, 487–94.

Baker, D. A. and B. H. Kirsop 1973b. A rapid procedure for reducing the diacetyl content of beer. *Journal of the Institute of Brewing* **79**, 43–4.

Bandyopadhyay, K. K. and T. K. Ghose 1982. Studies on immobilized *Saccharomyces cerevisiae*. III. Physiology of growth and metabolism on various supports. *Biotechnology and Bioengineering* **24**, 805–15.

Barbet, E. 1899. French Patent.

Baxter, W. 1976. Developments in fermentation and conditioning plant. *Brewers' Guardian* (November), 29–33.

Berdelle-Hilge, Ph. 1973. Wort under hydraulic pressure passed through a layer of yeast. US Patent, 3 737 323.

Birnbaum, S., R. Pendleton, P. O. Larsson and K. Mosbach 1981. Covalent stabilization of alginate gel for the entrapment of living whole cells. *Biotechnology Letters* **3**, 393–400.

Black, G. M., C. Webb, T. M. Mathews and B. Atkinson 1984. Practical reactor systems for yeast cell immobilization using biomass support particles. *Biotechnology and Bioengineering* **26**, 134–41.

Boinot, F. and J. Boige 1948. Yeast culture. US Patent 2 446 737.

Brohan, B. and A. J. McLoughlin 1984. The influence of yeast metabolism on dispersion in a fluidized yeast bed. *Applied Microbiology and Biotechnology* **20**, 146–9.

Bruens, L. J. 1966. Continuous fermentation and the tower fermenter. *Technical Quarterly of the Master Brewers Association of the Americas* **3**, 248–52.

Bu'Lock, J. D., M. C. Comberbach, C. Ghommidh and P. D. Williams 1984. A computer-controlled fermentation operating system for high-level alcohol production. *Chemical Industry* (London) **12**, 432–4.

Burrill, H. N., L. E. Bell, P. F. Greenfield and D. D. Do 1983. Analysis of distributed growth of *Saccharomyces cerevisiae* cells immobilized in polyacrylamide gel. *Applied Environmental Microbiology* **46**, 716–21.

Cheetham, P. S. J. 1979. Physical studies on the mechanical stability of columns of calcium alginate gel pellets containing entrapped microbial cells. *Enzyme and Microbial Technology* **1**, 183–8.

Chen, H. C. and R. R. Zall 1982. Continuous fermentation of whey into alcohol using an attached film expanded bed reactor. *Process Biochemistry* (January/February) 20–5.

Chiang, L. C., H. Y. Hsiao, M. C. Flickinger, L. F. Chan and G. T. Tsao 1982. Ethanol production from pentoses by immobilized microorganisms. *Enzyme Microbe Technology* **4**, 93–5.

Chiou, C. J. 1979. Studies on the principles and applications of immobilized yeast cells. *Chung-kuo Nung Yeh Hua Hsueh Hui Chih* **17** (3 & 4), 138–48.

Cho, G. H., C. Y. Choi, Y. D. Choi and M. H. Han 1981. Continuous ethanol production by immobilized yeast in a fluidized reactor. *Biotechnology Letters* **3**, 667–71.

Cho, G. H., C. Y. Choi, Y. D. Choi and M. H. Han 1982. Ethanol production by immobilised yeast and its $CO_2$ gas effects in a packed bed reactor. *Journal of Chemical Technology and Biotechnology* **32**, 959–67.

Chotani, G. K. and A. Constantinides 1984. Immobilized cell cross-flow reactor. *Biotechnology and Bioengineering* **26**, 217–20.

Comberbach, D. M. and J. D. Bu'Lock 1984. Continuous ethanol production in a gas-lift tower fermenter. *Biotechnology Letters* **6**, 129–34.

Corrieu, G., H. Blachere, A. Ramirez, J. M. Navarro, G. Durand, B. Duteutre and M. Moll 1976. An immobilized yeast fermentation pilot plant used for production of beer. *Abstracts. 5th International Fermentation Symposium, Berlin, 1976*, 294.

Cysewski, G. R. and C. R. Wilke 1976. Utilization of cellulosic materials through enzymatic hydrolysis. I. Fermentation of hydrolysate to ethanol and single-cell protein. *Biotechnology and Bioengineering* **18**, 1297–313.

Cysewski, G. R. and C. R. Wilke 1977. Rapid ethanol fermentations using vacuum and cell recycle. *Biotechnology and Bioengineering* **19**, 1125–43.

Daugulis, A. J., N. M. Brown, W. R. Cluett and D. B. Dunlop 1981. Production of ethanol by adsorbed yeast cells. *Biotechnology Letters* **3**, 651–6.

Day, D. F. and D. Sarkar 1982. An immobilized yeast cell column for the fermentation of molasses. In *Enzyme engineering*, I. Chibata, S. Fukushima and L. B. Wingard (eds), Vol. 6, 343–5. New York: Plenum Press.

Del Rosario, E. J., K. J. Lee and P. L. Rogers 1979. Kinetics of alcohol fermentation at high yeast levels. *Biotechnology and Bioengineering* **21**, 1477–82.

Delbruck, M. 1892. Schnellgarung und das Arbeiten mit gefesselfor Hefe. *Hochenscht. Brau* **9**, 645.

Dellweg, H. and W. Hartmeir 1983. The use of immobilized enzymes in the brewery. In *European Brewery Convention. Proceedings of the 19th Congress, London, 1983*, 369–76.

Deshpande, V., H. Sivaraman and M. Rao 1983. Simultaneous saccharification and fermentation of cellulose to ethanol using *Penicillium funiculosum* cellulase and free or immobilized *Saccharomyces uvarum* cells. *Biotechnology and Bioengineering* **25**, 1679–84.

Dias, S. M. M., J. M. Novais and J. M. S. Cabral 1982. Immobilization of yeasts on titanium activated inorganic supports. *Biotechnology Letters* **4**, 203–8.

Divies, C. 1977. Use of polymer matrix-embedded microorganisms in enzyme reactions. German Patent, 2 633 076.

Dostalek, M. and M. Haggstorm 1982. A filter fermenter-apparatus and control equipment. *Biotechnology and Bioengineering* **24**, 2077–86.

Emery, A. N. and J. P. Cardoso 1978. Parameter evaluation and performance studies in a fluidized-bed immobilized enzyme reactor. *Biotechnology and Bioengineering* **20**, 1903–29.

Fang, B. S., H. Y. Fang, C. S. Wu and C. T. Pan 1983. High productivity ethanol production by immobilized yeast cells. *Biotechnology and Bioengineering Symposium* No. 13, 457–64.

Faust, U., P. Prave and M. Schlingmann 1983. An integral approach to power alcohol. *Process Biochemistry* (May/June), 31–7.

Fricker, R. 1982. Industrial alcohol by continuous fermentation. *Proceedings of the Convention of the Institute of Brewing (Australia and New Zealand Section), 17th*, 162–6.

Fujimura, T. and I. Kaetsu 1981. Immobilization of yeast cells by radiation-induced polymerization. *Zeitschrift für Naturforschung* **37C**, 102–6.

Fukushima, S. and S. Hanai 1982. Pilot operation for continuous alcohol fermentation of molasses in an immobilized bioreactor. In *Enzyme engineering*, I. Chibata, S. Fukushima and L. B. Wingard (eds), Vol. 6, 347–8. New York: Plenum Press.

Fukushima, S. and H. Hatakeyama 1983. Novel immobilized bioreactor for rapid continuous ethanol fermentation of cane juice or fruit juices. *Annals of the New York Academy of Sciences* **413**, 483–5.

Furusaki, S., M. Saki and K. Fukimura 1983. Reaction characteristics of an immobilized yeast producing ethanol. *Biotechnology and Bioengineering* **25**, 2921–8.

Garlick, L. 1983. Biostil. Fermentation of molasses cane juice using continuous fermentation. *The Sugar Journal* (September), 13–16.

Gencer, M. A. and R. Mutharasan 1981. Ethanol fermentation in a yeast immobilized column fermenter. In *Advances in biotechnology*, M. Moo-Young (ed.), Vol. 1, 627–33. Toronto: Pergamon Press.

Gencer, M. A. and R. Mutharasan 1983. Ethanol fermentation in a yeast immobilized tubular fermenter. *Biotechnology and Bioengineering* **25**, 2243–62.

Genon, G., A. Gianetto and Y. Specchia 1983. Production of ethanol with *Saccharomyces cerevisiae* in a continuous reactor. Note III. An experimental, kinetic study with suspended, recycled biomass. *Chemical Engineering Communications* **23**, 245–58.

Ghose, T. K. and K. K. Bandyopadhyay 1980. Rapid ethanol fermentation in immobilized yeast cell reactor. *Biotechnology and Bioengineering* **22**, 1489–96.

Ghose, T. K. and K. K. Bandyopadhyay 1982. Studies on immobilized *Saccharomyces cerevisiae*. II. Effect of temperature distribution on continuous rapid ethanol formation in molasses fermentation. *Biotechnology and Bioengineering* **24**, 797–804.

Ghose, T. K. and R. D. Tyagi 1979a. Rapid ethanol fermentation of cellulose hydrolysate. II. Product and substrate inhibition and optimization of fermenter design. *Biotechnology and Bioengineering* **21**, 1401–20.

Ghose, T. K. and R. D. Tyagi 1979b. Rapid ethanol fermentation of cellulose hydrolysate. I. Batch versus continuous systems. *Biotechnology and Bioengineering* **21**, 1387–400.

Gianetto, A., V. Specchia and G. Genon 1983. Production of ethanol with *Saccharomyces cerevisiae* in a continuous reactor. Note I. Kinetics of immobilized yeast. *Chemical Engineering Communications* **23**, 215–31.

Godfredsen, S. E., M. Ottesen and B. Svensson 1981. Application of immobilized yeast and yeast co-immobilized with amyloglucosidase in the brewing process. In *European Brewery Convention Proceedings of the 18th Congress, Copenhagen, 1981*, 505–11.

Greenshields, R. N. and E. L. Smith 1974. The tubular reactor in fermentation. *Process Biochemistry* **9**(4), 11–17, 28.

Griffith, W. L. and A. L. Compere 1976. A new method for coating fermentation tower packing so as to facilitate microorganism attachment. *Developments in Industrial Microbiology* **17**, 241–6.
Grinbergs, M., R. P. Hildebrand and B. J. Clarke 1977. Continuous fermentations of glucose solutions. *Journal of the Institute of Brewing* **83**, 25–9.
Guidoboni, G. C. 1984. Continuous fermentation systems for alcohol production. *Enzyme and Microbial Technology* **6**, 194–200.

Hagerdal, B. 1980. Enzymes co-immobilized with microorganisms for the microbial conversion of non-metabolizable substrates. *Acta Chemica Scandinavica* **B34**, 611–13.
Hagerdal, H. and K. Mosbach 1979. The production of ethanol from cellobiose using baker's yeast coimmobilized with β-glucosidase. In *Food process engineering*, P. Linko and J. Larinkari (eds), Vol. 2, 129–32. London: Applied Science.
Hahn-Hagerdal, B. 1982. Enzymes co-immobilized with microorganisms. The bioconversion of whey permeate to ethanol with β-galactosidase and *Saccharomyces cerevisiae*. *Applied Biochemistry and Biotechnology* **7**, 43–5.
Hahn-Hagerdal, B. 1984. An enzyme coimmobilized with a microorganism: the conversion of cellobiose to ethanol using β-glucosidase and *Saccharomyces cerevisiae* in calcium alginate gels. *Biotechnology and Bioengineering* **26**, 771–4.
Hahn-Hagerdal, B. and B. Mattiasson 1982a. Azide sterilization of fermentation media: ethanol production from glucose using immobilized *Saccharomyces cerevisiae*. *European Journal of Applied Microbiology and Biotechnology* **14**, 140–3.
Hahn-Hagerdal, B. and B. Mattiasson 1982b. Shift in metabolism towards ethanol production in *Saccharomyces cerevisiae* by addition of metabolic inhibitors. *Biotechnology and Bioengineering*, Symposium No. 12, 193–5.
Haraldson, A. and C. G. Rosen 1982a. Studies on continuous ethanol fermentation of sugar cane molasses. I. A system for continuous fermentation. *European Journal of Applied Microbiology and Biotechnology* **14**, 216–19.
Haraldson, A. and C. G. Rosen 1982b. Studies on continuous ethanol fermentation of sugar cane molasses. II. Continuous alcohol fermentation and product removal in a laboratory scale plant. *European Journal of Applied Microbiology and Biotechnology* **14**, 220–4.
Hartmeier, W. 1981a. Basic trials on the conversion of cellulosic material to ethanol using yeast co-immobilized with cellulolytic enzymes. In *Advances in biotechnology*, M. Moo-Young (ed.), Vol. 3, 377–82. Toronto: Pergamon Press.
Hartmeier, W. 1981b. *Saccharomyces cerevisiae* coimmobilized with pepsin for wine making. In *Current developments in yeast research*, G. G. Stewart and I. Russell (eds), 105–10. Toronto: Pergamon Press.
Hartmeier, W. 1984. Co-immobilizates: new biocatalysts for fermentation technology. *Process Biochemistry* (February), 40–2.
Hoggan, J. 1977. Aspects of fermentation in conical vessels. *Journal of the Institute of Brewing* **83**, 133–8.
Holoberg, I. B. and P. Margalith 1981. Alcoholic fermentation by immobilized yeast at high sugar concentrations. *European Journal of Applied Microbiology and Biotechnology* **13**, 133–40.
Hospodka, J. 1966. Industrial applications of continuous fermentations. In *Theoretical and methodological basis of continuous culture of microorganisms*, I. Malek and Z. Fend (eds), 495–645. New York: Academic Press.
Hough, J. S. and A. H. Button 1972. Continuous brewing. *Progress in Industrial Microbiology* **11**, 89–132.
Hough, J. S. and T. P. Lyons 1972. Coupling of enzymes to microorganisms. *Nature* (London) **235**, 389.
Hough, J. S., C. W. Keevil, V. Maric, G. Philliskirk and T. W. Young 1976. Continuous

culture in brewing. In *Continuous culture 6: Applications and new fields*, A. C. R. Dean, D. C. Ellwood, C. G. T. Evans and J. Melling (eds), 226–37. Chichester: Ellis Horwood.

Hsiao, H. Y., L. C. Chang, C. M. Yang, L. F. Chen and G. T. Tsao 1983. Preparation and performance of immobilized yeast cells in column containing no inert carrier. *Biotechnology and Bioengineering* **25**, 363–75.

Inloes, D. S., D. P. Taylor, S. M. Cohen, A. S. Michaels and C. R. Robertson 1983. Ethanol production by *Saccharomyces cerevisiae* immobilized in hollow-fiber membrane bioreactors. *Applied Environmental Microbiology* **46**, 264–78.

Jones, R. P. and P. F. Greenfield 1982. Effect of carbon dioxide on yeast growth and fermentation. *Enzyme and Microbial Technology* **4**, 210–23.

Jones, R. P., N. Pamment and P. F. Greenfield 1981. Alcohol fermentation by yeasts – the effect of environmental and other variables. *Process Biochemistry* (April/May), 42–9.

Jones, S. T., R. A. Korus, W. Admassu and R. Heinsch 1984. Ethanol fermentation in a continuous tower fermenter. *Biotechnology and Bioengineering* **26**, 742–7.

Kierstan, M. and C. Bucke 1977. The immobilization of microbial cells, subcellular organelles and enzymes in calcium alginate gels. *Biotechnology and Bioengineering* **19**, 387–97.

Kierstan, M., A. McHale and M. P. Coughlan 1982. The production of ethanol from cellobiose using immobilized $\beta$-glucosidase coentrapped with yeast in alginate gels. *Biotechnology and Bioengineering* **24**, 1461–3.

King, V. A. E. and R. R. Zall 1983a. Ethanol fermentation of whey using polyacrylamide and kappa-carrageenan entrapped yeasts. *Journal of General Applied Microbiology* **29**, 379–93.

King, V. A. E. and R. R. Zall 1983b. Ethanol fermentation of whey using calcium alginate entrapped yeasts. *Process Biochemistry* (December), 17–20.

Kirsop, B. H. 1982. Developments in beer fermentation. In *Topics in enzyme and fermentation biotechnology*, A. Wiseman (ed.), Vol. 6, 80–131. Chichester: Ellis Horwood.

Klein, J. and B. Kreesdorf 1982. Immobilization of living whole cells in an epoxy matrix. *Biotechnology Letters* **4**, 375–80.

Klein, J., J. Stock and K. D. Vorlop 1983. Pore size and properties of spherical Ca-alginate biocatalysts. *European Journal of Applied Microbiology and Biotechnology* **18**, 86–91.

Klopper, W. J., R. H. Roberts, M. G. Royston and R. G. Ault 1966. Continuous fermentation in a tower fermenter. In *European Brewery Convention. Proceedings of the 10th Congress, Stockholm, 1965*, 238–59. Amsterdam: Elsevier.

Knudsen, F. B. and J. W. Larson 1974. Fermentation reproducibility – universal tanks vs. conventional tanks. *Technical Quarterly of the Master Brewers Association of the Americas* **11**, 1–8.

Knudsen, F. B. and N. L. Vacano 1972. The design and operation of uni-tanks. *Brewers' Digest* (July), 68–76.

Knudsen, F. B. and N. L. Vacano 1973. Economic aspects of uni-tanks. *Technical Quarterly of the Master Brewers Association of the Americas* **10**, 6–10.

Kolot, F. B. 1980. New trends in yeast technology. *Process Biochemistry* (October/November), 2–8.

Krouwel, P. G. and N. W. F. Kossen 1980. Gas production by immobilized microorganisms: Theoretical approach. *Biotechnology and Bioengineering* **22**, 681–7.

Krouwel, P. G. and N. W. F. Kossen 1981. Gas production by immobilized microorganisms: Calculation of theoretical maximum productivity. *Biotechnology and Bioengineering* **23**, 651–5.
Krouwel, P. G., A. Harder and N. W. F. Kossen 1982. Tensile stress–strain measurements of materials used for immobilization. *Biotechnology Letters* **4**, 103–8.
Kuu, W. Y. and J. A. Polack 1983. Improved immobilized biocatalysts by gel phase polymerization. *Biotechnology and Bioengineering* **25**, 1995–2006.
Kyung, K. H. and P. Gerhardt 1984. Continuous production of ethanol by yeast 'immobilized' in a membrane contained fermenter. *Biotechnology and Bioengineering* **26**, 252–6.

Larsson, M. and B. Mattiasson 1984. Novel process technology for biotechnological solvent production. *Chemical Industry* (London) **12**, 428–31.
Larsson, P. O. and K. Mosbach 1979. Alcohol production by magnetic immobilized yeast. *Biotechnology Letters* **1**, 501–6.
Lee, S. S. and H. Y. Wang 1982. Repeated fed-batch rapid fermentation using yeast cells and activated carbon extraction system. *Biotechnology and Bioengineering Symposium* No. 12, 221–31.
Lee, T. H., J. C. Ahn and D. D. Y. Ryu 1983. Performance of an immobilized yeast reactor system for ethanol production. *Enzyme and Microbial Technology* **5**, 41–5.
Lencki, R. H., C. H. Robinson and M. Moo-Young 1983. On-line extraction of ethanol from fermentation broths using hydrophobic adsorbents. *Biotechnology and Bioengineering Symposium* No. 13, 617–28.
Limtong, S., M. Nakata, H. Funahashi, T. Yoshida, T. Seki, J. Kumnuanta and H. Taguchi 1984. Continuous ethanol production by a concentrated culture of flocculating yeast. *Journal of Fermentation Technology* **62**, 55–62.
Lindsay, J. H. and J. W. Larson 1975. Large volume tanks – a review. *Master Brewers Association of America Technical Quarterly* **12**, 264–72.
Linko, P., S. Hakulin and Y. Y. Linko 1984. HTST-extrusion cooking in ethanol production from starchy materials. *Enzyme and Microbial Technology* **6**, 457–61.
Linko, P. and Y. Y. Linko 1982. Continuous ethanol fermentation by immobilized biocatalysts. In *Enzyme engineering*, I. Chibata, S. Fukui and L. B. Wingard (eds), Vol. 6, 335–42. New York: Plenum Press.
Linko, P. and Y. Y. Linko 1984. Industrial applications of immobilized cells. *CRC Critical Reviews in Biotechnology* **1**, 289–338.
Linko, P., M. Sorvari and Y. Y. Linko 1983. Ethanol production with immobilized cell reactors. *Annals of the New York Academy of Sciences* **413**, 424–34.
Linko, Y. Y. and P. Linko 1981. Continuous ethanol production by immobilized yeast reactor. *Biotechnology Letters* **3**, 21–6.
Linko, Y. Y., H. Jalanka and P. Linko 1981. Ethanol production from whey with immobilized living yeast. *Biotechnology Letters* **3**, 263–8.
Luong, J. H. T. and M. C. Tseng 1984. Process and technoeconomics of ethanol production by immobilized cells. *Applied Microbiology and Biotechnology* **19**, 207–16.
Lyons, T. P. 1984. Ethanol production in developed countries. *CRC Critical Reviews in Biotechnology* **1**, 189–207.

Maiorella, B., C. R. Wilke and H. Blanch 1981. Alcohol production and recovery. *Advances in Biochemical Engineering* **21**, 43–92.
Maiorella, B. L., H. W. Blanch and C. R. Wilke 1984. Economic evaluation of alternative ethanol fermentation processes. *Biotechnology and Bioengineering* **26**, 1003–25.

Maleszka, R., I. A. Veliky and H. Schneider 1981. Enhanced rate of ethanol production from D-xylose using recycled or immobilized cells of *Pachysolen tannophilus*. *Biotechnology Letters* **3**, 415–20.

Margaritis, A. 1975. *A study of the Rotorfermenter and the kinetics of ethanol fermentation*. Ph.D. Thesis, University of California at Berkeley, Berkeley, California, May 1975. Lawrence Berkeley Laboratory Report LBL-3278, July 1975.

Margaritis, A. and P. Bajpai 1981. Repeated batch production of ethanol from Jerusalem artichoke tubers using recycled immobilized cells of *Kluyveromyces fragilis*. *Biotechnology Letters* **3**, 679–82.

Margaritis, A. and P. Bajpai 1982. Continuous ethanol production from Jerusalem artichoke tubers. II. Use of immobilized cells of *Kluyveromyces marxianus*. *Biotechnology and Bioengineering* **24**, 1483–93.

Margaritis, A. and P. Bajpai 1983. Novel immobilized cell systems for the production of ethanol from Jerusalem artichoke. *Annals of the New York Academy of Sciences* **413**, 479–82.

Margaritis, A., P. Bajpai and M. A. Lachance 1983a. The use of free and immobilized cells of *Debaryomyces polymorphus* to produce ethanol from Jerusalem artichoke extracts. *Journal of Fermentation Technology* **61**, 533–7.

Margaritis, A., D. te Bokkel and M. El-Khashab 1983b. Pilot plant production of ethanol using immobilized yeast cells in a novel fluidized bioreactor system. Paper presented at the *186th ACS National Meeting*, Washington, DC, 28 August–2 September.

Margaritis, A. and F. J. A. Merchant 1984. Advances in ethanol production using immobilized cell systems. *CRC Critical Reviews in Biotechnology* **1**, 339–93.

Margaritis, A. and J. B. Wallace 1984. Novel bioreactor systems and their applications. *Bio/Technology* **2**, 447–53.

Margaritis, A. and C. R. Wilke 1972. Engineering analysis of the Rotorfermenter. *Developments in Industrial Microbiology* **13**, 159–76.

Margaritis, A. and C. R. Wilke 1978a. The Rotorfermenter. I. Description of the apparatus, power requirements and mass transfer characteristics. *Biotechnology and Bioengineering* **20**, 709–26.

Margaritis, A. and C. R. Wilke 1978b. The Rotorfermenter. II. Application to ethanol fermentation. *Biotechnology and Bioengineering* **20**, 727–53.

Martin, S., J. Bosch, J. Almenar and J. Posada 1975. Large outdoor sphero-conical tanks: a new realization on fermentation and lagering processes. In *European Brewery Convention Proceedings of the 15th Congress, Nice, 1975*, 301–10. Amsterdam: Elsevier.

Marwaha, S. S. and J. F. Kennedy 1984a. Ethanol production from whey permeate by immobilized yeast cells. *Enzyme Microbe Technology* **6**, 18–22.

Marwaha, S. S. and J. F. Kennedy 1984b. Alcohol production from whey permeate by immobilized and free cells of *Kluyveromyces marxianus* NCYC 179. *Process Biochemistry* **19** (April), 79–80.

Masschelein, C. A. and C. Francotte 1983. Possibilities et limites d'application des réacteurs a cellules de levure immobilisées en brasserie. *Cerevisiae* **3**, 135–42.

Mattiasson, B. and B. Hahn-Hagerdal 1982. Microenvironmental effects on metabolic behaviour of immobilized cells: a hypothesis. *European Journal of Applied Microbiology and Biotechnology* **16**, 52–5.

Maule, A. P. and P. D. Thomas 1973. Strains of yeast lethal to brewing yeasts. *Journal of the Institute of Brewing* **79**, 137–41.

Mehaia, M. A. and M. Cheryan 1984a. Ethanol production using membrane bioreactors. A comparison of membrane recycle and hollow fiber fermenters. Paper presented at *Biotech 84*, 10–12 September 1984, Washington, DC.

Mehaia, M. A. and M. Cheryan 1984b. Ethanol production in a hollow fiber bioreactor using *Saccharomyces cerevisiae*. *Applied Microbiology and Biotechnology* **20**, 100–4.
Mehaia, M. A. and M. Cheryan 1984c. Hollow fibre bioreactor for ethanol production: application to the conversion of lactose by *Kluyveromyces fragilis*. *Enzyme and Microbial Technology* **6**, 117–20.
Merchant, F. J. A. 1981. *The characteristics of* Saccharomyces cerevisiae *immobilized in calcium alginate gel performing ethanol production*. M.Sc. Thesis, Department of Chemical Engineering, University of Birmingham, England.
Michaux, M., M. Paquot, B. Baijot and Ph. Thonart 1982. Continuous fermentation: improvement of cell immobilization by zeta potential measurements. *Biotechnology and Bioengineering* Symposium No. 12, 475–84.
Minier, M. and G. Goma 1982. Ethanol production by extractive fermentation. *Biotechnology and Bioengineering* **24**, 1565–79.
Mitani, Y., Y. Nishizawa and S. Nagai 1984a. Ethanol production by immobilized cells with forced substrate supply. *Journal of Fermentation Technology* **62**, 249–53.
Mitani, Y., Y. Nishizawa and S. Nagai 1984b. Growth characteristics of immobilized yeast cells in continuous ethanol fermentation with forced substrate supply. *Journal of Fermentation Technology* **62**, 401–6.
Mohite, V. and H. Sivaraman 1984. Continuous conversion of sweet sorghum juice to ethanol using immobilized yeast cells. *Biotechnology and Bioengineering* **26**, 1126–7.
Moll, M. 1977. Continuous brewing of beer. U.S. Patent, 4 009 286.
Moo-Young, M., J. Lamptey and C. W. Robinson 1980. Immobilization of yeast cells on various supports for ethanol production. *Biotechnology Letters* **2**, 541–6.

Nagashima, M., M. Azuma, S. Noguchi, K. Inuzuka and H. Samejima 1984. Continuous ethanol fermentation using immobilized yeast cells. *Biotechnology and Bioengineering* **26**, 992–7.
Narziss, L. and P. Hellich 1971. A study on considerably accelerated fermentation and conditioning of beer. *Brauwelt* **111**, 1491–500.
Narziss, L. and P. Hellich 1972. Rapid fermentation and maturation of beer by means of the bio-reactor. *Brewers' Digest* (September), 106–18.
Nathan, L. 1930. Improvement in the fermentation and maturation of beers. *Journal of the Institute of Brewing* **36**, 538–50.
Navarro, A. R., C. Rubio and A. S. Callieri 1983. Production of ethanol by yeasts immobilized by pectin. *European Journal of Applied Microbiology and Biotechnology* **17**, 148–51.
Navarro, A., H. Marangoni, I. M. Plaza and D. Callieri 1984. Horizontal reactor for the continuous production of ethanol by yeasts immobilized in pectin. *Biotechnology Letters* **6**, 465–70.
Navarro, J. M. and G. Durand 1977. Modification of yeast metabolism by immobilization onto porous glass. *European Journal of Applied Microbiology and Biotechnology* **4**, 243–54.
Nishizawa, Y., Y. Mitani, M. Tamai and S. Nagai 1983. Ethanol production by cell recycling with hollow fibers. *Journal of Fermentation Technology* **61**, 599–605.
Nishizawa, Y., Y. Mitani, K. Fukunishi and S. Nagai 1984. Ethanol production by repeated batch culture with hollow fibers. *Journal of Fermentation Technology* **62**, 41–7.
Novais, J. M., S. M. M. Dias and J. M. S. Cabral 1983. Ethanol production from cheese whey by metal-chelated immobilized yeast cells. Paper presented at the *Fifth Symposium on Biotechnology for Fuels and Chemicals*, Gatlinburg, Tennessee, 10–13 May, 1983.

Oda, G., H. Samejima and T. Yamada 1983. Continuous alcohol fermentation technologies using immobilized yeast cells. Paper presented at *Biotech 83*. May 1983, London, England.

Pajunen, E. and V. Makinen 1975. The effect of pH and agitation on beer fermentation. In *European Brewery Convention Proceedings of the 15th Congress, Nice, 1975*, 525–38. Amsterdam: Elsevier.

Panushka, G. 1983. Continuous alcoholic fermentation for big-scale industrial plants. In *Energy from Biomass, 2nd E.C. Conference*, A. Strub, P. Chartier and G. Schleser (eds), 1045–9. London: Applied Science.

Pardonova, B., M. Polednikova, H. Sedova, M. Kehler and J. Ludvik 1982. Biocatalyst for beer production. *Brauwissenschaft* **35**, 254–8.

Paul, F. and P. M. Vignais 1980. Photophosphorylation in bacterial chromatophores entrapped in alginate gel: improvement of the physical and biochemical properties of gel beads with barium as the gel inducing agent. *Enzyme and Microbial Technology* **2**, 281–7.

Pines, G. and A. Freeman 1982. Immobilization and characterization of *Saccharomyces cerevisiae* in crosslinked prepolymerized polyacrylamide-hydrazide. *European Journal of Applied Microbiology and Biotechnology* **16**, 75–80.

Portno, A. D. 1970. Theoretical and practical aspects of continuous fermentation. *Wallerstein Laboratory Communications* **33**, 149–61.

Portno, A. D. 1979. Continuous fermentation in the brewing industry – the future outlook. In *European Brewery Convention on Fermentation and Storage, Symposium Monograph V*, Zoeterwoude, November 1978, 145–54.

Posada, J. 1978. Reasons to use sphero-conical fermenters. In *European Brewing Convention on Fermentation and Storage, Symposium Monograph V*, Zoeterwoude, November 1978, 207–18.

Prince, I. G. and J. P. Barford 1982a. Continuous tower fermentation for power ethanol production. *Biotechnology Letters* **4**, 263–8.

Prince, I. G. and J. P. Barford 1982b. Tower fermentation of sugar cane juice. *Biotechnology Letters* **4**, 469–74.

Prince, I. G. and J. P. Barford 1982. Induced flocculation of yeasts for use in the tower fermenter. *Biotechnology and Bioengineering* **4**, 621–6.

Purssell, A. J. R. and M. J. Smith 1968. Continuous fermentation. In *European Brewery Convention Proceedings of the 11th Congress, Madrid, 1967*, 155–65. Amsterdam: Elsevier.

Ramalingham, A. and R. K. Finn 1977. The vacuferm process: a new approach to fermentation alcohol. *Biotechnology and Bioengineering* **19**, 583–9.

Ramirez, A. and M. J. Boudarel 1983. Continuous production of ethanol on beet juice by a flocculent strain of *Saccharomyces cerevisiae*. *Biotechnology Letters* **5**, 659–64.

Righelato, R. C. 1980. Anaerobic fermentation: alcohol production. *Philosophical Transactions of the Royal Society of London* **B290**, 303–12.

Roffler, S. R., H. W. Blanch and C. R. Wilke 1984. *In situ* recovery of fermentation products. *Trends in Biotechnology* **2**, 129–36.

Rose, D. 1976. Yeast for molasses alcohol. *Process Biochemistry* **11**(3), 10–12, 36.

Rosen, K. 1978. Continuous production of alcohol. *Process Biochemistry* (May), 25–6.

Royston, M. G. 1966. Tower fermentation of beer. *Process Biochemistry* **1**, 215–21.

Russell, I. and G. G. Stewart 1979. Spheroplast fusion of brewer's yeast strains. *Journal of the Institute of Brewing* **85**, 95–102.

Ryu, D. D. Y., Y. J. Kim and J. H. Kim 1984a. Effect of air supplement on the performance of continuous ethanol fermentation system. *Biotechnology and Bioengineering* **26**, 12–16.

Ryu, D. D. Y., H. S. Kim and H. Taguchi 1984b. Intrinsic fermentation kinetic parameters of immobilized yeast cells. *Journal of Fermentation Technology* **62**, 255–61.

Ryu, Y. W., J. M. Navarro and G. Durand 1982. Comparative study of ethanol production by an immobilized yeast in a tubular reactor and in a multistage reactor. *European Journal of Applied Microbiology and Biotechnology* **15**, 1–8.

Schultz, J. S. and P. Gerhardt 1969. Dialysis culture of microorganisms: design, theory and results. *Bacteriology Reviews* **33**, 1–47.

Seddon, A. W. 1975. Continuous tower fermentation – experiences in establishing large scale commercial scale production. *Technical Quarterly of the Master Brewers Association of the Americas* **12**, 130–7.

Shardlow, P. J. 1972. The choice and use of cylindrico-conical fermentation vessels. *Technical Quarterly of the Master Brewers Association of the Americas* **9**, 1–5.

Shardlow, P. J. and C. C. Thompson 1971. The Nathan system. Conical fermenters. *Brewers' Digest* (August), 76–80.

Shiotani, T. and T. Yamane 1981. A horizontal packed-bed bioreactor to reduce $CO_2$ gas holdup in the continuous production of ethanol by immobilized yeast cells. *European Journal of Applied Microbiology and Biotechnology* **13**, 96–101.

Siess, M. H. and C. Divies 1981. Behaviour of *Saccharomyces cerevisiae* cells entrapped in a polyacrylamide gel and performing alcoholic fermentation. *European Journal of Applied Microbiology and Biotechnology* **12**, 10–15.

Sitton, O. C. and J. L. Gaddy 1980. Ethanol production in an immobilized cell reactor. *Biotechnology and Bioengineering* **22**, 1735–48.

Sivaraman, H., B. Seetaramarao, A. V. Pundle and C. Sivaraman 1982. Continuous ethanol production by yeast cells immobilized in open pore gelatin matrix. *Biotechnology Letters* **4**, 359–64.

Slininger, P. J., R. J. Bothast, L. T. Black and J. E. McGhee 1982. Continuous conversion of D-xylose to ethanol by immobilized *Pachysolen tannophilus*. *Biotechnology and Bioengineering* **24**, 2241–51.

Smart, N. J. 1984. Gaslift fermenters: theory and practice. *Laboratory Practice* **33**(7), 9–14.

Smith, E. L. and R. N. Greenshields 1973. Tower fermentation systems: their application in continuous brewing. *Biotechnology and Bioengineering* Symposium No. 4, 519–33.

Specchia, V., G. Genon and A. Gianetto 1983. Production of ethanol with *Saccharomyces cerevisiae* in a continuous reactor. Note II. Tests and modelling for a packed tubular reactor with immobilized yeasts. *Chemical Engineering Communications* **23**, 233–43.

Stewart, G. G. 1974. Some thoughts on the microbiological aspects of brewing and other industries utilizing yeast. *Advances in Applied Microbiology* **17**, 233–64.

Stewart, G. G. 1977. Fermentation – yesterday, today and tomorrow. *Technical Quarterly of the Master Brewers Association of the Americas* **14**, 1–15.

Suihko, M. L. and K. Poutanen 1984. D-xylulose fermentation by free and immobilized *Saccharomyces cerevisiae* cells. *Biotechnology Letters* **6**, 189–94.

Takayanagi, S. and T. Harada 1968. Experience with new types of freestanding fermentation and lagering tanks. In *European Brewery Convention Proceedings of the 11th Congress, Madrid, 1967*, 473–87. Amsterdam: Elsevier.

Tanaka, H., M. Matsumura and I. A. Veliky 1984. Diffusion characteristics of substrates in Ca-alginate beads. *Biotechnology and Bioengineering* **26**, 53–8.

Taniguchi, M., K. Wakamiya, M. Tsuchiya, R. Maksuno and T. Kanikubo 1983.

Continuous ethanol production by cell-holding culture of yeasts. *European Journal of Applied Microbiology and Biotechnology* **18**, 201–6.
Tyagi, R. D. and T. K. Ghose 1982. Studies on immobilized *Saccharomyces cerevisiae*. I. Analysis of continuous rapid ethanol fermentation in immobilized cell reactor. *Biotechnology and Bioengineering* **24**, 781–95.

Ulenberg, G. H., H. Gerritson and J. Huisman 1972. Experiences with a giant cylindrico-conical tank. *Technical Quarterly of the Master Brewers Association of the Americas* **9**, 117–22.
Updike, S. J., D. R. Harris and E. Shrago 1969. Microorganisms, alive and imprisoned in a polymer cage. *Nature* (London) **224**, 1122–3.

Varma, R., B. A. Baliga, S. K. Chatterjee, B. K. Ghosh, P. M. Dalal, A. K. Srivastava and N. G. Karanth 1984. Studies on the contamination of feed molasses for continuous ethanol production in an immobilized yeast cell reactor. *Journal of Chemical Technology and Biotechnology* **34B**, 111–15.
Veliky, I. A. and R. E. Williams 1981. The production of ethanol by *Saccharomyces cerevisiae* immobilized in polycation-stabilized calcium alginate gels. *Biotechnology Letters* **3**, 275–80.
Vick Roy, T. B., H. W. Blanch and C. R. Wilke 1983. Microbial hollow fiber bioreactors. *Trends in Biotechnology* **1**, 135–9.

Wada, M., J. Kato and I. Chibata 1979. A new immobilization of microbial cells: immobilized growing cells using carrageenan gel and their properties. *European Journal of Applied Microbiology and Biotechnology* **8**, 241–7.
Wada, M., J. Kato and I. Chibata 1980a. Electron microscopic observation of immobilized growing yeast cells. *Journal of Fermentation Technology* **58**, 327–31.
Wada, M., J. Kato and I. Chibata 1980b. Continuous production of ethanol using immobilized growing yeast cells. *European Journal of Applied Microbiology and Biotechnology* **10**, 275–87.
Wada, M., J. Kato and I. Chibata 1981. Continuous production of ethanol in high concentration using immobilized growing yeast cells. *European Journal of Applied Microbiology and Biotechnology* **11**, 67–71.
Wang, H. Y. and D. J. Hettwer 1982. Cell immobilization in K-carrageenan with tricalcium phosphate. *Biotechnology and Bioengineering* **24**, 1827–38.
Wang, H. Y., F. M. Robinson and S. S. Lee 1981. Enhanced alcohol production through on-line extraction. *Biotechnology and Bioengineering Symposium* No. 11, 555–65.
Weeks, M. G., P. A. Munro and P. L. Spedding 1982. Semi-continuous ethanolic fermentation using a novel yeast settling and recycling technique. *Biotechnology Letters* **4**, 85–90.
White, F. H. and A. D. Portno 1977. Continuous fermentation by immobilized brewer's yeast. *Journal of the Institute of Brewing* **84**, 228–30.
Wick, E. and K. Popper 1977. Continuous fermentation in slant tubes. *Biotechnology and Bioengineering* **19**, 235–46.
Wiesenberger, A. 1978. Wine making goes continuous. *Food Engineering International* **3**(2), 26–8.
Williams, D. and D. M. Munnecke 1981. The production of ethanol by immobilized yeast cells. *Biotechnology and Bioengineering* **23**, 1813–25.

Yamane, T. and S. Shimizu 1982. The minimum-sized ideal reactor for continuous alcohol fermentation using immobilized microorganisms. *Biotechnology and Bioengineering* **24**, 2731–7.

# 9 *Biochemical and genetic control of sugar and carbohydrate metabolism in yeasts*

G. G. STEWART and I. RUSSELL

## 9.1 Introduction

As a group of micro-organisms, yeast (i.e. unicellular fungi) are capable of utilizing a broad spectrum of carbohydrates and sugars. Nevertheless, none of the yeast species yet isolated from natural environs have been found to be capable of utilizing all of the readily available sugars and carbohydrates. Indeed, it is one of the objectives of many biotechnologists to manipulate a yeast strain genetically so that it will be capable of metabolizing a broader spectrum of carbohydrates and sugars in order to produce ethanol more efficiently (Stewart *et al.* 1982), biomass (Solomons 1984) and higher value-added products such as ascorbic acid (Cayle *et al.* 1986), vitamins (Trivedi 1986), glycerol (Vijaikishore & Karanth 1986), enzymes (Eveleigh 1981), flavourings and fragrances (Drawert & Barots 1978), lipids (Ratledge & Botham 1977) and proteins and peptides with industrial or therapeutic and pharmaceutical activity (Hitzeman *et al.* 1983; Bitter 1983; Chang & Cronan 1983; Smith *et al.* 1985). The major naturally occurring carbohydrates are starch and cellulose, together with the disaccharides sucrose (from sugar cane and sugar beet) and lactose (from milk and other dairy products). The hydrolysis products of these materials include glucose, fructose, galactose, maltose, maltotriose, maltodextrins, cellobiose, cellulodextrins, pentoses, xylans and lignin.

Some biochemists, molecular biologists and industrial microbiologists incorrectly employ the term 'yeast' as synonymous with *Saccharomyces cerevisiae*. Although most of the biochemical and genetic studies and industrial applications have been conducted on this species, there is a rich variety of yeast genera and species that await biological investigation. The more exotic species, particularly with regard to carbohydrate metabolism, may offer advantages for experimental study. Nevertheless, as previously discussed in Chapter 3, the genus *Saccharomyces* has often been referred to as 'the oldest plant cultivated by man'. Indeed, the history of beer-, wine- and breadmaking, with the fortuitous use of yeast, is as old as the history of man himself.

Their contribution to man has been based very largely on the capacity of certain yeasts to effect a rapid and efficient conversion of sugars into ethanol and carbon dioxide, and to effect an alcoholic fermentation of sugary liquids such as grain extracts, grape juice and milk. However, recent years have witnessed the transition of yeast from being solely the 'workhorse' in traditional food and beverage industries to being, in addition, one of the organisms of choice for gene manipulation involving the 'new genetics'.

Micro-organisms which metabolize glucose, or compounds that can give rise to glucose as a source of energy in the absence of oxygen, use a variety of different pathways to regenerate $NAD^+$. All of these pathways lead to the formation of so-called fermentation products, which include acids, such as lactic, acetic and propionic, and compounds such as acetone, butanol, isopropanol and acetylmethylcarbinol. They also include ethanol. Although many micro-organisms produce ethanol as an end product of their energy-generating metabolism, they are a minority of the total microbial population. However, pride of place in this minority group must be given to species of the genus *Saccharomyces* which have, for several thousands of years, been used to produce ethanol, both in a potable and non-potable form, from sugars. The main deficiencies of this ethanol-producing micro-organism are its inability to ferment pentose sugars, which can be made available in large quantity by hydrolysing wood, and its lack of depolymerizing enzymes which might yield mono- and disaccharides from cellulose. *Saccharomyces cerevisiae* is also unable to hydrolyse starch and dextrin but, as will be discussed later, there is a closely related species, *Saccharomyces diastaticus*, that can utilize starch by virtue of the production of glucoamylase (EC 3.2.1.3) which will hydrolyse the $\alpha$-1,4 linkages of starch and dextrin to glucose.

At present only *Saccharomyces cerevisiae* and related species are of major industrial importance. Yet some of the many yeast species could certainly assume a considerable role in industry, as a source of protein, for producing ethanol and for carrying out other chemical transformations, such as those involved in synthesizing precursors of important natural products. The following research programme would further such ends. (a) The carbohydrate-utilization capabilities of the more versatile yeasts should be characterized genetically and biochemically. (b) Artificial means of genetically manipulating yeasts should be developed further in order to make a wider range of yeasts available. It may no longer be essential to choose yeasts having an amenable, natural system of genetic recombination; the useful physiological characteristics found in various strains that have been little studied could be transferred, by gene cloning or fusion, to a commonly used but less versatile strain. (c) In order to assist this endeavour, more must be found out about genetic and physiological differences between yeast genera and species.

*Saccharomyces cerevisiae* has the ability to take up and ferment a wide range of sugars; for example, sucrose, glucose, fructose, galactose, maltose and maltotriose. In addition, the closely related species *Saccharomyces diastaticus* and *Saccharomyces uvarum* (*carlsbergensis*) are able to utilize dextrins and melibiose, respectively. The first step in the utilization of any sugar by yeast is usually either its passage intact across the cell membrane or initial hydrolysis outside the membrane followed by entry into the cell by some or all of the hydrolysis products.

Unlike some bacteria, which ferment a wide range of organic compounds, yeasts (with a few notable exceptions) only ferment those metabolized via the Embden–Meyerhof and Entner–Doudoroff pathways. Furthermore, cytochrome-deficient mutants apart, a yeast that uses sugar anaerobically also uses it aerobically (Barnett 1976, 1981).

*Saccharomyces cerevisiae* is constitutive for the ability to metabolize both glucose, mannose and fructose. With the exception of certain mutant yeast strains, all yeasts are able to utilize these sugars. In *Saccharomyces cerevisiae* a number of transport systems for sugars have been described: a constitutive system common for glucose, fructose and mannose, inducible systems for galactose, α-methyl-D-glucoside, maltose and maltotriose, respectively. The transport of hexoses by the constitutive and by the inducible galactose system is equilibrative and does not require energy. On the other hand, the α-methyl-D-glucoside, maltose and maltotriose systems have been characterized as active transport. For example, maltose transport, although independent of the intracellular ATP level, is coupled to the electrochemical gradient of protons (Serrano 1977). Sucrose is hydrolysed outside the cell membrane by the extracellular enzyme invertase (β-D-fructofuranoside fructohydrolase, EC 3.2.1.26), to glucose and fructose. Six genes *SUC*1, 2, 3, 4, 5, and *SUC*80 code for the production of invertase and a number of them have been mapped (Mortimer & Schild 1985). The disaccharide maltose and the trisaccharide maltotriose have independent uptake mechanisms (maltose or maltotriose permease), which transport the two sugars across the cell membrane into the cell. Once inside the cell, both sugars are hydrolysed to glucose units by the α-glucosidase system (Hautera & Lovgren 1975, Stewart *et al.* 1979).

As previously stated, there is a closely related species to *Saccharomyces cerevisiae* that is able to utilize dextrin material, dextrins being polymers of glucose, greater than three units of the monosaccharide. In 1952, Andrews and Gilliland isolated a strain of yeast capable of utilizing starch and dextrin by virtue of its ability to produce the extracellular enzyme glucoamylase or amyloglucosidase (α-1,4-glucan glucohydrolase, EC 3.2.1.3). Due to its dextrinase activity, it was defined as a new species of yeast – *Saccharomyces diastaticus*. The species *Saccharomyces uvarum* (*carlsbergensis*) is also differentiated from *Saccharomyces cerevisiae* on the basis of sugar-fermentation characteristics, *Saccharomyces uvarum* (*carlsbergensis*) being able to ferment the disaccharide melibiose by virtue of the production of the extracellular enzyme melibiase (α-galactosidase).

The disaccharide lactose consists of glucose and galactose units. High concentrations of lactose are found in milk and whey, the by-product of cheese manufacture. There are two species of yeast, *Kluyveromyces fragilis* and *Kluyveromyces lactis*, that can utilize lactose. The uptake mechanism is very similar to that of maltose. The lactose is transported across the cell membrane by means of a lactose permease system; once inside the cell it is hydrolysed by β-D-galactoside galactohydrolase (EC 3.2.1.23) into its constituent monosaccharides glucose and galactose, which then enter the common glycolytic pathways of the cell.

A yeast developing on a carbohydrate source must either be able to ferment it or to utilize it by respiration. It is customary in taxonomic work (Lodder 1970) to refer to the first process as 'fermentation' and to the second as 'assimilation'.

Although these terms are firmly established, they are somewhat misleading since actual assimilation of the carbohydrate, i.e. uptake by the cell, occurs both during the fermentative and the oxidative dissimilatory process.

The classical concept of the alcoholic fermentation by yeasts has been that of a process which takes place under strictly anaerobic conditions. The ability or inability to ferment carbohydrates to ethanol and carbon dioxide is a most useful characteristic for differentiating species. Although it finds less application in the definition of genera, it is nevertheless occasionally of some use. For example, the genus *Saccharomyces* is characterized by the vigorous fermentation of a number of sugars whereas other genera, e.g. *Lipomyces* and *Rhodotorula*, are strictly non-fermentative, and in other genera, e.g. *Hansenula*, an entire range from non-fermentative to strongly fermentative species may be found.

## 9.2 D-Galactose

The critical steps in the utilization of D-galactose by extracts of galactose-adapted yeast are described in the following reactions (Cardini & Leloir 1953). The first three are often referred to as the Leloir pathway.

$$\text{Galactose} + \text{ATP} \xrightarrow[\text{(EC 2.7.1.6)}]{\text{galactokinase}} \text{Gal-1-P} + \text{ADP} + \text{H}^+$$

$$\text{Gal-1-P} + \text{UDPG} \xrightarrow[\text{(EC 2.7.7.10)}]{\text{galactose-1-P uridylyltransferase}} \text{Gluc 1-P} + \text{UDP} - \text{galactose}$$

$$\text{UDP} - \text{galactose} \xrightarrow[\text{(EC 5.1.3.2)}]{\text{UDP-D-galactose 4-epimerase}} \text{UDP} - \text{glucose}$$

$$\text{Gluc-1-P} \xrightarrow[\text{(EC 2.7.5.1)}]{\text{phosphoglucomutase}} \text{Gluc-6-P}$$

The genetic system that encodes the enzymes responsible for the conversion of exogenous galactose to endogenous glucose 1-phosphate in *Saccharomyces* has been well studied (Douglas 1961). After transport across the cell membrane by the *GAL*2-encoded permease (Douglas & Hawthorne 1966), galactose is converted to glucose 1-phosphate by reactions catalysed by the products of the *GAL*1, *GAL*7 and *GAL*10 genes, i.e. galactokinase, galactose-1-P uridylyltransferase and UDP-D-galactose 4-epimerase, respectively. These three genes are located in a tightly linked cluster, six map units from the centromere of chromosome II (Douglas 1961). The genetic order of these genes has been determined by Bassel and Douglas (1968) as centromere–*GAL*7–*GAL*10–*GAL*1.

The enzymes encoded by the galactose gene cluster are co-ordinately induced by the addition of galactose to appropriately adapted cultures (Hartwell 1970). Broach (1979) has shown that the galactose enzymes are translated as separate polypeptides. This is consistent with the data of Hopper *et al*. (1978) which showed that the *GAL*1 and *GAL*7 genes specify distinct and separate mRNAs and are the result of *de novo* synthesis after induction.

Expression of the *GAL*7–*GAL*10–*GAL*1 gene cluster is known to be controlled primarily by two unlinked regulatory loci, *GAL*80 and *GAL*4 (Kew & Douglas 1976). These loci also affect the expression of the galactose permease and the extracellular enzyme $\alpha$-galactosidase (St John & Davis 1979). The *GAL*80 locus behaves genetically as a negative regulator. Both dominant and uninducible super-repressor mutations and recessive constitutive mutations (Matsumoto *et al*. 1980) have been isolated at this locus. The *GAL*4 locus is required for induction and expression of the galactose enzymes. It has been shown (Perlman & Hopper 1979) that the wild type *GAL*4 function is required for the expression of the messenger RNAs encoding at least two of the galactose gene cluster enzymes – the kinase (*GAL*1) and the transferase (*GAL*7). Dominant *GAL*80 mutations result in the constitutive expression of all of the galactose enzymes (Nogi *et al*. 1977). These mutations have been shown to drop within the *GAL*4 structural gene (Matsumoto *et al*. 1980).

## 9.3 Maltose and maltotriose

The disaccharide maltose and the trisaccharide maltotriose pass intact across the yeast plasma membrane. Both sugars are transported by inducible $\alpha$-glucoside permeases and hydrolysed by an $\alpha$-glucosidase (maltase, $\alpha$-D-glucoside glucohydrolase, EC 3.2.1.20), which is also inducible. The transport, hydrolysis and fermentation of maltose are particularly important in brewing and baker's yeast strains, since maltose is the main sugar component of both brewing wort and wheat dough. There are (at least) five unlinked polymeric genes that control the ability of yeast to produce $\alpha$-glucoside permease in response to maltose: *MAL*1, *MAL*2, *MAL*3, *MAL*4 and *MAL*6. Strains carrying an active allele at any one of these loci are inducible; strains carrying inactive alleles at all loci are uninducible but produce basal levels of $\alpha$-glucosidase (Khan *et al*. 1973). The role of the *MAL* genes is still not fully understood. It has been suggested that the *MAL* loci are structural genes for $\alpha$-glucosidase, regulatory genes controlling both $\alpha$-glucosidase and $\alpha$-glucoside permease or complex loci containing both regulatory and structural elements. The role of the *MAL* genes has been investigated genetically in two ways, either by analysing mutations at a single *MAL* locus (ten Berge *et al*. 1973a) or by comparing the effects of the different *MAL* loci (Tauro & Halvorson 1966). As an example of the former approach, ten Berge and his associates have analysed the effects of mutations in the *MAL*6 locus. They have found that both non-inducible and constitutive mutations map at the *MAL*6 locus (ten Berge *et al*. 1973b) and that two strains that are temperature sensitive for $\alpha$-glucosidase induction contain $\alpha$-glucosidase indistinguishable from that of the wild type.

These and similar results with other *MAL* loci (Khan *et al*. 1973) are consistent with the proposal that the *MAL* genes are regulatory loci that code for a protein required for $\alpha$-glucosidase induction (Zimmerman & Eaton 1974). These results do not rule out the possibility (discussed below) that the *MAL* genes are complex loci that also include structural genes (Naumov 1971) for the enzymes of maltose metabolism.

Halvorson and co-workers (Rudert & Halvorson 1962; Tauro & Halvorson 1966) have compared the effects of several different *MAL* loci on the synthesis of $\alpha$-glucosidase, and have found that strains carrying different active *MAL* loci produce different amounts of the enzyme. Mowshowitz (1979) has re-examined the role of the *MAL* loci and the influence of *MAL* gene dosage because: (a) it is generally assumed that regulatory genes do not exhibit gene dosage effects due to the fact that their products are present in excess; (b) structural genes for inducible and repressible enzymes do not necessarily show strictly additive gene dosage effects; and (c) it has not been demonstrated that the differences in $\alpha$-glucosidase levels are actually due to differences in the *MAL* gene dosage. Inbred strains of *Saccharomyces cerevisiae* carrying *MALI*, *MAL*2 or *MAL*6 in a common background were used to construct: (a) homozygous or heterozygous diploids carrying one or two active alleles of a single *MAL* locus (*MAL*1, *MAL*2 or *MAL*6); or (b) triploids carrying one, two or three active alleles of *MAL*2. The diploid and triploid strains were employed to investigate gene dosage effects on the different rates of $\alpha$-glucosidase synthesis. All three *MAL* loci exhibited a gene dosage effect on the rate of $\alpha$-glucosidase synthesis; *MAL*2 also exhibited a gene dosage effect on the enzyme kinetics of induction. The dosage effects of the *MALI* and *MAL*6 loci were additive, but the effects of the *MAL*2 locus were not of the same magnitude; the gene dosage effect decreased with increasing gene dosage. In a further study Mowshowitz (1981) has shown that the alleles, *MAL*1, *MAL*2 and *MAL*6 have different effects on the regulation of $\alpha$-glucosidase accumulation, and it has been documented that the *MAL* loci themselves regulate $\alpha$-glucosidase accumulation. However, it has not been demonstrated whether all *MAL*1, *MAL*2 or *MAL*6 alleles are identical, so it is possible that properties ascribed to each *MAL* locus are only attributable to a particular allele of that locus. It is therefore possible that there is as much variation between the different alleles of any one *MAL* locus as that observed between the alleles of three different *MAL* loci.

Fermentation studies with the *MAL*2 triploids (Stewart 1981) have been conducted in brewer's wort. In 12°Plato wort the strains containing the *MAL2/MAL2/mal* and *MAL2/MAL2/MAL*2 genotype fermented at a much faster rate than a production ale yeast strain (*Saccharomyces cerevisiae*). In 16°Plato wort the gene dosage effect was more pronounced and, again, a faster rate of fermentation with the multiple *MAL*2 strains was seen, this time compared with a production lager-brewing strain [*Saccharomyces uvarum* (*carlsbergensis*)].

It now appears that the five *MAL* loci (i.e. *MAL*1, *MAL*2, *MAL*3, *MAL*4 and *MAL*6) contain regulatory genes controlling the induction of the maltose fermentation enzymes. It has also been proposed that the *MAL* loci include the structural genes for $\alpha$-glucosidase and maltose permease. Naumov (1971, 1976) has analysed the progeny of crosses between various naturally occurring maltose-negative and maltose-positive isolates of *Saccharomyces*, and concluded that maltose fermentation requires the presence of two genes termed

*MALp* and *MALg*, and that each *MAL* locus is composed of these two linked genes. A maltose-negative strain could arise if it lacked either a functional *MALp* or a *MALg*, either at linked or at unlinked loci, allowing the cell to utilize maltose as a carbon source. Federoff *et al.* (1982) have constructed a clone bank from a donor yeast strain that carries the *MALp MALg* locus. The cloned yeast DNA was then amplified in *Escherichia coli* K-12 strain RR101. The recombinant plasmid DNA isolated from pools of 500 *Escherichia coli* (pYT) transformants was used to transform a $Mal^-$ yeast strain. $Mal^+$ colonies were identified as transformants. Most of the $Mal^+$ colonies were found to be highly unstable during mitotic growth on non-selective media, as they segregated mostly $Mal^-$ clones. DNA, prepared from several $Mal^+$ yeast transformants grown under selective conditions, was transferred to *Escherichia coli* and the plasmid *pMAL*26 identified by this approach. This plasmid was shown by positive hybridization translation, as well as Southern and Northern blot experiments, to carry a *MAL*6 structural gene for $\alpha$-glucosidase.

Cohen *et al.* (1984) have studied the *MALI* locus in some detail. The *MALI* locus is complex comprising at least three genes: *MALIR*, a gene involved in the coordinate regulation of the synthesis of $\alpha$-glucosidase and maltose transport; *MALIT*, a gene encoding a component of the maltose transport system; and *MALIS*, a likely candidate for the structural gene for $\alpha$-glucosidase. The same laboratory (Cohen *et al.* 1985) has physically and functionally identified three genes of the *MAL*6 locus, and the functional organization of this locus is identical to that of the *MALI* locus.

## 9.4 Sucrose

This disaccharide is first hydrolysed by the extracellular enzyme invertase into constituent monosaccharides, which are then taken up into the yeast cell. The specificity of the enzyme is absolute for the unsubstituted or unchanged $\beta$-fructofuranoside terminal:

$$R\text{–}\beta\text{-fructofuranoside} + H_2O \rightarrow \text{fructose} + R\text{OH}$$

Thus, $\alpha$-fructofuranosides, fructopyranosides, turanose and melezitose are not acted upon, whereas methyl-$\beta$-D-fructofuranoside, sucrose, raffinose, stachyose and verbascose are cleaved at the fructosidic linkage on the fructose side of the bridge-oxygen. The bulk of the literature is based on *Saccharomyces cerevisiae* as the source micro-organism and sucrose as the substrate, sucrose being the best substrate for the enzyme with the lowest $K_m$ and highest $V_{max}$.

In *Saccharomyces cerevisiae* the ability to hydrolyse sucrose is conferred by any one of six (or more) polymeric sucrose genes (denoted *SUC*1 to *SUC*6) which reside at loci distributed throughout the genome. Strains unable to ferment sucrose can arise either by segregation during crosses of strains containing different *SUC* genes or by mutation of a known *SUC* gene. Segregational negatives are termed *suc*0 to distinguish them from the mutational non-fermenters *suc*1, *suc*2, . . ., *suc*6 (Carlson *et al.* 1981).

The presence of any one of the *SUC* genes in the genome leads to the production of two forms of invertase: a large, secreted mannoprotein which is about 50% carbohydrate and has a molecular weight of 270 000, and a smaller intracellular non-glycosylated form of molecular weight 120 000. The production of these forms is controlled by catabolite repression and can vary widely, with the level of the large enzyme varying over a much greater range than that of the small enzyme. Invertase biosynthesis in yeast is not inducible by its substrates (e.g. sucrose and raffinose) but is subject to repression by the glucose concentration in the culture medium. It has been shown (Rodriguez *et al.* 1981) that external invertase increases proportionately much more than the internal invertase when cells are released from glucose repression.

Mormeneo and Sentandreu (1982) have reported that although invertase secretion requires protein synthesis it is independent of RNA formation. In addition, accumulation of invertase specific mRNA occurs in the absence of protein synthesis, but is expressed only after synthesis of protein. Invertase mRNA is continuously synthesized under repressive conditions and the level of this mRNA is regulated by the presence of glucose (Mormeneo & Sentandreu 1986). The hexose regulates the concentration of this mRNA at the level of transcription. The expression of the invertase mRNA present in the cells under repressive conditions is also regulated by glucose at the level of translation or secretion, or both. Consequently, under repressive conditions invertase is destroyed before secretion occurs.

The biosynthetic relationship between the two forms of invertase is not fully understood. Large invertase probably consists of two identical subunits, and the small enzyme is also probably a dimer. There are antigenic and kinetic similarities between the large and small enzymes, but attempts to demonstrate definitively a precursor–product relationship have been inconclusive. Tunicamycin, which inhibits the synthesis of external glycoproteins (Kuo & Lampen 1974), was found to have no effect on the formation of the small invertase, whereas the large form of the enzyme was not produced in the presence of this antibiotic. Furthermore, there was no accumulation of invertase activity with the properties of the small internal invertase. Similar results were also obtained by Kuo and Lampen (1974), using spheroplasts.

Mutagenesis of the sucrose-fermenting (*SUC*1) *Saccharomyces cerevisiae* strain 4059-358D yielded an invertase-negative mutant (D10) (Mizunaga *et al.* 1981). Although no mutant lacking only one form of invertase has been described, subsequent mutation tests of D10 gave a sucrose-fermenting invertase D10-ER1 which, under a variety of growth conditions, has a low level of small intracellular non-glycosylated invertase but a normal level of large invertase. It has been suggested that strain D10-ER1 carries a defect specific to a gene for small invertase. Limited genetic evidence indicated that these defects in D10 and D10-ER1 were allelic to the *SUC*1 gene. The large invertases from D10-ER1 and 4059-358D have been purified and compared. The two enzymes have similar specific activity and $K_m$ for sucrose, cross-react immunologically and show the same subunit molecular weight after removal of the carbohydrate with endo-$\beta$-N-acetyl glucosaminidase H. However, the large invertase from strain D10-ER1 is rapidly inactivated at 55°C, whereas that from the parent strain is relatively stable at 65°C. The small invertase in extracts of D10-ER1 is also heat sensitive compared with the small enzyme from the original parent strain. The low level of small invertase in mutant D10-ER1 may reflect increased intracellular degradation of this heat-labile form. In several crosses of D10-ER1 with strains carrying the *SUC*1 or *SUC*3 gene, the temperature sensitivity of the large and small invertases and the low cellular level of small invertase appeared to co-segregate. These findings are evidence that *SUC*1 is a structural gene for invertase, and that both the large and the small forms are encoded by a single gene.

The following model for the biosynthesis of invertase has been proposed by Rodriguez *et al.* (1981). The protein moiety for the external (large) enzyme is synthesized at the level of the rough endoplasmic reticulum. This protein carries a hydrophobic N-terminal peptide that permits its insertion into the lipid bilayer and its passage to the lumen of the rough endoplasmic reticulum. Glycosylation starts while the protein is still attached to the polysomes, and continues in the smooth endoplasmic reticulum. The addition of sugar residues could be either sequential or due to the transfer of a relatively bulky oligosaccharide. This latter mechanism would explain the changes in molecular weight and the ratio of carbohydrate to protein between the membrane-bound and the external isoenzymes. As far as the cytoplasmic or small enzyme is concerned, its synthesis might occur in free polysomes that release the protein into the cytoplasm or into an organelle. The organelle could disrupt during preparation of a cell-free extract, and would account for the cytoplasmic location of this invertase. In living cells the enzyme would be degraded at a rate similar to that of its synthesis. The metabolic significance of this small invertase is obscure.

The *SUC*2 gene has been cloned (Carlson & Botstein 1982) and sequenced (Taussig & Carlson 1983). The two forms of invertase are encoded by two differently regulated mRNAs which differ only at their 5′-ends (Carlson *et al.* 1983). A 1.9-kb RNA, which is glucose-repressible, encodes a signal peptide-containing precursor to the secreted invertase. However, the constitutively synthesized 1.8-kb RNA does not encode a complete signal sequence, and hence the translation product remains intracellular. Since the *SUC*2 gene carries a strong promoter as well as a signal sequence for secretion, it has been found to be useful in the construction of hybrid DNA sequences by fusing heterologous genes of interest with the promoter or signal sequence region, or both, of the *SUC*2 gene. Thus, secretion of calf prochymosin (the inactive precursor of chymosin or renin) has been achieved (Smith *et al.* 1985) by fusing the translational reading frame, the *SUC*2 transcriptional promoter regimen, the coding region for the amino-terminal secretion signal and codons for the first 11 amino-acid residues of mature invertase with a prochymosin-encoding DNA fragment. The fused genes were placed on a plasmid vector and transformed into an appropriate recipient yeast strain for expression. Experiments involving fusion of other heterologous genes with the *SUC*2 promoter have also been reported (Emr *et al.* 1984). Since depletion of glucose in the medium would activate the *SUC*2 promoter, the use of the *SUC*2 promoter-signal systems for expression and secretion of heterologous proteins such as glucoamylase is very appealing.

## 9.5 Melibiose

α-Galactosidase (α-D-galactoside galactohydrolase, EC 3.2.1.22) or melibiase is an inducible enzyme, found in a number of *Saccharomyces* spp., which hydrolyses the sugar melibiose to units of galactose and glucose. The enzyme is found mainly outside the cell membrane (Friis & Ottohenghi 1959); however, an intracellular form of the enzyme also exists (Lazo *et al.* 1977). Unlike invertase, the internal α-galactosidase is the same molecular size as the external enzyme. Similarly to most other extracellular enzymes in yeast, carbohydrate material has been found to be associated with the enzyme indicating that α-galactosidase is a glycoprotein (Lazo *et al.* 1977) where the carbohydrate portion accounts for 57% of the molecule (Lazo *et al.* 1978). Furthermore, α-galactosidase is similar in molecular weight to the extracellular enzymes invertase and acid phosphatase, being approximately 300 kda. When α-galactosidase was partly inactivated with 3 mol $l^{-1}$ urea, two proteins corresponding to the denatured enzyme resulted. The molecular weights of the two proteins were calculated to be 20 000 and 100 000 (Lazo *et al.* 1978).

As previously discussed in this chapter, the existence of extracellular hydrolytic enzymes in yeast (or any other micro-organism) can be attributed to the impermeability of the cell membrane to their corresponding substrates. One possibility to explain the induction of these enzymes is the appearance in the cytoplasm of hydrolysis products of the substrate, resulting from the presence of constitutive levels of enzyme on the cell surface. In the case of α-galactosidase this possibility is in agreement with the fact that galactose (a part of the melibiose molecule) is able to induce its synthesis (Kew & Douglas 1976). Moreover, when α-galactosides were used as inducers, the level of enzyme correlated with the number of galactose residues in the molecule, thus supporting the notion that the galactose, not the glucose, moiety is responsible for the enzyme induction. Indeed, it is logical that the galactose moiety is responsible for the induction, since the other half of the melibiose molecule, glucose, has been shown to be a good repressor of the enzyme. Further, co-ordinate induction of α-galactosidase and galactose-1-phosphate uridylyltransferase (EC 2.7.7.10), an enzyme involved in the utilization of galactose, has been reported by Kew and Douglas (1976). Indeed, it is difficult to envisage how the induction mechanism can be triggered when melibiose is hydrolysed at the cell surface, since an inducer, galactose, and a repressor, glucose, are the products of the reaction.

Lazo *et al.* (1981) have reported that constitutive levels of α-galactosidase can be detected in cells, but this activity is very clearly distinguishable from that of the cells growing in the presence of an inducer such as melibiose. The appearance of both enzymes (constitutive and induced) is efficiently repressed by glucose and fructose. A kinetic analysis of the repression has revealed both a transient and a permanent repression similar to that described for inducible enzymes in *Escherichia coli*.

The genetic control of α-galactosidase synthesis in *Saccharomyces* sp. has not, until recently, been the subject of detailed study. One gene, the *MEL*1 gene, has been identified to code for the production of this enzyme, but the presence of other polymeric *MEL* genes cannot be precluded. Indeed, the taxonomic difference between the two brewing yeast species *Saccharomyces cerevisiae* (ale yeast) and *Saccharomyces uvarum* (*carlsbergensis*) (lager yeast) is based on the fact that strains of *Saccharomyces uvarum* (*carlsbergensis*) produce this extracellular enzyme whereas strains of *Saccharomyces cerevisiae* are not able to produce β-galactosidase (Lodder 1970).

Recently the *MEL*1 gene from *Saccharomyces uvarum* (*carlsbergensis*) has been studied from the point of view of developing a secretion system for heterologous proteins. Consequently, a systematic examination of the molecular basis of the regulation and expression of extracellular melibiase has been conducted (Sumner-Smith *et al.* 1985; Ruohola *et al.* 1986).

The *MEL*1 gene has recently been cloned, and its expression, structure and gene product have recently been characterized in some detail. The molecular weight of the melibiase polypeptide calculated from the predicted amino-acid sequence is 52 000 and 50 000 for the mature polypeptide, which agrees with the molecular weight of the *in vivo* and *in vitro* synthesized protein. A plasmid containing the melibiase promoter region and the sequence encoding for the melibiase signal peptide can be used to express a bacterial endoglucanase as an extracellular yeast protein (Skipper *et al.* 1987).

## 9.6 Lactose

The disaccharide lactose consists of glucose and galactose units. It is the sugar found in milk, and the by-product of cheese manufacture, whey, contains lactose in high concentrations. A limited number of yeast species are able to utilize lactose (e.g. *Kluyveromyces fragilis*, *Kluyveromyces lactis* and *Candida pseudotropicalis*). The lactose-uptake mechanism in these yeast species is very similar to that of maltose in *Saccharomyces cerevisiae*. The lactose is transported across the cell membrane by means of lactose permease system; once inside the cell it is hydrolysed by β-galactosidase into its constituent monosaccharides, glucose and galactose, which then enter the common glycolytic pathways of the cell.

The disposal of cheese whey is a continuing and growing problem in the dairy industry, and a major hindrance to the utilization of it is its relatively high lactose content compared with its protein content (4–5% lactose and 0.6–0.8% protein). Over 40% of the whey produced in North America is not used as either human or animal food, but is discarded (Decleire *et al.* 1985). The biological oxygen demand (BOD) of raw whey is approximately 40 000 mg $l^{-1}$, making it one of the most potent food wastes. It is estimated that 100 kg liquid whey has the polluting strength equivalent to sewage reproduced by 45 people (Jelen 1979). The production of 1 kg cheese yields 9 kg whey, and cheese-makers, due to stricter environmental controls, have been seeking alternative disposal routes for whey. In 1983 there were $2.2 \times 10^9$ kg cheese produced in the USA, which translates to approximately $1.25 \times 10^9$ kg whey solids that must be disposed of in some manner.

The direct fermentation of lactose to ethanol is one possible route for disposal of this waste material. For this purpose selected strains of *Kluyveromyces fragilis*, *Kluyveromyces lactis* or *Candida pseudotropicalis* have been employed, and there are a number of reports describing both the use of immobilized cells and continuous fermentations with recycling of cells (Prenosil *et al.* 1984, Terrell *et al.* 1984). Another widely accepted method for lactose disposal is the enzymatic hydrolysis of lactose to glucose and galactose by $\alpha$-galactosidase (usually from a fungal source) followed by a traditional ethanol fermentation employing *Saccharomyces cerevisiae* or a related species. However, *Saccharomyces cerevisiae* is unable to ferment lactose directly because it lacks the cellular mechanisms responsible for lactose uptake and hydrolysis. Sreekrishna and Dickson (1985) have described the construction of strains of *Saccharomyces cerevisiae* that grow on lactose as sole carbon source. It is probably the first example of a eukaryotic membrane-bound permease being expressed in a heterologous organism. A plasmid containing a selectable marker that confers resistance to the antibiotic G418 and carrying a 13-kb region of the *Kluyveromyces lactis* genome was used to transform a haploid strain of *Saccharomyces cerevisiae*. In *Kluyveromyces lactis* the $\beta$-galactosidase and lactose permease genes have been found to be closely linked, and by using a plasmid that permitted a large *Kluyveromyces lactis* DNA insert expression of lactose utilization in *Saccharomyces cerevisiae* could be obtained. Integration of the permease gene into the host chromosome was necessary to obtain expression, and it is speculated that vector integration may serve to activate transcription of the permease gene.

From an economic viewpoint, a *Saccharomyces cerevisiae* strain capable of lactose utilization is an organism that will be invaluable to biotechnology, and specifically to the dairy industry. However, the transformed organism described by Sreekrishna and Dickson (1985) would not be suitable for immediate industrial use, because it is a slow-growing haploid harbouring nutritional deficiencies, and, most importantly, the stability of the $LAC^+$ phenotype requires further improvement. These are areas where the technology to solve the problems is rapidly becoming available. The most important stage, integration of the genes that encode functions for lactose uptake and utilization, has been accomplished.

Cayle *et al.* (1986) have evaluated a number of options for converting whey permeate to a value-added product. Lactose sweetener, single-cell protein, baker's yeast, alcohol and methane are examples of commercially viable products currently being considered. However, it was decided to seek a more profitable alternative. Although current commercial production of vitamin C is from D-glucose via the 'Reichstein synthesis' (Reichstein & Grussner 1934), yeasts are capable of synthesizing L-ascorbic acid from galactose (Bleeg & Christensen 1982). The final step in this biological synthesis is the oxidation of L-galactono-1,4-lactone to L-ascorbic acid and, in some yeasts, this is accomplished by L-galactonolactone oxidase (EC 1.1.3.24), an enzyme which uses oxygen as the oxygen acceptor:

$$\text{L-galactono-1,4-lactone} + O_2 \xrightarrow{\text{oxidase}} \text{L-ascorbic acid} + H_2O_2$$

A screening programme has revealed that the genus *Candida* includes many species that are capable of converting the lactose to ascorbate but, in most cases, very little of the ascorbate product is exported into the medium. However, a screening programme by Cayle *et al.* (1986) has uncovered a strain of *Candida norvegensis* CBS 2145 which produced 50 mg $l^{-1}$ ascorbate, and which exported a minimum of 50% of its product into the medium. During an ensuing strain improvement programme, the product yield of ascorbate increased from 50 mg $l^{-1}$ for CBS 2145 to 1.77 g $l^{-1}$.

As a result of the above research, a process for the conversion of whey to L-ascorbic acid has been devised: (1) whey is ultrafiltered and both the protein concentrate and the permeate are recovered; (2) permeate is continuously hydrolysed by an immobilized $\beta$-galactosidase reactor; (3) the D-glucose–D-galactose hydrolysate is fed to a continuous fermenter where a galactose-negative flocculent strain of *Saccharomyces cerevisiae* strain converts the glucose to ethanol; (4) the 'beer' is continuously withdrawn and the alcohol recovered by distillation; (5) the galactose is recovered by ion exclusion chromatography; (6) the dried galactose is oxidized with conventional inorganic catalysts to D-galacturonic acid; (7) the D-galacturonic acid is reduced with conventional inorganic catalysts to L-galactonic acid; (8) the L-galactonic acid is concentrated to remove water, thereby forming L-galactono-1,4-lactone; and (9) the L-galactono-1,4-lactone is converted to L-ascorbic acid via a mutant strain of *Candida norvegensis*, grown on the ethanol produced in step 4. Further research on *Candida norvegensis*, in order to increase ascorbic acid production from galactose, is currently on-going, particularly with regards to amplification of the gene for L-galactonolactone oxidase, thereby further enhancing the value of the process.

## 9.7 Dextrin and starch

Starch is a glucose polymer of $\alpha$-1,4 and $\alpha$-1,6 glucosidic bonds consisting of a mixture of amylose and amylopectin. It is one of the most widely available plant polysaccharides, and amylases are one of the most widely distributed enzymes in nature. The utilization of starch by yeasts to produce ethanol, biomass or other value-added products conventionally requires pretreatment of the substrate in order to produce fermentable sugars. This pretreatment consists of three steps: gelatinization, liquefaction and saccharification. Gelatinization requires heat and free water, and must precede liquefaction. Liquefaction, the dispersion of starch molecules into an aqueous solution, is accomplished by the use of heat and amylolytic enzymes. Heat-stable bacterial and fungal $\alpha$-amylase or malt enzymes may be employed. During liquefaction starch molecules are only partially hydrolysed, producing a form of carbohydrate which cannot be assimilated by ethanol-producing yeasts such as *Saccharomyces cerevisiae* (Stewart *et al.* 1983).

Nevertheless, there are a number of yeasts that are able to hydrolyse partially hydrolysed starch and dextrins, due to the production of extracellular amylolytic enzymes. Early work focused on *Saccharomyces* (*Endomycopsis*) *fibuligera* (Wickerham *et al.* 1944, Sanchez-Marroquin & Solorzano 1947, Gracheva *et al.* 1977), and this yeast species has been intensively studied with regard to amylase (Kato *et al.* 1976, Ueda & Saha 1983, Steverson *et al.* 1984) and biomass production (Lemmel *et al.* 1979, 1980, Admassu *et al.* 1981, 1983, 1984, Roeckel *et al.* 1985) from starch. Consequently, a dual fermentation system with *Saccharomyces fibuligera* and *Candida utilis* for the production of single-cell protein (SCP) from starchy wastes (Symba process) was developed (Jarl 1969, Skogman 1976, Litchfield 1983). The renewed interest in the potential use of amylolytic yeasts for starch conversion during the past decade (Spencer-Martins & van Uden 1977, Augustin *et al.* 1978, Oteng-Gyang *et al.* 1980a, b, Erratt and Stewart 1981a) has led to the identification of a number of starch-degrading yeasts (Moulin & Galzy 1978a, b, 1979, Moulin *et al.* 1982, Park *et al.* 1983). *Lipomyces kononenkoae* (Spencer-Martins & van Uden 1979, Spencer-Martins 1982, 1984), *Saccharomyces diastaticus* (Tamaki 1978, Erratt & Stewart 1981b, Searle & Tubb 1981, Laluce & Mattoon 1984, Yamashita *et al.* 1984, 1985b) and *Schwanniomyces* spp. (Oteng-Gyang *et al.* 1981, Calleja *et al.* 1982, Wilson & Ingledew 1982, Dhawele & Ingledew 1983, Sills *et al.* 1984a, b) turned out to be the most promising species for the production of ethanol or SCP from starchy materials, and these yeasts are becoming well characterized.

DeMot *et al.* (1984) have characterized the extracellular amylase systems of a number of yeast species by comparing their hydrolysis patterns with soluble starch as the substrate with those of 'well-known' amylolytic yeasts, namely *Saccharomycopsis fibuligera*, *Schwanniomyces alluvius* and *Torulopsis ingeniosa*. Essentially three types of hydrolysis patterns were observed. For *Crypotococcus flavus* and *Torulopsis ingeniosa* a rapid loss of iodine-staining was found, but relatively low levels of reducing sugars were detected. This type of reaction probably reflects the activity of $\alpha$-amylase. The patterns obtained for *Candida homilentoma*, *Candida silvanorum*, *Filobasidium capsuligenum* and *Trichosporon pullulans* were similar to those of *Saccharomycopsis fibuligera* and *Schwanniomyces alluvius*. The rapid loss of iodine-staining combined with a high final level of reducing sugars indicates the presence of both glucoamylase and $\alpha$-amylase. Starch hydrolysis by the culture supernatant of the strain of *Candida tsukubaensis* studied was characterized by a rapid initial production of reducing sugars, followed by a slow further hydrolysis of some residual iodine-staining fraction. The latter pattern can be expected when high glucoamylasc activity and low $\alpha$-amylase activity are present.

Because of its taxonomic proximity to *Saccharomyces cerevisiae* and the other brewing yeast species *Saccharomyces uvarum* (*carlsbergensis*), the amylolytic yeast species *Saccharomyces diastaticus* has been extensively studied (Stewart & Russell 1986) for its potential to produce fuel ethanol and low-carbohydrate (lite) beers. *Saccharomyces diastaticus* produces an extracellular glucoamylase, also called amyloglucosidase ($\alpha$-1,4 glucan glucohydrolase, EC 3.2.1.3). This enzyme possesses the ability to cleave $\alpha$-1,4 and in some cases $\alpha$-1,6 linkages, releasing glucose from the non-reducing end of starch chains. On the other hand, $\alpha$-amylase ($\alpha$-1,4 glucan 4-glucanohydrolase, EC 3.2.1.1) is an enzyme which hydrolyses random $\alpha$-1,4 glucosidic linkages by-passing $\alpha$-1,6 bonds. Three genes have been identified that are associated with glucoamylase production of *Saccharomyces diastaticus*: *DEX*1, *DEX*2 and *STA*3 (Erratt & Stewart 1978, Tamaki 1978). Recent studies by Erratt and Nasim (1986a) have ascertained allelism between *DEX*1 and *STA*2, and *DEX*2 and *STA*1, and it was proposed that in the future the two nomenclatures be consolidated. Based on the fact that the glucoamylase from *Saccharomyces diastaticus* can hydrolyse both dextrin and starch, it was proposed to retain *STA* as the designation for genes coding for glucoamylase production by this yeast species. However, the glucoamylase produced by *Saccharomyces diastaticus* only possesses $\alpha$-1,4 activity, and consequently this strain is incapable of complete utilization of native starch.

The production of glucoamylase by strains of *Saccharomyces diastaticus* is subject to carbon catabolite repression by glucose and other sugars (Sills & Stewart 1982). Subjecting two such strains to the 2-deoxyglucose (2-DOG) selection techniques (details are to be found in Ch. 3), stable spontaneous mutants derepressed for the formation of glucoamylase and mutants derepressed for maltose utilization have been isolated (Stewart *et al.* 1985). In brewer's wort with a 2-DOG starch mutant of *Saccharomyces diastaticus*, the level of glucoamylase was increased fivefold compared with the parental strain. The overall fermentation rates of the starch and maltose mutants were increased compared with the parental strain.

During distilled ethanol production, as previously discussed, the fermentation of starch conventionally requires the pretreatment of the substrate in order to produce fermentable sugars. This pretreatment produces a substrate that is only partially hydrolysed. Consequently, during fermentation conversion to lower-molecular weight sugars is accomplished by the addition of fungal glucoamylases to the fermentation vessel at the time of inoculation. This saccharifying glucoamylase represents a significant fraction of the total cost of producing ethanol. A reduction in the amount of added glucoamylase could significantly decrease the cost of the final product.

It has been found possible to decrease glucoamylase addition to starch mash fermentations by employing yeast which actively produce and secrete glucoamylase, e.g. strains of *Saccharomyces diastaticus* (Whitney *et al.* 1985, Russell *et al.* 1986). A genetically manipulated diploid strain of *Saccharomyces diastaticus* was studied as the glucoamylase-producing strain, and was compared with a strain of *Saccharomyces cerevisiae* which was unable to produce and secrete glucoamylase. The fermentation performance of the two strains was compared in a corn mash with and without added glucoamylase. When the *Saccharomyces diastaticus* culture was employed, added glucoamylase concentrations could be significantly decreased without reducing ethanol production or sugar uptake. Reduction of the added glucoamylase concentration from 0.1% by volume, based on the substrate weight, to 0.05% resulted in no significant decrease in ethanol yield or sugar uptake; the reduction of added glucoamylase being possible because the hybrid yeast is able to produce and secrete its own glucoamylase.

A number of laboratories are committed to achieving the objective of obtaining a strain of *Saccharomyces* sp. that possesses the ability to hydrolyse starch entirely, i.e. is able to synthesize and secrete $\alpha$-amylase and glucoamylase with debranching ability (Sills *et al.* 1983a, Tubb 1986). *Saccharomyces diastaticus* produces a thermostable glucoamylase, and no traces of $\alpha$-amylase or debranching ability can be detected. However, as previously discussed, starch is a polysaccharide composed of two polymers, 20–25% in the form of amylose (linear chains of $\alpha$-1,4 linked glucose residues) and 75–80% in the form of amylopectin (a highly branched polymer occurring by $\alpha$-1,6 linkages). Consequently, debranching activity is essential for complete hydrolysis of the polysaccharide. Two *Saccharomycopsis* (*Endomycopsis*) *fibuligera* strains were found to possess $\alpha$-amylase and glucoamylase activity but not debranching activity (Sills *et al.* 1983b). However, Yamashita *et al.* (1985a), have reported that the *Saccharomycopsis fibuligera* strain studied in their laboratory (strain HUT7212) did possess a glucoamylase with $\alpha$-1,4 and $\alpha$-1,6 activity. Indeed, this glucoamylase activity has been successfully cloned into a strain of *Saccharomyces cerevisiae* by these researchers.

The yeast genus *Schwanniomyces* is known to contain species that produce significant amounts of $\alpha$-amylase, glucoamylase and debranching activity. Although the genus was first described early in this century by Kloecker (1909) after the isolation of *Schwanniomyces oxidentalis* in the Antilles, it was not until the late 1970s and early 1980s that the industrial potential of this genus began to be appreciated (Augustin *et al.* 1978, Oteng-Gyang *et al.* 1980b, Wilson & Ingledew 1982, Sills *et al.* 1983b). *Schwanniomyces* spp. are able to ferment and assimilate a large number of carbon compounds in similar fashion to *Saccharomyces* spp. *Schwanniomyces* yeasts are normally haploid, but diploidization occasionally takes place by rare mating of individual haploid cells (Ferreira & Phaff 1959). An extensive study on the ability of this genus to undergo conjugation has supported the concept of *Schwanniomyces* being a monotypic genus. With no mating type and the absence of four-spore asci for tetrad analysis, conventional genetic analysis and manipulation, which is commonplace in the genus *Saccharomyces*, is not possible with *Schwanniomyces*. Consequently, spheroplast (protoplast) fusion has been studied for genetic manipulation of *Schwanniomyces* (Wilson *et al.* 1982). Although intraspecific fusion between a strain of *Schwanniomyces alluvius* was successful, intergeneric crosses between *Saccharomyces cerevisiae* and *Schwanniomyces alluvius* produced fusion products that were incapable of using starch. It is probable that the genetic diversity between the two genera may be too great to allow stable uninucleate hybrids to occur (Wilson *et al.* 1982).

One of the techniques for producing low-carbohydrate (low-calorie) 'lite' beers is to add fungal amylases to the wort during fermentation. Since 70–75% of the dextrins in wort are of the branched type, a debranching enzyme is essential for total hydrolysis of wort dextrins to fermentable sugars. The fungal glucoamylase used extensively in the production of lite-beer possesses debranching activity, therefore it can hydrolyse the dextrins. However, normal pasteurization of the final product employing a conventional temperature–time cycle, does not completely inactivate this enzyme. This is a major impediment, as the presence of active glucoamylase in the finished product means that the beer is likely to become sweeter due to the hydrolysis of residual dextrin to glucose, changing the beer's flavour characteristics and introducing a possible microbiological hazard.

An important characteristic of the amylases from *Schwanniomyces castellii* is their sensitivity to the normal pasteurization cycle employed in brewing (Sills *et al.* 1983a). As a consequence of this property, a two-stage system has been devised for the production of low carbohydrate beer (Sills *et al.* 1983a).

Amylases (gluco- and $\alpha$-amylase) from *Schwanniomyces castellii* are produced in a highly inducing medium. Subsequently, the cells are removed, the culture filtrate is concentrated and added to wort previously inoculated with a brewing-yeast strain. The beer produced from this process is very palatable, with a reduced carbohydrate content and, after pasteurization, it contains no detectable amylase activity.

As previously discussed, an objective of many research laboratories is to develop strains of *Saccharomyces* sp. that are able to synthesize and secrete $\alpha$-amylase and glucoamylase with debranching ability. One strategy has been to clone amylase activity from other yeast species into *Saccharomyces cerevisiae*, and examples of this have already been cited, e.g. *Saccharomycopsis fibuligera* (Yamashita *et al.* 1985b) and *Saccharomyces diastaticus* (Yamashita & Fukui 1983, Meaden *et al.* 1985, Erratt & Nasim 1986b). The alternative is to clone amylase activity into *Saccharomyces cerevisiae* from non-yeast species. Innis *et al.* (1985) have cloned and detected expression of *Aspergillus awamori* glucoamylase in an industrial strain of *Saccharomyces cerevisiae*. The glucoamylase gene is integrated into the genome of the yeast, and enzyme secretion is at a level sufficient to permit growth of the cloned yeast on liquefied starch. During 800 generations of continuous growth of the cloned yeast under non-selective conditions, there was no loss of glucoamylase activity.

Ashikari *et al.* (1985) have successfully cloned the glucoamylase from *Rhizopus* sp. into an auxotrophic haploid strain of *Saccharomyces cerevisiae*. Initial research employed *Hind*III-digested chromosomal DNA isolated from *Rhizopus* sp. inserted into the *Hind*III site of pBR322. However, expression of the chromosomal glucoamylase in the *Saccharomyces cerevisiae* strain was unsuccessful, probably due to the existence of intervening sequences. Construction of a cDNA library from *Rhizopus* total poly-(A)RNA produced a 2.3-kb *Hind*III fragment covering the glucoamylase gene. This was introduced into *Hind*III sites of yeast vector pYe227, which contained 2-μm origin and a *LEU2*-selectable marker. The resulting plasmid was successfully introduced into an auxotrophic haploid *Saccharomyces cerevisiae* strain. The transformant secreted glucoamylase into the medium when grown on starch media, and was stable over a number of generations.

Thomsen (1983) and Rothstein *et al.* (1984) have reported the cloning of mouse $\alpha$-amylase and wheat $\alpha$-amylase, respectively, into strains of *Saccharomyces cerevisiae*. Thomsen (1983) has shown that when pre-$\alpha$-amylase cDNA is inserted behind the yeast ADC1 promoter, it can direct the synthesis of approximately 0.1% of total yeast cell protein as amylase. This amount is similar to that obtained with other heterologous genes placed under the control of the yeast ADC1 promoter, e.g. human leukocyte interferon D (Hitzeman *et al.* 1981). In terms of fermentable ability, parental cells could not ferment high-molecular weight starch from a variety of sources (e.g. potato, wheat and corn). In similar fashion, Rothstein *et al.* (1984) have prepared and isolated cDNA for a wheat $\alpha$-amylase gene and inserted it into a yeast expression vector. This vector system has been successfully cloned into a strain of *Saccharomyces cerevisiae*. The processed protein is secreted into the medium in an active form and within the range of 30–60% of total $\alpha$-amylase produced.

## 9.8 Cellulosic materials

The search for renewable raw materials to replace fossil fuel has directed research towards the utilization of various cellulosic materials, particularly by yeasts. Cellulosic plant materials consist of three main components: cellulose, hemicellulose and lignin. In order to develop an economically feasible process it is necessary to utilize all of these materials. Cellulose and hemicellulose can be hydrolysed either by enzymatic or acid hydrolysis to yield sugars. In the case of both hardwoods and annual plants the main part of the hemicellulose consists of pentosans yielding xylose on hydrolysis. The total hydrolysate thus contains mainly glucose, cellobiose and xylose, which must be fermented to ethanol in order to obtain economic yields of ethanol. Glucose can be fermented by conventional yeasts or by some bacteria, e.g. *Zymomonas mobilis*, but xylose and cellobiose present a more difficult problem. Recently, there has been much interest in this field, and the results are most promising although no commercial process is yet available.

### *9.8.1 Pentoses*

Until recently it was believed that yeasts do not ferment aldopentoses, although many yeasts ferment ketopentoses. Thus *Saccharomyces cerevisiae* and *Schizosaccharomyces pombe* are able to produce ethanol from xylulose but are unable to utilize xylose. However, this fact has led to a novel approach which consists of employing microbial isomerases to transform xylose to xylulose, which can then be fermented (Gong *et al.* 1981, 1983).

Recently, Cheng Chan and Chan (1986) have reported the cloning of a strain of *Schizosaccharomyces pombe* with the xylose isomerase gene from *Escherichia coli*. Although the cloned strain could grow on xylose as sole carbon source, ethanol production was slow and xylitol production very active; apparently, the limiting step is the isomerization of xylose to xylulose.

Batt *et al.* (1986) have expressed the view that the inability of *Saccharomyces cerevisiae* to utilize xylose is not due to an absence of the enzymes necessary for catalysis or to an inability to induce their synthesis in the presence of xylose. Therefore, two possibilities exist: the enzymes are compartmentalized and the necessary substrates are not accessible to the enzymes, or the co-factors (NADPH for xylose reductase and $NAD^+$ for xylitol dehydrogenase) are not present at levels sufficient for the flow of metabolites. Although the former cannot be disproven, the sequestering of xylose and xylitol from their respective catalytic enzymes is an unlikely regulatory mechanism. The latter, co-factor regulation, appears to be the more likely explanation for the following reasons (Batt *et al.* 1986). Yeasts that are capable of xylose metabolism, do so under aerobic conditions (Slininger *et al.* 1982, Jeffries 1981, Margaritis & Bajpai 1982, Wang *et al.* 1980) which would enhance the recycling of NADH to NAD via respiration. Thus, this inability to recycle NADH to NAD would impede xylitol dehydrogenase activity, leading to a build-up of xylitol. The cellular location of the xylose metabolic enzymes may also be critical with respect to their partial relationship to the NADH dehydrogenase. Practically, strategies to genetically develop *Saccharomyces cerevisiae* strains to utilize xylose for ethanol production should attempt to circumvent the xylose reductase–xylitol dehydrogenase pathway. The cloning and introduction into *Saccharomyces cerevisiae* of these genes from a pentose-metabolizing yeast (e.g. *Pachysolen tannophilus*, *Candida shehatae* or *Pichia stipitis*) would not alleviate the problems of co-factor limitation. The obvious alternative is to introduce a xylose isomerase gene from another source, for direct co-factor free conversion of xylose to xylulose.

As previously discussed, there are species of yeast that are able to convert both D-glucose and D-xylose to ethanol that have been isolated and characterized. Some of the more efficient are *Pachysolen tannophilus* (Malesyka *et al.* 1982, Slininger *et al.* 1982, Mahmourides *et al.* 1983, 1985), *Candida shehatae* (du Preez *et al.* 1984, 1986, Wayman & Parekh 1985, Wayman & Tsuyuki 1985), *Pichia stipitis* (Slapack *et al.* 1984), *Kluyveromyces marxianus* (Margaritis & Bajpai 1982), *Candida tropicalis* (Jeffries 1981, 1982), *Candida* species X7217 (Gong *et al.* 1981), *Fusarium oxysporum* (Enari & Suihko 1984). Until recently, little was known about their physiology or biochemistry in general, and in particular about the factors that control the rate and yield of ethanol from D-xylose. Identification of these factors, and elucidation of the basis underlying their effects are of importance in efforts to develop suitably efficient processes.

*Pachysolen tannophilus*, until recently, has been thought to be the yeast with greatest industrial potential for converting ethanol rapidly and in good yield from D-xylose (Slininger *et al.* 1982, Schneider *et al.* 1981, Jeffries 1982, Mahmourides *et al.* 1985). Often it is assumed that *Pachysolen tannophilus* can convert each mole of D-xylose to D-xylulose and that each mole of D-xylulose follows the mechanism hypothesized by Gong *et al.* (1981), the overall conversion can be summarized as follows:

$$\underset{\text{D-xylose}}{C_5H_{10}O_5} \rightarrow \underset{\text{ethanol}}{\tfrac{5}{3}C_2H_5OH} + \underset{\text{carbon dioxide}}{\tfrac{5}{3}CO_2}$$

The theoretical ethanol yield based on this stoichiometry would be $0.51\ g\ g^{-1}$. However, Slininger *et al.* (1982) have reported that the ethanol yield was only 60% of this. Assuming that the yeast cell is composed of 50% carbon on a dry weight basis, 11% of the xylose was consumed to form biomass. This means that approximately 30% of the xylose was being used in cell maintenance or formation, or both, of metabolic products other than ethanol and carbon dioxide.

Slininger *et al.* (1982) have found that optimum ethanol production from D-xylose by *Pachysolen tannophilus* NRRLY-2460 was obtained when the circulation temperature was maintained at 32°C, the medium pH was controlled at 2.5, the substrate concentration was not higher than $50\ g\ l^{-1}$ and the ethanol concentration did not exceed $20\ g\ l^{-1}$. They have also reported upon the ability of *Pachysolen tannophilus* to convert D-xylose to ethanol. They have found that allowing limited access of air to the medium enhances the rate of conversion although this was not an absolute requirement.

Mahmourides *et al.* (1985) have reported that a requirement for a practical bioconversion of D-xylose to ethanol by *Pachysolen tannophilus* is high yield. Obtaining such yield necessitates minimization of the channelling of sugar to processes other than ethanol accumulation. At least one such process occurs during the early stages of aerobic batch culture of *Pachysolen tannophilus* in that D-xylose is utilized but ethanol is not formed. Ethanol yield was enhanced when wild-type cells were deliberately introduced into and maintained in a state where oxygen was used relatively slowly. If mutants with a low oxygen consumption potential could be isolated enhanced ethanol yield from D-xylose could also be obtained. Lee *et al.* (1986) have demonstrated that significant increases in ethanol yield from D-xylose can be achieved in *Pachysolen tannophilus* through mutations to reduce the ability of the organism to utilize ethanol as a carbon source. Nuclear gene mutations that reduce the ability to utilize ethanol occurred at 11 different loci, and the mutations at three of these loci conferred improved yields.

Malesyka and Schneider (1982) have surveyed 15 yeast strains, which are classified as being able to ferment D-glucose and to assimilate D-xylose, for the ability to convert pentose to ethanol. Included in this number were representatives of seven genera, different species within genera and variants within species. All grew aerobically on a defined medium consisting of yeast nitrogen base plus 4% D-xylose. However, some grow more rapidly on a rich medium consisting of yeast nitrogen base, casein amino acids and 4% D-xylose. The data indicate that conversion of D-xylose to ethanol by yeasts when air has access to the medium is probably a common phenomenon, since it occurred in six of the seven genera tested. The data also indicate that D-xylose fermentability cannot only vary among species within a genus, but also between variants of a single species. Thus, there are many candidates from which to choose in addition to the much researched *Pachysolen tannophilus*. Two such species with potential for the fermentation of D-xylose to ethanol are *Candida shehatae* and *Pichia stipitis*.

Wayman and Parekh (1985) have shown that *Candida shehatae* ATCC22984 will tolerate and ferment mixed hexose–pentose sugar solutions at concentrations up to 180 g $l^{-1}$ without inhibition other than a slowing of fermentation rate at the higher concentrations, with yields of approximately 0.45 g ethanol $g^{-1}$ sugar. Ethanol concentrations under these conditions were approximately 80 g $l^{-1}$ and the yeast is able to tolerate such high concentrations. However, when the ethanol concentration reached 100 g $l^{-1}$, fermentation of pentose sugars ceased. The yeast can apparently tolerate quite high ethanol concentrations, but not for pentose fermentation. However, when the ethanol was removed by low-temperature vacuum distillation, fermentation of pentose resumed, and very good yields of ethanol resulted.

The potential of D-xylose fermentation by *Candida shehatae* and *Pichia stipitis* has been studied by du Preez *et al.* (1986). Both yeasts were capable of fermenting over a relatively wide pH range. Although *Pichia stipitis* was somewhat more temperature-tolerant than *Candida shehatae*, both strains suffer from the disadvantage that they are not as temperature-tolerant as the *Saccharomyces cerevisiae* strains commonly employed for ethanol production, and their ethanol tolerance is also much lower. Of the two strains *Pichia stipitis* was more prone to inhibition by high xylose concentrations. However, the high xylose levels which caused the performance of *Pichia stipitis* to lag behind that of *Candida shehatae* are unlikely to be encountered in actual lignocellulosic hydrolysates. Of greater importance is the fact that the ethanol yield of *Pichia stipitis* (under optimal conditions usually 85–90% of the theoretical maximum) was markedly higher than that obtained with *Candida shehatae*. Furthermore, in contrast with *Candida shehatae*, the ethanol yield of *Pichia stipitis* was essentially unaffected by the xylose concentration until it exceeded 7 g $l^{-1}$. A further advantage is that *Pichia stipitis*, but not *Candida shehatae*, has no absolute requirements for growth factors (Dellweg *et al.* 1984). In addition, the strain of *Pichia stipitis* studied by du Preez *et al.* (1986) DSIR-Y644 can ferment all the major sugars produced by the hydrolysis of plant biomass, including D-cellobiose.

The pentose process technology can also be applied to grain alcohol production. For example, the combination of pentose fermentation with corn milling may permit conversion of both starch and fibre carbohydrates to ethanol in a single yeast process (Gong *et al.* 1983). This could increase the ethanol yield up to 20% (v/v). Pentose process technology could also facilitate the conversion of existing grain ethanol plants to cellulosic food stocks.

### *9.8.2 Cellobiose*

Enzymatic saccharification of cellulose to fermentable sugars is normally achieved by the concerted action of the cellulase complex produced by such mycelial fungi as *Trichoderma reesei* (Wood 1975). At least three different enzymes are involved in this hydrolysis: β-1,4 glucan glucanohydrolase (EC 3.2.1.4, endoglucanase; Cx-enzyme) which attacks the crystalline cellulose in a random fashion, generating oligosaccharides and free ends; β-1,4 glucan cellobiohydrolase (EC 3.2.1.91, exoglucanase, C1-enzyme) which removes cellobiose units from the non-reducing ends of the cellulose chains; and β-1,4 glucosidase (EC 3.2.1.21, cellobiase) which hydrolyses cellobiose and other cellodextrins into glucose (Shewale 1981).

The cellulose complex produced by *Trichoderma reesei* has been extensively characterized and mutants with increased cellulose activity have been isolated (Montenecourt & Eveleigh 1977, Sternberg *et al.* 1977, Bailey & Nevalainen 1981, Farkas *et al.* 1981, Labudova *et al.* 1981, Mishna *et al.* 1982, Sheir-Neiss & Montenecourt 1984). However, the deficiency of β-1,4 glucosidase activity remains a significant limitation, since accumulation of cellobiose occurs resulting in the inhibition of the endo- and exo-glucanases (Sternberg *et al.* 1977, Gong & Tsao 1979, Freer & Detroy 1983). Therefore simultaneous saccharification of cellulose and fermentation of cellobiose and glucose would be most desirable for ethanol production. *Saccharomyces cerevisiae* and related species are unable to ferment or oxidize cellobiose (Kreger-van Rij 1984). The structural genes coding for *Clostridium thermocellum*, β-1,4 glucosidase from *Aspergillus niger* (Penttila *et al.* 1984) and from *Kluyveromyces fragilis* (Raynal & Guerineau 1984) have been cloned into *Saccharomyces cerevisiae*; however, both types of transformants were unable to grow in media containing cellobiose as the sole source of carbon due to either low expression of the *Aspergillus niger* genes in the yeast or to the lack of a cellobiose permease in *Saccharomyces cerevisiae*, as the *Kluyveromyces fragilis* β-glucosidase is an extracellular enzyme.

Other yeasts, including *Brettanomyces* spp., *Candida* spp. and *Dekkera* spp., have been reported to ferment cellobiose to ethanol (Gonde *et al.* 1982, Blondin *et al.* 1982, Malesyka *et al.* 1982, Kilian *et al.* 1983a, b, Freer & Detroy 1983, Sills & Stewart 1985). A *Dekkera intermedia* strain was reported to produce up to 7.5% (w/v) ethanol at a theoretical yield of 80% under anaerobic conditions (Blondin *et al.* 1982); although with moderate aeration the fermentation rate increased, the theoretical yield decreased slightly. Food ethanol yields were also obtained from 12% (w/v) cellobiose, with *Brettanomyces claussenii* and *Brettanomyces custersii* under anaerobic conditions; however, increasing the disaccharide concentration to 18% (w/v) decreased the growth rate and ethanol yield, suggesting a synergistic inhibitory action of ethanol and osmotic pressure caused by the high sugar concentration (Gonde *et al.* 1982). *Candida wickerhamii* and *Candida lusitaniae* have been reported to have fast initial rates of ethanol production from cellobiose (Freer & Detroy 1983); however, they could not sustain ethanol concentrations higher than 5–6% (w/v).

In this laboratory (Sills & Stewart 1985) attempts have been made to select appropriate yeast species able to utilize cellobiose, which could be employed as DNA donors or fusion partners for transformation and spheroplast fusion with *Saccharomyces* spp. Initially over 40 yeast species were tested for their ability to produce β-glucosidase in the presence of cellobiose. Only four strains, *Brettanomyces anomalus*, *Brettanomyces claussenii*, *Dekkera intermedia* and *Candida wickerhamii* were able to ferment cellobiose to ethanol, the last of these strains being the best producer of β-glucosidase. Of cellobiose oxidizing species studied, significant levels of β-glucosidase activity were obtained under aerobic conditions with *Hansenula* and *Schwanniomyces* spp.

Studies on the regulation of β-glucosidase production have revealed that all of the strains tested exhibited a diauxic effect when grown in a glucose–cellobiose mixture, i.e. catabolite repression by glucose was observed. Attempts to isolate derepressed mutants from *Schwanniomyces castellii* and *Candida wickerhamii* strains were unsuccessful because, although initially mutants were isolated with 2-deoxyglucose selection, they reverted to their initial wild-type characteristics. The β-glucosidase system of *Saccharomyces cerevisiae* has been isolated, purified and characterized. The protein responsible for β-glucosidase activity was determined to be a 90 000-da glycoprotein with an approximate ratio of protein : carbohydrate of 4 : 1. The optimal temperature activity for this enzyme was determined to be about 50°C and the optimal pH 4.5.

### *9.8.3 Cellulose*

Although, as previously described, certain filamentous fungi can secrete large quantities of different cellulolytic enzymes into the culture medium and this mixture efficiently hydrolyses crystalline cellulose to glucose (Montenecourt 1983), no strain of yeast has been isolated with glucanase activity. Consequently, a number of laboratories have attempted to clone cellulase genes from filamentous fungi and bacteria into *Saccharomyces* spp. (Shoemaker *et al.* 1983, Teeri *et al.* 1983, Cantwell *et al.* 1985, Hinchliffe & Box 1985, Knowles *et al.* 1985).

Knowles *et al.* (1985) have conducted a detailed study on a number of genes coding for cellulolytic enzymes from filamentous fungi and the possibility of cloning these genes into *Saccharomyces* spp. Initially a gene bank for *Aspergillus niger* in a yeast cosmid vector was constructed and screened for the expression of five different cellulolytic genes in *Saccharomyces cerevisiae*. Only one of the genes, *Clostridium thermocellum* β-glucosidase, was expressed at a low level. These experiments illustrated that cloning of fungal genes by expression in yeast was not an efficient method for the isolation of specific fungal genes. For this reason differential hybridization of radioactive cDNA was employed to identify genes from *Trichoderma reesei* strongly expressed during cellulase synthesis. All four of the genes coding for the major cellulases of *Trichoderma reesei* were isolated from phage gene banks by differential hybridization using two populations of cDNA, made from either induced (cellulase-positive) or repressed (cellulase-negative) mRNA. These gene banks have been used to clone the cellulase genes successfully into a *Saccharomyces cerevisiae* strain and to obtain enzyme expression. The enzymes so produced are capable of degrading β-glucans. The amounts produced are, of course, relatively small compared with those produced by the original donor organism being of the order of 1% or less total cell protein, but for a number of commercial applications this may prove sufficient.

As well as employing mycelial fungi as potential sources of β-glucanase activity, a number of bacterial strains have been investigated. The first bacterial gene coding for an endo β-glucanase (endocellulase) was isolated by 'shotgun' cloning of *Cellulomonas firni* chromosomal DNA into *Escherichia coli* (Whittle *et al.* 1982). Later studies have involved the cloning of endocellulase genes from *Clostridium thermocellum* (Cornet *et al.* 1983) and *Thermomonospora YX* (Collmer & Wilson 1983). However, because of its GRAS status, most of the research on bacterial β-glucanase systems and their cloning into yeast has been conducted with strains of *Bacillus subtilis* (Bernier *et al.* 1983, Cantwell *et al.* 1985, Hinchliffe 1985). This Gram-positive bacterium produces an extracellular enzyme capable of hydrolysing alternating sequences of β-1,3 and β-1,4 linked glucan. The enzyme has been successfully introduced into strains of *Saccharomyces cerevisiae*, but only low levels of the biologically active enzyme have been detected to date, but with every new publication on this subject an increase in the efficiency of β-glucanase enzyme production is reported (Cantwell & McConnell 1987).

Cellulolytic yeasts will have a number of applications, but before constructing cellulolytic yeasts for a specific application, it is important to define the substrate in terms of the particular enzymes required. For example, cellulolytic yeasts could have application in the brewing industry. Roughly 10% of the total carbohydrate of barley grain is β-glucan. Much of this is hydrolysed during the malting and mashing processes by barley enzymes. Excess β-glucan in beer is undesirable because it can have a negative effect on beer filtration and may cause problems with beer clarity. The amount of β-glucan remaining in finished beer depends on factors such as the climatic and arable conditions under which the barley is grown and the precise malting and mashing procedures. In addition, the use of unmalted barley or wheat can significantly elevate the $\beta$-glucan content of the wort. For those reasons some breweries routinely add fungal or bacterial $\beta$-glucanases to the mash or the wort. A brewing yeast strain that possesses the ability to secrete $\beta$-glucanase could well find application in order to reduce beer $\beta$-glucan content. The other obvious application of cellulolytic yeast strains has already been discussed in this chapter, namely the simultaneous saccharification of cellulose and fermentation of cellobiose and glucose (SSF) to ethanol. However, for this application $\beta$-glucosidase as well as glucanase activity manipulated into the yeast would be required.

## 9.9 Conclusions

As a group of micro-organisms the yeasts are able to utilize most of the known sugars and carbohydrates. However, a single yeast species that is able to metabolize all of the sugars and carbohydrates that possess commercial potential has yet to be isolated or manipulated. The ideal yeast should possess the hexose-metabolizing abilities of *Saccharomyces cerevisiae* and its related species, the starch- and dextrin-utilizing ability of *Schwanniomyces castellii*, the pentose-utilizing ability of either *Pichia stipitis* or *Candida shehatae*, and the cellulolytic abilities of the mycelial fungus *Trichoderma reesei*.

## Acknowledgements

The authors wish to thank all of their colleagues from the Labatt Brewing Co. Ltd who have assisted in the writing and editing of this chapter. Part of the research from this laboratory that is discussed was funded by a PILP contract from the National Research Council of Canada and an ERDAF contract from Agriculture Canada, and these are gratefully acknowledged.

## References

Admassu, W., R. A. Korus, R. C. Heimsch and S. A. Lemmel 1981. Growth of *Saccharomycopsis fibuligera* in a continuous stirred tank fermenter. *Biotechnology and Bioengineering* **23**, 2361–6.

Admassu, W., R. A. Korus and R. C. Heimsch 1983. Two-stage continuous fermentation of *Saccharomycopsis fibuligera* and *Candida utilis*. *Biotechnology and Bioengineering* **25**, 2641–4.

Admassu, W., R. A. Korus and R. C. Heimsch 1984. Continuous fermentation of *Saccharomycopsis fibuligera* and *Candida utilis*. *Biotechnology and Bioengineering* **26**, 1511–16.

Andrews, J. and R. B. Gilliland 1952. A study of three organisms capable of causing abnormal attenuations. *Journal of the Institute of Brewing* **58**, 189–96.

Ashikari, T., N. Nakamura, Y. Tanaka, N. Kiucha, Y. Shibano, T. Tanaka, T. Amachi and H. Yoshizumi 1985. Cloning and expression of the *Rhizopus* glucoamylase gene in yeast. *Agricultural Biology and Chemistry* **49**, 2521–3.

Augustin, J., J. Zemek, A. Kockova-Kratochvilova and L. Kuniak 1978. Production of $\alpha$-amylase by yeasts and yeast-like organisms. *Folia Microbiologica* **23**, 353–61.

Bailey, M. J. and K. M. H. Nevalainen 1981. Induction, isolation and testing of stable *Trichoderma reesei* mutants with improved production of solubilizing cellulase. *Enzyme and Microbial Technology* **3**(2), 153–7.

Barnett, J. A. 1976. The utilization of sugars by yeasts. *Advances in Carbohydrate Chemistry and Biochemistry* **32**, 125–234.

Barnett, J. A. 1981. The utilization of disaccharides and some other sugars by yeast. *Advances in Carbohydrate Chemistry and Biochemistry* **39**, 347–401.

Bassel, J. and H. C. Douglas 1968. Osmotic remedial response in a galactose-negative mutant of *Saccharomyces cerevisiae*. *Journal of Bacteriology* **95**, 1103–10.

Batt, C. A., S. Carvallo, D. D. Easson, Jr, M. Akedo and A. J. Sinskey 1986. Direct evidence for a xylose metabolic pathway in *Saccharomyces cerevisiae*. *Biotechnology and Bioengineering* **28**, 549–53.

Bernier, R. Jr, H. Driguez and M. Desrochers 1983. Molecular cloning of a *Bacillus subtilis* xylanase gene in *Escherichia coli*. *Gene* **26**, 59–65.

Bitter, G. A. 1983. Transcription initiation in eukaryotes: analysis of heterologous *in vitro* systems utilizing components from mammalian and yeast cells. *Molecular and General Genetics* **191**, 434–41.

Bleeg, H. and F. Christensen 1982. Biosynthesis of ascorbate in yeast. *European Journal of Biochemistry* **127**, 391–6.

Blondin, R., R. Ratomahenina, A. Arnaud and P. Galzy 1982. A study of cellobiose fermentation by a *Dekkera* strain. *Biotechnology and Bioengineering* **24**, 2031–7.

Broach, J. R. 1979. Galactose regulation in *Saccharomyces cerevisiae*. *Journal of Molecular Biology* **131**, 41–53.

Calleja, G. B., S. Levy-Rick, C. V. Lusena, A. Nasim and F. Moranelli 1982. Direct and quantitative conversion of starch to ethanol by the yeast *Schwanniomyces alluvius*. *Biotechnology Letters* **4**, 543–8.

Cantwell, B. A. and D. J. McConnell 1987. Transformation of brewing yeast strains. In *Biological research on industrial yeasts*, G. G. Stewart, I. Russell, R. D. Klein and R. R. Hiebsch (eds). Boca Raton, Florida: CRC Press (in press).

Cantwell, B., G. Brazil, J. Hurley and D. McConnell 1985. Expression of the cloned endo-1,3-1,4-$\beta$-glucanase gene of *Bacillus subtilis* in *Saccharomyces cerevisiae*. In *European Brewery Convention Proceedings of 20th Congress, London*, 259–66. Oxford: IRL Press.

Cardini, C. E. and L. F. Leloir 1953. Enzymic phosphorylation of galactosamine and galactose. *Archives of Biochemistry and Biophysics* **45**, 55–64.

Carlson, M. and D. Botstein 1982. Two differentially regulated mRNA's with different 5′ ends encode secreted and intracellular forms of yeast invertase. *Cell* **28**, 145–53.

Carlson, M., B. C. Osmond and D. Botstein 1981. Mutants of yeast defective in sucrose utilization. *Genetics* **98**, 25–40.

Carlson, M., R. Taussig, S. Kustu and D. Botstein 1983. The secreted form of invertase in *Saccharomyces cerevisiae* is synthesized from mRNA encoding a surgical sequence. *Molecular and Cellular Biology* **3**, 439–43.

Cayle, T., J. Roland, D. Mehnert, R. Dinwoodie, R. Larson, J. Mathers, M. Raines, W. Alm, S. Ma'ayeh, S. Kiang and R. Saunders 1986. Production of L-ascorbic acid from whey. In *Biotechnology in food processing*, S. K. Harlander and T. P. Lubuza (eds), 157–69. New Jersey: Noyes Publications.

Chang, Y. Y. and J. E. Cronan, Jr 1983. Genetic and biochemical analyses of *Escherichia coli* strains having a mutation in the structural gene (poxB) for pyruvate oxidase. *Journal of Bacteriology* **154**, 756–62.

Cheng Chan, E. and L. Chan 1986. D-xylose fermentations to ethanol by *Schizosaccharomyces pombe* cloned with xylose isomerase gene. *Biotechnology Letters* **8**, 231–4.

Cohen, J. D., M. J. Goldenthal, B. Buchferer and J. Marmur 1984. Mutational analysis of the *MAL1* locus of *Saccharomyces*: identification and functional characterization of three genes. *Molecular and General Genetics* **196**, 208–16.

Cohen, J. D., M. J. Goldenthal, J. Chow, B. Buchferer and J. Marmur 1985. Organization of the *MAL* loci of *Saccharomyces*. Physical identification and functional characterization of the genes of the *MAL6* locus. *Molecular and General Genetics* **200**, 1–8.

Collmer, A. and D. B. Wilson 1983. Cloning and expression of a *Thermomonospora YX* endocellulase gene in *E. coli*. *Bio/Technology* **1**, 594–601.

Cornet, P., D. Tonik, J. Mallet and J. P. Aubert 1983. Characterization of two cell (cellulose degradation) genes of *Clostridium thermocellum* coding for endoglucanases. *Bio/Technology* **1**, 589–94.

Decleire, M., N. van Huyhn and J. C. Motte 1985. Hydrolysis of lactose solutions and wheys by whole cells of *Kluyveromyces bulgaricus*. *Applied Microbiology and Biotechnology* **21**, 103–7.

Dellweg, H., M. Rizzi, H. Methner and D. Debas 1984. Xylose fermentation by yeasts. 3. Comparison of *Pachysolen tannophilus* and *Pichia stipitis*. *Biotechnology Letters* **6**, 395–400.

DeMot, R., R. Andries and H. Verachtert 1984. Comparative study of starch degradation and amylase production by ascomycetous yeast species. *Systematic and Applied Microbiology* **5**, 106–11.

Dhawele, M. R. and W. M. Ingledew 1983. Starch hydrolysis by derepressed mutants of *Schwanniomyces castelli*. *Biotechnology Letters* **5**, 185–90.

Douglas, H. C. 1961. A mutation on *Saccharomyces* that affects phosphoglucomutase activity and galactose utilization. *Biochimica et Biophysica Acta* **52**, 209–11.

Douglas, H. C. and D. C. Hawthorne 1966. Regulation of genes controlling synthesis of the galactose pathway enzymes in yeast. *Genetics* **54**, 911–16.

Drawert, F. and H. Barots 1978. Production of monoterpenes by the yeast *Kluyveromyces lactis*. *Journal of Agricultural and Food Chemistry* **26**, 765–9.

du Preez, J. C., B. A. Prior and A. M. T. Monteiro 1984. The effect of aeration on xylose fermentation by *Candida shehatae* and *Pachysolen tannophilus*: A comparative study. *Applied Microbiology and Biotechnology* **19**, 261–6.

du Preez, J. C., M. Bosch and B. A. Prior 1986. Xylose fermentation by *Candida shehatae* and *Pichia stipitis*: effects of pH, temperature and substrate concentration. *Enzyme and Microbial Technology* **8**, 360–4.

Emr, S. D., I. Schauer, W. Hansen, P. Esmon and P. Schekman 1984. Invertase $\beta$-galactosidase hybrid proteins fail to be transported from the endoplasmic reticulum in *Saccharomyces cerevisiae*. *Molecular and Cellular Biology* **4**, 2347–51.

Enari, T.-M. and M.-L. Suihko 1984. Ethanol production by fermentation of pentoses and hexoses from cellulosic materials. *CRC Critical Reviews in Biotechnology* **1**, 229–40.

Erratt, J. A. and A. Nasim 1986a. Allelism within the *DEX* and *STA* gene families in *Saccharomyces diastaticus*. *Molecular and General Genetics* **202**, 255–6.

Erratt, J. A. and A. Nasim 1986b. Cloning and expression of a *Saccharomyces diastaticus* glucoamylase gene in *Saccharomyces cerevisiae* and *Schizosaccharomyces pombe*. *Journal of Bacteriology* **166**, 484–90.

Erratt, J. A. and G. G. Stewart 1978. Genetic and biochemical studies on yeast strains able to utilize dextrins. *Journal of the American Society of Brewing Chemists* **36**, 151–61.

Erratt, J. A. and G. G. Stewart 1981a. Fermentation studies using *Saccharomyces diastaticus* yeast strains. *Developments in Industrial Microbiology* **22**, 577–86.

Erratt, J. A. and G. G. Stewart 1981b. Genetic and biochemical studies on glucoamylase from *Saccharomyces diastaticus*. In *Current developments in yeast research*, G. G. Stewart and I. Russell (eds), 177–84. Toronto: Pergamon Press.

Eveleigh, D. E. 1981. The microbiological production of industrial chemicals. *Scientific American* **245**(9), 155–62.

Farkas, V., I. Labudova, S. Bauer and L. Ferenczy 1981. Preparation of mutants of *Trichoderma viride* with increased production of cellulase. *Folia Microbiologica* **26**, 129–32.

Federoff, H. J., J. D. Cohen, T. R. Eccleshall, R. B. Needleman, B. A. Buchferer, J. Giacalone and J. Marmur 1982. Isolation of a maltase structural gene from *Saccharomyces carlsbergensis*. *Journal of Bacteriology* **149**, 1064–72.

Ferreira, J. O. and H. J. Phaff 1959. Life cycle and nuclear behaviour of a species of the yeast genus *Schwanniomyces*. *Journal of Bacteriology* **78**, 352–9.

Freer, S. N. and R. W. Detroy 1983. Characterization of cellobiose fermentations to ethanol by yeasts. *Biotechnology and Bioengineering* **25**, 541–57.

Friis, J. and P. Ottohenghi 1959. Localization of melibiase in a strain of yeast. *Comptes Rendus des Travaux du Laboratoire Carlsberg* **31**, 272–83.

Gonde, P., B. Blondin, R. Ratomahenina, A. Arnaud and P. Galzy 1982. Selection of yeast strains for cellobiose alcoholic fermentation. *Journal of Fermentation Technology* **60**, 579–84.

Gong, Ch.-S. and G. T. Tsao 1979. Cellulase biosynthesis and regulation. In *Advances in biochemical engineering*, Vol. 6, A. Fiechter (ed.), 39–72. Berlin: Springer-Verlag.

Gong, Ch.-S., L. F. Chen, G. T. Tsao and T. C. Flickinger 1981. Conversion of hemicellulose carbohydrates. In *Advances in biochemical engineering*, Vol. 20, A. Fiechter (ed.), 93–118. Berlin: Springer-Verlag.

Gong, Ch.-S., T. A. Claypool, L. D. McCraken, C. M. Maun, P. D. Wang and G. T. Tsao 1983. Conversion of pentoses by yeasts. *Biotechnology and Bioengineering* **25**, 85–102.

Gracheva, I. M., T. A. Lushchik, Y. A. Trypsin and E. E. Pinchukova 1977. Purification and properties of glucoamylase from *Endomycopsis* sp. *Biokhimiya* **42**, 1258–64.

Hartwell, L. H. 1970. Biochemical genetics of yeast. *Annual Review of Genetics* **4**, 373–96.

Hautera, P. and T. Lovgren 1975. $\alpha$-Glucosidase, $\alpha$-glucosidase permease, maltose fermentation and leavening ability of baker's yeast. *Journal of the Institute of Brewing* **81**, 304–13.

Hinchliffe, E. 1985. $\beta$-Glucanase: the successful application of genetic engineering. *Journal of the Institute of Brewing* **91**, 384–9.

Hinchliffe, E. and W. Box 1984. Expression of the cloned endo-1,3-1,4-$\beta$-glucanase gene of *Bacillus subtilis* in *Saccharomyces cerevisiae*. *Current Genetics* **8**, 471–6.

Hitzeman, R. A., F. E. Hagie, H. L. Levine, D. V. Goeddel, G. Ammerer and B. D. Hall 1981. Expression of a human gene for interferon in yeast. *Nature* (London) **250**, 717–21.

Hitzeman, R. A., D. W. Leung, L. J. Perry, W. J. Kohr, H. L. Levine and D. V. Goeddel 1983. Secretion of human interferons by yeast. *Science* **219**, 620–3.

Hopper, J. E., J. R. Broach and L. B. Rowe 1978. Regulation of the galactose pathway in *Saccharomyces cerevisiae*: induction of uridyl transferase mRNA and dependency on *GAL4* gene function. *Proceedings of the National Academy of Sciences of the USA* **75**, 2878–82.

Innis, M. A., M. J. Holland, P. C. McCabe, G. E. Cole, V. P. Wittman, R. Tal, K. W. K. Watt, D. H. Gelfand, J. P. Holland and J. H. Meade 1985. Expression, glycosylation, and secretion of an *Aspergillus* glucoamylase by *Saccharomyces cerevisiae*. *Science* **228**, 21–5.

Jarl, K. 1969. Production of microbial food from low-cost starch materials and purification of industry's waste starch effluents through the Symba yeast process. *Food Technology* **23**, 1009–13.

Jeffries, T. W. 1981. Conversion of xylose to ethanol under aerobic conditions by *Candida tropicalis*. *Biotechnology Letters* **3**, 213–18.

Jeffries, T. W. 1982. A comparison of *Candida tropicalis* and *Pachysolen tannophilus* for conversion of xylose to ethanol. *Biotechnology and Bioengineering Symposium* No. 12, 103–10.

Jelen, P. 1975. Industrial whey processing technology: an overview. *Journal of Agricultural and Food Chemistry* **27**, 658–61.

Kato, K., K. Kuswanto, I. Banno and T. Harada 1976. Identification of *Endomycopsis fibuligera* isolated from ragi in Indonesia and properties of its crystalline glucoamylase. *Journal of Fermentation Technology* **54**, 831–7.

Kew, O. M. and H. C. Douglas 1976. Genetic co-regulation of galactose and melibiose utilization in *Saccharomyces*. *Journal of Bacteriology* **125**, 33–9.

Khan, N. A., F. K. Zimmerman and N. R. Eaton 1973. Genetic control of maltose formation in yeast II. Evidence for a gene regulating the level of maltose production. *Molecular and General Genetics* **124**, 365–9.

Kilian, S. G., B. A. Prior, H. J. Potgieter and J. C. du Preez 1983a. Utilization of glucose and cellobiose by *Candida wickerhamii*. *European Journal of Applied Microbiology and Biotechnology* **17**, 281–6.

Kilian, S. G., B. A. Prior, I. S. Pretorius, J. C. du Preez, J. J. Venter and H. J. Potgieter 1983b. Nutritional, temperature, pH, and oxygen requirements of *Candida wickerhamii*. *European Journal of Applied Microbiology and Biotechnology* **17**, 334–8.

Kloecker, A. 1909. Deux nouveaux genes de la famille des *Saccharomyces*. *Comptes Rendus des Travaux du Laboratoire Carlsberg* **7**, 273–89.

Knowles, J. K. C., M. Pentilla, T. T. Teeri, L. Andre, I. Salovuori and P. Lehtovaara 1985. Transfer of genes coding for fungal glucanases into yeast. In *European Brewery Convention Proceedings of 20th Congress, London*, 251–8. Oxford: IRL Press.

Kreger-van Rij, N. J. W. 1984. *The yeasts: a taxonomic study*. Amsterdam: Elsevier Science.

Kuo, S. C. and J. O. Lampen 1974. Tunicamycin – an inhibitor of yeast glycoprotein synthesis. *Biochemical and Biophysical Research Communications* **58**, 287–95.

Labudova, I., V. Farkas, S. Bauer, N. Kolarova and A. Brenyik 1981. Characterization of cellulolytic enzyme complexes obtained from mutants of *Trichoderma reesei* with enhanced cellulase production. *European Journal of Applied Microbiology and Biotechnology* **12**, 16–21.

Laluce, C. and J. R. Mattoon 1984. Development of rapidly fermenting strains of *Saccharomyces diastaticus* for direct conversion of starch and dextrins to ethanol. *Applied Environmental Microbiology* **48**, 17–19.

Lazo, P. S., A. G. Ochoa and S. Gascon 1977. $\alpha$-Galactosidase from *Saccharomyces carlsbergensis*: cellular localization, and purification of the external enzyme. *European Journal of Biochemistry* **77**, 375–82.

Lazo, P. S., A. G. Ochoa and S. Gascon 1978. α-Galactosidase (melibiase) from *Saccharomyces carlsbergensis*: structural and genetic properties. *Archives of Biochemistry and Biophysics* **191**, 316–21.

Lazo, P. S., I. G. Florez, A. G. Ochoa and S. Gascon 1981. Induction and catabolite repression of α-galactosidase from *Saccharomyces carlsbergensis*. *Cellular and Molecular Biology* **27**, 615–22.

Lee, H., A. P. James, D. M. Zahab, G. Mahmourides, R. Malesyka and H. Schneider 1986. Mutants of *Pachysolen tannophilus* with improved production of ethanol from D-xylose. *Applied Environmental Microbiology* **51**, 1252–8.

Lemmel, S. A., R. C. Heimsch and L. L. Edwards 1979. Optimizing the continuous production of *Candida utilis* and *Saccharomycopsis fibuligera* on potato processing waste-water. *Applied Environmental Microbiology* **37**(2), 227–32.

Lemmel, S. A., R. C. Heimsch and R. A. Korus 1980. Kinetics of growth and amylase production of *Saccharomycopsis fibuligera* on potato processing waste water. *Applied Environmental Microbiology* **39**(2), 387–93.

Litchfield, J. H. 1983. Single-cell proteins. *Science* **219**, 740–2.

Lodder, J. 1970. *The yeasts, a taxonomic study*, 2nd edn. Amsterdam: North–Holland.

Mahmourides, G., R. Malesyka, A. P. James and H. Schneider 1983. Mutants of *Pachysolen tannophilus* that produce enhanced amounts of acetic acid from D-xylose and other sugars. *Biotechnology Letters* **5**, 29–31.

Mahmourides, G., H. Lee, N. Maki and H. Schneider 1985. Ethanol accumulation in cultures of *Pachysolen tannophilus* on D-xylose is associated with a transition to a state of low oxygen consumption. *Bio/Technology* **3**, 59–62.

Malesyka, R. and H. Schneider 1982. Fermentation of D-xylose, xylitol and D-xylulose by yeasts. *Canadian Journal of Microbiology* **28**, 360–3.

Malesyka, R., P. Y. Wang and H. Schneider 1982. Ethanol production from D-galactose and glycerol by *Pachysolen tannophilus*. *Enzyme and Microbial Technology* **4**, 349–52.

Margaritis, A. and P. Bajpai 1982. Direct fermentation of D-xylose to ethanol by *Kluyveromyces marxianus* strains. *Applied Environmental Microbiology* **44**, 1039–41.

Matsumoto, K., T. Adachi, A. Toh-e and T. Oshima 1980. Function of positive regulatory gene *GAL4* in the synthesis of galactose pathway enzymes in *Saccharomyces cerevisiae*. *Journal of Bacteriology* **141**, 508–27.

Meaden, P., K. Ogden, H. Bussey and R. S. Tubb 1985. A *DEX* gene conferring production of extracellular amyloglucosidase on yeast. *Gene* **34**, 325–34.

Mishna, S., K. S. Gopalkrishnan and T. K. Ghose 1982. A constitutive cellulase-producing mutant of *Trichoderma reesei*. *Biotechnology and Bioengineering* **24**, 251–4.

Mizunaga, T., J. S. Tkacz, L. Rodriguez, R. A. Hackel and J. O. Lampen 1981. Temperature-sensitive forms of large and small invertase in a mutant derived from a *SUC1* strain of *Saccharomyces cerevisiae*. *Molecular and Cellular Biology* **1**, 460–8.

Montenecourt, B. S. 1983. *Trichoderma reesei* cellulases. *Trends in Biotechnology* **1**, 156–61.

Montenecourt, B. S. and D. E. Eveleigh 1977. Preparation of mutants of *Trichoderma reesei* with enhanced cellulase production. *Applied Environmental Microbiology* **34**, 171–82.

Mormeneo, S. and R. Sentandreu 1982. Regulation of invertase synthesis by glucose in *Saccharomyces cerevisiae*. *Journal of Bacteriology* **152**, 14–18.

Mormeneo, S. and R. Sentandreu 1986. Molecular events associated with glucose

repression of invertase in *Saccharomyces cerevisiae*. *Antonie van Leeuwenhoek* **52**, 15–24.
Mortimer, R. K. and D. Schild 1985. Genetic map of *Saccharomyces cerevisiae*, edition 9. *Microbiological Reviews* **49**, 181–206.
Moulin, G. and P. Galzy 1978a. Amylase activity of *Torulopsis ingeniosa* Di Menna. *Folia Microbiologica* **23**, 423–8.
Moulin, G. and P. Galzy 1978b. Remarks on the regulation of $\alpha$-amylase biosynthesis by *Pichia burtonii*. *Zeitschrift für Allgemeine Mikrobiologie* **18**, 329–33.
Moulin, G. and P. Galzy 1979. Study on an amylase and its regulation in *Lipomyces starkeyi*. *Agricultural Biology and Chemistry* **43**, 1165–9.
Moulin, G., H. Boze and P. Galzy 1982. Amylase activity in *Pichia polymorpha*. *Folia Microbiologica* **27**, 377–82.
Mowshowitz, D. B. 1979. Gene dosage effects on the synthesis of maltase in yeast. *Journal of Bacteriology* **137**, 1200–7.
Mowshowitz, D. B. 1981. The effects of three different *MAL* loci on the regulation of maltase synthesis in yeast. *Genetics* **98**, 713–28.

Naumov, G. I. 1971. Comparative genetics of yeasts. V. Complementation in maltose MA1 locus in hybrids of wild maltose-negative species of *Saccharomyces*. *Genetika* **7**, 141–8.
Naumov, G. I. 1976. Comparative genetics of yeast. XVI. Maltose fermentation in *Saccharomyces carlsbergensis* NCYC 74. *Genetika* **11**, 87–100.
Nogi, Y., K. Matsumoto, A. Toh-e and Y. Oshima 1977. Interaction of super-repressible and dominant constitutive mutations for the synthesis of galactose pathway enzymes in *Saccharomyces cerevisiae*. *Molecular and General Genetics* **152**, 137–49.

Oteng-Gyang, K., G. Moulin and P. Galzy 1980a. Influence of amylase excretion on biomass production by amylolytic yeasts. *Acta Microbiologica Academy of Science Hungary* **27**, 155–61.
Oteng-Gyang, K., G. Moulin and P. Galzy 1980b. Effect of medium composition on excretion and biosynthesis of the amylases of *Schwanniomyces castellii*. *European Journal of Applied Microbiology* **9**, 129–36.
Oteng-Gyang, K., G. Moulin and P. Galzy 1981. A study of the amylolytic system of *Schwanniomyces castellii*. *Zeitschrift für Allgemeine Mikrobiologie* **21**, 537–42.

Park, W. S., Y. J. Koo, D. H. Shin and B. Y. Min 1983. Study on the pattern of starch assimilation of *Sporobolomyces holsaticus*. *Korean Journal of Food Science and Technology* **15**, 177–9.
Penttila, M. E., K. M. H. Nevalaimen, A. Raynal and J. K. C. Knowles 1984. Cloning of *Aspergillus niger* genes in yeast. Expression of the gene coding *Aspergillus* $\beta$-glucosidase. *Molecular and General Genetics* **194**, 494–9.
Perlman, D. and J. E. Hopper 1979. Constitutive synthesis of the *GAL4* protein, a galactose pathway regulator in *Saccharomyces cerevisiae*. *Cell* **16**, 89–95.
Prenosil, J. E., E. Stuker, T. Hediger and J. R. Bourne 1984. Enzymatic whey hydrolysis in the pilot plant. *Bio/Technology* **2**, 441–3.

Ratledge, C. and P. A. Botham 1977. Pathways of glucose metabolism in *Candida* 107, a lipid-accumulating yeast. *Journal of General Microbiology* **102**, 391–5.
Raynal, A. and M. Guerineau 1984. Cloning and expression of the structural gene for $\beta$-glucosidase of *Kluyveromyces fragilis* in *Escherichia coli* and *Saccharomyces cerevisiae*. *Molecular and General Genetics* **195**, 108–15.
Reichstein, T. and A. Grussner 1934. A high-yield synthesis of L-ascorbic acid. *Helvetica Chimica Acta* **17**, 311–28.

Rodriguez, L., J. O. Lampen and V. L. MacKay 1981. *SUC1* gene of *Saccharomyces*: a structural gene for the large (glycoprotein) and small (carbohydrate-free) forms of invertase. *Molecular and Cellular Biology* **1**, 469–74.

Roeckel, M., R. Kummerlin and A. Gordon 1985. Growth limitation and cell production by *Saccharomycopsis fibuligera* on a starch medium. *Biotechnology Letters* **7**, 59–64.

Rothstein, S. J., C. M. Lazarus, W. E. Smith, D. C. Baulcombe and A. A. Gatenby 1984. Secretion of a wheat $\alpha$-amylase expressed in yeast. *Nature* (London) **308**, 662–6.

Rudert, F. and H. O. Halvorson 1962. The effect of gene dosage on the level of $\alpha$-glucosidase in yeast. *Bulletins of the Research Council of Israel II* **A4**, 337–44.

Ruohola, H., P. L. Liljestrom, T. Torkkeli, H. Kopu, P. Lehtinen, N. Kalkkinen and M. Korhola 1986. Expression and regulation of the yeast *MEL1* gene. *FEMS Letters* **34**, 179–85.

Russell, I., C. M. Crumplen, R. M. Jones and G. G. Stewart 1986. Efficiency of genetically engineered yeast in the production of ethanol from dextrinized cassava starch. *Biotechnology Letters* **8**, 169–74.

Sanchez-Marroquin, A. and M. Solorzano 1947. Estudios sobre la amilasa de *Endomycopsis fibuligera*. II. Poder amilolitico y fermentative en cultivos mixtos. *Anal. Esc. Nac. Cienc. Biol.* **4**, 311–16.

Schneider, H., P. Y. Wang, Y. K. Chang and R. Malesyka 1981. Conversion of D-xylose into ethanol by the yeast *Pachysolen tannophilus*. *Biotechnology Letters* **3**, 89–92.

Searle, B. A. and R. S. Tubb 1981. Regulation of amyloglucosidase production by *Saccharomyces diastaticus*. *Journal of the Institute of Brewing* **87**, 244–51.

Serrano, R. 1977. Energy requirements for maltose transport in yeast. *European Journal of Biochemistry* **80**, 97–102.

Sheir-Neiss, G. and B. S. Montenecourt 1984. Characterization of the secreted cellulases of *Trichoderma reesei* wild type and mutants during controlled fermentations. *Applied Microbiology and Biotechnology* **20**, 46–53.

Shewale, J. G. 1981. $\beta$-Glucosidase: its role in cellulase synthesis and hydrolysis of cellulose. *International Journal of Biochemistry* **14**, 435–43.

Shoemaker, S., V. Schweickart, M. Ladner, D. Gelfand, S. Kwok, K. Myambo and M. Innis 1983. Molecular cloning of exo-cellobiohydrolase I derived from *Trichoderma reesei* strain L27. *Bio/Technology* **1**, 691–6.

Sills, A. M. and G. G. Stewart 1982. Production of amylolytic enzymes by several yeast species. *Journal of the Institute of Brewing* **88**, 313–16.

Sills, A. M. and G. G. Stewart 1985. Studies on cellobiose metabolism by yeast. *Developments in Industrial Microbiology* **40**, 527–34.

Sills, A. M., I. Russell and G. G. Stewart 1983a. The production and use of yeast amylases in the brewing of low carbohydrate beer. In *European Brewery Convention Proceedings of the 19th Congress, London*, 377–84. Oxford: IRL Press.

Sills, A. M., C. J. Panchal, I. Russell and G. G. Stewart 1983b. Genetic manipulation of amylolytic enzyme production by yeasts. In *Gene expression in yeast*, M. Korhola and G. Varsänen (eds), 209–28. Helsinki: Foundation for Biotechnical and Industrial Fermentation Research.

Sills, A. M., M. E. Sauder and G. G. Stewart 1984a. Isolation and characterization of the amylolytic system of *Schwanniomyces castellii*. *Journal of the Institute of Brewing* **90**, 311–16.

Sills, A. M., P. S. J. Zygora and G. G. Stewart 1984b. Characterization of *Schwanniomyces castellii* mutants with increased productivity of amylases. *Applied Microbiology and Biotechnology* **20**, 124–8.

Skipper, N., R. Wong and J. E. Hopper 1987. Use of the melibiase promoter and signal peptide to express a bacterial cellulase from yeast. In *Biological research on industrial yeasts*, G. G. Stewart, I. Russell, R. D. Klein and R. R. Hiebsch (eds). Boca Raton, Florida: CRC Press (in press).

Skogman, H. 1976. The Symba process. *Starch/Staerke* **28**, 278–83.

Slapack, G., E. Edey, G. Mahmourides and H. Schneider 1984. Ethanol from hexoses and pentose in spent sulfite liquor. In *Biotechnology advances*, M. Moo-Young (ed.), 62–6. Oxford: Pergamon Press.

Slininger, P. J., R. J. Bothast, J. E. van Cauwenberge and C. P. Kurtzman 1982. Conversion of D-xylose to ethanol by the yeast *Pachysolen tannophilus*. *Biotechnology and Bioengineering* **24**, 371–84.

Smith, R. A., M. J. Duncan and D. T. Moir 1985. Heterologous protein secretion from yeast. *Science* **229**, 1219–21.

Solomons, G. L. 1984. Single cell protein. *CRC Critical Reviews in Biotechnology* **1**, 21–58.

Spencer-Martins, I. 1982. Extracellular isoamylase produced by the yeast *Lipomyces kononenkoae*. *Applied Environmental Microbiology* **44**, 1253–6.

Spencer-Martins, I. 1984. Production and properties of an extracellular cyclodextrin hydrolase from the yeast *Lipomyces kononenkoae*. *International Journal of Microbiology* **2**, 31–6.

Spencer-Martins, I. and N. van Uden 1977. Yields of yeast growth on starch. *European Journal of Applied Microbiology and Biotechnology* **4**, 29–34.

Spencer-Martins, I. and N. van Uden 1979. Extracellular amylolytic system of the yeast *Lipomyces kononenkoae*. *European Journal of Applied Microbiology and Biotechnology* **6**, 241–6.

Sreekrishna, K. and R. C. Dickson 1985. Construction of strains of *Saccharomyces cerevisiae* that grow on lactose. *Proceedings of the National Academy of Sciences of the USA* **82**, 7909–13.

St John, T. P. and R. W. Davis 1979. Isolation of galactose-inducible DNA sequences from *Saccharomyces cerevisiae* by differential plaque filter hybridization. *Cell* **16**, 443–52.

Sternberg, D., R. Vijayakumar and E. T. Reese 1977. β-Glucosidase: microbial production and effect on enzymatic hydrolysis of cellulose. *Canadian Journal of Microbiology* **23**, 139–47.

Steverson, E. M., R. A. Korus, W. Admassu and R. C. Heimsch 1984. Kinetics of the amylase system of *Saccharomycopsis fibuliger*. *Enzyme and Microbial Technology* **6**, 549–53.

Stewart, G. G. 1981. The genetic manipulation of industrial yeast strains. *Canadian Journal of Microbiology* **27**, 973–90.

Stewart, G. G. and I. Russell 1986. One hundred years of yeast research and development in the brewing industry. *Journal of the Institute of Brewing* **92**, 537–58.

Stewart, G. G., J. Erratt, I. Garrison, T. Goring and I. Hancock 1979. Studies on the utilization of wort carbohydrates by brewer's yeast strains. *Master Brewers Association of the Americas Technical Quarterly* **16**, 1–7.

Stewart, G. G., C. J. Panchal and I. Russell 1982. Advances in ethanol production from sugars and starch. In *Ethanol from biomass*, H. E. Duckworth and E. A. Thompson (eds), 4–53. Ottawa: Royal Society of Canada.

Stewart, G. G., I. Russell and A. M. Sills 1983. Factors that control the utilization of wort carbohydrates by yeast. *Master Brewers Association of the Americas Technical Quarterly* **20**, 1–7.

Stewart, G. G., R. Jones and I. Russell 1985. The use of derepressed yeast mutants in the fermentation of brewery wort. In *European Brewery Convention Proceedings of the 20th Congress, Helsinki*, 243–50. Oxford: IRL Press.

Sumner-Smith, M., R. P. Bozzato, N. Skipper, R. W. Davies and J. E. Hopper 1985.

Analysis of the inducible *MEL1* gene of *Saccharomyces carlsbergensis* and its secreted product, α-galactosidase (melibiase). *Gene* **36**, 333–40.

Tamaki, H. 1978. Genetic studies of ability to ferment starch in *Saccharomyces*. Gene polymorphism. *Molecular and General Genetics* **164**, 205–11.

Tauro, P. and H. O. Halvorson 1966. Effect of gene position on the timing of enzyme synthesis in synchronous cultures of yeast. *Journal of Bacteriology* **92**, 652–8.

Taussig, R. and M. Carlson 1983. Nucleotide sequence of the yeast *SUC2* gene for invertase. *Nucleic Acids Research* **11**, 1943–8.

Teeri, T., I. Salovuori and J. Knowles 1983. The molecular cloning of the major cellulase gene from *Trichoderma reesei*. *Bio/Technology* **1**, 696–8.

ten Berge, A. M. A., G. Zoutewelle and K. W. van de Poll 1973a. Regulation of maltose fermentation in *Saccharomyces carlsbergensis* I. Function of the gene *MAL6*, as recognized by *MAL6*-mutants. *Molecular and General Genetics* **123**, 233–46.

ten Berge, A. M. A., G. Zoutewelle, K. W. van de Poll and H. P. J. Bloemers 1973b. Regulation of maltose fermentation in *Saccharomyces carlsbergensis* II. Properties of a constitutive *MAL6*-mutant. *Molecular and General Genetics* **125**, 139–46.

Terrell, S. L., A. Bernard and R. B. Bailey 1984. Ethanol from whey: continuous fermentation with a catabolite repression-resistant *Saccharomyces cerevisiae* mutant. *Applied Environmental Microbiology* **48**, 577–80.

Thomsen, K. K. 1983. Mouse α-amylase synthesized by *Saccharomyces cerevisiae* is released into the culture medium. *Carlsberg Research Communications* **48**, 545–51.

Trivedi, N. 1986. Use of microorganisms in the production of unique ingredients. In *Biotechnology in food processing*, S. K. Harlander and T. P. Lubuza (eds), 115–32. New Jersey: Noyes Publications.

Tubb, R. S. 1986. Amylolytic yeasts for commercial applications. *Trends in Biotechnology* **4**, 98–104.

Ueda, S. and B. C. Saha 1983. Behaviour of *Endomycopsis fibuligera* glucoamylase towards raw starch. *Enzyme and Microbial Technology* **5**, 196–203.

Vijaikishore, P. and N. G. Karanth 1986. Glycerol production by fermentation – A review. *Process Biochemistry* **21**(2), 54–7.

Wang, P. Y., C. Shopsis and H. Schneider 1980. Fermentation of pentoses by yeast. *Biochemical and Biophysical Research Communications* **94**, 248–53.

Wayman, M. and S. Parekh 1985. Ethanol and sugar tolerance of *Candida shehatae*. *Biotechnology Letters* **7**, 909–12.

Wayman, M. and S. T. Tsuyuki 1985. Fermentation of xylose to ethanol by *Candida shehatae*. *Biotechnology and Bioengineering Symposium* No. 15, 167–77.

Whitney, G. K., C. R. Murray, I. Russell and G. G. Stewart 1985. Potential cost savings for fuel ethanol production by employing a novel hybrid yeast strain. *Biotechnology Letters* **1**(5), 349–54.

Whittle, D. J., D. G. Kilburn, R. A. Warner and R. C. Miller 1982. Molecular cloning of a *Cellulomonas firni* cellulase gene in *Escherichia coli*. *Gene* **17**, 139–45.

Wickerham, L. J., L. B. Lockwood, O. G. Pettijohn and G. E. Ward 1944. Starch hydrolysis and fermentation by the yeast *Endomycopsis fibuligera*. *Journal of Bacteriology* **48**, 413–18.

Wilson, J. J. and W. M. Ingledew 1982. Isolation and characterization of the amylolytic enzymes of *Schwanniomyces alluvius*. *Applied Environmental Microbiology* **44**, 301–6.

Wilson, J. J., G. G. Khachatourians and W. M. Ingledew 1982. Protoplast fusion in the yeast *Schwanniomyces alluvius*. *Molecular and General Genetics* **186**, 95–100.

Wood, T. M. 1975. Properties and mode of action of cellulases. *Biotechnology and Bioengineering Symposium* No. 5, 111–37.

Yamashita, I. and S. Fukui 1983. Molecular cloning of a glucoamylase-producing gene in the yeast *Saccharomyces*. *Agriculture Biology and Chemistry* **47**, 2689–92.

Yamashita, I., T. Hatano and S. Fukui 1984. Sub-unit structure of glucoamylase of *Saccharomyces diastaticus*. *Agriculture Biology and Chemistry* **48**, 1611–16.

Yamashita, I., T. Itoh and S. Fukui 1985a. Cloning and expression of the *Saccharomycopsis fibuligera* glucoamylase gene in *Saccharomyces cerevisiae*. *Applied Microbiology and Biotechnology* **23**, 130–3.

Yamashita, I., K. Suzuki and S. Fukui 1985b. Nucleotide sequences of the extracellular gene *STA1* in the yeast *Saccharomyces diastaticus*. *Journal of Bacteriology* **161**, 567–72.

Zimmerman, F. K. and N. R. Eaton 1974. Genetic and biochemical evidence of sucrose fermentation by maltase in yeast. *Molecular and General Genetics* **123**, 43–50.

# 10 *Substrate utilization, non-carbohydrate substrates*

TREVOR G. CARTLEDGE

## 10.1 Yeasts capable of growth on non-carbohydrate substrates

Although there are more than 50 recognized genera of yeasts, only a few genera are capable of utilizing a non-carbohydrate substrate as their sole source of carbon and energy; these genera include *Candida*, *Pichia*, *Kloeckera*, *Torulopsis*, *Hansenula*, *Rhodotorula*, *Saccharomyces* and *Saccharomycopsis*. Within these few genera, and particularly within the genus *Candida*, are found large numbers of species relevant to this chapter. Komagata *et al.* (1964) identified 56 isolates which utilized hydrocarbons, and all except one were members of the genus *Candida*. Lonsane *et al.* (1975) listed over 150 species which would grow on various hydrocarbons, especially *n*-alkanes. The picture is complicated as reports of new isolates are frequent and, inevitably, different groups of workers may use different names for the same organism. In addition, organisms are continually being renamed, for example *Kloeckera* 2201 first isolated by Ogata *et al.* (1969) is now classified as *Candida boidinii* by Lee and Komagata (1980).

Several yeasts, notably members of the genus *Saccharomyces*, have the almost unique ability amongst eukaryotes to grow readily either aerobically or anaerobically depending on substrates and environment. With all non-carbohydrate substrates mentioned in this chapter aerobic conditions are absolutely essential, most substrates being highly reduced, and their oxidation to yield energy utilizes ultimately the electron-transport chain and oxidative phosphorylation.

## 10.2 Non-carbohydrate substrates which will support growth of yeast

Methane, a major component of natural gas, is generally considered to be unable to support growth of yeasts, but *Rhodotorula glutinis* and *Sporobolomyces roseus* have been shown by Wolf *et al.* (1980) to grow very slowly on methane, with a generation time of at least two days. These yeasts will not grow on methanol, but they can utilize methionine and occasionally higher alkanes.

That yeasts grow on methanol was first shown by Shukla and Aroj (1967) using *Candida* spp. and Ogata *et al.* (1969) using *Kloeckera* 2201. A further important development was that of Cooney and Levine (1972), who first reported a methanol-using yeast (*Hansenula polymorpha*) able to grow at a temperature of 37°C or higher. Lee and Komagata (1980) reported that of 31 methanol-assimilating yeasts isolated, 27 were *Candida* spp., three *Pichia* spp. and one *Torulopsis* sp. These genera together with *Hansenula* represent the four most important genera of methanol-utilizing yeasts. Although methanol supports growth quite readily, its oxidation products, formaldehyde and formic acid, do not, and the latter particularly is inhibitory even at low concentrations (Pilat & Prokop 1975).

The relative ease with which alcohols are utilized falls dramatically with increasing chain length after ethanol. The corresponding amines will serve as a nitrogen source for many yeasts, though generally an alternative carbon source is required.

Although few yeasts will utilize ethane, some members of the genus *Candida* will utilize propane. Many yeasts accept alkanes of the $C_4$–$C_{20}$ range, and the most readily accepted are of the $C_{10}$–$C_{20}$ chain-length range. The organisms using $C_4$–$C_{10}$ compounds are fewer in number and tend to be very nutritionally demanding. Solubility of *n*-alkanes decreases with increasing chain length, $C_6$–$C_{16}$ *n*-alkanes being only soluble at 6–16 mg cm$^{-3}$ (Litchfield 1979). Alkanes above $C_{20}$ chain length are utilized slowly if at all, and are not important industrially as they are solids below 35°C (Evans 1969).

Alkenes are utilized by many *Candida* spp., and some yeasts will use other relatively simple carbon sources, for example alkynes (*Torulopsis*), cycloalkanes (*Saccharomyces*), isoalkenes, benzene or catechol (*Candida*) and phenol (*Torula*) (Lonsane *et al.* 1975). However, many of these substances are only metabolized very slowly by yeasts, and it is difficult to envisage any industrial value at present.

The most important carbon sources to be discussed in this chapter are methanol and ethanol (from natural gas and gas oils) and $C_4$–$C_{20}$ *n*-alkanes derived from gas oils. These substances are available either directly or as by-products of other processes, and their conversion into single-cell protein or other products is, or has been considered to be, commercially viable.

Other substrates, for example methylamine, may be useful in the future, not because they are available as cheap natural products but because they are troublesome waste products. Thus, the conversion of hazardous waste and production of useful food material may be accomplished in a single step.

## 10.3 Metabolism of methylotrophic yeasts

The oxidative metabolism of methanol-utilizing yeasts has been extensively studied, although gaps in our knowledge are still evident. Tani *et al.* (1972) first described a non-specific alcohol oxidase in methanol-oxidizing yeasts which has much higher activities than in yeasts grown on ethanol, glucose or glycerol (Sahm & Wagner 1973). Oxidation of methanol to formaldehyde is accompanied by formation of hydrogen peroxide; this being removed by catalase which has a specific activity many times higher in methanol-grown than in glucose-grown cells (Roggenkamp *et al.* 1974). Formaldehyde is then oxidized to carbon dioxide via formate using the inducible enzymes formaldehyde and formate dehydrogenases, respectively (Sahm & Wagner 1973, van Dijken *et al.* 1976). Regulation of these enzymes is by induction due to methanol and also catabolite repression and derepression. When growing on mixtures of glucose and methanol the two dehydrogenases are absent in *Hansenula polymorpha* (Eggeling & Sahm 1978) and *Candida boidinii* (Sahm 1977) until the glucose is depleted. Ethanol has a similar repressive effect to glucose on *Hansenula polymorpha* (Sahm 1979). Enzyme activities of formate and formaldehyde dehydrogenases are thought to be regulated by intracellular NADH and ATP concentrations (Sahm 1979).

The pathway shown in Figure 10.1 indicates that methanol could be oxidized via glyceraldehyde 3-phosphate, pyruvate and the TCA cycle. However, the vast increases in formate and formaldehyde dehydrogenase activities, together with lowered citrate synthase and aconitase activities during growth on methanol (Table 10.1), suggest that this is not an important route.

Methanol-assimilating yeasts do not operate by the bacterial ribulose monophosphate pathway, but by a xylulose monophosphate-dependent fixation of formaldehyde mediated by the enzyme DHA synthase and yielding dihydroxyacetone (Waites & Quayle 1980). A second enzyme, DHA kinase, is also required to phosphorylate DHA. These two enzymes are specific to this cycle, whereas all other enzymes in Figure 10.1 are normally functional in at

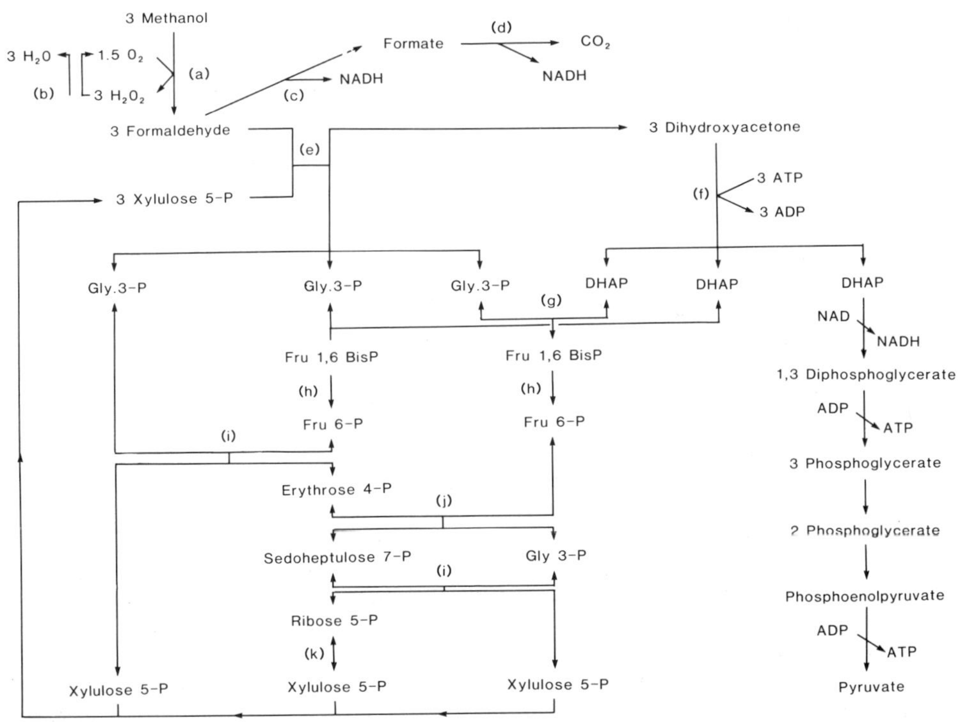


**Figure 10.1** Pathways of oxidation and assimilation of methanol in yeasts. (a) methanol oxidase, (b) catalase, (c) formaldehyde dehydrogenase, (d) formate dehydrogenase, (e) DHA synthase, (f) DHA kinase, (g) fructose 1,6-bisphosphate aldolase, (h) fructose 1,6-bisphosphate phosphatase, (i) transketolase, (j) transaldolase, (k) isomerase.

**Table 10.1** Specific activities of enzymes from cell extracts of methanol-grown *Candida boidinii* (Jenkins 1984).

| Enzyme | Specific activities (nmol substrate utilized or product produced $min^{-1}$ $mg^{-1}$ protein) | |
|---|---|---|
| | Glucose-grown cells | Methanol-grown cells |
| catalase | 14 | 719 |
| methanol oxidase | 0.1 | 240 |
| methanol dehydrogenase | 1.7 | 1.2 |
| ethanol dehydrogenase | 40 | 72 |
| formaldehyde dehydrogenase | 21 | 298 |
| formate dehydrogenase | 1.8 | 150 |
| DHA synthase | | 53 |
| cytochrome *c* oxidase | 240 | 97 |
| malate dehydrogenase | 3900 | 14 200 |
| citrate synthase | 200 | 59 |
| aconitase | 94 | 33 |
| isocitrate lyase | 1 | 1 |
| malate synthase | 1.2 | 1.5 |

least one other pathway in yeasts. Of three molecules of DHAP and three molecules of glyceraldehyde 3-phosphate, one DHAP is made available for assimilation into cellular material. The glycolytic enzymes fructose 1,6-bisphosphate aldolase and fructose 1,6-bisphosphatase convert two DHAP molecules and two glyceraldehyde 3-phosphates into two molecules of fructose 6-phosphate. Transaldolase and transketolase then rearrange these two hexoses, and the remaining triose to three molecules of xylulose 5-phosphate, thus completing the cycle. The net result of the cycle is:

$$3HCHO + 3ATP \rightarrow DHAP + 3ADP + 2P_i$$

and the rate-limiting enzymes appear to be one or more of DHA synthase, transketolase or transaldolase (Egli *et al.* 1983).

In addition to major changes in enzyme activities, profound structural changes are evident in methanol-grown cells. The major alteration is that microbodies, small and few in number in glucose-grown cells (van Dijken *et al.* 1975), are numerous forming a cluster which may occupy most of the interior of methanol-grown cells. Each microbody contains an electron-dense structure, the crystalloid, which contains methanol oxidase in *Hansenula polymorpha* (Veenhuis *et al.* 1976) but methanol oxidase and catalase in *Kloeckera* (Osumi *et al.* 1982). The microbody matrix contains catalase, D-amino acid oxidase and L-α-hydroxyacid oxidases, and urate oxidase if present in the cells. The role of microbodies in methanol-grown cells is complex, and elucidation has been further complicated by reports that two types of microbody may be present in the same cell (Jenkins 1984, Jenkins *et al.* 1985).

Formate and formaldehyde dehydrogenases are cytosolic, and it would be logical to assume that methanol assimilation via DHA synthase would have the same subcellular location. However, Jenkins *et al.* (1985) reported that 25–35% of the total DHA synthase activity sediments after fractionation of extracts in sucrose gradients to a density of 1.20 g $cm^{-3}$, a region containing mitochondria and vacuoles but not microbodies.

As mentioned above, formaldehyde could be oxidized via pyruvate and the TCA cycle, though the evidence suggests that the route is via the cytosolic formaldehyde and formate dehydrogenases. This extramitochondrial production of the majority of NADH is quite rare. The exact mechanism by which this NADH is used is not known, though there is a possibility that the maximum energy generation per NADH oxidized is only two, and not three, ATP molecules. This theory arises from the work of Ohnishi (1973), which showed that cytosolic NADH is oxidized by an outward-facing inner mitochondrial membrane-bound NADH dehydrogenase (yielding two ATP) whereas the inward-facing NADH dehydrogenase oxidizes NADH to yield three ATP. This lowered ATP yield together with the fact that no ATP is available in the oxidation of methanol to formaldehyde has obvious effects on the maximum possible cell yield per gram of methanol.

Methylamine may be used as a source of nitrogen by several yeasts and may supply some of the carbon requirements, although an alternative carbon source is needed. The initial reaction is the oxidation to the corresponding aldehyde, that is:

$$\underset{\text{methylamine}}{CH_3NH_2} \xrightarrow[H_2O \quad O_2 \quad NH_3 \quad H_2O_2]{\text{methylamine oxidase}} \underset{\text{formaldehyde}}{HCHO}$$

This enzyme has been shown to be located in peroxisomes, probably within the crystalloid (at least in *Hansenula polymorpha* grown on a mixture of methanol and methylamine; Veenhuis *et al.* 1981). Haywood and Large (1981) showed that *Candida boidinii* grown on glucose plus one of a variety of amines as nitrogen source had markedly increased activities of catalase and methylamine oxidase and, in certain cases, benzylamine oxidase was produced.

## 10.4 Metabolism of *n*-alkane-assimilating yeasts

Studies on the oxidative metabolism of *n*-alkane-assimilating yeasts have mainly utilized *n*-hexadecane as growth substrate. Although this is used more quickly than other alkanes, compounds within the range $C_4$–$C_{18}$ are generally accepted as sole source of carbon and energy.

*n*-Alkanes pass into the yeast cell and are accumulated in unchanged form (Davidova *et al.* 1977, Davidova & Rachinski 1981). The substrate has been shown to partition from the medium directly to the cell surface, and in a separate process is then translocated across the cell envelope (Kappeli & Fiechter 1981). Relative concentrations of alkane and cell protein together with pH, mixing efficiency and particularly temperature affect substrate uptake. Several workers have reported that hydrocarbon-utilizing yeasts release surfactants, possibly within vesicles, to the medium to solubilize the hydrocarbons and increase their adhesive properties to the cell (Goma *et al.*

1974). Petrikevich and Dovgun (1980) have suggested that the hydrocarbons are directed from the periplasmic space, through the membrane and into the cytoplasm in lamellar crystalline structures which may be collected into vacuolar type bodies until required.

The *n*-alkanes are oxidized first to the corresponding alkanol by a monooxygenase complex containing cytochrome $P_{450}$ and NADPH cytochrome $P_{450}$ reductase. These components together with a heat-labile lipid fraction are induced in response to the presence of *n*-alkane (Duppel *et al.* 1973) and are located in the microsomal fraction of *n*-alkane-grown *Candida tropicalis* (Kawamoto *et al.* 1977) and in the smooth endoplasmic reticulum of *Saccharomycopsis lipolytica* (Delaissé *et al.* 1981). The amount of cytochrome $P_{450}$ is markedly influenced by oxygen concentration, increasing three- to fourfold as the oxygen concentration is lowered (Mauersberger & Matyashova 1980). The alkanol is then oxidized via the alkanal to the corresponding fatty acid by NAD-linked long-chain alcohol and aldehyde dehydrogenases (Duppel *et al.* 1973) which are subject to derepression when *n*-alkane replaces glucose as the carbon source. The fatty acid is then activated to the CoA ester by acyl CoA synthase. Two such synthases, termed I and II, have been located in *Candida lipolytica*; I is present in glucose or *n*-alkane-grown cells whereas II is induced during growth on *n*-alkanes (Kamiryo *et al.* 1977, Mishina *et al.* 1978). The former is thought to produce acyl CoA derivatives for lipid biosynthesis, whereas the latter provides acyl CoA for $\beta$-oxidation. This acyl CoA is degraded by dehydrogenation and cleavage to acetyl CoA units by $\beta$-oxidation; the initial dehydrogenation reaction is catalysed by a flavoprotein, acyl CoA oxidase, which requires oxygen and generates hydrogen peroxide (Schimizu *et al.* 1979), the latter being degraded by catalase. Oxidation of acetyl CoA to yield energy is via the TCA cycle, and assimilation of two-carbon units is initially via the glyoxylate cycle to yield four-carbon units (Kawamoto *et al.* 1977).

The roles of the microbody in *n*-alkane oxidation have been of interest for many years, various strains of *Candida* have been shown to contain relatively few microbodies when grown on glucose, slightly more when grown on ethanol or acetate, and many more when grown on *n*-alkanes (Fukui & Osumi 1973, Osumi *et al.* 1974). Their numbers correlate closely with increase in total catalase activity within the cell. These organelles tend to lie alongside the smooth endoplasmic reticulum, making them spatially close to the site at which *n*-alkanes are processed to fatty acids.

These microbodies contain not only the conventional marker enzymes catalase and D-amino acid oxidase (Osumi *et al.* 1975), but also isocitrate lyase, malate synthase and NADP-linked isocitrate dehydrogenase (Kawamoto *et al.* 1977) in addition to the enzymes of $\beta$-oxidation (Kawamoto *et al.* 1978a). However, NAD-linked isocitrate dehydrogenase and the other TCA cycle enzymes are located in mitochondria. These data suggest that microbodies and mitochondria must co-operate to run the TCA and glyoxylate cycles efficiently (Fig. 10.2 and Tanaka *et al.* 1982). The acetyl CoA produced in the microbodies is required in the mitochondria, a process carried out by one or probably two carnitine acetyltransferases (Kawamoto *et al.* 1978b) which are located in mitochondria, microbodies and cytoplasm (Ueda *et al.* 1982). These enzymes

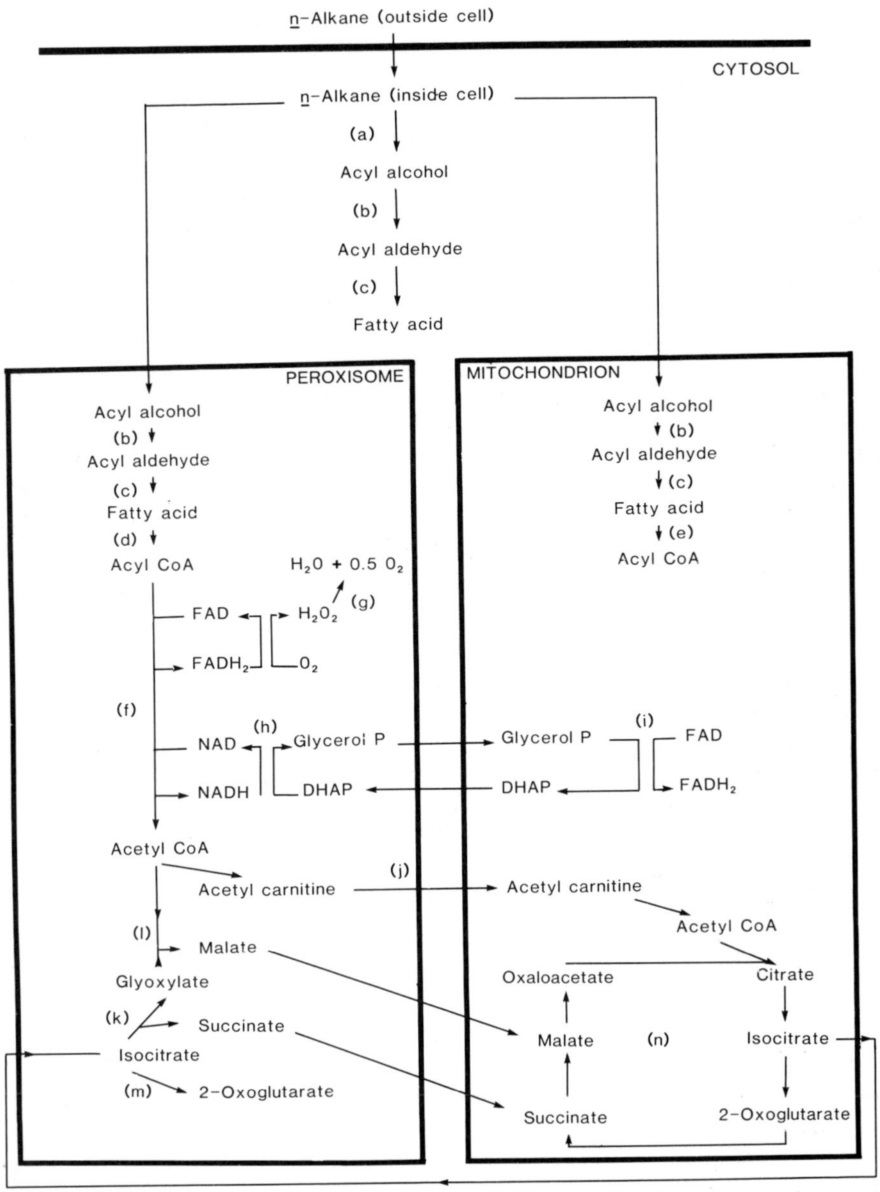


**Figure 10.2** Pathways of oxidation and assimilation of *n*-hexadecane in yeasts. (a) cytochrome $P_{450}$ plus NADPH cytochrome $P_{450}$ reductase, (b) NAD-linked long-chain alcohol dehydrogenase, (c) NAD-linked long-chain aldehyde dehydrogenase, (d) acyl-CoA synthase I and/or II, (e) acyl-CoA synthase I and/or II, (f) acyl CoA oxidase, (g) catalase, (h) NAD-linked glycerol-3-phosphate dehydrogenase, (i) FAD-linked glycerol-3-phosphate dehydrogenase, (j) carnitine acetyltransferase, (k) isocitrate lyase, (l) malate synthase, (m) NADP-linked isocitrate dehydrogenase, (n) the TCA cycle.

are specific for acetyl CoA and propionyl CoA, acetyl and propionyl being the sole products of $\beta$-oxidation of fatty acids (Ueda *et al.* 1982, 1983).

A further problem is that the NADH produced in the microbodies from $\beta$-oxidation is re-oxidized by the mitochondria, and thus the 'reducing equivalents' have to be relocated. Kawamoto *et al.* (1979) located two glycerol 3-phosphate dehydrogenases, an NAD-linked enzyme in microbodies and a mitochondrial FAD-linked enzyme. These authors proposed a DHAP ↔ glycerol 3-phosphate shuttle as shown in Figure 10.2. In terms of energy production, from each turn of the $\beta$-oxidation spiral one reduced FAD and one reduced NAD are produced. In theory five ATP molecules could be produced from these dinucleotides, though in these yeasts only two are available as discussed earlier.

The role of the peroxisome or at least of the microbody has been complicated by the findings of Jenkins *et al.* (1983) that *n*-hexadecane grown *Candida stellatoidea* contains two populations of microbodies.

The subcellular locations of long-chain alcohol and aldehyde dehydrogenases are not clear. Yamada *et al.* (1980) recovered them mainly from microbody fractions, with some activity evident in microsomal and mitochondrial fractions from *n*-alkane-grown *Candida tropicalis*. However, Jenkins *et al.* (1983) showed that although the enzymes were multilocational in alkane-grown *Candida stellatoidea*, they were primarily associated with mitochondria and not with microbodies.

In addition to profound structural changes, there are large differences in enzyme activities when comparing glucose- and *n*-hexadecane-grown cells (Table 10.2).

**Table 10.2** Specific activities of enzymes from cell extracts of *n*-hexadecane-grown *Candida stellatoidea* (Jenkins 1984).

| Enzyme | Specific activities (nmol substrate utilized or product produced $min^{-1}$ $mg^{-1}$ protein) | |
|---|---|---|
| | Glucose-grown cells | *n*-hexadecane-grown cells |
| catalase | 19 | 450 |
| long-chain alcohol dehydrogenase | 13 | 361 |
| long-chain aldehyde dehydrogenase | 32 | 70 |
| acyl CoA synthase II | 0.5 | 7 |
| palmitoyl CoA-dependent reduction of NAD | 0.8 | 36 |
| carnitine acetyl transferase | 45 | 330 |
| NADP-linked isocitrate dehydrogenase | 49 | 63 |
| citrate synthase | 120 | 440 |
| aconitase | 27 | 61 |
| *iso*-citrate lyase | 1 | 26 |
| malate synthase | 2 | 39 |
| malate dehydrogenase | 2800 | 4600 |
| NAD-linked *iso*-citrate dehydrogenase | 120 | 61 |

Inevitably fractions derived from petroleum contain both odd- and even-numbered carbon alkanes. Both are degraded in similar fashion with the final product from the last $\beta$-oxidation spiral of odd-chain length alkanes being propionyl CoA. This compound is degraded via the methyl TCA cycle (Tabuchi & Hara 1974), a cycle which appears to be mitochondrial as one of the key enzymes, methylcitrate synthase has been located in mitochondria of *Candida tropicalis* grown on odd-chain length alkanes (Tanaka *et al.* 1982). In this cycle the methylisocitrate formed is cleaved to succinate and pyruvate. *Candida tropicalis* growing on propionate as sole carbon source has relatively few very large microbodies, and contains an inducible propionyl CoA synthetase which is distinct from acetyl CoA synthetase (Ueda *et al.* 1983).

## 10.5 Metabolism of ethanol

Ethanol is metabolized by oxidation initially to acetaldehyde, then to acetate. The enzymes involved, NAD-dependent ethanol and NAD-dependent acetaldehyde dehydrogenases, are cytosolic enzymes and are of low or undetectable activity in glucose-grown cells, but are of much higher activity in ethanol-grown cells. The acetate may accumulate intra- or extracellularly, or is converted to acetyl CoA for entry into the TCA cycle. Entry is at one of two points as shown in Figure 10.3.

A build up in the concentration of acetate which occurs when the ethanol concentration is relatively high is economically undesirable in single-cell protein production. One of the major factors influencing whether acetate accumulates is the concentration of NADH. If the ethanol concentration is high, then a large quantity of NADH will be formed in the cytosol by the ethanol and aldehyde dehydrogenases. Acetate can either accumulate or be converted to acetyl CoA. The demand for acetyl CoA is determined partly by where it enters the TCA cycle and partly by the fate of malate. If oxaloacetate is to be formed, then NAD is required, though most of this co-factor may be present as NADH; if cell components are required ultimately NADH will be needed for anabolic processes. It is therefore necessary to keep the ethanol concentration high enough to produce sufficient acetate for growth and energy production, but not too high or acetate accumulates (Watteeuw *et al.* 1979). Using *Candida utilis* Watteeuw *et al.* showed that acetate accumulation has to be prevented, as acetate itself is inhibitory and even subsequent metabolism of this accumulated acetate still results in reduced cell yields.

## 10.6 Single-cell protein production

### *10.6.1 Introduction*

The term 'single-cell protein' is now widely accepted and is much preferred to terms such as fungal or bacterial protein. Although single-cell protein may be derived from plant material, the use of microbes gives many advantages; for example, shorter generation time of organisms, easy genetic modification,

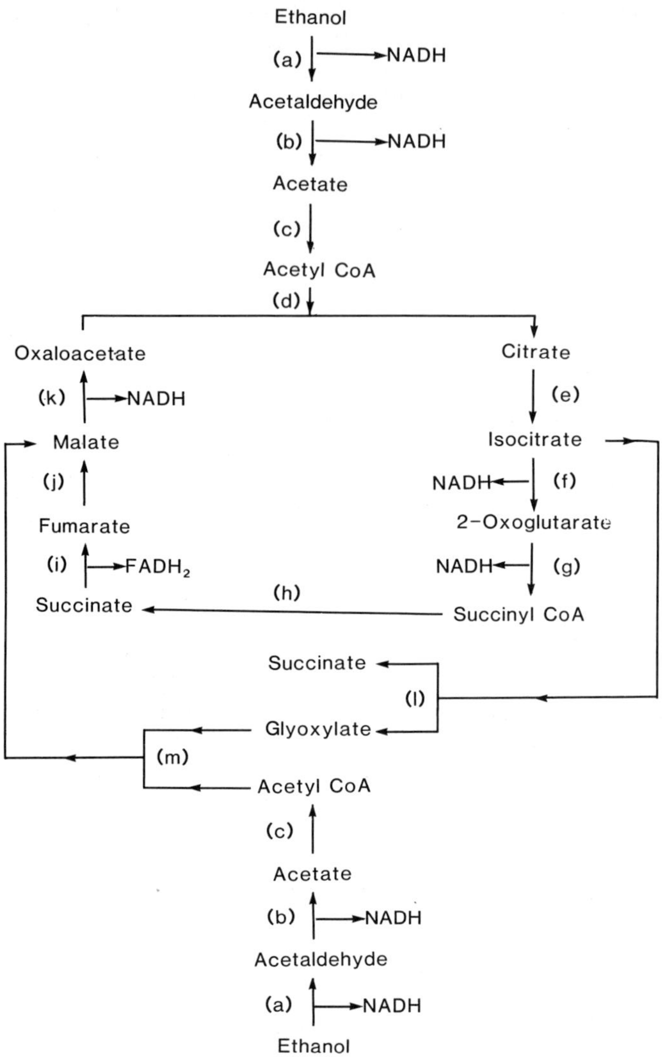


**Figure 10.3** Pathway of oxidation and assimilation of ethanol in yeasts. (a) ethanol dehydrogenase, (b) acetaldehyde dehydrogenase, (c) CoA transferase or thiokinase, (d) citrate synthase, (e) aconitase, (f) NAD-linked isocitrate dehydrogenase, (g) 2-oxoglutarate dehydrogenase, (h) succinate thiokinase, (i) succinate dehydrogenase, (j) fumarase, (k) malate dehydrogenase, (l) isocitrate lyase, (m) malate synthase.

versatility in that many substrates including waste materials may be utilized, no requirement for large areas of arable land and continuous production which may be carried out anywhere in the world (Kihlberg 1972).

The need for single-cell protein is obvious; already two-thirds of the world's population of 4000 million are said to be undernourished and, with a predicted doubling to 8000 million within 20 years, it would be impossible for agriculture alone to satisfy world food requirements (Cooney *et al.* 1980, Dixon 1984).

Over the last 20 years large amounts of money have been put into research and development projects investigating micro-organisms producing protein from non-carbohydrate substrates. British Petroleum led the field in hydrocarbon utilization by yeasts, and Shell investigated the growth of bacteria on methanol. Both groups subsequently withdrew from the field, and several other companies have also written off large sums of money (Sherwood 1984) as the financial returns have been so low. Relatively little interest has been generated in the USA, as its enormous agricultural base has meant the continuous production of relatively cheap soybean protein.

Single-cell protein production is capital-intensive and it is generally thought that plants producing less than 10 000 tonnes $year^{-1}$ would be uneconomic. Plants of 100 000 tonnes $year^{-1}$ capacity are economically much more viable (Kihlberg 1972), for example in 1975 BP estimated the cost of a 16 000-tonnes $year^{-1}$ plant at US$7 million, whereas a plant of capacity 100 000 tonnes $year^{-1}$ was estimated at $30 million (Humphrey 1975).

The cell yield (g dry weight cells $g^{-1}$ product) varies according to the substrate and the yeast species employed. In addition, it is affected by temperature, pH and dilution rate (Urakami *et al.* 1983). Once the yield is known, then an economic perspective can be obtained by calculating substrate cost per kilogram of single-cell protein (Cooney *et al.* 1980). This is important since misleading results are avoided, for example in 1980 *n*-paraffin cost over twice as much as methanol on a weight basis, but the final product cost less (Cooney *et al.* 1980).

The theoretical maximum yield can be calculated; in practice yields are lower due, in part, to maintenance energy requirements.

Capital investment is minimized by using continuous culture systems, preferably in which the carbon source is limiting so that maximum substrate utilization is obtained. This suggests that a low dilution rate is preferable, though caution is necessary as this rate influences both cell yield and the amount of substrate utilized. This rate may also affect the quality as well as the quantity of product (Cooney *et al.* 1980) although this does not appear to be the case with *Candida utilis* (Alroy & Tannenbaum 1973).

A high cellular-conversion yield (that is, substrate converted into cellular material) is desirable on economic grounds, but is also beneficial since the higher the value is, the lower the amount of heat produced, thus less cooling is required and also product purification may be easier.

Although economics are very important, absolute safety must be the first consideration if a product is intended for human consumption. Single-cell protein from yeasts has the disadvantage of indigestible cell walls, which can produce stomach pains and may contain levels of nucleic acids sufficient to cause the formation of kidney stones. Furthermore, toxins are produced by some yeasts, particularly when growing on ethanol (Litchfield 1979). It is also worth noting that problems of acceptability of product have arisen when *n*-alkane fractions, and particularly gas oil, have been used as substrate, due to the carry-over of unusable materials into the product.

### *10.6.2 From methanol*

Methanol is primarily obtained from natural gas, but also from the gas at oil wells, crude oils, coal or biogas. Efforts to obtain methanol from renewable raw materials are of increasing economic importance.

There are several advantages to using methanol compared with hydrocarbons including: (a) its high purity, and no pretreatment is needed; (b) it is completely miscible with water, and therefore there are no mass transfer problems and it is easy to remove residual methanol from the protein product; (c) it is of relatively restricted use by other micro-organisms, lessening the chance of contamination; (d) it is easy to handle and store; (e) it is partially oxidized, thus having a lower oxygen demand, and less cooling is needed as there is less heat generated (33% less than *n*-hexadecane); (f) only a relatively simple plant is needed; (g) using yeast at low pH together with (c) means sterilization may not be needed; and (h) a high-purity product is formed. This list is impressive, but there are two potential disadvantages; first, methanol is volatile and losses by evaporation may be significant, for example in shake flasks up to 20% of a 1% (w/w) methanol concentration may be lost (Tezuka *et al.* 1975). Secondly, the cost of substrate is particularly important in the case of methanol, as this carbon source typically represents 55–60% of the total cost compared with the more usual 35–45% for other substrates. In 1977 methanol was considered to be relatively cheap, but its price has since risen considerably. However, the cost of building a plant favours methanol, as such a plant would cost about two-thirds of the cost of one to utilize *n*-alkane.

Using growth on methanol as an example, it is appropriate at this point to compare the advantages of using yeast as opposed to bacteria for the synthesis of single-cell protein. Yeast has the advantage of being more acceptable as a food source; it grows at very low pH, possibly eliminating the need for sterilization of media and equipment; there are no bacteriophage problems; there are fewer scale-up problems (Urakami *et al.* 1983); the product contains lower levels of nucleic acids and is much easier to separate, as the cells are much larger; and, finally, according to Kuraishi *et al.* (1979), better results have been obtained after its use in animal feedstuffs. Bacteria, on the other hand, are faster growing; give a higher yield due to the production of energy from the oxidation of the methanol to formaldehyde (Harder 1979); have a higher protein content in the product (65% as opposed to 50%); and contain a higher percentage of methionine (Cooney & Levine 1973).

An ideal yeast would have the advantages mentioned above and the ability to grow at a reasonably high temperature, for example 37°C or above. These higher temperatures give faster growth, and usually easier recovery and better solid–liquid separation. A growth temperature of 37°C means that water at a temperature of 20–23°C is adequate for cooling, whereas growth temperatures of around 30°C mean extra cost due to a requirement for refrigeration. This is particularly important in countries near to the equator, for example Israel, where growth on methanol at 40°C rather than 30°C could save 16% of running costs due to reduced cooling requirements (Mateles 1979).

Several pilot- or larger-scale plants have been constructed for production of protein from methanol by yeasts. The Mitsubishi Chemical Company began research in 1969 and produced a pilot plant of 500 tonnes year$^{-1}$ capacity in 1974. From 1979 they were ready to scale up to 60 000 or 100 000 tonnes year$^{-1}$ plant, and still considered the process to be economically viable (Kuraishi *et al.* 1979). The pilot plant incorporated an air-lift fermenter having draft tubes with perforated plates, this type being less expensive than mechanically agitated fermenters. UHDE have a plant in Libya using the ICI methanol process and Hoechst-UHDE have a 1000 tonnes year$^{-1}$ large-loop fermenter producing protein from mixed cultures of bacteria and yeast. Here the process can be modified to give either fodder protein containing 80% protein and 12% nucleic acid or a protein concentrate containing 90% protein and 1% nucleic acid (Dimmling 1979). At least one, and possibly two, major plants in the USSR are due on-line shortly, and Phillips Petroleum (USA) still intend to construct a plant to produce protein dietary supplements for humans (Sherwood 1984). This would compete with the Amoco production of yeast protein from ethanol. Child and Suggitt (1980) patented a process to produce protein from methanol derived from coal. Briefly, carbon monoxide and hydrogen are produced from coal and are used to synthesize methanol. Ammonia is produced from atmospheric nitrogen and hydrogen. The off-gases, oxygen and carbon dioxide, are recycled to produce carbon monoxide again.

The maximum theoretical yield is 0.4–0.73 g (dry weight) biomass g$^{-1}$ methanol, depending on the amount of ATP produced. As yeasts do not obtain ATP from the methanol-to-formaldehyde oxidation step, the theoretical yield is then 0.41–0.54 g g$^{-1}$ (Cooney & Makiguchi 1977). In practice values of 0.3–0.45 g g$^{-1}$ are usual and values of at least 0.39 g g$^{-1}$ are considered essential for economic viability (Kuraishi *et al.* 1979). Table 10.3 shows examples of yeasts which have been or are considered potentially useful in single-cell protein production from methanol.

Although one aim must be to maximize yield, great care has to be taken when adjusting methanol concentrations, as even relatively small increases result in a drop in yield (Sahm 1979) and may result in a fall in specific growth rate. Several workers have also shown that formaldehyde and formate, oxidation products of methanol during energy formation, are very inhibitory. Formaldehyde is 100 times and formate 10 times more toxic than methanol (Pilat & Prokop 1975). The specific growth rate of *Candida boidinii* increases up to 0.8% methanol, is constant from 0.8 to 3.0%, and drops above this value (Pilat & Prokop 1975). *Candida lipolytica* 33 prefers 1–3% methanol for growth, and grows optimally in 2% methanol. Although high yields are obtained in 3 and 4% methanol concentrations, growth is inhibited before all of the methanol is utilized. This is thought to be due to formaldehyde and formate inhibition (Mesaric *et al.* 1981). These workers also considered that formate and not formic acid was inhibitory, and thus that the pH of the growth medium is important. Foda and El-Masry (1981) examined several *Candida* and *Hansenula* species and noted linear growth responses up to 2% methanol, although

**Table 10.3** Examples of yeasts considered useful for single-cell protein production from methanol.

| Organism | Growth temperature (°C) | Growth pH | Specific growth rate ($h^{-1}$) | Yield (g dry weight $g^{-1}$ methanol) | Maintenance (g methanol $g^{-1}$ cells $h^{-1}$) | Reference |
|---|---|---|---|---|---|---|
| *Candida boidinii* | 26.5 | 4 to 6 | 0.2 | 0.495 | | Tezuka *et al*. (1975) |
| *Candida boidinii* | 23–27 | 4.5 | 0.16 | 0.42 | | Kuvshinnikow *et al*. (1977) |
| *Candida intermedia* | 28.0 | 3.5–5 | | 0.5 | | Bachman and Pizlo (1977) |
| *Candida lipolytica* | | 7.0 | | 0.42 | | Mesaric *et al*. (1981) |
| *Candida melanii* | 30.0 | 5.6 | | 0.43 | | Hong and Jang (1976) |
| *Rhodotorula glutinis* | 30.0 | 5.2 | | 0.42 | | Hong and Jang (1976) |
| *Hansenula polymorpha* | 37.0 | 4.5–5 | 0.22 | 0.36 | | Cooney and Levine (1973) |
| *Hansenula polymorpha* | | | | 0.38 | 0.017 | van Dijken *et al*. (1976) |
| *Hansenula polymorpha* | | | | 0.36 | 0.016 | Mateles (1979) |
| *Hansenula polymorpha* | 38.0 | | 0.21 | 0.38 | 0.015 | Urakami *et al*. (1983) |
| *Hansenula polymorpha* | 37.0 | 3.5 | 0.215 | 0.41 | | Allais and Baratti (1983) |

up to 6% methanol would support growth, particularly if the inoculum was increased. However, Allais and Baratti (1983) suggested that the yield drops above 0.8% methanol, and suggested 0.4% as a suitable concentration.

Experiments to improve *Hansenula* strains genetically are being carried out in many laboratories. Yang *et al.* (1982) have isolated temperature-sensitive mutants which, being larger cells, aggregate (thus lowering the cost of recovery), and they also have a lowered ribonucleic acid content.

### *10.6.3 From ethanol*

Although production of single-cell protein from ethanol is very uncertain at present on economic grounds, several companies have recently been involved in research and development.

By 1976 Amoco (USA) had built a fully automated continuous culture plant requiring aseptic conditions to produce 7000 tonnes $year^{-1}$ single-cell protein from *Candida utilis*. Mitsubishi Petrochemicals used *Candida ethanothermophilum* to yield 1000 tonnes $year^{-1}$ and Chemopetrol (Czechoslovakia) envisaged using a yeast to produce up to 100 000 tonnes $year^{-1}$. In contrast, Exxon Nestle have produced a pilot plant producing protein from the bacterium *Acinetobacter calcoaceticus*. The Amoco product named 'Toruteen' was specifically produced for human consumption as a food and flavour additive and, along with the Mitsubishi and Chemopetrol plant products, contained about 55% protein and less than 10% nucleic acids.

The value of ethanol as a substrate for single-cell protein has been apparent for many years; the first patent for industrial purposes being granted in Czechoslovakia in 1964 (Vernerova & Rosa 1964). There are many advantages in using ethanol as substrate, including: (a) it is non-toxic and already accepted, in that it is itself consumed by humans; (b) it is easy to handle, store and transport; (c) it is soluble in water; (d) it is obtainable in high purity; (e) it is partially oxidized, and thus has a lowered oxygen demand; and, therefore, (f) it has relatively low heat production. Many of the advantages are in common with the advantages of methanol and, like methanol, the major disadvantage for ethanol is its relatively high cost. Ethanol, and particularly purified ethanol, is very expensive and its price is subject to considerable fluctuation. Except for the Mitsubishi plant, all of the industrial plants mentioned earlier were designed to run on pure ethanol. Ethanol also suffers the disadvantages of being volatile and inhibitory to growth and biomass production, even at low concentrations.

Many reports have been published regarding the actual concentration of ethanol which is inhibitory, though care has to be taken to specify the yeast and growth conditions. Omata *et al.* (1970) suggested that concentrations of ethanol greater than 3.2% were generally inhibitory to biomass production. Orlova *et al.* (1980) suggested that 1–1.5% ethanol gave optimal yields of *Saccharomyces cerevisiae*, 2% slightly lowered biomass yield and 2–3% gave low biomass and lowered key enzymes such as acetaldehyde dehydrogenase and isocitrate lyase. Kilian *et al.* (1981) noted that the yields and growth rate of *Candida krusei* were virtually unaffected by 0.43–3.6% ethanol, though above this concentration severe inhibition of both was found, particularly if growth was at 40°C rather than 30°C.

Much effort has gone into fermenter design, as careful choice of plant can virtually eliminate the problems of volatility and substrate inhibition. First, a continuous rather than a batch process is preferable; Nakanishi and Machida (1975) showed that the protein yield was of the order of 0.3 g biomass $g^{-1}$ ethanol for batch culture but up to 0.7 g $g^{-1}$ in continuous culture. If a batch culture system is employed, then a feed-batch system is essential, with careful addition of ethanol corresponding to the organism's expected growth rate (Machek *et al.* 1975).

Secondly, the continuous-culture system should be multistage, as indicated by the work of Watteeuw *et al.* (1979) referred to earlier.

A series of papers from Paca and co-workers (for example, Paca & Gregr 1979, 1980, Paca 1981, 1982) confirmed the findings of Watteeuw *et al.* After investigating the growth of *Candida utilis* under a variety of conditions, they concluded that 2–2.5% ethanol is optimal for yield; acetate inhibits the respiratory activity of yeasts growing on ethanol even at low concentrations; if the oxygen level drops below a critical level acetate will accumulate; the ability of high ethanol concentrations to inhibit depends on the partial pressure of oxygen; and care has to be taken in increasing oxygen concentration, as oxygen rapidly becomes toxic. Paca and co-workers concluded that a multistage tower fermenter with interstage mixing gave high biomass yield, complete ethanol conversion, low oxygen loss and little, if any, acetate formation.

One important economic saving mentioned earlier is the use of higher temperatures of fermentation. The examples mentioned in this section all employ 30°C, except the Mitsubishi plant which employs *Candida ethanothermophilum* at 40°C (Masuda *et al.* 1976). Kilian *et al.* (1981) reported growth of *Candida krusei* at an optimum of 37°C, whereas Tyi *et al.* (1982) and Chistyakova *et al.* (1983) have investigated growth of *Candida valida* at temperatures above 18°C. Optimal growth rate and biomass yield were at 36°C, with growth in particular being affected by temperature. The protein content of the biomass of *Candida utilis* grown at different rates differed between 55 and 65% (Machek *et al.* 1975) though Tyi *et al.* (1982) reported that the biomass of *Candida valida* grown at different temperatures showed no differences in carbon : hydrogen : nitrogen ratio.

The maximum theoretical biomass yield was calculated to be 0.826 (Suomalainen & Oura 1979). Published values include 0.68 for *Candida* species (Johnson 1969); 0.63 for *Candida utilis* (Mateles 1979); 0.67 for *Saccharomyces cerevisiae* (Suomalainen & Oura 1979); 0.65 for *Mycoderma vini* (Kharat'yan *et al.* 1975); 0.80 for *Candida valida* (Tyi *et al.* 1982) and 0.95 for *Candida ethanothermophilum* (Masuda *et al.* 1976). When calculating the potential biomass yield, the amount of ethanol needed for cell maintenance is required, and this has been estimated at 0.02 g ethanol $g^{-1}$ biomass $h^{-1}$ for *Candida valida* (Tyi *et al.* 1982).

As mentioned earlier, the possibility of using impure ethanol has been considered. Ethanol produced from ethylene contains many volatile compounds, most of which do not inhibit growth of *Candida utilis* (Sestakova *et al.* 1976). However, several compounds including allyl alcohol, crotonol and the major contaminant crotonaldehyde, inhibit growth and biomass yield severely even at concentrations below 10 mg $l^{-1}$. This inhibitory effect is reduced by increased aeration.

Finally, the use of ethanol as a carbon and energy supplement was reported by Suomalainen and Oura (1979). Since the Second World War, due to the shortage of raw materials, ethanol was added to molasses-wort in the ratio of 10 tonnes molasses to 700 kg ethanol during production of *Saccharomyces cerevisiae*. The authors suggest that if molasses continues to increase in price and a cheaper way of obtaining ethanol from, for example, cellulose is forthcoming, then the proportion of ethanol in the molasses-wort ethanol mixture could be increased. The use of ethanol to supplement yeast production from sulphite waste liquors (the liquor after extraction of cellulose from wood chips during manufacture of paper) has been studied (Rychtera 1979, Rychtera *et al.* 1982, Johnston & Oberman 1982). Although yield is much improved by ethanol addition, the economic viability of such processes for single-cell protein production is in doubt.

### *10.6.4 From hydrocarbons*

Microbial degradation of hydrocarbons was first reported by Miyoshi in 1895. British Petroleum became interested in the possibilities of single-cell protein production from *n*-alkanes in 1959, and the first industrial process and patent came in 1963 (Champagnat *et al.* 1963). In addition to gas oil and purified *n*-alkane fractions, which will be discussed in detail below, several other hydrocarbon-containing fractions from petrol have been investigated. Crude oil, fuel oil and kerosene support growth of one or more organisms, though none is promising enough to warrant scale-up. Ashy *et al.* (1980) reported that *Candida tropicalis* would grow on diesel fuel, but poorly on either gasoline or turbine fuel. Ammonium was the preferred nitrogen source and a product containing 68% protein was obtained.

Gas oil is a liquid petroleum fraction between diesel and lubricating oil (225–380°C) normally containing 10–25% *n*-alkane hydrocarbons of the $C_{10}$–$C_{18}$ range (Kihlberg 1972), though up to 40% *n*-alkane is possible (Cooney *et al.* 1980). The composition and percentage hydrocarbon depends upon the nature of the crude oil; for example, North African crude is high in *n*-alkane whereas Venezuelan crude is high in naphthalene and aromatic compounds. It is the *n*-alkane fraction which is used by yeasts, the remaining deparaffinized fraction, which may represent 90% of the original carbon mass, has a variety of fates as will be seen later. For use in any form of feed stuff the substrate must have less than 1 $\mu g\ kg^{-1}$ potential carcinogens such as 3,4-benzpyrene (Shacklady 1975). The importance of gas oil as a source of protein seems to have decreased, as no known commercial projects are operating or planned outside Commecon countries (Dimmling 1979).

*n*-Alkanes represent about 6% of crude oil, but are generally obtained from gas oil by a molecular-sieve process yielding a possible 99% $C_{10}$–$C_{25}$ *n*-alkanes (Laine *et al.* 1976). Aromatic impurities are relatively easily removed by treatment with fuming sulphuric acid (Kihlberg 1972). Carcinogenic impurities again need monitoring as indicated in the previous paragraph. The Western

European production of *n*-alkanes in 1979 was of the order of 1 million tonnes year$^{-1}$, although sufficient crude oil was processed to potentially yield 25 million tonnes year$^{-1}$ (Dimmling 1979).

A yield of over 1 g biomass g$^{-1}$ substrate is theoretically possible, and values for high-yielding strains are typically 0.88–1.1 g g$^{-1}$ (Litchfield 1979).

A comparison of the similarities and differences between gas oil and *n*-alkanes as carbon sources for single-cell protein production is shown in Table 10.4. The overall cost comparison is difficult, as carbon source prices vary considerably and it depends upon whether the commercial organization produces or buys the substrate. Gas oil represents about 36% of the total cost, whereas *n*-alkane represents 50% of its cost analysis. The plant investment and extraction costs are much higher for gas-oil systems, although the *n*-alkane system has to bear the cost of substrate purification. In terms of economics the fate of the deparaffinized gas oil becomes crucial. Bohlmann *et al.* (1979), when discussing the production of 'Fermosin', argue for gas oil for the following reasons: (a) the deparaffinized petroleum distillate is ready for use in diesel

**Table 10.4** Similarities and differences between gas oil and *n*-alkane as carbon sources for single-cell protein production in yeast.

| Gas oil | *n*-Alkane |
|---|---|
| 220–380°C fraction containing 10–25% *n*-alkane | up to 99% pure *n*-alkane |
| air-lift or agitated fermenter | air-lift or agitated fermenter |
| continuous process | continuous process |
| clean but non-sterile process requiring filtered air | aseptic (possibly non-sterile) |
| 10% of gas oil used, rest recycled | virtually all substrate used |
| pH 3–4, temperature 30°C | pH 3–4, temperature 30–32°C |
| ammonium as nitrogen source and pH control | ammonia as nitrogen source and pH control |
| generally run below maximum productivity, as the need for greater energy makes it no longer cost-effective | as opposite, but cost-effectiveness is less of a problem |
| collected yeast allowed to stand for 2 h to lower residual alkane concentration | collected yeast allowed to stand for 2 h to lower residual alkane concentration |
| protein separation complex due to remaining gas oil; phase separation followed by solvent extraction and washing | easier separation as substrate largely consumed; washing and centrifugation generally sufficient |
| very high capital investment | high capital investment |

motors and actually gives improved low-temperature characteristics; (b) during extended solvent extraction to purify 'Fermosin' a biolipid extract is obtained which may be marketed; and (c) the process is flexible in that it can be tailored to increase production of either 'Fermosin' or deparaffinized petroleum distillate as required.

Fermenters in the earlier research work were of the Waldhof type, aerated and agitated, although later more use was made of air-lift fermenters, particularly in the case of *n*-alkanes (Litchfield 1979). One of the Kamegafuchi Chemical group reported successful use of a modified air-lift fermenter which had an oxygen transfer rate at least as good as a conventional agitated fermenter and also employed a vacuum self-evaporated system for heat removal (Kanazawa 1975). Bohlmann *et al.* (1979), commenting on the 'Fermosin' production, noted that there are four phases in the fermenter – yeast, aqueous, vapour and oil – and the need for sophisticated, potentially expensive apparatus is obvious.

Some examples of plants which were either planned or built are given in Table 10.5. Most were designed and intended to come into operation by the mid-1970s. However, increasing fears of the effects of potentially carcinogenic impurities in the final products led several governments to ban their use, even after giving earlier encouragement to the industries concerned. By 1977 all plants outside Commecon countries were either shut down or not built.

The extraction of single-cell protein is much easier from *n*-alkane-grown yeast than from gas-oil-grown cells. In the former case the yeast utilizes virtually all of the *n*-alkane, such that it is sufficient to centrifuge, wash and dry the cells. The extraction of protein from gas oil is more complicated because about 90% of the gas oil remains after yeast growth. However, Bohlmann *et al.* (1979) considered that primary separation at least is very easy, as the biomass forms stable aggregates and the oil phase separates out. After low-speed centrifugation the aqueous phase containing some nutrient is recycled. The cells are extracted by polar and non-polar solvents to remove hydrocarbons and lipid; removal of the lipid is suggested to improve the keeping quality of the protein. The final product is generally odourless and tasteless, containing around 60% protein and 10% lipid (from *n*-alkane-grown cells) or 68% protein plus 2% lipid from gas-oil-grown cells (Shacklady 1975). The concentration of methionine and lysine are generally higher than from soy (Shacklady 1975) and would be adequate for animal feed, though low for poultry feed (Litchfield 1979). 'Fermosin' contains 60–65% protein and less than 1 part per $10^9$ of 3,4-benzpyrene, and it has given excellent results in feeding trials in hens and pigs (Bohlmann *et al.* 1979), although the nucleic acid concentration of around 7 is far too high for human foods.

One problem to consider is that although petroleum products contain 15–20% *n*-alkanes, they may contain up to 27% isohydrocarbons (Davidov *et al.* 1981). Bacteria and fungi utilize these iso-derivatives, but yeasts generally do not. Davidov *et al.* investigated the growth of *Candida guilliermondii* on octadecane and 2-methylhexadecane mixtures and found that only 5.5% of the cellular carbon comes from the latter carbon source. Most of the 2-methyl carbon is lost as $CO_2$, but that remaining in the cell is as isovaleric acid.

**Table 10.5** Examples of fermentation plants planned or built to produce single-cell protein from gas oil or *n*-alkane.

| Company and location | Organism | Capacity (tonnes year$^{-1}$) | Comments |
|---|---|---|---|
| *n*-alkane | | | |
| BP Grangemouth (Scotland) | *Candida* | 4000 | operational in 1965, though since shut down |
| Kamegafuchi Chem. Co. (Japan) | *Candida* | 60 000 | intended to start building in 1972–73, but both stopped on the advice of the Japanese government due to fear over carcinogens in product; government tried to ban each firm from exporting technology |
| Dai Nippon (Japan) | *Candida* | 60 000–100 000 | |
| BP ANIC (Sardinia) | *Candida lipolytica* | 100 000 | intended to sell product for animal feed as 'Toprina'; in 1976 'Toprina' accepted in eight European and many South American countries; Italian government withdrew permit and plant never actually fully functional |
| Petrom and Dai Nippon (Rumania) | Yeast | | on stream in 1980 |
| gas oil | | | |
| BP (France) | *Candida* | 16 000 | built in 1972, operational in 1973; closed in 1976 due to pressure on BP regarding impurities in product |
| GDR and USSR | *Candida* | 1000 (with scale-up planned) | product 'Fermosin'; permission to use it as an animal feed granted in GDR in 1973; plant producing by then |

### *10.6.5 Conclusion*

The preceding sections have indicated that many processes have reached technical feasibility at both pilot and full plant scale, although none appears to be economically viable at present. Probably the biggest loser financially has been ICI, with their plant to produce 'Pruteen' from methanol by the bacterium *Methylophilus methylotrophus*. Estimates of costs of over US$150 million have been made (see Sherwood 1984), although the plant is at least functioning and producing 'Pruteen', albeit at reduced levels. In addition to high costs of substrates there is also the problem that crude oil and natural gas will become increasingly scarce, probably in the 1990s.

Against this background of uncertainty there are one or two promising prospects. There is a growing vegetarian interest in single-cell protein additions to the diet which, while in the short term they will only ensure profit from low-volume high-value products (Sherwood 1984), could substantially increase the demand with major changes in attitude. In addition, the economics of food prices must alter as the increased demand due to population increases cannot be met by agriculture alone. A major area of interest must be the use of recyclable waste products as yeast substrates. For example, *Candida lipolytica* will grow on oxanone water (a mixture of organic acids from the manufacture of caprolactam, a raw material for nylon 6) to yield 4.5 kg $m^{-3}$ $h^{-1}$ yeast (Wiken 1972). The added benefit here is the removal of a potentially hazardous waste product.

Improvements to existing processes are continuing; for example, successful genetic manipulation of yeasts to improve methionine content, to produce higher protein : dry weight ratios and to produce yeasts with thinner cell walls (see Johnston & Oberman 1982).

## 10.7 Citrate production

Citrate has a wide variety of industrial applications, and as it is a metabolite of the TCA cycle (and therefore virtually non-toxic) its use as a human-food additive is allowed. It is used as the principal food additive in soft drinks, jams, jellies, sweets and wines. Citrate is also used in gelatin food products, artificial flavours, as a builder in detergent formulations and as a sequestering agent in cosmetics. It is added to blood for transfusions and, as a free acid, is available as an effervescent.

Citrate is produced as a free acid and marketed either as the crystalline monohydrate or sodium salt, both of which are fairly stable at temperatures up to 21°C.

Citrate was first shown to accumulate in micro-organisms in the 1890s by Wehmer, and from his work an industrial plant was actually built in France in the late 19th century to produce citrate from *Citromyces* (now *Penicillium*). The first fermentation process to yield citrate came in 1923, and by 1980 over 100 000 tonnes $year^{-1}$ were being produced in Western Europe and America, almost entirely by fermentation. Over the years the vast majority of citrate has been produced from plant carbohydrate by *Aspergillus niger*, although recently *n*-alkanes have partly replaced the plant carbon source. Yeasts were first reported to produce citrate from acetate in 1932 by Wieland and Sanderhoff, and since then the accumulation of citrate from glucose, acetate or *n*-alkane has been shown to be fairly common in *Candida* species (for example, see Nout 1972).

Citrate is not accumulated in the growth phase, but during the stationary phase in the presence of a carbon source and little if any available nitrogen. Under favourable conditions up to 200 g $l^{-1}$ is possible (Marchal *et al.* 1977a).

It is worth noting that citrate may be produced by yeasts growing on glucose, *n*-alkanes, ethanol, acetate or fatty acids. It is a TCA cycle intermediate, and each of the potential substrates are degraded or converted eventually to acetyl CoA which enters the TCA cycle (Figs 10.2 & 3).

Thus, it is necessary to influence the metabolism of the yeasts in such a way as to block the cycle at citrate, though if necessary allowing sufficient operation of the cycle to produce energy. As growth ceases due to nitrogen limitation the still-plentiful carbon source is metabolized, which increases the ATP concentration at the expense of the ADP and AMP concentrations. NAD-linked isocitrate dehydrogenase (which requires AMP for activity) is therefore slowed down or stopped. Citrate synthase is normally inhibited by high ATP concentration, but this effect appears to be overruled as the formation of citrate continues. *Iso*-citrate lyase is functional in these cells and is not inhibited. However, no build-up of succinate occurs, as this compound is used to provide the oxaloacetate necessary for citrate synthase activity. These cells therefore still produce varying amounts of NADH (depending on the substrate used), and this must be re-oxidized via the electron-transport chain linked to oxidative phosphorylation. As there is little demand for ATP, the continued recycling of ATP to yield ADP for oxidative phosphorylation is difficult to explain. One possible answer is that there is little or no coupling between the electron-transport chain and oxidative phosphorylation under these conditions (see Marchal *et al.* 1977a).

One obvious problem arising from the preceding paragraph is that due to aconitase activity unwanted isocitrate rather than citrate is likely to accumulate. Although aconitase is naturally much lower in stationary-phase cells than in log-phase cells, it is still necessary to produce yeast mutants with low activity of this enzyme, or to lower or interfere with any activity present. Much work has been carried out using the inhibitor monofluoroacetate which will combine in the cells with oxaloacetate to yield monofluorocitrate. This causes competitive inhibition of aconitase. Several workers have used the principle that strains or mutants which will not grow on citrate as sole source of carbon and energy or are monofluoroacetate-sensitive, or both, are likely to have low aconitase activity (Akiyama *et al.* 1972a). This assumption has been verified, as will be discussed later, in that several mutants giving high yields of citrate but not isocitrate have been produced. Akiyama *et al.* (1972a, b) produced such a strain of *Candida lipolytica* yielding 1.45 g biomass $g^{-1}$ *n*-hexadecane in a citrate to isocitrate ratio of 97:3 and an aconitase activity one-hundredth of the wild type. MacKenzie *et al.* (1980, 1981) produced a fluoroacetate mutant from *Candida lipolytica* yielding a 20:1 ratio of citrate to isocitrate, although with a relatively poor yield.

Ferrous ions are essential for aconitase activity (Dagley & Walker 1956) though Akiyama *et al.* (1972a) showed that concentrations below one part per million would be needed to stop *Candida* aconitase. To achieve such low ferrous-ion concentrations would probably be relatively expensive in practice.

The pH of the medium has to be monitored and not allowed to fall sharply during citrate accumulation. For example, Moresi *et al.* (1980) showed that a pH fall from 7 to 3 inhibited further citrate production by *Candida lipolytica*. Early strains required the presence of $CaCO_3$ as neutralizer, although precipitation of calcium citrate resulted in cloudy broth and made it difficult to aerate and agitate the culture and to separate the product from the cells. Furukawa *et al.* (1982) produced several mutants of *Candida lipolytica*, one of which preferred $NH_4OH$ as a neutralizer. This mutant gave a 1.5 g citrate plus isocitrate $g^{-1}$ *n*-paraffin in a ratio of 89:11. Limiting the phosphate concentration or adding cupric ions or borate improved the ratio by partially inhibiting isocitrate formation. This mutant had particularly low aconitase but very high citrate synthase activity.

Product formation occurs in stationary phase and not during growth, thus implying the use of batch cultures. Furukawa and Ogino (1982) attempted to get a semi-continuous cell-recycle system to shorten preproduction phase, lengthen production and to minimize product inhibition. One alternative test system used a non-aseptic 30-litre fermenter where yeast, after separation from the aqueous phase by centrifugation, was recombined with the *n*-paraffin layer and added to fresh medium. In a second method a 2-litre fermenter under aseptic conditions was allowed to stand for 1 h, during which the yeast and *n*-paraffin layer separated above the aqueous layer which contained citrate and spent medium. This lower layer was removed and replaced by fresh medium. Either of these methods produced around 0.92 g acid $l^{-1}$ $h^{-1}$, whereas batch cultures yielded only 0.58 g $l^{-1}$ $h^{-1}$. Runs of at least 160 h were made without contamination problems.

It has been reported that aeration levels are important and that high product yield requires high air consumption (Moresi *et al.* 1980), though MacKenzie *et al.* (1981) noted that only limited aeration was preferable as non-baffled flasks gave larger citrate yields than baffled flasks. Furukawa and Ogino (1982) suggested that aeration levels were not important as long as the cultures were agitated.

As mentioned previously, citrate may be produced by yeasts from a variety of substrates. Marchal *et al.* (1977a) grew *Candida lipolytica* in 1% potassium acetate, 1% ethanol, 7.5% glucose or 5% *n*-paraffin ($C_{13}$–$C_{18}$). In each case citrate was produced when growth stopped, and the *n*-paraffin gave by far the greatest yield. Ikeno *et al.* (1975) had earlier achieved 0.3–0.8 g acids $g^{-1}$ glucose, 0.3–0.7 g $g^{-1}$ coconut oil and 0.7–1.5 g $g^{-1}$ *n*-paraffin. The ratio of citrate to isocitrate varied from 2:1 in wild-type cultures to 44:1 for mutants growing on *n*-paraffin. Furukawa *et al.* (1982) showed that $C_{14}$ and $C_{15}$ *n*-paraffins were preferred from mixtures containing $C_{11}$–$C_{17}$ *n*-alkanes.

The theoretical maximum yield is 2.47 g citrate $g^{-1}$ hexadecane (Marchal *et al.* 1977b) assuming that all substrate is converted to product. However, these workers showed that a maximum of 84.5% of hexadecane is converted to product (Equation 1, below) and 15.5% is required for energy (Equation 2, below). Thus the maximum yield is 84.5% of 2.47, that is 2.09 g $g^{-1}$.

$$C_{16}H_{34} + 12.5O_2 \rightarrow 2.66C_6H_8O_7 \cdot H_2O + 4.66H_2O \quad (1)$$

$$C_{16}H_{34} + 24.5O_2 \rightarrow 16CO_2 + 17H_2O \quad (2)$$

Akiyama *et al.* (1972a) reported a yield of 1.45 g $g^{-1}$ *n*-paraffin within 72 h by *Candida lipolytica* and Nakanishi *et al.* (1972) obtained 1.5 g $g^{-1}$ *n*-paraffin using *Candida zeylanoides*, though citrate to isocitrate was in a 1:1 ratio. *Candida guilliermondii* showed a lower yield of 1 g $g^{-1}$ but a better ratio of 85:15. The ratio was increased by adding aconitase inhibitors, though it seemed to be unaffected by ferrous-ion deprivation. Whether low enough levels of ferrous ions were achieved is not clear.

Marchal *et al.* (1977b) showed that *Candida lipolytica* in a stirred nitrogen limited fermenter system grew for 20 h and produced citrate and isocitrate linearly over the next 70 h. Levels of 150–200 g acids $l^{-1}$ were possible.

## 10.8 L-Malic acid production

Malic acid is of interest industrially as a competitor to the role of citrate as a food additive. Sato *et al.* (1977a, b) investigated the potential usefulness of *Candidu brumptii*, a yeast known to produce organic acids from *n*-paraffins but relatively little acid from glucose. Growth proceeds for one to two days followed by malic acid production up to a maximum after a further four days. One problem is that malic acid and succinic acid are produced, and commercially it is only the malic acid that is required. Sato and co-workers found that growth at pH 6.2 and 30°C on 4% paraffin with 2 g $l^{-1}$ urea as nitrogen source yielded 25.4 g $l^{-1}$ malic acid, which is 0.8 g $g^{-1}$ *n*-paraffin. The specific activities of citrate synthase, aconitase, NADP-linked isocitrate dehydrogenase and 2-oxoglutarate dehydrogenase remain virtually constant, whereas during product production phase succinate dehydrogenase, fumarase and malate dehydrogenase show twofold increases and the two glyoxylate cycle enzymes show tenfold increases. NAD-linked isocitrate dehydrogenase is absent from this organism. Thus, during product formation the glyoxylate bypass must operate, channelling isocitrate and acetyl CoA to two malate molecules (see Fig. 10.3).

## 10.9 Other products

### *10.9.1 Fats*

In 1979 the world production of fats and oils, largely from plant sources and excluding dairy products, was over 50 million tonnes. An estimated extra 18 million tonnes $year^{-1}$ will be required by 1987, and only half is expected to come from agricultural sources, the shortfall possibly being derived from oleaginous micro-organisms (Ratledge 1982). Relatively few micro-organisms accumulate fat, though a few yeasts of the genera *Candida*, *Hansenula*, *Rhodotorula* and *Cryptococcus* can produce over 20% of their biomass as fat. One organism, *Cryptococcus terricolus*, can achieve 55% (dry weight) biomass as fat. In all cases fat accumulates after growth has finished and in the presence of excess carbon source and little or no nitrogen.

Several carbon sources may be used, including glucose, ethanol, *n*-paraffins and methanol. Methanol has generally been considered unsuitable for fat production by yeasts, as the yield rarely exceeds 7% (Rattray & Hambleton 1980). Jigami *et al.* (1979) recently showed that *Candida guilliermondii*, when grown under carefully limited nitrogen conditions and at 20°C, will accumulate 16% fat, though even with improved yield it is unlikely to be an economically feasible proposition.

The only current commercial enterprise utilizing non-carbohydrate carbon sources is that using gas oil to yield the protein 'Fermosin'. Biolipid is produced as a by-product.

### *10.9.2 Biochemicals and enzymes*

Although production is of a low-volume high-cost nature, it is to a definite and inevitably increasing market. Relevant products include nucleic acids, purines, pyrimidines, amino acids, vitamins, sterols and enzymes such as those of the TCA cycle and glycolytic pathway. Yeasts, and particularly *Saccharomyces cerevisiae* and *Candida utilis*, are the principal sources of these products which are used primarily for analytical and research purposes. The majority are derived from carbohydrate-grown yeasts, though change to other carbon sources results in spectacular increases of certain enzymes (for examples see Tables 10.1 & 2). Further fine control will allow increases in individual enzymes rather than levels of all enzymes in a pathway, for example careful monitoring of methanol concentration when growing yeast at low dilution rates can yield up to 20% of the soluble protein as methanol oxidase. Further improvement by manipulation at the genetic level using recombinant-DNA or protoplast-fusion techniques is also possible.

## 10.10 Final conclusion

Owing to the growing scarcity and increasing cost of naturally occurring non-carbohydrate carbon sources, their use for single-cell protein production by yeasts has become increasingly unattractive economically. On the other hand, the world population continues to expand and much available agricultural land is being diverted to other uses. Eventually agriculture alone will not be able to provide sufficient food material. More effective use of waste material is necessary, and it is effectively a double bonus when a material which is dangerous is not only neutralized, but is converted to something useful such as single-cell protein.

With the cessation of interest by organizations such as BP and Shell and the problems of the ICI methanol plant it is unlikely that large-scale fermentations outside the USSR and its associated countries will ever be attempted. Isolated examples are conceivable in which a country rich in a particular natural resource, for example methanol, may build such a fermenter, particularly if international trade or finance become difficult. However, these events would remove decision-making even further from the laboratory – further from scientists, industrialists and economists, and into the realms of national and international politics.

## References

Akiyama, S., T. Suzuki, Y. Sumino, Y. Nakao and H. Fukuda 1972a. Production of citric acid from *n*-paraffins by fluoroacetate sensitive mutant strains of *Candida lipolytica. Fermentation Technology Today. Proceedings of the 4th International Fermentation Symposium* 613–17.

Akiyama, S., T. Suzuki, Y. Sumino, Y. Nakao and H. Fukuda 1972b. Production of citric acid from *n*-paraffins by fluoroacetate sensitive mutant strains of *Candida lipolytica. Agricultural Biology and Chemistry Japan* **36**, 339–41.

Allais, J. J. and J. Baratti 1983. A new methanol assimilating thermotolerant yeast, *Hansenula polymorpha* with a high cell yield. *Journal of Fermentation Technology* **61**, 339–46.

Alroy, Y. and S. R. Tannenbaum 1973. The influence of environmental condition on the macromolecular composition of *Candida utilis. Biotechnology and Bioengineering* **15**, 239–56.

Ashy, M. A., M. Khalil, A. El-Galil and A. Abal-Zeid 1980. Utilisation of petroleum products in the production of single cell proteins. I Role of chemical ingredients of fermentation media on production of single cell proteins. *Indian Chemical Manufacture* **18**, 15–16.

Bachman, B. and A. Pizlo 1977. Low cost protein from methanol. *Polish Technical Reviews* **9**, 18.

Bohlmann, D., J. Bauch, L. I. Kozlova, G. I. Mescankin, M. I. Roskova, K. Triens, M. Ringpfeil and K. Sabek 1979. Process for the production of 'Fermosin®' – Fodder yeast from petroleum distillates. *Doechema Monographs. Proceedings of the XII International Congress on Microbiology* **83**, 1704–23.

Champagnat, A., C. Vernat, B. M. Lainc and J. Filosa 1963. Protein biosynthesis from petroleum. French Patent, 1 147 491.

Child, E. T. and R. M. Suggitt 1980. Biosynthesis of protein by fermentation of methanol obtained from the gasification of coal or residual oil. US Patent, 301 280.

Chistyakova, T. A., I. G. Minkevich and V. K. Eroshin 1983. Growth of the thermotolerant yeast, *Candida valida*, on the ethanol dependences of maximal growth rate and cell biomass yield on temperature. *European Journal of Applied Microbiology and Biotechnology* **18**, 225–8.

Cooney, C. L. and D. W. Levine 1972. Microbial utilisation of methanol. *Advances in Applied Microbiology* **15**, 337–65.

Cooney, C. L. and D. W. Levine 1973. Single cell protein production from methanol by yeast. In *Single cell protein II*, S. R. Tannenbaum and D. I. C. Wang (eds), 402–23. Cambridge, Massachusetts: MIT Press.

Cooney, C. L. and N. Makiguchi 1977. An assessment of single cell protein from methanol grown yeasts. In *Symposium on single cell protein from renewable and non-renewable resources*, E. L. Gaden and A. E. Humphrey (eds), Vol. 7, 65–76. New York: Wiley.

Cooney, C. L., C. Rha and S. R. Tannenbaum 1980. Single cell protein: engineering, economics and utilisation in foods. *Advances in Food Research* **26**, 1–52.

Dagley, S. and J. R. L. Walker 1956. Accumulation of citrate and pyruvate during growth of a *Vibrio*. *Biochimica et Biophysica Acta* **21**, 441–7.

Davidov, E. R., Y. I. Sokolov, N. F. Denanova and A. D. Gololobov 1981. Utilisation of 2-methyl hexadecane by the yeast *Candida guilliermondii*. *Applied Biochemistry and Microbiology* **17**, 236–47.

Davidova, E. G., V. V. Rachinski, N. F. Demanova, E. R. Davidov and A. Gololobov 1977. Kinetics of primary distribution of *n*-alkane oxidation products according to morphological components of the yeast cell. *Mikrobiologie* **45**, 1018–22.

Davidova, E. G. and V. V. Rachinski 1981. Determination of the uptake of liquid *n*-alkanes into the yeast cell by gas liquid radio chromatography. *Mikrobiologie* **50**, 248–51.

Delaissé, J. M., P. Martin, M. F. Verheyen-Bouvy and E.-J. Nyns 1981. Subcellular distribution of enzymes in the yeast *Saccharomycopsis lipolytica* grown on *n*-hexadecane with special reference to the omega hydroxylase. *Biochimica et Biophysica Acta* **676**, 7790–9.

van Dijken, J. P., R. Otto and W. Harder 1976. Growth of *Hansenula polymorpha* in a methanol limited chemostat. *Archives of Microbiology* **111**, 137–44.

van Dijken, J. P., M. Veenhuis, N. J. W. Kreger-van Rij and W. Harder 1975. Microbodies in methanol assimilating yeasts. *Archives of Microbiology* **102**, 41–4.

Dimmling, W. 1979. Substrates in fermentation technology – situation and economy. *Doechema Monographs. Proceedings of the XII International Congress in Microbiology* **83**, 1–26.

Dixon, B. 1984. Biotechnology and the population problem. *Biotechnology* **2**, 586.

Duppel, W., J.-M. Lebeault and M. J. Coon 1973. Properties of a yeast cytochrome $P_{450}$ containing enzyme system which catalyses the hydroxylation of fatty acids, alkanes and drugs. *European Journal of Biochemistry* **36**, 583–92.

Eggeling, L. and H. Sahm 1978. Derepression and partial insensitivity to carbon catabolite repression of the methanol dissimilating enzymes in *Hansenula polymorpha*. *European Journal of Applied Microbiology and Biotechnology* **5**, 197–202.

Egli, T., N. D. Lindley and J. R. Quayle 1983. Regulation of enzyme synthesis and variation of residual methanol concentration during carbon limited growth of *Kloeckera* sp. 2201 on mixtures of methanol and glucose. *Journal of General Microbiology* **129**, 1269–81.

Evans, G. H. 1969. In *4th Petroleum Symposium*. UN Economic Commission for Asia and Far East.

Foda, M. S. and H. G. El-Masry 1981. Parameters affecting methanol utilisation by yeasts. *Zentrabl Bakteriolparasitenkd Infektionsts Hyg Zweite Naturwiss Abt Microbiol. Landwirtsch Technol. Umweltschutzes* **136**, 427–32.

Fukui, S. and M. Osumi 1973. Some specific features of ultrastructure of *Candida* yeast cells growing on *n*-alkanes. *Proceedings of the 3rd Symposium on Techniques in Microbiology, Berlin*, 375–84.

Furukawa, T. and T. Ogino 1982. Citric acid production from *n*-paraffins by use of semicontinuous cell cycle system. *Journal of Fermentation Technology* **60**, 377–80.

Furukawa, T., T. Ogino and T. Matsuyashi 1982. Fermentative production of citric acid from *n*-paraffins by *Saccharomycopsis lipolytica*. *Journal of Fermentation Technology* **60**, 281–6.

Goma, G., A. Pareilleux and G. Durand 1974. Physical chemical aspects of the assimilation of hydrocarbons by *Candida lipolytica. Agricultural Biology and Chemistry (Japan)* **38**, 1273–80.

Harder, W. 1979. Metabolic diversity among methanol utilisers and its energetic implications. *Doechema Monographs. Proceedings of the XII International Congress on Microbiology* **83**, 83–93.

Haywood, G. W. and P. J. Large 1981. Microbial oxidation of amines. *Biochemical Journal* **199**, 187–201.

Hong, S.-W. and K.-Y. Jang 1976. Growth conditions for the biomass yield of two methanol utilising yeast spp. *Candida* spp. and *Rhodotorula* spp. *Proceedings of the Natural Sciences (Seoul National University)* **1**, 153–67.

Humphrey, A. E. 1975. Product outlook and technical feasibility of single cell protein. In *Single cell protein II*, S. R. Tannenbaum and D. I. C. Wang (eds), 1–23. Cambridge Massachusetts: MIT Press.

Ikeno, Y., M. Masuda, K. Tanno, I. Oomori and N. Takahashi 1975. Citric acid production from various raw materials by yeasts. *Hakko Kogaku Zasshi* **53**, 752–6.

Jenkins, R. O. 1984. An investigation into the role of subcellular organelles in various species of yeasts. Ph.D. thesis, Trent Polytechnic, Nottingham, UK.

Jenkins, R. O., T. G. Cartledge and D. Lloyd 1983. Subcellular fractionation of *Candida stellatoidea* after growth with glucose or *n*-hexadecane. *Journal of General Microbiology* **129**, 1171–85.

Jenkins, R. O., T. G. Cartledge and D. Lloyd 1985. Subcellular fractionation of *Candida boidinii* after growth on glucose or methanol. *Journal of General Microbiology* **131**, 335–44.

Jigami, Y., O. Suzuki and S. Nakasato 1979. Comparison of lipid composition of *Candida guilliermondii* grown on glucose, ethanol and methanol as the sole carbon source. *Lipids* **14**, 937–42.

Johnson, M. J. 1969. In *Fermentation advances*, D. Perlman (ed.), 833–42. New York: Academic Press.

Johnston, J. R. and H. Oberman 1982. Yeast genetics in industry. *Progress in Industrial Microbiology* **15**, 151–205.

Kamiryo, T., M. Mishina, S. Tashiro and S. Numa 1977. *Candida lipolytica* mutants defective in an acyl CoA synthetase: isolation and fatty acid metabolism. *Proceedings of the National Academy of Sciences of the USA* **74**, 4947–50.

Kanazawa, M. 1975. The production of yeasts from *n*-paraffins. In *Single cell protein II*, S. R. Tannenbaum and D. I. C. Wang (eds), 438–53. Cambridge Massachusetts: MIT Press.

Kappeli, O. and A. Fiechter 1981. Properties of hexadecane uptake by *Candida tropicalis. Current Microbiology* **6**, 21–6.

Kawamoto, S., A. Tanaka, M. Yamamura, Y. Teranishi, S. Fukui and M. Osumi 1977. Microbody of *n*-alkane grown yeast, enzyme localisation in the isolated microbody. *Archives of Microbiology* **112**, 1–8.

Kawamoto, S., C. Nozaki, A. Tanaka and S. Fukui 1978a. Fatty acid beta oxidation system in microbodies of *n*-alkane grown *Candida tropicalis. European Journal of Biochemistry* **83**, 609–14.

Kawamoto, S., M. Ueda, C. Nozaki, M. Yamamura, A. Tanaka and S. Fukui 1978b. Localisation of carnitine acetyltransferase in peroxisomes and in mitochondria of *n*-alkane grown *Candida tropicalis. Federation of the European Biochemical Society Letters* **96**, 37–40.

Kawamoto, S., T. Yamada, A. Tanaka and S. Fukui 1979. Distinct subcellular localisation of NAD linked glycerol-3-phosphate dehydrogenase EC 1.1.1.8 and FAD linked glycerol-3-phosphate dehydrogenase EC 1.1.99.5 in *n*-alkane grown *Candida tropicalis*. *Federation of the European Biochemical Society Letters* **97**, 253–6.

Kharat'yan, S. G., M. Y. A. Kalyuzhnyi, G. M. Petrushko, V. M. Belokov, A. I. Vol'Nova and T. V. Antonova 1975. Cultivation of the yeast *Mycoderma vini* on ethanol containing medium and nutritive evaluation of proteins isolated from them. *Prikladnaia Biokhimiia Mikrobiologiia* **11**, 35–8.

Kihlberg, R. 1972. The microbe as a source of food. *Annual Review of Microbiology* **26**, 427–66.

Kilian, S. G., B. A. Prior, P. M. Lategan and W. C. I. Kruger 1981. Temperature effects on ethanol and isoprenol utilisation by *Candida krusei*. *Biotechnology and Bio-engineering* **23**, 267–76.

Komagata, K., T. Nakase and N. Katsuya 1964. Assimilation of hydrocarbons by yeasts. II Determination of hydrocarbon assimilating yeasts. *Journal of General and Applied Microbiology* **10**, 323–31.

Kuraishi, M., I. Terao, H. Ohkauchi, N. Matsuda and I. Nagai 1979. Single cell protein – process development with methanol as substrate. *Doechema Monographs. Proceedings of the XII International Congress on Microbiology* **83**, 11–24.

Kuvshinnikow, V. D., M. Sobotka, I. G. Minkevich and V. K. Eroshin 1977. Dependence of cell yield and specific growth rate of *Candida boidinii* on temperature during continuous culture. *Folia Microbiologica* **22**, 189–97.

Laine, B. M., R. C. Snell and W. A. Peet 1976. Production of single cell proteins from *n*-paraffins. *Chemical Engineer* (London) **310**, 440–3, 446.

Lee, J. D. and K. Komogata 1980. Taxonomic study of methanol assimilating yeasts. *Journal of General and Applied Microbiology* **26**, 133–58.

Litchfield, J. H. 1979. Production of single cell protein for use in food or feed. In *Microbial technology, microbial processes*, H. J. Pepper and H. Perlman (eds), 2nd edn, Vol. 1, 93–105. New York: Academic Press.

Lonsane, B. K., H. D. Singh and J. N. Baruah 1975. Codification of petroleum hydrocarbon utilising microorganisms. *Journal of Scientific Industry Research* **34**, 614–24.

Machek, F., F. Stros, A. Prokop and L. Adamek 1975. Production and isolation of protein from synthetic ethanol. In *Continuous culture*, Vol. 6; *Applications and new fields*, A. C. R. Dean, D. C. Ellwood and C. G. T. Evans (eds), 135–45. Chichester: Ellis Horwood.

MacKenzie, C. R., S. J. Davids and G. R. Lawford 1980. Citric and isocitric acid production by *Saccharomycopsis lipolytica* grown on glucose and hexadecane. *Annual Meeting of the American Society of Microbiologists* **80**, 1484.

MacKenzie, C. R., S. J. Davids, G. R. Lawford and H. G. Lawford 1981. Citric and isocitric acid production from hydrocarbons by yeast fermentation. *Advances in Biotechnology (Proceedings of the 6th International Fermentation Symposium)* **2**, 57–62.

Marchal, R., J. P. Vandecasteele and M. Metche 1977a. Regulation of the central metabolism in relation to citric acid production in *Saccharomycopsis lipolytica*. *Archives of Microbiology* **113**, 99–104.

Marchal, R., C. Odile and M. Metche 1977b. Production of citric acid from *n*-paraffins by *Saccharomycopsis lipolytica*: kinetics and balance of the fermentation. *European Journal of Applied Microbiology* **4**, 111–23.

Masuda, Y., M. Nakanishi and Y. Sakakura 1976. Make single cell protein from ethanol. *Hydrocarbon Processing* **55**, 113–16.

Mateles, R. I. 1979. The physiology of single cell protein production. *Society of General Microbiologists Symposia* **29**, 29–52.

Mauersberger, S. and R. N. Matyashova 1980. The content of cytochrome $P_{450}$ in yeast cells growing on hexadecane. *Microbiology* **49**, 571–7.

Mesaric, J., V. Maric and A. Korculanin 1981. Biosynthesis of single cell protein from methanol by the yeast *Candida lipolytica* 33. *Prehrambene Technological Review* **19**, 3–6.

Mishina, M., T. Kamiryo, S. Tashiro and S. Numa 1978. Separation and characterisation of two long chain acyl CoA synthetases from *Candida lipolytica. European Journal of Biochemistry* **82**, 347–54.

Moresi, M., D. Cimarelli, G. Gasparrisi, G. Luizzo and R. Marinelli 1980. Kinetics of citric acid fermentation from *n*-paraffins by yeasts. *Journal of Chemical Technology and Biotechnology* **30**, 266–77.

Nakanishi, M. and T. Machida 1975. *Proceedings of the 1st Intersectional Congress of IAMS*, New York, **5**, 53.

Nakanishi, T., M. Yamamoto, K. Kimura and K. Tanaka 1972. Fermentative production of citric acid from *n*-paraffins by yeasts. *Hakko Kogaku Zasshi* **50**, 855–67.

Nout, M. J. R. 1972. A screening method for citric acid production by yeasts using glucose containing media and hydrocarbon containing media as substrates and *Candida* strains as test organisms. *Antonie van Leeuwenhoek Journal of Microbiology and Serology* **38**, 633–6.

Ogata, K., H. Nishikawa and M. Ohsugi 1969. A yeast capable of utilising methanol. *Agricultural Biology and Chemistry* **33**, 1519–20.

Ohnishi, T. 1973. Mechanisms of electron transport and energy conservation in a site I region of the respiratory chain. *Biochimica et Biophysica Acta* **301**, 105–28.

Omata, S., S. Murao and H. Terashima 1970. On yeast production from alcohol. Part 2, Yeast production from ethyl alcohol. *Journal of the Fermentation Association (Japan)* **26**, 317–21.

Orlova, V. S., N. M. Semikhatova and S. S. Rylkin 1980. Effect of ethanol concentration on physiological properties of *Saccharomyces cerevisiae* under periodic cultivation. *Microbiologicheskii Zhurnal* **42**, 718–22.

Osumi, M., N. Miwa, Y. Teranishi, A. Tanaka and S. Fukui 1974. Ultrastructure of *Candida* yeasts grown on *n*-alkanes. Appearance of microbodies and its relationship to high catalase activity. *Archives of Microbiology* **99**, 181–201.

Osumi, M., F. Fukuzumi, Y. Teranishi, A. Tanaka and S. Fukui 1975. Development of microbodies in *Candida tropicalis* during incubation of an *n*-alkane medium. *Archives of Microbiology* **103**, 1–12.

Osumi, M., M. Nagano, N. Yamada, J. Hosoi and Y. Mitsuhiro 1982. Three dimensional structure of the crystalloid in the microbody of *Kloeckera* sp: Composite crystal model. *Journal of Bacteriology* **151**, 376–83.

Paca, J. 1981. Effect of multistream ethanol feeding on growth and physiological characteristics of *Candida utilis* in a multistage tower fermenter. *Enzyme and Microbial Technology* **3**, 123–8.

Paca, J. 1982. Multistream ethanol and oxygen supply to a multistage tower fermentor during continuous yeast cultivations. *Journal of Fermentation Technology* **60**, 215–20.

Paca, J. and V. Gregr 1979. Effect of interstage mixing in multistage culture systems on continuous biomass production. *Biotechnology and Bioengineering* **21**, 1809–26.

Paca, J. and V. Gregr 1980. Growth characteristics of *Candida utilis* in a multistage culture system. *Enzyme Microbe Technology* **1**, 100–6.

Petrikevich, S. B. and L. I. Dovgun 1980. Ultrastructure of the yeast *Candida tropicalis*

during cultivation on *n*-alkanes: isolation in intracellular paraffin. *Mikrobiologie* **49**, 78–81.
Pilat, P. and A. Prokop 1975. The effect of methanol, formaldehyde and formic acid on growth of *Candida boidinii* 11BH. *Biotechnology and Bioengineering* **17**, 1717–28.

Ratledge, C. 1982. Microbial oils and fats: an assessment of their commercial potential. In *Progress in industrial microbiology*, M. J. Bull (ed.), Vol. 16, 119–206. Amsterdam: Elsevier.
Rattray, J. M. B. and J. E. Hambleton 1980. The lipid components of *Candida boidinii* and *Hansenula polymorpha* grown on methanol. *Canadian Journal of Microbiology* **26**, 190–5.
Roggenkamp, R., H. Sahm and F. Wagner 1974. Microbial assimilation of methanol induction and function of catalase in *Candida boidinii. Federation of the European Biochemical Society Letters* **41**, 283–6.
Rychtera, M. 1979. Single cell protein production from sulphite waste liquors on ethanol. *Doechema Monographs. Proceedings of the XII International Congress on Microbiology* **83**, 53–81.
Rychtera, M., G. Panuschlea, J. Vernerova and J. Mostecty 1982. Yeast biomass production on calcium bisulphite waste liquors with the addition of synthetic ethanol in a VB-12 fermenter. *Sbornik Vys. Sk. Chem Technol. Praze, Potraviny*, **E54**, 77–95.

Sahm, H. 1977. Metabolism of methanol by yeasts. *Advances in Biochemical Engineering* **6**, 77–103.
Sahm, H. 1979. Metabolic regulation in methanol utilisers. *Doechema Monographs. Proceedings of the XII International Congress on Microbiology* **83**, 95–109.
Sahm, H. and F. Wagner 1973. Microbial production from methanol. *Archives of Microbiology* **90**, 263–8.
Sato, S., T. Nakahara and Y. Minoda 1977a. Enzymatic studies on L-malic acid production from *n*-paraffins by *Candida brumptii* IFO-0731. *Agricultural Biology and Chemistry (Japan)* **41**, 1903–8.
Sato, S., T. Nakahara and Y. Minoda 1977b. L-Malic acid production from *n*-paraffins by *Candida brumptii* IFO-0731. *Agricultural Biology and Chemistry* **41**, 967–73.
Schimizu, S., K. Yasui, Y. Tani and H. Yamada 1979. Acyl CoA oxidase for *Candida tropicalis. Biochemical and Biophysical Research Communications* **91**, 108–13.
Sestakova, M., L. Adamek and F. Stros 1976. Effect of crotonaldehyde on the metabolism of *Candida utilis* during the production of single cell protein from ethanol. *Folia Microbiologica* **21**, 444–54.
Shacklady, C. A. 1975. Yeast growth on hydrocarbons. *International Biological Programme No. 4, Food protein sources*, 215–22. Cambridge: Cambridge University Press.
Sherwood, M. 1984. The case of the money hungry microbe. *Biotechnology* **2**, 606–9.
Shukla, J. P. and A. Aroj 1967. *35th Proceedings of the Annual Convention on Sugar Technology Association (India)*, 285.
Suomalainen, H. and E. Oura 1979. Ethanol as a substrate for baker's yeast. *Doechema Monographs. Proceedings of the XII International Congress on Microbiology* **83**, 43–51.

Tabuchi, T. and S. Hara 1974. Production of 2-methyl isocitric acid from *n*-paraffins by mutants of *Candida lipolytica. Agricultural Biology and Chemistry (Japan)* **38**, 1105–6.
Tanaka, A., M. Osumi and S. Fukui 1982. Peroxisomes of alkane grown yeast: fundamental and practical aspects. *Annals of the New York Academy of Sciences* 183–99.

Tani, Y., T. Miya and K. Ogata 1972. The microbial metabolism of methanol. Part 1. Properties of crystalline alcohol oxidase from *Kloeckera* sp. No. 2201. *Agricultural Biology and Chemistry (Japan)* **36**, 76–83.
Tezuka, H., T. Nakahara, Y. Minoda and K. Yamada 1975. Production of yeast cells from methanol. *Agricultural Biology and Chemistry (Japan)* **39**, 285–6.
Tyi, N. T. F., T. I. Chistyakova, I. G. Minkevich and V. K. Eroshin 1982. Growth of the thermotolerant yeast *Candida valida* during continuous cultivation on ethanol. *Mikrobiologie* **51**, 379–84.

Ueda, M., A. Tanaka and S. Fukui 1982. Peroxisomal and mitochondrial carnitine acetyltransferase in alkane grown yeast, *Candida tropicalis*. *European Journal of Biochemistry* **124**, 205–10.
Ueda, M., H. Okado, A. Tanaka, M. Osumi and S. Fukui 1983. Induction and subcellular localisation of enzymes participating in propionate metabolism in *Candida tropicalis*. *Archives of Microbiology* **136**, 169–76.
Urakami, T., I. Teroa and I. Nagai 1983. Isolation, identification and cultivation of methanol utilising yeasts. *Journal of Fermentation Technology* **61**, 221–32.

Veenhuis, M., J. P. van Dijken and W. Harder 1976. Cytochemical studies on the localisation of methanol oxidase and other oxidases in peroxisomes of methanol grown *Hansenula polymorpha*. *Archives of Microbiology* **111**, 123–35.
Veenhuis, M., K. B. Zwart and W. Harder 1981. Biogenesis and turnover of peroxisomes involved in the concurrent oxidation of methanol and methylamine in *Hansenula polymorpha*. *Archives of Microbiology* **129**, 35–41.
Vernerova, J. and M. Rosa 1964. Czechoslovak Patent, 109 658.

Waites, M. J. and J. R. Quayle 1980. Dihydroxyacetone: a product of xylulose 5′-phosphate dependent fixation of formaldehyde by methanol grown *Candida boidinii*. *Journal of General Microbiology* **118**, 321–7.
Watteeuw, C. M., W. B. Armiger, D. L. Ristroph and A. E. Humphrey 1979. Production of single cell protein from ethanol by fed batch process. *Biotechnology and Bioengineering* **21**, 1221–38.
Wiken, T. O. 1972. *Fermentation Technology Today. Proceedings of the 4th International Fermentation Symposium* 569–96.
Wolf, H. J., M. Christiansen and R. S. Hanson 1980. Ultrastructure of methanotrophic yeasts. *Journal of Bacteriology* **141**, 1340–9.

Yamada, T., H. Nawa, S. Kawamoto, A. Tanaka and S. Fukui 1980. Subcellular location of long chain alcohol dehydrogenase and aldehyde dehydrogenase in *n*-alkane grown *Candida tropicalis*. *Archives of Microbiology* **128**, 145–51.
Yang, S. S., C. C. Liao, F. I. Tang and H. H. Wang 1982. Temperature sensitive methanol utilising yeasts in single cell protein production. *American Society of Microbiology* 149.

# PART IV

# Product formation

# 11 *Production of organoleptic compounds*

D. R. BERRY and D. C. WATSON

## 11.1 Introduction

Yeasts are involved in the production of a variety of foods and beverages (Read & Peppler 1973, Wood 1985). In addition to the production of carbon dioxide and ethanol by fermentation, which is important in the production of both bakery products and alcoholic beverages, yeast also imparts a flavour to many of these products as a result of the production of a wide variety of organoleptic compounds. The production of these organoleptic compounds has been most extensively studied in the context of alcoholic beverages, especially beer. Most of this chapter is concerned with these processes; however, the role of yeast as a source of flavour compounds in the food industry should not be ignored.

*Saccharomyces cerevisiae* is grown aerobically on a large scale to produce baker's yeast. This is mixed with flour at a concentration of up to 6% by dry weight during the production of bread and can contribute a yeast flavour to the bread. The yeasty flavour does not appear to be associated with organoleptic compounds produced by the yeast during the raising of the dough. It has been reported that it is the thiamin content of the yeast which is responsible for the yeast flavour (Hohn *et al*. 1975, Margalith 1981). Aerobically grown yeast can also be modified to produce highly flavoured extracts using processes which involve either the natural autolysis of yeast or the hydrolysis of yeast by acid. Such products are widely used in the food industry to flavour preparations such as potato crisps, gravy and sauces, etc. (Peppler 1979).

Alcoholic beverages are produced throughout the world using a wide variety of carbohydrate substrates, such as grape juice, molasses, barley wort and starch from other sources such as maize, rice and cassava. In most of these processes the main alcoholic fermentation is carried out using *Saccharomyces cerevisiae*; however, a mixed population of yeasts is often used in wine production, and in some brewery fermentations. Fission yeast *Schizosaccharomyces pombe* is used in rum production. The production of ethanol by yeast is associated with the production of a wide variety of fermentation products which contribute to the final flavour of the beverage, either as organoleptic compounds, or precursors of organoleptic compounds which are produced in subsequent maturation or distillation processes.

## 11.2 Yeast organoleptic compounds

The organoleptic compounds produced by yeast can be classified into five categories: alcohols, esters, aldehydes and ketones (carbonyl compounds), sulphur-containing compounds, and organic acids. In addition, the yeast may affect flavour by altering compounds already present in the substrate. This is particularly true of the modification of carbonyls during alcohol fermentation. Several lists of organoleptic compounds produced during different alcoholic beverage processes have been published for beer (Engan 1981), whisky (Swan *et al.* 1981), wine and distilled beverages in general (Harrison & Graham 1970, Nykänen & Suomalainen 1983).

The amount of organoleptic compounds present is not, however, the only parameter to be considered in evaluating the flavour of a beverage, since different compounds have different sensory odour thresholds. A compound with a low threshold can be more important in influencing flavour than one with a higher threshold, even if present in smaller amounts. Odour thresholds and evaluation of the contribution of a compound to the overall odour of a product are difficult areas of study, since they can be influenced by other organoleptic compounds present. However, the list given in Table 11.1 gives a guide to the relative importance of different groups of compounds. It is clear that although the higher alcohols are the most abundant group of organoleptic compounds, they have odour thresholds ten times higher than the esters and 1000 times higher than the carbonyl compound diacetyl, so their contribution to the overall flavour of alcoholic beverages is not the most important.

Since the range of organoleptic compounds produced by yeast is similar in all alcoholic beverages, this chapter concentrates on the common features of the formation of organoleptic compounds produced by yeast; namely the biosynthetic pathways leading to their formation and the effect of different fermentation parameters on their rates of formation.

## 11.3 Higher alcohols

### *11.3.1 Biosynthesis of higher alcohols*

Some 45 alcohols have been reported in beer (Engan 1981); however, ethanol, *n*-propanol, isobutanol, 2-methyl-1-butanol (optically active amyl alcohol), 3-methyl-1-butanol (isoamyl alcohol) and phenethyl ethanol are the most important in beer and other alcoholic beverages. Most of these can be derived from the carbon skeletons of common amino acids. The relationship between higher alcohol formation and amino-acid metabolism was recognized as early as 1907 by Ehrlich. When a fermentation medium containing amino acids is used for yeast growth, the carbon skeletons of some of the amino acids are converted into higher alcohols by the catabolic sequence of reactions shown in Figure 11.1. The amino group is removed by transamination, then the resulting keto acid is decarboxylated to the corresponding aldehyde. This can be reduced by the same alcohol dehydrogenase enzyme which reduces acetaldehyde to ethanol, producing a higher alcohol with one less carbon atom than

**Table 11.1** Flavour thresholds of selected organoleptic compounds. The range of concentration found in beer is included for reference purposes (data obtained from Meilgaard (1975) and Engan (1981)).

| | Flavour threshold (ppm) | Range of concentration in beer (ppm) |
|---|---|---|
| alcohols | | |
| ethanol | 14 000 | — |
| 1-propanol | 800 | 7.5–13.8 |
| 2-propanol (*iso*-propanol) | 1500 | 0.2–2.4 |
| 1-butanol | 450 | — |
| 2-methyl propanol (*iso*-butanol) | 200 | 8.6–56.6 |
| 2-butanol | 16 | — |
| 2-methyl butanol (optically active amyl alcohol) | 65 | 7–23 |
| 3-methyl butanol (*iso*-amyl alcohol) | 70 | 27–122 |
| 2-phenethanol | 125 | 5–27 |
| acids | | |
| acetic | 175 | 150–280 |
| propionic | 150 | 5 |
| butyric | 2.2 | 0.6–3.3 |
| *iso*-butyric | 30 | 0.7–3.3 |
| caproic | 8 | 2.2–5.8 |
| caprylic | 13 | 3.3–8.2 |
| capric | 10 | 0.1–2.0 |
| phenylacetic | 2.5 | 0.93 |
| lactic (D + L) | 400 | 28–400 |
| esters | | |
| ethyl acetate | 33 | 8.2–47.6 |
| *n*-butyl acetate | 7.5 | 0.23 |
| *iso*-butyl acetate | 1.6 | 0.03–0.25 |
| phenethyl acetate | 3.8 | 0.1–1.17 |
| ethyl butyrate | 0.4 | 0.09 |
| ethyl caproate | 0.23 | — |
| ethyl caprylate | 0.9 | 0.08–0.01 |
| ethyl caprate | 1.5 | — |
| ethyl lactate | 250 | 0.1 |
| aldehydes and ketones | | |
| acetaldehyde | 25 | 2.5–24.4 |
| *iso*-butyraldehyde | 1.0 | 0–0.024 |
| acetone | 200 | 1 |
| pyruvate | 300 | 10–220 |
| diacetyl | 0.15 | 0.5–2.0 |
| sulphur compounds | | |
| dimethyl sulphide | 50 | 0–0.144 |

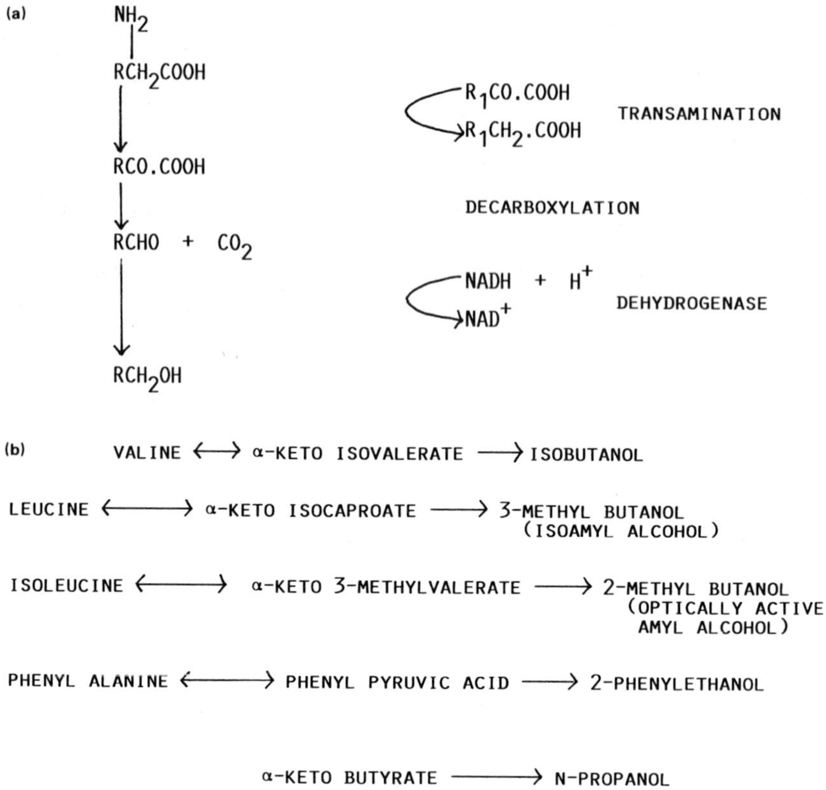


**Figure 11.1** Formation of higher alcohols from amino acids and the corresponding keto acids. (a) Biochemical sequence; (b) relationship between selected amino acids and the corresponding higher alcohols.

the original amino acid. However, this is not the only route for the biosynthesis of higher alcohols, since they can also be produced when there are no amino acids left in the substrate. In these conditions, the $\alpha$-keto acid containing one more carbon atom than the higher alcohol in question is synthesized by the same biosynthetic pathway that is used by the cell to produce the corresponding amino acid. The keto acid is then decarboxylated and reduced to the corresponding higher alcohol. Propanol appears to be produced from $\alpha$-keto butyrate by this biosynthetic route only, and is not derived from L-threonine, the amino acid produced from $\alpha$-keto butyrate. In the formation of the other higher alcohols the amount of alcohol produced and the relative importance of the two routes for their formation varies with the fermentation conditions.

### *11.3.2 Factors influencing higher alcohol formation*

Factors influencing the formation of higher alcohols in the beer fermentation have been reviewed by Engan (1981) and MacDonald *et al.* (1984), and in distilled beverages by Berry (1984) and Watson (1985). However, most of the work reported has been carried out in batch fermentations using complex media, so it is difficult to establish a direct relationship between a changed fermentation parameter and the change in higher alcohol levels. The addition of specific amino acids can stimulate the level of the corresponding higher alcohols; however, increasing the level of nitrogen by using amino acids such as glutamate and aspartate, which do not correspond to specific higher alcohols, or adding ammonium salts as the nitrogen source can also stimulate higher alcohol formation. However, the higher alcohols are a heterogeneous group, and the response to an increase in nitrogen varies with the higher alcohol being studied. Propanol in particular is stimulated by the addition of ammonia more than by increasing the $\alpha$-amino nitrogen content of the wort.

It is perhaps more surprising that changes in the carbohydrate composition of the growth medium caused by the addition of the readily metabolizable sugars such as glucose and sucrose to the wort as adjuncts, have been found to stimulate higher-alcohol formation. A stimulatory effect of glucose has also been observed when maltose was replaced by glucose in a malt whisky fermentation in which the $\alpha$-amino nitrogen level of the wort was held constant (Ramsay 1982).

Besides being influenced by medium composition, higher-alcohol formation is also influenced by fermentation conditions. Increased fermentation temperatures have been reported to increase higher-alcohol formation in both beer and whisky fermentations (Engan 1981, Ramsay & Berry 1984a). High pitching rates have not been reported to have a marked effect on the formation of higher alcohols in beer. However, oxygenation of worts, which leads to increased yeast growth in beer fermentations, has been reported to stimulate the production of higher alcohols (Geiger & Piendl 1973). In studies on a small-scale whisky fermentation, increasing inoculum and agitation in an open vessel only resulted in a small stimulation of higher-alcohol formation (Berry & Chamberlain 1983). This difference between beer and whisky fermentations can probably be attributed to the fact that the distilling yeast used in the whisky industry is grown aerobically and is not deficient in unsaturated fatty acids and sterols, as is much of the repitched yeast used in the brewing industry (Watson 1981, Berry 1984).

In general, fermentation procedures such as increased agitation, increased temperature and the use of continuous culture have been reported to increase higher alcohol formation. However, fermentation under pressure (0.8–2.0 atm) reduces higher-alcohol formation and can be used to control the excessive higher-alcohol formation which occurs at the higher fermentation rates obtained at higher temperatures. Experiments carried out using continuous culture in a defined medium indicated that when all of the parameters are held constant, higher-alcohol formation is actually reduced at higher growth rates (Berry & Chamberlain 1985).

Evidence has been presented that different brewing yeast strains produce different quantities of higher alcohols, both in laboratory-scale experiments and in full-scale plant trials (Engan 1981), and most breweries consequently take great care with their strain. On the other hand, the distilling industry rarely produces its own yeast but often purchases it either from food yeast companies or as spent brewer's yeast. Little attempt seems to be made to control the flavour of the spirit by choice of yeast strain (Watson 1985). Wine producers traditionally retain their own strain of yeast, and many of these have been shown to produce different quantities of higher alcohol (Kunkee & Goswell 1977). Recently an attempt has been made to reduce the level of higher alcohols in a wine fermentation by using a leucine auxotrophic mutant of a wine yeast (Snow 1983, Rous *et al.* 1983). One strain of wine yeast which was auxotrophic for leucine and lacked the enzyme $\alpha$-isopropylamate dehydratase, gave rise to a wine which had 20% less total higher alcohol and 50% less isoamyl alcohol than the parent strain. It is important to note that the wine produced by this strain did not exhibit any detectable decrease in flavour quality.

The biosynthetic pathways leading to isoleucine and valine are of considerable importance in the formation of higher alcohols and vicinal diketones such as diacetyl and 2,3-pentanedione. The genes coding for several of the enzymes involved in the biosynthesis of isoleucine and valine; namely threonine deaminase, aceto $\alpha$-hydroxy acid reductoisomerase, dihydroxy acid dehydratase and a branched-chain amino acid aminotransferase have been cloned (Holmberg 1984) and are being studied with a view to manipulating their regulatory mechanism to control the level of vicinal diketones in lager.

## 11.4 Organic acids

### *11.4.1 Organic acid formation*

Some 100 organic acids have been reported in alcoholic beverages. These arrive from three areas of yeast metabolism. Those such as acetate, succinate, $\alpha$-ketoglutarate, malate and citrate, which are derived from pyruvate via limited tricarboxylate acid cycle function (Fig. 11.2), and pyruvate itself constitute a qualitatively important group of acids. They may have a direct effect on flavour (e.g. the 'mouth feel' flavour of pyruvate), but they also contribute to the pH of the beer. Some, such as isobutyric and isovaleric acids, are probably derived from the amino-acid biosynthetic pathways, but the major group are produced from malonyl CoA by the fatty acid synthetase pathway (Lynen 1967; see Fig. 11.2). In yeast malonyl CoA is produced by a biotin-dependent acetyl CoA carboxylase. It is the substrate for the fatty acid synthetase, a single large multifunctional enzyme which is responsible for the production of saturated fatty acids in yeast and other organisms. Unsaturated fatty acids are produced from saturated fatty acids by an oxygen-requiring desaturase reaction. The main fatty acids produced by yeast are palmitic acid (16.0), palmitoleic acid (16.1), stearic acid (18) and oleic acid (18.1). However, shorter-chain fatty acids such as $C_6$ hexanoic (caproic) acid, $C_8$ octanoic (caprylic) acid and $C_{10}$ decanoic (capric) acid, are also produced by this route rather than the $\beta$-oxidation of longer-chain fatty acids (Taylor & Kirsop 1977). They have been considered to have been leaked from the main biosynthetic pathway. These medium-chain fatty acids are important flavour compounds in their own right; medium-chain fatty acids have been reported to give a caprylic, goaty, soapy or fatty flavour to beers, and when released by autolysis during the

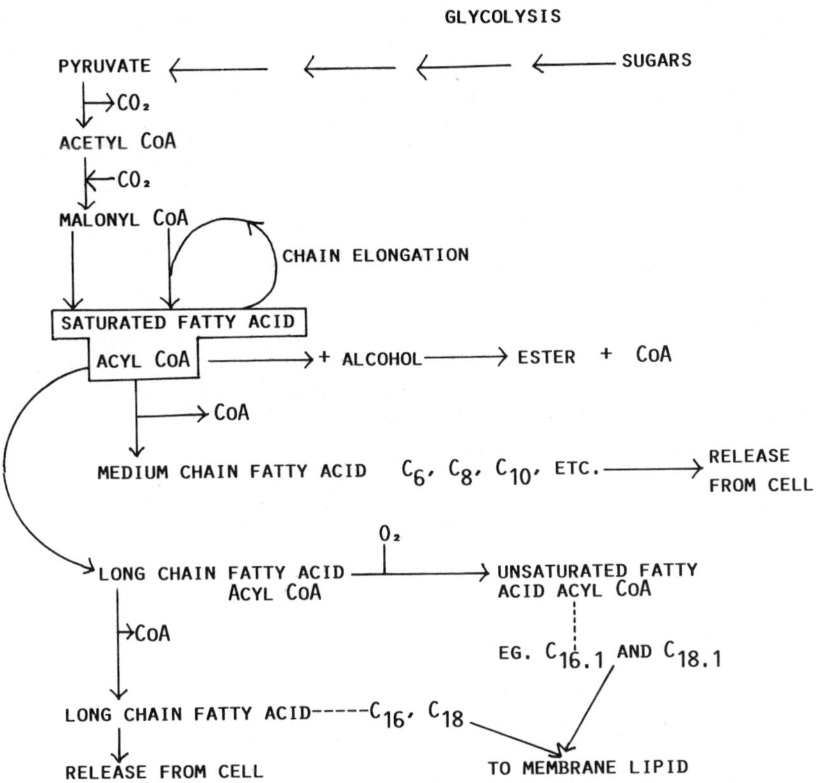


**Figure 11.2** Origin of fatty acids in yeast.

maturation of beer they have been associated with a yeast flavour (MacDonald *et al.* 1984). However, they are also important as precursors of esters in alcoholic beverage production. Most of the medium-chain and higher fatty acids produced by the yeast remain in the cell. Whereas these do not appear in beer and other non-distilled beverages to a great extent, they are released by distillation during the production of distilled spirits if the yeast is left in the wash during the distillation stage (Suomalainen & Nykanen 1966). Medium-chain fatty acids, particularly decanoic acid, have been reported to be released from the yeast cell during the maturation of beers and may contribute to yeast off-flavours which can develop during beer maturation (van der Meersshe *et al.* 1977).

### *11.4.2 Factors influencing organic acid formation*

The effect of fermentation conditions on the formation of fatty acids is complex and varies with the process parameter and the group of fatty acids being studied. The medium-chain fatty acids have been most intensively studied, and been found to be influenced by wort composition, yeast strain, inoculum level, fermentation temperature, pH, pressure and degree of agitation. In general, parameters which increase the rate of the fermentation, such as increasing pitching rate, increased fermentation temperature and increased mixing, lead to reduction in the level of fatty acids (Engan 1981, MacDonald *et al.* 1984). However, the most important parameter appears to be the level of aeration. In beer fermentations oxygenation of the wort is carried out to make good the deficiency of unsaturated fatty acids and steroids. This leads to a marked stimulation of growth. The demand for saturated fatty acids for membrane formation is increased and less acetyl CoA is diverted into the formation of medium-chain fatty acids. It has been reported (Sumper 1974) that when saturated fatty acids accumulate under fermentative conditions, they inhibit the activity of acetyl CoA carboxylase, and release of medium-chain fatty acids from the fatty acid synthetase complex is favoured. Utilization of the saturated fatty acids by resumed growth would reverse this process. One effect of increased agitation can be to increase the level of oxygenation of the wort. However, it also leads to increased mass transfer of substrates and increased growth rate, which leads to a limitation of the availability of free saturated fatty acids. Increased fermentation rates can also be achieved by using higher inoculum levels and higher fermentation temperatures. Both of these lead to a reduction in the level of medium-chain fatty acids produced in alcohol fermentations.

Wort is a complex medium and has diverse effects on acid formation. The level of nitrogen does not appear to have a direct effect on the level of most fatty acids, although the presence of high leucine levels can stimulate the formation of isovaleric acid and isobutyric acid. Wort contains a high level of saturated and unsaturated long-chain fatty acids. Many of these are lost during the boiling stage of beer wort productions, but are present in unboiled whisky worts. These long-chain fatty acids are assimilated by the yeast during the fermentation. When anaerobically grown yeast is used in an inoculum for beer fermentations, unsaturated fatty acids in the wort are essential for yeast growth. Dilution of the wort can lead to dilution of the unsaturated fatty acids and a limitation of yeast growth.

It has also been reported that increased levels of phosphate, magnesium and biotin stimulate fatty acid formation in beer fermentations (Forch *et al.* 1975). Several workers have reported that yeast strain is important in determining the level of medium-chain fatty acids in beer, and it is generally considered that lager yeasts produce more fatty acids than do ale yeasts (Clapperton & Brown 1978).

## 11.5 Esters

### *11.5.1 Ester biosynthesis*

Esters are numerically the largest group of organoleptic compounds in alcoholic beverages. Although some ester formation may occur during the distillation of spirits, the most common esters are produced by the yeast during the

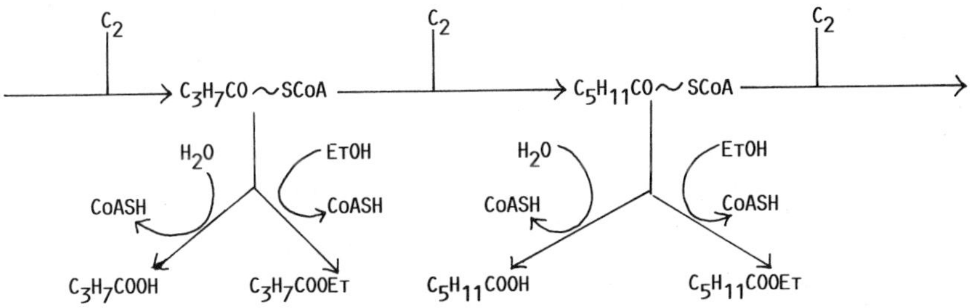


**Figure 11.3** Formation of esters and medium-chain length fatty acids by yeasts.

fermentation stage. Nordstrom (1964) proposed that esters were produced in the yeast cell by an enzymic reaction between acetyl CoA derivatives of fatty acids and free alcohols (Fig. 11.3) rather than by an extracellular chemical reaction. The presence of the enzyme responsible for the reaction was later demonstrated by Howard and Anderson (1976). Experiments involving the addition of individual acids and alcohols have indicated that there is competition between different alcohols and esters in ester formation such that the most abundant esters are derived from the most abundant acids and alcohols. Since ethanol is the most abundant alcohol the ethyl esters are the most abundant, followed by isoamyl and propyl esters. Acetate is the most abundant acid formed by yeast during fermentations, so acetate esters of ethanol and higher alcohols are the most abundant. Ethyl acetate at up to 50 mg $l^{-1}$ in beer and 175 mg $l^{-1}$ in certain whiskies is the most abundant ester in alcoholic beverages. However, since its flavour threshold is not high (25 mg $l^{-1}$ in beer) it is not necessarily the most important in terms of flavour.

### *11.5.2 Factors influencing the formation of esters*

The level of individual esters can be considered to be influenced by three independent factors: the availability of the corresponding acetyl CoA, the availability of the corresponding alcohol and the actual rate of ester formation. Since ethanol is present in high concentrations in anaerobic yeast fermentations, the rate of ethyl ester formation is related to the availability of the fatty-acid component of the ester. However, the rate of formation of acetate esters, other than ethyl acetate, is largely influenced by the availability of the corresponding higher alcohol since acetate is relatively abundant. In this situation changes in the availability of the higher alcohol do not influence the abundance of the alcohol ester relative to the free alcohol. Addition of specific amino acids to a small-scale whisky fermentation resulted in an increase in the level of the corresponding acetate esters, but the ratio of higher alcohols to higher alcohol esters remained unchanged (Ramsay 1982, Berry 1984). Varying the level of $\alpha$-amino nitrogen in the wort from a low level to 150 mg $l^{-1}$ resulted in an increase in acetate esters of higher alcohols, but not in the ratio of esters to higher alcohols. It would be anticipated that this ratio would be changed either by an increase in the level of acetate or possibly by a change in the availability of the enzyme controlling ester formation.

Different yeast strains have been reported to have characteristic patterns of acetate ester formation which relate to the capacity of the strain to produce higher alcohols (Engan 1981). A reduction in acetate esters has been reported when the inoculum level was increased in beer fermentations (Maule 1967). In studies on the whisky fermentation there was an initial increase in acetate esters as the inoculum level was increased (Ramsay & Berry 1984b). However, increasing inoculum levels caused a decrease in the level of ethyl esters of fatty acids. This was associated with a decrease in the level of medium-chain fatty acids produced by the yeast.

Wort composition and oxygen appeared to be the most important parameters affecting ethyl ester formation. High wort gravities lead to a dramatic increase in ester formation, giving rise to a problem in producing high-gravity beers. The relative levels of $\alpha$-amino nitrogen and carbohydrates do not appear to have much effect on ethyl ester formation; however, the level of unsaturated fatty acids in the wort are of critical importance. As was indicated in Section 11.4.1, yeasts grown in anaerobic conditions have an absolute requirement for unsaturated fatty acids and sterols. In general the presence of unsaturated fatty acids in the wort leads to a stimulation of yeast growth and a decrease in the formation of medium-chain fatty acids and their ethyl esters (Thurston *et al.* 1982). Saturated fatty acids and sterols are present in spent grain 'pressings', so some brewers have attempted to control ester formation by the addition of spent grain 'pressings' to the wort. Part of the stimulatory effect of unsaturated fatty acids may be attributable to a stimulation of growth; however, it has been proposed that linoleic acid may have a more direct effect on the membrane-bound ester synthesizing enzymes, since it had been reported to repress ester formation by up to 40% without having any effect on growth (MacDonald *et al.* 1984). However, the effect of unsaturated and saturated fatty acids on yeast growth are complex and very sensitive to the concentration of fatty acids supplied (Ramsay & Berry, unpubl. results). It seems possible that some of the effects obtained when wort was diluted by sugar or sugar mixtures may be attributable to dilution out of lipids in the wort. Aeration of worts had a similar effect on ester formation, as had the supply of unsaturated fatty acids. A continuous but low level of aeration has been found to be most effective in suppressing ester formation (Thurston *et al.* 1982). In these conditions it has been proposed that the continued growth of the yeast increases the demand for cellular lipids and decreases the availability of acetyl CoA for the formation of the fatty acids required for ester formation. Low levels of oxygen have also been reported to suppress the formation of selected fatty acids in whisky fermentations which have been inoculated with aerobically grown yeast not considered to be deficient in unsaturated fatty acids and sterols. In these conditions ethyl esters of unsaturated fatty acids were formed (Ramsay 1982, Ramsay & Berry 1984b).

Ester formation has also been reported to be influenced by fermentation pressures. Most results indicate that increased pressure leads to a reduction in growth and ester formation. However, increases in isoamyl acetate and phenyl acetate have been reported. The effect of pressure has been attributed to the effect of increased partial pressure of carbon dioxide on the decarboxylation of pyruvate and on malonyl CoA functions. Carbon dioxide has also been

**Table 11.2** Effect of glucose concentration on the formation of ethanol, amyl alcohol and amyl acetate at constant growth rate (from Berry & Chamberlain 1986).

| $D = 0.075\ h^{-1}$ glucose (g/l) | Dry weight ($g\ l^{-1}$) | Medium glucose ($g\ l^{-1}$) | Ethanol ($g\ l^{-1}$) | Amyl alcohol ($mg\ l^{-1}$ distillate) | Amyl acetate ($mg\ l^{-1}$ distillate) |
|---|---|---|---|---|---|
| 40 | 4.3 | 0.004 | 18.9 | 185 | — |
| 60 | 2.8 | 3.7 | 25.1 | 239 | 21 |
| 80 | 2.7 | 10.0 | 31.4 | 300 | 27 |
| 100 | 3.7 | 16 | 36.0 | 435 | 46 |
| 120 | 3.4 | 25 | 42.0 | 388 | 40 |

reported to affect the physical properties of the yeast cell membrane which may, in turn, influence ester formation. Removal of carbon dioxide by flushing the fermentation with nitrogen stimulates growth and ester formation (MacDonald *et al.* 1984).

Berry & Chamberlain (1986) have carried out a series of studies on the regulation of the formation of organoleptic compounds by yeast in defined media using continuous culture. In this way different fermentation parameters such as substrate level, pH, growth rate and temperature can all be controlled independently. It is clear that the relationship between growth and the formation of higher alcohols and esters is influenced by the level of glucose in the medium. High levels of esters were only obtained when there was an excess of glucose in the growth medium (Berry & Chamberlain 1986; Table 11.2).

Airflow rates were also very critical; whereas growth was repressed when no air was passed through the fermenter, even flow rates of 10 ml $min^{-1}$ were sufficient to give maximum cell yields. However, at this flow rate the level of ester formation was severely repressed (Berry & Chamberlain 1986). The regulation of the formation of flavour compounds by yeast is clearly complex. Investigation of the physiological parameters affecting the formation of higher alcohols, acids and esters in batch fermentation using complex media can give rise to misleading results, since the alteration of one parameter, e.g. the addition of adjunct, can affect several other characteristics of the fermentation. The fermentation conditions finally selected will inevitably be a compromise between different requirements. In these studies the aeration level which was required to give adequate yeast growth was too high to obtain maximum levels of esters.

## 11.6 Carbonyls

### *11.6.1 Introduction*

The carbonyl group of compounds consists of aliphatic and aromatic aldehydes and ketones. Acetals, which are formed by the stepwise addition of two alcohol molecules to an aldehyde in a chemical reaction catalysed by acids, and ketals, the ketone equivalents, are often included in the carbonyls. More than 200 members of this group have been detected in the range of alcoholic beverages (Nykänen & Suomalainen 1983). They are perceived as being extremely important flavour compounds, since some have very low thresholds; although this is not true of all of them, e.g. methyl ketones have similar thresholds to primary alcohols (Meilgaard 1975). The majority do not emanate from the action of yeast but instead result from chemical reactions, particularly in the pre- and post-fermentation stages of production processes. Saturated and unsaturated carbonyls are produced in a range of thermal reactions which occur during raw material and substrate processing, as well as in product pasteurization and distillation (Tressl *et al.* 1983, Watson 1985). Raw materials such as hops, fruits and their stones, can be another source of carbonyls. Wooden casks contribute a range of carbonyl compounds, particularly aromatic aldehydes, to matured beverages (Reazin 1981).

Although it cannot be classed as *de novo* synthesis, yeasts play an important role in transforming many of the highly flavour-active aldehydes and ketones present in fermentation substrates into compounds with higher thresholds or more desirable aromas, via reduction reactions. Aldehydes and vinyl ketones are reduced to their corresponding alcohols and acetate esters. Some of the saturated and non-conjugated unsaturated ketones are only partially reduced. It has been shown that the highly flavour-active carbonyls present in brewer's worts do not survive fermentation, and thus have little influence on beer flavour (Peppard & Halsey 1981, Peppard *et al.* 1981, Schreier 1979). Fermentation substrates derived from stone fruit can contain significant levels of benzaldehyde, some of which can be reduced to benzyl alcohol by the fermentation yeast (Albagnac & Vangheesdaele 1975).

### *11.6.2 Biosynthesis of aldehydes*

Aldehydes are synthesized by yeast as intermediates in the formation of alcohols through the decarboxylation of keto acids. The majority are further reduced by alcohol dehydrogenases, but a small amount may be oxidized to acids. During the active phase of fermentations, excess quantities can be excreted into the fermentation broth. The corresponding aldehydes to most of the alcohols formed by yeast have been detected in many alcoholic fermentations (Engan 1981, Nykänen & Suomalainen 1983). Acetaldehyde is thus quantitatively the most significant compound of this group. Generally, aldehydes have flavour thresholds two-to-three orders of magnitude below the alcohols. The aroma of the lower aldehydes is generally perceived as green leaves, grassy and even fruity. However, as the chain length increases they become more unpleasant, being cardboard-like and bitter (Meilgaard 1975). It is only in a few cases that aldehydes produced during fermentation have a significant flavour impact. Those produced outside the fermentation have greater importance, e.g. a stale flavour in beer is caused by oxidation reactions resulting in the formation of carbonyls, of which 2-nonenal is thought to be the major contributor to the off-flavour (Engan 1981, MacDonald *et al.* 1984).

### *11.6.3 Factors influencing the formation of aldehydes*

The majority of studies on parameters affecting aldehyde formation have concentrated on acetaldehyde. However, it can be accepted that many of the findings would hold for the group as a whole. As they are intermediates in the biosynthesis of alcohols, most of the parameters which influence the production of the latter, as discussed earlier, also influence aldehyde formation. In addition, excess quantities are produced when the reduction reaction, catalysed by alcohol dehydrogenases, is rate limiting. Parameters which increase the initial fermentation rate, such as aeration, readily utilizable sugars and other nutrients, higher temperatures, fast-fermenting yeast strains and higher pitching rates result in increased accumulation of aldehydes (Geiger & Piendl 1976, Engan 1981).

In the later stages of fermentation at least some of the excreted aldehyde can be reabsorbed and reduced or, to a lesser extent, oxidized by the yeast, if the latter is still highly active. Physiologically exhausted cells will be least able to further metabolize the aldehydes. The extreme would be cells which have died. Various factors influence the late fermentation activity. These include the condition of the yeast inoculum, the level of essential nutrients, including oxygen or unsaturated fatty acids, the alcohol concentration and the fermentation temperature. The latter two are generally associated with alcohol intolerance. However, low temperatures can inhibit the fermentation rate, such that the yeast can settle out prematurely, particularly if it is flocculent, thus restricting further fermentation (Geiger & Piendl 1976, MacDonald *et al.* 1984).

The final concentrations of aldehydes in a yeast fermentation are thus a balance between those which are formed in the initial stages of fermentation, and those which are re-utilized in the later stages. In addition, the presence of antioxidants which form complexes with aldehydes, such as sulphite ions and sulphur dioxide, can enhance the final concentrations (MacDonald *et al.* 1984). Although other parameters may have a more significant effect (Engan 1981), the yeast strain's inherent biochemistry will also influence aldehyde levels.

### *11.6.4 Biosynthesis of ketones*

It should be remembered that the keto acids are themselves ketones. Although they have low volatility and high odour thresholds, the most abundant, pyruvate, and to a lesser extent $\alpha$-ketobutyric acid, do influence beer and wine flavours by affecting the mouthfeel (Nykänen & Suomalainen 1983). Their role as intermediates in the higher alcohol production has been dealt with earlier. They are also involved in the formation of vicinal and 3-hydroxy-2-ketones, as outlined in Figure 11.4 (Wainwright 1973). The mono ketones are generally not counted as organoleptically important compounds (Meilgaard 1975).

The vicinal diketones, 2,3-butanedione (diacetyl) and 2,3-pentanedione are significant flavour compounds due to their low odour thresholds. In beers, the latter has a threshold value of about 1 mg $l^{-1}$, whereas the figure for diacetyl is in the range 0.1–0.2 mg $l^{-1}$ for lagers and approximately 0.4 mg $l^{-1}$ for ales. Both have a butterscotch-type aroma, which is perceived as being desirable in

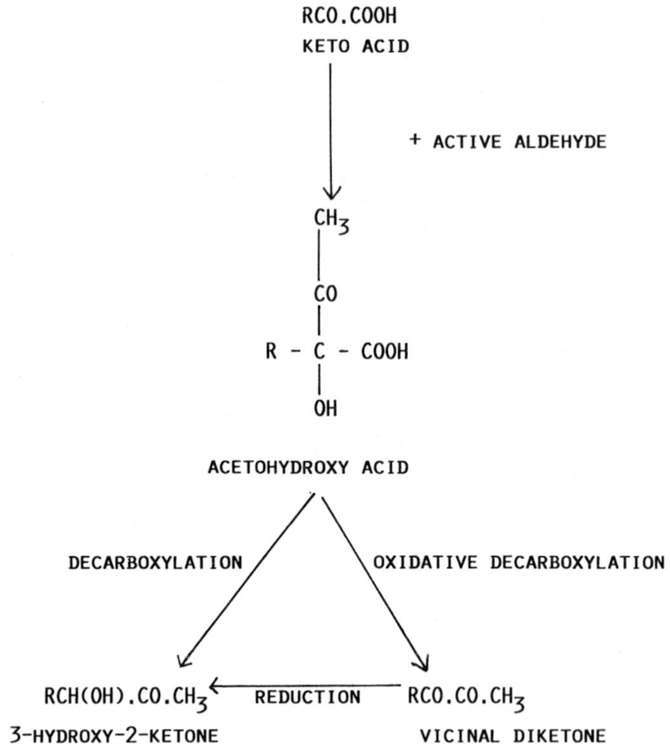


**Figure 11.4** The formation of vicinal diketones and 3-hydroxy-2-ketones.

small amounts, but offensive if overstrong, particularly in lagers (MacDonald *et al.* 1984). Similarly, in wines and distilled beverages the diacetyl-type aroma is perceived as desirable in small quantities in full bodied red wines, whiskies and rums, but offensive in most white wine, vodka and light rums (Nykänen & Suomalainen 1983).

The majority of diacetyl and 2,3-pentanedione is not truly synthesized by yeast. They are instead formed by non-enzymatic chemical reactions outside of the yeast cell, in which the corresponding $\alpha$-acetohydroxy acids, $\alpha$-acetolactate and $\alpha$-acetohydroxybutyrate, are oxidatively decarboxylated. These acetohydroxy acids leak from the yeast in the course of the fermentation. The yeast is, however, involved in the enzymatic reduction of the vicinal ketones to their corresponding 2,3-diols (Wainwright 1973). The concentration of vicinal ketones at the end of a fermentation is thus dependent on the balance between the synthesis and excretion of acetohydroxy acids, their oxidative decarboxylation and their further reduction by yeast, possibly involving alcohol dehydrogenase (MacDonald *et al.* 1984).

### 11.6.5 *Factors influencing the formation of ketones*

The organoleptically important keto acids, such as pyruvate, are influenced by similar factors to the aldehydes. They leak from the yeast cells during the highly active phase of fermentation and are readsorbed later if the yeast is still active. Similarly to acetaldehyde, the initial levels of pyruvate are controlled by factors influencing the rate of fermentation and yeast growth, while the readsorption is controlled by the factors influencing cell viability and biochemical activity (Coote *et al*. 1973, MacDonald *et al*. 1984).

The formation of the important vicinal diketones is influenced by other factors, but the final concentrations are similarly dependent on the activity of the yeast at the end of fermentation. Some recent studies have concentrated on the factors influencing acetohydroxy acid production. Inoue (1981) found that increases in yeast growth rates were accompanied by decreases in acetohydroxy acid formation, and suggested that the latter could be achieved by stirring, increased aeration and temperature, and possibly by reduced hydrostatic pressure. Nakatani *et al*. (1984a) found in studies on the kinetics of vicinal ketone formation, that there was a direct correlation with free amino nitrogen (FAN) levels and that there was a minimum FAN level which inhibited formation. $\alpha$-Acetolactate and $\alpha$-acetohydroxybutyrate are intermediates in the formation of valine and isoleucine, respectively. When these amino acids are present in excessive quantities they inhibit formation of the precursor acetohydroxy acid, and thus the vicinal diketone. Valine also has a suppressive effect on 2,3-pentanedione, but isoleucine does not affect diacetyl production (Nakatani *et al*. 1984b).

Although the levels of the acetohydroxy acids are important, the rate-limiting step is the oxidative decarboxylation reaction. This can be increased by the presence of oxygen, one of the factors which it is suggested could inhibit the acetohydroxy acid formation. However, increased oxygen could result in greater yeast activity at the end of fermentation, reducing the vicinal diketones (MacDonald *et al*. 1984). This example illustrates the complexity of the effects of individual parameters. Hoffmann (1985) summarized the most important factors in minimizing diacetyl concentrations in beer production. These were that: (a) the initial FAN concentration be greater than 200 mg $l^{-1}$; (b) the wort should be aerated to a $dO_2$ content of 8 mg $l^{-1}$ and that a second aeration stage is advisable; (c) the initial zinc content should be above 0.15 mg $l^{-1}$; (d) yeast strains produce varying concentrations; (e) the pitching rate should be 20–22 × $10^6$ cells $ml^{-1}$, as higher pitching rates increased diacetyl formation; (f) re-use of yeast should be restricted, possibly because of the effects of physiological exhaustion; and (g) temperature should be kept below 9°C, although higher temperatures do increase the rate of diacetyl reduction.

Since some of the suggested methods of control by different authors are contradictory, it is felt that the effects of certain parameters require further studies. It is likely that this confusion is due to differences in experimental systems and the fine balance between acetohydroxy acid formation, the oxidative decarboxylation reaction and the enzymatic reduction. However, it is clear that in practice ketone levels are controlled by a range of methods, which have been developed for the individual fermentation system.

## 11.7 Sulphur compounds

### *11.7.1 Introduction*

Over 50 volatile sulphur compounds, which could have an effect on flavour, have been detected in alcoholic beverages (Nykänen & Suomalainen 1983). The majority are not synthesized by yeast. Raw materials are a major source, particularly malt and hops in beer manufacture. In addition, some of the sulphur compounds, which the yeast does produce, can react outside the yeast cell with other metabolic products, e.g. mercaptans such as ethanethiol which can be formed by the reaction of hydrogen sulphide with alcohol (Amerine *et al.* 1966). The latter type of reaction occurs to a greater degree during distillation, and further reactions can occur during maturation, e.g. ethanethiol can be oxidized to diethyl disulphide. In addition, many sulphur compounds are formed from the sulphur-containing amino acids during yeast autolysis at the end of fermentations or during distillation (Suomalainen & Lehtonen 1979). As with other flavour compounds, abnormal quantities of sulphur compounds in a beverage can also be a result of the action of infecting bacteria.

Although small quantities may have a desirable effect on the characteristics of particular beverages, the sulphur compounds are generally perceived as off-flavours, if they are present in organoleptically significant quantities. They impart a range of aromas, such as cheesy, cooked vegetables, onion, garlic, rubbery, burnt, catty, rotten egg and rotten fish, all of which are grouped as putrefaction-type aromas. They have extremely low odour thresholds. In beer the threshold values for dimethyl sulphide and diethyl disulphide are 50 and 0.4 $\mu g\ l^{-1}$, respectively, with all of the mercaptans having values lower than 2 $\mu g\ l^{-1}$ (Meilgaard 1975).

### *11.7.2 Biosynthesis by yeast*

The major volatile sulphur compounds produced by yeast are sulphur dioxide and hydrogen sulphide. These are formed as by-products of the biosynthesis of the sulphur-containing amino acids, cysteine and methionine, from inorganic sulphate as outlined in Figure 16.5. During yeast autolysis they can be formed during the degradation of the amino acids. Organic sulphur sources are preferred by yeast, and the route from inorganic sulphate only becomes operative on their exhaustion. Sulphite in the substrate, from raw-material processing, can be a source of sulphur, being reduced to sulphide. This is an important source in winemaking. Elemental sulphur dusted or sprayed on to grapes before harvest can also be reduced to form hydrogen sulphide (Schütz & Kunkee 1977). Excessive quantities of either compound accumulate when their production exceeds their utilization, particularly during the active-growth phase. The production of both compounds corresponds with yeast growth and fermentation vigour. The pathway is under a complex system of control including feedback inhibition of ATP-sulphurylase and repression–derepression of enzyme synthesis (Engan 1981, Stewart & Russell 1981, MacDonald *et al.* 1984). S-methyl thioacetate, which can cause off-odours in some beers, is thought to be produced from methionine via methyl mercaptan.

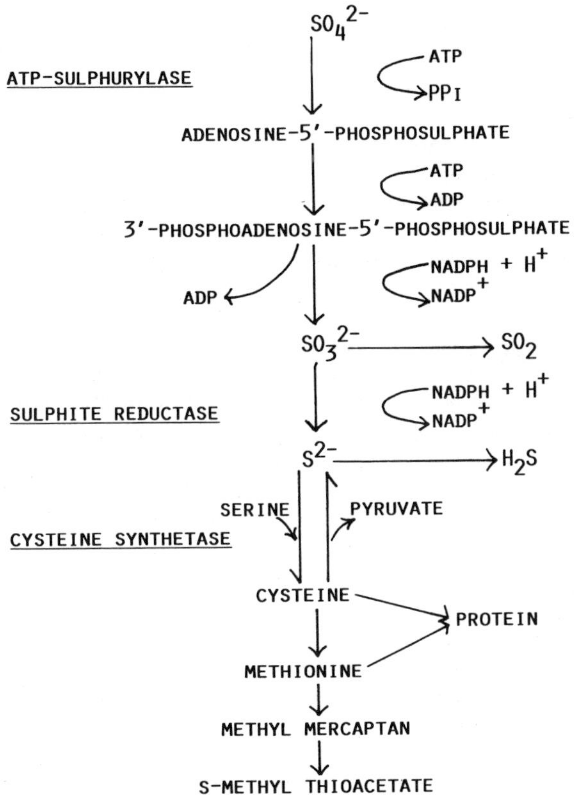


**Figure 11.5** The formation of important sulphur compounds as by-products of the sulphur-containing amino acid biosynthesis or degradation (Stewart & Russell 1981, MacDonald *et al.* 1984).

Yeast also produces dimethyl sulphide (DMS) if the precursors S-methylmethionine and dimethyl sulphoxide are present. Considerable amounts of DMS can emanate from malted cereal raw materials. In malt wort the presence of methionine sulphoxide can act as an inhibitor of the reduction of dimethyl sulphoxide, resulting in S-methylmethionine being the major precursor (Gibson *et al.* 1983). Stewart and Russell (1981) indicated that raw material and substrate processing were the major sources of DMS in beer, with yeast activity only contributing a minor amount.

### *11.7.3 Factors affecting biosynthesis*

There is considerable variation in sulphur dioxide and hydrogen sulphide formation between yeast strains. In natural wine fermentations some of the non-*Saccharomyces* wild yeasts, such as *Hansenula anomala*, produce quite high levels of $H_2S$ (Hernandez 1964). However, the major factors affecting production of both hydrogen sulphide and sulphur dioxide are the physiological state of the yeast and the substrate composition. Although higher quantities are produced with increased growth rate and fermentation vigour, these factors also result in loss of the sulphur compounds via the stripping action of evolved carbon dioxide. The various factors which affect growth and fermentation rates, as discussed in previous sections, will influence the production and loss of these compounds. Unhealthy cells are more likely to autolyse, resulting in breakdown products.

The levels formed are also dependent on the concentration of precursors in different substrates. Besides the general effects of the presence of organic sulphur sources, the sulphate-reducing pathway is strongly inhibited by methionine. Cysteine synthetase requires the vitamin pantothenate. When it is absent or at low levels, excessive quantities of hydrogen sulphide are formed. The lack of other growth factors, such as zinc, has similar effects. However, increases in other metal ions, such as iron and copper, have a stimulatory effect (Engan 1981, Stewart & Russell 1981, Nykänen & Suomalainen 1983, MacDonald *et al.* 1984).

The quantity of DMS formed by yeast is dependent on yeast strain, the physiological state and, most importantly, the presence of precursors. This compound has been investigated intensively in beer production. The methods of control of final levels have concentrated on the malt-kilning and wort-processing stages, because of their greater contribution to DMS production than yeast activity (MacDonald *et al.* 1984). Little is known about the factors which influence the biosynthesis of other sulphur compounds, but these are generally considered to be formed by non-enzymatic reactions, except for S-methyl thioacetate. Production of the latter is thought to be stimulated by excessive levels of methionine, although the intermediate, methyl mercaptan, has a greater effect (Stewart & Russell 1981).

## 11.8 Miscellaneous compounds

In addition to the major groups of congeners, there are a number of other flavour compounds metabolized by yeast. These are generally looked upon as being minor components of alcoholic beverages, and have not been investigated to the same extent as the major compounds. However, they can have significant effects on flavour if they are present at levels above their odour thresholds. Although they cannot be reviewed as thoroughly, a number of points can be made about some of them.

### *11.8.1 Phenols*

The majority of phenols present in alcoholic beverages originate from the raw materials and substrate processing. However, yeast can transform phenolic acids, via vinyl phenols, to volatile phenols. The example in Figure 11.6 illustrates that the initial step involves decarboxylation and the second step is reductional. *p*-Coumaric acid is similarly decarboxylated to *p*-vinyl phenol which is reduced to *p*-ethyl phenyl. Vanillin is decarboxylated to 4-methyl

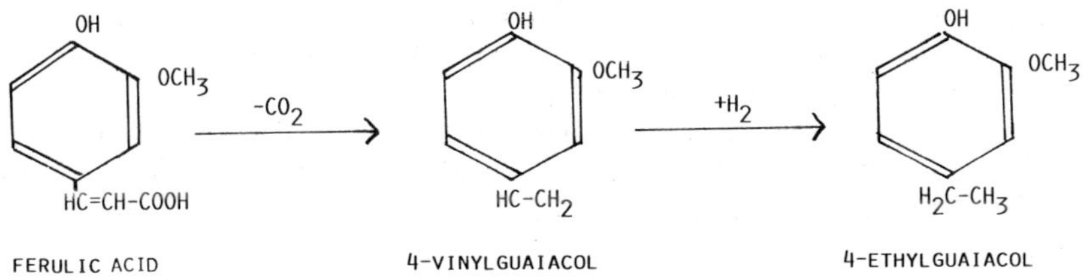


**Figure 11.6** An example of the transformation of a phenolic acid to more-volatile phenols.

guaiacol (Swan & Burtles 1978, Suomalainen & Lehtonen 1979, Jounela-Eriksson & Lehtonen 1981).

Little is known about the factors which control these reactions. However, in yeast breeding it was found that *Saccharomyces diastaticus* strains and hybrids with *Saccharomyces cerevisiae* produced beers with phenolic off-flavours. This was identified as being due to the decarboxylation of ferulic acid to *p*-vinyl guaiacol. It was found that a single dominant nuclear gene coded for the ferulic acid decarboxylating enzyme and that this could be bred out (Stewart *et al.* 1983). Since the majority of brewing strains lack the gene for the decarboxylation reaction, it is unlikely that yeasts are major contributors of phenolic flavours, unless significant levels of wild yeast, which do have the enzyme activity, are present.

### *11.8.2 Terpenes*

Under aerobic submerged conditions, *Kluyveromyces lactis* produces citronellol, linalool and traces of geraniol. When geraniol was added it was transformed to citronellol. Yields could be increased by the use of higher concentrations of asparagine as nitrogen source and by increasing the cultivation temperature (Drawert & Barton 1978). During autolysis of champagne yeast, *cis*- and *trans*-farnesol are among the flavour compounds released into the wine (Oura & Suomalainen 1981). In a simulated fino sherry, on which *Saccharomyces fermentati* was grown as a film, the yeast was found to produce low concentrations of linalool, *cis*- and *trans*-nerolidol and *trans,trans*-farnesol (Fagan *et al.* 1981).

Undoubtedly, the majority of terpenes present in alcoholic beverages emanate from the raw materials. However, these reports indicate that at least some yeasts do produce terpenes. Because of their results, Fagan *et al.* (1981) cautioned against the use of terpene concentrations in wines being the basis for differentiating among the original grape cultivars.

### *11.8.3 Amides and amines*

Very little has been reported on the production of amides and amines by yeast. These compounds can affect the flavour of wine and beer if present in quite small quantities. Amines are possibly formed by decarboxylation of amino acids in yeast fermentations (Nykänen & Suomalainen 1983). *Saccharomyces cerevisiae* produces the corresponding N-acetylamides from primary amines, such as 2-methyl-butylamine and 2-phenethylamine (Schreier *et al.* 1975). It is possible that some of the difference in levels in individual beers and wines is due to the action of yeast and the factors controlling it, as well as to raw materials and bacterial activity.

### *11.8.4 Lactones*

As with the other minor components, more attention has been given to the non-yeast sources of lactones, as these appear to be the primary sources in alcoholic beverages. Investigations into aroma compounds formed by the yeast *Sporobolomyces odorus* identified the lactones, 4-decanolide and *cis*-6-dodecen-4-olide. The latter has a peachy aroma (Tahera *et al.* 1973). However, in beverages the production of lactones by yeast is most noted in flor sherries and certain wines (Muller *et al.* 1973). Although the mechanisms have not been verified, it is likely that some yeast can produce $\gamma$-lactones from 4-hydroxy carboxylic acids and $\delta$-lactones from 5-hydroxy carboxylic acids. Due to their importance to sherry flavour, it is likely that further investigations will be made on the mechanisms of formation and the factors which influence the reactions.

## 11.9 Conclusions

During ethanol fermentations anaerobically grown yeasts always produce a large number of fermentation products which contribute to the flavour of alcoholic beverages. The biosynthesis of these compounds and the physiological factors which influence their formation are complex. However, since the level of organoleptic compounds produced can be influenced by yeast strain, medium composition and fermentation conditions, it is possible in principle for the brewer or the distiller to control the composition, and hence the flavour, of the beverage being produced. The main problem in this lies not with understanding the physiology and biochemistry of formation of different organoleptic compounds, but in knowing what combination of organoleptic compounds gives the desired flavour and in creating a set of growth conditions which satisfy the conflicting requirements of different types of organoleptic compounds.

In this chapter an attempt has been made to describe the main metabolic processes involved in the formation of organoleptic compounds by yeast, and to indicate those parameters which appear to be important in controlling their production in alcohol fermentations.

## References

Albagnac, G. and G. Vangheesdaele 1975. Etudes préliminaires sur l'origine de l'alcool benzylique présent dans les kirschs. *Industries alimentaire agricole* **92**, 1271–4.

Amerine, M. A., H. W. Berg and W. V. Cruess 1966. *The technology of wine making*, 2nd edn. Westport, Connecticut: AVI.

Berry, D. R. 1984. Physiology and microbiology of the malt whisky fermentation. In *Progress in industrial microbiology*, M. E. Bushell (ed.), 189–244. Amsterdam: Elsevier.

Berry, D. R. and H. Chamberlain 1986. Formation of organoleptic compounds by yeast grown in continuous culture on a defined medium. *Journal of the American Society of Brewing Chemistry* **44**(2), 52–6.

Berry, D. R. and C. M. Ramsay 1983. The whisky fermentation: past, present and future. In *Current developments in malting, brewing and distilling*, F. G. Priest and I. Campbell (eds), 45. London: Institute of Brewing.

Clapperton, J. F. and D. G. W. Brown 1978. Caprylic flavour as a feature of beer flavour. *Journal of the Institute of Brewing* **84**, 90–2.

Coote, N., B. H. Kirsop and G. K. Backee 1973. The concentration and significance of pyruvate in beer. *Journal of the Institute of Brewing* **79**, 298–304.

Drawert, F. and H. Barton 1978. Biosynthesis of flavour compounds by microorganisms 3: Production of monoterpenes by the yeast *Kluyveromyces lactis*. *Journal of Agricultural and Food Chemistry* **26**, 765–78.

Engan, S. 1981. Beer composition: volatile substances. In *Brewing sciences*, Vol. 2, J. R. A. Pollock (ed.), 98–105. London: Academic Press.

Fagan, G. L., R. E. Kepner and A. D. Webb 1981. Production of linalool, *cis*- and *trans*-nerolidol, and *trans,trans*-farnesol by *Saccharomyces fermentati* growing as a film on simulated wine. *Vitis* **20**, 36–92.

Forch, M. G., G. Krauss and H. Prokisch 1975. Wort composition and formation of lower free fatty acids (1). Effect of phosphorus, magnesium, calcium and biotin on the concentration of free fatty acids. *Journal of the American Society of Brewing Chemists* **33**, 148–50.

Geiger, E. and A. Piendl 1973. Technological influences on the formation of 2-phenylethanol during fermentation. *Proceedings of the Annual Meeting of the American Society of Brewing Chemists* **33**, 48–54.

Geiger, E. and A. Piendl 1976. Technological factors in the formation of acetaldehyde during fermentation. *MBAA Technical Quarterly* **13**, 51–63.

Gibson, R. M., P. J. Large, B. J. Anness and L. W. Bamforth 1983. The identity of an inhibitor in wort of dimethyl sulphoxide reductase from yeast. *Journal of the Institute of Brewing* **89**, 215–18.

Harrison, J. S. and J. C. J. Graham 1970. Yeasts in distillery practice. In *The yeasts*, Vol. 3, A. H. Rose and J. S. Harrison (eds), 283–348. London: Academic Press.

Haukeli, A. D. and S. Lie 1972. Production of diacetyl, 2-acetolactate and acetoin by yeasts during fermentation. *Journal of the Institute of Brewing* **78**, 229–32.

Hernandez, M. R. 1964. Production of $H_2S$ by wine yeasts grown with several sulfur containing compounds. *Sem Vitivinicola* **29**, 2359–64.

Hoffmann, S. 1985. Formation and degradation of diacetyl. *Brauwelt International* **1**, 59–61.

Hohn, E., J. Solms and H. R. Roth 1975. The aroma substances of yeast. II Sensory evaluation of thiamin and thiamin diphosphate. *Lebensm. Wiss & Technol.* **8**, 212–16.

Holmberg, S. 1984. Genetic improvement of brewer's yeast. *Trends in Biotechnology* **2**, 98–102.

Howard, D. and R. G. Anderson 1976. Cell-free synthesis of ethyl acetate by extracts from *Saccharomyces cerevisiae*. *Journal of the Institute of Brewing* **82**, 70–1.

Inoue, T. 1981. The relationship between the performance of yeast and acetohydroxy acid formation during wort fermentation. *MBAA Technical Quarterly* **18**, 62–5.

Jounela-Eriksson, P. and M. Lehtonen 1981. Phenols in the aroma of distilled beverages. In *The quality of foods and beverages, chemistry and technology*, Vol. 1, C. Charalambous and G. Inglett (eds), 167–82. London: Academic Press.

Kunkee, R. E. and R. Goswell 1977. Table wines. In *Economic microbiology*, Vol. 1: *Alcoholic beverages*, A. H. Rose (ed.). London: Academic Press.

Lynen, F. 1967. The role of biotin-dependent carboxylations in biosynthetic reactions. *Biochemical Journal* **162**, 381–400.

MacDonald, J., P. T. V. Reeve, J. D. Ruddlesden and F. H. White 1984. Current approaches to brewery fermentations. In *Progress in industrial microbiology*, Vol. 19: *Modern applications of traditional biotechnologies*, M. E. Bushell (ed.). Amsterdam: Elsevier.

Margalith, P. Z. 1981. *Flavour microbiology*. Springfield, Illinois: Charles C. Thomas.

Maule, D. R. 1967. Rapid gas chromatographic examination of beer flavour. *Journal of the Institute of Brewing* **73**, 351–61.

Meilgaard, M. C. 1975. Flavour chemistry of beer: Part II: Flavour and threshold of 239 aroma volatiles. *MBAA Technical Quarterly* **12**, 151–68.

Muller, J., E. Kepner and A. D. Webb 1973. Lactones in wines – a review. *American Journal of Enol. Viticult.* **24**, 5–9.

Nakatani, K., T. Takahashi, K. Nagami and J. Kumada 1984a. Kinetic study of vicinal diketones in brewing I: Formation of total vicinal diketones. *MBAA Technical Quarterly* **21**, 73–8.

Nakatani, K., T. Takahashi, K. Nagami and J. Kumada 1984b. Kinetic study of vicinal diketones in brewing II: Theoretical aspects for the formation of total vicinal diketones. *MBAA Technical Quarterly* **21**, 175–83.

Nordstrom, K. 1964. Studies on the formation of volatile esters in fermentations with brewer's yeast. *Svensk Kemisk Tidster.* **76**, 510–43.

Nykänen, L. and H. Suomalainen 1983. *Aroma of beer, wine and distilled alcoholic beverages*. Dordrecht: D. Reidel.

Oura, E. and H. Suomalainen 1981. Study of volatile substances produced during the autolysis of champagne yeast. *Acta Alimentaire* **10**, 27–36.

Peppard, T. L. and S. A. Halsey 1981. Malt flavour – transformation of carbonyl compounds by yeast during fermentation. *Journal of the Institute of Brewing* **87**, 386–90.

Peppard, T. L., S. A. Halsey and D. R. J. Laws 1981. Flavour constituents of malt. In *Flavour '81*, P. Schreier (ed.), 579–97. Berlin: Walter de Gruyter.

Peppler, H. J. 1979. Yeast extracts. In *Economic microbiology*, Vol. 7, A. H. Rose (ed.), 293–312. London: Academic Press.

Ramsay, C. M. 1982. PhD thesis, University of Strathclyde.

Ramsay, C. M. and D. R. Berry 1984a. The effect of temperature and pH on the formation of higher alcohols, fatty acids and esters in the malt whisky fermentation. *Food Microbiology* **1**, 117–21.

Ramsay, C. M. and D. R. Berry 1984b. The effect of inoculum level on the formation of higher alcohols, fatty acids and esters in the malt whisky fermentation. *Food Microbiology* **1**, 111–15.

Reazin, G. H. 1981. Chemical mechanisms of whiskey maturation. *American Journal of Enology and Viticulture* **32**, 283–90.

Reed, G. and H. J. Peppler 1973. *Yeast technology*. Westport, Connecticut: AVI.

Rous, C. V., R. Snow and R. E. Kunkee 1983. Reduction of higher alcohols by fermentation with a leucine auxotrophic mutant of wine yeast. *Journal of the Institute of Brewing* **89**, 274–8.

Schreier, P. 1979. The role of microorganisms in flavour formation. In *Progress in flavour research*, D. G. Land and H. E. Nurston (eds). London: Applied Science Publishers Ltd.

Schreier, P., F. Drawert and A. Junker 1975. Über die Biosynthese von Aromastoffen durch Mikroorganismen. I. Bildung von N-Acetyl-aminen durch *Saccharomyces cerevisiae. Zeitschrift für Lebensmittel-Untersuchungen und Forschung* **158**, 351–60.

Schütz, M. and R. E. Kunkee 1977. Formation of hydrogen sulfide from elemental sulfur during fermentation by wine yeast. *American Journal of Enology and Viticulture* **28**, 137–44.

Snow, R. 1983. Genetic improvement of wine yeast. In *Yeast genetics, fundamental and applied aspects*, J. E. T. Spencer, D. M. Spencer and A. W. Smith (eds), 439–59. New York: Springer-Verlag.

Stewart, G. G. and I. Russell 1981. The influence of yeast on volatile sulphur compounds in beer. *European Brewery Convention, Monograph VII, Flavour Symposium; Copenhagen*, 173–87. Amsterdam: Elsevier.

Stewart, G. G., I. Russell and A. M. Sills 1983. Factors that control the utilization of wort carbohydrates by yeast. *MBAA Technical Quarterly* **20**, 1–8.

Sumper, M. 1974. Control of fatty acid biosynthesis by long chain acyl CoA's and by lipid membranes. *European Journal of Biochemistry* **49**, 469–75.

Suomalainen, H. and M. Lehtonen 1979. The production of aroma compounds by yeast. *Journal of the Institute of Brewing* **85**, 144–56.

Suomalainen, H. and L. Nykanen 1966. The aroma components produced by yeast in a nitrogen free sugar solution. *Journal of the Institute of Brewing* **72**, 469–74.

Swan, J. S. and S. M. Burtles 1975. The development of flavour in potable spirits. *Chemical Society Review* **7**, 201–11.

Swan, J. S., D. Howie, S. M. Burtles and A. A. Williams 1981. Sensory and instrumental studies of Scotch whisky flavour. In *The quality of food and beverages*, G. Charalanbois and G. Inglett (eds), 201–23. New York: Academic Press.

Tahera, S., K. Fujinara and J. Mizutani 1973. Neutral constituents of volatiles in cultured broth of *Sporobolomyces odorus. Agricultural Biology and Chemistry* **37**, 2855–66.

Taylor, G. T. and B. H. Kirsop 1977. The origin of the medium chain fatty acids present in beer. *Journal of the Institute of Brewing* **83**, 241–3.

Thurston, P. A., P. E. Quain and R. S. Tubb 1982. Lipid metabolism and the regulation of volatile ester synthesis in *Saccharomyces cerevisiae. Journal of the Institute of Brewing* **88**, 90–4.

Tressl, R., D. Balivi and B. Helak 1983. Flavours of malt and other cereals. In *Flavour of distilled beverages: origin and development*, J. R. Piggott (ed.). Chichester: Ellis Horwood.

van der Meersshe, T., A. Devreux and C. A. Masschelein 1977. Le rôle de la levure dans la maturation du gout de la bière. *Proceedings of the European Brewery Convention, 16, Amsterdam*, 561–75. Amsterdam: Elsevier.

Wainwright, T. 1973. Diacetyl – a review. Part I – Analytical and biochemical considerations: Part II – Brewing experience. *Journal of the Institute of Brewing* **79**, 451–70.

Watson, D. C. 1981. The development of specialised yeast strains for use in Scotch malt whisky fermentations. In *Current developments in yeast research*, G. G. Stewart and I. Russell (eds), 57–63. Toronto: Pergamon Press.

Watson, D. C. 1985. Current developments in the potable distilling industry. *CRC Critical Reviews in Biotechnology* **2**, 147–92.

Wood, B. J. B. W. 1985. *Microbiology of fermented foods*, Vols 1 and 2. Amsterdam: Elsevier Applied Science.

# 12 *Hydrolytic enzymes*

WILFRED NIELS ARNOLD

## 12.1 Introduction

This chapter is concerned with a specific class of yeast products, hydrolytic enzymes, that are useful in analytical and industrial applications. The selection of subjects, and the depth of coverage, are based primarily on utility, but other hydrolases are included either for anticipated commercial value or as instructive examples of cell biology. Recent developments in biochemistry and genetics help to place the field on more than an empirical ground. Further developments will follow as a result of fundamental studies on species that have emerged from surveys.

Hydrolytic enzymes catalyse the transfer of specified chemical moieties from substrate molecules to acceptor water molecules. In some cases hydroxy compounds may act as alternative acceptors, but under most circumstances the end-result of enzyme action is simply hydrolysis of the substrate into two or more smaller compounds. This chapter addresses three subcategories: hydrolases acting on glycosyl compounds, ester bonds or peptide linkages. International Union of Biochemistry nomenclature is followed, and enzyme commission (EC) numbers are referenced. In some cases the systematic names are too unwieldy for frequent repetition. Trivial names are not only shorter, but they are so firmly entrenched in laboratory parlance that it would be a disservice to exclude them.

The biosynthesis of hydrolytic enzymes is vital to yeasts because they have evolved with plasma membranes that generally discriminate against any compound other than relatively small growth-supporting substances. For example, although glucose is avidly acquired from the environment by all yeast species, in contrast sucrose, raffinose, glucose phosphate, starch and casein are excluded from the cytoplasm. Many other disaccharides, oligosaccharides, phosphate esters and polymers of all kind are similarly debarred from the inner regions of the cell unless they are first degraded by secreted hydrolases. The extent of the complement of such enzymes is a prime factor in setting the range of substrates that a species can assimilate.

From the point of view of the industrial microbiologist the selection of a candidate species may be greatly influenced by both constitutive and inducible hydrolases, because they often determine the prospect for growth on alternative starting materials and the ultimate yield of useful products. Under special conditions selected species can be purposefully induced to produce massive amounts of a particular hydrolase for application elsewhere.

In this chapter the term extracellular is reserved for enzymes that are found in the culture medium, i.e. biosynthetic products that have departed the cell.

Specific examples may be the result of secretion by live cells or escape from dead cells due to autolysis. Some of the enzymes discussed are known to be located in cell envelopes, and are properly referred to as extracytoplasmic; they are nevertheless cellular, because the cell envelope is part of the cell. Notwithstanding this seemingly obvious distinction, the reader is cautioned against the occasional incorrect use of extracellular in the primary literature as an appellation for anything outside the yeast cell's plasma membrane. The distinction between extracellular and cellular enzymes also has commercial consequences, because isolation of the latter type requires disruption of the thick cell wall, which adds to the time and cost of production. Questions about enzyme location are thus both fundamental and practical. A brief discussion of the natural history of a secreted enzyme will help to place in context the specific examples offered later in the chapter.

## 12.2 Enzyme synthesis and secretion

It is accepted that all enzymes, even those destined for extracytoplasmic locales, are synthesized within the protoplasm. This follows from the observation that protoplasts, i.e. entities derived from cells by the artificial removal of the cell wall, retain the synthetic abilities of the parent cells (Arnold 1981a). Modifications of the primary structure may attend secreted enzymes as they are delivered to their final extracytoplasmic destination. It is also very likely that reversible or irreversible changes may occur in cell-envelope enzymes during the life of the cell, but no documented example is available at present.

The information for protein synthesis resides in nucleic acids, and for the most part (the known exception being some mitochondrial molecules) is organized in nuclear chromosomes, whose replication and migration enable the information to be passed from mother to daughter cells. The transcription and translation of that information into syntheses of specific enzymes by daughter cells is not infrequently affected by the environmental (cultural) conditions.

There are at least two characteristics that distinguish secreted enzymes from cytoplasmic enzymes. First, they have identifying sequences of amino acids which act as 'signals' for export (Blobel & Dobberstein 1975) and, secondly, these extracytoplasmic proteins are usually decorated with carbohydrate, i.e. the majority turn out to be glycoproteins with substantial amounts of covalently bound carbohydrate (Arnold 1981a). The polysaccharide moiety is frequently a mannan joined to asparagine via a chitobiose unit. It has become clear that the biosynthesis of glycoproteins involves first an indirect system of addition and subtraction of glucose residues (processing) that is energetically expensive, and then the addition of many individual mannose residues (Parodi 1981). The reasons behind this seemingly circuitous mechanism are far from explained. The end-result is certainly an enzyme that is much more bulky and it is usually more resilient than the unglycosylated precursor. Glycoproteins tend to be more resistant to heat inactivation and to proteolytic degradation (Arnold 1981a).

The following steps occur in the cytoplasm. (1) Copies of the mRNA for a specific enzyme are transcribed from the appropriate DNA sequence in the nucleus. (2) The message is translated into a polypeptide by the ribosomal apparatus in association with endoplasmic reticulum. (3) The nascent polypeptide is formed with an N-terminal signal sequence, and is coincidentally extruded into the lumen of the endoplasmic reticulum. (4) Vesiculation of reticulum leads to small, membrane-surrounded packages which are subsequently sorted, perhaps through a subcellular organelle like the Golgi apparatus. (5) The signal sequence is removed by a signal peptidase and glycosylation begins. (6) Processing of carbohydrate adducts prepares the now lightly glycosylated polypeptide for subsequent addition of mannose residues. (7) Mature glycoproteins in vesicles find their way to the plasma membrane.

The following steps occur at the plasma membrane. (1) There is a fusion between the membrane of the delivery vesicle and the plasma membrane. (2) The contents of the vesicle find themselves on the external side of the plasma membrane. (3) Some of the enlarged plasma membrane is depleted by reverse traffic of vesicles moving in the cytoplasmic direction.

The following steps occur in the cell envelope. (1) Enzymes diffuse through the cell wall and appear in the medium. (2) Alternatively, the enzyme molecule is too large to pass through the cell wall and is restrained in the periplasmic space. (3) In some cases there may be adsorption of secreted enzymes from the medium on to the surface of the cell wall; such appendages as a microfibrillar zone or a slime layer may be specifically involved.

The above is necessarily an overview of a large and very significant area of cell biology that is occupying investigators at present. The descriptions may turn out to require modification as information accrues. There is also the distinct possibility of mechanistic variations among individual enzymes and species. So, too, the question of retention in the periplasmic space or secretion all the way to the medium may depend on differences in the porosity of the cell wall from species to species, and the possibility of charge interactions between the cell-wall matrix and proteins. These questions have been addressed elsewhere in more detail (Arnold 1981a).

## 12.3 Glycoside hydrolases

Starch is a reserve polysaccharide of plants. In addition to being a direct foodstuff for man and domestic animals, it is a common starting material in the fermentation industry. The brief summary of starch chemistry that follows is a preamble to a discussion of the enzymology.

Starches are mixtures of two different polymers of $\alpha$-D-glucopyranose. The first, amylose, is a linear array with $\alpha$-1,4 links exclusively. The second, amylopectin, also contains linear $\alpha$-1,4-linked regions, but is more complex because branch points ($\alpha$-1,6 links) occur on 4–5% of the glucose residues. It can be mentioned in passing that the endogenous glycogens of yeasts and animals contain about 10% $\alpha$-1,6 linkages, but they are otherwise chemically similar to plant amylopectins. The starches of commerce vary in their proportion of amylose from about 20 to 28% (Kerr 1950).

Because it is uneconomical to separate amylose from amylopectin, some interest has accrued in the breeding of plants whose starches have different proportions. So-called waxy or glutinous starches (from specific varieties of corn and other cereals) contain not more than 6% amylose. At the other extreme special mutant varieties of corn make starch granules with as much as 77% amylose (Whistler & Corbett 1957).

Starches with a higher proportion of amylose have a greater propensity for 'retrogradation', i.e. precipitation from solution, which is undesirable. The phenomenon is due to hydrogen bonding among the linear molecules. Branched amylopectin molecules have less tendency to retrograde. Partly degraded starches (due to acid or enzymatic hydrolysis of some of the glucosidic bonds) are called dextrins. Because of the lower average chain length they are more soluble and exhibit a lower rate of retrogradation. Conversely, dextrins and amylopectin are less susceptible to enzymatic hydrolysis due to steric hindrance of the branch points.

Plant seeds are typically rich in $\alpha$-amylases (EC 3.2.1.1) and latent $\beta$-amylases (EC 3.2.1.2). Both classes of enzyme attack starch but are thwarted by $\alpha$-1,6 branch points. The products, as explained in more detail below, are glucose, maltose, maltotriose, maltotetraose, oligosaccharides containing one $\alpha$-1,6 branch point, and dextrins. Such dextrins may constitute as much as 20% of the partially digested starch in worts that are used as substrates for fermentations.

Brewer's yeasts ferment glucose, maltose and maltotriose although different permease systems for cellular uptake are involved (Kotyk & Horak 1981). Maltose and maltotriose are taken up intact by yeast cells, which is in contrast with the behaviour towards sucrose and raffinose to be discussed later. Typically all of the branched oligosaccharides and the dextrins remain in the final brewery product.

Although all yeasts appear to have the capability of endogenous glycogen turnover, nevertheless the secretion of glycogen- or starch-hydrolysing enzymes is relatively rare (DeMot *et al.* 1984). Those species that do may be used to increase the yield of single-cell protein from starch or to decrease the final carbohydrate content of potable beverages. We shall enumerate those species that can secrete $\alpha$-amylases, glucoamylases (EC 3.2.1.3) and enzymes that attack $\alpha$-1,6 links in dextrins. $\beta$-Amylases are not produced by yeasts.

### $\alpha$-1,4-Glucan 4-glucanohydrolase (EC 3.2.1.1)

The action pattern on amylopectin is shown in Figure 12.1. The trivial name for this class of enzyme is $\alpha$-amylase. Based on a wealth of studies on the mechanism of $\alpha$-amylases (all from non-yeast sources) it is established that $\alpha$-1,4-glycosidic linkages in amylose or amylopectin are broken by $C_1$–oxygen scission, and that the reducing ends of products have the $\alpha$-configuration (Mayer & Larner 1959). Maltotriose is slowly hydrolysed to maltose and glucose (Walker & Whelan 1960), but overall there is a very small liberation of glucose compared with a relatively large increase in reducing groups. (The term 'reducing' refers to the ability of a saccharide to reduce, for example, cupric ions to cuprous ions when heated under alkaline conditions; a glucose residue in any di-, oligo- or polysaccharide has the reducing property when its anomeric carbon, $C_1$, is not involved in glycosidic linkage.) The $\alpha$-amylases cannot split

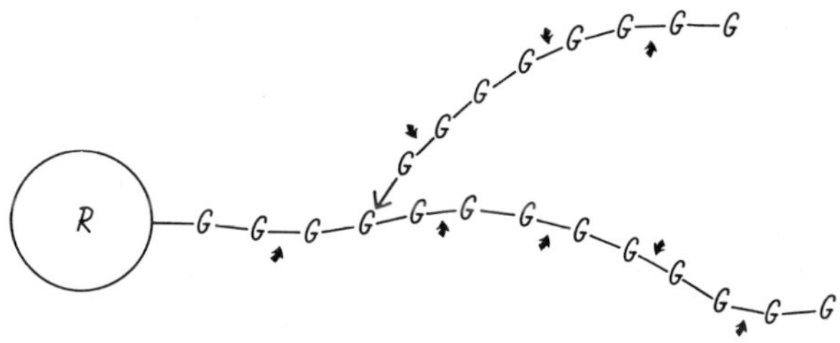


**Figure 12.1** Action pattern of $\alpha$-amylase. The schematic diagram represents a part of an amylopectin molecule (G, $\alpha$-D-glucopyranose; the interconnecting lines represent glycosidic 1,4 links; the interconnecting arrow indicates a 1,6 link; two non-reducing termini are shown; R, remainder of molecule, consisting of many other branched chains). The thick arrows indicate typical positions of hydrolytic attack by $\alpha$-amylase.

$\alpha$-1,6 branch linkages, and these reside in the dextrin products (Whelan 1960). The nature of the $\alpha$-limit dextrin varies with the source of the enzyme.

Augustin *et al.* (1978) surveyed 177 strains of yeast and yeast-like species, and found only five species capable of secreting reasonable amounts of $\alpha$-amylase, viz. *Endomycopsis capsularis*, *Schwanniomyces occidentalis*, *Schwanniomyces alluvius*, *Leucosporidium capsuligenum* and *Cryptococcus luteolus*. It is worth mentioning that the Czech workers employed a special gel-liquefaction method for survey purpose and they recommend an 8% (dry wt/vol.) amylose that has been cross-linked with 2-chloromethyloxirane, as described in more detail by Augustin *et al.* (1981). $\alpha$-Amylases are able to liquefy the gel because they can break internal $\alpha$-1,4 bonds, and hence the three-dimensional structure of the gel, whereas glucoamylases (see below), which attack from non-reducing termini, are soon thwarted by the artificial cross-links. Debranching enzymes are without effect because of the lack of $\alpha$-1,6 links in the gel. Accordingly, only those yeast colonies that secrete $\alpha$-amylases into the agar-medium overlay will produce zones of liquefaction in the underlying special gel.

Sills and Stewart (1982) used classical enzymology in exploring strains of five species and found that the best $\alpha$-amylase producers were strains of *Endomycopsis fibuligera*, *Schwanniomyces occidentalis* and *Schwanniomyces castellii*. Some selection among strains of *Schwanniomyces castellii* for higher $\alpha$-amylase activity was subsequently reported (Sills *et al.* 1983); Clementi *et al.* (1980) presented preliminary results on the same species. Spencer-Martins and van Uden (1979) indicate that $\alpha$-amylase is a part of the extracellular mixture of amylolytic enzymes produced by *Lipomyces kononenkoae*. Moulin and Galzy (1978) claim that *Torulopsis ingeniosa* produces an $\alpha$-amylase which, under their conditions of growth at least, is apparently adsorbed on the cell surface. This is quite exceptional in view of the majority of reports on other yeast species that the $\alpha$-amylases are found in the culture medium. However, it should be mentioned in passing that brewer's yeasts can adsorb barley amylases (Andrews & Gilliland 1952).

Wilson and Ingledew (1982) achieved about an 11-fold purification of the $\alpha$-amylase component from the spent medium of *Schwanniomyces alluvius*.

The enzyme was most active at pH 6.5 (50% as active at pH values of about 5.5 and 7.5).

A report by Wickerham *et al.* (1944) on starch hydrolysis by *Endomycopsis fibuliger* indicated production of α-amylase, but little enzymology was attempted. Clementi *et al.* (1980) extended the study of the effects of cultural conditions on the yields of crude amylolytic activity. Ebertova (1966a) worked on *Endomycopsis capsularis*, and found that if the pH of the culture medium is maintained above 6.0, then α-amylase predominates in the secreted mixture of enzymes. This enzyme was partially purified by chromatography on diethylaminoethyl-cellulose (DEAE-cellulose) and exhibited maximal activity at about pH 4.5 (50% as active at pH values of about 3.0 and 6.0).

**α-1,4-Glucan glucohydrolase (EC 3.2.1.2)**

The action pattern on amylopectin is shown in Figure 12.2. The trivial name for this class of enzymes is glucoamylase. They catalyse the hydrolysis of α-1,4-glucosidic bonds by stepwise attack from the non-reducing termini. Nascent glucose appears in the β-configuration (Weill *et al.* 1954).

It is claimed that some glucoamylases can bypass α-1,6 branch points of amylopectin and, with the purified enzyme of *Rhizopus delemar*, up to 90% conversion of starch into glucose is possible (Phillips & Caldwell 1951a, b, Tsujisaka *et al.* 1958). In other cases the branch points are a distinct barrier, and this certainly appears to be the case for the yeast enzymes. For example, work by Hopkins and Kulka (1957) has already indicated that *Saccharomyces diastaticus* secretes a glucoamylase and a debranching enzyme, the latter being inactivated preferentially at pH 6.4 and 55°C.

According to Manners (1962) the distribution of glucoamylases in nature is restricted to the fungi. *Aspergillus* and *Rhizopus* species are notable producers. The crude preparations that are commercially available probably contain debranching enzymes as fortuitous contaminants, so that quantitative yields of glucose from starch are possible. This is the basis of a specific method for starch or glycogen (see, for example, Thivend *et al.* 1972). The glucose can be assayed specifically, in the next step, by glucose oxidase reaction.

With regard to the yeasts, the early impetus for studying the glucoamylases came from observations that certain species can lower the ultimate specific

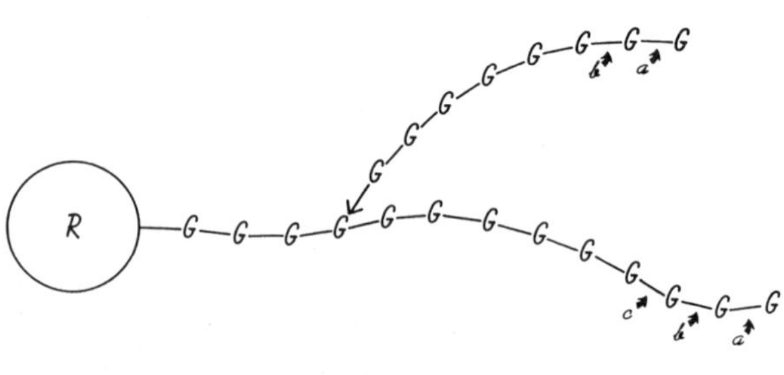


**Figure 12.2** Action pattern of glucoamylase. The heavy arrows indicate the positions of enzyme attack on amylopectin. Glucose residues are released in the order a, b, c, . . . . (See Fig. 12.1 for key to symbols.)

gravity of beers (see, for example, Andrews & Gilliland 1952). The lowest specific gravity that can be achieved with a particular wort and a particular yeast flora is called the attenuation limit, and is a reflection of the degree of removal of dextrins by enzyme systems that include a glucoamylase (Hopkins 1958). Interest in attenuation of beers still continues, with particular attention now being directed towards the production of palatable beers with lower total calorie content (see, for example, Erratt & Stewart 1981a).

Brief reports indicate that *Aureobasidium pullulans* (Federici & D'Elia 1983) and *Lipomyces kononenkoae* (Sa-Correia & van Uden 1981) convert starch into reasonable yields of biomass, apparently through the agency of secreted-enzyme systems that include glucoamylases. Preliminary evidence for glucoamylase production by *Pichia polymorpha* has appeared (Moulin *et al.* 1982). Brief communications have addressed other species, but the present interest in yeast glucoamylases seems to be concentrated on *Saccharomyces diastaticus*, *Endomycopsis* spp. (Hattori & Takeuchi 1961) and *Schwanniomyces* spp.

### *SACCHAROMYCES DIASTATICUS*

Tamaki (1980) purified three glucoamylase isoenzymes from genetically manipulated strains of *Saccharomyces diastaticus*; each preparation was deemed homogeneous on analytical ultracentrifugation, and molecular weights of 68 000, 77 000 and 84 000 were reported. The isoenzymes had slightly different mobilities under electrophoresis at pH 8 in polyacrylamide gels. Single bands were obtained in each case, and the coincidence of protein- and carbohydrate-staining in companion gels was offered as evidence that all three were glycoproteins. Enzymic parameters (including substrate specificity) are apparently not yet available, but will be of considerable interest.

Stewart's laboratory has given a preliminary report of their evidence for five isoenzymes, controlled by two genes in *Saccharomyces diastaticus*, that are separable by DEAE-cellulose chromatography. All five are mannoproteins that have similar biochemical properties (Erratt & Stewart 1981b). Of practical interest is the finding that one or more of these isoenzymes is not inactivated by pasteurization temperatures. This explains the slow but significant production of glucose that attended some bottled beers that had been given the secondary treatment with *Saccharomyces diastaticus* (Erratt & Stewart 1981a). It is worth emphasizing at this point that it is not sufficient to invoke a secondary fermentation to lower the dextrin content of a beverage; one must obviously be concerned with final taste and the storage performance of the product (Stewart 1981).

### *ENDOMYCOPSIS* spp.

The genus is marked by abundant development of true mycelium in addition to budding cells. Some previously-named *Endomyces* species are now included. Six of the ten species assigned to the genus by Kreger van Rij (1971) have the ability to assimilate starch, but in that respect the interest has centred upon *Endomycopsis fibuligera* and *Endomycopsis capsularis*.

Early studies on *Endomycopsis fibuligera* by Wickerham *et al.* (1944) and other investigators established assimilation rates for starches from diverse sources, and yields of biomass and alcohol, etc., but did not address the enzymology. Hattori (1961) and Fukumoto *et al.* (1960) used seemingly identical methods for partial purification of the secreted starch-degrading enzymes, and they are credited with first recognizing the 'saccharogenic' component (now identified as glucoamylase) in admixture with the 'dextrinogenic' component ($\alpha$-amylase). (Both reports were available to this reviewer only as English abstracts.) The glucoamylase was shown to exhibit maximal activity between pH 4.5 and 5.5, and was stable over a broad range (pH 4.5–9.0). Hattori (1961) and Hattori and Takeuchi (1962) showed that digestions of amylopectin (100%), glycogen (95%) and starches (70–80%) could be achieved with reasonable amounts of yeast enzyme in 24 h, and that the most prevalent product was free glucose.

Ebertova (1966b) employed acetone precipitation from spent medium and DEAE-cellulose chromatography of the proteins to achieve some purification of the glucoamylase from *Endomycopsis capsularis*, but the degree of purification is not stated. However, a pH optimum of 4.5 (50% as active at about pH 3.0 and 6.0) and relative rates of hydrolysis of amylose (100%), maltotetraose (83%) and maltose (47%) were established. The enzyme was active after exposure to pH extremes of 4.0 and 8.5 for 96 h, and after heating for 1 h at 60°C. Ebertova (1966a) found spent medium to contain at least an $\alpha$-amylase, a glucoamylase and a maltase. If the pH of the culture fell below 4.0, then the glucoamylase predominated.

Ruttloff *et al.* (1970) claim that three isoenzymes of glucoamylase are secreted by *Endomycopsis bisporus* cells growing on soluble starch. During log-phase, two isoenzymes occur in roughly 4 : 1 ratio, and they are reasonably assumed to be responsible for starch assimilation. However, after stationary-phase is achieved there is a great increase in total activity of which 90% is a third isoenzyme. The authors state that the function of this isoenzyme is unclear. This is probably another indication of derepression after glucose exhaustion, as has been found for a variety of similar enzymes (Sills & Stewart 1982).

### *SCHWANNIOMYCES* spp.

Phaff (1971) accepts four species in the genus, all of which can assimilate starch. Reports on the amylolytic enzymes from *Schwanniomyces* species are all comparatively recent.

Wilson and Ingledew (1982) subjected the crude glucoamylase of *Schwanniomyces alluvius* to anion-exchange chromatography, ultrafiltration and gel-permeation chromatography. No electrophoretic trials were conducted under non-denaturing conditions, so the true molecular weight is not known but a minimum molecular weight of 155 000 was estimated from electrophoretic mobility in the presence of sodium dodecyl sulphate (SDS).

The pH optimum of the *Schwanniomyces alluvius* glucoamylase is about 5.0 (50% as active at pH values of 3.5 and 6.5). The enzyme activity is quickly destroyed at 60°C (Wilson & Ingledew 1982). This is in contrast to the resistance offered by the glucoamylases of *Saccharomyces diastaticus* or *Endomycopsis capsularis* (see above). The increased lability of the *Schwanniomyces* enzyme is an advantage according to Stewart (pers. commun.)

because subsequent pasteurization of the treated beer stops dextrin degradation at the chosen point.

Dhawle and Ingledew (1983) have isolated mutants of *Schwanniomyces castellii* that produce three to four times the normal levels of $\alpha$-amylase and glucoamylase, but for some reason (perhaps debranching enzyme is limiting) this is still insufficient to elicit complete hydrolysis of starch in reasonable times. Sills and Stewart (1982) have reported some preliminary work on the glucoamylases of *Schwanniomyces castellii* and *Schwanniomyces occidentalis*. A recent abstract by Sills *et al.* (1984) described the application of a strain of *Schwanniomyces castellii* to lower the dextrin content of beer in combination with *Saccharomyces uvarum*.

**Amylopectin 6-glucanohydrolase (EC 3.2.1.9)**

The action pattern on amylopectin is shown in Figure 12.3. Maruo and Kobayashi (1951) characterized an interesting debranching enzyme from brewer's yeast autolysates. This enzyme (trivial names amylopectin 1,6-glucosidase or isoamylase) catalyses the hydrolysis of glutinous starch (average molecular weight 264 000) to linear polymers of glucose (average molecular weight 16 600) by attacking only $\alpha$-1,6 branch points; the $\alpha$-1,4 links are not modified. Gunja *et al.* (1961) confirmed these results, documented the occurrence of the enzyme in baker's yeast as well, and extended studies on the action pattern; they maintain that glycogen is not completely debranched in the presence of this enzyme. The function of this intracellular enzyme in baker's and brewer's yeasts is not established, but it supposedly plays a role in endogenous glycogen turnover.

Debranching enzymes of this type are somewhat analogous to pullulanase (Abdullah *et al.* 1966) and 'R-enzyme' (Hobson *et al.* 1951) from non-yeast sources. A crystalline enzyme prepared from *Pseudomonas* species by Harada *et al.* (1972) catalyses the hydrolysis of both inner and outer branching linkages of amylopectin, and should be the model for future comparisons; conversely, the pullulan 6-glucanohydrolase (pullulanase, EC 3.2.1.41) of *Aerobacter aerogenes* scarcely affects inner linkages (Wallenfels & Rachead 1966).

According to Sills and Stewart (1982) amylopectin 6-glucanohydrolase activity is an important constituent in any secreted mixture of enzymes that is

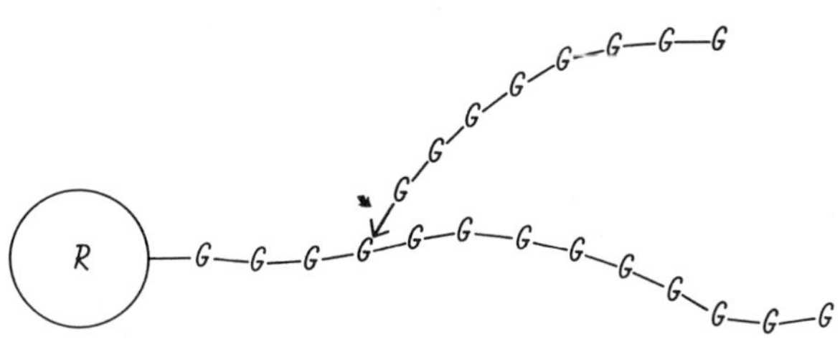


**Figure 12.3** Action pattern of debranching enzyme. The heavy arrow indicates the position of enzyme attack. Other 1,6 bonds are similarly attacked in the remainder (R) of the molecule of amylopectin. (See Fig. 12.1 for key to symbols.)

to have a reasonable potential for starch degradation, but surprisingly few results are available. Sills and Stewart (1982) measured glucose release from pullulan, a linear polymer of $\alpha$-glucan containing $\alpha$-1,4 and $\alpha$-1,6 links in 2:1 ratio (Bender *et al.* 1959, Marshall 1980), to reveal reasonable debranching activity in the crude extracts of *Schwanniomyces occidentalis* and *Schwanniomyces castellii*. It is worth noting that a purified debrancher should be monitored via formation of reducing groups rather than glucose, because the $\alpha$-1,4 links would supposedly not be cleaved in the absence of amylases.

**Others**

TRANSGLUCOSYL-AMYLASE

Sawai and Hehre (1962) described a novel starch-degrading enzyme from *Candida tropicalis*. This hydrolase is intracellular, stable under toluene autolysis, and has an unusual propensity for $\alpha$-D-glucosyl transfer from starch and dextrins to polyols. There is retention of configuration and glycerol surpasses water in acceptor capacity. The almost complete failure to attack glycogen distinguishes this enzyme from other amylases and glucoamylases of fungal origin. The biological function remains a puzzle. Only three species of *Candida* among 28 tested displayed this activity; nearly all strains of *Candida tropicalis* and *Candida albicans* were positive (Sawai & Hehre 1963).

CLASSICAL DEBRANCHING ENZYME

Yeast glycogen can also be debranched by an intracellular enzyme system (Lee *et al.* 1967) that is analogous to the animal model, i.e. a combination of oligo-1,4 $\rightarrow$ 1,4-glucantransferase (EC 2.4.1.25) and amylo-1,6-glucosidase (EC 3.2.1.33). Both activities reside in a single protein, and in the presence of glycogen phosphorylase (EC 2.4.1.1) (Fosset *et al.* 1971) and inorganic phosphate the complete degradation of glycogen to a mixture of glucose 1-phosphate and glucose (about 10%) is realized (Lee *et al.* 1970).

### *12.3.1 Acting on hemicellulose*

The waste hemicelluloses from the wood pulping industry represent a potential substrate of 'negative cost', i.e. if single-cell protein can be produced from it, then the expense of getting rid of a nuisance product could be turned to profit. The most abundant hemicelluloses in higher plants are pentosans, and in particular xylans that are typically linear chains of D-xylopyranose in $\beta$-1,4 linkage (see Whistler & Tu 1953, and references therein).

**$\beta$-1,4-Xylan xylanohydrolase (EC 3.2.1.8)**

Recent work by Biely and associates has focused on *Cryptococcus albidus* as a source of xylan-degrading enzymes. This yeast produces an extracellular $\beta$-xylanase (EC 3.2.1.8) that catalyses the hydrolysis of xylan to xylose and xylo-oligosaccharides. The oligosaccharides are apparently transported by a specific, inducible permease, and are then hydrolysed by an intracellular $\beta$-xylosidase (EC 3.2.1.37) into free xylose (Kratky & Biely 1980, Peciarova & Biely 1982). Xylobioses are the best inducers of the xylan-degrading system

(Biely & Petrakova 1984). This is in keeping with the principles elucidated by Reese *et al.* (1969), viz. that glycanases are induced by the products of their action on polymeric substrates, as a result of small levels of constitutive enzymes.

### *12.3.2 Acting on fructans*

The fructans of the plant world fall roughly into two categories. First, the levans contain D-fructofuranose residues mainly in $\beta$-2,6 linkage, but also with $\beta$-2,1-linked branch points. Although the enzymology is of considerable interest in bacterial systems (see, for example, Dedonder *et al.* 1963, Rapoport & Dedonder 1963a, b) it will not be discussed further in the present context. Secondly, the inulins are major reserve polysaccharides in some tuberous plants of which the most notable are *Dahlia rosea* (dahlia), *Taraxacum officinale* (dandelion), *Helianthus tuberosus* (Jerusalem artichoke) and *Chicorium intybus* (chicory) (Horton & Wolfrom 1963). This polymer is linear, with an average of 35 D-fructofuranose units held in $\beta$-2,1-glycosidic linkage. The otherwise reducing terminus is joined to the anomeric carbon of D-glucopyranose through a sucrose type linkage (Fig. 12.4). A few yeast species are able to assimilate inulins.

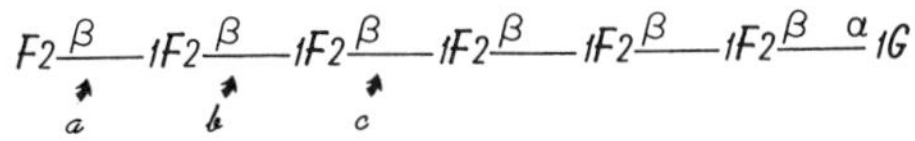


**Figure 12.4** Schematic structure of inulin (F, D-fructofuranose; G, D-glucopyranose; the connecting lines represent glycosidic linkages between the carbon atoms indicated; the configuration ($\alpha$ or $\beta$) about anomer carbons is given). The action pattern of inulinase is represented by heavy arrows and occurs in the order a, b, c, . . . .

**$\beta$-2,1-Fructan fructohydrolase (EC 3.2.1.7)**

Fructose residues are released one at a time as indicated in Figure 12.4. Although the $\beta$-fructofuranosidase fructohydrolase (EC 3.2.1.26) of baker's yeast catalyses the hydrolysis of lower-molecular weight inulins, the larger fructans are hardly affected (Arnold 1981a). Snyder and Phaff (1960) characterized an enzyme from *Kluyveromyces fragilis* that is secreted into the growth medium and acts on sucrose, raffinose and inulin. They believe that an inulinase (EC 3.2.1.7) is involved. Hewitt and GrootWassinck (1984) apparently embrace the evidence for an inulinase, but incorrectly append the systematic name and number of invertase. The latter authors have suggested that *Kluyveromyces fragilis* has potential as an industrial source of inulinase. Given the ease of acid hydrolysis of inulins and the other routes to fructose syrups, it is difficult to conceive anything more than a small regional interest in the inulins compared with starch.

### *12.3.3 Acting on mannans*

The mannans are important constituents of yeasts. They play at least three roles: first, as constituents of the cell wall (Bacon 1981); second, as antigenic determinants of yeast species (Suzuki 1981); and third, as components of several extracytoplasmic enzymes (Arnold 1981a). The yeast mannans are homopolymers of α-D-mannopyranose, with great diversity of structure among species (Spencer & Gorin 1973), but they usually include -1,6-, -1,2- and -1,3-linkages (see, for example, Ballou and Raschke 1974); some contain peripheral phosphodiester linkages (Cawley *et al.* 1972). Much, if not all, of the mannan in the cell wall appears to be associated with protein in massive macromolecules (Eddy & Longton 1969). Bacon (1981) should be consulted for an interesting discussion of the disposition of mannoproteins and other polymers within the yeast-cell envelope.

In the normal scheme of things one would expect some turnover of mannoproteins in the growing yeast cell. However, according to Kratky *et al.* (1975) this is not the case, or at least the rate of release of radio-labelled mannan, once inserted into the fabric of the wall, is very low. On the other hand, about 10% of the mannan seems to escape into the growth medium during the multiplication of cells.

The bulk of the published investigations on yeast mannans has dealt with structure and synthetic mechanisms (see, for example, Parodi 1981). All of the structure proof studies have involved enzymes from bacterial sources; only a few papers have appeared on the degradative enzymes from the yeasts.

**α-D-Mannoside mannohydrolase (EC 3.2.1.24)**

Kaya *et al.* (1973) purified an α-mannosidase from a toluene-induced autolysate of *Saccharomyces cerevisiae*, about 90-fold. The pH optimum is 6.8. A molecular weight of 300 000 was estimated from gel-permeation elution position. This enzyme catalyses the hydrolysis of aryl-α-D-mannosides at a reasonable rate; attack on yeast mannans is said to be slow but significant (Kaya *et al.* 1973), supposedly by stepwise removal of mannose units from the non-reducing termini. Barratt-Bee *et al.* (1982) reported briefly on changes in α-mannosidase activity (as measured by hydrolysis of *p*-nitrophenyl mannoside) during a 3-day culture of *Candida albicans*. Total enzyme activity followed the growth curve. The amount of mannanase activity (measured by hydrolysis of radio-labelled mannan) showed a different time course.

**α-D-Mannanase**

Augustin *et al.* (1980) have surveyed some 57 species of yeasts for mannanase activity. An artificially cross-linked mannan gel was used in a manner analogous to that employed by the same authors (Augustin *et al.* 1981) for α-amylase, as described above. Some growth on the modified mannan as sole carbon source was observed in 47 cases and suggested the presence of both a mannanase (to liquefy the gel) and an α-mannosidase (to release free mannose). A preliminary examination of products by paper chromatography indicated a prevalence of mannotriose and mannotetraose, some of which were presumably branched. The nature of the products was thus in keeping with earlier work by Maddox and Hough (1971) on a mannanase from *Saccharomyces uvarum*.

### 12.3.4 *Acting on β-glucans*

The mechanical strength of yeast cell walls is due to β-glucan polymers (Bacon 1981). It is generally accepted that turnover of β-glucans attends wall expansion in growing cells, and a large body of information is available on β-glucanases that are intimately associated with the cell wall (see Arnold 1981a, and references therein).

During the process of autolysis, no matter how it is induced, the action of endogenous β-glucanases are most important (Arnold 1979). Neuberg and Roberts (1946) reviewed the early literature; Arnold (1981b) has written the only current review on this industrially important method for releasing useful products from yeast. The enzymology definitely includes two separate glucanases that catalyse the hydrolyses of β-1,3 and β-1,6 links, and there is also an indication of a proteinase (Arnold 1972b).

Villa *et al.* (1979) showed that two β-glucanases were secreted by cultures of *Candida utilis*, but secretion is the exception rather than the rule even for *Candida* spp. Other examples of detection in spent media involve *Hanseniaspora valbyensis* and *Hanseniaspora uvarum* (Abd-El-Al & Phaff 1969) and *Kluyveromyces fragilis* and *Hansenula anomala* (Abd-El-Al & Phaff 1968), although in most cases it is not certain how much enzyme is contributed from dead cells. Villa and Phaff (1980) have described a method for recovery of β-glucanases from waste broths coming out of the baker's yeast industry. One or more of the enzymes mentioned above might find application in protoplast generation procedures, but to date the favoured sources have been snail digestive juice and bacteria (Arnold 1981c, Eddy & Longton 1969). Some yeast glucan preparations have been offered as food extenders, but there has been no great interest in the fungal glucans as potential fermentation substrates.

### 12.3.5 *Recalcitrant polysaccharides*

The cellulose of higher plants, which is a linear polymer of D-glucopyranose uniformly linked by β-1,4-glycosidic bonds, is not attacked by yeasts or yeast-like organisms. Dextrans, which are bacterial polymers containing D-glucopyranose units mainly in α-1,6 linkage (but with some α-1,4, and α-1,3 bonds, depending upon the source) are likewise not degraded by yeasts. (Note that dextrans are quite distinct from dextrins, the latter being partially degraded starches and glycogens.) The mannans of higher plants, linear β-1,4-linked D-mannopyranose units, are also refractory to yeasts. One can mention the pectic substances, mixtures of three polysaccharides (galactan, methyl ester of galacturonan, and araban) as generally not being substrates for yeasts. In all four cases well-documented examples do exist among *Aspergillus* species and other non-yeast fungi (see, for example, Fogarty & Kelly 1979) for production of significant amounts of the requisite hydrolases.

Chitin, a β-1,4-linked polymer of N-acetyl glucosamine, is a major polysaccharide in insects, Crustacea and some higher fungi. In the budding yeasts it is specifically located in the bud scars. Yeasts are not notable producers of chitinase (EC 3.2.1.14), but such an enzyme from *Saccharomyces cerevisiae* has been characterized (Elango *et al.* 1982, Correa *et al.* 1982) as a mannoprotein of the periplasmic space. Although of much potential interest in terms of chitin regulation, in conjunction with chitin synthetase (Kang *et al.* 1984), the chitinase exists in small amounts and is of no commercial interest.

### 12.3.6 *Acting on di- and trisaccharides*

Several disaccharides are important carbon sources for yeast fermentations. These include sucrose, maltose (indirectly from starch) and lactose. Trehalose is of academic interest. The trisaccharides raffinose and maltotriose will also be addressed in this section. A summary of the important structures and the bond specificities of the requisite enzymes are given in Figure 12.5.

**β-D-Fructofuranoside fructohydrolase (EC 3.2.1.26)**
Many authors shorten this to β-fructofuranosidase, and acceptable trivial names are sucrase and invertase. The natural substrate is sucrose, but others include methyl-β-fructofuranoside, raffinose, stachyose and verbascose, all of which are cleaved at the fructosidic linkage, on the fructose side of the bridge-oxygen (see Fig. 12.5). In the case of raffinose the products are melibiose and fructose. The $K_m$ for sucrose is 26 mmol $l^{-1}$ at pH 5 compared with 150 mmol $l^{-1}$ for raffinose; the $V_{max}$ value for the trisaccharide is about 50% of that for sucrose. The ubiquity of invertases among the yeasts probably reflects the native habitats of these fungi; plant nectars and exudates are rich in sucrose.

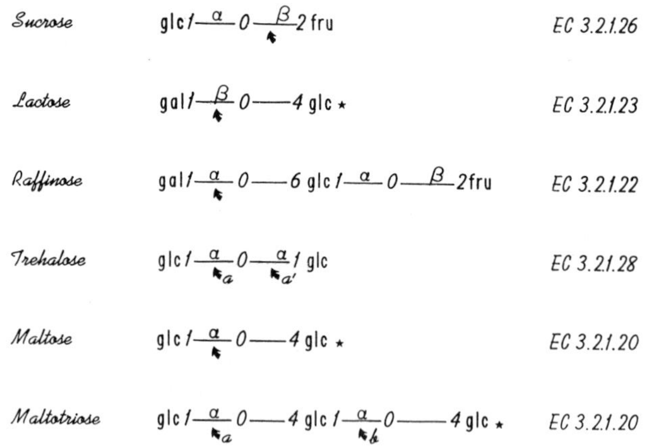


**Figure 12.5** Schematic structures of oligosaccharides (glc, D-glucopyranose; fru, D-fructofuranose; gal, D-galactopyranose). Glycosidic linkages are represented by lines from the numbered carbons through bridging oxygen (–O–). The configuration (α or β) about anomeric carbons is given. Reducing groups are indicated with a star (*). The arrows give the positions of hydrolyses by glycoside hydrolases whose EC numbers are presented on the right.

Yeast invertase was one of the earliest of enzymes to be investigated. Thus, Persoz reported on the inverting action of yeast in 1833. (The term 'inverting' and thence 'invertase' derives from the change in sign of optical rotation that attends the hydrolysis of sucrose to glucose and fructose.) Numerous investigations in the next 150 years, and indeed several hallmark studies in enzymology, involved yeast invertase. The early literature was summarized by Neuberg and Roberts (1946). Several reviews have appeared at approximately regular intervals since then. The most recent (Arnold 1981a) includes a more exhaustive bibliography than is appropriate here. Accordingly, the features will be summarized (with a minimum of primary references) together with an assessment of the results of the last few years as they illuminate the biology of this enzyme. The great majority of investigations has been conducted on *Saccharomyces cerevisiae*, and that species will be understood unless otherwise specified.

Yeast invertases are bulky glycoproteins (molecular weight about 200 000) with an average of 50% of their mass being mannan. Arnold (1969a, b) showed that purified enzyme is polydisperse with regard to carbohydrate content, and that a continuum between 39 and 54% can be demonstrated for active enzyme molecules (polydispersity has been rediscovered several times since then). The size of the enzyme is largely determined by its location in the periplasmic space, where it is soluble, but restrained by the cell wall from diffusion into the medium. In addition, it is established that the covalently bound carbohydrate affords increased enzyme stability at elevated temperatures, and increased resistance to proteolytic enzyme attack on the polypeptide moiety.

Because of its location, periplasmic invertase is naturally subjected to a range of media constituents that can penetrate the cell wall, and some fluctuations in pH, all of which are readily endured. The optimum for long-term stability is pH 5.5; the optimum range for enzyme activity is between pH 4.5 and 5.5. Toluene autolysates of baker's yeast have been known to retain their $\beta$-fructofuranosidase activity for decades. Partially purified preparations with a specific activity greater than 2600 IU $mg^{-1}$ protein can be dialysed against distilled water, lyophilized and stored as a stable powder at 4°C (Arnold 1969a). The absence of water improves stability (Neuberg & Mandl 1950).

The presence of a small amount of cytoplasmic invertase in *Saccharomyces cerevisiae* cells has been of academic interest. This is a non-glycosylated variety with no precursor-to-product relationship to the extracytoplasmic glycoprotein. The current understanding is that distinct messenger ribonucleic acids (mRNAs) for the two enzymes are encoded by the same gene (Perlman & Halvorson 1981, Carlson & Botstein 1982). Messenger RNAs for extracytoplasmic invertase include a sequence that has been deemed a signal for secretion (Perlman *et al.* 1982, Carlson *et al.* 1983).

The multiplicity of genes for sucrose assimilation has recently been rationalized, to the effect that only one of the six previously-identified structural genes is usually expressed by any particular strain, and that these genes have been transposed to different chromosomal locations in closely related strains (Carlson & Botstein 1983). The results of Ottolenghi (1971) show a remarkable constancy in the kinetic parameters for $\beta$-fructofuranosidase from five strains of *Saccharomyces cerevisiae*, each containing a different non-allelic gene for sucrose assimilation, and this is in agreement with the most recent genetic statements mentioned above.

The biosynthesis of invertase is repressed by high concentrations of glucose in the growth medium. The titre of enzyme in baker's yeast can be increased by overnight aeration in a peptone-containing medium to which sucrose is slowly and continuously added (see, for example, Arnold 1972a).

Yeast invertase is used for the analysis of sucrose in food products, and commercial preparations from *Saccharomyces cerevisiae* and *Candida utilis* are available. A useful product of much higher specific activity is easily prepared by the method of Myrback and Schilling (1965). Neuberg and Roberts (1946) discussed the earlier literature on autolysis methods of isolation, and a more recent review is given by Arnold (1981b).

Invertase is also used in the food industry in the manufacture of soft-centred chocolates (see, for example, Janssen 1964) and baked goods with stable texture (see, for example, Youngquist & Brables 1982). In each case the principle rests on the conversion of crystalline sucrose into the more soluble invert sugars. These applications represent a significant, but rather small, market for commercial invertase. High-sucrose diets in man cause an increase in dental caries, and along those lines a mouthwash containing invertase has been claimed as beneficial in affording some protection from dental caries and gingivitis (Herai *et al.* 1981).

From the point of view of the brewer, the various strains provide no limitation on the use of sucrose as a substrate. Baker's and brewer's yeast strains have been selected for their high invertase potential. Arnold (1982) found two orders of magnitude more invertase activity in commercial baker's yeast than in genetically defined laboratory strains grown under the same conditions. In countries such as Australia with an overproduction of cane sugar, some sucrose is used in the brewing industry. However, the increased cost of sucrose on the world market in the last decade or so probably precludes much competition with starch as a carbon source.

**β-D-Galactoside galactohydrolase (EC 3.2.1.23)**

The trivial name is lactase, after the natural substrate lactose which is split into galactose and glucose (Fig. 12.5b). Interest in the enzymatic hydrolysis of lactose from milk products comes from at least two avenues. First, there is a reasonably high incidence of lactose intolerance in adult humans. For those individuals it is desirable to hydrolyse the disaccharide (before ingestion) in such a manner that neither the nutritional value nor the taste of the treated product are compromised. A specific enzyme method is thus indicated. Secondly, there is a surplus of milk solids from the cheese industry; methods to convert whey into single-cell protein or ethanol would be attractive if they were economical. However, relatively few yeast species are able to assimilate lactose, a major constituent of whey. The yeast species that have attracted the most interest to date are *Kluyveromyces fragilis* (syn. *Saccharomyces fragilis*) and *Kluyveromyces lactis* (syn. *Saccharomyces lactis*, *Zygosaccharomyces lactis*). *Candida pseudotropicalis* is now considered to be an imperfect form of *Kluyveromyces fragilis* (van der Walt 1971).

*KLUYVEROMYCES FRAGILIS*

The ability of this species to assimilate lactose is due to a permease system for the intact disaccharide (de la Fuente & Sols 1962) and a cytoplasmic $\beta$-galactosidase (Caputto *et al.* 1948). Uwajima *et al.* (1972) started with a toluene-induced autolysate and purified the enzyme to homogeneity. They found a pH optimum of 6.8 (50% as active at pH values of 5.8 and 7.4) and moderate heat inactivation susceptibility (90% enzyme activity destroyed by 50°C for 10 min). The $K_m$ for lactose is 21 mmol $l^{-1}$, and the estimated molecular weight (analytical ultracentrifugation) is 203 000. The unusual requirement for $K^+$ and a divalent cation ($Mn^{2+}$, $Mg^{2+}$ or $Co^{2+}$) was first described by Caputto *et al.* (1948) and further documented by Uwajima *et al.* (1972) and Mahoney and Whitaker (1977). The enzyme contains two subunits and is not a glycoprotein (Mahoney & Whitaker 1978). Maintenance of sulphydryl status is required for enzymatic activity (Uwajima *et al.* 1972, Mahoney & Whitaker 1978).

Growth conditions for *Kluyveromyces fragilis* that favour production of lactase have been explored by Wendoff *et al.* (1970), Mahoney *et al.* (1975) and Gomez and Castillo (1983). Galactose and lactose induce synthesis of the $\beta$-galactosidase; glucose acts as a repressor.

Releasing active $\beta$-galactosidase from cells adds to the expense. Fenton (1982) has addressed autolysis methods for this specific application. Pretreatment of cells with 80% ethanol for 1.5 h followed by autolysis at pH 6.6 and 28°C for 15 h was one of the recommended regimens for which 90% yield is claimed.

*KLUYVEROMYCES LACTIS*

An intracellular $\beta$-galactosidase is induced in this species by lactose or galactose (Tingle & Halvorson 1972). Bierman and Glantz (1968) purified this enzyme through a four-step procedure, and the final product (19% yield) was shown to be homogeneous by electrophoresis on polyacrylamide gels. The pH optimum was 7.2 (50% as much activity at about pH 6.6 and 8.1). The purified enzyme required 2 mmol $l^{-1}$ $Mg^{2+}$ for optimum activity. Mercaptoethanol (20 mmol $l^{-1}$) was included in assays, but the sulphydryl status of the enzyme was not addressed. Dickson *et al.* (1979) used a very similar purification regimen and confirmed the pH parameters. These authors estimated a $K_m$ for lactose of 17 mmol $l^{-1}$ and a minimum molecular weight of 135 000. A requirement for 40 mmol $l^{-1}$ $Na^+$ or $K^+$ and a preference for $Mn^{2+}$ over $Mg^{2+}$ were reported. Neither Bierman and Glantz (1968) nor Dickson *et al.* (1979) quote the metal requirements of the *Kluyveromyces fragilis* enzyme, data that were already available in 1948 (Caputto *et al.* 1948).

Dickson and Markin (1980) found that the maximum rate of $\beta$-galactosidase induction in *Kluyveromyces lactis* was achieved when the lactose concentration in the growth medium was maintained above 2 mmol $l^{-1}$. Gonzalez and Berry (1982) found glucose repression at concentrations of 1 mmol $l^{-1}$ or greater; on the other hand, stimulation of $\beta$-galactosidase synthesis occurred at lower levels of glucose.

The great majority of *Saccharomyces* species cannot utilize lactose as a carbon source. A provocative paper by Puglisi *et al.* (1970) indicates that

*Saccharomyces cerevisiae* can be induced to synthesize a cytoplasmic β-galactosidase by exposure to galactose in the growth medium. That growth on lactose still does not occur is attributed to failure to develop a lactose permease system. Further work on the genetics of the lactose permease system might be instructive. Elegant fluorometric methods for β-galactosidase activity measurement in single yeast cells are now available (Yashphe & Halvorson 1976).

**α-Galactoside galactohydrolase (EC 3.2.1.22)**

The trivial names are α-galactosidase and melibiase. The latter derives from the substrate melibiose (6-O-α-D-galactopyranosyl-D-glucose), a rare sugar in nature but readily prepared from raffinose (see Fig. 12.5c).

Although the α-galactosyl group is widely distributed in both plants and animals as a moiety of glycoproteins, glycolipids and oligosaccharides, the compound of immediate interest is raffinose. After sucrose, raffinose is probably the most abundant oligosaccharide in the plant world (French 1954). This non-reducing disaccharide has long been associated with the beet-sugar refining industry and, in fact, the name is inspired by the French verb *raffiner* (= to refine). It exists in small concentrations (0.05%, compared with 16% sucrose) in sugar beets, but accumulates in the mother liquors during sucrose crystallization (de Whalley & Gross 1954). Nevertheless, the amount of raffinose available as a by-product is not sufficient to compete with other carbohydrates for commercial fermentations, and it remains a speciality chemical.

Early interest in yeast α-galactosidases arose from the observation that brewer's bottom yeasts (*Saccharomyces carlsbergensis*, now *Saccharomyces uvarum*) can ferment raffinose completely whereas top yeasts (*Saccharomyces cerevisiae*) can only ferment the fructose moiety. Both species have a β-fructofuranosidase (see above), but the latter lacks an α-galactosidase so that melibiose accumulates in the test medium. The fermentation of raffinose by selected *Saccharomyces* spp. has been discussed in more detail by Arnold (1981a). Some preliminary work has appeared on α-galactosidases from *Cryptococcus laurentii* (Bhanot & Brown 1980) and from *Pichia guilliermondii* (Church *et al.* 1980).

The α-galactosidase of *Saccharomyces uvarum* is an inducible enzyme that is located in the periplasmic space (de la Fuente & Sols 1962, Friis & Ottolenghi 1960, see Arnold 1981a for a summary). Lazo *et al.* (1977) have purified the enzyme to homogeneity as judged by electrophoresis. It is a glycoprotein (about 60% carbohydrate) with a molecular weight of about 300 000; the pH optimum is broad (between 4.0 and 5.0) (Lazo *et al.* 1978).

Cartledge and Lloyd (1972) and Herrero *et al.* (1980) have studied cytoplasmic vesicles in *Saccharomyces uvarum* and their possible role in export of α-galactosidase, as well as other extracytoplasmic glycoproteins. Florez *et al.* (1981) have repeated the earlier work on enzyme induction by galactose, melibiose and raffinose, and they have surveyed other sugars and derivatives.

There is a demand for α-galactosidases (that are free from other glycosidase activities) for analytical purposes, e.g. in the elucidation of glycoprotein structure. Enzymes from *Aspergillus niger* and green coffee beans are commercially available, but the yeast enzymes have not been used as such. An $\alpha$-galactosidase from the mould *Mortierella vinacea* has been thoroughly studied by Suzuki *et al.* (1970), and their paper is a good model for this type of application.

**$\alpha,\alpha'$-Glucoside 1-glucohydrolase (EC 3.2.1.28)**

The trivial name is trehalase, after the naturally occurring substrate (see Fig. 12.5d); 6-substituted trehaloses are also substrates. An interesting account of the discovery of trehalose has been compiled by Harding (1923). This non-reducing disaccharide is a constituent of most fungi, some bacteria, a few higher plants and possibly all insect lymphs. Man is equipped with a trehalase to deal with dietary sources of trehalose.

Trehalose and glycogen are the common reserve carbohydrates of fungi. In yeast there are cytoplasmic trehalases whose activity is increased by some physiological signal preceding the onset of budding (see Arnold & McLellan 1975). The cytoplasmic trehalases are difficult to purify because of their susceptibility to inactivation (Arnold 1981a), but they elicit considerable interest for their role in controlling the flow of endogenous carbon that attends new growth.

A lag in the uptake of trehalose by *Saccharomyces cerevisiae* is due to the time required for the induction of a specific transport system (Kotyk & Michaljanicova 1979). It should be mentioned in passing that inducible sugar carriers in general are turned over; the half-lives of some permeases under non-induction conditions are a matter of 1–2 h (Alonso & Kotyk 1978).

Relatively few yeast species assimilate trehalose without lag, one example is *Torulopsis glabrata* which has an additional, constitutive trehalase located in the periplasmic space (Arnold 1979). Trehalases are apparently not secreted by live yeast cells. Although trehalose is a ubiquitous sugar there are no major sources that could be used for fermentations, so in the present context the trehalases remain of fundamental interest rather than practical import.

**$\alpha$-D-Glucoside glucohydrolase (EC 3.2.1.20)**

An $\alpha$-glucosidase has been purified from *Saccharomyces italicus* autolysates. It has a broad specificity with respect to alkyl- and aryl-$\alpha$-D-glucopyranosides, but supposedly maltose (Fig. 12.5e) and maltotriose (Fig. 12.5f) are the natural substrates (Halvorson & Ellias 1958). The enzyme (trivial name maltase) is active against sucrose (not due to $\beta$-fructofuranosidase contamination, according to the authors) but not against trehalose. A very similar protein seems to have been isolated by Phillips (1959) from *Saccharomyces cerevisiae*. Gottschalk (1951) surveyed the considerable earlier literature on yeast $\alpha$-glucosidases, some of which is now known to have been based on a mixture of enzymes. Matsusaka *et al.* (1977) and Needleman *et al.* (1978) purified to homogeneity $\alpha$-glucosidases from *Saccharomyces carlsbergensis* strains. This class of enzyme is functional in the intracellular hydrolysis of maltose and maltotriose which are taken across the plasma membrane by an induced permease (Kotyk & Horak 1981). Similar activities found in the spent culture medium by various workers may be due to autolytic release from dead cells.

## 12.4 Phosphoric monoester hydrolases

Phosphorus is an essential element of all living cells, and the availability of, and competition for, orthophosphate ions in nature is well documented. The biology, physiology and biochemistry of phosphate uptake and assimilation were recently reviewed by Beever and Burns (1981) for fungi in general. Phosphate esters of carbohydrates, lipids, nucleosides and other organic compounds are not transported across microbial plasma membranes. These compounds are first hydrolysed by non-specific phosphatases located extracytoplasmically. In bacteria the enzymes are alkaline phosphatases (EC 3.1.3.1), in yeast they are acid phosphatases (EC 3.1.3.2), an obvious reflection of preferred growth environments.

The levels of acid phosphatase produced by most yeast species are controlled by the concentration of orthophosphate in the growth medium; millimolar concentrations repress synthesis, which is in keeping with good cellular economy. An increase in the production of acid phosphatase can usually be achieved by depleting the medium of orthophosphate by pretreatment with magnesia.

**Orthophosphoric monoester phosphohydrolase (EC 3.1.3.2)**
The trivial name of this enzyme is acid phosphatase:

$$RO\text{PO}_3\text{H}^- + \text{H}_2\text{O} = R\text{OH} + \text{H}_2\text{PO}_4^-$$

The optimum pH for activity is in the vicinity of 4.0. This enzyme is ubiquitous among the yeasts. A wide spectrum of phosphomonoesters, including $\alpha$- and $\beta$-glycerophosphates, hexose phosphates and aryl phosphates are substrates. Adenosine triphosphate is attacked stepwise, with the liberation of orthophosphate. Diesters of phosphate are not hydrolysed. A more complete summary of substrates is given in Arnold (1981a).

Highly purified preparations from *Saccharomyces cerevisiae* (Boer & Steyn-Parve 1966), *Schizosaccharomyces pombe* (Dibenedetto 1972) and *Rhodotorula glutinis* (Trimble *et al.* 1981) are all glycoproteins of molecular weight about 200 000. In the great majority of cases acid phosphatase is located in the periplasmic space; claims for additional activity on the cell wall of *Saccharomyces cerevisiae* have been dismissed as artefacts by Arnold and Garrison (1981). Cracking the cell wall releases 75% of the periplasmic enzyme, and the remainder is associated with the plasma membrane (Arnold 1972a). A more extensive evaluation of the evidence is given in Arnold (1981a).

There are exceptional species with respect to localization of acid phosphatase. The cellular component of *Saccharomyces rouxii* is in the periplasmic space, but an approximately equal amount is found in the spent medium (Arnold & Garrison 1979). *Saccharomyces bisporus* var. *mellis* secretes acid phosphatase into a medium fortified with 0.25 mol $l^{-1}$ KCl (Weimberg & Orton 1965). *Hansenula holstii* also produces extracellular acid phosphatase (San Blas & Cunningham 1974). There is some evidence that secreted enzyme can be adsorbed on the slime layer of *Cryptococcus neoformans* (Mahvi *et al.* 1974) and microfibrils of *Sporothrix schenckii* (Garrison & Arnold 1983).

The only commercial opening for acid phosphatase at present is as an analytical reagent, and that is currently met by preparations from wheat germ, potato and bovine semen. Isolation from baker's yeast is not economical, in part because mechanical disruption is required; the enzyme is unstable to autolysis (Arnold 1972a). Reasonably pure preparations are easily produced from *Saccharomyces rouxii* by induced secretion of previously washed cells (Arnold & Garrison 1979).

According to Thill *et al.* (1983) and Arima *et al.* (1983) an N-terminal region of 17 amino acids in *Saccharomyces cerevisiae* acid phosphatase is a signal for secretion. The physiological control of repressible acid phosphatase by orthophosphate is apparently mediated at the gene transcription level (Bostian *et al.* 1983).

**Phosphoprotein phosphohydrolase (EC 3.1.3.16)**

The primary role of acid phosphatase (see above) in yeasts is the hydrolysis of external phosphate esters (Heredia *et al.* 1963), and until recently this was the only postulated role. Although only a working hypothesis at present, the acid phosphatases may also be involved in the control of plasma membrane or periplasmic space enzymes by dephosphorylation of these other proteins. The trivial name for such an enzyme is protein phosphatase, and several examples from animal sources are available (Fischer & Brautigan 1982). Unpublished results from experiments in the author's laboratory indicate that acid phosphatases from *Saccharomyces rouxii* and *Sporothrix schenckii* can catalyse the hydrolyses of phosphate esters of serine, threonine and tyrosine, and that phosphoproteins such as phosvitin and casein are acted upon to a significant extent.

## 12.5 Carboxylic ester hydrolases

**Carboxylic-ester hydrolase (EC 3.1.1.1)**

$$\text{A carboxylic ester} + \text{water} = \text{an alcohol} + \text{a carboxylate}$$

These enzymes catalyse the approach to equilibrium as written above; a wide substrate specificity is common among members of this class. Several communications (e.g. Suomalainen 1969) address the turnover of esters during brewing procedures. The specificities of the intracellular and periplasmically located esterases are different (Parkkinen & Suomalainen 1982). In all of the industrial yeasts that have been examined, the esterases are not secreted into the medium. The periplasmic activities are demonstrable in washed, live-cell suspensions. Spaepen and Verachtert (1982) reported on the esterase from *Brettanomyces* spp. (so-called secondary yeasts in brewing), which is active against a number of esters normally encountered in Belgian beers. Yoshioka and Hashimoto (1981) can be consulted for the esterases of Japanese brewer's yeasts (*Saccharomyces uvarum* and *Saccharomyces cerevisiae* strains) while Peel (1951) and Laurema and Erkama (1968) worked with the esterases from *Hansenula anomala*.

**Glycerol-ester hydrolase (EC 3.1.1.3)**
The yeasts are not active secretors of lipases even though such enzymes must be active in the internal turnover of triglycerides. Publications by Jensen (1974) on the mould *Geotrichum candidum*, and Federici (1982) on a yeast-like organism, *Aureobasidium* (*Pullularia*) *pullulans* should be consulted for lipase characterization and survey, respectively.

**Phosphatide acyl-hydrolase (EC 3.1.1.4)**
Nurminen and Suomalainen (1970) provided evidence for the facile elution of phospholipase A from baker's yeast plasma membranes. Kokke *et al.* (1963) described a lysolecithin acyl-hydrolase (EC 3.1.1.5) that was easily eluted from intact cells of baker's yeast.

## 12.6 Peptide hydrolases

Some di- and tripeptides are taken up intact by *Saccharomyces cerevisiae* cells (Marder *et al.* 1977) and are acted upon internally by peptidases. Frey and Rohn (1978) claim that there is also a periplasmic aminopeptidase (EC 3.4.1.1).

Proteolysis and its control are important processes of the yeast cytoplasm (Holzer & Heinrich 1980), and have much to do with the turning on and off of alternative metabolic pathways. Conversely, the yeasts are not noted for secreting proteinases. Small amounts are found in the growth medium of brewer's yeast according to Maddox and Hough (1970). A survey by Ahearn *et al.* (1968) lists several yeasts as having some degradative effect on gelatine or casein. Martini and Federici (1980) reported on a milk-coagulating enzyme from *Cryptococcus albidus* var. *aerius* that they think may have industrial use.

## 12.7 Concluding remarks

Oligosaccharides and esters of moderate size can penetrate the yeast cell wall (Arnold & Lacy 1977) and, as described above, they may be acted on by periplasmic hydrolases. It has been argued elsewhere (Arnold 1981a) that a species that retains the requisite enzymes in the periplasmic space is more highly evolved than its competitors that share their energetically expensive proteins with the environment. Conversely, enzymes that are secreted into the medium are certainly easier for us to handle for secondary applications.

In the case of bulky substrates (starch, inulin, proteins, etc.) the fungi have no alternative but to secrete the particular hydrolases into the medium. It should be clear from this chapter that the yeasts, as a group, are not prolific producers of extracellular enzymes. Notwithstanding the possible results of future surveys, it would be surprising if yeast species came to light as industrial candidates in some of the categories we have briefly discussed.

In other cases the extent of the coverage herein is approximately proportional to the present depth of understanding and the industrial interest that has evolved. One anticipates that there will be continued interest in starch-degrading enzymes, lactase, invertase, and possibly others. From the fundamental aspect the starch debranching enzymes produced by certain yeast species seem to be the least studied, yet perhaps they are the most likely to be limiting in terms of rate and extent of starch breakdown. There is obviously a need for further research in that area.

From the brewer's perspective there is not much point in demonstrating alternative substrates, or more-complete utilization of a particular substrate, if the end-product is compromised in taste or odour. The addition of enzyme extracts from other yeast (or non-yeast) sources may turn out to be more practical than mixed species fermentations. Otherwise, some rather specific genetic engineering is called for, as discussed in other chapters.

On the other hand, production of single cell protein may call for mixed or tandem fermentations. The 'waste' polysaccharides from the food, paper and other industries are attractive substrates for this, and further processes involving yeasts are to be anticipated.

## Acknowledgements

Literature retrieval in connection with this chapter was greatly facilitated by the assistance of Larry C. Mann. The author's research was supported by grants from the US Department of Health and Human Services.

## References

Abd-El-Al, A. T. V. and H. J. Phaff 1968. Exo-$\beta$-glucanases in yeast. *Biochemical Journal* **109**, 347–60.

Abd-El-Al, A. T. V. and H. J. Phaff 1969. Purification and properties of endo $\beta$-glucanase in the yeast *Hanseniaspora valbyensis*. *Canadian Journal of Microbiology* **15**, 697–701.

Abdullah, M., B. J. Catley, E. Y. C. Lee, J. Robyt, K. Wallenfels and W. J. Whelan 1966. The mechanism of carbohydrase action. XI. Pullulanase, an enzyme specific for hydrolysis of $\alpha$-1,6-bonds in amylaceous oligo- and polysaccharides. *Cereal Chemistry* **43**, 111–18.

Ahearn, D. G., S. P. Meyers and R. A. Nichols 1968. Extracellular proteinases of yeasts and yeastlike fungi. *Applied Microbiology* **16**, 1370–4.

Alonso, A. and A. Kotyk 1978. Apparent half-lives of sugar transport proteins in *Saccharomyces cerevisiae*. *Folia Microbiologia* **23**, 118–25.

Andrews, J. and R. B. Gilliland 1952. Superattenuation of beer: a study of three organisms capable of causing abnormal attenuations. *Journal of the Institute of Brewing* **58**, 189–96.

Arima, K., T. Oshima, I. Kubota, N. Nakamura, T. Mizunaga and A. Tohe 1983. The nucleotide sequence of the yeast PHO 5 gene: a putative precursor of repressible acid phosphatase contains a signal peptide. *Nucleic Acids Research* **11**, 1657–72.

Arnold, W. N. 1969a. Heat inactivation kinetics of yeast $\beta$-fructofuranosidase. A polydisperse system. *Biochimica et Biophysica Acta* **178**, 347–53.

Arnold, W. N. 1969b. Chromatography of a polydisperse enzyme, yeast $\beta$-fructofuranosidase on DEAE-cellulose. *Anal. Letters* **2**, 547–52.

Arnold, W. N. 1972a. Location of acid phosphatase and $\beta$-fructofuranosidase within yeast cell envelopes. *Journal of Bacteriology* **112**, 1346–52.

Arnold, W. N. 1972b. The structure of the yeast cell wall. Solubilization of a marker enzyme, β-fructofuranosidase, by the autolytic enzyme system. *Journal of Biological Chemistry* **247**, 1161–9.

Arnold, W. N. 1979. Trehalose assimilation and turnover by *Torulopsis glabrata*. *Current Microbiology* **2**, 109–12.

Arnold, W. N. 1981a. Enzymes. In *Yeast cell envelopes: biochemistry, biophysics, and ultrastructure*, W. N. Arnold (ed.), Vol. II, Chapter 1, pp. 1–46. Boca Raton: CRC Press.

Arnold, W. N. 1981b. Autolysis. In *Yeast cell envelopes: biochemistry, biophysics, and ultrastructure*, W. N. Arnold (ed.), Vol. II, Chapter 7, 129–37. Boca Raton: CRC Press.

Arnold, W. N. 1981c. Protoplasts. In *Yeast cell envelopes: biochemistry, biophysics, and ultrastructure*, W. N. Arnold (ed.), Vol. II, Chapter 5, 93–103. Boca Raton: CRC Press.

Arnold, W. N. 1982. The cryptic β-fructofuranosidase of *Saccharomyces rouxii*. *Molecular and Cellular Biochemistry* **45**, 59–64.

Arnold, W. N. and R. G. Garrison 1979. An Fe-activated acid phosphatase from *Saccharomyces rouxii*. *Journal of Biological Chemistry* **254**, 4919–24.

Arnold, W. N. and R. G. Garrison 1981. Kinetic limitations on the trapping of nascent phosphate for cytochemical localization of yeast acid phosphatase. *Current Microbiology* **5**, 57–60.

Arnold, W. N. and J. S. Lacy 1977. Permeability of the cell envelope and osmotic behavior in *Saccharomyces cerevisiae*. *Journal of Bacteriology* **131**, 564–71.

Arnold, W. N. and M. N. McLellan 1975. Trehalose and glycogen levels during the initial stages of growth of *Candida albicans*. *Physiological Chemistry and Physics* **7**, 369–80.

Augustin, J., J. Zemek, A. Kockova-Kratochvilova and L. Kuniak 1978. Production of α-amylase by yeasts and yeast-like organisms. *Folia Microbiologia* **23**, 353–61.

Augustin, J., J. Zemek, L. Kuniak and A. Kockova-Kratochvilova 1980. Mannan hydrolyzing enzymes of yeasts and yeastlike organisms. *Folia Microbiologia* **25**, 301–5.

Augustin, J., J. Zemek, O. Fassatiova and L. Kuniak 1981. Production of α-amylase by microscopic fungi. *Folia Microbiologia* **26**, 142–6.

Bacon, J. S. D. 1981. Nature and disposition of polysaccharides within the cell envelope. In *Yeast cell envelopes: biochemistry, biophysics, and ultrastructure*, W. N. Arnold (ed.), Vol. I, Chapter 5, 65–84. Boca Raton: CRC Press.

Ballou, C. E. and W. C. Raschke 1974. Polymorphism of the somatic antigen of yeast. *Science* **184**, 127–34.

Barrett-Bee, K. J., J. Lees and W. Henderson 1982. Variation in the activities of enzymes associated with cell wall metabolism during a growth cycle of *Candida albicans*. *FEMS Microbiological Letters* **15**, 275–8.

Beever, R. E. and D. J. W. Burns 1981. Phosphorus uptake, storage and utilization by fungi. In *Advances in botanical research*, K. W. Woolhouse (ed.), Vol. 8, 127–219. London: Academic Press.

Bender, H., J. Lehmann and K. Wallenfells 1959. Pullulan, an extracellular glucan from *Pullularia pullulans*. *Biochimica et Biophysica Acta* **36**, 309–16.

Bhanot, P. and R. G. Brown 1980. Effect of 3-O-methyl-D-glucose on the production of glycosidases by *Cryptococcus laurentii* and *Saccharomyces cerevisiae*. *Canadian Journal of Microbiology* **26**, 1289–95.

Biely, P. and F. Petrakova 1984. Novel inducers of the xylan-degrading enzyme system of *Cryptococcus albidus*. *Journal of Bacteriology* **160**, 408–12.

Biermann, L. and M. D. Glantz 1968. Isolation and characterization of $\beta$-galactosidase from *Saccharomyces lactis*. *Biochimica et Biophysica Acta* **167**, 373–7.

Blobel, G. and B. Dobberstein 1975. Transfer of proteins across membranes. 1. Presence of proteolytically processed and unprocessed nascent immunoglobulin light chains on membrane bound ribosomes of murine myeloma. *Journal of Cell Biology* **67**, 835–51.

Boer, P. and E. P. Steyn-Parve 1966. Isolation and purification of an acid phosphatase from baker's yeast (*Saccharomyces cerevisiae*). *Biochimica et Biophysica Acta* **128**, 400–2.

Bostian, K. A., J. M. Lemire and H. O. Halvorson 1983. Physiological control of repressible acid phosphatase gene transcripts in *Saccharomyces cerevisiae*. *Molecular and Cellular Biology* **3**, 839–53.

Caputto, R., L. F. Leloir and R. E. Trucco 1948. Lactase and lactose fermentation in *Saccharomyces fragilis*. *Enzymologia* **12**, 350–5.

Carlson, M. and D. Botstein 1982. Two differentially regulated mRNA's with different 5′ ends encode secreted and cytoplasmic forms of yeast invertase. *Cell* **28**, 145–54.

Carlson, M. and D. Botstein 1983. Organization of the SUC gene family in *Saccharomyces*. *Molecular and Cellular Biology* **3**, 351–9.

Carlson, M., R. Taussig, S. Kustu and D. Botstein 1983. The secreted form of invertase in *Saccharomyces cerevisiae* is synthesized from mRNA encoding a signal sequence. *Molecular and Cellular Biology* **3**, 351–9.

Cartledge, T. G. and D. Lloyd 1972. Subcellular fractionation of particles containing acid hydrolases from *Saccharomyces carlsbergensis*. *Biochemical Journal* **126**, 755–7.

Cawley, T. N., M. G. Harrington and R. Letters 1972. A study of the phosphate linkages in the phosphomannan in the cell walls of *Saccharomyces cerevisiae*. *Biochemical Journal* **129**, 711–20.

Church, F. C., S. P. Meyers and V. R. Srinivasen 1980. Isolation and characterization of $\alpha$-galactosidase from *Pichia guilliermondii*. *Developments in Industrial Microbiology* **21**, 339–48.

Clementi, F., J. Rossi, L. Costamagna and J. Rossi 1980. Production of amylase(s) by *Schwanniomyces castellii* and *Endomycopsis fibuligera*. *Antonie van Leeuwenhoek Journal* **46**, 379–405.

Correa, J. U., N. Elango, I. Polacheck and E. Cabib 1982. Endochitinase, a mannan-associated enzyme from *Saccharomyces cerevisiae*. *Journal of Biological Chemistry* **257**, 1392–7.

de la Fuente, G. and A. Sols 1962. Transport of sugars in yeasts. II. Mechanisms of utilization of disaccharides and related glycosides. *Biochimica et Biophysica Acta* **56**, 49–62.

Dedonder, R., E. Jozon, G. Rapoport, Y. Joyeux and A. Fritsch 1963. La saccharose-levane $\beta$-D-fructofuranosyl-transferase (levane-sucrase) de *Bacillus subtilis*. I. Obtention de l'enzyme pur. *Bulletin de la Société de Chimie Biologique* **45**, 477–91.

DeMot, R., K. Andries and H. Verachtert 1984. Comparative study of starch degradation and amylase production by ascomycetous yeast species. *Systematic and Applied Microbiology* **5**, 106–18.

de Whalley, H. C. S. and D. Gross 1954. Determination of raffinose and kestose in plant products. *Methods in Biochemical Analysis* **1**, 307–39.

Dhawle, M. R. and W. M. Ingledew 1983. Starch hydrolysis by derepressed mutants of *Schwanniomyces castellii*. *Biotechnology Letters* **5**, 185–90.

Dibenedetto, G. 1972. Acid phosphatase in *Schizosaccharomyces pombe*. I. Regulation and preliminary characterization. *Biochimica et Biophysica Acta* **286**, 363–74.

Dickson, R. C. and J. S. Markin 1980. Physiological studies of $\beta$-galactosidase induction in *Kluyveromyces lactis*. *Journal of Bacteriology* **142**, 777–85.
Dickson, R. C., L. R. Dickson and J. S. Markin 1979. Purification and properties of an inducible $\beta$-galactosidase isolated from the yeast *Kluyveromyces lactis*. *Journal of Bacteriology* **137**, 51–61.

Ebertova, H. 1966a. Amylolytic enzymes of *Endomycopsis capsularis*. I. Formation of the amylolytic system in cultures. *Folia Microbiologica* **2**, 14–20.
Ebertova, H. 1966b. Amylolytic enzymes of *Endomycopsis capsularis*. II. A study of the properties of isolated $\alpha$-amylase, amyloglucosidase and maltase-transglucosidase. *Folia Microbiologia* **2**, 422–38.
Eddy, A. A. and J. Longton 1969. A large mannan molecule released from isolated yeast cell walls by the action of snail gut juice. *Journal of the Institute of Brewing* **75**, 7–9.
Elango, N., J. U. Correa and E. Cabib 1982. Secretory character of yeast chitinase. *Journal of Biological Chemistry* **257**, 1398–400.
Erratt, J. A. and G. G. Stewart 1981a. Fermentation studies using *Saccharomyces diastaticus* yeast strains. In *Developments in industrial microbiology*, L. A. Underkoffler and M. L. Wulf (eds), Vol. 22, Chapter 50, 577–87. Arlington: Society for Industrial Microbiology.
Erratt, J. A. and G. G. Stewart 1981b. Genetic and biochemical studies on glucoamylases from *Saccharomyces diastaticus*. In *Current developments in yeast research*, G. G. Stewart and I. Russell (eds), 279–84. Toronto: Pergamon Press.

Federici, F. 1982. Extracellular enzymatic activities in *Aureobasidium pullulans*. *Mycologia* **74**, 738–43.
Federici, F. and M. D'Elia 1983. Growth and amylolytic activity of *Aureobasidium pullulans* in starch-limited culture. *Enzyme and Microbial Technology* **5**, 225–6.
Fenton, D. M. 1982. Solvent treatment for $\beta$-galactosidase release from yeast cells. *Enzyme and Microbial Technology* **4**, 229–32.
Fischer, E. H. and D. L. Brautigan 1982. A phosphatase by any other name: from prosthetic group removing enzyme to phosphorylase phosphatase. *Trends in Biochemical Sciences* **7**, 3–4.
Florez, I. G., P. S. Lazo, A. G. Ochoa and S. Gascon 1981. The specificity of induction of $\alpha$-galactosidase from *Saccharomyces carlsbergensis*. *Biochimica et Biophysica Acta* **674**, 71–7.
Fogarty, W. M. and C. T. Kelly 1979. Developments in microbial extracellular enzymes. In *Topics in enzyme and fermentation biotechnology*, A. Wiseman (ed.), Vol. 3, 12–102. Chichester: Ellis Horwood.
Fosset, M., L. W. Muir, L. D. Nielsen and E. H. Fischer 1971. Purification and properties of yeast glycogen phosphorylase a and b. *Biochemistry* **10**, 4105–13.
French, D. 1954. The raffinose family of oligosaccharides. *Advances in Carbohydrate Chemistry* **9**, 149–84.
Frey, J. and K. H. Rohn 1978. Subcellular localization and levels of aminopeptidases and dipeptidases in *Saccharomyces cerevisiae*. *Biochimica et Biophysica Acta* **527**, 31–41.
Friis, J. and P. Ottolenghi 1960. Localization of melibiase in a strain of yeast. *Compte Rendu des Travaux du Laboratoire Carlsberg* **31**, 272–81.
Fukumoto, J., Y. Tsujisaka and M. Araki 1960. Studies on amylase of *Endomycopsis*. *Kagakuto Kogyo (Osaka)* **34**, 423–7. *Chemical Abstracts* **55**, 24917b.

Garrison, R. G. and W. N. Arnold 1983. Cytochemical localization of acid phosphatase in the dimorphic fungus *Sporothrix schenckii*. *Current Microbiology* **9**, 253–8.

Gomez, A. and F. J. Castillo 1983. Production of biomass and $\beta$-D-galactosidase by *Candida pseudotropicalis* grown in continuous culture on whey. *Biotechnology and Bioengineering* **25**, 1341–57.

Gonzalez, J. B. and D. R. Berry 1982. Induction of $\beta$-galactosidase in the yeast *Kluyveromyces lactis*. *Biotechnology Letters* **4**, 369–74.

Gottschalk, A. 1951. $\alpha$-Glucosidases. In *The enzymes*, J. B. Sumner and K. Myrback (eds), Vol. 1, Part 1, Chapter 15, 551–82. New York: Academic Press.

Gunja, Z. H., D. J. Manners and K. Maung 1961. Studies on carbohydrate-metabolizing enzymes. 7. Yeast isoamylase. *Biochemical Journal* **81**, 392–8.

Halvorson, H. and L. Ellias 1958. The purification and properties of an $\alpha$-glucosidase of *Saccharomyces italicus* Y-1225. *Biochimica et Biophysica Acta* **30**, 28–40.

Harada, T., A. Misaki, H. Akai, K. Yokobayashi and K. Sugimoto 1972. Characterization of *Pseudomonas* isoamylase by its action on amylopectin and glycogen: comparison with *Aerobacter* pullulanase. *Biochimica et Biophysica Acta* **268**, 497–505.

Harding, T. S. 1923. The sources of the rare sugars. IX. History of trehalose, its discovery and methods of preparation. *Sugar* **25**, 476–8.

Hattori, Y. 1961. Amylolytic enzymes produced by *Endomyces* species. II. Purification and general properties of amyloglucosidase. *Agricultural Biology and Chemistry* **25**, 895–901.

Hattori, Y. and I. Takeuchi 1961. Raw starch digestibility of the amylase of *Endomycopsis fibuligera*. *Rika Gaku Kenkysusho Kokoku* **37**, 37–42. *Chemical Abstracts* **55**, 26132f.

Hattori, Y. and I. Takeuchi 1962. Amylases produced from *Endomyces*. III. Hydrolysis of starch and glucosyl carbohydrates with amyloglucosidase. *Agricultural Biology and Chemistry* **26**, 316–22.

Herai, T., Y. Watanabe, K. Kamoi, K. Sato and K. Ishibashi 1981. Chemical effect of invertase mouthwash. *Nippon Shishubyo Gakkai Kaishi* **23**, 451–7.

Heredia, C. F., Y. Yen and A. Sols 1963. Role and formation of the acid phosphatase in yeast. *Biochemical and Biophysical Research Communications* **10**, 14–18.

Herrero, P., F. Moreno and S. Gascon 1980. Role of vesicles in the transport and secretion of extracytoplasmic enzymes by yeast. *Cellular and Molecular Biology* **26**, 485–92.

Hewitt, G. M. and J. W. D. GrootWassinck 1984. Simultaneous production of inulase and lactase in batch and continuous cultures of *Kluyveromyces fragilis*. *Enzyme and Microbial Technology* **6**, 263–70.

Hobson, P. N., W. J. Whelan and S. Peat 1951. Enzymic synthesis and degradation of starch. XIV. R enzyme. *Journal of the Chemical Society* 1451–9.

Holzer, H. and P. C. Neinrich 1980. Control of proteolysis. *Annual Review of Biochemistry* **49**, 63–91.

Hopkins, R. H. 1958. Amylase systems in culture and wild yeasts. *Wallerstein Communications* **21**, 309–19.

Hopkins, R. H. and D. Kulka 1957. The glucoamylase and debrancher of *Saccharomyces diastaticus*. *Archives of Biochemistry and Biophysics* **69**, 45–55.

Horton, D. and M. L. Wolfrom 1963. Polysaccharides. In *Comprehensive biochemistry*, M. Florkin and E. H. Stotz (eds), Vol. 5, Chapter VIIb, 189–232. Amsterdam: Elsevier.

Janssen, F. 1964. Effect of high temperatures on the efficiency of invertase. *Manufacturing Confectioner* **44**, 63.

Jensen, R. G. 1974. Microbial lipolytic enzymes; characterization of the lipase from the mold *Geotrichum candidum*, a review. *Lipids* **9**, 149–57.

Kang, M. S., E. Narayanasamy, E. Mattia, J. Au-Young, P. W. Robbins and E. Cabib 1984. Isolation of chitin synthetase from *Saccharomyces cerevisiae*. Purification of an enzyme by entrapment in the reaction product. *Journal of Biological Chemistry* **259**, 14966–72.

Kaya, T., M. Shibano and T. Kutsumi 1973. Isolation and characterization of $\alpha$-D-mannosidase from baker's yeast. *Journal of Biochemistry* **73**, 181–2.

Kerr, R. W. 1950. *Chemistry and industry of starch*, 2nd. edn. New York: Academic Press.

Kokke, R., G. J. M. Hooghwinkel, H. L. Boij, H. van den Bosch, L. Zelles, E. Milder and L. L. M. van Deenan 1963. Metabolism of lysolecithin and lecithin in a yeast suspension. *Biochimica et Biophysica Acta* **70**, 351–4.

Kotyk, H. and J. Horak 1981. Transport processes in the plasma membrane. In *Yeast cell envelopes: biochemistry, biophysics and ultrastructure*, W. N. Arnold (ed.), Vol. I, Chapter 4, 49–64. Boca Raton: CRC Press.

Kotyk, A. and D. Michaljanicova 1979. Uptake of trehalose by *Saccharomyces cerevisiae*. *Journal of General Microbiology* **110**, 323–32.

Kratky, Z. and P. Biely 1980. Inducible $\beta$-xyloside permease as a constituent of the xylan-degrading enzyme system of the yeast *Cryptococcus albidus*. *European Journal of Biochemistry* **112**, 367–73.

Kratky, Z., P. Biely and S. Bauer 1975. Wall mannan of *Saccharomyces cerevisiae*. Metabolic stability and release into growth medium. *Biochimica et Biophysica Acta* **404**, 1–6.

Kreger van Rij, N. J. W. 1971. Genus 5, *Endomycopsis*. In *The yeasts, a taxonomic study*, 2nd edn, J. Lodder (ed.), 166–208. Amsterdam: North–Holland.

Laurema, S. and J. Erkama 1968. Formation of ethyl acetate in *Hansenula anomala*. *Acta Chemica Scandinavica* **22**, 14482–6.

Lazo, P. S., A. G. Ochoa and S. Gascon 1977. $\alpha$-Galactosidase from *Saccharomyces carlsbergensis*. Cellular localization and purification of the external enzyme. *European Journal of Biochemistry* **77**, 375–82.

Lazo, P. S., A. G. Ochoa and S. Gascon 1978. $\alpha$-Galactosidase (melibiase) from *Saccharomyces carlsbergensis*. Structural and kinetic properties. *Archives of Biochemistry and Biophysics* **191**, 316–24.

Lee, E. Y. C., L. D. Nielsen and E. D. Fischer 1967. A new glycogen-debranching system in yeast. *Archives of Biochemistry and Biophysics* **121**, 245–6.

Lee, E. Y. C., J. H. Carter, L. D. Nielsen and E. H. Fischer 1970. Purification and properties of yeast amylo-1,6-glucosidase and oligo-1,4 → 1,4-glucan transferase. *Biochemistry* **9**, 2347–55.

Maddox, I. S. and J. S. Hough 1970. Proteolytic enzymes of *Saccharomyces carlsbergensis*. *Biochemical Journal* **117**, 843–8.

Maddox, I. S. and J. S. Hough 1971. Yeast glucanase and mannanase. *Journal of the Institute of Brewing* **77**, 44–7.

Mahoney, R. R. and J. R. Whitaker 1977. Stability and enzymatic properties of $\beta$-galactosidase from *Kluyveromyces fragilis*. *Journal of Food Science* **1**, 327–50.

Mahoney, R. R. and J. R. Whitaker 1978. Purification and physicochemical properties of $\beta$-galactosidase from *Kluyveromyces fragilis*. *Journal of Food Science* **43**, 584–91.

Mahoney, R. R., T. A. Nickerson and J. R. Whitaker 1975. Selection of strain, growth conditions, and extraction procedures for optimum production of lactase from *Kluyveromyces fragilis*. *Journal of Dairy Science* **58**, 1620–9.

Mahvi, T. A., S. S. Spicer and N. J. Wright 1974. Cytochemistry of acid mucosubstance and acid phosphatase in *Cryptococcus neoformans*. *Canadian Journal of Microbiology* **20**, 833–8.

Manners, D. J. 1962. Enzymic synthesis and degradation of starch and glycogen. *Advances in Carbohydrate Chemistry* **17**, 371–430.
Marder, R., J. M. Becker and F. Naider 1977. Peptide transport in yeast. Utilization of leucine and lysine containing peptides by *Saccharomyces cerevisiae. Journal of Bacteriology* **131**, 906–16.
Marshall, J. J. 1980. A novel glucoamylase from *Cladosporium resinae*. In *Mechanisms of saccharide polymerization and depolymerization*, J. J. Marshall (ed.), 119–24. New York: Academic Press.
Martini, A. and F. Federici 1980. Partial purification of a yeast extracellular acid protease. *Journal of Dairy Science* **63**, 1397–402.
Maruo, B. and T. Kobayashi 1951. Enzymic scission of the branch links in amylopectin. *Nature* (London) **167**, 606–7.
Matsusaka, K., S. Chiba and T. Shimomura 1977. Purification and substrate specificity of brewer's yeast $\alpha$-glucosidase. *Agricultural Biology and Chemistry* **41**, 1917–23.
Mayer, F. C. and J. Larner 1959. Substrate cleavage point of the $\alpha$- and $\beta$-amylases. *Journal of the American Chemical Society* **81**, 188–93.
Moulin, G. and P. Galzy 1978. Amylase activity of *Torulopsis ingeniosa* di Menna. *Folia Microbiologia* **23**, 423–7.
Moulin, G., H. Boze and P. Galzy 1982. Amylase activity in *Pichia polymorpha. Folia Microbiologia* **27**, 377–81.
Myrback, K. and W. Schilling 1965. Studies on yeast $\beta$-fructofuranosidase. Part XVI. Partial purification. *Enzymologia* **29**, 306–14.

Needleman, R. B., H. J. Federoff, T. R. Eccleshall, B. Buchferer and J. Marmur 1978. Purification and characterization of an $\alpha$-glucosidase from *Saccharomyces carlsbergensis. Biochemistry* **17**, 4657–61.
Neuberg, C. and R. Mandl 1950. Invertase. In *The enzymes*, J. B. Sumner and K. Myrback (eds), Vol. 1, Part 1, Chapter 14. New York: Academic Press.
Neuberg, C. and I. S. Roberts 1946. Invertase. *Sugar Research Foundation Science Reports Series* **4**, 1–62.
Nurminen, T. and H. Suomalainen 1970. The lipolytic activities of the isolated cell envelope fractions of baker's yeast. *Biochemical Journal* **118**, 759–63.

Ottolenghi, P. 1971. A comparison of five genetically distinct invertases from *Saccharomyces. European Journal of Biochemistry* **18**, 544–52.

Parkkinen, E. and H. Suomalainen 1982. Esterases of baker's yeast. II. Substrate specificities towards esters formed during sugar fermentations. *Journal of the Institute of Brewing* **88**, 34–8.
Parodi, A. J. 1981. Biosynthetic mechanisms for cell envelope polysaccharides. In *Yeast cell envelopes: biochemistry, biophysics, and ultrastructure*, W. N. Arnold (ed.), Vol. II, Chapter 2, 47–64. Boca Raton: CRC Press.
Peciarova, A. and P. Biely 1982. $\beta$-Xylosidases and a novel non-specific wall bound $\beta$-glucosidase of the yeast *Cryptococcus albidus. Biochimica et Biophysica Acta* **716**, 391–9.
Peel, J. L. 1951. Ester formation by yeasts. 2. Formation of ethyl acetate by washed suspensions of *Hansenula anomala. Biochemical Journal* **449**, 62–7.
Perlman, D. and H. O. Halvorson 1981. Distinct repressible mRNA's for cytoplasmic and secreted yeast invertase are encoded by a single gene. *Cell* **25**, 525–36.
Perlman, D., H. O. Halvorson and L. E. Cannon 1982. Presecretory and cytoplasmic invertase polypeptides encoded by distinct mRNA's derived from the same structural gene differ by a signal sequence. *Proceedings of the National Academy of Sciences of the USA* **79**, 781–5.

Phaff, H. J. 1971. Genus 20, *Schwanniomyces*. In *The yeasts, a taxonomic study*, 2nd edn, J. Lodder (ed.), 756–66. Amsterdam: North–Holland.

Phillips, A. W. 1959. The purification of a yeast maltase. *Archives of Biochemistry and Biophysics* **80**, 346–52.

Phillips, L. L. and M. L. Caldwell 1951a. A study of the purification and properties of a glucose-forming amylase of *Rhizopus delemar*, gluc amylase. *Journal of the American Chemical Society* **73**, 3559–63.

Phillips, L. L. and M. L. Caldwell 1951b. A study of the action of gluc amylase, a glucose-producing amylase, formed by the mold *Rhizopus delemar*. *Journal of the American Chemical Society* **73**, 3563–8.

Puglisi, P. P., I. Ferrero and A. Algeri 1970. An inducible $\beta$-galactosidase in *Saccharomyces cerevisiae*. *Giorn. Microbiol.* **18**, 57–67.

Rapoport, G. and R. Dedonder 1963a. La levane-sucrase de *Bacillus subtilis*. II. Hydrolyse et transfer à partir des levans. *Bulletin de la Société de Chimie Biologique* **45**, 493–513.

Rapoport, G. and R. Dedonder 1963b. La levanesucrase de *Bacillus subtilis*. III. Réactions d'hydrolyse de transfert et d'échange avec des analogues du saccharose. *Bulletin de la Société de Chimie Biologique* **45**, 515–35.

Reese, E. T., J. E. Lola and F. W. Parrish 1969. Modified substrates and modified products as inducers of carbohydrases. *Journal of Bacteriology* **100**, 1151–4.

Ruttloff, H., A. Taeufel, R. Friese and F. Zickler 1970. Secretion of glucoamylase isoenzymes by a strain of the genus *Endomycopsis*. *Zeitschrift für Allgemeine Mikrobiologie* **10**, 335–40.

Sa-Correia, I. and N. van Uden 1981. Production of biomass and amylases by the yeast *Lipomyces kononenkoae* in starch-limited continuous culture. *European Journal of Applied Microbiology and Biotechnology* **13**, 24–8.

San Blas, G. and W. L. Cunningham 1974. Production of an extracellular acid phosphatase by the yeast *Hansenula holstii* NCYC 560. *Biochimica et Biophysica Acta* **343**, 208–10.

Sawai, T. and E. J. Hehre 1962. A novel amylase (*Candida* transglucosyl-amylase) that catalyzes glucosyl transfer from starch and dextrins. *Journal of Biological Chemistry* **237**, 2047–52.

Sawai, T. and E. J. Hehre 1963. The distribution of trans glucosyl-amylase among *Candida* yeasts. *Antonie van Leeuwenhoek Journal* **29**, 377–85.

Sills, A. M. and G. G. Stewart 1982. Production of amylolytic enzymes by several yeast species. *Journal of the Institute of Brewing* **88**, 313–16.

Sills, A. M., I. Russell and G. G. Stewart 1983. The production and use of yeast amylases in the brewing of low-carbohydrate beer. *Proceedings of the Congress of the European Brewery Convention, 19th*, 377–84. *Chemical Abstracts* **101**, 5515 (1984).

Sills, A. M., I. Russell and G. G. Stewart 1984. The production and use of yeast amylases in the brewing of low-carbohydrate beer. *Chemical Abstracts* **101**, 5515.

Snyder, H. E. and H. J. Phaff 1960. Studies on beta-fructofuranosidase (inulinase) produced by *Saccharomyces fragilis*. *Antonie van Leeuwenhoek Journal* **26**, 433–52.

Spaepen, M. and H. Verachtert 1982. Esterase activity in the genus *Brettanomyces*. *Journal of the Institute of Brewing* **88**, 11–17.

Spencer, J. F. T. and P. A. J. Gorin 1973. Mannose-containing polysaccharides of yeasts. *Biotechnology and Bioengineering* **15**, 1–12.

Spencer-Martins, I. and N. van Uden 1979. Extracellular amylolytic system of the yeast *Lipomyces kononenkoae*. *European Journal of Applied Microbiology and Biotechnology* **6**, 241–50.

Stewart, G. G. 1981. The genetic manipulation of industrial yeast strains. *Canadian Journal of Microbiology* **27**, 973–90.

Suomalainen, H. 1969. Trends in physiology and biochemistry of yeasts. *Antonie van Leeuwenhoek Journal* **35** (suppl.), 83–111.

Suzuki, S. 1981. Antigenic determinants. In *Yeast cell envelopes: biochemistry, biophysics, and ultrastructure*, W. N. Arnold (ed.), Vol. I, Chapter 6, 85–96. Boca Raton: CRC Press.

Suzuki, H., S. C. Li and Y. T. Li 1970. $\alpha$-Galactosidase from *Mortierella vinacea*. Crystallization and properties. *Journal of Biological Chemistry* **245**, 781–6.

Tamaki, H. 1980. Purification of glucoamylase isoenzymes produced by *Saccharomyces diastaticus*. *Doshisha Joshi Daigaku Gki jatsu Kinkyu Nenpo* **31**, 166–82.

Thill, G. P., R. P. Kramer, K. T. Turner and K. A. Bostian 1983. Comparative analysis of the 5′-end regions of two repressible acid phosphatase genes in *Saccharomyces cerevisiae*. *Molecular and Cellular Biology* **3**, 570–9.

Thivend, P., C. Mercier and A. Guilbot 1972. Determination of starch with glucoamylase. In *Methods in carbohydrate chemistry*, R. L. Whistler and J. N. BeMiller (eds), Vol. VI, Chapter 14, 100–5. New York and London: Academic Press.

Tingle, M. and H. O. Halvorson 1972. Mutants in *Saccharomyces lactis* controlling both $\beta$-glucosidase and $\beta$-galactosidase activities. *Genetical Research* **19**, 27–32.

Trimble, R. B., F. Maley and W. Watorek 1981. Subunit structure and carbohydrate composition of the extracellular acid phosphatase of *Rhodotorula glutinis*. *Journal of Biological Chemistry* **256**, 10037–43.

Tsujisaka, Y., J. Fukumoto and T. Yamamoto 1958. Specificity of crystalline amylase of moulds. *Nature* (London) **181**, 770–1.

Uwajima, T., H. Yagi and O. Terada 1972. Purification, crystallization and some properties of $\beta$-galactosidase from *Saccharomyces fragilis*. *Agricultural Biology and Chemistry* **36**, 570–7.

van der Walt, J. P. 1971. Genus 8, *Kluyveromyces*. In *The yeasts, a taxonomic study*, 2nd edn, J. Lodder (ed.), 316–78. Amsterdam: North–Holland.

Villa, T. G. and H. J. Phaff 1980. Recovery of invertase and laminarinase from industrial waste broth of baker's yeast. *European Journal of Applied Microbiology and Biotechnology* **9**, 9–14.

Villa, T. G., V. Notario and J. R. Villanueva 1979. Occurrence of an endo-1,3-$\beta$-glucanase in culture fluids of the yeast *Candida utilis*. Purification and characterization of the enzyme activity. *Biochemical Journal* **177**, 107–14.

Walker, G. J. and W. J. Whelan 1960. The mechanism of carbohydrase action. 7. Final stages in the salivary $\alpha$-amylolysis of amylose, amylopectin and glycogen. *Biochemical Journal* **76**, 257–63.

Wallenfels, K. and I. R. Rachead 1966. Crystallization and stereospecificity of pullulanase from *Aerobacter aerogenes*. *Biochemical Journal* **344**, 524–6.

Weill, L. E., R. J. Burch and J. W. van Dyck 1954. An $\alpha$-amyloglucosidase that produces $\beta$-glucose. *Cereal Chemistry* **31**, 150–8.

Weimberg, R. and W. L. Orton 1965. Elution of acid phosphatase from the cell surface of *Saccharomyces mellis* by potassium chloride. *Journal of Bacteriology* **90**, 82–94.

Wendoff, W. L., C. H. Amundsen and N. F. Olsen 1970. Nutrient requirements and growth conditions for production of lactase enzyme by *Saccharomyces fragilis*. *Journal of Milk Food Technology* **33**, 451–4.

Whelan, W. J. 1960. The action patterns of $\alpha$-amylases. *Die Starke* **12**, 358–64.

Whistler, R. L. and W. M. Corbett 1957. Polysaccharides. In *The carbohydrates, chemistry, biochemistry, physiology*, W. Pigman (ed.), Chapter XII, Part I, 641–708. New York: Academic Press.

Whistler, R. L. and C. C. Tu 1953. Crystalline xyloheptaose. *Journal of the American Chemical Society* **75**, 645–7.

Wickerham, L. J., L. B. Lockwood, D. G. Pettijohn and G. E. Ward 1944. Starch hydrolysis and fermentation by the yeast *Endomycopsis fibuligera*. *Journal of Bacteriology* **48**, 413–27.

Wilson, J. J. and W. M. Ingledew 1982. Isolation and characterization of *Schwanniomyces alluvius* amylolytic enzymes. *Applied Environmental Microbiology* **44**, 301–7.

Yashphe, J. and H. O. Halvorson 1976. β-D-Galactosidase activity in single cells during cell cycle of *Saccharomyces lactis*. *Science* **191**, 1283–4.

Yoshioka, K. and N. Hashimoto 1981. Ester formation by alcohol acetyltransferase from brewer's yeast. *Agricultural Biology and Chemistry* **445**, 2183–90.

Youngquist, R. W. and W. J. Brables 1982. Single-dough cookies having stable-storage texture. US Patent. *Chemical Abstracts* **97**, 180619.

# 13 *Expression and secretion of foreign polypeptides in yeast*

A. R. GOODEY, S. DOEL,
J. R. PIGGOTT, M. E. E. WATSON
and B. L. A. CARTER

## 13.1 Introduction

The yeast *Saccharomyces cerevisiae* is a unicellular eukaryote that exhibits many features of cell structure normally associated with higher organisms such as a nucleus, nucleolus, golgi body and endoplasmic reticulum. In addition, many elements of macromolecular synthesis in yeast are similar to those observed in higher eukaryotes.

Yeast has only about five times the amount of DNA found in the *Escherichia coli* chromosome, but this is distributed amongst 17 chromosomes in a haploid cell. These chromosomes are similar to those of higher eukaryotes; the DNA is associated with protein to form chromatin. In addition, yeast chromosomes have centromeres which associate with microtubules in the mitotic spindle during chromosome segregation; telomeres are present at the ends of chromosomes to ensure the proper resolution of the replication of linear DNA structures.

Yeast has been extensively analysed biochemically and physiologically, primarily because of its early availability from the baking and brewing industries. It has been susceptible to genetic analysis for several decades. Since 1978, when it was shown that yeast cells could be transformed with exogenous DNA (Hinnen *et al*. 1978) it has been possible to apply recombinant DNA technology to yeast.

The possibility of studying yeast physiologically, biochemically and by both classical and molecular genetical techniques has led to an upsurge of interest in yeast as an organism for the study of fundamental problems in eukaryotic biology. It has also led a number of groups to explore the possibility of using yeast as a host for the expression and secretion of commercially useful peptides and proteins.

There are several reasons why yeast is being considered as an alternative to *Escherichia coli*. These include the fact that yeast has no endotoxins; indeed, it has been a major contributor to the food and beverage industries for centuries. Yeast expression offers the potential for removal of the initiating methionine codon from mammalian proteins; a difficult thing to achieve using *Escherichia coli*. Yeast can glycosylate and secrete into the medium foreign polypeptides, neither of which is possible in *Escherichia coli*.

Secretion in yeast is very similar to that observed in mammalian cells. Proteins destined for secretion have an amino-terminal hydrophobic extension termed a leader or signal, which is responsible for translocation of the protein to the endoplasmic reticulum. Once in the secretory pathway, proteins are transferred to the golgi body and thence via secretory vesicles across the plasma membrane and out of the cell.

Secretion has several attractions. It is likely that many proteins when expressed internally in the cell will be, to a greater or lesser extent, toxic to the cell. This will result in selective pressure for a reduction in expression; either through a reduction in copy number (possibly even loss of plasmid) or through plasmid rearrangement. It may be possible to avoid these problems by rapid and efficient secretion from yeast. Secretion is also important, because if glycosylation is desired it can only occur if a protein is directed through the secretion pathway as glycosylation occurs in the endoplasmic reticulum and golgi body. Secretion offers possible purification advantages; it may be easier to purify a protein secreted into medium containing only the ten or so proteins naturally secreted by yeast than from the thousands of proteins present in a yeast or *Escherichia coli* cell-free extract.

It is a common experience that high expression of a mammalian gene in *Escherichia coli* results in the production of a protein in an insoluble and aggregated form that is difficult to renature to full biological activity. Secretion of soluble active material may avoid this problem.

Glycosylation in yeast is similar to, but not identical with, that observed in mammalian cells. N-glycosylation occurs at asparagine residues where they are followed by any amino acid and then threonine or serine. Glycosylation occurs in two stages in two different places in the cell. In both yeast and mammalian cells inner core glycosylation occurs via the addition of the dolichol-linked oligosaccharide shown in Figure 13.1 to asparagine via two N-acetylglucosamine residues (Parodi & Leloir 1979, Robbins 1982, Ballou 1982). Subsequently, in both yeast and mammalian cells the protein is transferred to the golgi body where outer core glycosylation occurs. In yeast this is

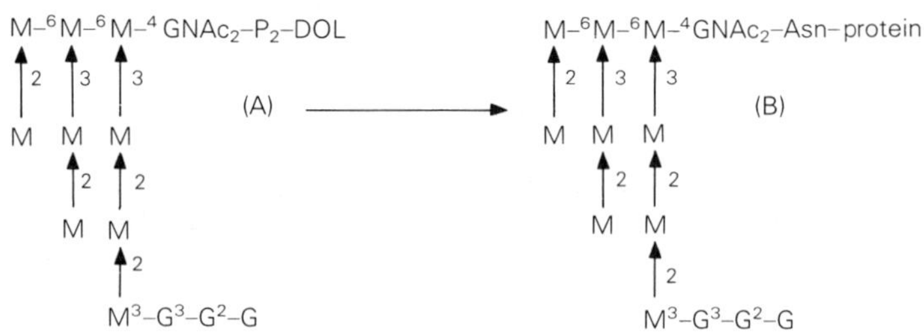


**Figure 13.1** Inner core N-glycosylation in yeast. The dolichol-linked oligosaccharide is attached via two N-acetylglucosamine (GNAC) residues to asparagine (ASN). M, mannose; G, glucose.

accomplished by the stepwise addition of mannose residues (Ballou 1976, Sheckman & Novick 1982). In mammalian cells the inner core is subject to trimming and subsequently complex, sometimes branched, oligosaccharides are added to the inner core; carbohydrates other than mannose can be included in the oligosaccharide structure (Standoni & Leloir 1982).

How important is it that yeast will not glycosylate in exactly the same way as mammalian cells? This depends on the function of glycosylation, how important a particular glycosylation pattern is to that function, and of course the possibility of antigenicity problems. Glycosylation may not be important for the biological function of some proteins. On the other hand, there will be instances where a specific glycosylation pattern may be vital for biological activity or for recognition and targeting. If the latter is the case yeast-derived glycoproteins may not be suitable. Glycosylation may also be important for half-life in the serum or to ensure proper folding of the proteins. It may also ensure that cleavage of the leader–signal occurs efficiently without aberrant processing of potential cleavage sites in the mature protein sequence. The effect of yeast-specific glycosylation on these possible functions is not clear. However, since protein folding occurs in the endoplasmic reticulum (ER) and inner-core glycosylation in the ER is similar in yeast and mammalian cells, glycosylation in yeast should not affect folding.

## 13.2 Yeast transformation

In order to achieve efficient transformation of *Saccharomyces cerevisiae* the cell wall must be removed. This is accomplished by treating cells with a mixture containing, in an osmotically stabilized medium, a reducing agent (mercapto-ethanol or dithiothreitol) to break down the disulphide bridges in the yeast cell wall and an enzyme preparation (zymolyase or $\beta$-glucuronidase) to break down the cell wall constituents which include glucan and mannan. Having removed part or all of the cell wall to produce sphaeroplasts or protoplasts, respectively, plasmid DNA is then added to the osmotically sensitive cells. The DNA uptake is aided by the addition of calcium ions and polyethylene glycol (PEG), which results in aggregation of protoplasts, localized membrane fusion and DNA uptake.

Aggregated protoplasts are then embedded in a hypertonic selective agar. Embedding in a solid selective medium is necessary for efficient cell-wall regeneration in *Saccharomyces cerevisiae*. This embedding procedure is not necessary for *Schizosaccharomyces pombe*; potential transformants are spread on top of agar (Beach & Nurse 1981). As an alternative to this lengthy procedure, techniques have been developed for the transformation of intact cells that do not involve cell fusion. These techniques rely on alkali cations (typically lithium) in a procedure analogous to *Escherichia coli* transformation (Ito *et al*. 1983).

Procedures based on those described here have been successfully applied to the transformation of other yeast species.

## 13.3 Plasmid systems

A number of vector systems are available for the transformation of *Saccharomyces cerevisiae*.

The first steps towards *Saccharomyces cerevisiae* plasmid systems were taken by Ratzkin and Carbon (1977) who complemented the *Escherichia coli leu*B mutant with a yeast DNA fragment. This was then shown by Hinnen *et al.* (1978) to complement the yeast *leu*2 gene. Such auxotrophic selection techniques have proved to be the most common method to select for yeast transformants. Many other auxotrophic genes have been directly selected from the yeast genome by their ability to complement specific mutations. In general yeasts are relatively insensitive to antibiotics. However, some dominant selectable drug resistant markers are available for yeast transformation such as G418 (Jimenez & Davis 1980), chloramphenicol acetyltransferase (Cohen *et al.* 1980) and hygromycin (Gritz & Davis 1983). The CUP1 gene provides copper resistance and is an alternative dominant selectable marker (Karin *et al.* 1984). The expression of the HSV1 thymidine kinase gene in yeast (McNiel & Friesen 1981) gives resistance to folate antagonists; this particular selection is described in more detail below.

Plasmid vectors available for yeast transformation generally include an *Escherichia coli* plasmid origin of replication and drug-resistance marker allowing propagation and manipulation in *Escherichia coli* before their introduction into yeast. Such plasmids are referred to as shuttle vectors, as they can be 'shuttled' between two organisms. The *Escherichia coli* plasmid sequences can be removed from these plasmids so that they can be propagated solely in yeast. This is an important point to note in the case of industrial processes.

The first yeast-cloning vector described by Hinnen *et al.* (1978) was a yeast integrating vector, or YIp, which consists of a bacterial cloning vector carrying a yeast gene (in this case *leu*2). These vectors lack any yeast origin of replication, and can only cause phenotypic change by integration into the yeast genome at a point of homology. Such plasmids have a very low transformation efficiency; 1–10 transformants $\mu g^{-1}$ DNA. Efficiency can be dramatically increased by restricting and consequently linearizing the plasmid within yeast sequences generating highly recombinogenic ends. This results in a 100–1000-fold enhanced transformation efficiency (Orr-Weaver *et al.* 1981). YIp vectors can also be 'retrieved' from the genome (see review by Rothstein 1985) to generate vectors with mutant or wild-type alleles, and can be used to generate specific lesions in genes.

Some plasmids carrying yeast chromosomal DNA sequences gave elevated transformation levels. It was first demonstrated by Stinchombe *et al.* (1979) that these plasmids were maintained extrachromosomally and that they contained an 'autonomously replicating sequence' (ARS) that functioned as a replication origin for the plasmid. Such sequences were thought to be chromosomal origins of DNA replication.

These plasmids are known as yeast replicating plasmids (YRp) and are capable of attaining high copy number in yeast (100 copies cell$^{-1}$) (Stinchombe *et al.* 1979), but are exceedingly unstable at mitosis and meiosis. Consequently, these plasmids are rapidly lost from cells when grown under non-selective conditions and the average copy number is low (Zakian & Kupfer 1982, Murray & Szostak 1983).

DNA sequences from a range of eukaryotes permitting autonomous plasmid replication in yeast have been isolated from a list of organisms that includes *Physarum* (Gorman *et al.* 1981), *Drosophila*, *Zea mays*/*Neurospora crassa* (Stinchombe *et al.* 1980) and *Xenopus* (Zakian 1981). It is not known whether these sequences promote DNA replication in the organisms from which they are derived. It has been shown that a DNA sequence obtained from *Pichia pastoris* functions as an ARS in *Saccharomyces cerevisiae*, but does not do so in *Pichia pastoris* itself (Cregg *et al.* 1985).

The unstable nature of YRp plasmids was exploited by Clarke and Carbon (1980) to identify clones which contained yeast centromeres. These centromeric fragments were seen to stabilize YRp vectors in *cis* to produce YRpcen vectors. YRpcen vectors are stable, but present at a low copy number of 1–2 $cell^{-1}$.

More recently Szostak and Blackburn (1982) have cloned yeast telomeres (ends of chromosomes). These have been combined with a centromere and ARS on a linear plasmid. Such linear vectors have been shown to behave as synthetic mini-chromosomes, the stability of which appears to be a function of their length (see review by Blackburn 1985).

The class of vectors that were originally studied by Beggs (1978) and are most commonly used for the expression of heterologous protein in *Saccharomyces cerevisiae* are the YEp vectors or 'yeast episomal plasmids'. The yeast episome that is included in these plasmids is the 2-μm native plasmid, an autonomously replicating high-copy number plasmid (50–100 $cell^{-1}$) found in the majority of *Saccharomyces cerevisiae* strains and inherited in a non-Mendelian manner (for review see Broach 1981). The plasmid contains two regions of length 600 bp that are precise repeats of each other and separate two unique regions of dissimilar size. It has four extended open-coding regions. The 2-μm plasmid has a single origin of replication. Although autonomous replication is conferred by the 2-μm origin of replication, propagation at high copy number requires three proteins encoded on the plasmid itself. These are encoded by REP1 and REP2, which may be produced in trans by native 2 μm and REP3 that acts in *cis* and must be present on the recombinant plasmid itself, along with the origin of replication to ensure high copy number (Kikuchi 1983). In addition to its replication, 2-μm circle has a recombinational system. In yeast it readily undergoes recombination between the inverted repeat sequences. The 2-μm plasmid can convert between two forms that differ in the orientation of one unique region with respect to another. Since recombination is efficient, both forms are present in approximately equal proportions in the cell. This recombination is catalysed in part by a protein encoded by the 2-μm FLP gene. Various combinations of 2-μm sequences have been used in yeast–*Escherichia coli* shuttle vectors. Hybrid plasmids containing the entire 2-μm DNA sequence, the yeast *leu*2 gene and bacterial sequences can transform strains of yeast lacking native 2 μm ($cir^0$) with high efficiency and have a high copy number such as pJDB219 (Beggs 1981). Other smaller plasmids, for instance pJDB207, which contain a fragment of 2 μm containing the replication origin and REP3 will also transform yeast at high frequency, and be maintained at high copy number. Plasmids like pJDB207 will only transform strains provided the recipient contains 2 μm to provide REP1 and REP2 functions in *trans* (Beggs 1981).

Plasmid copy number and stability are interrelated in the case of 2-μm based vectors. For example, plasmids like pJDB219 and derivatives of pJDB219 lacking bacterial sequences are capable of very high copy number and stability in cir$^0$ host yeast cells (Futcher & Cox 1984, Walmsley *et al.* 1983). This is presumably because (a) they suffer no competition for replication from native 2-μm plasmid and (b) the particular yeast *leu*2 gene they contain has a deletion in its promoter region, so it is not efficiently expressed (Erhart & Hollenberg 1983). The plasmid has to be present at high copy number to produce sufficient gene product for survival on leucine-minus medium.

It has been shown that 2-μm plasmid sequences act as ARS elements in *Saccharomyces pombe* (Beach & Nurse 1981), but do not stabilize plasmids. Indeed 2 μm has not been shown to be the basis of a stable plasmid in species of yeast or fungus other than *Saccharomyces cerevisiae*. Other yeast species do have their own native plasmids. For instance, *Zygosaccharomyces baillii* (Toh-e *et al.* 1982), *Zygosaccharomyces rouxii* (Toh-e *et al.* 1984), *Zygosaccharomyces bispora* (Toh-e & Utatsu 1985), have plasmids which in gross terms show resemblance to the structure of 2 μm but show no precise homology. The yeast *Kluyveromyces lactis* has two plasmids one of which is associated with the production of a secreted toxin (Gunge *et al.* 1981). The presence of an accessible DNA plasmid in this yeast may permit the development of a host–vector system that provides efficient heterologous protein expression.

The choice of plasmid system for heterologous expression is clearly an important one; plasmid copy number and stability must be taken into account. For high-level production and short fermenter runs 2-μm based vectors offer high copy number and reasonable stability. However, for long-term or continuous fermentation plasmid stability will be all-important and it may be advisable to use an integrating vector. In our experience a major influence on stability of heterologous protein expression is the nature of the heterologous protein itself and its effect on the yeast cell, which can be exceedingly deleterious. Yeast plasmid systems with a degree of copy number control similar to those which are known in *Escherichia coli* (Twigg & Sherratt 1980) would be useful for the conditional expression of proteins that are deleterious to cell health. Clearly a high-copy number plasmid that is stable without selection is desirable; the native *Saccharomyces cerevisiae* 2-μm plasmid is just such a plasmid. However, manipulating its sequences and the inclusion of heterologous sequences appears to produce some degree of instability.

There is a paucity of good information on sequences involved in plasmid copy number control and stability.

## 13.4 Expression vectors

A typical yeast expression vector is shown in Figure 13.2. This utilizes the alcohol dehydrogenase (ADH1) promoter and terminator which are separated by a unique *Hin*dIII site for the insertion of a structural gene encoding a foreign

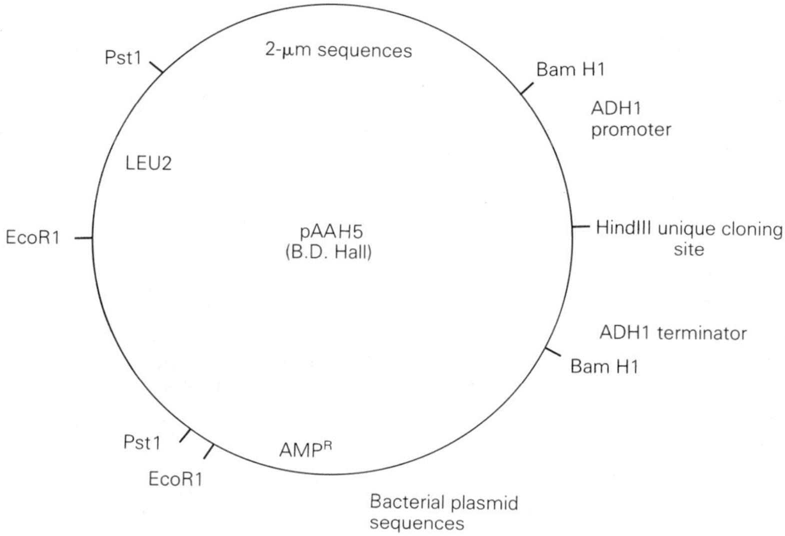


**Figure 13.2** A typical yeast expression vector. The plasmid incorporates several features: 2-μm sequences and LEU2 selection for replication in yeast, bacterial plasmid sequences and ampicillin selection for replication in bacteria, the ADH1 promoter and terminator sequences with a unique HindIII cloning site between them for the insertion of a heterologous gene of interest.

protein whose expression is desired. The ADH1 promoter was the first yeast promoter used to express a human protein, viz. leukocyte interferon (Hitzeman *et al*. 1981).

A number of glycolytic genes have been cloned and their promoters used for foreign protein production. The interest in glycolytic promoters occurred because glycolytic enzymes are amongst the best represented proteins in a yeast cell; each can represent between 1 and 5% of total cell protein, even though they are often encoded by single copy genes. It was reasonably assumed that high-level expression was due to the association of strong promoters with these genes. A list of glycolytic promoters that have been used for foreign gene expression is shown in Table 13.1.

Although everyone has their favourite promoter – usually the one they are working on – all of the promoters probably meet the requirement for promoter strength, such that other factors determine foreign protein expression. It is notable that expression levels of foreign proteins produced by a variety of promoters are more protein dependent than promoter dependent. We will return to this point later.

The major difficulty with the glycolytic promoters is that they are most active at the start of fermentation, when the sugar concentration is highest. High expression at the start of fermentation is at best an energy drain, and at worst results in the production of a protein that is toxic to a cell. The effect of this is to produce a selective pressure for a reduction in expression. The result is a

**Table 13.1** Glycolytic promoters used for heterologous protein expression in *Saccharomyces cerevisiae*.

| Promoter | Protein | Reference |
|---|---|---|
| GAPDH | Thaumatin | Edens *et al.* (1984) |
| GAPDH | $\alpha$-IFN | Bitter and Egan (1984) |
| GAPDH | HBsAg | Bitter and Egan (1984) |
| GAPDH | hEGF | Urdea *et al.* (1983) |
| PYK | $\alpha$-IFN | Goodey *et al.* (unpubl.) |
| PYK | $\beta$-IFN | Goodey *et al.* (unpubl.) |
| PYK | $\gamma$-IFN | Goodey *et al.* (unpubl.) |
| TPI | PDGF | Kelly *et al.* (1985) |
| PGK | $\alpha$-IFN | Tuite *et al.* (1982), Hitzeman *et al.* (1981) |
| PGK | HBsAg | Hitzeman *et al.* (1983) |
| PGK | $\gamma$-IFN | Derynck (1983) |
| PGK | Chymosin | Mellor *et al.* (1983) |
| PGK | Mouse immunoglobin | Wood *et al.* (1985) |
| PGK | Wheat amylase | Rothstein *et al.* (1984) |

reduction in plasmid copy number or loss of plasmid; alternatively plasmid rearrangement resulting in a loss of foreign gene expression can occur (Jabbar *et al.* 1985, Miyajima *et al.* 1985).

If yeast cells containing a multi-copy plasmid in which the pyruvate kinase promoter is used to express a foreign gene, or for that matter the pyruvate kinase structural gene, are constantly restreaked and the streaks are used as an inoculum for cultures in which protein expression is monitored, then the level of heterologous and homologous expression declines with successive inocula. On Petri dishes strong-growing papillae are observed against the background lawn of the streak, and these have markedly reduced expression of the protein of interest.

There are basically two routes to the production of foreign proteins in yeast. One is batch culture and the other (particularly if expression is allied to protein secretion into the medium) is continuous-flow culture. A suitable promoter for batch culture is one that is regulated such that expression is turned off during the initial stages of growth and switched on when biomass accumulation is essentially complete. The use of a single-stage continuous-flow culture system dictates that the promoter is active throughout. Clearly, the glycolytic promoters have this property, but the instability of expression with time precludes their use in chemostat culture. It is vital (to maximize the benefits of continuous operation) to use a system that gives no diminution of yield for generation after generation. It may be necessary to sacrifice to some extent the levels of expression as a means to the attainment of stability of expression and rely on the greater productivity of continuous systems to compensate for this reduction.

We have attempted to identify systems that are effective at foreign gene expression in several ways, one of which will be discussed now. A selection system has been devised that can be used to isolate directly yeast promoters of unknown origin that are effective at heterologous gene expression. A guiding principle behind this work has been the recognition that selection methods provide systems that work, even if in the first instance one does not understand why. An additional advantage of selection systems is that they provide flexibility; by altering selection conditions it is possible to obtain constitutive or regulated promoters.

The selection system relies on the fact that yeast cells do not have a thymidine kinase. They rely on thymidylate synthetase to convert dUMP to TMP, a precursor for DNA synthesis (Fig. 13.3). Endogenous production of TMP can be blocked by the folate antagonists sulphanilamide and amethopterin. Yeast cells cannot grow in medium containing these folate antagonists, even if thymidine is added to the medium (Little & Haynes 1979). A shuttle vector was constructed that contained the herpes simplex virus (HSV-1), thymidine kinase structural gene and a unique *Bgl*II site 5′ to the gene which permitted the insertion of yeast chromosomal DNA fragments. A gene bank was prepared in *Escherichia coli* and used to transform yeast cells. Transformants were selected on medium containing sulphanilamide, amethopterin and thymidine. Although yeast cells transformed with a plasmid containing the HSV-1 thymidine kinase structural gene and its own promoter do not grow on this medium (Kiss *et al.* 1980), they do if a yeast promoter is substituted for the HSV-1 thymidine kinase promoter (McNiel & Friesen 1981). It was therefore reasoned that yeast transformants carrying DNA fragments which had promoter activity would

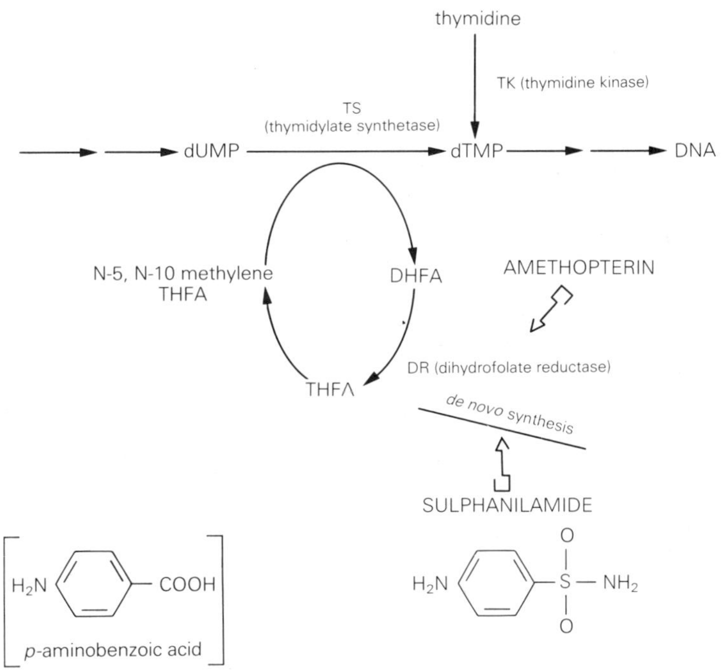


**Figure 13.3** TMP synthesis in yeast and its inhibition by folate antagonists.

```
(Pst 1)

5' C A G T G G C T T A T C A C G C C A G G A C T G C G G G A G T G G

 C G G G G C A A A C A C A C C C G C G A T A A A G A G C G C G A
     -118                                                   -90
 T G A A T A T A A A A G G G G C C A A T G T T A C G T C C C G T

 T A T A T T G G A G T T C T T C C C A T A C A A A C T T A A G A G T

         Alu 1
 C C A A T T A G C T T C A T C G C C A A T A A A A A A A C A A A*C

 A A A C C T A A T T C T A A C A A G C A A A G A T G A A G T T A
                                                Met   Lys   Leu

 G A A A A T A C T C T A T T T A C A C T C G G T G C C C T A
 Glu   Asn   Thr   Leu   Phe   Thr   Leu   Gly   Ala   Leu

 G G G A G C A T C T C T G C T G C T T T G G T C A T C C C A A A T
 Gly   Ser   ile   Ser   Ala   Ala   Leu   Val   Ile   Pro   Asn

 C T T G A A A A T G C C G C C G A C C A C C A C G A A C T G A T T
 Leu   Glu   Asn   Ala   Ala  Asp   His   His  Glu  Leu   Ile

 A A C A A G G A A G A T C A C C A C G A G A G A C C C A G A
 Asn   Lys   Glu   Asp   His  His   Glu   Arg   Pro   Arg
                                                      EcoR1
```

**Figure 13.4** The sequence of the '246' promoter and surrounding sequences.

**Table 13.2** Stability test of 246 and PK driving the expression of IFN SB 430. Ten-millilitre cultures of YEP 2% glucose were inoculated from yeast-defined media plates without leucine and supplemented with 1.2 mol $l^{-1}$ Sorb and grown for 24 h. The cells were counted, 50 μl was transferred to fresh 10-ml YEPD and the remaining culture spun down pellets or media was frozen as appropriate. This was repeated for 9 days and the pellets were all lysed in PBS–0.1% BSA and assayed for INF. The results are given in IU IFN $l^{-1}$ $10^{-11}$ cells.

| Day | SD 942 (246 expression) | pAYE 115 (PYK expression) |
|---|---|---|
| 1 | $6.4 \times 10^8$ | $4.4 \times 10^8$ |
| 2 | $4.9 \times 10^8$ | $9.4 \times 10^7$ |
| 3 | $5.9 \times 10^8$ | $5.9 \times 10^7$ |
| 4 | $8.1 \times 10^8$ | $1.7 \times 10^7$ |
| 7 | $5.2 \times 10^6$* | $6.9 \times 10^5$* |
| 8 | $3.0 \times 10^6$ | $1.2 \times 10^6$ |
| 9 | $2.7 \times 10^8$ | $7.0 \times 10^6$ |

* These cultures grew to stationary phase over the weekend.

express TK, produce TMP and grow on this medium. A number of such yeast transformants were identified, and promoter strength was assessed by an *in vitro* assay of thymidine kinase activity. The sequence of one promoter selected in this way is shown in Figure 13.4.

The promoter has been incorporated into a yeast expression vector and used to express a modified IFN-beta (SB 430) in which amino acids 36–48 were replaced by the corresponding amino acids of IFN-alpha.

A comparison of expression levels using either the pyruvate kinase promoter or the 246 promoter (Table 13.2) indicates that the latter permits maintenance of productivity for many more generations and may provide a system capable of continuous operation.

## 13.5 Regulated systems

Although it may be possible to operate continuous-flow culture systems for some heterologous proteins, the strong selective pressure for faster-growing variants encountered in these systems make it likely that for many heterologous proteins batch cultures will be the method of choice for production. Clearly, regulated expression is desirable so that selective pressure can be minimized. The ideal regulation would be such that biomass accumulation occurred in the absence of heterologous protein production, with expression being switched on towards the end of the growth phase.

A number of regulated systems have been described that require an alteration in medium composition. Regulation of acid phosphatase has been investigated for a number of years, and it is known that a number of genes including PHO2, PHO80, PHO81, PHO85 and a gene cluster PHO82, PHO4 (Toh-e *et al.* 1973, Toh-e & Oshima 1974, Ueda *et al.* 1975) are involved in the expression of repressible acid phosphatase (PHO5) in response to inorganic-phosphate levels in the medium.

Induction of acid phosphatase in medium depleted of inorganic phosphate is regulated at the transcription level (Bostian *et al.* 1980, 1983). A number of laboratories have used the acid-phosphatase promoter to achieve regulated expression of heterologous proteins in yeast (Hinnen *et al.* 1983, Miyanhara *et al.* 1983, Kramer *et al.* 1984, Rosenberg *et al.* 1984, Lemontt *et al.* 1985). The acid-phosphatase system is not completely on–off. Basal levels, whether of acid phosphatase or heterologous proteins, are produced in the presence of inorganic phosphate, but the levels can be induced up to 200-fold by removal of inorganic phosphate from the medium.

Changes in the composition of the medium are not convenient in large-scale fermenters, and this consideration led Kramer *et al.* (1984) to combine a molecular and classical genetical approach to achieve temperature-regulated expression of IFN-alpha$_1$. The pho80 gene product is a repressor of acid phosphatase; pho80 mutants lead to constitutive synthesis of acid phosphatase in high and low inorganic phosphate. PHO4 encodes a positive regulator of acid phosphatase; pho4 mutants do not produce acid phosphatase, irrespective of medium conditions. Temperature-sensitive pho4 mutants exist. The double mutant pho4ts pho80 results in the production of acid phosphatase at the permissive temperature (24°C), but not at the restrictive temperature (37°C); the effect of temperature is independent of the inorganic-phosphate concentration in the medium. When yeast plasmids utilizing the acid-phosphatase promoter to drive IFN-alpha are used to transform a pho4ts pho80 host, interferon expression can be induced 50-fold by a drop in temperature from 36 to 24°C. It is possible that the full induction potential (200-fold) is not achieved because temperature-sensitive gene products are often not fully effective, even at the permissive temperature. Nevertheless, these experiments suggest the potential importance of combining both classical and molecular genetical approaches.

Another system that is regulated and benefits from a background knowledge derived from classical genetics is the galactose system. The basis of our understanding of the system stems from the pioneering work of Douglas and Hawthorne (1966), although their original model has required some refinement (see review by Oshima 1983).

Certain enzymes involved in galactose metabolism are encoded by the GAL7–GAL10–GAL1 cluster on chromosome II. GAL1 is transcribed in a divergent direction from GAL7 and GAL10. Douglas and Hawthorne (1966) proposed that the GAL4 gene product is an essential positive regulator for induction of the galactose pathway enzymes, and that the GAL80 product is a negative regulator. It is thought to inhibit the action of GAL4 in the absence of an inducer, possibly by a protein–protein interaction (Matsumoto *et al.* 1981).

The GAL7–10–1 genes are subject to glucose repression. Growth in galactose plus glucose represses the level of the galactose-utilizing enzymes (Adams 1972, Matsumoto *et al.* 1981). The GAL80 gene product is not involved in this repression (Torchia *et al.* 1984). The enzymes are co-ordinately induced about 1000-fold at the level of transcription by growth on galactose (Hopper *et al.* 1978, St John & Davis 1981). Overproduction of the GAL4 gene product is necessary for maximum induction on multicopy plasmids (Johnston & Hopper 1982).

The galactokinase (GAL1) promoter has been used to direct expression of calf prochymosin (Goff *et al.* 1984) and human insulin (Stepien *et al.* 1983). In neither instance was any foreign protein detectable when yeast cells were grown in glucose medium. The amount of foreign protein produced on galactose induction was rather low. It is possible that a single genomic copy of the GAL4 gene is insufficient to provide sufficient inducer for both foreign-gene expression from a multicopy plasmid and expression of the galactose utilization enzymes. Stepien *et al.* (1983) presumed that poor growth of transformants on galactose medium is explicable in terms of poor expression of the GAL7–10–1 gene cluster. An alternative explanation is that cells do not grow well on galactose because proinsulin is inimical to cell health. The system is amenable to an investigation of which, if either, of these two factors is limiting the expression levels.

Use of the alpha-factor promoter permits conditional expression of proteins. Alpha-factor is a tridecapeptide that is secreted into the medium at the end of logarithmic growth (Scherer *et al.* 1974). It appears from this result that the alpha-factor promoter (MF$\alpha$1) is regulated and only active at the end of growth phase. We favour an alternative explanation, based on internal and secreted estimates of heterologous protein production (Goodey *et al.* unpubl. obs.); the alpha-factor promoter is active at all stages of batch fermentation, but secretion is favoured towards the end of batch growth. Nevertheless, Brake *et al.* (1984) have elegantly demonstrated conditional expression of heterologous proteins using the MF$\alpha$1 promoter. Alpha-factor is expressed in *a* mating type haploid cells but not in *a*/$\alpha$ diploids. Mutations in SIR genes result in a non-mating phenotype, because both *Mata* and *Mat*$\alpha$ sequences are expressed. The result is that such mutants behave like *a*/$\alpha$ diploids. When a *Mat*$\alpha$,Sir3$^{ts}$ mutant was used as a host for the production of human epidermal growth factor (hEGF) and grown at 36°C, the host behaved as an *a*/$\alpha$ diploid and very little hEGF was produced. After a shift down to the permitted temperature the strain behaved like a *Mat*$\alpha$ strain, and hEGF production rose from 20 ng $l^{-1}$ to 4 mg $l^{-1}$ (Brake *et al.* 1984).

Although the examples that have been discussed show that regulated expression of heterologous proteins is possible, and illustrate the strength of a combined classical and molecular genetical approach, they involve either media alteration or temperature change. It would be preferable to exploit the natural changes in the physiology of yeast cells that accompany batch culture on glucose. Initially, when the glucose concentration is high, the cells grow fermentatively; subsequently, as the glucose concentration falls, the cells switch to a respiratory metabolism as various genes are released from catabolite repression. By harnessing heterologous gene expression to a promoter that is switched on late in batch culture, one avoids the necessity of intervening in the environmental conditions. There are two approaches to the isolation of such promoters. It is possible to identify genes that are so regulated, and to clone the appropriate genes and their promoters. Such a gene, iso-1-cytochrome C (CYC1) has been isolated (M. Smith *et al.* 1979). This protein is five to seven times more abundant in cells grown on non-fermentative carbon sources than in media in which glucose is the carbon source. Similar regulation is observed when the CYC1 promoter is used to express foreign proteins (Guarente & Ptashne 1981, Rymond *et al.* 1983). Unfortunately, a factor of seven is inadequate. A promoter with the pattern of regulation observed in CYC but the induction potential of GAL1 is desirable. The thymidine kinase selection technique offers a route to the isolation of such promoters. Yeast transformants that carry such promoters could be identified by their ability to grow on folate antagonists plus thymidine when glycerol : ethanol is used to mimic the respiratory conditions at the end of fermentation, but not when glucose is the carbon source. An added advantage of this system is that the selection technique can be modified to discriminate on promoter strength; stronger promoters can grow on lower thymidine concentrations (Goodey *et al.* 1986).

A rather different approach to regulated expression is to assemble desirable elements from different promoters into a single promoter. This is possible because of the finding that yeast promoters transcribed by RNA polymerase II are made up of at least two functional elements, upstream activation sites (UAS) and TATA elements (Guarente 1984), together with sequences at the transcription initiation site. The latter provide the information specifying the transcription start site. The former (UAS) are responsible for the pattern of regulation that is characteristic of a particular promoter. If UAS elements from one promoter are combined with the TATA element and initiation region of the other, the transcripts have 5′-ends identical to the promoters from which the latter are derived and the pattern of regulation observed in the promoters from which the UAS are derived.

The upstream region of one promoter can confer regulation that is characteristic of the UAS in experiments where heterologous gene expression is being examined. Sarokin and Carlson (1985) found that an upstream region of the SUC2 gene conferred glucose-repressible expression to a heterologous gene, a LEU2–lacZ fusion, that is not normally regulated by glucose repression. It has also been observed that UAS of the CYC1 gene confer regulation by catabolite repression and heme to a LEU2–lacZ fusion (Guarente *et al.* 1984). UAS can function when inverted (Guarente & Hoar 1984), and in this respect resemble the enhancer elements observed in higher eukaryotes.

It is likely that greater attention will be paid to the use of hybrid promoters for the regulated production of commercially useful proteins. These will initially be rather unsophisticated, but as we gain more knowledge concerning sequences responsible for regulation and their interaction with *trans* acting elements, together with a knowledge of the factors determining promoter strength, they could form the basis for promoters which are regulated so that the induction capacity is greater than can be achieved from any natural promoter.

Regulated expression is the method of choice for the production of heterologous proteins that are toxic to a yeast. However, it does not follow that regulated expression will permit the production in yeast of commercially useful amounts of all such proteins. Once expression is turned on there is likely to be only a short time before the toxic effects of the protein result in a cessation of cell growth and lysis. The period may be too short to achieve sufficient production of the products for commercial exploitation.

## 13.6 Factors affecting heterologous protein production

Many yeast promoters, when associated with the homologous yeast gene on a multicopy plasmid, result in the production of up to 50% of total cell protein as the protein under investigation. When heterologous genes are associated with the same yeast promoter, the yield of heterologous protein varies considerably, but rarely surpasses 5% of total cell protein. Since heterologous is used most often to describe non-yeast proteins, there has been a tendency to consider the disparity in production levels of homologous and heterologous proteins as equating to differences between yeast and non-yeast proteins. It is much more likely that the different protein levels are simply protein effects. It is probable that, for instance, the use of the pyruvate-kinase promoter to drive pyruvate kinase (a homologous gene) may result in different protein levels than that achieved using the same promoter to drive another yeast gene (a heterologous gene).

We have observed that the levels of expression of IFN-beta are at least 100 times lower than that achieved from the identical plasmid construct when a modified IFN-beta in which amino acids 36–48 are substituted by equivalent regions of IFN-alpha (Watson *et al.* unpubl. obs.). The reasons for this are not obvious, but this experiment illustrates clearly that the level of expression is protein (or gene) dependent rather than promoter dependent. A similar conclusion can be concluded from experiments relating to homologous gene expression. The level of pyruvate kinase achieved from the full-yeast pyruvate kinase promoter is the same as that achieved from a promoter in which all but 670 bp of the promoter upstream of the ATG are deleted. The copy number of plasmids incorporating the full promoter is less than ten, whereas that containing the deleted promoter is more than 100 (Cafferkey unpubl. obs.). This shows that, whereas the protein levels are identical, the deleted promoter achieves that level from at least ten times more gene copies, presumably because it has reduced promoter efficiency. What determines the level of protein obtained? It is as if the yeast cell only tolerates a certain level of pyruvate kinase, and the plasmid copy number is adjusted downwards to accomplish this. How this is achieved is not clear. It may be an active adjustment or a reflection of selective pressures on this population.

A corollary of this observation is that experiments that are designed to explore the effect of codon bias usage or sequences immediately upstream of the ATG are irrelevant if, in fact, it is the protein itself that is limiting expression levels. Even if sequences immediately upstream of the ATG were altered to increase translation efficiency, there would be no increase in pyruvate kinase levels, and a false conclusion would be drawn.

Factors that affect expression can often only be investigated properly under conditions where an enhancement of protein can be accommodated by the cell. This may require the use of weak promoters and single copy vectors, and means that in practice many elements that do affect gene expression do not limit heterologous gene expression in a commercial setting.

## 13.7 Secretion

Proteins that are secreted from yeasts and other eukaryotic cells have a hydrophobic amino-terminal extension that is responsible for translocation of the protein into the endoplasmic reticulum, the first stage in the secretion pathway. In general the extension can be of two types; signals that are made up of approximately 20 amino acids are removed by a signal peptidase after translocation to the endoplasmic reticulum, and longer leaders that are not removed from the mature protein until the precursor reaches the golgi body. Yeast acid phosphatase and invertase have signal sequences, as do proteins such as interferon in mammalian cells. Alpha-factor and killer toxin which are both secreted from yeast rely on a golgi-associated cleavage step to produce the appropriate secreted form, as do certain pituitary hormones.

Most of the proteins that are of interest to a pharmaceutical company are proteins that are secreted from one cell type and exert their effect on another cell type. In general the mammalian proteins whose secretion has been attempted from yeast possess signal sequences cleaved in the endoplasmic reticulum.

Three approaches have been used to secrete foreign proteins from yeast; the use of the native mammalian signal sequence, the use of a yeast signal sequence and the use of a yeast leader sequence.

When a yeast promoter has been used to express a foreign gene and signal sequence, secretion of the foreign protein into the medium has been observed in some, but not all, cases. No secretion into the medium was observed with human alpha-antitrypsin (Cabezon *et al.* 1984) or IFN-beta (Watson pers. commun.). In contrast, the use of the signal sequences of mouse alpha-amylase (Thomson 1983), human leukocyte interferon (Hitzeman *et al.* 1983), mouse immunoglobulin light and heavy chains (Wood *et al.* 1985) and tissue inhibitor of metalloproteases (Goodey unpubl. results) result in the secretion of some of the respective proteins into the medium. Although Mellor *et al.* (1983) reported no secretion into the medium using the prochymosin signal, R. A. Smith *et al.* (1985) did when the GAL1 promoter was used in place of the PGK promoter. Wheat alpha-amylase was secreted from yeast cells, even when the signal was preceded by an amino-terminal fusion consisting of the first 12 amino acids of phosphoglycerate kinase and 12 amino acids derived from linker DNA (Rothstein *et al.* 1984). The alpha-amylase secreted into the medium had a molecular weight consistent with cleavage at the signal–mature gene junction. It is possible that the secreted material derived from proteins in which initiation of translation occurred at an internal methionine codon adjacent to the alpha-amylase signal.

The different results achieved with foreign signals may occur because of variations in the efficiency with which foreign signals are recognized and processed properly by yeast. Alternatively, the proteins themselves may carry information necessary for secretion, and this information may be appropriate for yeast in some, but not all, foreign proteins.

Although IFN-beta is not secreted when its own signal is used, analysis of material recovered from the cell indicates that the signal is cleaved off and the IFN is glycosylated (Piggott *et al.* 1986). This indicates that the signal is recognized and that the IFN is translocated to the ER, where the signal is cleaved, but that the yeast is unable to direct the protein through the secretion pathway such that it ends up in the medium.

An alternative approach has been to use yeast signal sequences to ensure translocation to the endoplasmic reticulum. Both the acid phosphatase (PHO5) and invertase (SUC2) signal sequences have been used to obtain secretion of prochymosin (R. A. Smith *et al.* 1985). A hybrid signal in which the first amino acids of the acid phosphatase signal was fused to the cleavage site of a human leukocyte interferon resulted in the expression and secretion of interferon (Hinnen *et al.* 1983). Ten per cent of the interferon produced was secreted into the medium. A similar approach in which the first 22 amino acids of the alpha-factor leader (to achieve translocation to the endoplasmic reticulum) were fused to the last eight amino acids of the IFN-alpha$_2$ signal sequence (to provide a cleavage site that is processed in the endoplasmic reticulum) resulted in 10–50 mg $l^{-1}$ of IFN being secreted into the medium (Carter *et al.* 1986, Piggott *et al.* 1986). A construct which was identical other than the replacement of the mature IFN-alpha$_2$ gene by IFN-beta did not result in secretion of the latter into the medium.

**Table 13.3** A comparison of activable prochymosin expressed and secreted using the GAL1 promoter and a range of secretion signals. ssc1 and ssc2 are supersecreting mutants, ssc1,ssc2 is a double mutant. Data from R. A. Smith *et al.* (1985).

| | | Level of activable prochymosin (μg $g^{-1}$) | | | | |
|---|---|---|---|---|---|---|
| | | | Secreted | | | |
| Promoter | Secretion signal | Internal wild-type | Wild-type | ssc1 | ssc2 | ssc1,ssc2 |
| GAL1 | 0 | 10 | 0 | 0 | 0 | 0 |
| GAL1 | SUC2 | 18 | 27 | 65 | 141 | 238 |
| GAL1 | pre chymosin | — | <6 | 117 | 46 | 275 |
| GAL1 | PHO5 | — | <6 | 268 | 129 | 390 |

Although there are few studies that have compared the effect of native and yeast signal sequences on foreign protein secretion, R. A. Smith *et al.* (1985) have compared the secretion of prochymosin using the natural prochymosin signal and the signal sequences of invertase (SUC2) and acid phosphatase (PHO5) (Table 13.3). The constructs were not strictly comparable, as the use of the SUC2 signal necessitated a 33-bp inframe fusion of the yeast SUC2 gene between the signal and prochymosin. The level of activable enzyme secreted into the medium was highest using the SUC2 signal in wild-type cells. When two supersecreting mutants *ssc*1 and *ssc*2 were used as hosts the results were less clearcut, and in the double mutant ssc1, ssc2 the highest level was achieved using the PHO5 signal, and the prochymosin signal was slightly better than the SUC2 signal (Table 13.2).

There is still insufficient information to permit firm conclusions about the efficiency of secretion of IFN-beta using native mammalian signals compared with yeast signals. It is clear that in the case of prochymosin either will work, but this is not the case for all proteins. For instance, neither the natural signal of IFN-beta nor a yeast signal lead to secretion of IFN-beta into the medium. It is probable that either a mammalian signal or yeast signal is sufficient for the first stage of secretion (namely translocation to the ER, where the signal is removed), but is insufficient to direct the cell proteins to the exterior of the cell. Thus, signal sequences from whatever source cannot be used for general secretion vectors, but may be useful in particular instances.

The third approach to the secretion of heterologous proteins from yeast cells has been to use yeast leader sequences that are cleaved in the golgi apparatus. Leader sequences from the precursor to the yeast pheromone alpha-factor (Kurjan & Herskowitz 1982) and from the killer-toxin precursor (Bostian *et al.* 1984) have been used. Most investigators have concentrated on the former because of its earlier availability.

Alpha-factor is a pheromone involved in mating. Haploid yeast cells of the *α* mating type secrete into the medium alpha-factor, a tridecapeptide that arrests a mating type cell (*Mata*) in the G1 phase of the cycle before mating. *Mata* cells secrete a-factor which arrests alpha-cells in G1. Thus, synchronization before mating is affected. Alpha-factor is processed from a precursor of 165 amino acids, known as prepro alpha-factor, which contains four copies of mature alpha-factor (Kurjan & Herskowitz 1982). There is a hydrophobic sequence at the amino terminus that ensures translocation to the endoplasmic reticulum (ER). It is unlike conventional signals, since the initial cleavage does not occur in the ER. The first cleavage occurs in the golgi apparatus after the lysine and arginine at positions 84 and 85; signal sequences are shorter (usually about 20 amino acids).

The processing of alpha-factor has been extensively studied (Kurjan & Herskowitz 1982, Julius *et al.* 1983, 1984a, b). The picture that emerges is shown in Figure 13.5. There is a convenient HindIII restriction endonuclease site immediately preceding the first repeat of mature alpha-factor. This permits removal of the alpha-factor repeats and the substitution of any foreign gene of interest. Such a tactic has been successfully used to express and secrete human epidermal growth hormone (Brake *et al.* 1984), $\beta$-endorphin and a synthetic human alpha-interferon (Bitter & Egan 1984). However, in a proportion of the secreted material the sequence glu–ala–glu–ala was found at the amino terminus, indicating incomplete processing by the dipeptidyl amino peptidase encoded by STE13. This is not an altogether surprising result, as it has been

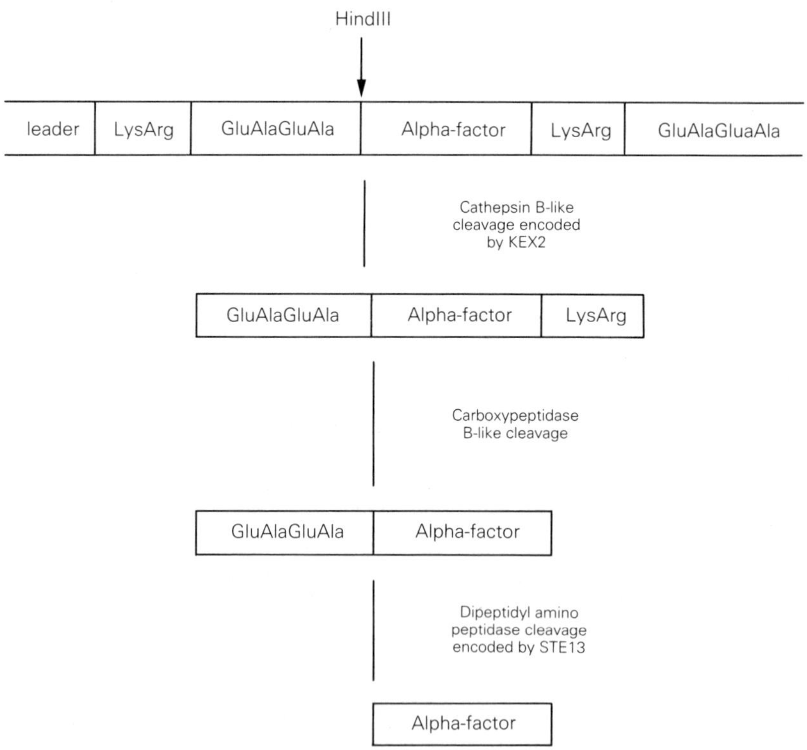


**Figure 13.5** The processing of prepro alpha factor. Only the first of the four copies of mature alpha-factor is shown.

observed that when $\alpha$ mating type cells are transformed with a multicopy plasmid bearing the alpha-factor gene the normal pathway of synthesis and processing is overloaded; cleavage of the glu–ala repeats by the dipeptidyl amino peptidase becomes limiting, and a large proportion of secreted alpha-factor retains the glu–ala repeats at its amino terminus (Julius *et al.* 1983). There are several possible ways of overcoming this problem. A wild-type copy of the STE13 gene encoding the dipeptidyl amino peptidase can be incorporated on a multicopy plasmid (Julius *et al.* 1983). Alternatively, a DNA synthetic linker can be used to interpose a KEX2 encoded endopeptidase site (lys–arg) between the glu–ala repeats and the mature foreign gene (Brake *et al.* 1984). When such a tactic was employed to produce human epidermal growth factor, 80% of the secreted material was processed correctly but 20% retained the glu–ala repeats at the amino terminus (Brake *et al.* 1984). This result indicated that the endopeptidase could cleave at either of the two Lys–Arg sequences, but that at least a proportion was not cut at both.

A rather neater solution to this problem is to delete the glu–ala repeats such that the mature gene follows the $lys^{84}$–$arg^{85}$ sequence in the alpha-factor leader. To facilitate this approach, *in vitro* mutagenesis is used to create a restriction site in the leader before the lys–arg sequence (Brake *et al.* 1984, Carter *et al.* 1986), and a synthetic DNA linker is used to reconstitute the leader up to and including the lys–arg followed by the gene of interest. This approach has been used to secrete human epidermal growth factor (Brake *et al.* 1984). Sequence analysis indicated that 100% of the secreted material was cleaved after the lys–arg sequence. Unfortunately, deleting the glu–ala repeats does not always result in proper processing. Piggott *et al.* (1986) found that when this tactic was employed to secrete fibroblast interferon from yeast the vast majority of the material secreted was a hyperglycosylated fusion protein. This indicated that a lys–arg sequence, while necessary, is not always sufficient for proper cleavage. The lys–arg must be presented in such a way that it is recognized by the endopeptidase.

When this problem arises there are two possible solutions. Correct processing may be achieved if the mature gene of interest is ligated to the HindIII site in the leader, and a STE 13 gene encoding the dipeptidyl peptidase is included on the multicopy plasmid in an attempt to provide sufficient enzyme to cleave the glu–ala repeats. An alternative approach is to alter sequences preceding the lys–arg (in a glu–ala deleted construct) in an attempt to provide proper presentation of the lys–arg sequence; changing sequences downstream from the lys–arg may result in proper processing, but will result in a change in the amino-acid sequence of the protein of interest.

It is important to obtain proper processing, so that the mature protein is secreted without N-terminal additions, but it is equally important that no processing occurs at sites within the mature protein sequence peptide. Many proteins have potential KEX2 cleavage sites (Lys–Arg, Arg–Arg, Lys–Lys), and it has been shown that at least a proportion of B-endorphin when secreted using the alpha-factor leader is cleaved on the carboxyl side of two lysines within the B-endorphin peptide (Bitter & Egan 1984).

These problems suggest that it will not be possible to construct a general yeast secretion vector, based on the alpha-factor leader, that will be successful

```
              10         20         30         40          50         60         70          80          90
pppGAAAAATAAAGAA ATG ACG AAG CCA ACC CAA GTA TTA GTT AGA TCC GTC AGT ATA TTA TTT TTC ATC ACA TTA CTA CAC CTA GTC GTA GCG
                 Met Thr Lys Pro Thr Gln Val Leu Val Arg Ser Val Ser Ile Leu Phe Phe Ile Thr Leu Leu His Leu Val Val Ala
                     "δ"                         10                            20

         100         110 PstI     120       130        140        150        160   BglII 170         180
CTG AAC GAT GTG GCC GGT CCT GCA GAA ACA GCA CCA GTG TCA TTA CTA CCT CGT GAA GCG CCG TGG TAT GAC AAG ATC TGG GAA GTA AAA
Leu Asn Asp Val Ala Gly Pro Ala Glu Thr Ala Pro Val Ser Leu Leu Pro Arg Glu Ala Pro Trp Tyr Asp Lys Ile Trp Glu Val Lys
 (P1)?  30                                 40                        α Toxin               50

         190        200        210        220        230        240        250        260        270
GAT TGG CTA TTA CAG CGT GCC ACA GAT GGC AAT TGG GGC AAG TCG ATC ACC TGG GGT TCA TTC GTA GCG AGC GAT GCA GGT GTA GTA ATC
Asp Trp Leu Leu Gln Arg Ala Thr Asp Gly Asn Trp Gly Lys Ser Ile Thr Trp Gly Ser Phe Val Ala Ser Asp Ala Gly Val Val Ile
        60                                 70                                 80

         280        290        300        310        320        330        340        350        360
TTT GGT ATC AAT GTG TGT AAG AAC TGC GTG GGT GAG CGT AAG GAT GAT ATC AGT ACG GAC TGC GGC AAG CAA ACA CTT GCT TTA CTA GTC
Phe Gly Ile Asn Val [Cys] Lys Asn [Cys] Val Gly Glu Arg Lys Asp Asp Ile Ser Thr Asp [Cys] Gly Lys Gln Thr Leu Ala Leu Leu Val
        90                                  100                                 110

         370        380 HpaII  390        400        410  HpaII 420        430        440        450
AGC ATT TTT GTA GCA GTT ACA TCC GGC CAT CAT CTT ATA TGG GGT GGT AAT AGG CCG GTG TCG CAG TCA GAT CCT AAT GGC GCT ACC GTT
Ser Ile Phe Val Ala Val Thr Ser Gly His His Leu Ile Trp Gly Gly Asn Arg Pro Val Ser Gln Ser Asp Pro Asn Gly Ala Thr Val
        120                                 130  γ Immunity (?)                 140

         460        470        480        490        500        510        520        530        540
GCT CGT CGT GAC ATT TCT ACT GTC GCA GAC GGG GAT ATT CCA CTC GAC TTT AGT GCG TTG AAC GAC ATA TTA AAT GAA CAT GGT ATT AGT
Ala Arg Arg Asp Ile Ser Thr Val Ala Asp Gly Asp Ile Pro Leu Asp Phe Ser Ala Leu Asn Asp Ile Leu Asn Glu His Gly Ile Ser
        150                                 160                                 170
```

```
             550                 560                 570                 580                 590                 600                 610                 620                 630
              •                   •                   •                   •                   •                   •                   •                   •                   •
ATA CTC CCA GCT AAC GCA TCA CAA TAT GTC AAA AGA TCA GAC ACA GCC GAA CAC ACG ACA AGT TTT GTA GTG ACC AAC AAC TAC ACT TCT
Ile Leu Pro Ala [Asn] Ala Ser Gln Tyr Val Lys Arg Ser Asp Thr Ala Glu His Thr Thr Ser Phe Val Val Thr Asn [Asn] Tyr Thr Ser
                180                                     190                                     200

             640       HinfI     650                 660                 670                 680                 690                 700                 710                 720
              •                   •                   •                   •                   •                   •                   •                   •                   •
TTG CAT ACC GAC CTG ATT CAT CAT GGT AAT GGA ACA TAT ACC ACG TTT ACC ACA CCT CAC ATT CCA GCA GTG GCC AAG CGT TAT GTT TAT
Leu His Thr Asp Leu Ile His His Gly [Asn] Gly Thr Tyr Thr Thr Phe Thr Thr Pro His Ile Pro Ala Val Ala Lys Arg Tyr Val Tyr
                210                                     220                                     230            β-Toxin →

             730                 740                 750                 760                 770                 780                 790                 800                 810
              •                   •                   •                   •                   •                   •                   •                   •                   •
CCT ATG TGC GAG CAT GGT ATC AAG GCC TCA TAC TGT ATG GCC CTT AAT GAT GCC ATG GTG TCG GCT AAT GGT AAC CTG TAT GGA CTA GCA
Pro Met [Cys] Glu His Gly Ile Lys Ala Ser Tyr [Cys] Met Ala Leu Asn Asp Ala Met Val Ser Ala Asn Gly Asn Leu Tyr Gly Leu Ala
                240                                     250                                     260

             820                 830                 840                 850                 860                 870                 880                 890                 900
              •                   •                   •                   •                   •                   •                   •                   •                   •
GAA AAG CTG TTT AGT GAG GAT GAG GGA CAA TGG GAG ACG AAT TAC TAT AAA TTG TAT TGG AGT ACT GGC CAG TGG ATA ATG TCG ATG AAG
Glu Lys Leu Phe Ser Glu Asp Glu Gly Gln Trp Glu Thr Asn Tyr Tyr Lys Leu Tyr Trp Ser Thr Gly Gln Trp Ile Met Ser Met Lys
                270                                     280                                     290

             910                 920                 930                 940                 950              960  BglI       970          980          990
              •                   •                   •                   •                   •                •                •            •            •
TTT ATT GAG GAA AGT ATT GAT AAC GCC AAT AAT GAC TTT GAA GGC TGT GAC ACA GGC CAC TAG GGCATCGTGTCTGACCTCTGATGCGATAACTC
Phe Ile Glu Glu Ser Ile Asp Asn Ala Asn Asn Asp Phe Glu Gly [Cys] Asp Thr Gly His stop
                300                                     310

   1000       1010
    •          •
GACCCTACAGAGCA
```

**Figure 13.6** Complete nucleotide sequence of the $M_1$ dsRNA pre-protoxin gene. Restriction endonuclease sites are marked and the $M_1$-P1 preprotoxin open reading frame starting with AUG at 14–16 is shown. The in-frame $\alpha$- and $\beta$-toxin component N termini are indicated by arrows. Possible processing sites for cleavage within $\delta$, by a leader peptidase, and between $\alpha$ and the central glycosylated region ($\delta$) are also indicated. Cysteine residues and potential asparagine glycosylation sites are boxed (Bostian *et al*. 1984).

in producing properly processed protein in all cases. Each protein may require a different vector. However, it has been our experience that mammalian proteins which are not secreted using a mammalian or yeast signal can be secreted using the alpha-factor leader, although in some instances proper processing remains a problem. Examples include beta-interferon, tissue inhibitor of metalloprotease and tissue plasminogen activator.

Various methods may be used to secrete foreign proteins across the cell membrane and into the periplasmic space. The factors that determine whether the proteins will subsequently cross the cell wall and enter the medium are not established. However, it does not appear that the yeast cell wall acts as a sieve, since recombinant derived material with a molecular weight in excess of 100 000 can be secreted into the medium.

Mammalian glycoproteins are glycosylated when they are secreted from yeast. Although there are no reports analysing the composition of the oligosaccharide components of these proteins, the outer core of the oligosaccharide chains of yeast-derived glycoproteins will be of the high-mannose type (Ballou 1982) and, as such, different in many instances from mammalian-derived material.

An attempt has also been made to use the killer toxin leader to secrete a foreign protein (bacterial cellulase) from yeast (Skipper *et al.* 1985). The pre protoxin includes a pre region responsible for co-translational insertion of protoxin into the endoplasmic reticulum. The N-terminal pre region is followed by an alpha- and beta-subunit of the toxin separated by a glycosylated gamma segment (Fig. 13.6) which is thought to confer immunity to the toxin (Bostian *et al.* 1984).

Secretion and processing of alpha-factor and killer toxin have certain features in common. It is thought that release of the beta-toxin subunit is accomplished by cleavage after the $Lys^{232}$–$Arg^{233}$ residues by the same KEX2 encoded endopeptidase that is involved in prepro alpha-factor processing. Indeed, KEX2 was originally identified as a mutant of a killer strain that failed to produce active toxin but produced the immunity function. Such mutants are also sterile, because of a failure in prepro alpha-factor processing (Leibowitz & Wickner 1976).

The similarity of processing involved in alpha-factor and killer factor production led Skipper *et al.* (1985) to perform an experiment in which the ADH1 promoter was used to express a protein composed of the first 233 amino acids of the prepro toxin sequence followed by a bacterial endo-1,4,$\beta$-D-glucanase (carboxymethyl cellulase), which according to the authors may lack the sequence encoding the $NH_2$-terminal region of the carboxymethyl cellulase. This fusion effectively substitutes carboxymethyl cellulase for the beta toxin subunit that follows the $lys^{232}$–$arg^{233}$ in the prepro toxin sequence.

Little, if any, active carboxymethyl cellulase was secreted into the medium from yeast cells transformed with this construct on a multicopy vector. It is possible that no protein is secreted, or that an inactive protein is secreted. The authors suggest that substitution of the carboxymethyl cellulase polypeptide for the beta-toxin subunit may result in a precursor conformation that interferes with normal processing. The results discussed earlier concerning the use of the alpha-leader indicate that processing does not occur at all lys–arg sequences, and that the context in which this sequence appears is important. More work on the alpha-factor and killer toxin systems is necessary for a better understanding of KEX2 processing. It would be useful to have a test protein that was only active if precise cleavage occurred; such a system would permit both classical and *in vitro* mutagenesis studies of processing.

## 13.8 Products

No attempt will be made to list all of the heterologous protein products of yeast that are being considered or achieved; as such, a list would be outdated by the time of publication.

One of the earliest commercial products of heterologous expression in yeast will be a vaccine utilizing the hepatitis B surface antigen (HBsAg). This is a product that we would like to highlight in this section. HBsAg is not only expressed in yeast, but is also complexed into viral-like particles which are similar to those found in serum. The production of such particles was not a predictable result. These particles are antigenic. A vaccine preparation has been made from them, which is safe and immunogenic in humans (Jilg *et al.* 1984, Scholnick *et al.* 1984) and is effective in preventing hepatitis infection in chimpanzees (McAleer *et al.* 1984). Unlike HBsAg derived from human plasma, the yeast antigen is held together by non-covalent interactions and can be dissociated by 2% sodium dodecyl sulphate without the use of reducing agents. During *in vitro* purification of the yeast antigen some disulphide bonds form spontaneously between the antigen subunits, resulting in a particle. This disulphide-bonded particle resembles the naturally occurring plasma-derived surface antigen particle (Wampler *et al.* 1985). Such polymeric particles overcome the problem of antigen presentation. It is a polymer of high molecular weight, which is both very stable and highly immunogenic. It will therefore be possible to use the membrane insertion and *in vivo* assembly and presentation of other antigens to form polyvalent vaccines. As a model system a genetically engineered hybrid particle containing HBsAg with the glycoprotein D(gD) of *Herpes simplex* virus (HSV) has been expressed in yeast. The HSV-1 gD coding sequence was inserted into the pre-S or pre-surface region of hepatitis, just outside the HBsAg coding region. Yeast extracts were shown in a sandwich ELISA assay using bonded anti-HBsAg antibodies to capture the hybrid antigen, and a second soluble anti-HSV-1 glycoprotein antibody was used to detect the complex. Moreover, the evidence indicates that the hybrid antigen is associated with a protein of a molecular weight close to the expected fusion of both HBsAg and HSV-1 gD proteins, and that the hybrid antigen is assembled into particles (Valenzuela *et al.* 1985).

HBsAg and its potential polyvalent derivatives are currently the best example of how cloning in yeast has overcome problems with potential vaccines made by recombinant DNA as well as chemically synthesized epitopes, i.e. antigen presentation.

It is likely that further examples will follow. An advantage of yeast expression, particularly when allied to secretion, is that it permits the production of soluble active proteins. It is important to optimize the yield of recombinant proteins from yeast. This is likely to come from advances in molecular genetics concerning expression, permitting enhanced expression per gene copy and vectors in which the copy number of the gene can be manipulated, advances in host genetics to increase the yield and advances in fermentation technology. A difficulty with the use of yeast as a host for human therapeutic agents is whether the glycosylation observed with yeast will present problems, either with regard to biological function or immunogenicity. The latter is difficult to predict; it is a potential problem with long-term therapy, but is unlikely to be critical with proteins such as tissue plasminogen activator where repeated doses over a long period are not anticipated.

## References

Adams, B. G. 1972. Induction of galactokinase in *Saccharomyces cerevisiae*: kinetics of induction and glucose effects. *Journal of Bacteriology* **111**, 308–15.

Ballou, L. E. 1976. Structure and biosynthesis of the mannan component of the yeast cell envelope. *Advances in Microbial Physiology* **14**, 93–158.

Ballou, L. E. 1982. The yeast cell wall and cell surface. In *Molecular biology of the yeast* Saccharomyces: *metabolism and gene expression*, J. N. Strathern, E. W. Jones, J. R. Broach (eds), New York: Cold Spring Harbor Laboratory.

Beach, D. and P. Nurse 1981. High-frequency transformation of the fission yeast *Schizosaccharomyces pombe. Nature* (London) **290**, 140–2.

Beggs, J. D. 1978. Transformation of yeast by a replicating hybrid plasmid. *Nature* (London) **275**, 104–9.

Beggs, J. 1981. Multiple-copy yeast plasmid vectors. In *Molecular genetics in yeast*, D. von Wettstein, J. Friis, M. Keilland-Brandt, A. Stenderup (eds). Copenhagen: Alfred Benzon. Symposium 16.

Bitter, G. A. and K. A. Egan 1984. Expression of heterologous genes in *Saccharomyces cerevisiae* from vectors utilizing the glyceraldehyde-3-phosphate dehydrogenase gene promoter. *Gene* **32**, 263–74.

Blackburn, E. H. 1985. Artificial chromosomes in yeast. *Trends in Genetics* **1**, 8–12.

Bostian, K. A., J. M. Lemire, L. E. Cannon and H. O. Halvorson 1980. *In vitro* synthesis of repressible yeast acid phosphatase; identification of multiple mRNAs and products. *Proceedings of the National Academy of Sciences of the USA* **77**, 4504–8.

Bostian, K. A., J. M. Lemire and H. O. Halvorson 1983. Physiological control of repressible acid phosphatase gene transcripts in *Saccharomyces cerevisiae. Molecular and Cellular Biology* **3**, 839–53.

Bostian, K. A., Q. Elliot, H. Bussey, V. Bum, A. Smith and D. J. Tipper 1984. Sequence of the prepro-toxin dsRNA gene of type 1 killer yeast: multiple processing events produce two component toxin. *Cell* **36**, 741–51.

Brake, A. J., J. P. Merryweather, D. G. Coit, V. A. Heberlain, F. R. Masiarz, G. T. Mullenbach, M. S. Urdea, P. Valenzuela and P. J. Barr 1984. $\alpha$-Factor-directed synthesis and secretion of mature foreign proteins in *Saccharomyces cerevisiae. Proceedings of the National Academy of Sciences of the USA* **81**, 4642–6.

Broach, J. R. 1981. The yeast 2 microncircle. In *The molecular biology of the yeast* Saccharomyces, J. N. Strathern, E. W. Jones, J. R. Broach (eds), Vol. 1. New York: Cold Spring Harbor Laboratory.

Cabezon, T., M. De Wilde, P. Herion, R. Lorain and A. Bollen 1984. Expression of human alpha-antitrypsin cDNA in the yeast *Saccharomyces cerevisiae*. *Proceedings of the National Academy of Sciences of the USA* **81**, 6594–8.

Carter, B. L. A., S. Doel, A. R. Goodey, J. R. Piggott and M. E. W. Watson 1986. Secretion of mammalian polypeptides from yeast. *Microbiological Sciences* **3**, 23–7.

Clarke, L. and J. Clarkson 1980. Isolation of a yeast centromere and construction of a functional small circular chromosome. *Nature* (London) **287**, 504–9.

Cohen, J. D., T. R. Eccleshall, R. B. Needleman, H. Federoff, B. A. Buchferer and J. Marmur 1980. Functional expression of the *Escherichia coli* plasmid gene coding for chloramphenicol acetyltransferase. *Proceedings of the National Academy of Sciences of the USA* **77**, 1078–82.

Cregg, J. M., K. J. Barringer, A. Y. Hessler and K. R. Madden 1985. *Pichia* as a host system for transformations. *Molecular and Cellular Biology* **5**, 3376–85.

Derynck, R., A. Singh and D. V. Goeddel 1983. Expression of the human interferon $\gamma$ cDNA in yeast. *Nucleic Acids Research* **11**, 1819–37.

Douglas, H. C. and D. C. Hawthorne 1966. Regulation of genes controlling synthesis of the galactose pathway enzymes in yeast. *Genetics* **54**, 911–16.

Edens, L., I. Bom, A. M. Ledeboer, Y. T. Marjolain, C. Visser and C. T. Verrips 1984. Synthesis and processing of the plant protein thaumatin in yeast. *Cell* **37**, 629–33.

Erhart, E. and C. P. Hollenberg 1983. The presence of a defective *leu 2* gene on 2 μm DNA recombinant plasmids of *Saccharomyces cerevisiae* is responsible for curing and high copy number. *Journal of Bacteriology* **156**, 625–35.

Futcher, A. B. and B. S. Cox 1984. Copy number and the stability of 2 μm circle based artificial plasmids of *Saccharomyces cerevisiae*. *Journal of Bacteriology* **157**, 283–90.

Goff, C. G., D. T. Moir, T. Kohno, T. C. Gravius, R. A. Smish, E. Yamaski and A. Taunton-Rigby 1984. Expression of calf prochymosin in *Saccharomyces cerevisiae*. *Gene* **27**, 35–46.

Goodey, A. R., S. M. Doel, J. R. Piggott, M. E. Watson, G. R. Zealey, R. Cafferkey and B. L. A. Carter 1986. The selection of promoters for the expression of heterologous genes in the yeast *Saccharomyces cerevisiae*. *Molecular and General Genetics* **204**, 505–11.

Gorman, J. A., W. F. Dove and N. Warren 1981. Isolation of *Physarum* DNA segments that support autonomous replication in yeast. *Molecular and General Genetics* **183**, 306–13.

Gritz, L. and J. Davies 1983. Plasmid encoded hygromycin B resistance: the sequence of hygromycin B phosphotransferase and its expression in *Escherichia coli* and *Saccharomyces cerevisiae*. *Gene* **25**, 179–88.

Guarente, L. 1984. Yeast promoters: positive and negative elements. *Cell* **36**, 799–800.

Guarente, L. and E. T. Hoar 1984. Upstream activation sites of the CYC1 gene of *Saccharomyces cerevisiae* are active when inverted but not when downstream to the TATA box. *Proceedings of the National Academy of Sciences of the USA* **81**, 7860–4.

Guarente, L. and M. Ptashne 1981. Fusion of *Escherichia coli* Lac Z to the cytochrome c gene of *Saccharomyces cerevisiae*. *Proceedings of the National Academy of Sciences of the USA* **78**, 2199–203.

Guarente, L., B. Lalonde, P. Gifford and E. Alani 1984. Distinctly regulated tandem upstream activation sites mediate catabolic repression of the CYC 1 gene of *S. cerevisiae*. *Cell* **36**, 503–11.

Gunge, N., A. Tamaru, F. Ozawa and K. Sakaguchi 1981. Isolation and characterisation of linear deoxyribonucleic acid plasmids from *Kluyveromyces lactis* and the plasmid associated killer character. *Journal of Bacteriology* **148**, 988–90.

Hinnen, A., J. B. Hicks and G. R. Fink 1978. Transformation of yeast. *Proceedings of the National Academy of Sciences of the USA* **75**, 1929–33.

Hinnen, A., B. Meyhack and R. Tsapis 1983. High expression and secretion of foreign proteins in yeast. In *Gene expression in yeast. Proceedings of the Alko Yeast Symposium*, M. Korhula and E. Vaisanen (eds), Vol. 229, 1219–24. Helsinki: Foundation for Biotechnical and Industrial Fermentation Research.

Hitzman, R., F. Hagie, H. Levine, D. Goeddel, G. Ammerer and B. Hall 1981. Expression of a human gene for $\alpha$ interferon in yeast. *Nature* (London) **293**, 717–22.

Hitzman, R. A., C. Y. Chen, F. E. Hagie, E. G. Patzer, C. C. Lin, P. A. Estell, J. V. Miller, A. Yaffe, D. G. Kleid, A. D. Levinson and H. Opperman 1983. Expression of Hepatitis B virus surface antigens in yeast. *Nucleic Acids Research* **11**, 2745–63.

Hopper, J. E., J. R. Broach and L. B. Rowe 1978. Regulation of the galactose pathway in *Saccharomyces cerevisiae*: induction of uridyltransferase mRNA and dependency on GAL4 gene function. *Proceedings of the National Academy of Sciences of the USA* **75**, 2878–82.

Ito, H., Y. Fukuda, K. Murata and A. Kimura 1983. Transformation of intact yeast cells treated with alkali cations. *Journal of Bacteriology* **153**, 163–8.

Jabbar, M. A., N. Sivasubramarian and D. P. Nayak 1985. Influenza viral (A/WSN/33) hemagglutinin is expressed and glycosylated on the yeast *Saccharomyces cerevisiae*. *Proceedings of the National Academy of Sciences of the USA* **82**, 2019–23.

Jilg, W., M. Schmidt, G. Zoulek, B. Lorbeer, B. Wilske and F. Deinhardt 1984. Clinical evaluation of a recombinant hepatitis B vaccine. *Lancet* **ii**, 1174–5.

Jiminez, A. and J. Davis 1980. Expression of a transposable antibiotic resistance element in *Saccharomyces*. *Nature* (London) **287**, 869–71.

Johnston, S. A. and J. E. Hopper 1982. Isolation of the yeast regulatory gene Gal 4 and analysis of its dosage effects on the galactose/melibiose regulon. *Proceedings of the National Academy of Sciences of the USA* **79**, 6971–5.

Julius, D. J., L. C. Blair, A. J. Brake, G. F. Sprague and J. Thorner 1983. Yeast alpha-factor is processed from a larger precursor polypeptide: the essential role of a membrane-bound dipeptidyl amino peptidase. *Cell* **32**, 839–52.

Julius, D., R. Schekman and J. Thorner 1984a. Glycosylation and processing of prepro alpha-factor through the yeast secretory pathway. *Cell* **36**, 309–18.

Julius, D., A. Brake, L. Blair, R. Kunisawa and J. Thorner 1984b. Isolation of the putative structural gene for the lysine–arginine-cleaving endopeptidase required for processing of yeast prepro-alpha factor. *Cell* **37**, 1075–85.

Karin, M., R. Najorian, A. Haslinger, P. Valenzuela, J. Welch and S. Fogel 1984. Primary structure and transcription of an amplified genetic locus. The CUP1 locus of yeast. *Proceedings of the National Academy of Sciences of the USA* **81**, 337–42.

Kelly, J. D., E. W. Raines, R. Ross and M. J. Murray 1985. The B chain of PDGF alone is sufficient for mitogenesis. *EMBO Journal* **4** (13a), 3399–405.

Kikuchi, Y. 1983. Yeast plasmid requires a *cis*-acting locus and two plasmid proteins for its stable maintenance. *Cell* **35**, 487–93.

Kiss, G. B., K. V. Cornish, P. E. Pearlman and J. D. Friesen 1980. The Herpes simplex virus thymidine kinase gene is not expressed in *Saccharomyces cerevisiae*. *Recombinant Technical Bulletin* **3**, 21–4.

Kramer, R. A., T. M. Chiara, M. D. Schaber and S. Hilliker 1984. Regulated expression of a human interferon gene in yeast; control by phosphate concentration or temperature. *Proceedings of the National Academy of Sciences of the USA* **81**, 367–70.

Kurjan, J. and I. Herskowitz 1982. Structure of a yeast pheromone (MF alpha). A putative alpha factor precursor contains four tandem copies of mature alpha-factor. *Cell* **30**, 933–43.

Leibowitz, M. J. and R. B. Wicker 1976. A chromosomal gene required for killer plasmid expression, mating, and sporulation in *Saccharomyces cerevisiae*. *Proceedings of the National Academy of Sciences of the USA* **73**, 2061–5.

Lemontt, J. L., C.-M. Wei and W. R. Dackowski 1985. Expression of active human uterine tissue plasminogen activator in yeast. *DNA* **4**, 419–28.

Little, J. G. and R. H. Haynes 1979. Isolation and characterization of yeast mutants 2′-deoxythymidine 5′ monophosphate. *Molecular and General Genetics* **168**, 141–51.

McAleer, W. J., E. G. Buynak, R. Z. Maigetter, D. E. Wampler, W. J. Miller and M. R. Hilleman 1984. Human Hepatitis B vaccine from recombinant yeast. *Nature* (London) **307**, 178–80.

McNiel, J. and J. Friesen 1981. Expression of the Herpes simplex virus thymidine kinase gene in *Saccharomyces cerevisiae*. *Molecular and General Genetics* **184**, 386–93.

Matsumoto, K., A. Toh-e and Y. Oshima 1981. Isolation and characterization of dominant mutations resistant to carbon catabolite repression of galacto-kinase synthesis in *Saccharomyces cerevisiae*. *Molecular and Cellular Biology* **1**, 83–93.

Mellor, J., M. J. Dobson, N. A. Roberts, M. F. Tuite, J. S. Emtage, S. White, P. A. Lowe, T. Patel, A. J. Kingsman and S. M. Kingsman 1983. Efficient synthesis of enzymically active calf chymosin in *Saccharomyces*. *Gene* **24**, 1–14.

Miyajima, A., M. W. Bond, K. Otsu, K. Aria and N. Aria 1985. Secretion of mature mouse interlukin-2 by *Saccharomyces cerevisiae*: use of a general secretion vector containing promoter and leader sequences of the mating pheromone $\alpha$-factor. *Gene* **37**, 155–61.

Miyanohaa, A., A. Toh-e, C. Nozaki, F. Hamudu, M. Ohtomo and U. Matsubaru 1983. Expression of hepatitis B surface antigen in yeast. *Proceedings of the National Academy of Sciences of the USA* **80**, 1–5.

Murray, A. W. and J. W. Szostak 1983. Pedigree analysis of plasmid segregation in yeast. *Cell* **34**, 961–70.

Orr-Weaver, T., J. W. Szostak and R. J. Rothstein 1981. Yeast transformation: a model system for the study of recombination. *Proceedings of the National Academy of Sciences of the USA* **78**, 6354–8.

Oshima, Y. 1983. Regulatory circuits for gene expression. In *The molecular biology of the yeast*, J. H. Strathern, E. W. Jones and J. R. Broach (eds), 159–80. New York: Cold Spring Harbor Laboratory.

Parodi, A. J. and L. F. Leloir 1979. The role of lipid intermediates in the glycosylation of proteins in the eukaryotic cell. *Biochimica et Biophysica Acta* **559**, 1–15.

Piggott, J. R., M. E. E. Watson, S. M. Doel, A. R. Goodey and B. L. A. Carter 1986. The secretion and post-translational modification of interferons from *Saccharomyces cerevisiae*. (In preparation.)

Ratzkin, B. and J. Cabon 1977. Functional expression of cloned yeast DNA in *Escherichia coli*. *Proceedings of the National Academy of Sciences of the USA* **2**, 487–91.

Robbins, D. W. 1982. The processing of cell surface glycoproteins. In *Molecular genetic neurosche*, F. O. Schmitt, S. J. Bird and F. E. Bloom (eds), 161–9. New York: Raven Press.

Rosenberg, S., P. J. Barr, R. C. Najorian and R. A. Hallewell 1984. Synthesis in the yeast *Saccharomyces cerevisiae* of a functional oxidation resistant mutant of human $\alpha$ 1-antitrypsin. *Nature* (London) **312**, 77–80.

Rothstein, R. 1985. Cloning in yeast. In *DNA cloning*, Vol. ii, J. M. Glover (ed.). Oxford and Washington, DC: IRL Press.

Rothstein, S. J., C. M. Lazarus, W. E. Smith, D. C. Baulcombe and A. A. Gatenby 1984. Secretion of a wheat $\alpha$-amylase expressed in yeast. *Nature* (London) **308**, 633–5.

Rymond, B. C., R. S. Zitomer, D. Schumperli and M. Rosenberg 1983. The expression in yeast of the *Escherichia coli* gal K gene on CYC 1. *Gene* **25**, 249–62.

St John, T. P. and R. W. Davis 1981. The organisation and transcription of the galactose gene cluster of *Saccharomyces*. *Journal of Molecular Biology* **152**, 285–316.

Sarokin, L. and M. Carlson 1985. Upstream region of the SUC2 gene confers regulated expression to a heterologous gene in *Saccharomyces cerevisiae*. *Molecular and Cellular Biology* **5**, 2521–6.

Schekman, R. and P. Novick 1982. The secretory process and yeast cell surface assembly. In *Molecular biology of the yeast* Saccharomyces. *Metabolism and gene expression*, J. H. Strathern, E. W. Jones and J. R. Broach (eds), Vol. 42, 361–93. New York: Cold Spring Harbor Laboratory.

Scherer, G., G. Haag and W. Durtze 1974. Mechanism of $\alpha$-factor biosynthesis in *Saccharomyces cerevisiae*. *Journal of Bacteriology* **119**, 386–93.

Scholnick, E. M., A. A. McLean, D. J. West, W. J. McAleer, W. J. Miller and E. B. Buynak 1984. Clinical evaluation in healthy adults of a hepatitis B vaccine made by recombinant DNA. *Journal of the American Medical Association* **251**, 2812–15.

Skipper, N., M. Sutherland, R. W. Davis, D. Kilburn, R. C. Miller, A. Warren and R. Wong 1985. Secretion of a bacterial cellulase by yeast. *Science* **230**, 958–60.

Smith, M., D. W. Leung, S. Gillam and C. R. Astell 1979. Sequence of the gene for iso-1-cytochrome c in *Saccharomyces cerevisiae*. *Cell* **16**, 753–61.

Smith, R. A., M. J. Duncan and D. T. Moir 1985. Heterologous protein secretion from yeast. *Science* **229**, 1219–24.

Standoni, R. J. and L. F. Leloir 1982. The biosynthetic pathway of the asparagine-linked oligosaccharides of glycoproteins. *Critical Reviews in Biochemistry* **12**, 298–326.

Stepien, P. P., R. Brousseau, R. Wu, S. Narang and D. Y. Thomas 1983. Synthesis of a human insulin gene VI. Expression of the synthetic pro-insulin gene in yeast. *Gene* **24**, 289–97.

Stinchcombe, D., K. Struhl and R. Davis 1979. Isolation and characterization of yeast chromosomal replicator. *Nature* (London) **282**, 39–43.

Stinchcombe, D. T., M. Thomas, J. Kelly, E. Selker and R. W. Davis 1980. Eukaryotic DNA segments capable of autonomous replication in yeast. *Proceedings of the National Academy of Sciences of the USA* **77**, 4559–68.

Szostak, J. W. and E. H. Blackburn 1982. Cloning yeast telomeres on a yeast linear plasmid. *Cell* **29**, 245–55.

Thomson, K. K. 1983. Mouse alpha-amylase synthesized by *Saccharomyces cerevisiae* is released into the culture medium. *Carlsberg Research Communications* **48**, 545–55.

Toh-e, A. and Y. Oshima 1974. Characterisation of a dominant constitutive mutation phoO for repressible acid phosphatase synthesis in *Saccharomyces cerevisiae*. *Journal of Bacteriology* **120**, 608–17.

Toh-e, A. and I. Ututsu 1985. Physical and functional structure of a yeast plasmid pSB3 isolated for *Zygosaccharomyces bisporus*. *Nucleic Acids Research* **13**, 4267–83.

Toh-e, A., Y. Ueda, S. Kakimoto and Y. Oshima 1973. Isolation and characterization of acid phosphatase mutants in *Saccharomyces cerevisiae*. *Journal of Bacteriology* **113**, 727–38.

Toh-e, A., S. Toda and Y. Oshima 1982. 2 μm DNA-like plasmids in the osmophilic yeast *Saccharomyces rouxii*. *Journal of Bacteriology* **151**, 1380–90.
Toh-e, A., H. Araki, I. Utatsu and Y. Oshima 1984. Plasmids resembling 2 μm DNA in the osmotolerant yeasts *Saccharomyces bailii* and *Saccharomyces bisporus*. *Journal of General Microbiology* **130**, 2527–34.
Torchia, T. E., R. W. Hamilton, C. L. Cano and J. E. Hopper 1984. Disruption of regulatory gene Gal-80 in *Saccharomyces cerevisiae*. Effects of carbon-controlled regulation of the galactose–melibiose pathway genes. *Molecular and Cellular Biology* **4**, 1521–7.
Tuite, M. F., M. J. Dobson, N. A. Roberts, R. M. King, D. C. Burke, S. M. Kingsman and A. J. Kingsman 1982. Regulated high efficiency expression of human interferon-alpha in *Saccharomyces cerevisiae*. *EMBO Journal* **1**, 603–8.
Twigg, A. and D. Sherratt 1980. Transcomplementable copy-number mutants of plasmid ColE1. *Nature* (London) **283**, 216–18.

Ueda, Y., A. Toh-e and Y. Oshima 1975. Isolation and characterization of recessive, constitutive mutations for repressible acid phosphatase synthesis in *Saccharomyces cerevisiae*. *Journal of Bacteriology* **122**, 911–22.
Urdea, M. S., J. P. Merryweather, G. T. Mullenbach, D. Coit, U. Heberlain, P. Valenzuela and P. J. Barr 1983. Chemical synthesis of a gene for human epidermal growth factor urogastrone and its expression in yeast. *Proceedings of the National Academy of Sciences of the USA* **80**, 7461–5.

Valenzuela, P., D. Coit, M. A. Medina-Selby, C. H. Kuo, G. V. Nest, R. L. Burke, P. Bull, M. S. Urdea and P. V. Graves 1985. Antigen engineering in yeast: synthesis and assembly of hybrid hepatitis B surface antigen–herpes Simplex 1 gD particles. *Biotechnology* **3**, 323–6.

Walmsley, R. M., D. C. Gardener and S. G. Oliver 1983. Stability of a cloned gene in yeast grown in chemostat culture. *Molecular and General Genetics* **194**, 361–5.
Wampler, P. E., E. D. Lehman, J. Boger, W. J. McAleer and E. M. Scolnick 1985. Multiple forms of hepatitis B surface antigen produced in yeast. *Proceedings of the National Academy of Sciences of the USA* **82**, 6830–4.
Wood, C. R., M. A. Boss, J. M. Kerton, J. E. Calvert, N. A. Roberts and J. S. Emtage 1985. The synthesis and *in vivo* assembly of functional antibodies in yeast. *Nature* (London) **314**, 446–9.

Zakian, V. A. 1981. Origin of replication from *Xenopus laevis* mitochondrial DNA promotes high-frequency transformation of yeast. *Proceedings of the National Academy of Sciences of the USA* **78**, 3128–32.
Zakian, V. A. and D. M. Kupfer 1982. Replication and segregation of an unstable plasmid in yeast. *Plasmid* **8**, 15–28.

# PART V

# Downstream processing

# 14 *The isolation and purification of protein and peptide products*

ALAN WISEMAN, DAVID J. KING
and MICHAEL A. WINKLER

## 14.1 Introduction

Downstream processing is a very significant factor in the economics of any product from fermentation, and is especially important in the production of proteins and peptides. These products are often delicate enzymes which nevertheless require extensive purification. Almost all of the bacterial enzymes of commercial importance are secreted from the cell into the growth medium, but most enzymes located inside the cell are not secreted. The study of protein secretion in *Saccharomyces cerevisiae* has focused, until recently, primarily on invertase and acid phosphatase, which are secreted into the periplasmic space, hence facilitating their subsequent isolation and purification. Secretion of the important new genetically-engineered proteins from the yeast cell would obviously be desirable, to simplify their subsequent purification. Some progress has been made in this direction (see Section 14.7). It is not yet known whether all such proteins and peptides will be able to be secreted, and the rate of secretion may be too slow. Also, although the secreted product would be contaminated with fewer other proteins (in the medium), in some cases secretion may not be desirable. Secretion can introduce the need for a substantial concentration step to reduce processing volumes to those which can be used for intracellular proteins (probably by ultrafiltration or by a selective binding method). In addition, proteases are often excreted which could cause product loss, and the environment in the fermenter may itself cause denaturation of secreted proteins. If the desired products cannot be produced extracellularly or are then unstable, then they must be produced intracellularly, the cells being disrupted for product release. Cell breakage is often the stage of downstream processing at which most product loss occurs, so the disruption method should be chosen with care and investigated for efficacy. The methods available for the disruption of yeast cells can be broadly divided into the mechanical and the enzymic. They must all be considered in relation to product release rather than other criteria of 'cell breakage' involving total enzyme assay or microscopic appearance (Wiseman & Jones 1971).

## 14.2 Techniques of yeast disruption

### *14.2.1 Mechanical methods of yeast cell disruption – laboratory scale*

Yeast cells are relatively hard to disrupt compared with many other microbial cells, so a disruption method must be used which is vigorous enough to break the cells and yet gentle enough not to destroy the desired product. In the laboratory several methods of mechanical cell disruption are routinely used. These include agitation with glass beads and similar abrasives, solid-shear disruption of frozen cell pastes (freeze-pressing), liquid-shear methods in which a cell suspension is passed through a restricted orifice under pressure such as the French press, and sonication.

**Sonication**

Cell disruption by ultrasound is dependent upon the mechanical effects of gaseous cavitation in the suspending medium. The mechanical stresses which cause cell disruption by cavitation may arise from shock waves as well as from flows generated by pulsating bubbles (Doulah 1978). Cell disruption is largely independent of frequency, and most modern ultrasonic disruptors work at low frequency, i.e. 15–25 kHz (Hughes *et al.* 1971). James *et al.* (1972) found that the rate of release of protein from brewer's yeast was unaffected by cell concentration up to 60% suspensions [60 g (wet weight) yeast per 100 ml]. Protein release was inversely proportional to the volume of the treatment vessel (in the range 75–450 ml) and almost proportional to the input acoustic power. These workers also developed a flow system in which 60% of the protein was released at a flow rate of 40 ml $min^{-1}$, but they made no attempt to measure enzymatic activities. In general sonication is not widely used as a yeast-cell disruption method, as the power required to break the cells often results in enzyme denaturation. For example, yeast alcohol dehydrogenase (sonicated in purified form) was lost completely after sonication for 15 s (Wiseman *et al.* 1973). Also, sonication has not been applied to large-scale preparations because of the difficulties of transmitting enough power to large volumes of liquid. Neppiras and Hughes (1964) have developed a flow system for sonication of yeast cells, although they found poor reproducibility in the activity of alcohol dehydrogenase liberated.

**The French press**

Liquid-shear disruption methods break cells by passage through a restricted orifice under pressure. This is a very popular method of yeast-cell disruption for enzyme purification, and both small- and large-scale equipment is available. On a large scale the Manton–Gaulin homogenizer (APV Co. Ltd, Crawley, UK) is widely used [see Section 14.2.2(d)] and on a laboratory scale (up to 1–2 kg cell paste) the French press is widely used. Several models of the French press are available, and an automatic model is produced by Aminco (Silver Springs, Maryland, USA).

The French press consists of a steel cylinder with a small orifice and needle valve at the bottom. A piston fits into the cylinder with a pressure-tight washer, and pressure is applied to the piston with a hydraulic press. Before operation, the press is precooled to approximately 0°C and filled with the suspension of yeast cells. Care is taken to exclude air during operation, as compressed air may squirt out explosively at the end. Cell disruption is more effective at high cell concentrations. Hughes *et al.* (1971) have reported up to 70% release of total cell nitrogen from baker's yeast by a single passage through a French press at 20 000 lb $in^{-2}$ (140 MPa) with a cell concentration of 60–70 g (wet weight) per 100 ml. The choice of disruption buffer obviously depends upon the nature of the product, but enzyme stabilizers and protease inhibitors are often added. Also, it is often necessary to use more than one passage though the press.

Eisenstein *et al.* (1984) used a French press for disruption of *Saccharomyces cerevisiae* in their purification of ornithine transcarbamosylase (EC 2.1.3.3). These workers used 150 g (wet weight) of cells suspended in 450 ml ice-cold cell-breakage buffer (in this case 100 mmol $l^{-1}$ potassium phosphate buffer pH 7.0 containing 50 mmol $l^{-1}$ ornithine, 10 mmol $l^{-1}$ 2-mercaptoethanol, 2 mmol $l^{-1}$ EDTA, 2 μmol $l^{-1}$ leupeptin and 1 mmol $l^{-1}$ phenylmethylsulfonyl fluoride). Clumps of cells were dispersed by briefly mixing in a high-speed blender, and after filtering through cheesecloth the cells were disrupted by passage (twice) through a French press at 18 000–20 000 lb $in^{-2}$ (125–140 MPa). At least 75% cell disruption was attained with two passes through the French press. Another example of the use of a French press for yeast enzyme purification is in the purification of glucokinase (Maitra 1970). Glucokinase (ATP:D-glucose-6-phosphotransferase, EC 2.7.1.2) was purified from 65 g (wet weight) of *Saccharomyces cerevisiae* suspended in 55 ml 50 mmol $l^{-1}$ potassium phosphate buffer, pH 7.4, containing 2 mmol $l^{-1}$ EDTA and 2 mmol $l^{-1}$ 2-mercaptoethanol. This suspension was then passed through a French press to release the enzyme for subsequent purification.

**Freeze-pressing**

Freeze-pressing is a disruption method in which the sample is cooled to approximately −20°C or less, then forced through a small orifice under high pressure. Disruption occurs through solid shear by compression of ice crystals with the cells. The Hughes' press and the X-press both work on this principle (Hughes *et al.* 1971). The X-press has been used for the disruption of *Saccharomyces cerevisiae* by Magnusson and Edebo (1976a, b). The X-press is precooled in an alcohol bath to −25°C for at least 1 h. The suspension of cells is precooled into the press, and when the sample is frozen the X-press is returned to the alcohol bath for another 2 h before freeze-pressing. The frozen material is then extruded through the orifice (1–2 mm in diameter) with a piston propelled by an hydraulic pump. The pressure required for flow through the orifice is related to the temperature and to the phase boundaries of water. Magnusson and Edebo (1976a) found that the presence of salts and gelatin in the suspending medium had a marked effect on the extent of disruption of the yeast cells. Gelatin helped the frozen cell suspension to flow smoothly, and this promoted disruption. The same workers also found that if a frozen cell paste of baker's yeast of 270 mg $g^{-1}$ (rather than the dilute suspension used before of 5.4 mg $g^{-1}$) was used the extent of disruption was greatly increased (Magnusson & Edebo 1976b). The disruption was improved further by using a lower temperature (−35°C) and a high velocity of flow through the orifice. Under these conditions more than 95% of the yeast was disrupted in one passage through the X-press. This method, although time consuming, may be particularly good for the isolation of highly heat-labile enzymes, as disruption occurs at very low temperatures.

The standard X-press allows the disruption of approximately 25 ml cell paste in each run. However, a larger-scale version of the X-press has also been developed which allows the disruption of up to 10 kg yeast cells $h^{-1}$ (Magnusson & Edebo 1976c). In this version the material to be disrupted is prefrozen in cylindrical rods which fit into the pressure chamber. These rods are then freeze-pressed at −20°C, resulting in approximately 90% disruption.

**Agitation with glass beads**

The most widely used mechanical method for yeast breakage is rapid agitation with glass beads, or similar abrasives (sometimes called milling techniques). In a simple form glass beads can be agitated with yeast cells in a laboratory blender (Kopetzki & Entian 1982), although this leads to heating of the sample as a result of the agitation, and this may lead to extensive product loss. This problem is largely overcome by the use of a commercially available bead-mill with a cooling system. There are several of these available, but the most widely used are the Braun shaker (Braun Melsungen, West Germany) and the Vibro-mill (E. Buhler, Tübingen, West Germany) for small-scale work and the Dyno-mill for larger-scale work [see Section 14.2.2(e)].

The Braun shaker consists of a chamber for the disruption flask which is shaken in an elliptical pattern at 4000 rev $min^{-1}$ and cooled by liquid $CO_2$. For yeast cell disruption Shetty and Kinsella (1978) used equal volumes of a cell suspension (5% in water) and glass beads (0.45–0.50 mm in diameter). These were placed in the disruption flask and shaken for 3 min while liquid $CO_2$ was passed through the shaker for cooling. After 3 min 100% disruption of the cells was achieved. These workers have used this same method of disruption in their purification of ribonuclease (Shetty *et al.* 1980).

Needleman and Tzagoloff (1975) developed an adaptor for the Braun shaker which allows the processing of 16 small samples at once. The yeast samples were suspended in 20 mmol $l^{-1}$ Tris–acetate buffer, pH 7.5, containing 200 mmol $l^{-1}$ mannitol and 1 mmol $l^{-1}$ EDTA: 0.4 ml of cell suspension ($7.4 \times 10^8$ cells $ml^{-1}$) was used for each sample, which was shaken with 0.45-mm diameter glass beads. After 3 min of shaking over 90% of the cells were broken with no appreciable heating of the samples. The percentage breakage of cells was independent of cell density from $3.6 \times 10^8$ to $3.7 \times 10^9$ cells $ml^{-1}$. Log-phase cells were easier to disrupt than stationary-phase cells, probably due to cell-wall thickening in stationary phase. This procedure could enable the disruption of several hundred samples per hour, and the cell-free preparations obtained contained active ATPase, cytochrome oxidase and NADH–cytochrome c reductase.

The Braun shaker has been used for many yeast enzyme preparations, for example in the purification of glycerol-3-phosphate dehydrogenase (EC 1.1.1.8) from *Saccharomyces cerevisiae* (Merkel *et al.* 1982), where a cell suspension in 10 mmol $l^{-1}$ Tris–HCl buffer, pH 7.2, containing 10 mmol $l^{-1}$

EDTA and 2 mmol $l^{-1}$ 2-mercaptoethanol, was disrupted with 0.45–0.50-mm glass beads.

The Vibro-mill disruptor works by shaking glass beads with a yeast-cell suspension or paste, and is cooled by cold water circulating through a water jacket. Continuous-flow heads are available for the Vibro-mill disruptor. Welch and Scopes (1981) used a flow head on a Vibro-mill for the disruption of *Saccharomyces cerevisiae* during their purification of phosphofructokinase (EC 2.7.1.11). These workers used 300 g (wet weight) of yeast dispersed in 600 ml cold buffer (10 mmol $l^{-1}$ potassium phosphate, pH 7.0, containing 1 mmol $l^{-1}$ EDTA, 5 mmol $l^{-1}$ 2-mercaptoethanol and 1 μmol $l^{-1}$ pepstatin A). Phenylmethylsulphonylfluoride (50 mg in 1 ml acetone) was then added and the ice-cold slurry pumped through the Vibro-mill, producing a homogenate at 20 ml $min^{-1}$.

Extracellular invertase ($\beta$-fructofuranosidase, EC 3.2.1.26) was easily released by Vibro-mill disruption of *Saccharomyces cerevisiae* (Williams and Wiseman 1973). All of the invertase activity was recovered in the cell-free fraction after 10 min, after which time 100% of the cells were disrupted. The Vibro-mill has been used successfully for the disruption of *Saccharomyces cerevisiae* in the purification of cytochrome P-450 (Azari & Wiseman 1982, King *et al.* 1984). The steel Vibro-mill vessel was filled with approximately one-third yeast cell paste and two-thirds glass beads (1.00–1.05 mm in diameter) and milled for a total of 6 min, with cooling by a circulating cold-water system. An improved enzyme yield was obtained by using the Vibro-mill in 1-min bursts with 1-min intervals for cooling. This reduced the heating of the sample and increased the enzyme yield by approximately 30% (Azari *et al.*, unpubl. results).

Recently, human interferon-$\gamma$ expressed in *Saccharomyces cerevisiae* after transformation with a plasmid containing its cDNA, has been isolated from yeast by disruption with glass beads, as well as by osmotic shock of yeast spheroplasts (see Section 14.3; Derynck *et al.* 1983). Cells from 5-ml cultures were pelleted, washed and resuspended in PBS (20 mmol $l^{-1}$ $NaH_2PO_4$, pH 7.4, 150 mmol $l^{-1}$ NaCl) containing 1% bovine serum albumin. The cells were disrupted by a 2-min shaking with 0.3 ml of glass beads (0.45–0.50 mm in diameter) at 4°C. The level of interferon-$\gamma$ activity was dependent upon the disruption procedure. The milder method of lysing spheroplasts by osmotic shock gave three- to fivefold higher interferon levels than the mechanical method, but since the mechanical method was much easier and less time-consuming, it was used for routine analysis.

### *14.2.2 Mechanical methods of yeast disruption – pilot-scale and large-scale*

In considering the release of protein products from yeasts by disruption, most attention has been given to enzyme recovery, as is evident from Section 14.2. Apart from their intrinsic value as useful products, enzyme activities provide a convenient indicator for assessing protein denaturation in the disruption process. Where the intracellular location of particular enzymes is known, their release during disruption can, in principle, provide clues as to the detailed mechanisms involved in disruption (Follows *et al.* 1971, Dunnill & Lilly 1972, Wang *et al.* 1979).

The following enzymes have been released from yeasts by disruption on the pilot scale or larger scale:

- acid and alkaline phosphatases (EC 3.1.3.1 and 3.1.3.2), invertase (EC 3.2.1.26), alcohol dehydrogenase (EC 1.1.1.1), fumarase (EC 4.2.1.2), glucose-6-phosphate dehydrogenase (EC 1.1.1.49) and 6-phosphogluconate dehydrogenase (EC 1.1.1.44) from baker's yeast (Follows *et al.* 1971) and principally from baker's yeast (Mogren *et al.* 1974)
- maltase (EC 3.2.1.20), acid phosphatase (EC 3.1.3.1), malate dehydrogenase (EC 1.1.1.37), glucose-6-phosphate dehydrogenase (EC 1.1.1.49) and 6-phosphogluconate dehydrogenase (EC 1.1.1.44) from *Saccharomyces carlsbergensis* (Marffy & Kula 1974)
- Alcohol dehydrogenase (EC 1.1.1.1) from baker's yeast (Limon-Lason *et al.* 1979)
- Cytochrome P-450 from *Saccharomyces cerevisiae* (brewer's yeast) (Sadler *et al.* 1983, 1985, Winkler & Wiseman 1984)
- glucose-6-phosphate dehydrogenase from alkane-grown *Candida lipolytica* (Whitworth 1974).

Large-scale disruption of food yeasts is also of interest as a stage in separating yeast protein from nucleic acids to reduce its toxicity (Lawford *et al.* 1979, Gierhart & Potter 1979) and from cell-wall material to improve digestibility (Lee *et al.* 1979, Whitworth 1974, Machek *et al.* 1976).

The recovery of a mitochondrial fraction from yeast has been reported (Deters *et al.* 1976, Mason *et al.* 1973).

**Susceptibility to disruption**

The susceptibility of yeast in general to disruption by different methods has been rated (Edebo 1969) as less than that of animal cells, Gram-negative bacilli and cocci and Gram-positive bacilli, greater than that of spores and similar to that of Gram-positive cocci. Charm and Matteo (1971) noted that the breakage of cells such as micrococci, streptococci, yeast and lactobacilli is more difficult than the easily ruptured Gram-negative bacteria. However, the susceptibility of a particular yeast to disruption depends upon the conditions under which it was grown, as well as its species.

In comparative disruption trials, fresh baker's yeast, *Saccharomyces cerevisiae*, and fresh brewer's yeast, *Saccharomyces cerevisiae*, showed similar behaviour (Rehacek & Schaefer 1977), as did fresh baker's yeast, dried baker's yeast and fresh brewer's yeast, *Saccharomyces carlsbergensis* (Mogren *et al.* 1974). In both studies *Candida utilis*, frozen (Mogren) and grown on sulphite waste liquor (Rehacek) was distinctly more resistant to disruption. The frozen *Candida utilis* in particular showed only 40% disintegration, compared with 85% for brewer's and baker's yeast under the same conditions. *Candida utilis* has a susceptibility to disruption which increases with increasing growth rate, attributed to concomitant decreasing cell age (Engler & Robinson 1981) or increasing cell-size (Machek *et al.* 1976). In addition, *Candida utilis* grown in continuous culture with ethanol as the principal substrate showed increased resistance to disruption when grown under conditions of nitrogen limitation (Machek *et al.* 1976). Few disruption procedures proved successful for yeasts grown on alkanes, and, in particular, agitation with glass beads proved unsuitable with *Candida lipolytica* (Whitworth 1974).

**Criteria of cell disruption**

As pointed out in Section 14.1, from the process point of view the most useful criterion of cell-disruption efficiency is the release of the desired intracellular product. However, from the investigative point of view it can be useful to distinguish between cell breakage and product release, particularly in assessing the extent of protein denaturation occurring during disruption. Most of the investigators whose work is discussed here have followed cell disruption by means of release of total protein or of particular enzymes. Some assessed cell breakage microscopically (Mogren *et al.* 1974, Rehacek & Schaefer 1977, Deters *et al.* 1976), and Magnusson and Edebo (1974, 1976a, b, c) assessed the number of intact cells remaining after disruption, using a Coulter counter. Lee *et al.* (1979) correlated visual observations of cell breakage with a scanning electron microscope with protein release, determined as total nitrogen. Hughes *et al.* (1971) consider the Coulter counter method unreliable in the presence of cell envelopes or other large particles, and regarding microscopic counting, point out that damaged yeast cells may take up basic stains more intensely than undamaged cells. Evaluating cell disintegration by means of protein release involves certain problems. Estimation of the protoplasmic solids released requires a correction for the corresponding increase in the aqueous-phase fraction (Engler & Robinson 1979) and may give variable and inconclusive results with extensively disrupted samples. Dilution of samples which may well contain partially denatured macromolecules can cause changes in solubility, resulting in inconsistencies in comparative determinations of soluble protein. Engler and Robinson suggest that the Kjeldahl method for nitrogen determination is therefore preferable for protein determination, as it is less sensitive to solubility changes. Yeast biomass subjected to a prolonged disruption process may release protein or peptide material associated with the cell wall (Mosqueira *et al.* 1981), which may distort the released soluble-protein determination. A problem in assessing the efficiency of enzyme release is in determining the true, pre-disruption level of the enzyme in the intact cell, and thus the true fraction of enzyme released by disruption. This can lead to misleading yield values in subsequent operations (Darbyshire 1981). An enzyme which may be of value in this respect is the subject of investigation by the present authors. Cytochrome P-450 of particular strains of brewer's yeast, *Saccharomyces cerevisiae*, is unusual for an intracellular enzyme, since it can be assayed directly in the intact cell (Sadler *et al.* 1985), so that enzyme yields after disruption can be determined rapidly and accurately. For example, preliminary results from disruption experiments suggest that, although more-severe disruption conditions release higher proportions of enzyme, they also cause more extensive denaturation, so that the net enzyme yield appears to be almost independent of the disruption conditions (Winkler & Wiseman 1984).

**Choice of process**

The choice of disruption procedure should ideally be made only after careful evaluation of the available methods, as to their suitability to the organism and the product to be released from it, as discussed in Section 14.1. However, some disruption processes are unsuitable for scaling-up. As for large-scale operation, the choice is restricted by the need to deal rapidly with large quantities of yeast biomass while effecting the extent of disruption found to be necessary on the laboratory scale. In practice the equipment used is likely to be chosen on the 'Mount Everest' principle – 'because it is there' – so that availability may override ideality. In a summary of 24 large-scale purification procedures (Scawen *et al.* 1980), a liquid shear method was used in 13 cases with the other available methods fairly evenly distributed, and while it is an effective technique guaranteeing some degree of cell breakage, its selection may well have been influenced by availability (Darbyshire 1981). Such situations are not unusual in the biological industries where, for example, the design of fermenter selected for a process is not necessarily optimal, but is chosen for reliability and predictability of performance (Winkler 1983). While in general it is necessary to effect rapid processing at low temperatures, intracellular enzymes have the advantage that they are usually stable provided the integrity of the parent cell is maintained, and in many cases can be stored in the cell mass for 2 years or more at −20°C (Darbyshire 1981). However, some enzymes are cold-labile (Bucke 1983). Prompt cell disruption has the advantage that it reduces the bulk of biomass solids, as lysed biomass solids occupy a much smaller volume than the same quantity of whole cells, as much as 60–80% reduction in some cases (Atkinson & Mavituna 1983).

Much of the fundamental investigation of pilot-scale processes for disruption of micro-organisms has been carried out using yeasts as test organisms. This is partly because yeasts are recognized as being more resistant to disruption than most microbial species, but also because baker's yeast and spent brewer's yeast are cheap, readily available and consistent in their properties. There is thus a considerable body of quantitative information in respect of yeast disruption.

Reviews of microbial disruption processes have been presented by Hughes *et al.* (1971), Charm and Matteo (1971), Darbyshire (1981), Scawen *et al.* (1980), Melling and Phillips (1975a, b), Bucke (1983) and Wang *et al.* (1979).

For yeast disruption the two systems that have received most attention are liquid shear (homogenization) and agitation with solid particles (bead milling). Other procedures with reported or potential large-scale use are discussed briefly in Section 14.2.2(f).

**Disruption by liquid shear**

Liquid shear, usually in a Manton–Gaulin homogenizer, has proved one of the most widely used cell-disruption methods (Scawen *et al.* 1980). The Manton–Gaulin homogenizer comprises a high-pressure, positive-displacement pump with an adjustable, restricted-orifice discharge valve. The cell suspension is drawn through a check valve into the pump cylinder, and on the pressure stroke is forced through the discharge-valve assembly. During discharge the cell suspension is forced between the valve and its seat, and impinges on an impact ring, whence it passes through an annular channel between the valve and its housing to the discharge port. Two types of valve seat have been used, a flat seat of tungsten carbide (known as a 'WC seat') and a 'knife-edge' seat made of stellite. The discharge-valve clearance, and thus the operating pressure, is controlled by a hand-wheel screw assembly through a spring-loaded valve rod. Pressures up to 60 MPa (600 bars) are easily obtainable, and the unit works well with yeasts and some bacteria, but not filamentous organisms. Its main problem is heat generation, and the feed and the output must be cooled. The suspension feed should be homogeneous, without frozen lumps or solid foreign bodies.

Hetherington *et al.* (1971) measured the protein release from baker' yeast with two different versions of the Manton–Gaulin homogenizer, a single-cylinder model, capacity 54 l $h^{-1}$, and a triple-cylinder model, capacity 280 l $h^{-1}$, both with a nominal pressure range of 0–56 MPa. The heating effect was a 4°C temperature rise in the single-cylinder model and 8°C in the triple-cylinder model. Under the same operating conditions there was no significant difference between the performances of the two models. Protein release was found to be markedly temperature dependent and pressure dependent, but independent of the yeast concentration in the slurry. For example, at 30°C and with a pressure of 55 MPa, 62% of the water-soluble protein was released in a single pass. Between 5 and 30°C protein release was found to follow a first-order kinetic relationship of the form

$$\ln\left(\frac{R_m}{R_m - R}\right) = KNP^{2.9} \tag{14.1}$$

where $R_m$ is the maximum protein available for release, $R$ is the actual protein release, $N$ is the equivalent number of discrete passes through the homogenizer, $P$ is the working pressure and $K$ is the rate constant, found to be a function of temperature. The high exponent of pressure indicates the importance of using as high a pressure as possible, and protein release rate increased $1\frac{1}{2}$ times when the temperature was increased from 5°C to 30°C. Above 40°C denaturation occurred. At higher pressures, above about 30 MPa, and with a high yeast concentration (0.75 g $ml^{-1}$), the disruption rate was reduced and no longer followed the first-order rate equation. While relationships such as Equation 14.1 may be used to describe protein release under carefully defined conditions, in practice appropriate operating conditions must be determined empirically (Melling & Phillips 1975a). The release of seven different enzymes (Section 14.3; Follows *et al.* 1971) in the process was also studied. For each enzyme the rate of release correlated with the rate of protein release from yeast disrupted at several concentrations, pressures and temperatures. Invertase and acid phosphatase were released faster than the overall protein release rate, alkaline phosphatase and fumarase more slowly, and alcohol dehydrogenase, glucose-6-phosphate dehydrogenase and 6-phosphogluconate dehydrogenase at about the same rate, in accordance with their reported intracellular location. However, these differences were insufficient for enzyme fractionation, and the rates of enzyme release, relative to protein release, were independent of the operating conditions.

The first-order relationship was followed in later work on the disruption of baker's yeast (Mosqueira *et al.* 1981), and the ratio of the output volumetric solids content to the initial volumetric solids content of the yeast suspension, usually between 0.49 and 0.64, was used as an indicator of homogenizer wear. The temperature rise was counteracted by passing the outflow through a heat-exchanger cooled with fluid at −10°C.

Disruption of alkane-grown *Candida lipolytica* using a Manton–Gaulin homogenizer was reported as satisfactory, where agitation with glass beads had not been (Whitworth 1974). An enzyme, glucose-6-phosphate dehydrogenase, was present at high levels, and was released rapidly and progressively by homogenization. However, protein release did not follow a first-order relationship, which was attributed to the yeast being a heterogeneous mixture of filamentous and elongated ovoid cells.

Homogenization also greatly increased the extractability of protein from dried *Candida lipolytica* (BP) grown on hydrocarbon, from 20 to 93% in alkali (Lee *et al.* 1979).

High-quality undenatured protein was isolated from *Candida utilis*, grown on molasses to give yeast biomass containing 50–55% protein, homogenized as a 10% (dry weight) cream at about 60 MPa (Lawford *et al.* 1979). A Manton–Gaulin homogenizer was also used in disrupting brewer's yeast, *Saccharomyces cerevisiae*, at 60 MPa, in the preparation of a microsomal fraction containing cytochrome P-450 (Sadler *et al.* 1983, 1985). Particular enzymes in eukaryotic cells are bound to the endoplasmic reticulum, which, during disruption is pinched-off to form small vesicles, commonly known as 'microsomes', or after high-speed centrifugation, as a 'microsomal fraction'. Cytochrome P-450 in yeast is such an enzyme (Winkler *et al.* 1985) and, as it can be assayed directly in the intact cell, it would appear to be a suitable indicator for investigating enzyme yields on disruption (see Section 14.3.2). However, preliminary results show no correlation between enzyme yield and homogenization pressure, suggesting that, as cytochrome P-450 is highly susceptible to denaturation, increasing enzyme release with increasing pressure is counteracted by increasing enzyme denaturation (Winkler & Wiseman 1984).

Mason *et al.* (1973) recovered a mitochondrial fraction from yeast using a Manton–Gaulin homogenizer.

Other investigations of yeast-cell disruption in high-shear systems have provided some insight as to the mechanisms of disruption in high-shear processes. Engler and Robinson (1981) disrupted *Candida utilis* suspensions in a high-velocity jet impinging on to a static plate. This enabled the effects of shear, normal and impingement stresses to be evaluated independently. They showed that impingement was the most important effect, with normal stress calculated as only 20% as effective in disruption as impingement, and calculated shear stresses three orders of magnitude lower than impingement stresses for the same operating pressure. In addition, although a first-order relationship was followed at relatively low pressures, the exponent of pressure (2.9 in Equation 14.1) was much lower and also depended on the yeast growth rate, and therefore on average cell age: 1.77 for 10-h-old yeast, 1.17 for 2-h-old yeast.

Brookman and Davis (1973) and Brookman (1975) investigated the disruption of baker's yeast in suspensions forced through a needle valve and through a fine-bore tube at pressures up to 5000 atm (500 MPa). Protein release increased logarithmically with increasing pressure up to the maximum available for release, but with the microbore tube attachment protein release decreased with increasing tube length. From this, Brookman concluded that yeast disruption was probably not due to shear, and suggested that a pressure-shock mechanism was the cause, with the magnitude and rate of pressure drop as the suspension left the tube the key factors. This accorded with the disruption of baker's yeast in a shock-tube disruptor (Edwards & Wiseman 1971). Doulah *et al.* (1975) developed a mathematical model for yeast disruption in a homogenizer, by analogy with the break-up of bubbles, which suggested that turbulence is the controlling factor in disruption. From Brookman's results, Scully (1975) suggested that the breakage of yeast cells was due to cell-wall expansion during rapid decompression, when compressive strain converts to tensile strain. However, the intracellular water cannot sustain tensile strain, and its compressive strain converts into kinetic energy, causing an impulsive blow to the cell wall and further increasing its tensile strain.

**Agitation with solid particles**

Organisms suspended in a liquid medium and agitated with solid particles, such as glass beads, are disrupted by grinding between beads, collisions between beads and organisms and from shear forces resulting from liquid velocity gradients generated by movement of the beads. The method has the advantage of disrupting notoriously resistant cells for which other large-scale methods are unsuited (Darbyshire 1981), and is capable of dealing with filamentous mycelia which 'blind' an homogenizer (Dunnill & Lilly 1972). It has the disadvantage of requiring the separation of the particles from the disrupted biomass, and of material adhering to the particles (Charm & Matteo 1971). As with homogenization, heat generation is a considerable problem and some means of cooling the process is usually necessary. Disruption efficiency depends upon at least five parameters: agitator speed, proportion of beads, bead size, cell-suspension flow rate and cell concentration.

A widely used machine is the Bachofen 'Dyno-mill', in which discs or impellers are mounted on a horizontal axis in a water-jacketed cylinder partly filled with glass beads. Mogren *et al.* (1974) used a 'Dyno-mill' with five impellers rotating generally at 2340 rev $min^{-1}$ to release enzymes from fresh and dried baker's yeast, fresh *Saccharomyces carlsbergensis*, frozen *Candida utilis* and lyophilized green algae, *Scenedesmus obliquus*. Baker's yeast was used for most of the experiments. With a 5.8-kg charge of beads of diameter 0.5–0.75 mm, leaving a net volume of 2.2 litres, yeast disintegration was found to follow a first-order relationship:

$$\ln\left(\frac{1}{1 - D}\right) = k/f \qquad (14.2)$$

where $D$ is the fractional cell disruption, $f$ is the flow rate of cell suspension through the mill and $k$ is a rate constant. The mean residence time of cells in the mill is inversely proportional to the flow rate $f$, and at 200 l $h^{-1}$, 85% disruption of baker's yeast was obtained with a single pass through the mill. The degree of disintegration was determined by microscopic counting before and after disintegration. There was a considerable rise in temperature in the suspension passing through the mill, being 10°C at a flow rate of 300 l $h^{-1}$, but as much as 50°C at 100 l $h^{-1}$, representing about 70% of the electrical energy supplied.

Marffy and Kula (1974) recovered enzymes from brewer's yeast (*Saccharomyces carlsbergensis*) using a 'Dyno-mill', and also found that enzyme release followed first-order kinetics approximately, with the mill operating in a continuous, single-pass mode. Agitator speed was varied from 2000 to 6000 rev $min^{-1}$, and beads in three different size-ranges were used, 0.1–0.25, 0.25–0.5 and 0.75–1 mm. The first-order rate constant was found to increase with agitator speed and cell density, and decrease with increasing flow rate, and the middle range of bead sizes was markedly better than the other two.

A comprehensive investigation of the performance of a Bachofen mill was carried out by Limon-Lason *et al.* (1979), comparing batch and continuous disruption of baker's yeast in a laboratory-scale, 600-ml, four-impeller mill and a pilot-scale, 5-litre, five-impeller mill, using beads of diameter 0.42–0.5 mm. Variables investigated were impeller tip speed, temperature and yeast concentration in the range 15–75% by weight of packed yeast. Disruption approximately followed first-order kinetics, but with the rate constant a function of impeller tip speed and yeast concentration. Temperature rise was a problem in the 5-litre mill, and possibly explained the deviation from true first-order kinetics. In continuous operation, the mills behaved as a series of continuous stirred-tank reactors, with the impeller design affecting the extent of back-mixing in the system and the consequent effect on disruption efficiency. The effect of yeast concentration was attributed to the change in rheology as the Newtonian whole-yeast suspension was converted into the non-Newtonian pseudoplastic disruptate. It was also suggested that the assessment of protein release was confused by the release of complexed protein from the break-up of cell debris, as with Mosqueira *et al.* (1981) [Section 14.2.2(b)].

Deters *et al.* (1976) investigated the breakage of *Saccharomyces cerevisiae* using a 'Dyno-mill' in the separation of mitochondria. With a continuous flow of 2.5 kg yeast $h^{-1}$ and 80% cell breakage (monitored by microscopic examination) the mitochondrial yield was 5–7 g protein $kg^{-1}$ pressed yeast, about five- to eight-times higher than with any other large-scale method. The size of beads used was found to be important, as small beads broke 30–50% fewer cells and sub-mitochondrial particles were heavily contaminated by other intracellular material. An optimal impeller speed of 3000 rev $min^{-1}$ was determined. Higher speeds raised the temperature, which damaged the mitochondria and reduced the yield, and lower speeds gave poorer cell breakage.

The Netzsch mill, comprising a horizontal shaft on which circular impellers are mounted normally and alternately with obliquely mounted ellipsoidal impellers, was used by Rehacek and Schaefer (1977) to investigate the disruption of fresh baker's yeast, fresh brewer's yeast (*Saccharomyces cerevisiae*) and *Candida utilis* grown on sulphite waste liquor. The degree of disintegration was determined microscopically. The mill was cooled by means of water circulating through a cooling jacket and through the hollow shaft and impellers. Horizontal mills had previously been shown as more efficient for disruption than vertical mills (Rehacek *et al.* 1969). In this case the angle at which the oblique impellers were mounted was significant at low speeds; it became progressively less important as the speed increased. The degree of disintegration fell with increasing flow rate of yeast suspension, and increased as the bead charge was increased, as did the temperature rise and the power consumption. All of the yeasts showed high degrees of disintegration at low flow rates of cell suspension, but that of *Candida utilis* fell more rapidly with increasing flow rate than those of the other yeasts. With a residence time of 2.5 min a 15% suspension of cells was 90% disrupted with *Saccharomyces cerevisiae* and 85% with *Candida utilis*. Currie *et al.* (1972) investigated the disruption of baker's yeast with a vertical Netzsch–Molinex variable-speed agitator mill, which contained a central shaft on which circular impellers were mounted eccentrically to form a helical array. The effects of six variables were studied: temperature, agitator speed, bead size, weight of beads used, upward flow rate of yeast suspension and yeast concentration. Protein release was found to follow a first-order relationship, and the rate constant was dependent upon all six parameters investigated. Increased disruption efficiency was obtained with higher agitator speeds, greater bead loading and reduced upward yeast recycle rates. Disruption efficiency was reduced by about 20% when the temperature was raised from 5 to 42°C. The first-order disruption rate constant increased rapidly with increasing agitator speed and decreased almost linearly with increasing bead size and yeast slurry flow rate, and increased rapidly as the weight of beads charged is increased. With suitable operating conditions, the mill could effect 90% disruption to produce 5.32 kg $h^{-1}$ soluble protein.

**Other disruption methods**

While several of the disruption procedures commonly used in the laboratory are inherently difficult to scale-up, there are some which would appear to be worthy of development as large-scale processes. For example, enzymic lysis of cell walls, using snail-gut enzyme from *Helix pomatia*, has been successful in producing protoplasts from *Saccharomyces cerevisiae* and *Saccharomyces carlsbergensis* and intact mitochondria from *Candida albicans* (Hughes *et al.* 1971; Section 14.2.3). Invertase has been released from yeast by digestion with papain, after pre-treatment with warm toluene (Darbyshire 1981). The principal disadvantage of enzymic lysis is that the cost is prohibitive for large-scale operations, as well as the problem of contamination of the disruptate with the enzyme. The use of immobilized enzymes had the problem of steric restriction imposed by the insoluble supports (Lilly & Dunnill 1972). However, the use of soluble polymer supports for lytic enzymes is now practicable as a result of the availability of microporous filters which retain the enzyme–polymer complexes while allowing products of lysis to be removed (Dunnill 1983). Dunnill also pointed to the development of temperature-sensitive microbial mutants which lyse above a particular temperature. This would require careful engineering in an industrial-scale fermenter, where temperature changes take some time to achieve, so as to avoid adventitious premature lysis. A more traditional method which is easily adapted to large-scale operation is autolysis, where a cell suspension is maintained at 23–37°C for several hours. The cell extract is collected by centrifugation. This method has the disadvantage that the required protein may undergo thermal inactivation or be destroyed by cellular proteases, but, for example, aldolase has been obtained from *Candida utilis*.

UDP-Gal-4-epimerase from lyophilized *Candida pseudotropicalis* and tripeptide synthetase from air-dried baker's yeast (Charm & Matteo 1971). The problem in scaling-up sonication [Section 14.2.1(a)] for use in large-scale disruption is the necessity of removing the heat generated by the high energy input, a familiar limitation in large-scale disruption processes.

### *14.2.3 Enzymic methods of yeast disruption*

**Introduction**

The ease and convenience of dissolving the tough cell wall of yeast rapidly with suitable enzymes would present an attractive alternative to mechanical methods of disruption. This is, indeed, a useful method in small-scale work, but problems exist on the large scale. Suitable enzyme preparations are currently rather expensive, but no doubt they could be made available at a cost-effective price if there was a large enough demand. Recovery of such enzymes after use would greatly improve the cost position, and attempts have been made to use soluble immobilized enzymes. Nevertheless, some problems of steric inaccessibility still remain, even with soluble immobilized enzymes and the yeast cell wall. Another problem is that the polymer-bound immobilized enzyme must be retained on a suitable microporous filter. This itself must not impede the dissolution of the cell wall by entrapment of high-molecular weight products released during lysis (for a general review, see Dunnill 1983).

There is considerable current interest in the use of enzymes to lyse yeast, in some cases via the preparation of the yeast protoplast, which is easily disrupted by a variety of gentle procedures. Controlled autolysis is another important method.

**Uses of enzymes to degrade the yeast cell wall**

Snail-gut enzyme preparations have been used for many years to produce small quantities of yeast protoplasts (for a general review of earlier literature, see Wiseman 1969). A highly concentrated and complex mixture of enzymes is employed to dissolve the yeast cell wall in about 60 min at 35°C. However, large-scale work requires the use of enzymes from micro-organisms which are capable of being produced in quantity (see Knorr *et al.* 1979).

Many enzymes have been described from micro-organisms that can dissolve the cell wall of living yeast. Some of these enzymes have been patented. For example, such extracellular enzymes have been isolated from strains of *Coprinus*, namely *Coprinus macrorrhizus* var. *microsporus*, ATCC 20120, *Coprinus radians*, ATCC 20014 and *Coprinus micaceus*, ATCC 20122 (Kawai & Mukai 1972). The lytic enzyme accumulates in the culture media during aerobic submerged culture at 20–35°C for up to 5 days. The lytic enzyme in phosphate buffer at pH 7.3 was reported to effect the lysis of *Saccharomyces cerevisiae*, which had previously been heated for 15 min at 80°C. An enzyme preparation from *Rhizoctonia* (probably *Rhizoctonia solani*) has been described which increases the release of cell-wall invertase in subsequent use of the Dyno-mill (Kobayashi *et al.* 1982).

Some success has been claimed with lytic enzymes from *Bacillus* spp., including a mixed culture of *Bacillus circulans* and *Phaffia rhodozyma*

(Okagbue & Lewis 1983). The earlier patent literature claims the isolation of lytic enzymes from *Bacillus circulans*, ATCC 21367, *Bacillus coagulans*, ATCC 21366 and *Bacillus stearothermophilus*, ATCC 21365 (Suekane & Satake 1972). These enzymes remove yeast cell wall.

An important source of bacterial enzyme for lysis of yeast is found in the *Cytophaga* bacteria. Such enzymes have been patented (e.g. Craveri *et al.* 1976). Culture fluids of these bacteria can be used directly to lyse live yeast biomass without previous treatment. These bacteria can also be grown in mixed culture with the yeast to achieve lysis *in situ*, if preferable. Two important papers have been published on these enzymes, involving the growth of *Cytophaga* sp. in batch culture (Assenjo *et al.* 1981) and the isolation of yeast lytic enzymes (Assenjo & Dunnill 1981).

**Controlled autolysis of yeast**

Yeast can be autolysed quickly with sodium chloride, although most enzymes are likely to be lost. Plasmolysis and autolysis can be readily achieved with non-polar organic solvents such as toluene, chloroform, ether, ethyl acetate or amyl acetate. Some enzymes survive this treatment. Many examples of such autolysis with chemicals can be found in the early literature, and more recently some interest is again seen.

For example, an important paper by D. M. Fenton describes the use of solvents to release 90% of $\beta$-galactosidase from yeast on a large scale (Fenton 1982). The $\beta$-galactosidase (lactase) isolation is from a mutant (97.1) of *Kluyveromyces fragilis*. It was grown in a 14-litre fermenter containing 3% lactose, phosphate buffer, magnesium sulphate and yeast extract, at 30°C and pH 4.5 for 60 h. Various solvents were used at high concentration (e.g. 80%, calculated as weight of solvent divided by weight of solvent plus dry weight of yeast). The contact time was 90 min at temperatures between 20°C and 30°C. A slow extraction of $\beta$-galactosidase into 0.1 mol $l^{-1}$ potassium phosphate buffer, pH 6.6, was then needed. This required the use of a gyratory shaker for 15 h using 25 ml buffer $g^{-1}$ yeast in the continued presence of the solvent. The most effective solvents were reported to be ethanol, methanol, *t*-butanol and isopropanol. The solvent prevents losses of enzyme due to microbial degradation, and seems also to prevent yeast proteases from degrading the required enzyme during the subsequent slow extraction with the buffer. The author reports the successful scale-up of this process to pilot-plant volume of 300 gallons (1350 litres). The mechanism of enzyme release is not fully understood in this solvent 'permeabilization' technique, as little or no lysis occurs, as shown by scanning electron microscopy.

**Use of spontaneous cell rupture of yeast**

There is considerable current interest in biological mechanisms of cell lysis, for example in the use of temperature-sensitive mutants or inducible phages. For *Saccharomyces cerevisiae*, important papers have appeared on spontaneous cell rupture (Al-Shahwani *et al.* 1978, 1982). Release of several enzymes has been reported by these workers, including invertase, $\beta$-1,3-glucanase, acid phosphatase, chitin synthetase and alkaline phosphatase. In a related manner, the industrial interest in this phenomenon has been discussed (Boudrant *et al.*

1979) with reference to the process characteristics of such cell lysis mutants of *Saccharomyces cerevisiae*.

## 14.3 Isolation of intracellular enzymes

### *14.3.1 Introduction to general theory of large-scale isolation and purification*

Well-established laboratory-scale procedures have been fairly successfully scaled up to cope with the large volumes of materials to be processed. Some of the main considerations will be briefly discussed in this short introductory section. For previous general reviews see Darbyshire (1981) and Scawen and Melling (1985).

**Continuous or batch processing**

The best arrangement would seem to be the continuous isolation of several enzymes from yeast disruptate. It would be conceivable to link together the fermentation and isolation (and subsequent immobilization) steps.

The advantages of continuous processing would seem to be in the rapid and controlled throughput, which may lead to less loss of enzymes by denaturation (see below). The process could be achieved in a relatively small size of plant, carrying out the necessary 'unit operations', with a low cost for labour and for energy.

However, there may be real disadvantages in setting up the process in continuous mode, in addition to the research and development cost in setting up or conversion from batch processing. Extremely reliable equipment is needed, and this must be kept solely for the particular process if a fully operational continuous process is to be operated. It is now likely that some or all of the process will be microprocessor controlled. A properly established process can be remarkably effective, but other problems, including initial cost, may be generated by the introduction of highly sophisticated equipment, especially in the event of component failure or damage.

**Assay and loss of product**

It is most valuable to have a quick test for activity, and therefore recovery, of the required protein or peptide product. This should allow the recovery and loss of, for example, an enzyme to be followed closely at every step. Particular enzyme activity is easily tested for, although a quick test may be sufficient using a relatively crude method for an industrial enzyme. Very specific and sensitive tests are available now for particular proteins and peptides based upon a range of commercially available antibody reagents.

Unexpected loss of enzyme at any stage in the isolation and purification process must be investigated. Commonly, losses are due to loss of coenzyme (try adding back a dialysate!), contact with several well-known inhibitors such as copper ions or mercury ions, destruction due to excesses of pH, temperature, foaming or dilution, and activation of destructive enzymes (proteases) present in the yeast. The addition of protective agents such as ethylenediaminetetra-acetic acid (EDTA) is useful to remove the heavy metal ions noted above. With enzymes that contain cysteine in the active site it is usually beneficial to add thiol-protective agents such as cysteine itself or dithiothreitol. Another useful protective agent is glycerol. This subject has been considered in detail by Wiseman (1978).

**Removal of nucleic acids from yeast disruptates**

It is often necessary first to remove nucleic acids from yeast disruptates. This can be carried out by precipitation with, for example, protamine sulphate, but it is usually convenient to employ nucleases instead. These cannot, of course, be used if the required product must necessarily be guaranteed to be free of extraneous nuclease contamination. Restriction endonucleases, exonucleases, DNA ligase or topoisomerase products from yeast would fall into this category of product.

**Large-scale precipitation of enzymes**

It is most important that fast efficient mixing is achieved, to avoid the build-up of local excesses of pH, temperature, organic solvents and salts. Frothing due to turbulent agitation must also be avoided, as surface denaturation of enzymes easily occurs. Turbulent agitation may also prevent the aggregation of protein required for easy filtration or centrifugation.

There are several good precipitating agents for proteins and enzymes, but a careful choice must be made based on the nature of the desired enzymes, especially if they are lipid-bound (subcellular fraction) or contain bound-carbohydrate moieties. In general, it is better to precipitate first in a few broad fractions and then to dissolve each precipitate in a smaller volume. Ammonium sulphate is a very soluble and self-cooling precipitating salt, and it achieves a rapid salting out of most proteins (and enzymes). Use of polyethylene glycol is discussed below. It can be used at room temperature, and it dissolves to high final concentration in water, with rapid equilibration (which avoids problems of late precipitates forming).

**Large-scale fractionation of enzymes**

Some of the main problems of large-scale work are due to the difficulties of solid–liquid separations, for example, in large centrifuges. Only relatively low **g**-values are obtainable here, and the common laboratory procedures using ultracentrifuges to separate, for example, the 'microsomal fraction' at around 100 000 × **g** is not possible. We discuss below the use of polyethylene glycol coupled with rather slower centrifugation, to solve this problem in the recovery of one microsomal enzyme from yeast, cytochrome P-450.

Most methods of fractionation and purification require the use of chromatography, usually column chromatography, often in novel developments associated with specific ligand binding (affinity chromatography). Many problems exist in scale-up here, including zoning edge effects in wide columns and compacting of absorbents under their own weight (decreased by using several columns stacked vertically). High flow rates, if so achieved, can cause overheating and other problems, so that careful temperature control is often necessary.

**Conclusions on scale-up problems**

The problems associated with the batch or continuous handling of large quantities, often for long periods, can be solved with the right equipment and its control, including the correct personnel considerations. Local variation in process liquid temperature due to poor mixing remains a problem. Unit operations can be made satisfactory, but with some reduced efficiency and yields in preliminary fractionations, and in solid–liquid separations. Column chromatography can be effective and economic. Concentration of solutions is usually achieved satisfactorily by ultrafiltration (reverse osmosis), followed by freeze-drying.

### *14.3.2 Microsomal enzymes and their precipitation with poly(ethylene glycol)*

During disruption of eukaryotic cells, the endoplasmic reticulum is broken up to form small vesicles, known as 'microsomes' or as a 'microsomal fraction' when separated from cell disruptate by high-speed centrifugation [Section 14.2.2(d)]. Enzymes bound to the intracellular membrane in the intact cell which remain in the microsomal fragments after disruption are thus known as 'microsomal enzymes'. The microsome contains a range of enzymes and other proteins, as well as the membrane fragment to which they are bound, and is thus several orders of magnitude larger than an individual enzyme molecule. A microsomal enzyme is thus ensconced in the same protective microenvironment that it had in the original intact cell. To be susceptible to most of the available techniques for purification and concentration of enzymes (Section 14.3.1), however, an enzyme has to be in solution. A microsomal enzyme must therefore be detached from its membrane fragment and brought into solution by 'solubilization' with a surface-active agent, such as sodium cholate. A microsomal enzyme becomes very much more susceptible to denaturation when removed from its microenvironment in the microsome, so that procedures for purifying microsomal enzymes should maintain the enzyme in its microsomal form until as late a stage as possible. This can be advantageous in laboratory-scale purification, as in centrifugation the intracellular location of an enzyme can simplify a separation process by confining the enzyme to a particular fraction defined by the size and shape of the organelle with which it is involved. Microsomes have a typical dimension of less than 0.1 μm and, although they are very much larger than an enzyme molecule, their separation by sedimentation requires centrifugation at high speeds or for long periods, or both. High-speed centrifuges have a low volume capacity, and cause a serious hold up in treating the product of any reasonably large-scale process. The standard procedure for separating the microsomal fraction of brewer's yeast disruptate containing cytochrome P-450, for example, involves centrifugation for 1 h using a relative centrifugal acceleration ('RCF') of $10^5 \times \mathbf{g}$. Much less demanding centrifugation conditions may be used if the required protein can be induced to agglomerate into large aggregates by one of the standard techniques such as salting out or precipitation with metal ions or 'organic' solvents. Microsomal cytochrome P-450 from *Candida tropicalis* has been successfully precipitated with calcium ions (Kappeli *et al*. 1982), and cytochrome P-450 has been precipitated with ammonium sulphate from a microsomal preparation after solubilization with sodium cholate (Azari & Wiseman 1982). However, some enzymes are apparently sensitive to the conditions of high ionic strength or low dielectric constant which occur in these standard precipitation methods. For example, extensive enzyme loss was experienced in attempting to precipitate cytochrome P-450 from brewer's yeast, disrupted by a Manton–Gaulin homogenizer and centrifuged in batch in a pilot-scale operation, using ammonium sulphate, calcium ions or acetone (Sadler *et al.* 1985). A precipitation method which does not involve extreme values of ionic strength and dielectric constant is the use of non-ionic polymers. As they do not form ions in solution, they do not interact electrostatically with solute proteins. Their mechanism of precipitation is widely accepted to be one in which proteins are made to aggregate and thence precipitate as a result of steric exclusion from the solution by the non-ionic polymer.

The steric exclusion mechanism was first proposed by Laurent (1963) from the results of investigating the effect on protein solubility of dextran. Polson *et al.* (1964) examined several high-molecular weight non-ionic polymers to compare their suitability for protein precipitation. Of those examined, viz. poly(ethylene glycol), dextran, nonyl phenyl ethoxylate, poly(vinyl alcohol) and poly(vinyl pyrrolidone); poly(ethylene glycol) was found to be most suitable, as it caused virtually no denaturation of protein at room temperature and its solutions were less viscous than those of the other polymers examined. Juckes (1971) reported poly(ethylene glycol) precipitation as a gentle process giving high recovery levels of different proteins, and Knoll and Hermans (1983) investigated the effect of the concentration of poly(ethylene glycol) on the solubility of bovine serum albumin. Using poly(ethylene glycol) of different average molecular weights between 1000 and 100 000, they concluded that the protein precipitation effects observed were accounted for satisfactorily by the steric exclusion mechanism.

A preliminary study of the precipitation of microsomal cytochrome P-450 from centrifuged brewer's yeast disruptate with poly(ethylene glycol) having an average molecular weight of 6000 ('PEG 6000') was carried out by Sadler *et al.* (1983). About 90% recovery of enzyme was obtained using a PEG 6000 concentration of only 6% and centrifugation at 2300 × **g** for 20 min. Some degree of fractionation of the enzyme also appeared to take place but, as the preparation was intended for immobilization, production of the enzyme in a concentrated form was more important than its purity. Further investigation using different concentrations of PEG 6000 and centrifuge speeds showed that PEG 6000 could be considered as effectively substituting for centrifuge speed in obtaining a given level of enzyme recovery (Sadler *et al.* 1985). It was concluded that precipitation with PEG 6000 at low concentrations followed by low-speed short-time centrifugation was very effective in separating a high-concentration, low-purity microsomal enzyme preparation in a pilot-scale operation.

Several semi-empirical quantitative models of the precipitation of protein by non-ionic polymers have been presented. From a consideration of the increase in chemical potential of one solute in the presence of another, Edmond and Ogston (1968) developed an expression very similar in form to the standard Cohn salting out equation:

$$\ln S = K_G - BG \tag{14.3}$$

where $S$ is the concentration of protein in a solution containing a concentration $G$ of non-ionic polymer, and $K_G$ and $B$ are coefficients characteristic of the protein and the polymer. A similar expression, actually based on the Cohn salting out equation and using the assumption that the precipitation mechanism was one of steric exclusion, was developed by Juckes (1971) and fitted well with data obtained with the precipitation of brome grass virus, haemoglobin, ovalbumin and bovine serum albumin with poly(ethylene glycol). The precipitation of plasma protein with poly(ethylene glycol) was described by a semi-empirical relationship, developed by Hasko *et al.* (1982), expressing the solubility of plasma protein as a function of the concentration of poly(ethylene glycol):

$$\ln (S - S_f) = K_g + b \ln (1 - G) \tag{14.4}$$

where $S$ is the protein concentration in a solution containing a concentration $G$ of poly(ethylene glycol), $S_f$ is the final residual protein concentration, which is unaffected by further increases in $G$, and $K_g$ and $b$ are characteristic coefficients.

Sadler *et al.* (1985) proposed a model to describe the precipitation of microsomal cytochrome P-450 from a centrifuged disruptate of *Saccharomyces cerevisiae* with poly(ethylene glycol) with an average molecular weight of 6000 ('PEG 6000'). This was developed from the idea that as the microsomal protein is forced to agglomerate and then precipitate by the addition of the non-ionic polymer; the concentration of PEG 6000 used affects both the average size of the protein agglomerates so formed and their size distribution. As increasing concentrations of PEG 6000 are used, increasing proportions of the microsomal protein are forced out of solution to form agglomerates of increasing average sizes and increasing breadth of size ranges. If such suspensions are centrifuged, then for a given centrifuge configuration and centrifugation time it can be shown that the diameter of the largest agglomerate that can escape sedimentation is proportional to the reciprocal of the square root of the centrifugal acceleration $C$ used. There was also evidence that the protein agglomerates had an approximately Normal size distribution, so that the percentage enzyme recovery, $P$ was given by

$$P = 100[1 - \phi(h)] \tag{14.5}$$

where

$$h = \frac{1}{\sigma}(C^{-1/2} - C_{50}{}^{-1/2}) \tag{14.6}$$

$C_{50}$ is the centrifugal acceleration (RCF) giving 50% enzyme recovery at a given concentration of PEG 6000, $\phi(h)$ is the cumulative Normal distribution function of $h$, and $\sigma$ its standard deviation.

Results also indicated that $C_{50}$ and the standard deviation $\sigma$ are functions of the concentration $G$ of PEG 6000 used, and

$$C_{50} = AG^m \quad (14.7)$$

and

$$\sigma = q(A^{-1/2})G^{-m/2}$$

where $A$ and $q$ are characteristic coefficients and $m$ is a characteristic exponent. Equation 14.6 then becomes

$$h = q^{-1}A^{1/2}G^{m/2}(C^{-1/2} - A^{-1/2}G^{-m/2}) \quad (14.8)$$

Plots of percentage enzyme recovery for different concentrations of PEG 6000 and centrifugal accelerations used were constructed using the above relationships, and gave a reasonable description of the experimental results. The expression is semi-empirical, as the values of $q$, $A$ and $m$ have to be determined from the experimental data. Their values may give some insight to the mechanisms involved in the precipitation of microsomal protein by non-ionic polymers.

### *14.3.3 General examples of isolation and purification of yeast enzymes*

The techniques involved in the isolation and purification of yeast enzymes can be easily understood when considered in a series of examples. Table 14.1 lists the examples covered in this section.

**Table 14.1** Examples of protein purifications.

| Enzyme | Purification steps | Reference |
|---|---|---|
| alcohol dehydrogenase (EC 1.1.1.1) | ethanol precipitation $(NH_4)_2SO_4$ precipitation, recrystallization | Scopes *et al.* (1981) |
| formaldehyde dehydrogenase (EC 1.2.1.1) | DEAE-cellulose (×2), hydroxyapatite | Schütte *et al.* (1976) |
| formate dehydrogenase (EC 1.2.1.2) | DEAE-cellulose (×2), hydroxyapatite | Schütte *et al.* (1976) |
| invertase (EC 3.2.1.26) | acid and heat treatment, ethanol precipitation, $(NH_4)_2SO_4$ precipitation, SE-Sephadex, DEAE-Sephadex | Neumann and Lampen (1967) |
| | concanavalin A-Sepharose | Gallili (1975) |
| trehalase (EC 3.2.1.28) | ethanol precipitation, $(NH_4)_2SO_4$ precipitation, hydroxyapatite, concanavalin A-Sepharose 4B | Kelly and Catley (1976) |

*Continued*

**Table 14.1** *Continued.*

| Enzyme | Purification steps | Reference |
|---|---|---|
| aldehyde dehydrogenase (EC 1.2.1.5) | acid and heat treatment $(NH_4)_2SO_4$ precipitation, DEAE-Sephadex, $NAD^+$-Sepharose 4B | Bostian and Betts (1978) |
| phosphoglycerate kinase (EC 2.7.2.3) | $(NH_4)_2SO_4$ precipitation, heat treatment, DEAE-cellulose, Cibacron blue F3G-A Sepharose 4B, Sephadex G-100 | Kulbe and Schuer (1979) |
| phosphofructokinase (EC 2.7.1.1) | PEG precipitation, Cibacron blue affinity partitioning, DEAE-cellulose, Sepharose 4B-CL (gel-filtration) | Kopperschlager and Johansson (1982) |
| phosphofructokinase (EC 2.7.1.1) | acetone precipitation, acid precipitation, protamine sulphate precipitation, Matrex blue A, $(NH_4)_2SO_4$ precipitation | Welch and Scopes (1981) |
| cytochrome P-450 (EC 1.14.14.1) | subcellular fractionation, solubilization, $(NH_4)_2SO_4$ precipitation, 8-amino-*n*-octyl Sepharose, hydroxyapatite, CM-Sephadex C50, DEAE-Sephacel | Azari and Wiseman (1982) King *et al.* (1984) |
| NADPH:Cytochrome P-450 reductase (EC 1.6.2.4) | subcellular fractionation, solubilization, $(NH_4)_2SO_4$ precipitation, 8-amino-*n*-octyl Sepharose, 2′,5′-ADP agarose | Azari and Wiseman (1982) King *et al.* (1984) |
| cytochrome $b_5$ | subcellular fractionation, solubilization, $(NH_4)_2SO_4$ precipitation, 8-amino-*n*-octyl Sepharose, Sephadex G-200 | Azari and Wiseman (1982) King *et al.* (1984) |
| NADH:Cytochrome $b_5$ reductase (EC 1.6.2.2) | subcellular fractionation, solubilization, $(NH_4)_2SO_4$ precipitation, 8-amino-*n*-octyl Sepharose, hydroxyapatite, DEAE-Sephacel | Azari and Wiseman (1982) King *et al.* (1984) |

**Examples of patents on enzyme isolation from yeast**

Many such examples of industrial interest are described in the patent literature.

*Saccharomyces* ATCC 20488 produces large amounts of $\alpha$-glucosidase (maltase) when grown in a complex medium containing maltose (and glucose at no greater than 10% concentration; an unusually high resistance to glucose repression). Autolysis of the yeast (see Section 14.2.1) was achieved with ethyl acetate at room temperature for 45 min (Hall 1978). The maltase was recovered in 81% yield after a six-step purification process involving protamine sulphate, Sephadex G-25, DEAE-cellulose, hydroxyapatite, Sephadex G-25 and DEAE-cellulose, respectively.

Another patented example is for the production of cholesterol esterase and lipase from *Candida rugosa* (Esders *et al.* 1977). Separation of these two enzymes is achieved by hydrophobic chromatography (see below), using a stepwise salt gradient, on columns of carboxylated polysaccharides with attached alkylamine, e.g. butylamine-CH-Sepharose.

Superoxide dismutase has been isolated from yeast after plasmolysis with small amounts of organic solvents, much like the well-known use of hot toluene in the isolation of invertase from yeast. In the superoxide dismutase patent (Johansen 1982) the required enzyme is purified by fractional precipitation of unwanted proteins followed by ion-exchange chromatography on CM-cellulose or DEAE-cellulose. Further purification is achieved by gel-filtration or fractional precipitation with alcohol followed by a further column step using CM-cellulose.

**Examples of yeast protein purification (especially by affinity chromatography)**

A simple and very rapid purification scheme has been devised for alcohol dehydrogenase (EC 1.1.1.1) from *Saccharomyces cerevisiae* (Scopes *et al.* 1981). This method exploits the unusually high stability of this enzyme to high ethanol concentrations. The protease inhibitors phenylmethylsulphonyl fluoride (PMSF) and pepstatin A were added to a disrupted yeast suspension (prepared by one of several methods) at pH 6.0. One half-volume of ethanol was added and the temperature allowed to rise to 25°C for 15 min. Much of the protein in the extract precipitated and was removed by centrifugation. A further 0.7 volume of ethanol was added and the mixture was stirred at 0–2°C. This time the precipitate (which contained the alcohol dehydrogenase activity) was collected and redissolved in potassium phosphate buffer, pH 7.0, and ammonium sulphate was added to 45% saturation. The precipitate was removed and the ammonium sulphate concentration adjusted to 60% of saturation, which precipitated alcohol dehydrogenase. The precipitate was collected and redissolved in a minimal volume of buffer, followed by the addition of buffer containing 50% saturated ammonium sulphate to 0.1 times the original extract volume. This resulted in the rapid crystallization of alcohol dehydrogenase, and after standing for 1 h at room temperature the crystals were collected, redissolved and recrystallized.

This purification procedure thus exploits the stability of this enzyme to high ethanol concentrations, and uses fractionation with ammonium sulphate. The yeast glycolytic enzymes, phosphoglycerate mutase, phosphoglucose isomerase and triose phosphate isomerase are also stable to these concentrations of ethanol, but are left in solution during ammonium sulphate precipitation. These enzymes could be collected by adding more ammonium sulphate, and be further purified using ion-exchange chromatography with DEAE-cellulose (Scopes *et al.* 1981).

The purification of more than one enzyme from a single preparation can obviously lead to great savings in time and resources. Schütte *et al.* (1976) purified both formaldehyde dehydrogenase (EC 1.2.1.1) and formate dehydrogenase (EC 1.2.1.2) from the same preparation of cells of the yeast *Candida boidinii*. These workers used methanol as the sole carbon source for *Candida boidinii* which was grown up in an 80-litre fermenter. After disruption with glass beads in a Dyno-mill and removal of cell debris by centrifugation, streptomycin sulphate was added (final concentration 1%) and the resulting precipitate was removed. Ion-exchange chromatography on a DEAE-cellulose column was then performed, enabling the resolution of formaldehyde and formate dehydrogenases. The enzymes were applied to the column in 10 mmol $l^{-1}$ potassium phosphate (pH 7.5), and eluted as discrete bands with a stepwise buffer change to (in mmol $l^{-1}$) 50, 100, 100 + 100 NaCl and 100 + 300 NaCl. After concentration and dialysis to reduce the buffer concentration to 10 mmol $l^{-1}$ phosphate, formaldehyde dehydrogenase was further purified by DEAE-cellulose chromatography, with the enzyme being eluted with a linear gradient between 21 and 50 mmol $l^{-1}$ phosphate and 21–50 mmol $l^{-1}$ phosphate + 100 mmol $l^{-1}$ NaCl. The enzyme was again concentrated and dialysed and subjected to the final purification step, chromatography on a hydroxy-apatite column. The formaldehyde dehydrogenase preparation was applied to the column in 10 mmol $l^{-1}$ phosphate and eluted with a stepwise increase in buffer concentration of 10, 20, 50 and 100 mmol $l^{-1}$. The enzyme eluted at 100 mmol $l^{-1}$ buffer. This purified formaldehyde dehydrogenase was obtained in 75% yield from the crude extract (282 mg from 2 litres of crude extract) with a purification factor of 129.5-fold, and was electrophoretically homogeneous.

Formate dehydrogenase was also purified to homogeneity by a similar procedure. The eluent from the first DEAE-cellulose column containing formate dehydrogenase was concentrated and dialysed to a buffer concentration of 5 mmol $l^{-1}$ phosphate. The enzyme was then subjected to a second DEAE-cellulose chromatography step, in which the enzyme was eluted with a linear gradient of 5–100 mmol $l^{-1}$ phosphate. The enzyme which eluted at 45–55 mmol $l^{-1}$ phosphate was again dialysed against 5 mmol $l^{-1}$ buffer and was then applied to a hydroxy-apatite column, and eluted again with a linear gradient of 5–100 mmol $l^{-1}$ phosphate. The enzyme eluted from the hydroxy-apatite column at 60 mmol $l^{-1}$ phosphate. This enzyme, obtained from the same cell extract as formaldehyde dehydrogenase, was obtained in 23% yield (575 mg from 2 litres of crude extract) with a purification factor of 18.5-fold.

Invertase (EC 3.2.1.26) exists in two forms in *Saccharomyces cerevisiae*. Extracellular invertase is located in the periplasmic space, is a glycoprotein comprising 50% mannan and two identical subunits of molecular weight 60 000 (Trimble & Maley 1977), and accounts for about 96–99% of the cell's invertase activity. There is also intracellular invertase, which is not a glycoprotein and is not secreted. Both extracellular and intracellular invertases have been purified by many workers (for a review, see Lampen 1971). Extracellular invertase for commercial use is usually isolated by prolonged autolysis, and as a result has often led to invertase preparations with widely differing protein carbohydrate ratios. Neumann and Lampen (1967) have used a single pass through the French press for the disruption of *Saccharomyces* strain FH4C, a high invertase-yielding strain. These workers then obtained a homogeneous preparation of invertase in 22% yield. The crude disruptate was centrifuged to remove cell debris. Streptomycin sulphate was then added (final concentration 10 mg $ml^{-1}$) to precipitate nucleic acid and after centrifugation the enzyme preparation was adjusted to pH 5.0 with acetic acid and stored overnight at 0°C. The solution was then heated to 50°C for 30 min and cooled. The precipitated protein was removed by centrifugation, and the invertase precipitated with an equal volume of ethanol. The precipitate was collected and redissolved in 100 mmol $l^{-1}$ sodium acetate, pH 5.0. Ammonium sulphate was then added at a concentration of 0.56 g $ml^{-1}$. At this concentration invertase was not precipitated, and after centrifugation the remaining ammonium sulphate was removed from the supernatant by gel-filtration on Sephadex G-25. After this, two ion-exchange chromatographic steps were used to purify the enzyme to homogeneity. The enzyme preparation was applied to an SE-Sephadex column and eluted with a linear sodium-chloride gradient of 0–300 mmol $l^{-1}$. The invertase-containing fractions were pooled, desalted with Sephadex G-25, and further purified by chromatography on DEAE-Sephadex A-50. The invertase was eluted with a linear gradient of sodium chloride (0–500 mmol $l^{-1}$), and again desalted, resulting in a homogeneous preparation with a specific activity of 2700–3000 μmol sucrose hydrolysed $min^{-1}$ $mg^{-1}$ protein at 30°C.

More recently, affinity chromatography using concanavalin A bound to Sepharose has been used as the key step in the purification of extracellular invertase (Gallili 1975). As concanavalin A binds glucans, mannans and fructans (Goldstein *et al.* 1974), it can be used to isolate glycoproteins such as invertase.

Affinity chromatography with concanavalin A-Sepharose has also been used to purify the enzyme trehalase (EC 3.2.1.28) from *Saccharomyces cerevisiae* (Kelly & Catley 1976). Trehalase which catalyzes the hydrolysis of trehalose (α-D-glucopyranosyl-α-D-glucopyranoside) to glucose, was purified 733-fold with a yield of 38% of the original activity. Cells of *Saccharomyces cerevisiae* were broken by agitation with glass beads, and after the removal of cell debris the enzyme was precipitated with ethanol at 0°C. The precipitated trehalase was resuspended in acetate buffer, pH 5.6, and after the removal of insoluble material ammonium sulphate was added to 70% of saturation. The trehalase remained in solution and the precipitate was removed by centrifugation. After dialysis against 50 mmol $l^{-1}$ phosphate buffer, pH 6.8, the enzyme was applied to a hydroxy-apatite column and eluted with a stepwise phosphate gradient (50–500 mmol $l^{-1}$). Fractions containing trehalase activity were pooled, dialysed against 50 mmol $l^{-1}$ phosphate, pH 7.0, and applied to a column of concanavalin A-Sepharose 4B, which was washed and eluted with 1 mmol $l^{-1}$ α-methylglucoside. The final step of affinity chromatography resulted in an increase in the purification factor from 345-fold to 733-fold, and removed contaminating invertase activity which remained bound to the column.

Affinity chromatography is a very powerful technique which is now widely used in enzyme purification and increasingly on a large scale (for reviews, see Janson 1984, Wilchek *et al.* 1984). Several types of affinity chromatography ligands are widely used. Some affinity ligands are specific for an individual enzyme, such as the enzyme's substrate, a substrate analogue or a specific competitive inhibitor. Other ligands will bind several closely related enzymes, e.g. co-factors and co-factor analogues, with the separation of individual enzymes being achieved by differential displacements using, for example, co-factor gradients or the formation of a ternary complex with the co-factor and substrate. Also, other general ligands are often used, such as dyes or hydrocarbon chains.

Bostian and Betts (1978) have used affinity chromatography with the co-factor $NAD^+$ as ligand, to purify aldehyde dehydrogenase (EC 1.2.1.5) from *Saccharomyces cerevisiae*. These workers used baker's yeast disrupted by shaking with glass beads in the presence of mercaptoethanol and PMSF. This preparation was subjected to heat treatment at 54°C for 15 min, acid treatment (pH 4.3) and a second heat treatment, and the resulting precipitates were removed. The ammonium sulphate precipitate between 35 and 60% saturation was collected and re-dissolved in buffer for ion-exchange chromatography on a DEAE-Sephadex column. The aldehyde dehydrogenase was eluted with 1 mol $l^{-1}$ potassium phosphate containing mercaptoethanol, PMSF and glycerol, and applied to an $NAD^+$-Sepharose 4B column. After washing with buffer the enzyme was eluted with 10 mmol $l^{-1}$ $NAD^+$. A final preparation with a specific activity of 34 IU $mg^{-1}$ protein was obtained.

Bostian and Betts (1978) also demonstrated that yeast aldehyde dehydrogenase can bind to the dye Cibacron blue, and that this could be used as the basis of an affinity chromatography purification. For large-scale affinity chromatography the cost of the ligand becomes very important. The use of reactive dyes as affinity ligands is a great advantage, as these are available in large quantities at low cost. Also, these dyes often have a higher protein binding capacity than conventional affinity ligands and are easily coupled to matrix materials. Mono- or dichlorotriazine dyes, for example Cibacron dyes from Ciba–Geigy and Procion dyes from ICI, are very useful in the purification of a number of enzymes (Kopperschlager *et al.* 1982, Lowe & Pearson 1984). Initially dyes of this type were thought to bind to the enzyme via a dinucleotide binding site, and thus were used to purify many nucleotide-dependent enzymes. Now it is recognized that a wide range of enzymes can be purified using this technique. The enzyme is eluted from the dye column by changing either the pH or ionic strength, or with a specific co-factor or substrate. Table 14.2 gives some examples of yeast enzymes which have been purified by affinity chromatography with dyes as the affinity ligand.

Affinity chromatography with Cibacron blue–Sepharose 4B is a key step in the large-scale preparation of phosphoglycerate kinase (EC 2.7.2.3) from *Saccharomyces cerevisiae* reported by Kulbe and Schuer (1979). From an initial 3.5 kg (wet weight) yeast, 1.7 g homogeneous phosphoglycerate kinase was obtained. The yeast was disrupted by autolysis and the ammonium sulphate precipitate between 45 and 80% saturation was prepared. The enzyme preparation was then re-dissolved and subjected to a heat treatment at 50°C for

**Table 14.2** Yeast enzymes purified by affinity chromatography with dyes.

| Enzyme | Dye used as affinity ligand | Reference |
|---|---|---|
| alcohol dehydrogenase (EC 1.1.1.1) | Cibacron blue F3G-A | Easterday and Easterday (1974) |
| glycerol 3-phosphate dehydrogenase (EC 1.1.1.8) | Cibacron blue F3G-A | Merkel *et al.* (1982) |
| malate dehydrogenase (EC 1.1.1.37) | Cibacron blue F3G-A | Hagele *et al.* (1978) |
| 6-phosphogluconate dehydrogenase (EC 1.1.1.43) | Procion red HE-3B | Watson *et al.* (1978) |
| glucose-6-phosphate dehydrogenase (EC 1.1.1.49) | Cibacron blue F3G-A | Easterday and Easterday (1974) |
| flavocytochrome $b_2$ (EC 1.1.2.3) | Cibacron blue F3G-A | Pompon and Lederer (1978) |
| aldehyde dehydrogenase (EC 1.2.1.5) | Cibacron blue F3G-A | Tamaki *et al.* (1977) |
| glyceraldehyde-3-phosphate dehydrogenase (EC 1.2.1.12) | Cibacron blue F3G-A | Easterday and Easterday (1974) |
| orotate phosphoribosyl-transferase (EC 2.4.2.10) | Cibacron blue F3G-A | Reyes and Sandquist (1978) |
| hexokinase (EC 2.7.1.1) | Cibacron blue F3G-A | Easterday and Easterday (1974) |
| | Procion green H-4G | Lowe and Pearson (1984) |
| phosphofructokinase (EC 2.7.1.11) | Cibacron blue F3G-A | Böhme *et al.* (1982)<br>Diezel *et al.* (1973)<br>Tamaki and Hess (1975)<br>Kopperschlager and Johansson (1982) |
| pyruvate kinase (EC 2.7.1.40) | Cibacron blue F3G-A | Roschlau and Hess (1972) |
| phosphoglycerate kinase (EC 2.7.2.3) | Cibacron blue F3G-A | Kulbe and Schuer (1979) |
| orotidylate decarboxylase (EC 4.1.1.23) | Cibacron blue F3G-A | Reyes and Sandquist (1978) |
| aminoacyl-t-RNA synthetases (EC 6.1.1) | Cibacron blue F3G-A | Nikodem *et al.* (1977) |

10 min, followed by cooling and removal of the precipitated material by centrifugation. After dialysis against 10 mmol $l^{-1}$ triethanolamine–HCl, pH 7.5, the enzyme solution was applied to a DEAE-cellulose column. The column was washed and phosphoglycerate kinase was eluted with a linear gradient of sodium chloride (0–150 mmol $l^{-1}$) in starting buffer. Phosphoglycerate kinase was eluted at 60–80 mmol $l^{-1}$ sodium chloride. The enzyme was then applied to a Cibacron blue F3G-A-Sepharose 4B column pre-equilibrated with 10 mmol $l^{-1}$ triethanolamine HCl, pH 7.5. After washing with the equilibration buffer, phosphoglycerate kinase was eluted with 0.4 mol $l^{-1}$ sodium chloride in equilibration buffer or with a linear gradient of 0–1 mol $l^{-1}$ sodium chloride, again in the equilibration buffer. The enzyme preparation was concentrated and further purified by gel-filtration on a Sephadex G-100 column. In some cases the enzyme was not homogeneous (by disc– and SDS–polyacrylamide gel electrophoresis) after this procedure, and in these cases an extra step was used to remove trace impurities. This was either hydroxyapatite chromatography, affinity chromatography on ATP-Sepharose 4B or reversible salting out on Sepharose CL-4B. Reversible salting out was achieved by first bringing the enzyme solution to 47% saturation of ammonium sulphate and adjusting the pH to 7.5. The enzyme suspension was then applied to a column of Sepharose CL-4B equilibrated with 55% saturated ammonium sulphate in 10 mmol $l^{-1}$ triethanolamine, pH 7.5. The column was eluted with a linear gradient of decreasing ammonium sulphate concentration, with phosphoglycerate kinase being recovered almost quantitatively at 44.5% saturation. This procedure resulted in a homogeneous enzyme preparation of 890 IU $mg^{-1}$ protein in 52% yield from the crude extract.

Cibacron blue F3G-A has also been used in the purification of phosphofructokinase (EC 2.7.1.1) by affinity chromatography by several workers (Böhme *et al.* 1982, Diezel *et al.* 1973, Tamaki & Hess 1975). Kopperschlager and Johansson (1982) described a large-scale method for the purification of phosphofructokinase from baker's yeast, which uses Cibacron blue F3G-A affinity partitioning. Affinity partitioning is a technique in which separation is achieved by the preferential distribution of a soluble affinity ligand–polymer complex to one of two phases in an aqueous polymer two-phase system. The affinity ligand (in this case Cibacron blue F3G-A) is attached to poly(ethylene glycol) of molecular weight 6000. Most proteins, nucleic acids and cell debris prefer the polar phase, which is based on high-molecular weight dextran. Proteins with binding affinity for the dye ligand will thus be selectively extracted into the poly(ethylene glycol) phase.

The purification scheme devised by Kopperschlager and Johansson (1982) used a cell homogenate prepared from 1 kg of baker's yeast. The first step was a fractionation by precipitation with poly(ethylene glycol). Solid PEG was added to the homogenate to a concentration of 4% and the resulting precipitate removed by centrifugation. The supernatant was then made 8% in PEG and the precipitate, which now contained phosphofructokinase, was collected and redissolved in 120 ml of 25 mmol $l^{-1}$ sodium phosphate buffer, pH 7.1, containing 2-mercaptoethanol (5 mmol $l^{-1}$), EDTA (0.5 mmol $l^{-1}$) and PMSF (0.5 mmol $l^{-1}$). This solution was then added to a mixture of dextran (104.9 g, 20%), PEG (33.4 g, 40%), buffer (6.67 ml, 0.5 mol $l^{-1}$), PMSF (67 μl of 1% in dioxane) and 2-mercaptoethanol (100 μl). After mixing carefully the two-phase system was centrifuged to separate the phases. The upper phase was removed and the lower phase (containing the enzyme) washed with more upper phase. Phosphofructokinase was then extracted from the lower phase by adding 135 ml of an upper phase containing 8.4% Cibacron blue–PEG in buffer also containing PMSF, 2-mercaptoethanol and EDTA. After washing the upper phase 15 g solid $KH_2PO_4 : K_2HPO_4$ (1 : 1.2 molar ratio) was dissolved in it. Within 5 min a colourless salt-containing phase separated from the mixture which contained the phosphofructokinase. This salt-containing phase was collected and the enzyme further purified by ion-exchange chromatography with DEAE-cellulose and gel-filtration on a Sepharose 4B-CL column.

A rapid method for the purification of phosphofructokinase has also been described by Welch and Scopes (1981). This method was the adsorption of phosphofructokinase to a protamine–nucleic acid precipitate as the key step in the purification, although a dye affinity column was also used. The yeast was disrupted using a Vibro-mill, and to the crude extract 0.3 volume of acetone was added. After standing at 0°C for 15 min the precipitate was removed and the supernatant pH adjusted to 5.8. The precipitate was collected after standing for 10 min and re-dissolved in a MOPS (4-morpholine propane sulphonic acid) buffer, pH 7.0, containing $10^{-7}$ mol $l^{-1}$ pepstatin A. A solution of protamine sulphate (20 mg $ml^{-1}$) was added to the preparation at 90 mg $g^{-1}$ protein. The resulting precipitate was removed by centrifugation. All phosphofructokinase activity remained in the supernatant, but was then precipitated by the addition of a second equal amount of protamine sulphate. This precipitate was then re-dissolved and applied to a Matrex blue column, and after washing, eluted with 2 mmol $l^{-1}$ ATP. The enzyme was precipitated with ammonium sulphate at 60% of saturation, collected, re-dissolved and then crystallized or trace impurities removed by gel-filtration on a Sephacryl S-300 column.

Another form of affinity chromatography relies on the interaction of hydrophobic regions on proteins and a hydrophobic group attached to a suitable matrix, and is thus sometimes called hydrophobic chromatography (for a review, see Shaltiel 1984). This procedure is a key step in the purification of cytochrome P-450 (EC 1.14.14.1), NADPH:cytochrome P-450 reductase (EC 1.6.2.4), cytochrome $b_5$ and NADH:cytochrome $b_5$ reductase (EC 1.6.2.2) from the same preparation of yeast microsomes (Azari & Wiseman 1982, King *et al.* 1984). These enzymes are all bound to the microsomal membrane fraction and thus, after cell disruption with a Vibro-mill, and the preparation of microsomes by one of several techniques (see Section 14.2), need to be solubilized before further purification can take place. This was achieved by stirring at 4°C with 1% sodium cholate for 1 h. Non-solubilized material was then removed by centrifugation and the precipitate between 35 and 65% saturation of ammonium sulphate prepared. This precipitate was re-dissolved, dialysed to remove ammonium sulphate and applied to a column of 8-amino-*n*-octyl Sepharose 4B pre-equilibrated with 10 mmol $l^{-1}$ potassium phosphate, pH 7.0, containing 20% glycerol, 1 mmol $l^{-1}$ EDTA and 0.3% sodium cholate. After washing with the same buffer, cytochrome P-450 and NADH:cytochrome $b_5$ reductase were eluted with the same buffer also containing 0.1% of the detergent emulgen 911. These two enzymes could not be resolved on this column. However, NADPH:cytochrome P-450 reductase remained on the column and was eluted by adding 0.2% deoxycholate and 2 μmol $l^{-1}$ FMN to the elution buffer. Finally, cytochrome $b_5$ was eluted by changing the buffer to 10 mmol $l^{-1}$ phosphate containing 20% glycerol, 0.35% deoxycholate and 0.15% sodium cholate. Cytochrome $b_5$ was further purified by gel-filtration on Sephadex G-200. NADPH:cytochrome P-450 reductase was purified to homogeneity by a second affinity-chromatography step using a column of 2′,5′-ADP agarose. The enzyme was applied to this column after pre-equilibration with 0.3 mol $l^{-1}$ phosphate, pH 7.7, containing 20% glycerol and 0.1% emulgen 911, and after washing the pure NADPH:cytochrome P-450 reductase was eluted with buffer containing 2′-AMP (5 mmol $l^{-1}$). Cytochrome P-450 and NADH:cytochrome $b_5$ reductase were resolved by chromatography on a hydroxy-apatite column. This column was equilibrated with 30 mmol $l^{-1}$ phosphate, pH 7.0 (containing 20% glycerol and 0.2% emulgen 911) and the enzyme mixture was applied. After washing, NADH:cytochrome $b_5$ reductase was eluted with 50 mmol $l^{-1}$ phosphate buffer and then cytochrome P-450 was eluted with 100 mmol $l^{-1}$ buffer. The NADH:cytochrome $b_5$ reductase was further purified by ion-exchange chromatography on DEAE-Sephacel, and the cytochrome P-450 was further purified by two ion-exchange steps, using carboxymethyl-Sephadex C50 and DEAE-Sephacel. This purification scheme is summarized in Figure 14.1.

Hydrophobic chromatography has also been used for the purification of a yeast sex pheromone, the secreted peptide $\alpha$-factor from *Saccharomyces cerevisiae* culture fluid. Ciejek *et al.* (1977) used a column of octyl-Sepharose as the final step in the purification of the $\alpha$-factor peptide from up to 200 litres of culture fluid. The technique of reverse-phase high-performance liquid chromatography has also been used for the purification of yeast $\alpha$-factor (Julius *et al.* 1983). In this technique a hydrophobic column is used and peptides are separated on the basis of differences in the hydrophobic interaction with the column, the peptides being eluted with a solvent system, which can be designed to the desired degree of polarity, or can use a gradient of increasing hydrophobicity. High-performance liquid chromatography is becoming widely used for protein and peptide purification, and can use the principles of gel-filtration, ion-exchange chromatography or affinity chromatography, depending on the nature of the column chosen. It has the advantage of very fast separations, but is difficult to use on a large scale (for a review of HPLC techniques in protein purification, see Jakoby 1984).

Several non-yeast peptides and proteins have now been produced in *Saccharomyces cerevisiae*. Hitzeman *et al.* (1981) cloned a DNA sequence for human leukocyte interferon D into yeast, which was expressed in the yeast cell. These workers used an affinity chromatography technique for the purification of the interferon based on its binding to antibodies prepared to it, which were attached to a support for packing in an immuno-affinity column. As mentioned in the introduction to this chapter, secretion of products into the growth medium may greatly facilitate their subsequent purification. This has recently been achieved for human interferons (Hitzeman *et al.* 1983) and wheat $\alpha$-amylase (Rothstein *et al.* 1984). There is now much interest in using the $\alpha$-factor promoter for enzyme secretion, for example in the production of

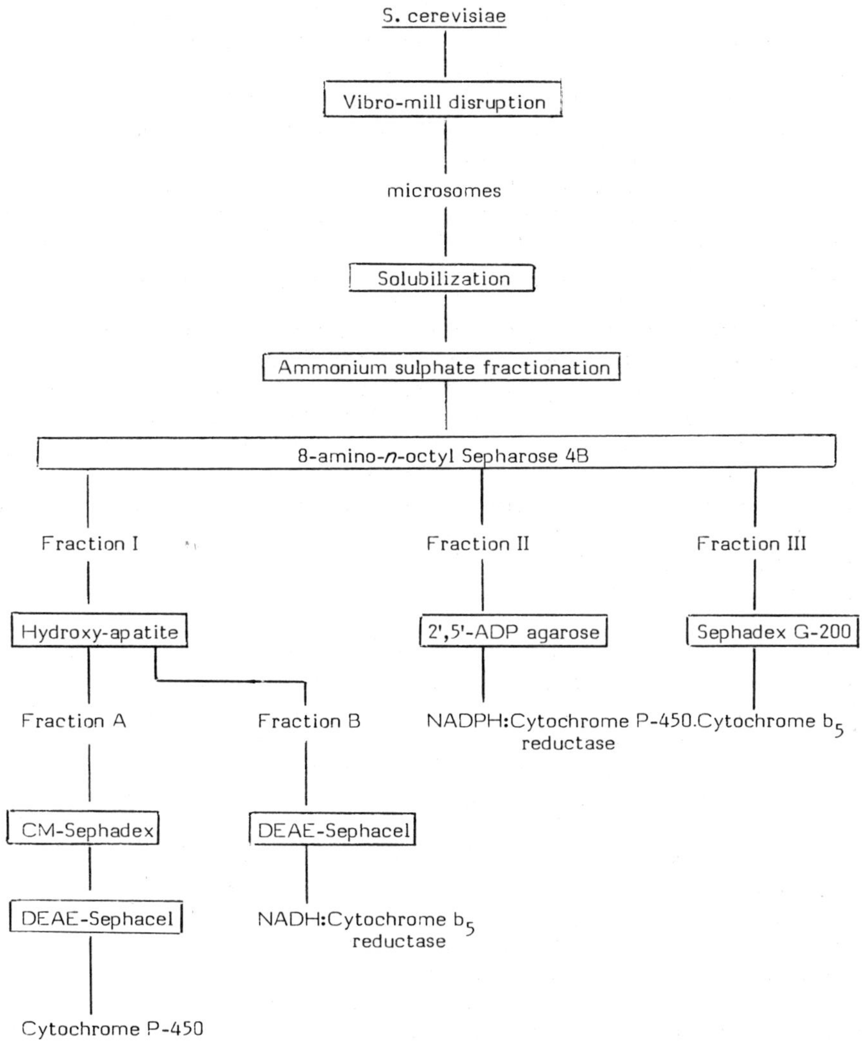


**Figure 14.1** Purification scheme for components of the microsomal electron transfer system from *Saccharomyces cerevisiae*.

human epidermal growth factor (Brake *et al.* 1984). Goff *et al.* (1984) have partially purified calf prochymosin, a milk-coagulating enzyme used in cheese-making, which was expressed in *Saccharomyces cerevisiae*. Here the yeast cells were disrupted by agitation with glass beads, and after removal of cell debris by centrifugation the enzyme was partially purified by ion-exchange chromatography on a Whatman DE-52 column, pre-equilibrated with 100 mmol $l^{-1}$ sodium phosphate, pH 5.8. The prochymosin was eluted with a linear gradient of 100 mmol $l^{-1}$ sodium phosphate, pH 5.8, and 300 mmol $l^{-1}$ sodium phosphate, pH 4.3. This resulted in a seven-fold purification of the prochymosin.

Several other foreign proteins have been expressed in yeast; for example, $\beta$-glucosidase from *Aspergillus niger* (Penttila *et al.* 1984), chicken ovalbumin which was found to be present in engineered yeast after disruption with glass beads (Mercereau-Puijalon *et al.* 1980), hepatitis B surface antigen which was detected after the osmotic lysis of spheroplasts prepared from yeast cells with zymolase (Miyanohara *et al.* 1983), and *Herpes simplex* virus thymidine kinase found in yeast cells after disruption in a French press (Zhu *et al.* 1984). However, in none of these cases has extensive purification of the product been reported. Further research is necessary on the optimization of yeast disruption procedures in the isolation of membrane-bound human proteins and glycoproteins genetically engineered into yeast. Model work in this area is being pursued in our laboratory.

## References

Al-Shahwani, M. F., E. A. Berry and D. R. Berry 1978. Growth conditions inducing spontaneous cell rupture in *Saccharomyces cerevisiae*. *Transactions of the British Mycology Society* **70**, 257–63.

Al-Shahwani, M. F., E. A. Berry and D. R. Berry 1982. Enzymatic studies during spontaneous cell rupture of *Saccharomyces cerevisiae*. *European Journal of Applied Microbiology and Biotechnology* **15**, 153–5.

Assenjo, J. and P. Dunnill 1981. The isolation of lytic enzymes from *Cytophaga* and their application to the rupture of yeast cells. *Biotechnology and Bioengineering* **23**, 1045–56.

Assenjo, J., P. Dunnill and M. D. Lilly 1981. Production of yeast-lytic enzymes by *Cytophaga* species in batch culture. *Biotechnology and Bioengineering* **23**, 97–109.

Atkinson, B. and F. Mavituna 1983. Product recovery processes and unit operations. In *Biochemical engineering and biotechnology handbook*, 934–68. London: Macmillan.

Azari, M. R. and A. Wiseman 1982. Purification and characterisation of the cytochrome P-448 component of a benzo(a)pyrene hydroxylase from *Saccharomyces cerevisiae*. *Analytical Biochemistry* **122**, 129–38.

Böhme, H. J., G. Kopperschulager, J. Schulz and E. Hofmann 1982. Affinity chromatography of phosphofructokinase using Cibacron blue F3G-A. *Journal of Chromatography* **69**, 209–14.

Bostian, K. A. and G. F. Betts 1978. Rapid purification and properties of potassium-activated aldehyde dehydrogenase from *Saccharomyces cerevisiae*. *Biochemical Journal* **173**, 773–86.

Boudrant, J., J. De Angelo, A. J. Sinsleey and S. R. Tannenbaum 1979. Process characteristics of cell lysis mutants of *Saccharomyces cerevisiae*. *Biotechnology and Bioengineering* **21**, 659–70.

Brake, A. J., J. P. Merryweather, D. G. Coit, U. A. Heberlein, F. R. Masiarz, G. T. Mullenbach, M. S. Urdea, P. Valenzuela and P. J. Barr 1984. $\alpha$-Factor-directed synthesis and secretion of mature foreign proteins in *Saccharomyces cerevisiae*. *Proceedings of the National Academy of Sciences of the USA* **81**, 4642–6.

Brookman, J. S. G. 1975. Further studies on the mechanism of cell disruption by extreme pressure extrusion. *Biotechnology and Bioengineering* **17**, 465–79.

Brookman, J. S. G. and M. Davis 1973. An extreme pressure pump for continuous cell disintegration. *Biotechnology and Bioengineering* **15**, 693–705.

Bucke, C. 1983. The biotechnology of enzyme isolation and purification. In *Principles of biotechnology*, A. Wiseman (ed.), 151–71. Surrey University Press.

Charm, S. E. and C. C. Matteo 1971. Scale-up of protein isolation. In *Methods in enzymology*, S. P. Colowick and N. O. Kaplan (eds), Vol. 22, 476–556. New York: Academic Press.

Ciejek, E., J. Thorner and M. Geier 1977. Solid phase peptide synthesis of α-factor, a yeast making pheromone. *Biochemical and Biophysical Research Communications* **78**, 952–61.

Craveri, R., C. Colia and V. Cavazzoni 1976. Lysis with enzyme complexes from *Cytophaga bacterium* assigned to Society Italiana Resine Spa Italy. US Patent 3 997 367.

Currie, J. A., P. Dunnill and M. D. Lilly 1972. Release of protein from baker's yeast (*Saccharomyces cerevisiae*) by disruption in an industrial agitator mill. *Biotechnology and Bioengineering* **14**, 725–36.

Darbyshire, J. 1981. Large scale enzyme extraction and recovery. In *Topics in enzyme and fermentation biotechnology*, A. Wiseman (ed.), Vol. 5, 147–86. Chichester: Ellis Horwood.

Derynck, R., A. Singh and D. V. Goeddel 1983. Expression of the human interferon-γ cDNA in yeast. *Nucleic Acids Research* **11**, 1819–37.

Deters, D., U. Muller and H. Homberger 1976. Breakage of yeast cells: large scale isolation of yeast mitochondria with a continuous-flow disintegration. *Analytical Biochemistry* **70**, 263–7.

Diezel, W., H.-J. Bohme, K. Nissler, R. Freyer, W. Heilmann, G. Kopperschlager and E. Hofmann 1973. A new purification procedure for yeast phosphofructokinase minimizing proteolytic degradation. *European Journal of Biochemistry* **38**, 479–88.

Doulah, M. S. 1978. Application of the statistical theory of reliability to yeast cell disintegration in ultrasonic cavitation. *Biotechnology and Bioengineering* **20**, 1287–9.

Doulah, M. S., T. H. Hammond and J. S. G. Brookman 1975. A hydrodynamic mechanism for the disintegration of *Saccharomyces cerevisiae* in an industrial homogenizer. *Biotechnology and Bioengineering* **17**, 845–58.

Dunnill, P. 1983. Trends in downstream processing of proteins and enzymes. *Process Biochemistry* **18** (October), 9–13.

Dunnill, P. and M. D. Lilly 1972. Continuous enzyme isolation. In *Enzyme engineering, biotechnology and bioengineering symposium* No. 3, 97–113.

Easterday, R. L. and J. M. Easterday 1974. Affinity chromatography of kinases and dehydrogenases on Sephadex and Sepharose dye derivatives. *Advances in Experimental Medicine and Biology* **42**, 123–33.

Edebo, L. 1969. Disintegration of cells. In *Fermentation advances*, D. Perlman (ed.), 249–71. New York: Academic Press.

Edmond, E. and A. G. Ogston 1968. An approach to the study separation in ternary systems. *Biochemical Journal* **109**, 569–76.

Edwards, D. G. and A. Wiseman 1971. Shock tube for disrupting baker's yeast. *Process Biochemistry* **32**, 32–4.

Eisenstein, E., J. C. Osborne, I. M. Chaiken and P. Hensley 1984. Purification and characterisation of ornithine transcarbamylase from *Saccharomyces cerevisiae*. *Journal of Biological Chemistry* **259**, 5139–45.

Engler, C. R. and C. W. Robinson 1979. New method for measuring cell-wall disruption. *Biotechnology and Bioengineering* **21**, 1861–9.

Engler, C. R. and C. W. Robinson 1981. Disruption of *Candida utilis* cells in high pressure flow devices. *Biotechnology and Bioengineering* **23**, 765–80.

Esders, T. W., J. R. Schaeffer and H. W. Harris 1977. Assigned to Eastman Kodak Company. Cholesterol esterase from *Candida rugosa*. US Patent 4 042 461.

Fenton, D. M. 1982. Solvent treatment for $\beta$-galactosidase release from yeast cells. *Enzyme and Microbial Technology* **4**, 229–32.

Follows, M., P. J. Hetherington, P. Dunnill and M. D. Lilly 1971. Release of enzymes from baker's yeast by disruption in an industrial homogenizer. *Biotechnology and Bioengineering* **13**, 549–60.

Gallili, G. E. 1975. Synthesis of small and large invertases and the yeast cell cycle. PhD Thesis, Rutgers University, New Jersey, USA.

Gierhart, D. L. and N. N. Potter 1979. Effects of ribonucleic acid removal methods on proteolytic activity and protein solubility in *Candida utilis*. *Biotechnology and Bioengineering* **21**, 1963–80.

Goff, C. G., D. T. Moir, T. Kohno, T. C. Gravius, R. A. Smith, E. Yamasaki and A. Taunton-Rigby 1984. Expression of calf prochymosin in *Saccharomyces cerevisiae*. *Gene* **27**, 35–46.

Goldstein, I. J., C. M. Reichert and A. Misaki 1974. Interaction of concanavalin A with model substrates. *Annals of the New York Academy of Sciences* **234**, 283–96.

Hagele, E., J. Neeff and D. Mecke 1978. The malate dehydrogenase isoenzymes of *Saccharomyces cerevisiae*. *European Journal of Biochemistry* **83**, 67–76.

Hall, L. M. 1978. Assigned to The Board of Trustees at the University of Alabama. Maltase from *Saccharomyces*. US Patent 4 071 407.

Hasko, F., R. Vaszileva and L. Halasz 1982. Solubility of plasma proteins in the presence of poly(ethylene glycol). *Biotechnology and Bioengineering* **24**, 1931–9.

Hetherington, P. J., M. Follows, P. Dunnill and M. D. Lilly 1971. Release of protein from baker's yeast (*Saccharomyces cerevisiae*) by disruption in an industrial homogenizer. *Transactions of the Institute of Chemical Engineers* **49**, 142–8.

Hitzeman, R. A., F. E. Hagie, H. L. Levine, D. V. Goeddel, G. Ammerer and B. D. Hall 1981. Expression of a human gene for interferon in yeast. *Nature* (London) **293**, 717–22.

Hitzeman, R. A., D. W. Leung, J. Perry, W. J. Kohr, H. L. Levine and D. V. Goeddel 1983. Secretion of human interferons by yeast. *Science* **219**, 620–5.

Hughes, D. E., J. W. T. Wimpenny and D. Lloyd 1971. The disintegration of micro-organisms. In *Methods in microbiology*, Vol. 5B, J. R. Norris and D. W. Ribbons (eds), 1–54. London: Academic Press.

Jakoby, W. B. 1984. Enzyme purification and related techniques. In *Methods in enzymology*, Vol. 104, S. P. Colowick and N. O. Kaplan (eds). New York: Academic Press.

James, C. J., W. T. Coakley and D. E. Hughes 1972. Kinetics of protein release from yeast sonicated in batch and flow systems at 20 kHz. *Biotechnology and Bioengineering* **14**, 33–42.

Janson, J. C. 1984. Large scale affinity purification – state of the art and future prospects. *Trends in Biotechnology* **2**, 31–8.

Johansen, J. T. 1982. Assigned to De Forenede Bryggerier A/S Denmark. Superoxide dismutase recovered from yeast. US Patent 4 340 675.

Juckes, I. R. M. 1971. Fractionation of proteins and viruses with poly(ethylene glycol). *Biochimica et Biophysica Acta* **229**, 535–46.

Julius, D., L. Blair, A. Brake, G. Sprague and J. Thorner 1983. Yeast $\alpha$ factor is processed from a larger precursor polypeptide: the essential role of a membrane-bound dipeptidyl aminopeptidase. *Cell* **32**, 839–52.

Kappeli, O., M. Sauer and A. Fiechter 1982. Convenient procedure for the isolation of highly enriched cytochrome P-450 containing fraction from *Candida tropicalis*. *Analytical Biochemistry* **126**, 179–82.

Kawai, M. and N. Mukai 1972. Assigned to Kyowa Hakko Kogyo Co. Ltd, Japan. Production of cell lytic enzymes from *Coprinus* strains. US Patent 3 682 778.

Kelly, P. J. and B. J. Catley 1976. A purification of trehalase from *Saccharomyces cerevisiae*. *Analytical Biochemistry* **72**, 353–8.

King, D. J., M. R. Azari and A. Wiseman 1984. Studies on the properties of highly purified cytochrome P-448 and its dependent activity benzo(a)pyrene hydroxylase from *Saccharomyces cerevisiae*. *Xenobiotica* **14**, 187–206.

Knoll, D. and J. Hermans 1983. Polymer-protein interactions. *Journal of Biological Chemistry* **258a**, 5710–15.

Knorr, D., K. Shetty and J. Kinsella 1979. Enzymatic lysis of yeast cell walls. *Biotechnology and Bioengineering* **21**, 2011–21.

Kobayashi, R., T. Miwa, S. Yamamoto and S. Nagasaki 1982. Preparation and evaluation of an enzyme which degrades yeast cell walls. *European Journal of Applied Microbiology and Biotechnology* **15**, 14–19.

Kopperschlager, G. and G. Johansson 1982. Affinity partitioning with polymer-bound Cibacron blue F3G-A for rapid, large-scale purification of phosphofructokinase from baker's yeast. *Analytical Biochemistry* **124**, 117–24.

Kopperschlager, G., H.-J. Bohme and E. Hofman 1982. Cibacron blue F3G-A and related dyes as ligands in affinity chromatography. *Advances in Biochemical Engineering* **25**, 101–38.

Kopetzki, E. and K.-D. Entian 1982. Purification of yeast hexokinase isoenzymes using affinity chromatography and chromatofocusing. *Analytical Biochemistry* **121**, 181–5.

Kulbe, K. D. and R. Schuer 1979. Large scale preparation of phosphoglycerate kinase from *Saccharomyces cerevisiae* using Cibacron Blue-Sepharose 4B pseudoaffinity chromatography. *Analytical Biochemistry* **93**, 46–51.

Lampen, J. O. 1971. Yeast and *Neurospora* invertases. In *The enzymes*, 3rd edn, P. D. Boyer (ed.), Vol. 5, 291–305. New York and London: Academic Press.

Laurent, T. C. 1963. The interaction between polysaccharides and other macromolecules. *Biochemical Journal* **89**, 253–7.

Lawford, G. R., A. Kligerman and T. Williams 1979. Production of high-quality edible protein from *Candida* yeast grown in continuous culture. *Biotechnology and Bioengineering* **21**, 1163–74.

Lee, C.-H., S. K. Tsang, R. Urakabe and C. K. Rha 1979. Disintegration of dried yeast cells and its effect on protein extractability, sedimentation property and viscosity of the cell suspension. *Biotechnology and Bioengineering* **21**, 1–17.

Lilly, M. D. and P. Dunnill 1972. Engineering aspects of enzyme reactors. *Biotechnology and Bioengineering Symposium* No. 3, 221–7. New York: Wiley.

Limon-Lason, J., M. Hoare, C. B. Orsborn, D. J. Doyle and P. Dunnill 1979. Reactor properties of a high-speed bead mill for microbial cell rupture. *Biotechnology and Bioengineering* **21**, 745–74.

Lowe, C. R. and J. C. Pearson 1984. Affinity chromatography on immobilized dyes. In *Methods in enzymology*, W. B. Jakoby (ed.), Vol. 104, 97–113. New York: Academic Press.

Machek, F., F. Stros, A. Prokop and L. Adamek 1976. Production and isolation of protein from synthetic ethanol. In *Continuous culture 6: Applications and new fields*, A. C. R. Dean, D. C. Ellwood, C. G. T. Evans and J. Melling (eds), 135–45. Chichester: Ellis Horwood/SCI.

Magnusson, K.-E. and L. Edebo 1974. Estimation of the disruption in freeze-pressed

*Saccharomyces cerevisiae* by an electronic particle counter. *Biotechnology and Bioengineering* **16**, 1273–82.

Magnusson, K. E. and L. Edebo 1976a. Influence of salts and gelatin on disintegration of *Saccharomyces cerevisiae* by freeze-pressing. *Biotechnology and Bioengineering* **18**, 449–63.

Magnusson, K. E. and L. Edebo 1976b. Influence of cell concentration, temperature and press performance on flow characteristics and disintegration in the freeze-pressing of *Saccharomyces cerevisiae* with the X-press. *Biotechnology and Bioengineering* **18**, 865–83.

Magnusson, K. E. and L. Edebo 1976c. Large-scale disintegration of micro-organisms by freeze-pressing. *Biotechnology and Bioengineering* **18**, 975–86.

Maitra, P. K. 1970. A glucokinase from *Saccharomyces cerevisiae*. *Journal of Biological Chemistry* **245**, 2423–31.

Marffy, F. and M.-R. Kula 1974. Enzyme yields from cells of brewer's yeast disrupted by treatment in a horizontal disintegrator. *Biotechnology and Bioengineering* **16**, 623–34.

Mason, T. L., R. O. Poynton, D. C. Wharton and G. Schatz 1973. Cytochrome C oxidase from baker's yeast. *Journal of Biological Chemistry* **246**, 1346–54.

Melling, J. and B. W. Phillips 1975a. Large-scale extraction and purification of enzymes. In *Handbook of enzyme biotechnology*, A. Wiseman (ed.), Part 1, 58–88. Chichester: Ellis Horwood.

Melling, J. and B. W. Phillips 1975b. Practical aspects of large-scale enzyme purification. In *Handbook of enzyme biotechnology*, A. Wiseman (ed.), Part 2, 181–202. Chichester: Ellis Horwood.

Mercereau-Puijalon, O., F. Lacroute and P. Kourilsky 1980. Synthesis of a chicken ovalbumin-like protein in the yeast *Saccharomyces cerevisiae*. *Gene* **11**, 163–7.

Merkel, J. R., M. Straume, S. A. Sajer and R. L. Hopfer 1982. Purification and some properties of $\alpha$-glycerol-3-phosphate dehydrogenase from *Saccharomyces cerevisiae*. *Analytical Biochemistry* **122**, 180–5.

Miyanohara, A., A. Toh-e, C. Nozaki, F. Hamada, N. Ohtomo and K. Matsubara 1983. Expression of hepatitis B surface antigen gene in yeast. *Proceedings of the National Academy of Sciences of the USA* **80**, 1–5.

Mogren, H., M. Lindblom and G. Hedenskog 1974. Mechanical disintegration of micro-organisms in an industrial homogenizer. *Biotechnology and Bioengineering* **16**, 261–74.

Mosqueira, F. G., J. J. Higgins, P. Dunnill and M. D. Lilly 1981. Characteristics of mechanically disrupted baker's yeast in relation to its separation in industrial centrifuges. *Biotechnology and Bioengineering* **23**, 335–43.

Needleman, R. B. and A. Tzagoloff 1975. Breakage of yeast: a method for processing multiple samples. *Analytical Biochemistry* **64**, 545–9.

Neppiras, E. A. and D. E. Hughes 1964. Some experiments on the disintegration of yeasts by high-intensity ultrasound. *Biotechnology and Bioengineering* **6**, 247–70.

Neumann, N. P. and S. O. Lampen 1967. Purification and properties of yeast invertase. *Biochemistry* **6**, 468–75.

Nikodem, V. M., R. C. Johnson and J. R. Fresco 1977. Interaction between blue dextran and aminoacyl t-RNA synthetases from baker's yeast. *Federation Proceedings* **36**, 822.

Okagbue, R. N. and M. J. Lewis 1983. Mixed culture of *Bacillus circulans* WL-12 and *Phaffia rhodozyma* on different carbon sources: yeast wall lytic enzyme production and extractability of astaxanthin. *Biotechnology Letters* **5**, 731–6.

Penttila, M. E., K. M. H. Nevalainen, A. Raynal and J. K. C. Knowles 1984. Cloning of *Aspergillus niger* genes in yeast. Expression of the gene coding *Aspergillus* $\beta$-glucosidase. *Molecular and General Genetics* **194**, 494–9.

Polson, A., G. M. Potgeiter, J. F. Largier, G. R. F. Mears and F. J. Joubert 1964. The fractionation of protein mixtures using linear polymers of high molecular weight. *Biochimica et Biophysica Acta* **104**, 463–75.

Pompon, D. and F. Lederer 1978. Binding of Cibacron blue F3G-A to flavocytochrome $b_2$ from baker's yeast. *European Journal of Biochemistry* **90**, 563–9.

Rehacek, J. and J. Schaefer 1977. Disintegration of micro-organisms in an industrial horizontal mill of novel design. *Biotechnology and Bioengineering* **19**, 1523–34.

Rehacek, J., K. Beran and V. Bicik 1969. Disintegration of micro-organisms and preparation of yeast cell walls in a new type of disintegrator. *Applied Microbiology* **17**, 462–6.

Reyes, P. and R. B. Sandquist 1978. Purification of orotate phosphoribosyltransferase and orotidylate decarboxylase by affinity chromatography on Sepharose dye derivatives. *Analytical Biochemistry* **88**, 522–31.

Roschlau, P. and B. Hess 1972. Affinity chromatography of yeast pyruvate kinase with Cibacronblau bound to Sephadex G-200. *Hoppe-Seyler's Zeitschrift für Physiologische Chemie* **353**, 441–3.

Rothstein, S. J., C. M. Lazarus, W. E. Smith, D. C. Baulcombe and A. A. Gatenby 1984. Secretion of a wheat $\alpha$-amylase expressed in yeast. *Nature* (London) **308**, 662–5.

Sadler, A. M., M. A. Winkler and A. Wiseman 1983. Recovery of cytochrome P-450 from *Saccharomyces cerevisiae* using poly(ethylene glycol) precipitation. *Biochemical Society Transactions* **11**, 402.

Sadler, A. M., M. A. Winkler and A. Wiseman 1985. Recovery of microsomal cytochrome P-450 from yeast using low-speed centrifugation. *Chemical Engineering Journal* **30**, 1343–9.

Scawen, M. and J. Melling 1985. Large-scale extraction and purification of enzymes. In *Handbook of enzyme biotechnology*, 2nd edn, A. Wiseman (ed.), 15–53, 247–73. Chichester: Ellis Horwood.

Scawen, M. D., A. Atkinson and J. Darbyshire 1980. In *Applied protein chemistry*, R. A. Grant (ed.), 281–324. London: Applied Science Publishers.

Schütte, H., J. Flossdorf, H. Sahm and M.-R. Kula 1976. Purification and properties of formaldehyde dehydrogenase and formate dehydrogenase from *Candida boidinii*. *European Journal of Biochemistry* **62**, 151–60.

Scopes, R. K., K. Griffiths-Smith and D. G. Millar 1981. Rapid purification of yeast alcohol dehydrogenase. *Analytical Biochemistry* **118**, 284–5.

Scully, D. B. 1975. Cell disintegration by rapid decompression. *Biotechnology and Bioengineering* **17**, 1545–50.

Shaltiel, S. 1984. Hydrophobic chromatography. In *Methods in enzymology*, W. B. Jakoby (ed.), Vol. 104, 69–96. New York: Academic Press.

Shetty, K. J. and J. E. Kinsella 1978. Effect of thiol reagents on extractability of protein from yeast. *Biotechnology and Bioengineering* **20**, 755–66.

Shetty, J. K., R. C. Weaver and J. E. Kinsella 1980. A rapid method for the isolation of ribonuclease from yeast *Saccharomyces carlsbergensis*. *Biochemical Journal* **189**, 363–6.

Suekane, M. and C. Satake 1972. Assigned to C.P.C. International Inc., Japan. Lysis of yeast with *Bacillus* or *Thermoactinomyces* enzymes. US Patent 3 681 195.

Tamaki, N. and B. Hess 1975. Purification and properties of phosphofructokinase (EC

2.1.11) of *Saccharomyces cerevisiae. Hoppe-Seyler's Zeitschrift für Physiologische Chemie* **356**, 399–415.

Tamaki, N., M. Nakamura, K. Kimura and T. Hama 1977. Purification and properties of aldehyde dehydrogenase from *Saccharomyces cerevisiae. Journal of Biochemistry* **82**, 73–9.

Trimble, R. B. and F. Maley 1977. Subunit structure of external invertase from *Saccharomyces cerevisiae. Journal of Biological Chemistry* **252**, 4409–12.

Wang, D. I. C., C. L. Cooney, A. L. Demain, P. Dunnill, A. E. Humphrey and M. D. Lilly 1979. In *Fermentation and enzyme technology*, Chapter 12. New York: Wiley.

Watson, D. H., M. J. Harvey and P. D. G. Dean 1978. The selective retardation of $NADP^+$-dependent dehydrogenases by immobilized procion red HE-3B. *Biochemical Journal* **173**, 591–6.

Welch, P. and R. K. Scopes 1981. Rapid purification and crystallization of yeast phosphofructokinase. *Analytical Biochemistry* **112**, 154–7.

Whitworth, D. A. 1974. Hydrocarbon fermentation: protein and enzyme solubilization from *Candida lipolytica* using an industrial homogenizer. *Biotechnology and Bioengineering* **16**, 1399–406.

Wilchek, M., T. Miron and J. Kohn 1984. Affinity chromatography. In *Methods in enzymology*, W. B. Jakoby (ed.), Vol. 104, 3–55. New York: Academic Press.

Williams, N. J. and A. Wiseman 1973. Ease of invertase ($\beta$-fructofuranosidase) solubilization in disruption of brewer's yeast by Vibro Mill. *Biochemical Society Transactions* **1**, 1299–301.

Winkler, M. A. 1983. Application of the principles of fermentation engineering to biotechnology. In *Principles of biotechnology*, A. Wiseman (ed.), 94–143. Surrey University Press.

Winkler, M. A. and A. Wiseman 1984. Polyethylene glycol precipitation of microsomal enzymes. Paper presented at SCI Meeting on Discovery and Isolation of Microbial Products, SCI London, 23 October.

Winkler, M. A., A. Wiseman and A. M. Sadler 1985. Poly(ethylene glycol) precipitation of a microsomal enzyme. In *Discovery and isolation of microbial products*, M. Verrall (ed.), 84–97. SCI/Horwood.

Wiseman, A. 1969. Enzymes for breakage of micro-organisms. *Process Biochemistry* **4**(5), 63–5.

Wiseman, A. 1978. Stabilization of enzymes. In *Topics in enzyme and fermentation biotechnology*, A. Wiseman (ed.), Vol. 2, 280–303. Chichester: Ellis Horwood.

Wiseman, A. and P. R. Jones 1971. Assay increment and solubilization of $\alpha$-glucosidase and invertase in disruption of brewer's yeast. *Journal of Applied Chemistry and Biotechnology* **21**, 26–8.

Wiseman, A., P. H. Leslie and G. C. Granville 1973. Ultrasonic inactivation of yeast alcohol dehydrogenase: protection by oxidized nicotinamide-adenine dinucleotide. *Biochemical Society Transactions* **1**, 606–8.

Zhu, X.-L., C. Ward and A. Weissbach 1984. Control of *Herpes simplex*, virus thymidine kinase gene expression in *Saccharomyces cerevisiae* by a yeast promoter sequence. *Molecular and General Genetics* **194**, 31–41.

# 15 *Production of baker's yeast*

KNUT ROSEN

## 15.1 Introduction

Yeast has been used for brewing and baking for several thousand years, but it was not until about 1850 that the specific production of a yeast intended especially for bread-making was initiated. Until that time, the bakers would take away a small portion of the raised dough and seed the next dough batch with it. This seeding dough was later called leaven, and was the principal raising means until the 19th century. This method is still used in several countries for special types of bread.

Eventually, the bakers in Europe went over to using the waste yeast from distilleries and breweries, but often the quality of the yeast was not good enough and the keeping quality, in particular, was poor.

Around 1800 a modification of the normal procedure for alcohol production was developed in Holland. In the breweries, where the chief raw material was cereal which was saccharified by malt, they began to separate the husks and sprouts from the mash before the fermentation by leaving the mash to sediment. A modest amount of yeast was then added to the clear liquid. During the fermentation alcohol was generated, and also some yeast propagation – although modest – took place. The yeast gathered partly in the bottom and partly as a foam layer on the surface. The two layers of yeast were united and the yeast layer was freed from the liquid by being pressed in linen bags, hence the name 'pressed yeast'.

In 1846 this method was improved by an Austrian, Ad. Ign. Mautner. The raw materials were still cereal and malt, but the husks were not strained off. By choosing a suitable dilution of the mash the result was that, at a certain point during the fermentation, the yeast rose to the surface and could be skimmed off. The yeast was washed with cold water and freed from the husks by straining through silk-gauze strainers. The method, called the skimmed yeast or the Vienna method, gradually found common use until the air yeast method gained a footing at the turn of this century (Kiby 1912).

In 1876 Louis Pasteur published the work *Etudes sur la biere* containing a section with the title 'des Rapports de l'oxygène avec la Levure'. In this is a series of experiments describing the influence of oxygen on the amount of yeast generated in a sugar solution. The conclusion was that the growth of the yeast increased considerably when air was added to the fermenting liquid and, furthermore, the experiments showed that where the yeast yield was large, a proportionally small part of the sugar had been fermented and a proportionally large part assimilated.

This observation laid the foundation of the air yeast method. The first to exploit this principle was the Dane, Eusebius Bruun in 1881, and in the years to come the method was further developed, in particular by German chemists. Around 1900 the method has spread all over the world. By straining an ordinary cereal mash, a clear wort was made which was fed to a yeast vat, and to this was added an appropriate amount of seed yeast. At the bottom of the yeast vat was a system of perforated tubes through which large amounts of air were blown into the wort. The husks had been filtered off by hot water, and the thin wort was fed to the yeast vat 5–7 h after the fermentation had started. Thus, the fermentation took place in a still thinner nutrient liquid. The strong starting concentration, concurrently with the yeast propagation, resulted in a considerable amount of alcohol being generated. In order to reduce this the yeast producers gradually began to dilute the wort essentially with water from the very beginning of the fermentation and to blow more air through, but of course this modification made still heavier demands on the size of the yeast vats and the capacity of the air compressors.

In 1915–19 a Dane, Soren Sak, and F. F. Hayduck in Germany, simultaneously succeeded in developing a new method: the Z-method or the Danish method (German Patent No. 300662 1915, Danish Patent No. 28507 1919).

The principle of the Z-method is that the fermentation is differentially controlled, i.e. the nutrient substances necessary for the growth of the yeast are added to the fermenting liquid gradually as they are consumed by the yeast cells. The fermentation is started by adding the seed yeast to a thin diluted wort containing only a small amount of the total sugar quantity. Air is blown into the vat in the same way as by the air yeast method, but concentrated wort is added simultaneously at an increasing rate concurrent with the yeast propagation. In this way a constant, very low concentration of nutrient materials in the wort is ensured during the entire fermentation.

Through this method a much larger yeast yield was achieved than was earlier possible. The formation of alcohol was practically completely avoided (the 'Z' in the name of the method alludes to zero-alcohol) and, what was at least equally important, the yeast had a better keeping quality. The yield development is illustrated in Table 15.1.

The remarkable increase of the yeast yield by the Z-method, and by using the same total dilution in the yeast vats as by the air yeast method, permitted a marked increased production in the existing plants and contributed to the concentration of yeast production in few plants.

The method was soon adapted to molasses supplemented by inorganic ammonia and phosphate-containing salts, and in a few years was introduced worldwide. The timing was very fortunate, because a prohibition movement started all over the world after World War I, resulting in a drastic decline of alcohol consumption, and a complete separation of yeast production and alcohol production became necessary. The method outlined still forms the

**Table 15.1** Yield development by various methods.

| | Yield from 100 g cereal | |
|---|---|---|
| | Yeast (kg) | Alcohol (l) |
| Dutch method | 6 | 25 |
| skimmed yeast method | 14 | 25 |
| air yeast method | 32 | 15 |
| air yeast method with dilution | 40 | 10 |
| Z-method | 85–100 | 0–1 |

basis of modern yeast production, which will be further explained later in this chapter.

## 15.2 Review of yeast production

Normally, yeast for bread-making has a dry matter content of 27–30% and a water content of 70–73%. The percentage composition of the dry matter is as follows: 47.0 C, 6.3 H, 33.0 O, 8.0 N, 1.2 P, 4.5 salts.

Based on this, the following gross formula for yeast dry matter can be computed:

$$C_{3.92}H_{6.30}O_{2.06}N_{0.57}P_{0.04} + 4.5\% \text{ salts}$$

or, rounded off,

$$C_8H_{12}O_4NP_{0.1}$$

Assuming that the yeast is produced from sugar in molasses and that 1 kg molasses contains 480 g saccharose, and that this quantity yields 950 g yeast with 27% dry matter, or 245 g, in accordance with the gross formula (the figures stated conform to the actually achieved yields) we get the following:

$$0.55C_{12}H_{22}O_{11} + 0.57NH_3 + 2.51O_2 + 0.02P_2O_5 \rightarrow$$

$$C_{3.92}H_{6.3}O_{2.06}N_{0.57}P_{0.04} + 2.68CO_2 + 3.85H_2O$$

or

1 kg molasses (480 g sugar) + 205 g $O_2$ (= 0.72 $Nm^3$ air) +
7.25 g $P_2O_5$ gives 950 g yeast with 27% dry matter +
301 g $CO_2$ ( = 0.15 $Nm^3$) + 172.26 g $H_2O$

From these assumptions the following can be concluded. For production of yeast we need: carbohydrate (molasses), N, P, oxygen from air, and various other chemicals which will be further described in the following section. Apart from yeast, carbon dioxide, water and energy in the form of heat are produced (Harrison 1967, Oura 1972).

The $CO_2$ produced results from the decomposition of sugar, and from the equation it appears that approximately 40% of the sugar is used as an energy source, the rest being present in the yeast, partly as carbohydrates (approximately 25%) and partly as building materials in the protein and nucleic acids (approximately 35%).

The energy liberated has been measured to be approximately 3.87 kcal $g^{-1}$ yeast (dry matter) or 1000 kcal $kg^{-1}$ newly produced pressed yeast with 27% dry matter (Harrison 1967) (1 kcal ≈ 4.2 kJ). This quantity is more than the energy arising from combustion of 40% of the sugar, and must therefore derive also from other exothermic processes during the synthesis where partly decomposed compounds are built into other, more-complicated combinations. Furthermore, it is seen that a considerable amount of oxygen in the form of air is needed to maintain the aerobic conditions during fermentation.

By the process outlined the assumption is, of course, that all of the carbohydrate is used as raw material for organic material and as an energy source, and that completely aerobic conditions prevail, i.e. that no alcohol is generated. This is the basis for the above procedure, which is now normally called fed-batch fermentation.

Based on the above mass balance, the following principle diagram for yeast production can be set up (Fig. 15.1).

The main raw material, molasses, is diluted with water and sterilized to kill foreign micro-organisms. Various impurities are precipitated during the heating process and are removed by separation. The molasses solution is then ready for yeast manufacture.

Baker's yeast is a micro-organism that propagates by proliferation. To ensure that the culture used is the correct one and a pure one, the cultivation is initiated in the laboratory. From a pure culture on an agar slant, single colonies are produced on Petri dishes, and from these is produced an adequate quantity of pure culture grown on the basis of one single yeast cell. This quantity is transferred to the propagating plant, where the cultivation is continued.

Apart from molasses, nitrogen and phosphorus must be added to the yeast. The inorganic compounds are added in the form of aqueous solutions such as ammonium sulphate, ammoniacal water or urea as a nitrogen source, and diammonium phosphate or phosphoric acid as a phosphorus source.

The cultivation of the yeast is continued in the large fermenters by adding molasses, nitrogen and phosphate compounds according to a fixed scheme. At the same time large quantities of filtered sterilized air are blown through the substrate to keep the yeast in an aerobic condition, perhaps controlled by the oxygen solubility potential in the liquid. In some cases agitation is also used.

When the fermentation has ended the mixture of yeast and liquid is fed to the separators, where the yeast is concentrated into a yeast cream, and simultaneously washed with water. The yeast cream is subsequently cooled to 2–3°C and fed to the yeast-cream storage vat.

From here the yeast cream is pumped to filters where excess liquid is removed. Thus, the yeast is of the consistency necessary to permit the yeast to be extruded in blocks and packaged.

After filtration the yeast can be dried in a special drying plant, where the yeast is treated gently so that the greater part of the yeast cells keep their

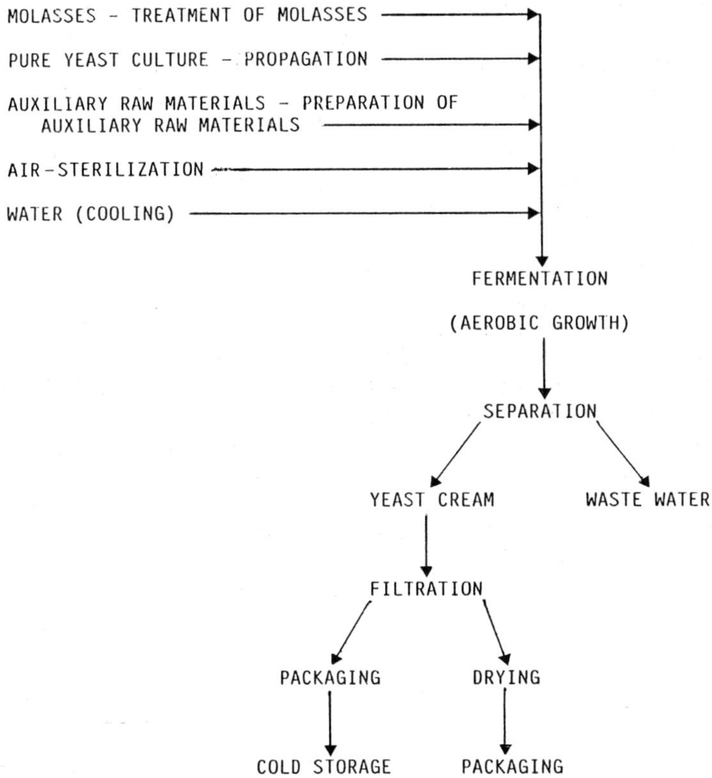


**Figure 15.1** Flowchart of the yeast production process.

activity. The dried yeast can be packaged and then stored for more than 12 months.

## 15.3 Raw materials

### *15.3.1 Carbon source*

As mentioned previously, the most important source of raw material is sucrose, which is normally employed in the form of molasses, a by-product of sugar production. Obviously, the composition of a product like molasses will vary greatly, and generally the yeast manufacturers can not control this problem, they can only mix various qualities (Olbrich 1956).

There are two main types of molasses, beet and cane, and as can be seen from Table 15.2, they differ in various areas (Baker 1976).

Beet molasses has a low content of invert sugar, a high nitrogen content, generally a lower calcium content and low biotin, whereas cane molasses has a

**Table 15.2** Composition of molasses.

| | Beet | Cane |
|---|---|---|
| dry matter (%) | 74–78 | 75 |
| sugar, total (%) | 48–52 | 48–56 |
| invert (%) | 0.2–1.2 | 15–20 |
| raffinose (%) | 0.5–2.0 | |
| fermentable (%) | 45–47 | 46–52 |
| unfermentable (%) | | 2–4 |
| organic non-sugars (%) | 12–17 | 9–12 |
| N-containing comp. (%) | 6–8 | 2–3 |
| betain (%) | 3–4 | |
| glutamic acid (%) | 2–3 | |
| organic acids (%) | 6–8 | 3 |
| gums, etc. (%) | | 4 |
| ash (%) | 10–12 | 10–15 |
| Na (%) | 0.3–0.7 | 0.1–0.4 |
| K (%) | 2–7 | 1.5–5 |
| Ca (%) | 0.1–0.5 | 0.4–0.8 |
| Cl (%) | 0.5–1.5 | 0.7–3.0 |
| P (%) | 0.02–0.07 | 0.6–2.0 |
| pH | 7–9 | 5–6 |
| vitamins (ppm) | | |
| biotin | 0.04–0.13 | 1.2–3.2 |
| inositol | 6000–8000 | 6000 |
| pantothenic acid | 50–100 | 54–64 |
| thiamine | 1.3 | 1.8 |
| nicotinic acid | 20–45 | 30–80 |

high content of invert sugar, low nitrogen content but a very significant biotin content. A high calcium content may cause problems during the processing, and it may be necessary to remove this as calcium sulphate, particularly if the calcium content is higher than 0.5%.

Beet molasses sometimes contains so much sulphur dioxide that this substance can act as a poison to the yeast and reduce the yield. With a sulphur dioxide content higher than 0.05–0.1% it is necessary to reduce the sulphur dioxide content; for instance, through oxidation with hydrogen peroxide or potassium chlorate.

Beet molasses always contains minor amounts of acetic acid, and occasionally propionic acid, butyric acid, and valeric acid are also found, these being caused by infection during the sugar production process (Dierssen 1956, Andersen 1956). The latter acids are poisonous to the yeast in quantities exceeding 0.1%. No easy way exists to remove them, and in some cases the only solution is to mix such 'infected' molasses with molasses from a well-functioning sugar factory.

Nitrite is also poisonous to the yeast, and the maximum content allowed is 0.005%.

Sometimes formic acid, probably caused by the excessive use of formaldehyde in sugar production, is found. Formic acid is poisonous to the yeast, and particularly affects the keeping quality.

In cane molasses considerable amounts of hydroxymethylfurfural are sometimes present. This compound may have an inhibitory effect on yeast metabolism under anaerobic conditions, whereas under aerobic conditions it can be tolerated in quantities up to 0.4% of the molasses quantity (Rose & Harrison 1970).

Crystals of a rather insoluble potassium salt, i.e. potassium imido disulphonate [$HN(SO_3K)_2$], are sometimes found in beet molasses. The compound has only a slightly inhibitory effect on yeast formation, but it might deceive the user as these are often mistaken for sugar crystals (Veibel & Brieghel-Muller 1946).

In some countries the beet-sugar manufacturers employ an ion exchange of sodium by magnesium; this offers increased sugar yield, but the resulting molasses (the so-called 'Quentin' molasses) is less suitable for yeast production.

As is seen from Table 15.2, there is rather a significant amount of betain $(CH_3)_3N^+CH_2COO^-$ in beet molasses. This material cannot be assimilated by the yeast and is found in the waste water from yeast production. As it has a high biological oxygen demand (BOD), it may sometimes be of interest to use debetainized molasses. This reduces the BOD by almost 50%. The betain is removed by a liquid chromatographic method, and the sale of the betain pays the cost of the process (Penttila 1984).

As mentioned above, molasses is a by-product from the sugar factories, and therefore it is unavoidable that the molasses contains materials which have been used as auxiliary agents such as anti-foaming agents, humidifying agents, pesticides, etc., by the sugar production process (Olbrich 1979). Many of these compounds have a harmful effect on yeast production, and it is becoming a major problem for the yeast manufacturers to obtain raw material of good quality which does not contain these compounds.

As mentioned, molasses is the chief carbohydrate source but, as described in the introduction to this chapter, cereal has previously been used and, in some countries where it is difficult to procure a satisfactory quality of molasses, it may be necessary to use, for example, starch hydrolysates containing large amounts of glucose and maltose, and minor amounts of oligosaccharides and protein.

Sulphite waste liquor has also been used as a raw material for yeast production and, until 1975, a yeast plant in Finland produced yeast from this material. Ethanol has also been used together with molasses as raw material for yeast production.

Recently experiments have been conducted in West Germany and Austria on the application of various sugar juices derived from the juice in a sugar factory. The so-called thick-juice, as well as green syrup, have been used and have yielded satisfactory results alone or mixed with molasses (Bronn 1985). Hydrol can also be used, but should be mixed with molasses to give a suitable yeast quality (Klaushofer 1979).

### *15.3.2 Nitrogen sources*

The amino acids in beet molasses make up approximately one-third of the nitrogenous compounds that can be utilized by the yeast. The rest – and, when cane molasses is used, the whole quantity – of the nitrogen required must be added in the form of easily assimilable compounds such as ammonia, ammonium sulphate or urea.

### *15.3.3 Minerals*

Phosphate must be added, and this is usually in the form of diammonium hydrogen phosphate or phosphoric acid. Previously, superphosphate was sometimes used, as this was less expensive, and it is still used in some regions (Bergander 1959). However, superphosphate often contains considerable amounts of fluorine, which is poisonous to the yeast, so care must be taken that the amount of this compound is not excessive.

For all chemicals used, the rule prevails that the maximum content of toxic impurities allowed is: fluorine 400 mg $kg^{-1}$, arsenic 20 mg $kg^{-1}$ and lead 2 mg $kg^{-1}$.

Magnesium, being of importance for the glycolytic enzyme system, must sometimes be added if it is not present in the molasses or the water used in sufficient quantities. If magnesium is lacking, then pyruvic acid is converted into acetaldehyde and purged away with the outlet air. This causes reduced yeast yield and inferior yeast quality.

Apart from the above, the yeast needs a series of elements such as potassium, calcium, sodium, zinc, iron, manganese, copper and cobalt; but these materials are normally supplied through the molasses.

When employing other substrates, or in laboratory tests with synthetic substrate, these minerals must be added in sufficient quantities. The quantities required are listed in Table 15.3 (Bronn 1985).

**Table 15.3** Required minerals for baker's yeast production.

| Minerals | Yeast with 27% dry matter (mg $kg^{-1}$) |
|---|---|
| potassium | 5000 |
| magnesium | 500 |
| sodium | 250 |
| zinc | 50 |
| manganese | 5 |
| copper | 2 |
| cobalt | 1 |

**Table 15.4** Comparison of beet and cane molasses for biotin, pantothenic acid and inositol.

| | Beet molasses | Cane molasses | Yeast 27% dry matter |
|---|---|---|---|
| biotin activity (ppm) | 0.04–0.13 | 1.2–3.2 | 0.25 |
| pantothenic acid activity (ppm) | 50–100 | 54–64 | 44 |
| inositol activity (ppm) | 6000–8000 | 6000 | 1000 |

### *15.3.4 Vitamins*

The strains most often used for producing baker's yeast usually require the addition of biotin, pantothenic acid and inositol to obtain optimum yields. Molasses contains some of these substances, and Table 15.4 shows the content in beet and cane molasses compared with the demand (White 1954).

A microbiological method has been used for determination of these vitamin activities, and the test organism used is a production baker's yeast strain. In this way the activities shown will also include precursors and compounds of an analogous composition with the same growth-promoting effect.

If the molasses does not contain a sufficient amount of these growth factors it may become necessary to supplement with vitamins from other sources. In such cases precursors or combinations identical to vitamins may be sufficient. It is also common yeast practice to use a mixture of beet and cane molasses (for instance, 10% cane molasses) and, through this, the lack of biotin in beet molasses can be outweighed by the very significant content usually found in cane molasses. Also DL-desthiobiotin can be used instead of D-biotin, and in lieu of pantothenic acid, β-alanine can be used. If the nitrogen source is urea one-third more biotin must be added. Aspartic acid and oleic acid can replace biotin in a synthetic substrate (Suomalainen & Keranen 1963).

The yeast strains used are normally able to synthesize thiamine, but in order to obtain optimal growth and yeast quality it is often advantageous to add 10 mg thiamine $kg^{-1}$ pressed yeast.

### *15.3.5 Water and air*

The process water must be of potable water quality according to WHO, i.e. must fulfil the demands shown in Table 15.5.

After having been compressed to working pressure, the process air is filtered to sterilize the air. The filter material most often used is glass fibres, and the control of the efficiency of the filters is carried out according to the so-called sodium flame test, where salt particles are atomized in front of the filter and the reaction is registered by a flame behind the filter.

**Table 15.5** Demand on water used in process.

| | |
|---|---|
| total hardness | 15°dH = 150 mg $l^{-1}$ CaO = 26.8 fH |
| iron max. | 0.1 mg $l^{-1}$ |
| manganese max. | 0.05 mg $l^{-1}$ |
| oxygen demand | 10 mg $l^{-1}$ |
| chloride | 250 mg $l^{-1}$ |
| nitrate | 50 mg $l^{-1}$ |
| sulphate | 500 mg $l^{-1}$ |
| the water must be free of organic substances | |
| microbiological factors: | |
| total count, 21°C | 100 bacteria $ml^{-1}$ |
| fluorescents | 5 bacteria $ml^{-1}$ |
| total count, 37°C | 10 bacteria $ml^{-1}$ |
| total coliforms | 0 per 100 ml |
| total *Escherichia coli* | 0 per 100 ml |
| total *Clostridium perfringens* | 0 per 100 ml |

## 15.4 Yeast

### *15.4.1 Yeast strains*

Baker's yeast belongs to the species *Saccharomyces cerevisiae* and is a top-fermenting yeast. There are many strains in this group, all having somewhat different qualities, and, of course, the yeast manufacturers take great pain to find the strains most suited to the various local conditions.

As mentioned above, the propagation is vegetative in industrial yeast production, but *Saccharomyces cerevisiae* belongs to the sporogenous yeast types and new strains have been constructed from different spores to give a yeast with better qualities (French Patent No. 1 336 363 1961).

The yeast strains used today can be divided into two main groups. One has the enzyme system for fermentation of maltose, and is therefore well suited for sugar-free doughs. The other must first adapt to this, and therefore has a lag phase, fermenting in a dough made from wheat flour and yeast without any other fermentable carbohydrates. However, this yeast often has a better osmotolerance and a better keeping quality, and is therefore used in a number of countries.

### *15.4.2 Yeast-propagation mathematics*

The propagation of yeast cells follows the pattern also known from other micro-organisms: a lag phase where no growth takes place, an exponential phase where the growth follows a first-order reaction scheme, a steady phase and then a declining phase (Pirt 1975).

When the lag phase is ended the yeast cells propagate unhampered as long as the substrate contains sufficient amounts of all necessary nutrients and the incremental growth of yeast (d$x$) in the period (d$t$) can be expressed by:

$$dx = \mu x \, dt$$

where $\mu$ is a constant, called the specific growth rate. As long as $\mu$ is constant the differential equation above can be integrated giving, with the initial concentration $x_0$,

$$x = x_0 e^{\mu t}$$

In some cases the incremental growth factor $H$ is used, being equal to the ratio between the yeast quantity after 1 h of growth divided by the yeast quantity at the start of the hour:

$$\frac{x_{T+1}}{x_T} = e^{\mu} = H$$

Another expression often used is the generation time, i.e. the time it takes for the yeast quantity to double. From the above equation, it is seen that

$$2 = e^{\mu T} \quad \text{or} \quad T = (\ln 2)/\mu = 0.693/\mu$$

For 'active' yeast production, $\mu$ is in the range 0.05–0.25, and in this range the best yield is obtained as the total amount of carbohydrate in the substrate is used for yeast propagation. By higher values of the specific growth rate, an essential part of the carbohydrate will be fermented into alcohol because a surplus of carbohydrate will cause a repression of the respiration, the so-called 'Crabtree effect'.

It has been proven that the constant $\mu$ is substrate dependent and can be expressed by the Monod equation:

$$\mu = \mu_m \frac{S}{K + S}$$

with $\mu_m$ being the maximum value for the specific growth rate, $S$ is the concentration of the growth-inhibiting substrate and $K$ is a constant. (Note that the expression is completely analogous with the Michaelis–Menten equation for enzyme processes.) The value $K$ corresponds to the substrate concentration for which the specific growth rate is half of the maximum value.

Only a few values of $\mu_m$ and $K$ have been referred to for yeast growth. Burrows (1970) states that at a 30°C fermentation temperature, pH 4.0, and with glucose as inhibiting substrate:

$$\mu_m = 0.37 \pm 0.03 \text{ h}^{-1} \quad \text{and} \quad K = 3.6 \pm 0.5 \times 10^{-4} \text{ mol l}^{-1}$$

Pirt and Kurowski (1970) state the following value also for glucose:

$$K = 1.4 \times 10^{-4} \text{ mol l}^{-1}$$

In practice, use is made of a constant value of the specific growth rate during the first part of the propagation, and in this period carbohydrate is added according to an exponential feeding curve. At a certain time the oxygen supply to the yeast cells becomes the inhibiting factor, and the addition of carbohydrate is then changed to a constant, or even a decreasing addition.

## 15.5 Production

### *15.5.1 Molasses treatment*

In yeast factories the viscous raw molasses is stored in large storage vats which can be heated in order to reduce the viscosity. It is normal to have stocks to cover several months of production, but it is important that the temperature of the molasses in the vats does not become too high. Many incidents have been described where molasses has been spoiled by heating. At temperatures of 100°C and higher decomposing processes begin, and these generate heat, resulting in ruined vats because the contents expand (Olbrich 1979). To keep the molasses relatively fluid it should be heated with hot water, and not with steam. Also, cane molasses has proved to be unstable on storage, in particular in such cases where the N-content is higher than normal. There have been reports of foaming molasses due to the generation of $CO_2$, of sugar destruction, and of reactions between the amino acids and the hexoses. The last of these processes seems to be the factor that starts the process of the destruction of the molasses (Chen 1970).

Before the molasses is used it undergoes an initial treatment, the purpose of which is to remove any colloids, firm particles, sand, etc., and to kill unwanted micro-organisms.

In the normal procedure the molasses is diluted with water to 35–45°Brix, and the pH is regulated if needed. Then the molasses is either heated to the boiling point and kept at this point for a short period (a couple of hours) after which the sterile liquid is drawn off from the settled sediment, cooled to at least 75°C and stored until required; or a high-temperature, short-time heating (HTST) (140°C for 4 s) is effected by direct steam injection with subsequent flashing. Precipitated matter can be removed either by simple decanting or by separation in special sludge separators. Figure 15.2 shows a plant capable of treating 5000 litres of molasses solution per hour.

The clean, sterile molasses solution is kept in vats from which it is fed to the yeast fermenter through a pipe with a control valve for flow regulation. The molasses solution is kept either hot (75°C) or is cooled to 10–12°C. If the molasses has a very high content of calcium it may be necessary to precipitate a large part of this as calcium sulphate.

**Figure 15.2** Plant capable of treating 5000 litres of molasses solution per hour.

### *15.5.2 Propagating plant*

The basis for the cultivation of yeast is the single-cell culture method developed by Emil Chr. Hansen at the Carlsberg Laboratories some 100 years ago. The starting material is a pure culture on the laboratory scale. The culture is inoculated in sterile substrates and left to grow until a suitable amount of absolutely pure yeast is achieved. This is taken from the laboratory to the first fermenter in the propagating plant, for instance a 0.5-$m^3$ tank. Before that a substrate is prepared in the laboratory, consisting of a molasses solution to which various salts and anti-foam agents have been added. The substrate is sterilized and cooled before addition of the laboratory culture.

Under light aeration the mixture is left for about 20 h and is then transferred to the next propagation stage (e.g. a fermenter of capacity 15 000 litres). Here also the substrate has been sterilized and cooled in the fermenter before the finished first stage is pumped into it. Again, light aeration takes place and after approximately 16 h the fermentation is complete and the content is transferred to one of the large fermenters.

The two propagating tanks mentioned above are normally placed in a separate room, where all precautions can be taken to avoid microbial contamination, and the substrates (i.e. the water and the air) can be sterilized before use. This results in a fermented liquid from stage 2 which is normally without infection. Figure 15.3 shows an example of a yeast propagation plant.

### *15.5.3 Cultivation tanks*

Further cultivation of the yeast takes place in large, closed fermenters with air injection and continuous substrate feeding during the entire cultivation period. The tanks are usually up to 200 $m^3$ and are made from stainless steel. Two-thirds of the volume can be used, the rest being filled with foam during cultivation.

The design of the yeast fermenters must take the following into consideration: dimension (height/diameter), aeration and cooling. Height/diameter is of importance with regard to air consumption. With a higher height the amount of air per kilogram of new yeast is reduced but, on the other hand, the air must be compressed to a higher pressure to overcome the liquid pressure in the vat, and thus the energy consumption per kilogram of yeast is generally unchanged.

The aeration systems most often employed are the rotating and the stationary ones. The advantage of the rotating systems is the significant stirring effect obtained in the liquid by the rotation and, as the air is atomized to a high degree, the consumption of air per kilogram of new yeast is less, but again energy is needed for the rotation (Austrian Patent No. 199 170 1958). The disadvantages are that a centrally placed air pipe must be fed into the tank, there must be a bottom bearing, and the drive shaft is usually fed through the bottom and on to a motor placed outside and under the bottom of the vat. All of these factors make the cleaning of the vat difficult.

Other rotating aeration systems that should be mentioned are Frings (German Patent No. 1 667 042 1966) and Inferator (German Patent No. 1 071 024 1957). In addition, a series of systems have been developed especially for

**Figure 15.3** Yeast propagation plant.

fodder yeast production, combining air inlet, substrate feeding and removal of the heat generated by the propagation (Reiff *et al.* 1962).

In the stationary aeration system the bottom is covered by a horizontal, diametrically placed main air pipe containing many side pipes with a great number of holes of diameter 1.5–2.5 mm. Through these the air is pressed as fine bubbles into the liquid. The consumption of air for rotating systems is 2–4 N $m^3$ $kg^{-1}$ newly formed yeast with 27% dry matter, while for stationary systems the consumption is 5–8 N $m^3$ $kg^{-1}$ for 27% dry matter.

Cooling of the cultivation vats is necessary, as approximately 4000 kJ are generated per kilogram of newly formed yeast with 27% dry matter. Depending on the local climatic conditions and the availability of suitable cooling water, cooling can be effected by internal cooling coils or an outside cooling mantel, or via a plate heat exchanger.

The use of cooling coils should generally be avoided, as these complicate the cleaning and sterilization of the cultivation vat. Cooling via a plate heat exchanger puts a great demand on the cleaning of the plate aggregate, pipes and pump. It must be taken into account that the content of the vat will be circulated through the cooling system about three times per hour. With a working volume of 150 $m^3$ the pipes and pumps must therefore be prepared for 450 $m^3$ $h^{-1}$.

External cooling through a cooling mantel is the preferred method. If there is ample cold water available, then direct spraying can be used, otherwise a secondary system of circulating cold water must be established, keeping the in-coming water at a suitably low temperature (i.e. less than 10°C) by means of cooling towers or cooling machines. Finally, the heat generated can be removed by letting the cooling medium from the cooling plant evaporate directly into a cooling mantel.

As mentioned above, the cultivation vats must be made of stainless steel, all internal weldings must be ground so that they are level with the remaining surface, and all of the connections, also made from stainless steel, must be manufactured with a view to the highest imaginable hygiene. Figure 15.4 shows the inner side of a cultivation vat provided with a stationary aeration system and a cleaning turbine for cleaning-in-place (CIP).

The outlet tube for the air blown through the liquid in particular requires careful execution, as condensate will gather at this place during the cultivation and provide fertile soil for unwanted micro-organisms.

### *15.5.4 Production of seed yeast*

The yeast culture produced in the propagation plant, which is practically pure, is fed to one of the vats in the plant, and to this is added substrate and dilution water. The substrate is fed during the entire propagation period and, after 24–26 h a 200-$m^3$ vat, for example, will contain 20 tonnes of yeast. The yeast cells are separated from the liquid, and the resulting yeast cream, approximately 27 $m^3$, can now be used as seed material for six sales* fermentations, each resulting in approximately 20 tonnes of yeast. The cultivation can be summed up in Table 15.6.

* The term 'sales yeast' is used in this context for yeast destined for the market place.

**Figure 15.4** Inner side of a cultivation vat provided with a stationary aeration system and a cleaning turbine.

The cultivation scheme shown, based on a 200-$m^3$ vat, should of course only be considered as an example; however, it does show the procedure normally used by yeast factories to cultivate from a single colony on the laboratory Petri dish to the finished product. Particularly, it should be noted that it is only in the first two steps that alcohol is generated; after that the molasses addition in step C is regulated so that the alcohol fed to the fermenter, with the liquid from the fermented step B, is assimilated rather quickly and then the rest of the fermentation is implemented as a pure fed-batch fermentation.

In some factories it may be desirable to produce alcohol together with yeast. In that case step C is split into two steps: in the first the main part of the molasses

**Table 15.6** Yeast cultivation scheme.

| Stage | Molasses consumption (kg) | Yeast produced (kg) | Alcohol generated (l) | Fermentation time (h) |
|---|---|---|---|---|
| laboratory | 1 | 0.1–0.2 | | |
| step A | 40 | 9 | 9 | 20 |
| step B | 1000 | 200 | 250 | 16 |
| step C | 22 000 | 20 000 | 0 | 26 |
| 6 × step D | 6 × 18 000 | 6 × 20 000 | 0 | 6 × 18 |

131 041 kg molasses yields 120 000 kg yeast (27% dry matter) for sale.

is fed at the beginning of the fermentation and aeration is light. When the substrate has been consumed, approximately 5% alcohol has been produced plus the yeast, which is separated and used for the second part of step C while the wort with the alcohol is fed to the distilling column. By varying the amount of molasses to the first part of step C the quantities of alcohol desired at any time can be obtained (Butscher & Kautzmann 1962).

Another method which has gained some ground, particularly in the Far East, is the so-called Deloffre-method (French Patent No. 885959 1943). According to this method alcohol is also produced in the 1st step of the cultivation, but yeast and alcohol are not separated, the liquid is diluted, the aeration is increased and the molasses inlet is started. The system is often employed in small factories under unfavourable climatic conditions, and often in such a way that the yeast obtained from the 2nd part of step C is used as the sales yeast. This results in simplified equipment, and the anaerobic steps A, B, and C1 may result in a reduction of the infection level.

### *15.5.5 Production of sales yeast*

Production of sales yeast takes place in one of the cultivation vats described above, and the procedure will be briefly described.

Into the vat, which has been sterilized with steam and then cooled, are pumped an adequate amount of clean water, various chemicals (phosphate, magnesium and ammonia compounds), a small amount of the prepared molasses and the desired amount of seed yeast. In this context it should be mentioned that the seed yeast from step 3 is usually stored as a cream yeast in cooled tanks (at 0–4°C) for 1 or 2 weeks, and that the necessary yeast is pumped from these tanks to the fermenters at the beginning of the fermentations. If necessary a short acid treatment to a pH value of approximately 2 is conducted. This will not harm the yeast, but most of the contaminating micro-organisms are diminished to such a degree that they will not cause problems during the subsequent fermentation.

The air-supply pipe is opened and then the feeding of molasses and chemicals is started in quantities corresponding to the actual consumption by yeast propagation. During the first hours these inlets can take place according to an exponential curve, as discussed in Section 15.4.2.

At a certain point, 6–8 h after the beginning, one of the factors necessary for optimal growth will have been fully exploited and will then start to have an inhibitory effect on the growth, which will be constant for the next hours, until the salt concentration or the viscosity of the liquid becomes inhibitory. During the fermentation the temperature is preferably kept in the region 25–35°C, or just below 30°C.

The pH value in the fermenting liquid is maintained at 4.0–5.5. These lower values give the yeast a somewhat darker colour, but at the same time these values have an inhibitory effect on the growth of foreign micro-organisms (Butscher & Kautzmann 1962). Often the procedure chosen is to maintain a comparatively low pH value during the main part of the fermentation and to increase the pH towards the upper limit during the last hours, depending on the local possibilities of carrying out the fermentation in an aseptic manner.

By the method described the molasses feeding is, according to experience, controlled so that practically no alcohol is produced. This control can be carried out automatically.

According to the Vogelbusch system, where aeration in the vat takes place through a constant flow of air through the rotating aeration system, the alcohol content is measured in the outlet air. The aim is to keep alcohol at a constant low level and the molasses inlet to the vat is controlled by the signal from the alcohol sensor (US Patent No. 3 010 881 1959).

According to another method the alcohol is measured in the wort or the oxygen potential is measured and the amount of air is regulated so that a slight oxygen surplus is constantly prevailing.

As the existing alcohol and oxygen potential sensors have often proven not to be well suited on an industrial scale (they cannot withstand sterilization or cleaning *in situ* with turbines), it is possible to measure instead the removed oxygen and generated $CO_2$ and to aim at a respiration quotient of 1.0. Today, however, developments in the area of sensors are rapid and the control of the yeast formation will be automated and computer controlled to a much greater extent in the near future.

During the constant inlet the number of cell divisions is reduced but, in lieu of carbohydrate reserves here, glycogen and trehalose are generated in the cells. The yeast concentration in the wort at the end of a cultivation may be up to 200 g yeast with 27% (dry matter) $l^{-1}$, corresponding to 5.4% dry matter. By employing high concentrations at the end of the fermentation the yeast becomes accustomed to working in a high-salt concentration and the yeast produced will be less osmosensitive, and therefore suitable for doughs containing high levels of sugar and salt.

When the addition of carbohydrates has ended, the air supply continues for another 0.5–1 h to mature the yeast, i.e. make it more stable and with better keeping qualities. Yeast cells in the process of division get time to finalize the process.

Further processing of the yeast consists of the following steps.

(1) Separation, by which the yeast is separated from the main part of the liquid resulting in a yeast cream.
(2) Washing, by which the yeast cream is diluted with ample water. This process is often carried out in an ejector where the water flow pulls the yeast cream with the flow.
(3) Separation, by which the wash water is separated.

The washing may be repeated, and in one particular method the wash water from the second washing is used again as wash water for the first washing.

As a result of these operations, a yeast cream containing 650–800 g yeast $l^{-1}$ can be obtained. The yeast cream is cooled, in plate heat exchangers, in counterflow with cold brine or in an alcohol–water mixture, is collected in closed, insulated tanks with slow-working agitators, and is provided with appropriate cooling so that the yeast cream can be kept at approximately 3–4°C.

An interesting separation variant was proposed a few years ago (Muller 1983). The fermented wort is led to a filter consisting of a series of candles with cotton stocking, and the liquid is pulled through these while the yeast cells build up a layer on the outside. When the layer is of an appropriate thickness the feeding of wort is stopped, wash water is let in and pressure is put on from the inside. The yeast falls off in lumps and can be removed from the bottom of the filter with a dry matter content of 29–30% and is now ready for packaging. The system has not yet been used in practice, but it should be applicable under certain suitable conditions.

### 15.5.6 *Yeast yield*

By using an ordinary beet molasses with the addition of the necessary growth factors and chemicals the actual yield is 90–95 kg pressed yeast with 27% dry matter from 100 kg molasses with 48% fermentable sugar, corresponding to approximately 0.5 kg yeast (dry matter) $kg^{-1}$ fermentable sugar. This compares well with a number of theoretical views. In this connection it should be remembered that the content of fermentable sugar is often many percentage points below the total sugar content.

### 15.5.7 *Alternative types of fermentation*

The literature on patents presents many modifications and proposals for the amendment of the normal procedure for yeast propagation described above, but only very few – i.e. the formation of alcohol during seed yeast propagation and the Deloffre method – seem to have achieved some acceptance.

Continuous methods have also been described. The basic principles of the practical performance have been presented in a patent by Olsen 1927 (British Patent No. 299 336 1927), but it was only by the end of the 1950s that the techniques on aseptic growth and on regulation were mastered to an extent that it was feasible to put the theory into practice (Olsen 1961).

The English yeast factory in Dovercourt, belonging to the Distillers Company Ltd, was the first to employ the continuous method. It should be mentioned that the theoretical basis for this type of fermentation had been elaborated by Monod in France (Monod 1950), Herbert *et al.* (1956) in England and by a number of Czechoslovakian scientists under the direction of Malek (Malek *et al.* 1966). The last of these groups has worked particularly with the conditions existing for the continuous yeast production, and through this they have contributed essentially to the understanding of the processes taking place during the growth of the cells.

One of the reasons why there are not more factories converting to the continuous process is, of course, that all the equipment is intended for the batch process and that the continuous methods put essentially increased demands on aseptic working methods. Another important problem is the yeast quality, which depends upon the growth parameters and the maturation of the yeast. This causes a reduction of the advantages of the continuous system.

### 15.5.8 *Filtration of yeast*

The excess water was previously removed by filtration in filter presses, but in most areas these have been replaced by vacuum filter drums. These consist of a

**Figure 15.5** Filtration of yeast.

horizontally placed cylinder or drum rotating around its shaft. A screen of cotton cloth or a wire netting, with a layer of starch, has been stretched on the surface and the drum dips down in a vat with yeast cream or the yeast cream is fed through a box put on the side of the filter (see Fig. 15.5).

By means of a vacuum pump, an appropriate vacuum is established on the entire surface, and in this way a thin layer of yeast is pulled on to the part of the cloth dipped down or at the passage to the box. While the rotation continues, the main part of the water is removed from this layer so that the yeast, after one revolution, can be scraped off in a consistency which is, in many countries, suitable for packaging.

In those cases where a drier yeast is required, this can be obtained by the so-called salt method of Kuestler and Rokitansky (US Patent No. 2 947 668 1960). To the yeast cream is added a sodium chloride solution, by means of which water is drawn out from between the cells until an osmotic balance is established. When the yeast layer has been sucked relatively dry on the filter drum, water is sprayed on, replacing the salt solution between the yeast cells. In the yeast scraped off the osmotic balance has not yet been completely re-established, but this will happen rather quickly because water is taken into the cells from the ambient water. Through this a yeast is achieved with a higher consistency than normal.

The filtered yeast is extruded, divided and packed in 1- or 0.5-kg packages for bakers, or in smaller units of 50 g, 42 g or similar for home-baking and, eventually, the yeast may be granulated and packaged in 25-kg bags for large-scale consumers.

The packaged yeast should be kept in cold storage at 2–6°C and provided prompt distribution and appropriate treatment at the bakers (i.e. dry and cool storage) occurs, the yeast normally has a shelf-life of 3–4 weeks. If, on the other hand, the yeast is stored at higher temperatures, autolysis will quickly take place, ruining the yeast, and the leavening power will be lost.

### 15.5.9 *Infection*

As mentioned several times above, all efforts are taken during the entire production process to reduce any infection of the yeast caused by unwanted micro-organisms such as wild yeast, moulds and bacteria. A large content of these in a lump of yeast will not only reduce the raising power of the yeast, but will also result in reduced keeping quality, and both of these are what the yeast manufacturer wishes to avoid.

Therefore, the propagation, separation, packaging, etc., are performed as aseptically as possible. The molasses is sterilized before use, the fermenters are cleaned by mechanical cleaners consisting of rotary turbines capable of over-spraying the entire interior of the vat with cleaning liquid. The vats are then steamed, if necessary. Tubes, pumps and separators are rinsed with cleaning liquid (cleaning-in-place). Valves with a top section in which infection can gather must be avoided, where possible. The working rooms must be kept well cleaned, and spilled yeast or nutrient solutions must be removed as quickly as possible. In short, the best defence against infection is obtained when all of these factors have been taken into consideration from the first stage of designing the factory.

Another few points must be emphasized: chemical solutions must be concentrated to such a degree that no bacteria can grow in them (this applies in particular to diammonium hydrogen phosphate). The air to the fermenters must be cleaned before it is piped in. In this context it should be mentioned that many of the aseptic precautions adopted for antibiotic production are applied to yeast production to an increasing extent, but it must be kept in mind that there is an essential difference in the selling price for antibiotics and for yeast, which puts limits to the yeast producer's potential.

Careful control in filtration and packaging must be taken so that the surface is not contaminated by mould spores or other micro-organisms which might develop on the surface of the yeast during storage and lend it an unpleasant appearance and, in the worst case, allows aflatoxin to form. It is therefore necessary for air to the packaging rooms to be filtered before being blown in, and that floors, walls and ceilings be kept clean and that contact with the yeast be avoided.

In this chapter it has only been possible to mention a few of the many factors that have to be taken into consideration in a modern yeast production facility in order to secure that the yeast, to the widest possible extent, is kept free from infection by foreign micro-organisms.

### 15.5.10 *Active dry yeast*

As mentioned above, the keeping quality of compressed yeast is limited to 3–4 weeks in cold storage. Consequently, there has for many years been a desire to be able to dry the yeast with the purpose of maintaining the leavening power and prolonging the keeping quality.

The first experiments were conducted by drying in trays. The yeast was spread out in small lumps or granulates on a tray with a perforated bottom through which hot air was blown. The method was further developed through the use of continuous conveyer belt drying plants. The yeast was extruded on to a conveyer consisting of a perforated stainless-steel plate moving slowly through drying zones with different temperatures and degrees of humidity. The air is blown through the layer alternately, from below and from above, and after the yeast has passed through the various sections, for up to 10 h at temperatures ensuring that the yeast particles are not heated beyond 30–40°C, the product has been dried to approximately 8% humidity and is ready for packaging.

By this procedure, which is still in common use, the yeast strain used has to be a robust one, and therefore relatively slow at dough raising. In particular, it will be noticed that the capability of the yeast to adapt to maltose fermentation has been reduced, i.e. it takes much longer before the yeast starts to ferment the maltose in a dough without sugar.

The dry yeast has an excellent keeping quality, i.e. 1–2 years at room temperature, and antioxidants may be added to increase the stability on storage.

When the yeast is going to be used it must first be reconstituted or rehydrated. The dry yeast is mixed with lukewarm water, at 40–43°C, and allowed to stand for 5–10 min. Then the yeast can be stirred. If the temperature of the water is lower the re-establishment of the cell membranes will take longer and more of the cell contents may leak out. If too high temperatures are employed the yeast will be killed (Herrera *et al*. 1956).

Among the materials that may leak during the rehydration is glutathione, a tripeptide. This compound activates the proteolytic enzymes in the flour, and the enzymes affect the gluten membranes in the dough. If the protein membranes slacken, then they cannot hold back the carbon dioxide generated. The consequence is that the finished loaves become flat and 'fluid' (Jorgensen 1945). Therefore the aim was to find special yeast strains with a low content of glutathione; this has, to some extent, been successful.

If the drying of the yeast is carried out in a drum dryer, then the result is a dry yeast in the form of beads with a hard surface. This type must be rehydrated as described above, but may require an additional few minutes.

By the end of the 1960s a new type of dry yeast was developed: instant active dry yeast (Greup 1974). By using fluid-bed drying and extrusion into 0.5-mm threads the drying period was reduced to less than 1 h. By adding an emulsifier, such as sorbitan monostearate or citric acid ester of monoglycerides, the water content can be reduced to approximately 5%, and by using a very active yeast a dry yeast product can be obtained which for dough preparation can be mixed immediately with the flour without rehydration (instant) and which has, in spite of the very large surface, an acceptable keeping quality of up to 1 year in vacuum or in an inert atmosphere at room temperature. The yeast has a further advantage: the glutathione content is very low, resulting in very satisfactory baking results.

**Figure 15.6** A fluid bed dryer employed as a continuously working dryer.

This type of dry yeast, introduced by the Dutch yeast factory Gist-Brocades, quickly gained a foothold on most of the important markets for dry yeast, and is now produced by almost every dry-yeast manufacturer.

A fluid-bed dryer can be used either as a batch dryer or a continuously working dryer. The latter is, of course, much less labour-intensive, and the drying is more energy-efficient. Figure 15.6 shows the latter type.

Because of the short drying period and the lenient drying conditions, at least 85% of the yeast cells survive the treatment, so the amount of yeast of this type needed is less than what is needed of the first type, where only approximately two-thirds of the cells survive the drying process.

The dry yeast is normally packaged in laminated aluminium foil in vacuum or inert gas. If $CO_2$ is used this gas will be adsorbed to the yeast particles and thereby cause a vacuum (Amsz *et al.* 1956). However, the most essential factor is the requirement to bring the oxygen content in the package down to the lowest level possible (0.1–0.2% $O_2$), thus improving the storage stability of the dry yeast.

### *15.5.11 Quality control of the yeast produced*

The yeast produced is monitored in the laboratory. Table 15.7 illustrates the most commonly tested parameters which comprise a chemical, microbiological and a physiological analysis.

**Table 15.7** Laboratory analysis of yeast.

| | |
|---|---|
| chemical | dry matter<br>nitrogen<br>phosphorus<br>carbohydrates, in particular trehalose |
| microbiological | total count of living cells<br>total count of bacteria<br>coliform bacteria<br>*Escherichia coli*<br>*Streptococcus faecalis*<br>wild yeasts<br>moulds |
| physiological | raising power<br>keeping quality (leavening power after storage of the yeast at 30°C for 2 or 4 days, 20°C for 7 days, 5°C for 4 weeks)<br>autolysis (number of days at 35°C before the yeast is soft) |

With regards to the physiological examination it is important to use methods, and in particular doughs, conforming to those used in practice in the bakeries in the region in question. A yeast may completely fulfil the demands in a lean dough without sugar and fat but, if it is not sufficiently osmostable, it will not suffice in sweet doughs. When examining the yeasts in the laboratory of the yeast plant it is common practice to use different dough types, ranging from lean to medium to heavy doughs, containing up to 30% of the flour quantity as sugar and also often fat.

The leavening power is determined in a simple apparatus consisting of a container in which the dough is placed, and this is connected to a gasometer where the carbon dioxide generated is measured, a so-called Fermentograph (Fig. 15.7). The dough is placed in a baking mould, which is placed in a thermostatically controlled chamber. As the dough is leavening the air is pressed out of the chamber and up into a bell flowing into an oil bath. On the top of the bell is a pen marking the position on a rotating paper trace which is activated by clockwork. In this way a curve is produced correlating carbon dioxide development and time.

It should be noted that such a result cannot immediately be transferred to bakeries, as the yeast may affect the gluten membranes in the dough (as described in Section 15.5.10), and laboratory results must therefore always be combined with examinations in a pilot bakery.

Another method for determining $CO_2$ generated is the use of an apparatus where the displaced gas is measured as a total amount, and alternately after passage through an alkaline absorber which removes the $CO_2$ (system Chopin Zymotachograph; Kent-Jones & Amos 1967). This method, to a great extent, compensates for the uncertainty mentioned above, as it allows the total gas generation to be measured, as well as the carbon dioxide escaping from the dough when the gluten membranes are no longer able to retain the $CO_2$.

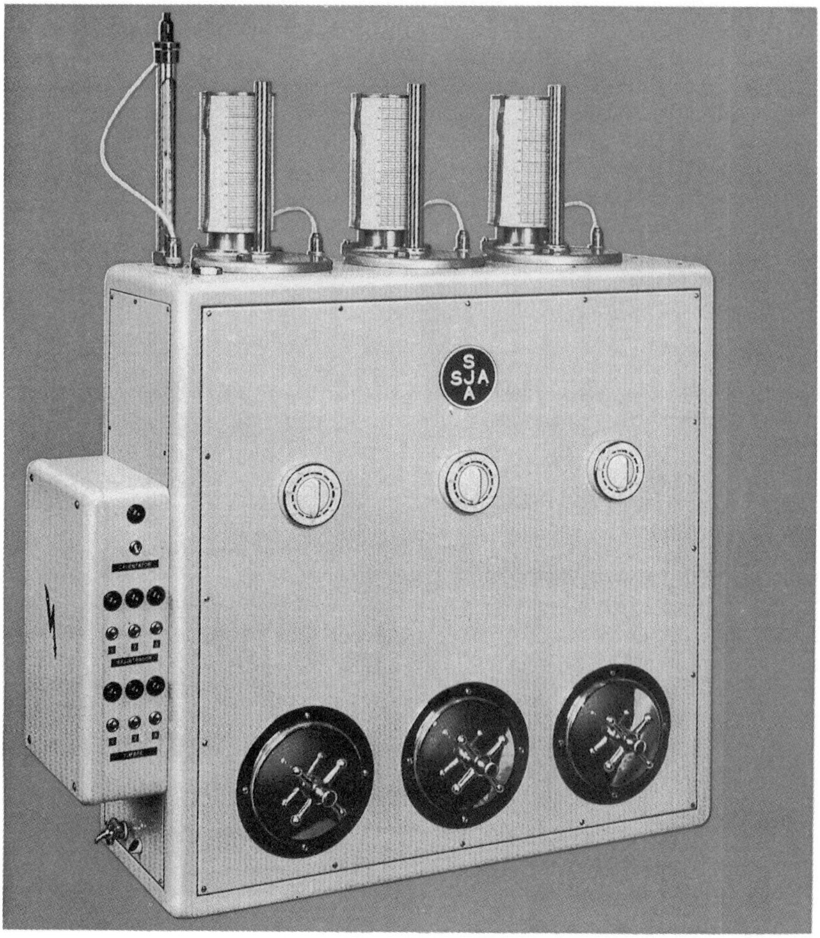

**Figure 15.7** Fermentograph for leavening power determination.

### *15.5.12 Waste water*

After the removal by yeast of carbohydrates and assimilable N from the raw material molasses, there is still much organic material left in the waste water, especially in the separated wort and also in the water from the cleaning of the plant and the equipment.

If beet molasses is used as the raw material, then the waste water contains approximately 0.2 kg BOD $kg^{-1}$ yeast or 0.3 kg COD $kg^{-1}$. Using cane molasses the corresponding figures are 0.1 and 0.3 kg $kg^{-1}$ yeast.

In Table 15.8 the discharge from a factory producing 20 000 tons $year^{-1}$ of compressed yeast has been compared with the waste water from a town with 200 000 inhabitants. It appears that the organic material from the factory

**Table 15.8** Comparison of factory and town waste water discharges.

| | |
|---|---|
| 1 kg beet molasses used for yeast production gives: | 0.2 kg BOD<br>0.01 kg N<br>0.001 kg P |
| A factory producing 20 000 t compressed yeast per year discharges: | 4800 t BOD<br>240 t N<br>24 t P |
| A town with 200 000 inhabitants discharges per year: | 4400 t BOD<br>1100 t N<br>400 t P |

corresponds to approximately 220 000 inhabitants, while the nitrogen outlet corresponds to the discharge from about 50 000 inhabitants and phosphorus to the discharge from 12 000 inhabitants. Consequently, it is normally necessary to subject the waste water to purification.

The most simple method uses the waste water to irrigate fields, thus returning the potassium, nitrogen and phosphorus to nature. The problem is that this cannot be done in regions with frosty winters, or if the soil already contains abundant amounts of potassium. In tropical areas the effluent can be sent to a lagoon for evaporation.

The waste water can also be fed to the city purification plant if it is capable of treating the heavy organic charge. Household waste water often has an excess of nitrogen and phosphorus, and it may be advantageous to mix the yeast waste water and the household waste water.

Using anaerobic purification the organic compounds can be decomposed into methane and carbon dioxide, and the gas produced can be used to operate the factory boilers (Skogman 1979).

A reduction of the BOD of approximately 80% is obtained. An aerobic treatment may follow the anaerobic. The disadvantage of the anaerobic treatment method is that the sulphate present is reduced to hydrogen sulphide, which will result in a bad smell if there is a gas leak to the atmosphere, and may cause heavy corrosion by combustion, and that nitrogen will be reduced to ammonia. By a subsequent aerobic treatment this can be oxidized into nitrate.

If beet molasses is the main raw material, then it may be advantageous in some cases to evaporate the effluent, giving condensed molasses solubles (CMS) or vinasse containing about 65% dry matter consisting of 20–22% crude protein (betains, etc.), 20–25% other organic material and 20% inorganic salts (normally potassium sulphate). The CMS may be used for cattle feed. If necessary, the potassium content may be reduced before or during the evaporation process (Belgian Patent No. 654 148 1964).

As the initial expenditure for waste-water treatment may be of a considerable size, this factor must be taken into account when a new project is started, and with existing plants it must be taken into consideration that now and in the years to come the local authorities may demand that the factory undertakes cleaning of the waste water to protect the environment.

## 15.6 Concluding remarks

In the production of yeast for baking, the yeast manufacturers constantly seek to exploit the progress developing within the biotechnology area, but with due respect to the economic aspects of production of a living micro-organism which is sold at only a fraction of the price obtainable for antibiotics.

It must be admitted that, although we know a great deal today about the microbiology and biochemistry of yeast, there are still many points to be cleared up, in particular with regard to the demands put on the yeast by the practical application for bread-making, where the yeast must be able to work under very different conditions.

The research and development carried out by the yeast manufacturers are aimed at this, and there will certainly be some very interesting new achievements in the future.

## References

Amsz, J., R. F. Dale and H. J. Peppler 1956. Carbon dioxide sorption by yeast. *Science* **123**, 463.

Andersen, E. 1956. Volatile acids in molasses. *International Sugar Journal* **58**, 133–5.

Austrian Patent No. 199 170, 25 August 1958. Patentauswertung Vogelbusch GmbH.

Baker, B. P. 1976. Composition, properties and uses of molasses and related products. *UMT Co. Ltd*. London: United Molasses Trading Company Ltd.

Belgian Patent No. 654 148, 1964. Lesaffre et Cie.

Bergander, E. 1959. *Biochemie und Technologie der Hefe*. DDR.

British Patent No. 299 336, 1927. Olsen, A. J. C.

Bronn, W. K. 1985. Moglichkeiten der Substitution von Melasse durch andere Rohstoffe für die Backhefeherstellung. *Die Branntweinwirtschaft* **125**, 228–33.

Burrows, S. 1970. *The yeasts*, Vol. III: *Yeast technology – baker's yeast*, A. H. Rose and J. S. Harrison (eds), 349–413. London: Academic Press.

Butscher, G. and R. Kautzmann 1962. In *Die Hefen*, Vol. II. Nuremberg: Verlag Hans Carl.

Chen, J. C. P. 1970. Rapid destruction of final molasses in storage. *The Sugar Journal* (December), 19–22.

Danish Patent No. 28 507, 1919. Soren Sak.

Dierssen, G. A., K. Holtegaard, B. Jensen and K. Rosen 1956. Volatile carboxylic acids in molasses. *International Sugar Journal* **58**, 35–9.

French Patent No. 885 959, 1943. Deloffre, M. C. H.

French Patent No. 1 336 363, 1961. Gist-Brocades (NGSF).

German Patent No. 300 662, 1915. Verein der Spiritusfabrikanten in Deutschland.
German Patent No. 1 071 024, 1957. Hefefabrik Weingarten GmbH.
German Patent No. 1 667 042, 1966. Heinrich Frings GmbH.
Greup, D. H. 1974. Aktive Trockenbackhefe. *Getreide, Mehl und Brot* **28**, 259–63.

Harrison, J. S. 1967. Aspects of commercial yeast production. *Process Biochemistry* **2**, 41–5.
Herbert, D., R. Ellsworth and R. C. Telling 1956. The continuous culture of bacteria – theoretical and experimental studies. *Journal of General Microbiology* **14**, 601–22.
Herrera, T., W. H. Peterson, E. J. Cooper and H. J. Peppler 1956. Loss of cell constituents on reconstruction of active dry yeast. *Archives of Biochemistry and Biophysics* **63**, 131–43.

Jorgensen, J. 1945. *Studien über die Natur der Bromat-Wirkung*. Leipzig: Schafer.

Kent-Jones, D. W. and A. J. Amos 1967. *Modern cereal chemistry*. London: Food Trade Press.
Kiby, W. 1912. *Handbuch der Presshefenfabrikation*. Vieweg: Braunschweig.
Klaushofer, H. 1979. *Proceedings 'Problems with Molasses in the Yeast Industry'*, 67–76. Helsinki: Alko.

Malek, I. and Z. Fencl 1966. Theoretical and methodological basis of continuous culture of micro-organisms. *Czechoslovak Academy of Sciences, Prague*.
Monod, J. 1950. La technique de culture continué. *Annales de l'Institut Pasteur* **79**, 390–410.
Muller, H. 1983. Ein neues Trennverfahren für die Backhefeaufbereitung. *Die Branntweinwirtschaft* **123**, 398–402.

Olbrich, H. 1956. *Die Melasse*. Berlin: Institut für Garungsgewerbe.
Olbrich, H. 1979. Schadensfall durch Explotion eines geschlossenen Melassetanks. *Die Branntweinwirtschaft* **119**, 111–12.
Olsen, A. J. C. 1961. Manufacture of baker's yeast by continuous fermentation. *Society of the Chemical Industry (London) Monograph* **12**, 81–93.
Oura, E. 1972. *The effect of aeration on the growth energetics and biochemical composition of baker's yeast*. Helsinki: Alko-Foundation for Biotechnical and Industrial Fermentation Research.

Penttila, L. 1984. Alternative raw materials for baker's yeast production. *Nordic Yeast Producers Symposium*. Helsinki: OY Alko.
Pirt, S. J. 1975. *Principles of microbe and cell cultivation*. Oxford: Blackwell.
Pirt, S. J. and W. M. Kurowski 1970. An extension of the theory of the chemostat with feed-back of organisms. *Journal of General Microbiology* **63**, 357–66.

Reiff, F., R. Kautzmann, H. Luers and M. Lindemann 1962. *Die Hefen*, Vol. II: *Backhefefabrikation*, G. Butschek and M. Kautzman (eds), 501–610. Nuremberg: Hans Carl.
Rose, A. H. and J. S. Harrison 1970. *The yeasts*, Vol. III: *Yeast technology – introduction*, A. H. Rose and J. S. Harrison (eds), 349–420. London: Academic Press.

Skogman, H. 1979. Effluent treatment of molasses based fermentation wastes. *Process Biochemistry* **14**, 5–6.

Suomalainen, H. and A. J. A. Keranen 1963. The effect of biotin deficiency on the synthesis of fatty acids by yeast. *Biochimica et Biophysica Acta* **70**, 493–503.

US Patent No. 2 947 668, 1960. Kuestler, E. and K. Rokitansky.
US Patent No. 3 010 881, 1959. Patentauswertung Vogelbusch GmbH.

Veibel, S. and A. Brieghel-Muller 1946. Isolation of potassium-iminodisulphonate from beet molasses. *Socker* **2**, 373–7.

White, J. 1954. *Yeast technology*. London: Chapman and Hall.

# 16 *Alcohol recovery*

VIDAR F. LARSEN

## 16.1 Introduction

The method traditionally used to recover alcohol throughout the ages has been distillation. Its history is lost in antiquity, but there are indications that the art was known to the Chinese as early as 1000 years BC (Rose 1977). It is thought that it was brought to Europe through Spain by the Moors in the 12th century (Ray 1973). The initial use of distillation was in the production of 'natural medicines' from naturally fermented wine, aqua vitae or infusions of herbs (Hastie 1949). The enjoyable aspects of aqua vitae were soon realized, and the *Annals of Clonmacnoise* recorded in 1405 that the 'Chieftain of Moyntyreolas died at Christmas by taking a surfeit of aqua vitae' (Ross 1970). The more modern 'ills' of government regulation were introduced at an early stage to the industry. Tax on distilled spirit was introduced as early as 1360 in Germany (Daiches 1969), whereas the requirement to possess a licence to operate a still legally was introduced in Ireland in 1556, although peers, gentlemen of property and borough freemen were exempt (Lyons & Rose 1977).

In the distillation process the separation of two or more components is achieved due to the different volatility of the individual components. If we consider the case of an ethanol–water binary mixture, and if this mixture is brought to its boiling point, the vapour so formed would be richer in ethanol than the original liquid (Fig. 16.1). Thus, separation is achieved if the vapour is removed. The concentration obtained in the vapour phase, and the boiling point of the liquid depend on the concentration of ethanol in the liquid. Although pure ethanol boils at 78.3°C, it can be seen (Fig. 16.2) that a 5% ethanol solution boils at 95°C, and that the boiling point temperature increases rapidly towards 100°C as the ethanol concentration decreases to zero. Unfortunately, binary mixtures of ethanol and water deviate from ideal behaviour to such a large extent that a minimum boiling point azeotrope is formed. At 95.57% ethanol the boiling point of liquid and the dew point of vapour coincide, thus no further separation is possible without special means.

Other problems also arise due to the presence of the congeners, or secondary products. These products, with the exception of methanol and some aldehydes, are normally less volatile than ethanol, and therefore are not expected to be present in any significant quantities in the distilled vapour. However, the presence of water does change this. The majority of these compounds also form non-ideal mixtures with water, and many of the mixtures deviate to a greater extent than the ethanol–water mixture from ideality. This increased deviation will render these compounds more volatile than ethanol when water is present. Such increases in relative volatilities have been observed for the higher alcohols

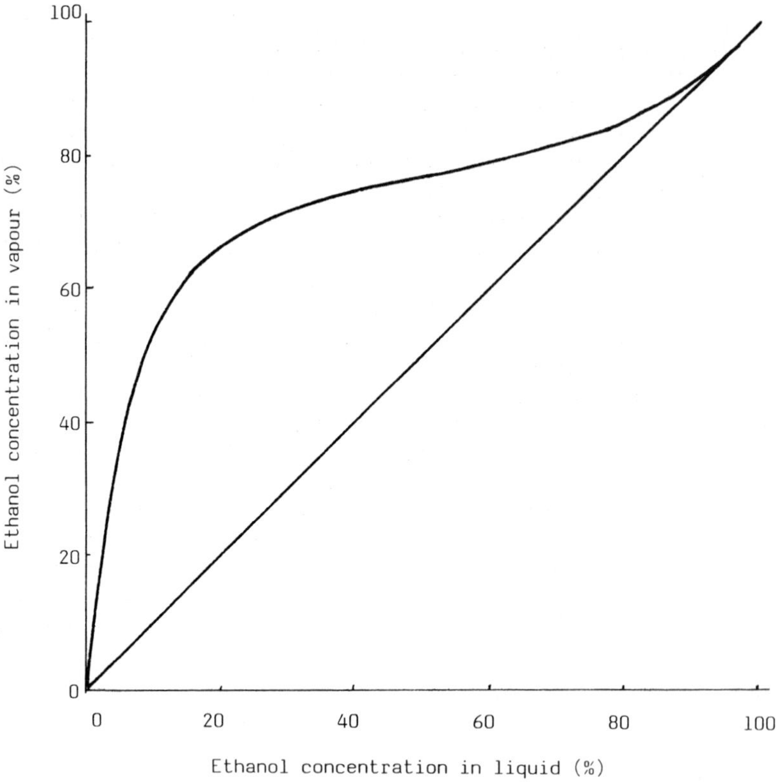


**Figure 16.1** Equilibrium composition diagram for ethanol–water mixtures (Perry 1963).

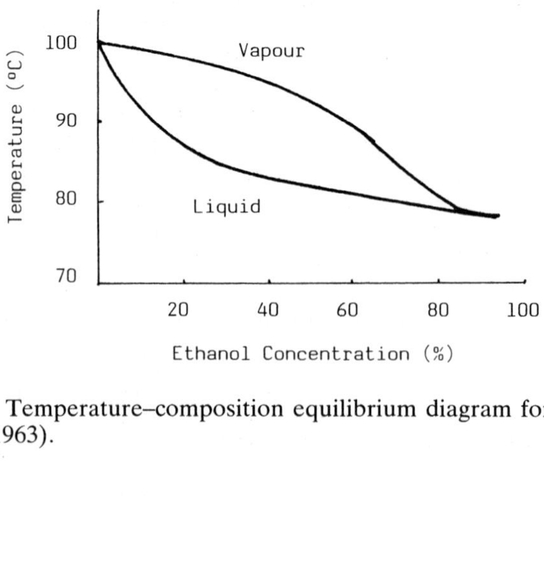


**Figure 16.2** Temperature–composition equilibrium diagram for ethanol–water mixtures (Perry 1963).

found in fusel oil, for esters, and for those aldehydes which normally are less volatile than ethanol (Unger & Coffey 1975, Carlson *et al*. 1954). Furthermore, these increases in relative volatilities depend upon the concentration of water present. As the concentration decreases, the volatilities decrease, thus at low ethanol concentration these compounds are more volatile than ethanol, and at high ethanol concentration they are less volatile. The ethanol concentration at which this changeover in volatility occurs is for all compounds greater than the ethanol concentration in the wash. Thus, for the conditions experienced with the majority of fermented broths these components are more volatile than ethanol.

For methanol the situation is different, in that the greater non-ideal behaviour of the higher alcohols renders it less volatile than these alcohols up to an ethanol concentration of 33.4%* (Unger & Coffey 1975), and less volatile than ethanol up to ethanol concentrations of 52.1% (Carlson *et al*. 1954). The volatile fatty acids are also another exception, in that the degree of deviation is insufficient to render them more volatile than ethanol (Unger & Coffey 1975).

The exact details of the process for the production of ethanol depend to a large extent upon the end use of the product. The production may be carried out in traditional batch stills, as is the case for cognac and malt whisky, or it may utilize continuous stills employing substances like pentane or benzene if anhydrous ethanol is to be produced. The distillation of ethanol can be carried out in a variety of equipment, ranging from kitchen utensils, as often seen in the production of moonshine, to large continuous stills employing special corrosion resistant materials.

## 16.2 Pot stills

The pot stills employed today are a natural extension of the stills used in the Middle Ages. They are simple in construction and operation, and consist of a kettle which is heated either directly by open fire or indirectly by steam. The vapour is withdrawn from the kettle by means of a long pipe, the swan neck or lyne arm, and is condensed in a coiled tube, the worm, situated in a water bath, the condenser (Figs 16.3 & 4). The same basic design is in operation today, although the size may vary from 135 $m^3$ for a wash still in Irish whiskey production (Court & Bowers 1970) to around 0.9 $m^3$ for small, low wine stills used in Cognac production (Ray 1973). The construction material of such stills is copper, as is all parts of the equipment which come in contact with the spirit or its vapour. The reason for selecting copper is partly due to its resistance to corrosion, but it has also been found that the copper plays an active part in the quality of the spirit. It renders volatile fatty acids non-volatile (Simpson 1971), and fixes obnoxious sulphur-containing compounds (Thorne *et al*. 1971), thus greatly improving the aroma of the final product. Copper also catalyses reactions between ethanol and fatty acids to produce esters, and the reduction

* All percentages quoted in this chapter are by weight unless otherwise stated. Conversion from mole % has been done by calculation, whereas that from volume % by a published conversion table (Herstein & Jacobs 1948). In converting from energy to steam consumption, it was assumed that the steam had a latent heat of 2230 kJ $kg^{-1}$. This is the heat of evaporation of water at 110°C.

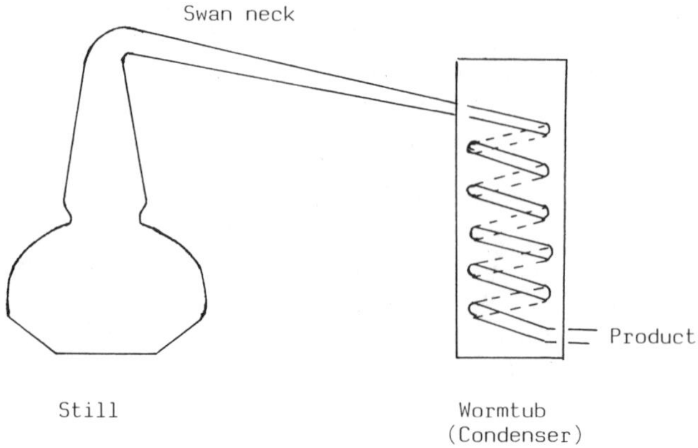


**Figure 16.3** Schematic diagram of a pot still.

**Figure 16.4** Pot stills in a Scotch malt distillery (courtesy of Chivas Brothers Ltd).

of aldehydes to alcohols (Watson 1983). These reactions appear to influence the organoleptic qualities of the product to such an extent that when malt whisky was produced in a laboratory-scale glass pot still, the product had the same graininess as a North American whisky produced in a continuous still. When copper wires were introduced into the still, the still head and the condenser on both wash and low wine stills, the whisky produced had the same organoleptic balance as a malt whisky (Watson 1983).

Improvement in the design of the still has also been made over the years. All wash stills, which receive the total contents of the fermenters ('washbacks' in distillery terminology), are fitted with a mechanical stirrer (the rummager). In a whisky still this is four rotating arms dragging a copper chain mesh around the bottom of the still to prevent build up of solids on the heating surface (Ross 1970).

The means of heating the stills were traditionally by open fire, using either wood or coal. More-modern stills use either direct heating with gas or indirect heating with steam. It has been reported that directly heated stills give rise to a larger quantity of secondary products, due to the thermal degradation of insoluble materials caused by local superheating (Graham 1940a, b, c). This is said to enhance flavour, although others argue that indirectly heated stills are better due to the absence of furfural from the finished product (Simpson 1968).

The output stream from a pot still is usually separated into three or four fractions. The first fraction to come over from the wash still – the head, feints or foreshot fraction – contains the aldehydes and some of the esters. The second fraction, the low wine, is sent to the low wine still for further distillation. It contains virtually all of the ethanol. In some processes this fraction is split into two, a weak low wine and a strong low wine fraction. In such cases the low wine fraction is usually recycled, and only the high wine fraction is sent to the next distillation unit. The third fraction from the wash still, the pot ale, contains virtually no alcohol or other important by-products, and is either dumped, used as fertilizer or used in a dry form as animal fodder.

The product from the low wine still is also divided into several fractions, usually four. The first and third fractions, the heads and tails, are recycled, whereas the fourth, the spent lees, is dumped. The second fraction, the spirit fraction, is collected in the spirit receiver for subsequent storage and bottling.

The continual recycling of the heads and tails fractions should, over the months of operation, increase the concentration of aldehydes and fatty acids in the spirit. This is not the case, possibly due to the catalytic effect of the copper in removing these components (Watson 1983).

## 16.3 Brandy

Brandy is a term applied to the product obtained by the distillation of wine or fermented juice of any fruit (Valaer 1939). As such it has a widespread 'origin', and different production methods exist from country to country. Although continuous stills are frequently used in the production of Californian brandy, the brandies of France, Cognac and Armagnac are distilled exclusively by pot stills (Ray 1973). The wine used in Cognac is characterized by a low concentration of alcohol, 6.0–6.8% on average (Herstein & Jacobs 1948), and a high acidity (Valaer 1939). The size of the stills used are small, varying from 0.9 to 2.5 $m^3$, which is the maximum legal limit for the low wine stills (Ray 1973). Cognac distillation is a slow process taking between 10 h (Herstein & Jacobs 1948) and 18 h (Graham 1940a, b, c). This slow distillation is said to ensure that a larger fraction of the congeners (the secondary products) are included in the final product (Valaer 1939).

The head fraction in such a distillation is low, varying from 1 to 5% of the charge to the still (Simpson 1971, Graham 1940a, b, c). The low wine fraction from the wash still and the spirit fraction from the low wine still are approximately 30–35% of the charge to the stills (Graham 1940c). The concentration of ethanol in the low wine fraction is 22.1–23.0% (Simpson 1971), and the concentration in the spirit fraction varies from 62.4% at the start of the runoff to 42.5% at the end (Graham 1940a, b, c, Valaer 1939).

## 16.4 Whisky

Whisky is the generic name for the spirit obtained by the distillation of fermented extracts of malted barley or other cereals. It is produced worldwide, and the character of the whisky produced depends to a large extent upon the geographical location of the producer. Scotch whisky is produced solely within the country of Scotland from malted barley and distilled in pot stills (malt whisky), or from other cereals and distilled in continuous stills (grain whisky). The Scotch is marketed either as malt whisky, single malt or blends of two or more malts, or a blend of malt and grain whiskies. In the last of these cases the blend may contain as many as 20–30 individual whiskies, of which 30–40% are malt whiskies (Lyons & Rose 1977).

The production of Scotch malt is somewhat simpler than that of Cognac, in that only one product stream is taken off the wash still. During the distillation of the wash which contains on average 6.8% ethanol, no heads or tails fractions are separated. All of the ethanol is accumulated over a period of 5–6 h in the low wine fraction. This fraction is approximately 30–35% of the charge to the wash still, and it contains on average 21% ethanol (Lyons & Rose 1977). Only the low wine still heads and tails fractions are recycled back to the same still. The runoff for the spirit fraction starts at approximate 65% ethanol, and this fraction contains on average 56–62% ethanol.

Irish whisky, or whiskey as the preferred spelling is, is produced solely within the Republic of Eire or in the province of Northern Ireland. It is also traditionally produced in pot stills, and is sold as such (Irish pot still whiskey), or in blends with grain whisky (Lyons & Rose 1977). The pot stills used in Irish whiskey production differ from those used in Scotch whisky or Cognac production, in that a horizontal length of pipe is inserted into the swan neck, or lyne arm (Fig. 16.5). This pipe is a low point in the lyne arm, and is placed in the lyne tank which is filled with water. In this way the pipe acts as a partial condenser of the vapour. The condensate is returned to the still by means of a pipe. This arrangement has the effect of increasing the concentration of ethanol in the vapour phase, as partial condensation preferentially removes water. The

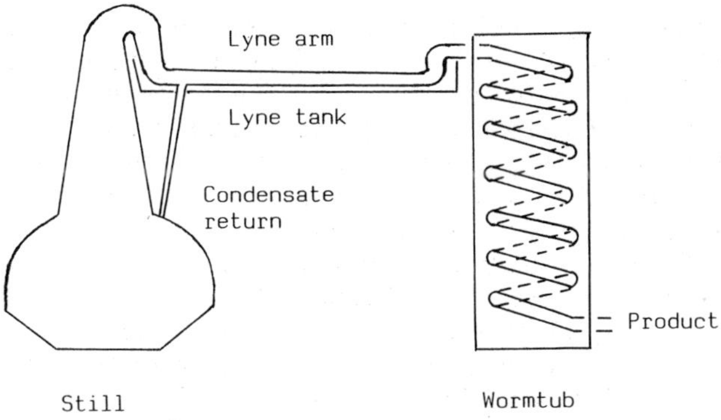


**Figure 16.5** Schematic diagram of Irish pot still.

return of the condensate to the still also ensures a more uniform concentration of ethanol in the vapour during the distillation.

In the distillation of Irish whiskey the low wine fraction is split into two, a weak low wine and a strong low wine. The weak low wine is sent on to the low wine still for further processing. Here a weak and a strong feints fraction are produced. The weak feints fraction is recycled, and the strong one is passed on to the spirit still together with the strong low wine from the wash still (Lyons 1974). In the spirit still the head and tail fractions are recycled, whereas the middle fraction is withdrawn as the spirit fraction. The ethanol concentration of the wash is approximately 6.8%, and the ethanol concentration in the first vapours 49%. This high concentration in the vapour indicates that a small amount of rectification does take place, as the vapour will only contain 46% if no rectification occurs. The strength of the spirit in this process is 80% (Lyons 1974).

## 16.5 Rum

According to a British definition going back to 1906, rum is a spirit distilled directly from sugar-cane products in sugar-producing countries. Pot distillation of rum is decreasing in importance as the demand for rum is switching to the light-bodied rums. However, such an operation is still associated with the darker, more heavy bodied, navy type rums (Lehtonen & Suomalainen 1977). A typical pot still operation contains three stills in series. The wash still, of 2.25–31.5 $m^3$ in size, produces low wine which is processed in the low-wine retort. This still then produces a high-wine fraction which is processed into spirits in the high-wine retort (I'Anson 1971). Some pot stills operate with a rectification column attached to the high-wine retort (Lehtonen & Suomalainen 1977). In all stills the heads and tails fractions are recycled.

A survey was made of the steam consumption of pot stills in Jamaica. The total steam consumption was found to vary with the size of the still, but not with the type of still (Floro 1945). A 1500-gal (6750-litre) still required between

40.78 and 43.53 lb (18.35 and 19.59 kg) steam per proof gallon, whereas a 2000-gal still required 32.02 lb. It was further concluded that the fraction of steam used in each part of the process was independent of the size and type of the still. The amount of steam required to heat the wash from the fermentation temperature to boiling was 36% of the total. The production of low wine required a further 30%, and the remainder split evenly between the production of high wine and rum.

## 16.6 Continuous stills

The development of continuous distillation columns can be traced back to the early 19th century. Although several designs were put forward at that time, the most successful design was that of Aeneas Coffey, a former Inspector-General of Excise in Ireland. This still is often called the Coffey or patent still. It was partly developed at the Islay distillery of Port Ellen in Scotland (Daiches 1969), and was very similar in design to present columns. It differed in one respect, in that reflux was not used to achieve rectification, but the same effect was obtained by passing the cold feed down through a pipe in the rectification column. By arranging the pipe to make one or two passes along each plate in the column, partial condensation of the rising vapour was achieved. As the liquid so obtained contained a larger concentration of water than the original vapour, the water was preferentially removed from the vapour. The vapour was rectified.

The simplest form of a continuous distillation column is a single-stage column consisting of two sections. The stripping section, below the feed point, strips the ethanol from the liquid phase, leaving an ethanol-free bottom product. The top section, the rectification section, concentrates the vapour coming from the stripping section, thus leaving a top product which is close to the ethanol–water azeotropic concentration. The column is fitted with devices which generate good contact between the liquid and vapour phases. For ethanol distillation the contact device is usually a tray or a plate, as it is sometimes known. This can be either a bubble cap tray (Fig. 16.6a), or a sieve tray (Fig. 16.6b), the latter being more suitable if no solid separation is performed on the fermented broth before the distillation.

The process occurring on each plate is very simple. The liquid flowing down to a plate on its journey from the top of the column is contacted on the plate by the hotter vapour coming up from the bottom. On achieving thermal equilibrium the vapour is partially condensed and the liquid partially evaporated. Both of these processes increase the concentration of ethanol in the vapour phase, leaving the liquid phase with an increased concentration of water. This process is repeated on each plate in the column. The concentrated vapour is withdrawn from the top of the column, condensed and split into two streams. One stream is the final product, the other is refluxed back into the column to provide a liquid stream for the rectification process.

The vapour required to drive the rectification is provided at the bottom of the column in the form of steam. The steam can either be directly injected into the column, or it can be used to generate a vapour in the reboiler from the watery

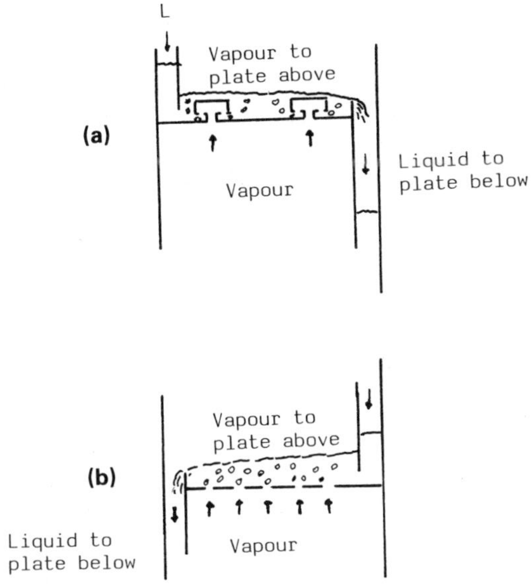


**Figure 16.6** Schematic diagram of (a) a bubble cap plate and (b) a sieve tray.

liquid leaving the column at the bottom. The former is often used in ethanol distillation, although columns operating with direct steam injection require a larger number of plates than columns with a reboiler. However, this additional cost is usually more than offset by removing the need to purchase a reboiler, and by removing the heat-transfer problems normally associated with boiling liquids containing proteins and salts.

The presence of the congeners has implications for the design of the continuous distillation system. If beverage spirit is to be produced, the still is often a single column with sufficient plates in the rectification section to yield a vapour of the same ethanol concentration as required for the spirit. No further separation of the different components is performed.

If a neutral potable or a high grade industrial spirit is required, then several columns are used. The number of columns depends upon the purity of the product required, and upon the amount of ethanol to be recovered from the waste streams. Congeners which are more volatile than ethanol over the entire ethanol concentration range are withdrawn at the top of the rectification column. The vapour, composed mainly of aldehydes and ethanol, is condensed, and the majority of the condensate used as a reflux to the column. The remainder may be passed on to an aldehyde-concentrating column for recovery of the ethanol. The main product stream is in this case withdrawn as a liquid sidestream from a point near the top of the rectification column. The bottom product of the column is the spent wash, which may be dried for animal fodder.

The product stream from the rectification column may then be passed on to an extractive distillation column. The product is diluted with water to give an ethanol concentration of around 10% by volume, and this is fed to the column at a point near the middle. The dilution is sometimes performed in the column itself, with the water and ethanol streams being fed separately. The presence of water renders the congeners more volatile than the ethanol, and the remaining aldehydes present are removed at the top and refluxed to the column. Takeoff points at the side of the column are available for the withdrawal of fusel oil and esters. Several such points are normally provided above the feed point, so that the streams can be withdrawn where the concentration of the appropriate component is the greatest (Fig. 16.7). It is normal to have separate takeoff points for fusel oil and for low- and high-boiling point esters. Where no ester takeoff points are installed the takeoff points for fusel oil may be used, as they concentrate at approximately the same location in the column. The fusel oil is usually sent to a concentrating column for recovery of the ethanol. The bottom product from the extractive distillation column is a dilute but pure ethanol–water mixture. This stream is then concentrated in a second rectification column, yielding a pure concentrated ethanol top product, and a bottom product of high-quality water.

In the aldehyde-concentrating column water is again used to separate the aldehydes from the ethanol. Sufficient plates are used above the feed point to achieve concentration of the aldehydes. The bottom section of the column contains sufficient plates to strip the aldehydes from the ethanol–water

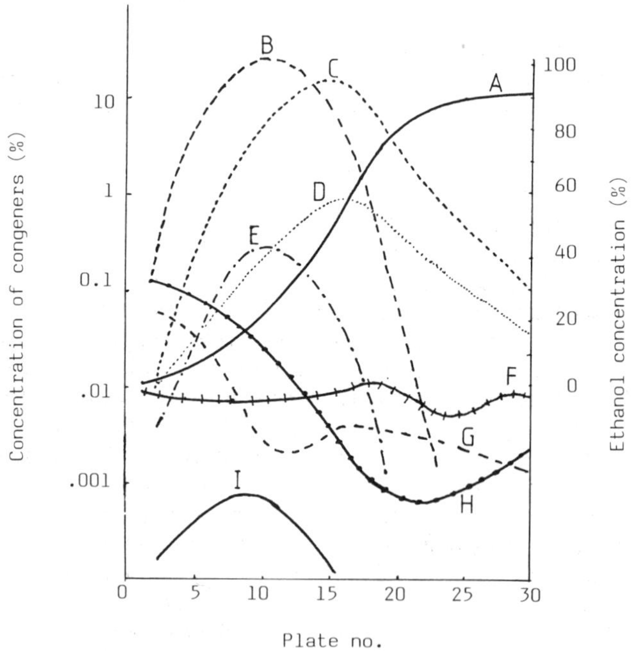


**Figure 16.7** Changes in vapour concentration in a rectifier. A, ethanol; B, amyl alcohol; C, isobutanol; D, *n*-propanol; E, *n*-butanol; F, total aldehydes; G, total acids; H, total esters; I, furfural (Pyke 1965).

mixtures, leaving a dilute mixture as a bottom product. This product is then recycled to the extractive distillation column.

The fusel oil concentrating column separates the fusel oil–water–ethanol mixture so that the ethanol is given off as the top product and is recycled to the extractive distillation column. The bottom product from the fusel oil column is water, while a mixture of fusel oil and water is withdrawn as a side product. This product is sent to a decanter for the separation of oil and water.

## 16.7 Production of anhydrous ethanol

The production of anhydrous ethanol requires special techniques, such as extractive or azeotropic distillation. Both methods require the addition of a third component to the system, but the requirements of these components vary between the distillation techniques. Furthermore, an additional column may be required to recover the third component.

### *16.7.1 Azeotropic distillation*

In azeotropic distillation the role of the third component, the entrainer, is to reduce the chemical attraction between the molecules of the different chemical species in the original mixture (Holland *et al.* 1981). Thus, a non-polar chemical may be added to a mixture of two polar components in order to 'break' the azeotrope. It is a further requirement of the entrainer that it should form an azeotrope with at least one of the original components. This azeotrope must be more volatile than the two original components, and form two liquid phases on condensation.

For ethanol–water distillation several entrainers are suitable, benzene, trichloroethane, pentane, diethyl ether, cyclohexane, carbon tetrachloride, butyl chloride and ethyl acetate (Scheibel 1950). The process for the dehydration of ethanol using benzene was developed in Dublin at the turn of the century by Young (Wentworth & Othmer 1940). He showed that benzene forms a ternary azeotrope with water and ethanol, and that this azeotrope is more volatile than ethanol, with a boiling point temperature of 64.85°C (Norman 1945). The composition of the azeotrope is 18.5% ethanol, 74.1% benzene and 7.4% water, and this is split into two liquids on condensation, a benzene layer containing 12.7% ethanol and 1.3% water, and a water layer containing 4.8% benzene and 52.1% ethanol (Othmer 1978). As the azeotrope contains a higher proportion of ethanol than water, preconcentration of the fermented broth must be performed. The concentrated product, 95% ethanol, from a rectification column is fed to the azeotropic column. The entrainer is introduced into the column through the decanter, and fed to the top of the column as a benzene layer saturated with water (Fig. 16.8). The bottom product from the column is anhydrous ethanol containing traces of water (Norman 1945). The vapour taken off at the top of the column has a composition near to the azeotropic composition, and is condensed and fed to the decanter. The benzene layer is used as reflux to the azeotropic column, whereas the water layer is passed to a stripping column to recover the small amount of

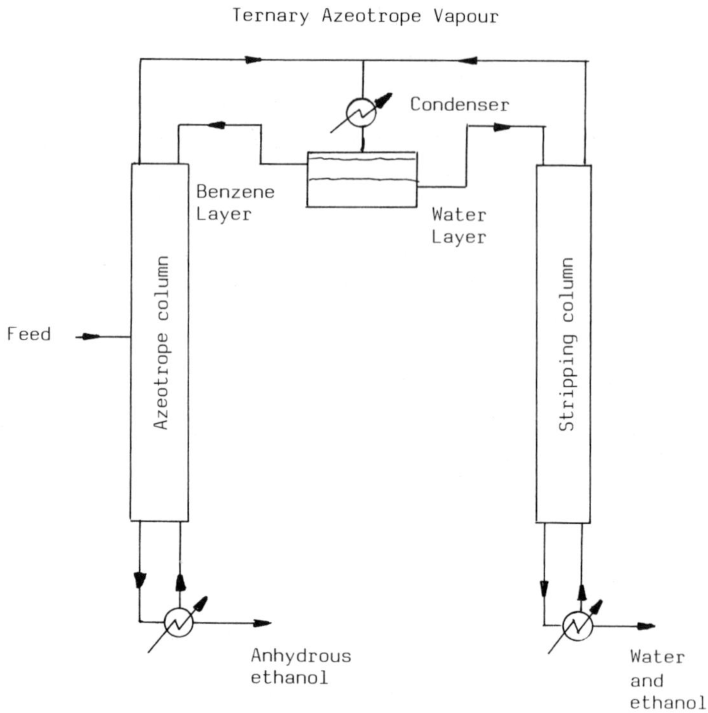


**Figure 16.8** Schematic diagram of azeotropic distillation system.

benzene in this layer. The vapour from the stripping column, being pure benzene, is recycled to the decanter via the condenser. The bottom product from the stripper is a water–ethanol mixture containing typically 46% ethanol. This stream is returned to the rectification column as a second feed to this column.

If pentane is used as an entrainer a similar system is used. The concentration of water in the ethanol product is considerably less, and contains no entrainer (Black 1980). The vapour from the azeotropic column contains 8% ethanol and 61.3% water, the remainder being entrainer. The bottom product from the stripper column is in this case pure water, and the top product a mixture of water and alcohol virtually free of entrainer.

The use of diethyl ether as an entrainer gives rise to a slightly different process flowsheet, in that this entrainer forms a binary azeotrope with water only. This azeotrope is highly volatile, with a boiling point of 34.15°C at 1 atm pressure. Thus, entrainer and water are removed in the overhead vapour, and anhydrous ethanol is withdrawn as the bottom product. This has the advantage that only the component present in the smallest concentration in the azeotropic ethanol is evaporated. However, problems occur in that the water–entrainer azeotrope contains only 1.25% water. This, coupled with the high water solubility in the liquid ether refluxed to the column, 1.15% at 15°C, means that only small quantities of water are removed in the decanter. Operation under increased pressure increases the concentration of water in the azeotrope. At 7.8 atm absolute pressure the water content is 3.76%, thus approximately two-thirds of the water in the azeotrope is removed in the decanter. For such an operation 40 kg ether are required per kilogram of water removed. This may appear to be a large quantity to evaporate for the removal of such a small quantity of water, but this is offset by the low heat of evaporation of the ether, only 12% that of water.

A comparison has been made between the three different entrainers for the azeotropic distillation. In a comparison between benzene, diethyl ether and pentane, it concluded that pentane gave the best operating performance. It gave the driest ethanol product, less than 0.0008% water, as water forms the greatest non-ideal mixture with this entrainer. Diethyl ether gave a water content in the product of between 0.0035 and 0.0047%. Benzene yielded a water content of around 0.0012%. There was very little difference in the energy input requirements between benzene and diethyl ether, whereas for pentane only 50–65% of the energy required for benzene and diethyl ether was used. The exact amount depended on the entrainer to ethanol mole ratio. This ratio could be as low as 2.5 for pentane and benzene, whereas for diethyl ether the ratio had to be above 4. The use of pentane further led to smaller column diameters, as the largest flow on any tray in the column was only between 70 and 75% that of comparative conditions with the other entrainers. Shorter columns were also indicated by the use of pentane, with the number of equilibrium stages being approximately 20 with pentane, 30–40 with benzene and 50–55 with diethyl ether (Black *et al.* 1972).

### *16.7.2 Extractive distillation*

In extractive distillation the third component is added to increase the relative volatility between the two original components. The third component, the solvent, behaves as indicated earlier, in that it forms non-ideal solutions with each of the original components, but the degree of this non-ideality differs for each of these two components. Thus, their relative volatility changes. The requirements of the solvent are different from that of the entrainer in azeotropic distillation. The solvent should be less volatile than either of the original components in the binary mixture, and it should be easily separated from the component with which it deviates less from ideality. Furthermore, it should not form an azeotrope with either of these components. There should therefore be a greater number of possible solvents available than entrainers, although this does not appear to be the case for ethanol–water systems.

Ethylene glycol has been used extensively as a solvent for the extractive distillation of ethanol. As this solvent associates chemically with water, it causes the volatility of ethanol to increase relative to that of water. The water content of the absolute alcohol will depend on the solvent to feed ratio (s:f) and the reflux to feed ratio (r:f). A small s:f ratio (2.5) will give a relatively high water content (130 ppm) at the optimum r:f ratio, whereas a high s:f ratio (4) will give a low water content (20 ppm) (Black & Ditsler 1972). However, this increase in the s:f ratio will increase the steam requirement by nearly 40%. The

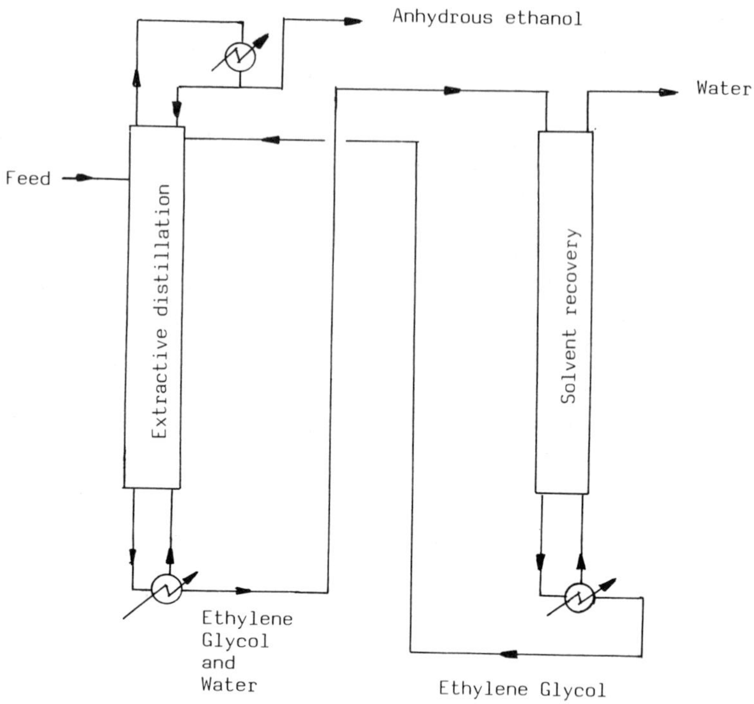


**Figure 16.9** Schematic diagram of extractive distillation system.

bottom product from the extractive distillation column is a solvent–water mixture which is passed to a solvent recovery column (Fig. 16.9). In this column water is distilled from the solvent, and the solvent recycled to the extractive distillation column.

In a comparison between extractive distillation using ethylene glycol and azeotropic distillation using pentane, Black and Ditsler (1972) concluded that the azeotropic distillation gave a better performance. It required fewer trays in the dehydration column, 18 compared with 43 for a given separation. It required a thinner column, column diameter <5 ft (1.5 m) compared with 5.3 ft. Furthermore the steam requirement using pentane was only 45% of that using ethylene glycol.

### *16.7.3 Use of salt in extractive distillation*

The use of dissolved salt as a solvent in extractive distillation has been extensively researched by Furter and co-workers (Johnson & Furter 1957, 1960, 1965, Meranda & Furter 1966, 1971, 1972, 1974, Cook & Furter 1968, Jaques & Furter 1972, Furter 1968, 1972). The dissolved salt affects the activities of the two original components present by forming association complexes with one of them. Such preferential attraction will usually be made with the more polar component in order to decrease the activity of water in the mixture. In this manner the relative volatility of ethanol to water is increased. This is sometimes called salting out (Furter 1972). Some salts, particularly potassium nitrate and mercury chloride, have the effect of salting-in the ethanol, thus making the separation more difficult.

Several salts have been found to enhance the separation of ethanol from water. However, it appears that potassium acetate is the best because only small concentrations of salt are required to 'break the azeotrope'. Utilizing a salt concentration of only 5% (mole) the azeotrope can be completely removed (Cook & Furter 1968). It can also be seen that the relative volatility of the ethanol has increased over the part of the concentration range where separation is difficult, i.e. where the equilibrium curve is close to the diagonal (Fig. 16.10).

Several problems have been identified in using dissolved salt as a solvent. These are mainly associated with preventing the vapour from the boiling reflux from entering the salt-handling and metering systems. Should this occur the efficiency of the distillery would decrease through increased ethanol losses. Vapour in the solid-handling system can also lead to salt-bridging, and therefore loss of salt flow to the system (Cook & Furter 1968). However, these problems have not deterred the large-scale operation of such plants, as is evident from the literature on the subject (Furter & Cook 1967).

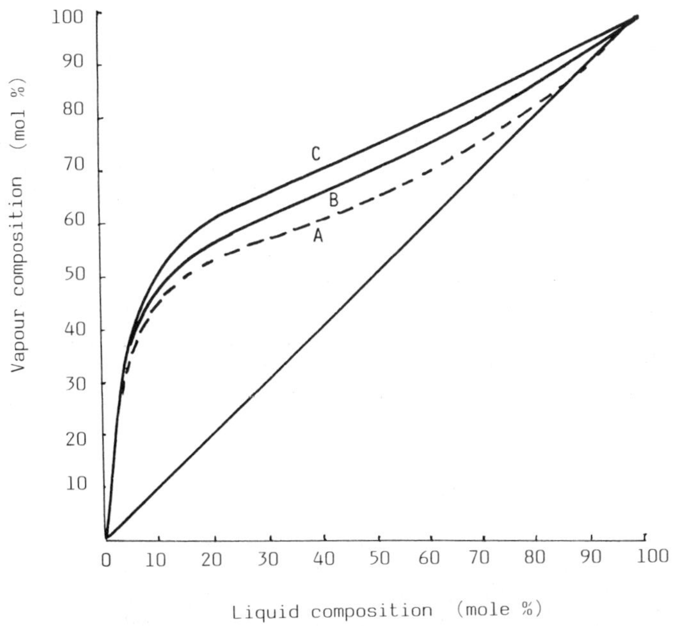


**Figure 16.10** Effect of potassium acetate on the vapour–liquid equilibrium for ethanol–water mixtures. A, pure ethanol–water systems, B, 5 mole % potassium acetate, C, 10 mole % potassium acetate (Cook & Furter 1968).

## 16.8 Energy requirements

The energy requirement of ethanol distillation is often given special attention, particularly when the final product is industrial or fuel alcohol. This is due to the influence of the steam cost on the final production cost. For a grain distillery the steam cost for distillation alone can be as high as 40% of the total cost when excluding the sugar cost (Cysewski & Wilke 1978). As the sugar contributes between 55 and 70% of the total cost, the steam will contribute some 5–15%, depending upon the circumstances, and is one of the major contributors to the production cost.

The minimum energy requirement is dependent upon the feed concentration and the final product concentration. The more concentrated the feed is, the less steam is required, and the more concentrated the final product is, the more steam is required per litre of product. For ethanol distillation there is a further complication; as the product concentration approaches the azeotropic composition, so the energy requirement increases sharply (Fig. 16.11). This is due to the small differences between the equilibrium concentration of ethanol in the vapour and liquid phases at these high concentrations. Increased reflux is required to achieve any practical differences between these concentrations on the trays, and this is obtained at the expense of increased vapour flowrate, and hence increased steam consumption. For normal fermentation broths of 6.8% ethanol, this sharp increase occurs at a final product concentration of 94.0%. The minimum steam requirement per litre of distilled product increases nearly linearly with the product composition up to this value.

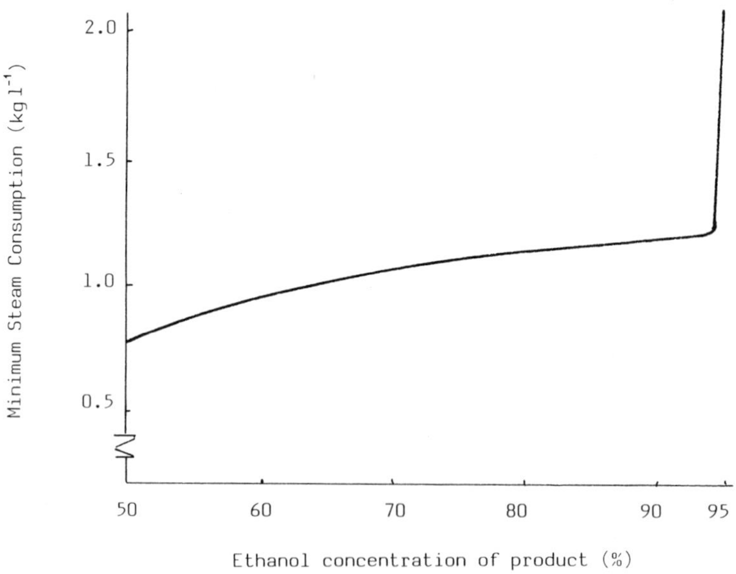


**Figure 16.11** Minimum steam consumption for the distillation of broths containing 6.8% ethanol.

The actual energy requirement will be higher than the minimum, and there are large variations from plant to plant. Plants installed when energy was relatively inexpensive use considerably more energy than the minimum, whereas the more-modern plants will be highly energy efficient. Some old plants may use up to 21.9 kg steam $l^{-1}$ of product (Marion 1979), whereas it has been claimed that 'an undisclosed rearrangement of conventional process techniques' only requires 0.72 kg $l^{-1}$ (Anonymous 1981). The factor governing the steam requirement is the reflux ratio, the amount of liquid returned to the column at the top to the amount withdrawn as product. If we can assume that the feed to the column is at its boiling point, the steam consumption per unit distillate can be equated to this ratio by

$$V/D = R_D + 1$$

The economical optimum $R_D$ will vary according to the relative magnitudes of investment and operating costs, but if the cost of energy increases more rapidly than the investment cost, then the optimum $R_D$ will move close to the minimum.

If the ethanol losses in aldehyde- and fusel oil-concentrating columns can be assumed to be of the same order as the increase in flowrates due to the presence of these compounds, the flow diagram for the rectification column may be represented by Figure 16.12. It is noted that for this system the steam consumption is 1.67 kg $l^{-1}$ of product, and that cooling water is required at a rate of 18.2 kg $l^{-1}$ for the condensation of the overhead vapour. Cooling-water is also required for vent condensers and product cooling, but the amounts required for these operations are considerably less than that needed in the product condenser. Nearly 96% of the energy in the steam is transferred to the cooling-water in the condenser, and this heat is often lost, as distilleries frequently have a surplus of hot water. A large amount of energy could be saved if the heat removed in the condenser could be upgraded to be used in the reboiler.

For beverage spirit the energy requirements are considerably less. A continuous whisky still will consume between 1.4 and 1.8 kg steam $l^{-1}$ of product. This lower steam requirement is due to the lower concentration of ethanol in the product (Packowski 1978).

If anhydrous ethanol is to be produced, additional energy is required to drive the dehydration and solvent and entrainer recovery columns. It has been estimated that when using pentane as an entrainer to produce essentially pure ethanol from 95% ethanol, steam is required at a rate of between 0.3 and 0.55 kg $l^{-1}$ of product for the dehydration column alone. If benzene is used, the steam requirement for the same column would be between 0.55 and 0.7 kg $l^{-1}$, and for diethyl ether between 0.58 and 0.68 kg $l^{-1}$ (Black *et al.* 1972). The additional requirement for the pentane recovery column would be approximately 0.07 kg $l^{-1}$ (Black & Ditsler 1972), whereas it has been estimated that a total of 1.0–1.1 kg $l^{-1}$ steam is required for both the dehydration and solvent recovery columns when using benzene (Packowski 1978, Maiorella *et al.* 1981). These values compare with an estimated 0.71 and 0.32 kg $l^{-1}$ for the dehydration and solvent recovery columns when using ethylene glycol (Black & Ditsler 1972).

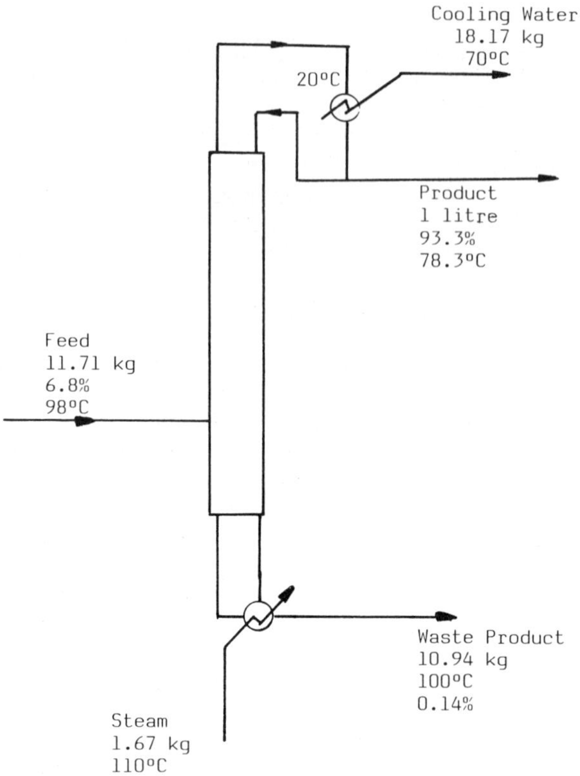


**Figure 16.12** Mass flow-rates for a conventional ethanol distillation system. (Basis 1 litre product of 93.3% ethanol; feed 6.8% ethanol; operating reflux 1.25 × minimum; recovery 95%.)

An estimate of the steam requirement when using dissolved potassium acetate as the solvent, indicates that approximately 1.2 kg $l^{-1}$ steam is required when producing 99% ethanol from 6.2% ethanol feed, and using 5% (mole) salt. Although no further improvement is obtained by using 10% salt, the energy requirement compares very favourably with conventional anhydrous ethanol production. However, there is the additional cost of the salt to be considered, and if a salt recovery system is used then 12.2 kg water $l^{-1}$ of product must be evaporated. Additional cost will also be imposed by the need to remove any solids present in the liquid before the addition of the salt.

## 16.9 Energy conservation

The problem in using the energy from the overhead vapour to drive the reboiler is that the energy is available at a lower temperature, 78°C compared with the 100–110°C required in the reboiler. To upgrade this energy the vapour must be available at a higher pressure, as the boiling point and dew point temperatures of saturated liquid and vapour increase with pressure. This increase in vapour pressure can be achieved in two ways. Othmer (1936) advocated that the beer still, the stripping section of the rectification column, should be operated at a pressure of 4.3 atm absolute. The overhead vapour could then be used in the reboilers of the rectifier, and the aldehyde-concentrating columns, when these columns were operated at atmospheric pressure. The same technique can be applied to the distillation of ethanol itself, and it has been estimated that such a multi-effect distillation system would reduce the steam consumption to 1.0 kg $l^{-1}$ (Busche 1983).

In the azeotropic distillation using diethyl ether, the overhead vapour temperature is above 100°C. This makes this vapour reuse method possible, and a small industrial plant was found to require 3.1 kg $l^{-1}$ steam, when producing anhydrous ethanol from an aqueous solution containing 7.4% ethanol. It was further estimated that a full-scale plant would require between 2.6 and 3.3 kg $l^{-1}$ steam (Wentworth *et al.* 1943). A similar effect can be achieved with a beer column operating at atmospheric pressure and the subsequent columns under vacuum (Geyer 1979). Vacuum distillation is of particular interest to ethanol recovery in that at a pressure below 0.12 atm, ethanol and water do not form an azeotrope (Busche 1983). However, it has been shown that to dehydrate a 10% ethanol mixture under vacuum, the steam requirement would be greater than for conventional dehydration, and the columns would be excessively large (Black 1980).

The alternative method to make the energy in the vapour available is to compress the vapour, and then use the reboiler as a condenser for this vapour. The flow diagram for the plant given in Figure 16.12 would then be as shown in Figure 16.13. To obtain the required temperature the vapour is compressed to 3 atm, at which pressure the temperature is approximately 110°C. The vapour is then condensed in the reboiler, and cooled under pressure to 78°C. At this temperature the pressure is released, and part of the now saturated liquid is used as reflux. The remainder is passed to the extractive distillation column or to a product cooler. The available heat in the compressed vapour is equivalent to 1.53 kg steam $l^{-1}$ of distillate. This is further increased by the work done by the compressor, and if all the theoretical work done on the vapour ends up as heat in the vapour, then an additional equivalent of 0.15 kg $l^{-1}$ steam is available. Thus, by use of vapour recompression the distillation column is self-sufficient in heat energy at the expense of an additional 0.14 kWh electrical energy $l^{-1}$ of distillate. It should also be noted that the cooling-water requirement has been reduced significantly from 18.2 to 0.6 kg $l^{-1}$.

The advantage of such a system will depend on the relative costs of steam, electricity, cooling-water and capital investment, but it should be economically viable where the steam cost is relatively high compared with the cost of electricity. For New Zealand conditions, where the cooling-water cost can often be taken as virtually zero, and with a relative cost of 1 kWh to 1 kg steam of 4:1, a saving of 60% can be made on the utility cost. It has been estimated that the relative investment between distillation with vapour recompression and conventional distillation is a factor of 1.4 (Essien & Pyle 1983). Such an increase in the investment should only have a minor effect on the total

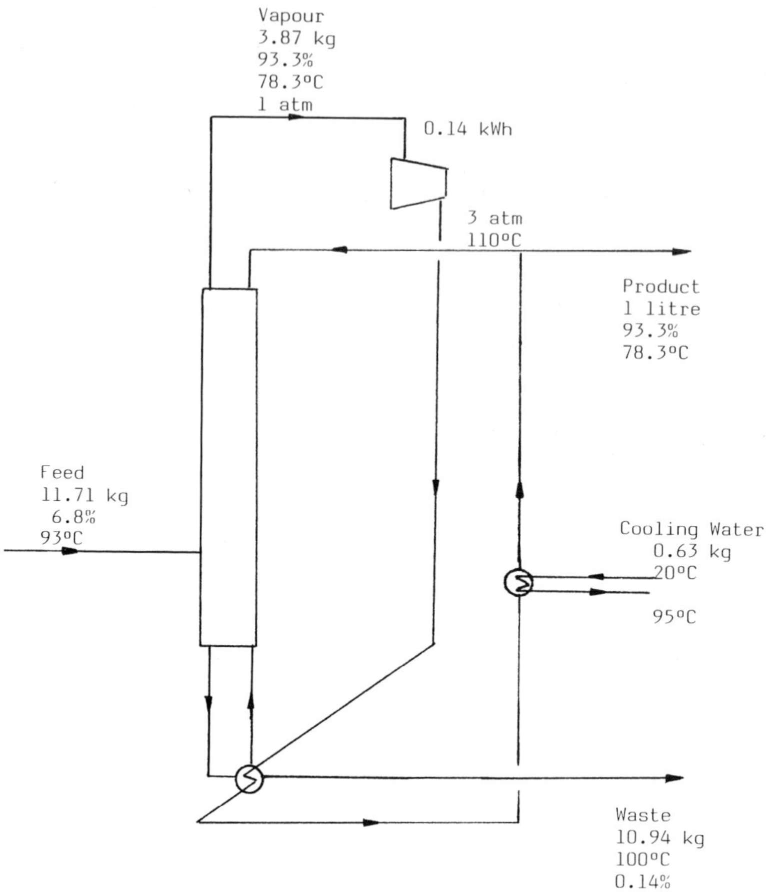


**Figure 16.13** Mass flow-rates for ethanol distillation systems with vapour recompression. (Basis, see Fig. 16.12.)

production cost, as the ethanol recovered accounts for some 12% of total investment for large-scale ethanol plants. It therefore contributes only 3–4% of the total annual production cost (Cysewski & Wilke 1978). This compares with a value of between 5 and 15% for the steam cost.

For older plants operating at a large reflux ratio, the retrofit of extra plates may be of considerable benefit. This will enable the reflux ratio, and hence the steam consumption to be reduced, and the desired separation to be maintained (Mix *et al.* 1978).

## 16.10 Alternatives to distillation

Due to the high demand for energy by a conventional distillation process, the different alternatives to distillation have received considerable interest.

Although this interest is not a new phenomenon, it was the energy crisis in the 1970s that started the real interest. Processes like freeze-concentration, adsorption, liquid-extraction and the membrane processes of reverse osmosis and pervaporation have all been explored. In the vast majority of experiments with these 'novel' methods, the ethanol has been recovered from a pure ethanol–water mixture, thus no allowance or consideration has been given to the presence of congeners. Furthermore, there is often a considerable lack of technical data, making a full evaluation of these processes difficult.

### *16.10.1 Solvent extraction*

Solvent, or liquid–liquid, extraction is a potential alternative to the distillation process in recovering a solute from a solution. Instead of relying on the addition and removal of heat, and on a difference in volatilities, solvent extraction relies on the use of an additional phase, and on a difference in solubilities. This operation has traditionally been used when the recovery by distillation has been difficult or impossible. The recovery of the solute is achieved by extracting it into a solvent and then separating it from this solvent by traditional means. Thus, two processes are required in the separation compared with one in distillation.

For the extraction of ethanol from an aqueous solution, the solvent must have a high affinity for ethanol, due to its low concentration in the feed. This requirement is said to preclude most common solvents (Essien & Pyle 1983), and even the more suitable solvents do not have the affinity to produce ethanol at a high concentration directly from a fermented broth. In summarizing their own work and that of Roddy (1981) and Souissi and Thyrion (1981), Munson and King (1984) showed that chlorinated hydrocarbons and aromatic compounds gave the best compromise between selectivity and low solvent requirements. They did, however, specify that a selectivity of 610 would be required if the ethanol concentration in the product were to exceed that of the azeotropic composition. This compared with the highest value quoted of 150 for tetrachloroethane. However, it is unknown whether this value of 610 was estimated for extraction systems with or without reflux.

A process proposed by Essien and Pyle (1983) used ethyl trichloroacetate as the solvent. This process yielded an extract phase containing 80% ethanol on a solvent-free basis. This extract was then further processed in two distillation columns, one for the recovery of the solvent, the second for the further concentration of the ethanol to 95%. The steam requirement for this process was estimated to be 1.5 kg $l^{-1}$, although half of this was required at a pressure twice that required in the distillation. This ethanol recovery system also required four times the capital investment of that of the distillation alone.

The use of a proprietary solvent is said to eliminate the need for further concentration of the ethanol, when producing anhydrous ethanol from grain. This process consists of a solvent extractor and one distillation column to separate the ethanol and the solvent. It is claimed that the total energy requirement for the plant is only 1.44 kg $l^{-1}$ steam (Anonymous 1980b).

The solvent-extraction process has also been evaluated for the production of anhydrous ethanol from azeotropic ethanol. Scheibel (1950) proposed a process using fractional solvent extraction to dehydrate the ethanol. In this process two solvents were used, one with a high affinity for ethanol and one with a high affinity for water. Ethylene glycol was used as the solvent with affinity for water, whereas both methyl *n*-amyl ketone and *o*-xylene were tested as the solvent with high affinity for ethanol. The proposed plant produced ethanol with a water content of 0.4%. The optimum volumetric ratio of ketone to glycol was 11.7, whereas the optimum *o*-xylene to glycol ratio was 143. The two solvents were fed in at either end of the extraction column, with the feed added at an appropriate point in between. The two extract phases were then passed to two distillation columns for separation of the solvents from ethanol and water, respectively. The use of *o*-xylene gave rise to the lower steam consumption with 1.4 kg $l^{-1}$ steam being used to separate ethanol and *o*-xylene and 0.11 kg $l^{-1}$ to separate water and glycol. The values obtained for the steam consumption when using the ketone were 1.6 and 0.2 kg $l^{-1}$, respectively.

The use of solvent extraction to dehydrate ethanol for gasohol production has also been proposed. This has the advantage that if petrol can be used as the solvent, and if the ethanol concentration in the extract exceeds 10%, then the need for further separation of ethanol and solvent would be eliminated. In such a process it was found that to achieve the required extract concentration the ethanol had to be preconcentrated to approximately 90% (Leeper & Wankat 1982). By using unleaded petrol an extract of 13% ethanol could be obtained. However, the ethanol recovery was only 57%, with a solvent concentration of 15% and a solvent loss of 1.8% in the raffinate. These values were obtained with a solvent to feed ratio (s:f) of 3.1, but it was shown that they could be improved by using higher s:f ratios. Using naphtha to simulate petrol, a recovery of 68% was achieved with a solvent loss of 0.8%, at an s:f ratio of 5.9. Despite this low solvent loss it was estimated that solvent recovery from the raffinate phase would be required. The energy requirement for the dehydration process was equivalent to 0.31 kg $l^{-1}$ steam.

### *16.10.2 Dense-gas extraction*

As a chemical engineering process, dense- or supercritical-gas extraction is still in its development stage. It lies somewhere between distillation and solvent extraction, and differs from both in important process details. The separation factor for the process does not only depend upon the temperature, as in distillation, but also on the pressure. The extraction process differs from solvent extraction in that a reduction in pressure is used to separate solvent from solute and not heat. The process relies on the large variation observed in the physical properties of the fluid for small changes in pressure or temperature near the critical point (Irani & Funk 1977). By careful manipulation of the temperature or pressure around the critical point, the solvent power of the fluid can be varied from that of a gas (low), to that of a liquid (high) (Williams 1981). The additional advantage of using dense-gas extraction is that the diffusivity of the solute in the fluid is greater than in a liquid and that the density is lower. This means that the mass transfer rates are increased, and that the separation of the two resulting phases, extract and raffinate, is easier (Filippi & Moses 1982).

Carbon dioxide has been used as a solvent in dense-gas extraction in coffee decaffeination (Filippi & Moses 1982), and in the extraction of hops for the brewing industry (Scott *et al.* 1981). It has also been proposed as a solvent for ethanol extraction. Equilibrium data for the ternary system of water–ethanol–$CO_2$ at 100 atm and 35°C have been published by Paulaitis (1982). Using $CO_2$ as a supercritical fluid (SCF) at 100 atm and 35°C, this diagram indicates that 80–85% ethanol is the maximum possible concentration of ethanol on a solvent-free basis. This is also only possible when the extraction is employing reflux, as a feed of 8% ethanol will give rise to an extract of approximately 70% on a solvent-free basis. A similar situation exists when using a pressure near the critical (73 atm), but with a temperature below the critical (25°C). Another aspect of the process is that the extract will contain in excess of 90% carbon dioxide. Thus, large quantities of solvent are required, and this solvent must be stripped from the extract and recompressed for reuse. Although the energy usage of such a process is lower than conventional distillation, 0.26 kWh $l^{-1}$ and 1.25 kg steam $l^{-1}$, the capital cost of such a system is high (Busche 1983).

### *16.10.3 Adsorption*

The use of solids to dry ethanol was demonstrated more than 40 years ago with calcium oxide (Ladisch & Dyke 1979). Since then several solids have been tested for their ability to separate ethanol and water. They vary in nature from inorganic compounds such as silicalites to organic compounds such as activated carbon, starch and cellulose.

Silicalite was found to adsorb ethanol preferentially, and a 2% ethanol–water liquid mixture could be concentrated to 35% ethanol. The process required 14 kg adsorbent $kg^{-1}$ ethanol adsorbed at 20°C, and this had to be heated to approximately 130°C to recover the ethanol (Milestone & Biddy 1981).

Some commercially available resins were found to adsorb water preferentially. However, their adsorption capacity was found to be low, with the best value being 9.1 kg resin $kg^{-1}$ ethanol adsorbed. Modified polystyrene beads also performed the same function, but with no significant improvement in the capacity. Activated carbon was found to have a significantly higher adsorption capacity, at 5.6 kg solids $kg^{-1}$ ethanol adsorbed. However, this solid could not be used as the pressure drop experienced in an adsorption column was too high (Malik *et al.* 1983). The ethanol concentration of the eluted product was found to be as high as 43%. A two-stage process could produce a 94% product when the starting concentration was 10% ethanol, with a recovery of 90%.

Liquid-phase dehydration of azeotropic ethanol has also been performed using modified starch as the adsorbent. The success of this process relied on the concurrent production of gasohol, since mixtures with high ethanol concentrations would not completely dehydrate in the absence of petrol. If enough petrol were added to yield a 10% gasohol mixture, then the process would yield a gasohol of sufficient dryness to meet the gasohol specification (Fanta *et al.* 1980).

Vapour-phase dehydration of high-strength ethanol has also been shown to be a feasible proposition. Ladisch and Dyke (1979) tested several inorganic and organic solids for their adsorption capability. Several solids were capable of producing ethanol with a water content of less than 2%, but corn starch and CM cellulose were found to give the driest ethanol. The adsorption capacity of several biomass adsorbents were tested by Hong *et al.* (1980). Their capacity was found to vary in descending order from potato and corn starch > xylan > cornmeal > Avicel > bagasse, corn residue and wheat straw.

The use of cornmeal and corn grist as the adsorbent was tested at both laboratory and pilot plant scale. Ground corn was capable of producing a vapour containing 99% ethanol from a vapour containing 49%. The dryness of the vapour increased with increasing feed concentration. The adsorption capacity of the starch was 16.7 kg starch $kg^{-1}$ water adsorbed. The process was tested over a 3-month period on a farm scale using a two-stage system containing 18 kg corn grist. The system was found to be satisfactory over this period, and consistently produced a vapour containing 99.8% ethanol.

The starch was regenerated by passing hot air (80–95°C) down through the column, and the energy requirement for regeneration and heat losses from the insulated column was 530 kJ $kg^{-1}$, or the equivalent of 0.19 kg $l^{-1}$ steam (Ladisch *et al.* 1984). The reason for this low heat requirement in the regeneration is the low heat of adsorption of starch. It has also been estimated that such an adsorption system would add 4 cents to the cost of 1 gal ethanol if the corn could be used for 20 adsorption and desorption cycles (Anonymous 1980a).

### *16.10.4 Reverse osmosis*

The membrane processes of reverse osmosis and pervaporation depend on the use of selective porous membranes to achieve molecular separation. For reverse osmosis this separation is entirely a liquid-phase operation, thus there is no need to supply heat for evaporation. In reverse osmosis low-molecular weight species are retained by the membrane, so there is a considerable osmotic pressure to overcome. As the osmotic pressure increases with solute concentration, increased pressure energy is required in order to obtain high product concentrations (Parulekar *et al.* 1982). Although reverse osmosis can in general claim to be energy efficient, its use in separating ethanol and water has until now been considered impracticable, due to the high osmotic pressure of such mixtures. A 12.1% ethanol solution has an osmotic pressure of 65 atm, and a 42.5% solution has a pressure of 250 atm. As the upper practical limit of a reverse osmosis operation is close to 65 atm, the use of this process for other than preconcentration has not been considered practicable (Lee *et al.* 1984).

However, it has been shown that a process using countercurrent reverse osmosis reduced the need for such extremely high pressures. By recycling the concentrated ethanol stream to the permeate side, small ethanol concentration differences were established across the membrane. This reduced the osmotic pressure and led to an increased flow of water through the membrane when operating at normal pressures. Although such a process was shown to be technically feasible, the ethanol concentration could not be taken above 42.5%, due to the instability of the membranes tested at high ethanol concentrations. The final permeate also had to be recycled back into the process, as this contained ethanol at a concentration just below that of the feed (Lee *et al*. 1984).

The use of reverse osmosis to preconcentrate dilute ethanol solutions has also been proposed (Dick & Mavel 1982). In order to achieve high ethanol recoveries (97%) a permeate flux of no more than $100\,l\,m^{-2}\,day^{-1}$ was needed, so such systems would only be practical on small- to medium-sized plants, due to the membrane area required.

### *16.10.5 Pervaporation*

Pervaporation is a membrane process which can be said to be a mixture of distillation and reverse osmosis, as the species permeating through the membrane is also evaporated. The separation of a mixture depends only upon the selectivity of the membrane, and the driving force for this separation is the difference in chemical potential between the liquid on the feed side and the vapour on the permeate side. To maintain large differences in this potential the vapour must be removed either by vacuum pump, gas sweeping or by cold-wall condensation. The chemical nature of the membrane greatly affects the selectivity of the process. Some membranes show a selectivity for water, others show a reversibility in the selectivity as the solute concentration changes, and some show high selectivity for ethanol.

Medical-grade silicone rubber was found to have a high selectivity for ethanol and high permeate flux of ethanol. This selectivity increased as the ethanol concentration decreased, whereas the permeability showed a minimum around 85.6% ethanol. The permeation rate was highest at very low ethanol concentrations with a value of $50 \times 10^{-9}$ mol m $(s\ m^2\ kPa)^{-1}$ or, if assuming a membrane thickness of 1 mm and a pressure difference of 710 mmHg (94.6 kPa) the value was 780 g $(m^2\ h)^{-1}$. Using a countercurrent pervaporation system consisting of two columns an ethanol stream of 18.2% was concentrated to 52.2%. However, problems did arise in that the ethanol recovery was low (76.3%), with an ethanol concentration in the waste stream of 9%. Although the recovery may have been limited by the available area in the stripping column, they proposed that such a pervaporation system was best suited to be used in an extraction system in the fermentation (Hoover & Hwang 1982). Other membranes were also found to have high selectivity for the component which was present in the smallest quantity. A polystyrene membrane was shown to have high selectivity for water when the feed was azeotropic ethanol (95.6%). It was claimed that this membrane should be able to produce a 99% ethanol product with a recovery of approximately 95% (Schissel & Orth 1984). In their extensive work on pervaporation membranes Mulder *et al*. (1983) proposed a two-stage pervaporation system for extracting and concentrating ethanol from a fermentation broth. By use of membranes selective for the component present in the smallest quantity, a 95% ethanol product could be produced directly from the broth with the minimum use of energy for evaporation. However, they stated that such a system would not be commercially attractive today.

### *16.10.6 Freeze-concentration*

When considering the basic principles, it may be argued that the freeze-concentration process resembles that of distillation. Both processes rely on the difference in concentration of solute between two phases, solids and liquid for freeze-concentration and liquid and vapour for distillation. Heat is also the means of driving both processes, although in freeze-concentration it is withdrawn, and not added as in distillation. However, the process differs from distillation in that, when sufficient heat has been removed to lower the temperature to below the freezing point of the solution, the phase now formed, ice, is free of solute.

Several large-scale plants using this concentration technique have been installed worldwide. It is presently being used to purify salt water, fractionate xylene (Heist 1979) and to concentrate vinegar (Anonymous 1961), fruit juices and beer (Wagner 1983). It has also been tested on coffee and tea extracts, sugar solutions (Wagner 1983) and wine (Bergeret 1962).

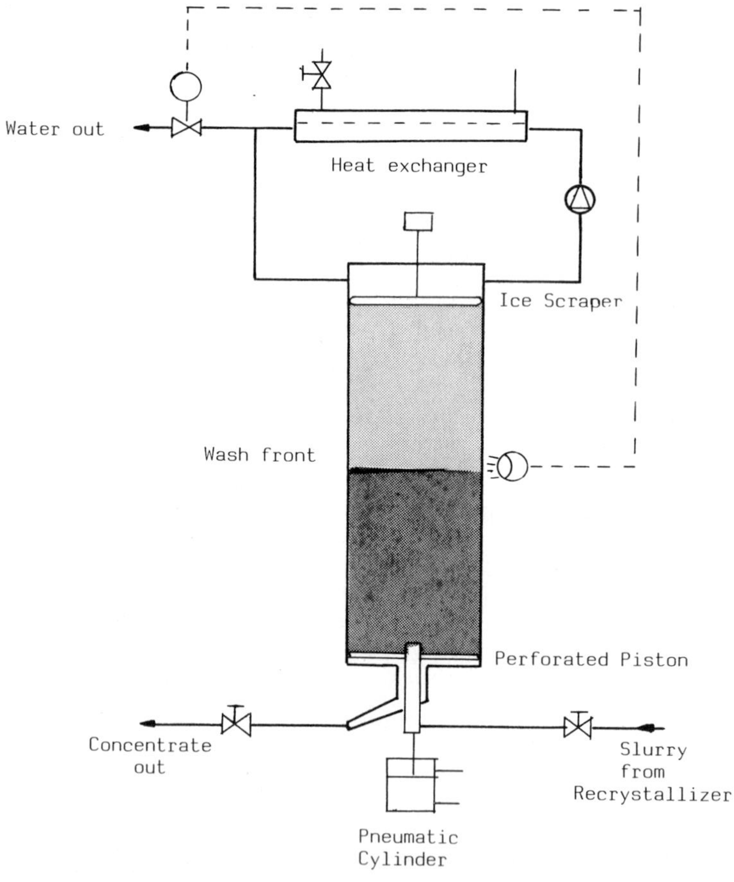


**Figure 16.14** Schematic diagram of the Grenco wash column (van Pelt 1975).

The claimed benefit of freeze-concentration is that food liquids can be concentrated without appreciable loss of aroma, taste, nutrients or colour. Such advantages have implications on the design of plants in that smaller storage capacities are required, and products can be concentrated before transportation, thus satellite plants can be used to feed larger plants. For chemical processing the advantage is that many of the corrosion problems occurring when salts are present at moderate to high temperatures are removed. The claimed disadvantage of the process is that it is more costly than distillation, due to its high capital cost (Muller 1966).

However, the cost of freeze-concentration is being reduced by the replacement of the centrifuges by wash columns. There are two different designs for these columns, that of Grenco (van Pelt 1975; Fig. 16.14) and that of Phillips (Muller 1966). The basic principles of the columns are the same. The slurry, containing pure water crystals (ice) and concentrate, is fed into the column. In the column the slurry is subjected to an oscillating pressure which forces the free concentrate out of the column, and draws wash water into the column. The wash water is obtained by melting the ice, and recycling part of this back to the column. The water washes the ice crystals countercurrently, thus removing any concentrate adhering to them. Thus, relatively little concentrate is lost with the ice when melted.

The application of freeze-concentration to the separation of ethanol and water has also been considered. Complete separation is not possible, as this mixture forms an eutectic point at −118°C and 93.5% ethanol (Mellan 1977). However, the process is less energy intensive than that of distillation due to the lower heat of fusion than heat of evaporation. The freezing point of aqueous ethanol solutions decreases sharply above 70% ethanol (Fig. 16.15), thus

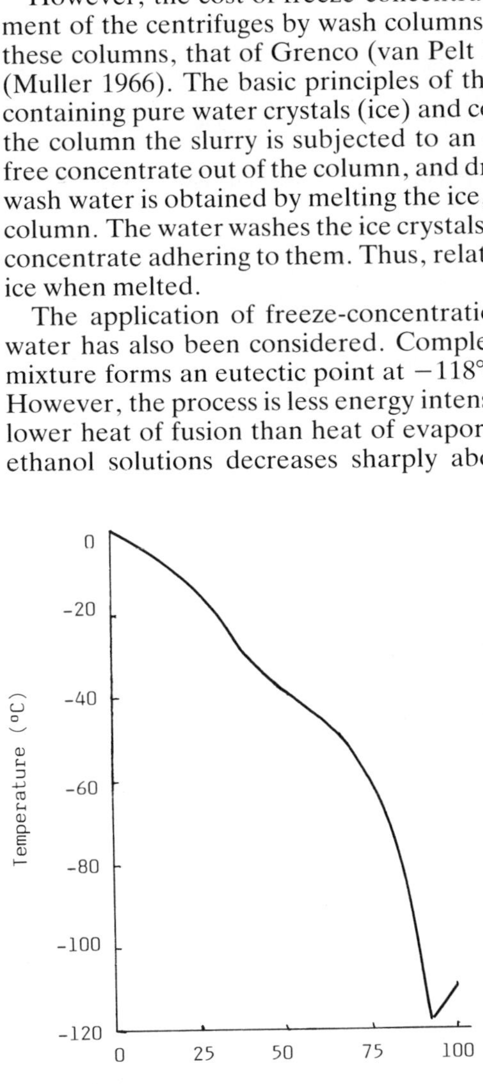


**Figure 16.15** Temperature composition diagram for liquid–solid ethanol–water mixtures (Essien & Pyle 1983).

making heat transfer that much more difficult when attempting to concentrate to these high concentrations.

A process using two-stage freeze-concentration to 70% and distillation to 95% was evaluated by Essien and Pyle (1983). Although the process utilized less energy than straight distillation, 1.0 kg $l^{-1}$ steam and 0.47 kWh $l^{-1}$ electrical energy, it was found that the capital investment required had increased by a factor of 8.5. Such a process is therefore unlikely to compete economically with distillation.

## References


Anonymous 1961. Continuous process freeze-concentrates vinegar. *Food Engineering* **33**(3), 82–3.

Anonymous 1980a. New ethanol route wears a low-energy label. *Chemical Engineering* **87**(23), 103.

Anonymous 1980b. Processing twists to further cut grain-derived ethanol costs continue to burgeon. *Chemical Engineering* **87**(18), 19.

Anonymous 1981. Energy required for distillation of ethanol in biomass-conversion facilities is reduced. *Chemical Engineering* **88**(3), 17.

Bergeret, J. 1962. Applications du froid en oenologie. Vinification-stabilisation et concentration des vins. *Le Vide* **101**, 43–9.

Black, C. 1980. Distillation modeling of ethanol recovery and dehydration processes for ethanol and gasohol. *Chemical Engineering Progress* **76**(9), 78–85.

Black, C. and D. E. Ditsler 1972. Dehydration of aqueous ethanol mixtures by extractive distillation. *Advances in Chemistry Series* **115**, 1–15.

Black, C., R. A. Golding and D. E. Ditsler 1972. Azeotropic distillation results from automatic computer calculations. *Advances in Chemistry Series* **115**, 64–92.

Busche, R. M. 1983. Recovering chemical products from dilute fermentation broths. *Biotechnology and Bioengineering Symposium* No. 13, 597–615.

Carlson, C. S., P. V. Smith, Jr and C. E. Morrell 1954. Separation of oxygenated organic compounds by water extractive distillation. *Industrial and Engineering Chemistry* **46**, 350–5.

Cook, R. A. and W. F. Furter 1968. Extractive distillation employing a dissolved salt as a separating agent. *Canadian Journal of Chemical Engineering* **46**, 119–23.

Court, R. E. and V. H. Bowers 1970. Irish whiskey. *Process Biochemistry* **5**(10), 17–20.

Cysewski, G. R. and C. R. Wilke 1978. Process design and economic studies of alternative fermentation methods for the production of ethanol. *Biotechnology and Bioengineering* **20**, 1421–44.

Daiches, D. 1969. *Scotch whisky. Its past and present.* London: Deutsch.

Dick, R. and G. Mavel 1982. R O technology for preconcentrating alcoholic solutions. *Proceedings of the 5th International Alcohol Fuel Technology Symposium*, Auckland, New Zealand, 13–18 May, Vol. 1, 151–8.

Essien, D. and D. L. Pyle 1983. Energy conservation in ethanol production by fermentation. *Process Biochemistry* **18**(4), 31–7.

Fanta, G. F., R. C. Burr, W. L. Orton and W. M. Doane 1980. Liquid-phase dehydration of aqueous ethanol–gasoline mixtures. *Science* **210**, 646–7.

Filippi, R. P. de and J. M. Moses 1982. Extraction of organics from aqueous solutions using critical-fluid carbon dioxide. *Biotechnology and Bioengineering Symposium* No. 12, 205–19.
Floro, M. B. 1945. Steam consumption of pot stills. *International Sugar Journal* **47**, 246–7.
Furter, W. F. 1968. Salt effect in distillation: a technical review. *The Chemical Engineer* **219**, CE 173–7.
Furter, W. F. 1972. Extractive distillation by salt effect. *Advances in Chemistry Series* **115**, 35–45.
Furter, W. F. and R. A. Cook 1967. Salt effect in distillation: a literature review. *International Journal of Heat and Mass Transfer* **10**, 23–36.

Geyer, C. R. 1979. Conserving energy in a distillation train. *Chemical Engineering Progress* **75**(1), 41–5.
Graham, W. O. 1940a. The influence of distillation methods on brandy composition. *Australian Brewing and Wine Journal* **58**(6), 40–2.
Graham, W. O. 1940b. The influence of distillation methods on brandy composition. *Australian Brewing and Wine Journal* **58**(7), 31–3.
Graham, W. O. 1940c. The influence of distillation methods on brandy composition. *Australian Brewing and Wine Journal* **58**(8), 26, 29, 31.

Hastie, D. F. S. 1949. The manufacture of Scotch whisky. *Journal of the Incorporated Brewers' Guild* **35**(415), 170–89.
Heist, J. A. 1979. Freeze crystallization. *Chemical Engineering* **86**(10), 72–82.
Herstein, K. M. and M. B. Jacobs 1948. *Chemistry and technology of wines and liquors*, 2nd edn. New York: Van Nostrand.
Holland, C. D., S. E. Gallun and M. J. Lockett 1981. Modeling azeotropic and extractive distillations. *Chemical Engineering* **88**(6), 185–200.
Hong, J., M. Voloch, M. R. Ladisch and G. T. Tsao 1982. Adsorption of ethanol–water mixtures by biomass materials. *Biotechnology and Bioengineering* **24**, 725–30.
Hoover, K. C. and S.-T. Hwang 1982. Pervaporation by a continuous membrane column. *Journal of Membrane Science* **10**, 253–71.

I'Anson, J. A. P. 1971. Rum manufacture. *Process Biochemistry* **6**(7), 35–9.
Irani, C. A. and E. W. Funk 1977. Separations using supercritical gases. In *Recent developments in separation science*, N. Norman (ed.), Vol. 3, Part A, 171–93. Cleveland, Ohio: CRC Press.

Jaques, D. and W. F. Furter 1972. Prediction of vapor composition in isobaric vapor–liquid systems containing salts at saturation. *Advances in Chemistry Series* **115**, 159–68.
Johnson, A. I. and W. F. Furter 1957. Salt effect in vapor–liquid equilibrium Pt 1. *Canadian Journal of Technology* **34**, 413–24.
Johnson, A. I. and W. F. Furter 1960. Salt effect in vapor–liquid equilibrium Pt 2. *Canadian Journal of Chemical Engineering* **38**, 78–87.
Johnson, A. I. and W. F. Furter 1965. Vapor–liquid equilibrium in systems containing dissolved salts. *Canadian Journal of Chemical Engineering* **43**, 356–7.

Ladisch, M. R. and K. Dyck 1979. Dehydration of ethanol: new approach gives positive energy balance. *Science* **205**, 898–900.
Ladisch, M. R., M. Voloch, J. Hong, P. Blenkowski and G. T. Tsao 1984. Cornmeal adsorber for dehydrating ethanol vapors. *Industrial Engineering and Chemical Process Design Development* **23**, 437–43.

Lee, E. K. L., W. C. Babcock and P. A. Bresnahan 1984. Ethanol–water separation by countercurrent reverse osmosis. *Polymer Material Science* **50**, 251–5.

Leeper, S. A. and P. C. Wankat 1982. Gasohol production by extraction of ethanol from water using gasoline as solvent. *Industrial Engineering and Chemical Process Design Development* **21**, 331–4.

Lehtonen, M. and H. Suomalainen 1977. Rum. In *Economic microbiology*, A. H. Rose (ed.), Vol. 1: *Alcoholic beverages*, 595–633. New York: Academic Press.

Lyons, T. P. 1974. Irish whiskey. *The Brewer* **60**, 634–7.

Lyons, T. P. and A. H. Rose 1977. Whisky. In *Economic microbiology*, A. H. Rose (ed.), Vol. 1: *Alcoholic beverages*, 635–89. New York: Academic Press.

Maiorella, B., C. R. Wilke and H. W. Blanch 1981. Alcohol production and recovery. *Advances in Biochemical Engineering* **20**, 43–92.

Malik, R. K., P. Ghosh and T. K. Ghose 1983. Ethanol separation by adsorption–desorption. *Biotechnology and Bioengineering* **25**, 2277–82.

Marion, L. 1979. In the spotlight: ethanol. *Chemical Engineering* **86**(5), 78–9.

Mellan, I. 1977. *Industrial solvents handbook*, 2nd edn. Park Ridge, New Jersey: Noyes Data.

Meranda, D. and W. F. Furter 1966. Vapor–liquid equilibrium data for system: ethanol–water saturated with potassium acetate. *Canadian Journal of Chemical Engineering* **44**, 298–300.

Meranda, D. and W. F. Furter 1971. Vapor–liquid equilibrium in alcohol–water systems containing dissolved acetate salts. *American Institute of Chemical Engineers Journal* **17**, 38–42.

Meranda, D. and W. F. Furter 1972. Vapor–liquid equilibrium in alcohol–water systems containing dissolved halide salts and salt mixtures. *American Institute of Chemical Engineers Journal* **18**, 111–16.

Meranda, D. and W. F. Furter 1974. Salt effects on vapor–liquid equilibrium: some anomalies. *American Institute of Chemical Engineers Journal* **20**, 103–8.

Milestone, N. B. and D. M. Biddy 1981. Concentration of alcohols by adsorption on silicalite. *Journal of Chemical Technology and Biotechnology* **31**, 732–6.

Mix, T. J., J. S. Dweck, M. Weinberg and R. C. Armstrong 1978. Energy conservation in distillation. *Chemical Engineering Progress* **74**(4), 49–55.

Mulder, M. H. V., J. O. Hendrikman, H. Hegeman and C. A. Smolders 1983. Ethanol–water separation by pervaporation. *Journal of Membrane Science* **16**, 269–84.

Muller, J. G. 1966. Recent developments in crystallization of ice as applied to freeze concentration. *Bulletin of the International Institute of Refrigeration* Annexe-3, 15–46.

Munson, C. L. and C. J. King 1984. Factors influencing solvent selection for extraction of ethanol from aqueous solutions. *Industrial Engineering and Chemical Process Design Development* **23**, 109–15.

Norman, W. S. 1945. The dehydration of ethanol by azeotropic distillation. *Transactions of the American Institute of Chemical Engineers* **23**, 66–74.

Othmer, D. F. 1936. Vapor re-use process. Separation of mixtures of volatile liquids. *Industrial and Engineering Chemistry* **28**, 1435–8.

Othmer, D. F. 1978. Azeotropic and extractive distillation. In *Kirk–Othmer encyclopedia of chemical technology*, 3rd edn, Vol. 3, 352–77. New York: Wiley.

Packowski, G. W. 1978. Beverage spirits, distilled. In *Kirk Othmer encyclopedia of chemical technology*, 3rd edn, Vol. 3, 830–63. New York: Wiley.

Parulekar, S. J., M. M. Sharma, J. B. Joshi and Y. T. Shah 1982. Separation processes for recovery of valuable chemicals from aqueous effluents. *Journal of Separation Process Technology* **3**(2/3), 2–29.

Paulaitis, M. E. 1982. Separation processes utilizing supercritical fluids. *Proceedings of the Scientific Conference of the Corn Refiners Association Inc.*, Lincolnshire, Illinois, 16–18 June, 168–97.
Perry, R. H. (ed.) 1963. *Chemical engineering handbook*, 4th edn. New York: McGraw-Hill.
Pyke, M. 1965. The manufacture of Scotch grain whisky. *Journal of the Institute of Brewing* **71**, 209–18.

Ray, C. 1973. *Cognac*. London: Peter Davies.
Roddy, J. W. 1981. Distribution of ethanol–water mixtures to organic liquids. *Industrial Engineering and Chemical Process Design Development* **20**, 104–8.
Rose, A. H. 1977. History and scientific basis of alcoholic beverage production. In *Economic microbiology*, A. H. Rose (ed.), Vol. 1: *Alcoholic beverages*, 1–41. New York: Academic Press.
Ross, J. 1970. *Whisky*. London: Routledge & Kegan Paul.

Scheibel, E. G. 1950. Dehydration of ethyl alcohol by fractional liquid extraction. *Industrial and Engineering Chemistry* **42**, 1497–508.
Schissel, P. and R. A. Orth 1984. Separation of ethanol–water mixtures by pervaporation through thin, composite membranes. *Journal of Membrane Science* **17**, 109–20.
Scott, R. W., P. D. Theaker and A. S. Marsh 1981. Use of extracts prepared with liquid carbon dioxide as a substitute for dry hops. *Journal of the Institute of Brewing* **87**, 252–8.
Simpson, A. C. 1968. Manufacture of Scotch malt whisky. *Process Biochemistry* **3**(1), 9–12.
Simpson, A. C. 1971. Manufacture of brandy. *Process Biochemistry* **6**(2), 25–7.
Souissi, A. and F. C. Thyrion 1981. Liquid–liquid extraction of ethanol from aqueous solutions. *Proceedings of the 2nd World Congress on Chemical Engineering, Montreal, Canada*, Vol. 4, 443–6.

Thorne, R. S. W., E. Helm and K. Svendsen 1971. Control of sulphury impurities in beer aroma. *Journal of the Institute of Brewing* **77**, 148–53.

Unger, E. D. and T. R. Coffey 1975. Production of light-bodied rum by an extractive distillation process. *Annals of Technology in Agriculture* **24**, 469–95.

Valaer, P. 1939. Brandy. *Industrial and Engineering Chemistry* **31**, 339–53.
van Pelt, W. H. 1975. Dutch process cuts cost of freeze concentration. *Food Engineering* **47**(11), 77–9.

Wagner, J. 1983. Improvement in freeze concentration. *Food Engineering* **55**(2), 98–100.
Watson, D. C. 1983. A laboratory apparatus for distillation research. In *Current developments in malting, brewing and distilling*, F. G. Priest and I. Campbell (eds), 249–55. London: Institute of Brewing.
Wentworth, T. O. and D. F. Othmer 1940. Absolute alcohol, an economical method for its manufacture. *Transactions of the American Institute of Chemical Engineers* **36**, 785–99.
Wentworth, T. O., D. F. Othmer and G. M. Pohler 1943. Absolute alcohol, an economical method for its manufacture. II Plant data. *Transactions of the American Institute of Chemical Engineers* **39**, 565–78.
Williams, D. F. 1981. Extraction with supercritical gases. *Chemical Engineering Science* **36**, 1769–88.

# Index